Legumes Biofortification

Muhammad Azhar Nadeem
Faheem Shehzad Baloch
Sajid Fiaz • Muhammad Aasim
Ephrem Habyarimana • Osman Sönmez
Nusret Zencirci
Editors

Legumes Biofortification

Springer

Editors
Muhammad Azhar Nadeem
Faculty of Agricultural Sciences
and Technologies
Sivas University of Science and Technology
Sivas, Türkiye

Faheem Shehzad Baloch
Faculty of Agricultural Sciences
and Technologies
Sivas University of Science and Technology
Sivas, Türkiye

Sajid Fiaz
Department of Plant Breeding and Genetics
The University of Haripur
Haripur, Pakistan

Muhammad Aasim
Faculty of Agricultural Sciences
and Technologies
Sivas University of Science and Technology
Sivas, Türkiye

Ephrem Habyarimana
International Crops Research Institute for
the Semi-Arid Tropics
Patancheru, Telangana, India

Osman Sönmez
Department Soil Science and Plant
Nutrition
Erciyes University
Kayseri, Türkiye

Nusret Zencirci
Biology Department
Abant Izzet Baysal University
Golkoy, Bolu, Türkiye

ISBN 978-3-031-33956-1 ISBN 978-3-031-33957-8 (eBook)
https://doi.org/10.1007/978-3-031-33957-8

This Springer imprint is published by the registered company Springer Nature Switzerland AG
The registered company address is: Gewerbestrasse 11, 6330 Cham, Switzerland

Contents

Chapter 1
Introduction to Biofortification and Challenges for Nutrition Security

Asima Rasheed, Sabir Hussain, Muhammad Abdul Rehman Rashid, Ijaz Rasul, and Farrukh Azeem

Abstract By the middle of the century, the global population will have surpassed 9 billion people, increasing the demand for food, water, and space. Maintaining food security and sustainability presents several significant hurdles like nutritional deficits, postharvest losses, and inconsistent regulation. Micronutrient deficiency is one of the major concerns of the time that imparts negative health impacts on millions of people and is also referred to as "hidden hunger." To deal with the deficiency impacts, biofortification is presented as the most effective strategy that enhances the micronutrients in staple crops. This technique can also increase bioavailability by removing antinutrients from plants. The cultivars developed by biofortification are tagged as ideal for nutritional security which shows a positive response in vulnerable countries. The ability of biofortification to improve crop micronutrient levels has been demonstrated through research; the next step is effective execution and public consumption. Current chapter highlights various approaches for food biofortification and challenges related to food nutritional security.

1 Introduction

Food insecurity makes it more difficult for people to get the amount of food they need to meet their caloric needs. Due to the resulting deficiencies, a person may not be able to work properly or have enough strength to do daily duties, which also

A. Rasheed · M. A. R. Rashid · I. Rasul · F. Azeem (✉)
Department of Bioinformatics and Biotechnology, Government College University Faisalabad, Faisalabad, Pakistan
e-mail: farrukh@gcuf.edu.pk

S. Hussain
Department of Environmental Sciences, Government College University Faisalabad, Faisalabad, Pakistan

M. A. Nadeem et al. (eds.), *Legumes Biofortification*,
https://doi.org/10.1007/978-3-031-33957-8_1

lowers their ability and productivity to earn (Banerjee and Duflo 2011). The world population is expected to reach 9 billion people in 2050, which would bring numerous issues for the sustainability of food due to the rising demand for food. The economy of the target nation is also impacted by ongoing global population expansion in the absence of technological and environmental policy measures (Tian et al. 2016). A growing problem in the world's expanding population is malnutrition. Malnutrition affects 792.5 million people worldwide, with developing countries bearing the brunt of the problem. In underprivileged nations, the majority of people either go hungry or eat food that is lacking in nutrients. Almost, 24,000 people a day die from hunger-related causes worldwide. A third of the population on average is facing "hidden hunger." They lack one or more essential macronutrients or micronutrients in their diets, e.g., Zn, Fe, Se, I, folic acid, lysine, vitamin A, vitamin B12, vitamin C, and Vitamin D (Malik and Maqbool 2020). Vitamin A, iron, and zinc deficiencies are the three most typical nutritional deficiencies worldwide. In regions of the world such as Sub-Saharan Africa, the Caribbean, and East and West Asia, it is a significant public health issue (Siwela et al. 2020). Malnutrition and poverty have a crucial connection. Unstable and poor conditions brought on by poverty may exacerbate the malnutrition issue. People who live in poverty frequently experience financial constraints, which makes it difficult for them to obtain sufficient, wholesome meals (Peña and Bacallao 2002). Short-term micronutrient deficiency is harmless, but persistent deficiency can cause a variety of illnesses, including anemia (iron deficiency), beriberi (vitamin B deficiency), pellagra (niacin deficiency), rickets (vitamin D deficiency), and scurvy (vitamin C deficiency), some of which can be fatal (Ratajczak et al. 2021). To boost the nutrient output of farming systems, a variety of agricultural instruments (such as crop diversification, crop selection, fertilizers, cropping systems, soil amendments, etc.) could be used. The first agricultural strategy now being used to combat micronutrient deficiencies is biofortification (Bouis and Welch 2010).

As compared to other traditional approaches, biofortification is considered as most economical one due to its tremendous outcomes in short period. Even though initial investments are substantial, numerous researchers have already examined the cost-effectiveness of biofortification in numerous studies (Kumar and Pandey 2020). Because of its several advantages over food diversification and artificial food fortification, it has been demonstrated to be an effective strategy in many industrialized countries and produced germplasms can be shared internationally.

The first aim of this chapter is to highlight the challenges in nutrition security and food sustainability and, second, to highlight the different biofortification strategies for the development of improved and easily approachable foods while maintaining the food chain.

2 Current Challenges for Nutrient Security

A key problem at the moment is the need for more food and fiber sources to feed future generations. Various risks and constraints, some of which are addressed below, can severely limit the capacity to develop and maintain a sustainable global agri-food system to satisfy these demands.

2.1 Global Population

Food security is seriously threatened by the rise in global population, climate change, and the shrinking amount of arable land. The world population increased from 4.4 billion to 6.1 billion people between 1980 and 2000, and food production increased by 50% during that time. Since 2000, the global population has increased by about 2% per year, reaching 7.3 billion people in 2014. The global population is expected to increase to reach 9.7 billion people by 2050 (McCarthy et al. 2018). Increased population directly impacts the environment which in turn affects the production rate of food for people. This will happen directly by changing the land structure that is available and suitable for farming, as well as indirectly by preventing the formation of clouds that are driven by volatile organic compounds (Tian et al. 2016). The Food and Agriculture Organization (FAO) predicts that emerging countries' urbanized areas would see the majority of the world's population growth. It has also been shown that there is enough food production to feed the entire world's population, but due to socioeconomic hurdles, harsh environmental conditions, and a lack of social safety nets, this food supply cannot be spread equally throughout all continents, especially in developing countries (Sunderland et al. 2013). Hunger and malnutrition are two persistent problems in these emerging nations. More individuals are expected to experience inadequate nutrition as a result of the ongoing population growth, especially youngsters, making them more vulnerable to chronic illnesses and even mortality.

2.2 Climate Change

Climate change is defined as an increase in atmospheric temperature, increased carbon dioxide levels, and changes in precipitation. According to the FAO, these factors will all have an impact on agriculture and food production, leading to drought and more extreme temperature swings in many areas where food is produced. Accuracy and precision in estimating climate sensitivity are essential to any climate predictions (Franzke et al. 2015). It is the primary duty of the authorities dealing with food security to strictly consider and follow the predictions of climate. The ignorance of these predictions could impact the dimensions of food security (Burke et al. 2015). People being hungry or perhaps starving is a particular challenge faced by those projecting climate change's effects on the agri-food industry. It is possible to focus too much on some outcomes (such as the worst, most extreme temperature increases scenarios) and neglect to prepare and plan for a wider range of potential future climate trends. Furthermore, the established nonuniform regional climate trends that take place within an overall changing climate hinder the applicability of global climate change projections to food security. To accommodate for climatic changes in crop and nutritional policies and practices, regional decisions must be founded on research. This is crucial for ensuring global food security (Chandio et al. 2020). Extreme heat stress brought on by rising global temperatures can have a severe impact on crop yield. Crop yield and world food supply are predicted to suffer from projected changes in the frequency and intensity of extreme climatic events, especially at lower latitudes (Porter et al. 2014). More than just a risk, climate change is a challenge that calls for quick and decisive actions. To improve climate change mitigation and adaptation in developing countries, new approaches are required. Climate change is expected to have a significant impact on food costs, which will make goods unaffordable and heavily reliant on imports, particularly in developing nations (Arora 2019). One hundred and ninety-five nations adopted the first global climate pact to address climate change during the December 2015 United Nations Conference on Climate Change, which was held in Paris. The goal of this agreement was to keep global warming in the twenty-first century to less than 2 °C compared to preindustrial levels. According to estimates, the world's greenhouse gas emissions must fall by 40–70% by 2050 to achieve carbon neutrality, which would increase the sustainability of food production (Tian et al. 2016).

2.3 *Water-Related Issue*

Water being an essential need in agriculture, household activities, and other industrial infrastructures, is utilized extensively which results in a shortage of water around the globe. 1 kilogram of rice requires 3500 liters of water and 1 kg of beef needs 15,000 liters of water to produce, which are examples of the intense usage of water (Tian et al. 2016). Water use has increased at a rate that is more than twice as fast as population growth over the past century. Geographically, water is distributed unevenly throughout the world, and a lot of it is wasted, polluted, and handled in an unsustainable manner (Premanandh 2011). According to statistics, a quarter of global population faces water shortages, and the one-fifth population lives in areas with limited water resources. When climate change hits, two-thirds of the world's population would have insufficient access to clean water (Shan et al. 2020). Proper water management and greater access to fresh water are required to cope with the demands of food production and agricultural operations.

2.4 Postharvest Food Losses

Every year, around 1.3 billion tons of food have been wasted due to mishandling of whole supply chain, especially postharvest handling. In contrast, over 870 million people experience daily hunger, which accounts for more than one-third of the food produced globally. Food loss can also be taken place by using polluted water, which has additional financial expenses. Agricultural sector, food security, and food supply chain should be placed at the utmost priority in the development of ecologically friendly crop protection measures. To decrease food waste and postharvest losses, FAO developed a toolbox to give awareness among population on how to minimize and recycle trash (Tian et al. 2016). Large amounts are lost and squandered, which results in losses of not only food and nutrition but also of the natural resources utilized to manufacture and handle these goods, including land, water, chemicals, energy, and labor. Additionally, the losses and waste of agricultural products also have a great contribution to the development of environmental issues, such as the emission of greenhouse gases. Minimizing postharvest losses and wastes can only be an operative way to upsurge food availability in the food system, ensuing in shortened food insecurity, improved nutrition, improved income ratio, and reduced wasting of several resources such as land, water, chemicals, and energy (Yahia et al. 2019).

3 Application of Biofortification to Improve the Nutritional Profile of the Diet

For a long time, developing countries have focused to upgrade their agriculture-based research to develop improved and nutritious cereals. Besides the development in research, every third person is facing hidden hunger, which might be due to the hindrance in availability of mineral-based foods or foods with low quality. Recently, plant scientists have developed a new policy to produce varieties with an elaborated nutrient profile instead of focusing on increasing production rates to reduce hidden hunger (Christou and Twyman 2004). A sustainable and cost-effective process that enhances the nutrient ratio by adopting different techniques in the diet is called biofortification. Biofortified food may not provide as many micronutrients per day as fortified or commercially available foods, but they can provide an adequate daily intake of nutrients or vitamins throughout the individual's life (Riaz et al. 2020). Biofortification provides micronutrients or treats nutrient deficiencies more sustainably and cheaply to fewer resources community instead of eliminating malnutrition from their life. Once established, the biofortified crop system is extremely sustainable, even if government interest and finance for solving micronutrient issues wane, nutritionally enhanced seeds will continue to be planted and consumed for a longer time period (Saltzman et al. 2017). Furthermore, biofortification provides a feasible method of supplying naturally fortified meals to undernourished communities in relatively remote rural areas that have limited access to commercially marketed fortified foods, which are more easily accessible in urban areas (Singh et al. 2016). Many organizations have been working hard to eliminate malnutrition from the world and introduced many varieties of different crops into the markets By the end of 2016, more than 20 million people in 30 countries were eating biofortified crops, and 150 varieties of 10 crops were available (Jha and Warkentin 2020). Marketed surpluses of these crops may make their way into retail outlets, reaching consumers in first rural and then urban areas, in contrast to complementary interventions, such as fortification and supplementation, that begin in urban centers. Biofortification research is ongoing in the hope of minimizing or eradicating malnutrition caused by micronutrient deficiency. Biofortification research continues in the hope of reducing or eliminating malnutrition caused by micronutrient deficiency.

Many crops like wheat, maize, rice, and soybean, etc. are broadly consumed by the population that is unable to provide recommended daily doses of essential nutrients which in turn causes health issues. Biofortification successfully enhances Fe, Zn, Ca, Se, or different vitamins in highly consumed staple crops and grows improved crops for the population. Biofortification makes sure to enhance mineral absorption, their delivery to edible parts, and bioavailability. Many factors in crops require special attention to provide people with balanced diets (Kumar and Pandey 2020). The inadequacies that have drawn the attention of responsible authorities and governments for biofortification are briefly described in the section that follows.

3.1 Micronutrients

In basic foods, Fe, Zn, and vitamin A are the most deficient ones that are directly involved in the hidden hunger around the globe. The deficiency of these components causes anemia, color blindness, and bone diffraction and also affects mental development (Majumder et al. 2019). For humans, micronutrients are also considered important candidates as they work as cofactors of many functioning enzymes in the body that are involved in the metabolism and regulatory functions of the body. Wheat, rice, and maize are the primary sources of nutrients for people living in underdeveloped countries (Shahzad et al. 2021). Unfortunately, these agriculture-based foods contain deficient amounts of different nutrients especially Fe, Zn, and vitamin A which enable them to fulfill the daily nutrient requirements of the body and result in many major disorders in the body. Underaged children and pregnant women are the main targets of these deficiencies and about 40% population around the globe is labeled as anemic (Rempel et al. 2021). Wheat is the red meat of poor communities that must be biofortified with Fe, Zn selenium, iodine, vitamin, and other micronutrients for a healthy population. Maize is found with antioxidants, vitamin A and E (tocopherol) deficiencies that affect the quality of proteins in maize. The quality of zein proteins can be enhanced by increasing the concentration of tocopherol and lysin. As opposed to wild-typed lines of maize, varieties with opaque-2 (o2) mutant form maize have the potential to increase tocopherol and lysin content (Grover et al. 2020). Other crops like barley, sorghum chickpea pigeon pea, etc. are also deficient in essential micronutrients and required great attention to avoid hidden hunger.

3.2 Antinutrients

The chemical substances that exert harmful effects on human health by the reduction of nutrient absorption are labeled as antinutrients. These factors are mostly found in the edible part of crops. Within plant material, these substances have critical impacts on nutrients like minerals, vitamins, and proteins and their amount and intensity vary in each plant depending upon the chemical fertilizers used for growth purposes, method of propagation, as well as storage conditions, also affect their influence (Ugwu and Oranye 2006). Some major antinutrients are alkaloids, lectins, phytases, oxalates, and tannins. These substances are gifts to plants by nature for their defense against fungi, insects, and predators. The continued consumption of these compounds by the human population causes serious health issues as these substances interfere with the metabolic process which in turn reduces the bioavailability of essential nutrients (Sinha and Khare 2017). Some antinutrients are briefly discussed below.

Lectins Lectins are also known as phytohemagglutinins involved in the agglutination of red blood cells and have many active carbohydrates binding sites. These antinutrient substances are widely distributed in plants especially in grain products by nature, involved in the protection mechanism of plants (Mishra et al. 2019) but when these lectin-rich products are consumed by humans, several disorders like nausea, gastroenteritis, and diarrhea have caused. Some severe disorders such as destruction of the epithelial tissue of the gastrointestinal tract, local hemorrhage, and damage to the kidney, heart, or liver have also been reported by the toxicity of lectin (Vasconcelos and Oliveira 2004). Because lectins directly affect digestive enzymes, they decrease the digestibility of nutrients.

Phytase Phytases present in plants, animals, and soil are the salted form of phytic acid and are also known as Inositol hexakisphosphate (InsP6) (Desai et al. 2014). In monocotyledon plants, it can be removed easily during milling as it is present in the aleurone or bran layers. In dicotyledon plants, it is directly linked with the protein that can be removed along with phytase during processing and added adverse impacts on the nutritional value of the concerned food. Phytases are the storage house of phosphates and play important role in the germination of seeds and also serve as the energy house of plants (Gibson et al. 2018). Despite the beneficial role of phytates for plants, it is crucial for humans and animals because it forms complexes with Fe, Zn, Ca, and Mn in the digestive tract and lessens the bioavailability of these minerals (Schlemmer et al. 2009).

Tannins The plant kingdom also has one of the highest molecular weighted antinutrients named tannins and is involved in the defensive properties of plants. This type of antinutrient has chelated properties with Fe and Zn. Condensed tannins and hydrolyzable tannins are the two types of tannins. The first one is the ester of gallic acid while the second one is the polymer of polyhydroxy flavan-3-ol monomers and also known as proanthocyanidins (Balasundram et al. 2006). The tannin-rich foods are considered as least nutritious because these tannin compounds are responsible for a low growth rate, indigestibility of proteins, and lack of appetizers (Ozcan et al. 2014). According to Chung et al. (1998), tannins can contribute to the acceleration of esophageal and cheek cancers. However, some tannin molecules also contain antiviral, antifungal, and antioxidant (Pizzi 2019).

Saponins These are the widely distributed antinutrients mostly found in soybean, peanuts, spinach, broccoli, potatoes, apples and eggplants (Kregiel et al. 2017). The presence of a very low concentration of saponins in foods can build up soapy-nature constituents that can be distinguished by their bitter taste. Saponins are water-soluble agents with a major portion of nonsugary aglycone which term as sapogenin (Góral and Wojciechowski 2020). By the chemical nature of sapogenin, saponins are differentiated into steroidal and triterpenoid saponins. The minute concentration of saponins can decrease the absorption of glucose and cholesterol in the gut region via intraluminal physicochemical interaction and also destroy the red blood cells in the human body (Sinha and Khare 2017).

4 Biofortification Approaches

The basic goal of biofortification (Shahzad et al. 2021) can be achieved through three routes: conventional plant breeding, agronomic approach, and genetic engineering (Fig. 1.1). These routes of biofortification involved agriculturalists, economics, and nutritionists working together (Garcia-Casal et al. 2017) for the welfare of mankind by producing improved and safe staple crops like wheat, rice, maize, pulses, potato, sweet potato, tomatoes, etc. The three routes of biofortification are briefly described in the below section.

4.1 Agronomic Approach

The agronomic approach of biofortification uses fertilizer, soil, and beneficial microorganisms to enhance the nutrients of edible parts of plants (Philipo et al. 2021). This technique is economic and easy to temporally enrich the plants with nutrient contents. The adoption of this technique requires much knowledge about environmental factors as well as soil microorganisms. The accessibility, transmission, and consumption of nutrients by plant parts are the primary objectives of

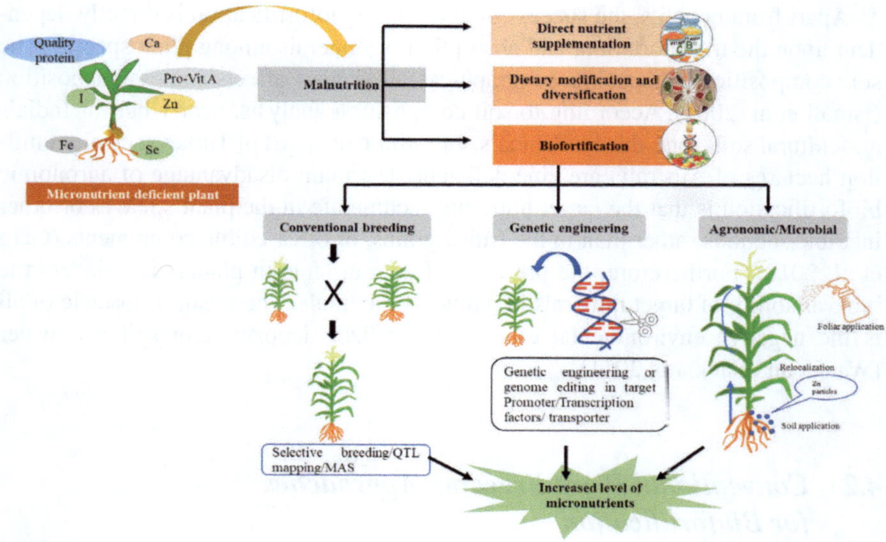

Fig. 1.1 Graphical representation of different biofortification approaches to overcome malnutrition. (Sourced from Sheoran et al. 2022)

agronomic biofortification (Shahzad et al. 2021). Naturally, the soil is rich with essential nutrients absorbed by the plants but sometimes plants are enabled to absorb nutrients and undergo nutrient deficiencies. In this condition, different nutrients are given to the oil. Nitrogen, phosphorus and potassium (NPK) are the frequently used nutrients in this approach (Rashid et al. 2020). Other than NPK, Zn and Fe are also reported as deficient the in the human diet, so these two are also applied under this approach to overcome their deficiencies. In China, a 75% increase in Zn content has been reported in wheat grain after Zn fertilizer application (Wang et al. 2016). Adequate attention related to environmental factors, ways of application and nature of nutrients required and adoption of an agronomic approach to biofortification. There are many options to enrich the plant nutrients under this strategy of biofortification. The target nutrient usually NPK can be directly added to the soil bed either during the preparation of soil for sowing or after the germination of plants. Thus, nutrients are taken up by plants and integrated into the food chain. Micronutrients in the liquid form are sprayed out on the aerial parts or on the reproductive parts of the plant that are absorbed by stomatal opening and resulted in healthier foods for consumers. Thirdly, micronutrients can also be flooded alongside irrigation that is absorbed by plants through root uptake and accumulated in edible parts of the plant (Shahzad et al. 2021). Several environmental factors like humidity, wind speed, temperature and time of application directly affect the efficiency of foliar and soil application. The moist or warm conditions of weather directly affect the permeability of plant tissues which in turn disturbs the efficiency of foliar and soil application. Before adopting these options for biofortification, sufficient research about the location and environmental conditions is necessary (De Valença et al. 2017).

Apart from benefits, the success of agronomic biofortification is directly dependent upon the transportation, and absorption of minerals among plant species. The soil composition of specific geographical areas also affects mineral deposition (Ismail et al. 2007). According to soil composition analysis, nearly half of India's agricultural soils, one-third of China's, 14 million hectares of Turkey, and eight million hectares of Australia are zinc deficient. The main disadvantage of agronomic biofortification is that the target nutrients accumulate in the plant's leaves or other inedible sections rather than in the fruits, grains, or other edible components (Garg et al. 2018). Furthermore, the presence of antinutrients in plants also hinders the bioavailability of target minerals in humans or animals. The greatest obstacle of all is the negative environmental effects of fertilizer deposition in soil and water (Waters and Sankaran 2011).

4.2 Conventional Plant Breeding Approaches for Biofortification

The main purpose of breeding activities is to enhance nutrient concentration to the target level in the edible parts of staple foods to ensure a healthy lifestyle for the population and to overcome malnutrition. In this method, plant breeders, nutritionists, and food technologists have collaborated to produce results that are long-lasting and sustainable while taking into consideration any gene screening or nutrient-related points (Kumar et al. 2022). Plant breeding not only focuses on improving the micronutrient content of edible plant parts but also takes into account how readily the body can absorb the targeted nutrients after cooking or preparation (Bouis and Welch 2010). Around the globe, several crops enriched with different nutrients have been developed and released under conventional plant breeding. Efficacy trials for vitamin A-rich OSP, provitamin A-fortified orange maize, provitamin A-fortified yellow cassava, iron pearl millet, and iron beans all provide promising evidence that biofortification improves micronutrient intake and deficiency status among target populations (Saltzman et al. 2017). To run fast and feasible breeding of plants with improved micronutrient and vitamins profile, the presence of adequate genotypic variation is mandatory. Parent lines with enhanced nutrient concentration are crossed with other lines several times to get a perfect product with desirable nutrients and agronomic traits. Contrary to agronomic biofortification, conventional breeding needed genetic diversity in the gene pool for the trait of interest (TOI). Sometimes, breeders need to cross lines with distant relative lines due to the limited genetic variations of TOI in the gene pool to get lines with the desired trait which slowly moved into commercial cultivars. Mutagenesis is another approach to transfer new traits into commercial varieties (Garg et al. 2018). Discovery of genetic differences that impact heritable mineral properties, evaluation of their stability under various situations, and determination of their breeding viability for higher mineral content in edible tissues without fluctuating yields or other quality attributes are some breeding approaches (Kaur et al. 2020). Traditional breeding depends on successful selection based on additive genetic effects, the heterosis phenomenon in F1 progeny, and transgressive segregation in later generations once a suitable genetic variation is available (Welch and Graham 2005). By lowering the amounts of antinutrients, the breeding strategy can be used to boost the bioavailability of nutrients. Antinutrients served as metabolites in plants and are directly related to plant metabolism and help out in biotic and abiotic resistance. The decrease in antinutrients required more precautions as it may impose negative impacts on crop health if not done carefully (Singh et al. 2016; Siwela et al. 2020). Since breeding programs are probably the quickest approach to improving plants, many international organizations have launched attempts to raise the nutritional content of crops. The European Union's Health Grain Project (2005–2010), funded with over £10 million and 44 partners from 15 countries, aimed to create high-quality cereal foods and components that promote good health. Numerous crops have been targeted for biofortification through crop breeding due to their improved acceptance (Bouis and Welch 2010).

Biofortification via conventional breeding also has several limitations. The major one is the limited genetic variations of traits in the gene pool of plants which might be solved by breeding with distant relatives but plant breeders also face difficulty to search for the gene of interest in distant relatives. This scenario makes it impossible for breeders to introduce crops with improved or desired traits through conventional breeding, e.g., Se improvement in wheat (Lyons et al. 2005) and oleic and linoleic acid improvement in soybean (Sarwar et al. 2020; Yeom et al. 2020). Moreover, only a small number of crops are enhanced through traditional biofortification, and the population grows solely dependent on these crops for greater nutrients, destroying the diversity of the environment and food. Diet diversification is crucial since some nutrients can be found in foods other than the basic crops that have been biofortified. Additionally, because existing biofortification solutions do not take into account the effects of "normal" meals on micronutrient deficits, they frequently fail over time (Lewis 2021).

4.3 *Transgenic Approach*

When agronomic and conventional breeding approaches fail to elevate the desired nutrient level in crops then transgenic biofortification becomes the only choice to plant scientists to enhance or introduce new nutrients in crops under limited or no genetic variation of traits in the plant's gene pool (Zhu et al. 2007). The transgenic modification approach can also be used to transfer genes of the desired trait from one species of plant to another which are unable to carry such genes naturally or when antinutrients present in a crop affect the uptake of nutrients (Jha and Warkentin 2020). It is also possible via a transgenic approach to introduce bacterial or other organisms' genes into crops to develop the desired trait or to explore pathways for metabolic engineering. The only key to developing a transgenic plant is the identification, characterization and utilization of the desired gene function (M. Vasconcelos et al. 2003). When a gene is identified and stamped as useful for a specific function, then it can be utilized in multiple crops under transgenic engineering biofortification and this is one of its major advantages over conventional breeding. For example, phytoene synthase (*PSY*), carotene desaturase, and nicotinamide synthase genes have been reported in many crops for essential mineral biofortification. By comparing the transgenic approach with breeding, the former proved as the most researched and emphasized one whereas later labeled as the most successful for commercial cultivar release (Garg et al. 2018). This approach has been broadly used on oilseeds biofortification due to the availability of limited genetic variation. Several crops such as golden rice with provitamin A, cassava with iron and vitamin A, and maize with high lysine content have been released as genetically modified crops (Vasconcelos et al. 2003).

There are also many barriers to the success rate of genetically modified crops for biofortification. The prominent one is that GM crops are not easily accepted by farmers, masses or communities (Vasconcelos et al. 2003). Secondly, it is an expensive approach that makes it beyond the access of poor people. Furthermore, extensive and detailed research regarding the identification, assessment and utilization of target genes is a time-consuming thing that lowers its success in terms of release. The other limitation is the poor regulatory and commercial releasing system of transgenic crops (Lewis 2021).

5 HarvestPlus Program

HarvestPlus program was launched with the contribution of CGIAR, International Center for Tropical Agriculture (CIAT), and International Food Policy Research Institute to develop nutrient-rich staple crops. The inputs of the HarvestPlus program along with agricultural and nutritional organizations to fight micronutrient deficiencies and hidden hunger are much appreciable around the globe. This program worked on strategies to control deficiencies among more vulnerable communities and to provide solutions based on food ingredients (La Frano et al. 2014). The World Health Organization (WHO) outlined Fe, Zn, and vitamin A as the three most underutilized micronutrients in the diet that require special attention. In this regard, the HarvestPlus Program plays a crucial and defined role in developing crop varieties rich in the aforementioned micronutrients by using practical, affordable methods like conventional breeding (Pfeiffer and McClafferty 2007). Currently, this program worked in Pakistan, India, Bangladesh, Nigeria, Rwanda, and Zimbabwe intending to provide nutrient-enriched foods to 1 billion people by 2030 (Bouis and Welch 2010). The main task of the HarvestPlus Program is to identify suitable nutrients to be biofortified because not all micronutrients are suitable for biofortification approaches due to their low concentrations or low absorption property in staple crops. Recently, maize (*Zea mays* L.), rice (*Oryza sativa* L.), wheat (*Triticum* spp.), beans (*Phaseolus vulgaris* L.), cassava (*Manihot esculenta* Crantz), sweet potatoes

(*Ipomoea batatas* L.), and pearl millet (*Pennisetum glaucum*) have been successfully biofortified by conventional breeding technique under this program (Ortiz-Monasterio et al. 2007).

HarvestPlus program comprises three phases: the discovery phase (2003–2008), the development phase (2009–2013), and the delivery phase (2014 onward). In the discovery phase, highly vulnerable populations, their dietary habits, resources, and studies were identified. In the development phase, the development of Fe, Zn, and vitamin A-rich crops in concerned countries, stability tests at several locations, and assessment of developed varieties in terms of nutrient ratio and planning to deliver them around the globe were all the aims focused by the researchers. Lastly, the establishment of consumer demand for biofortified crops to reach maximal populations was the main goal of the delivery phase. Researchers are engaged in estimating the area occupied by biofortified crops and working toward ensuring their long-term viability (La Frano et al. 2014). To date, the HarvestPlus program has released around 243 biofortified varieties of different crops in 30 countries. This single program has provided resources to nine million low-income farmers to grow developed varieties for the 42.4 million population (Shahzad et al. 2021).

6 Conclusion

It can be concluded that the use of both current and emerging technologies (coupled with the implementation of a wise policy) will be the most effective and long-lasting answer to issues with food security. The work of crop biofortification is difficult. The main objectives of many plant breeding efforts are to increase productivity, stress tolerance, food taste, and biotic and abiotic stress tolerance. Enhancing nutritional quality has been included as a new breeding objective in recent years. With the aid of regional, international, and domestic initiatives like the HarvestPlus program, this goal is being realized. To accomplish this goal, coordination between nutrition scientists and plant breeders is essential. Furthermore, because there is insufficient genetic variation for the micronutrients in the germplasm, several biofortification initiatives cannot be put into practice. Utilizing genetic engineering methods is necessary for these circumstances, and collaboration between molecular biologists and plant breeders is essential. Although there is more emphasis on transgenic methods, breeding-based approaches have much higher success rates because transgenically fortified crop plants must overcome barriers such as consumer acceptance issues and various pricey and time-consuming regulatory approval adopted by different countries. Despite these challenges, biofortified crops have a bright future since they have the potential to end micronutrient malnutrition among billions of poor people, especially in developing countries.

References

Arora NK (2019) Impact of climate change on agriculture production and its sustainable solutions. In: Environmental sustainability, 2, 95–96. https://doi.org/10.1007/s42398-019-00078-w

Balasundram N, Sundram K, Samman S (2006) Phenolic compounds in plants and agri-industrial by-products: antioxidant activity, occurrence, and potential uses. Food Chem 99(1):191–203. https://doi.org/10.1016/j.foodchem.2005.07.042

Banerjee A, Duflo E (2011) More than 1 billion people are hungry in the world. Foreign Policy 186:66–72

Bouis HE, Welch RM (2010) Biofortification—a sustainable agricultural strategy for reducing micronutrient malnutrition in the global south. Crop Sci 50(S1):S-20–S-32. https://doi.org/10.2135/cropsci2009.09.0531

Burke M, Dykema J, Lobell DB, Miguel E, Satyanath S (2015) Incorporating climate uncertainty into estimates of climate change impacts. Rev Econ Stat 97(2):461–471

Chandio AA, Jiang Y, Rehman A, Rauf A (2020) Short and long-run impacts of climate change on agriculture: an empirical evidence from China. Int J Clim Change Strategies Manage 12(2):201–221

Christou P, Twyman RM (2004) The potential of genetically enhanced plants to address food insecurity. Nutr Res Rev 17(1):23–42. https://doi.org/10.1079/NRR200373

Chung K-T, Wei C-I, Johnson MG (1998) Are tannins a double-edged sword in biology and health? Trends Food Sci Technol 9(4):168–175. https://doi.org/10.1016/S0924-2244(98)00028-4

De Valença AW, Bake A, Brouwer ID, Giller KE (2017) Agronomic biofortification of crops to fight hidden hunger in sub-Saharan Africa. Glob Food Sec 12:8–14

Desai M, Rangarajan P, Donahue JL, Williams SP, Land ES, Mandal MK, Phillippy BQ, Perera IY, Raboy V, Gillaspy GE (2014) Two inositol hexakisphosphate kinases drive inositol pyrophosphate synthesis in plants. Plant J 80(4):642–653

Franzke CLE, O'Kane TJ, Berner J, Williams PD, Lucarini V (2015) Stochastic climate theory and modeling. Wiley Interdiscip Rev Clim Chang 6(1):63–78

Garcia-Casal MN, Peña-Rosas JP, Giyose B, Groups CW (2017) Staple crops biofortified with increased vitamins and minerals: considerations for a public health strategy. Ann N Y Acad Sci 1390(1):3–13

Garg M, Sharma N, Sharma S, Kapoor P, Kumar A, Chunduri V, Arora P (2018) Biofortified crops generated by breeding, agronomy, and transgenic approaches are improving lives of millions of people around the world. Front Nutr 5:12. https://doi.org/10.3389/fnut.2018.00012

Gibson RS, Raboy V, King JC (2018) Implications of phytate in plant-based foods for iron and zinc bioavailability, setting dietary requirements, and formulating programs and policies. Nutr Rev 76(11):793–804

Góral I, Wojciechowski K (2020) Surface activity and foaming properties of saponin-rich plants extracts. Adv Colloid Interf Sci 279:102145. https://doi.org/10.1016/j.cis.2020.102145

Grover K, Arora S, Choudhary M (2020) Development of quality protein product using biofortified maize to combat malnutrition among young children. Cereal Chem 97(5):1037–1044

Ismail AM, Heuer S, Thomson MJ, Wissuwa M (2007) Genetic and genomic approaches to develop rice germplasm for problem soils. Plant Mol Biol 65(4):547–570

Jha AB, Warkentin TD (2020) Biofortification of pulse crops: status and future perspectives. Plan Theory 9(1):73

Kaur S, Kumari A, Singh P, Kaur L, Sharma N, Garg M (2020) Biofortification in pulses, pp 85–103. https://doi.org/10.1007/978-981-15-2874-3_4

Kregiel D, Berlowska J, Witonska I, Antolak H, Proestos C, Babic M, Babic L, Zhang B (2017) Saponin-based, biological-active surfactants from plants. Appl Characterization Surfactants 6(1):184–205

Kumar S, Pandey G (2020) Biofortification of pulses and legumes to enhance nutrition. Heliyon 6(3):e03682. https://doi.org/10.1016/j.heliyon.2020.e03682

Kumar S, Dikshit HK, Mishra GP, Singh A, Aski M, Virk PS (2022) Biofortification of staple crops: present status and future strategies. In: Kumar S, Dikshit HK, Mishra GP, Singh A (eds) Biofortification of staple crops. Springer Singapore, pp 1–30. https://doi.org/10.1007/978-981-16-3280-8_1

La Frano MR, de Moura FF, Boy E, Lönnerdal B, Burri BJ (2014) Bioavailability of iron, zinc, and provitamin A carotenoids in biofortified staple crops. Nutr Rev 72(5):289–307. https://doi.org/10.1111/nure.12108

Lewis E (2021) Biofortification as a sustainable solution for 'hidden hunger': evaluating its impact on the environment, nutrition, culture, and ethics

Lyons G, Ortiz-Monasterio I, Stangoulis J, Graham R (2005) Selenium concentration in wheat grain: is there sufficient genotypic variation to use in breeding? Plant Soil 269(1):369–380. https://doi.org/10.1007/s11104-004-0909-9

Majumder S, Datta K, Datta SK (2019) Rice biofortification: high iron, zinc, and vitamin-A to fight against "hidden hunger.". Agronomy 9(12):803

Malik KA, Maqbool A (2020) Transgenic crops for biofortification. Front Sustain Food Syst 4:571402

McCarthy U, Uysal I, Badia-Melis R, Mercier S, O'Donnell C, Ktenioudaki A (2018) Global food security–issues, challenges and technological solutions. Trends Food Sci Technol 77:11–20

Mishra A, Behura A, Mawatwal S, Kumar A, Naik L, Mohanty SS, Manna D, Dokania P, Mishra A, Patra SK, Dhiman R (2019) Structure-function and application of plant lectins in disease biology and immunity. Food Chem Toxicol 134:110827. https://doi.org/10.1016/j.fct.2019.110827

Ortiz-Monasterio JI, Palacios-Rojas N, Meng E, Pixley K, Trethowan R, Peña RJ (2007) Enhancing the mineral and vitamin content of wheat and maize through plant breeding. J Cereal Sci 46(3):293–307. https://doi.org/10.1016/j.jcs.2007.06.005

Ozcan T, Akpinar-Bayizit A, Yilmaz-Ersan L, Delikanli B (2014) Phenolics in human health. Int J Chem Eng Appl 5(5):393–396. https://doi.org/10.7763/IJCEA.2014.V5.416

Peña M, Bacallao J (2002) Malnutrition and poverty. Annu Rev Nutr 22(1):241–253

Pfeiffer WH, McClafferty B (2007) HarvestPlus: breeding crops for better nutrition. Crop Sci 47(S3):S-88–S-105. https://doi.org/10.2135/cropsci2007.09.0020IPBS

Philipo M, Ndakidemi PA, Mbega ER (2021) Importance of common bean genetic zinc biofortification in alleviating human zinc deficiency in sub-Saharan Africa. Cogent Food Agric 7(1):1907954. https://doi.org/10.1080/23311932.2021.1907954

Pizzi A (2019) Tannins: Prospectives and actual industrial applications. Biomol Ther 9(8):344

Porter JR, Xie L, UK AC, Howden M, Iqbal MM, Lobell D, Travasso MI, Garrett K, UK JI, Lipper L (2014) Food security and food production systems. Methods 7:1

Premanandh J (2011) Factors affecting food security and contribution of modern technologies in food sustainability. J Sci Food Agric 91(15):2707–2714. https://doi.org/10.1002/jsfa.4666

Rashid A, Ram H, Zou C, Guilherme LRG, Corguinha APB, Guo S, Kaur C, Naeem A, Yamuangmorn S, Ashraf MY (2020) Simultaneous biofortification of rice with zinc, iodine, iron and selenium through foliar treatment of a micronutrient cocktail in five countries. Front Plant Sci 11:589835

Ratajczak AE, Rychter AM, Zawada A, Dobrowolska A, Krela-Kaźmierczak I (2021) Do only calcium and vitamin D matter? Micronutrients in the diet of inflammatory bowel diseases patients and the risk of osteoporosis. Nutrients 13(2):525

Rempel J, Grover K, El-Matary W (2021) Micronutrient deficiencies and anemia in children with inflammatory bowel disease. Nutrients 13(1):236

Riaz U, Aziz H, Anum W, Mehdi SM, Murtaza G, Jamil M (2020) Biofortification technologies used in agriculture in relation to micronutrients. In: Plant micronutrients. Springer, pp 225–239

Saltzman A, Birol E, Oparinde A, Andersson MS, Asare-Marfo D, Diressie MT, Gonzalez C, Lividini K, Moursi M, Zeller M (2017) Availability, production, and consumption of crops biofortified by plant breeding: current evidence and future potential. Ann N Y Acad Sci 1390(1):104–114. https://doi.org/10.1111/nyas.13314

Sarwar N, Akhtar M, Kamran MA, Imran M, Riaz MA, Kamran K, Hussain S (2020) Selenium biofortification in food crops: key mechanisms and future perspectives. J Food Compos Anal 93:103615. https://doi.org/10.1016/j.jfca.2020.103615

Schlemmer U, Frølich W, Prieto RM, Grases F (2009) Phytate in foods and significance for humans: food sources, intake, processing, bioavailability, protective role and analysis. Mol Nutr Food Res 53(S2):S330–S375

Shahzad R, Jamil S, Ahmad S, Nisar A, Khan S, Amina Z, Kanwal S, Aslam HMU, Gill RA, Zhou W (2021) Biofortification of cereals and pulses using new breeding techniques: current and future perspectives. Front Nutr 665:721728

Shan, V., Singh, S.K., Haritash, A.K. (2020). Water crisis in the Asian countries: status and future trends. In: Kumar, M., Munoz-Arriola, F., Furumai, H., Chaminda, T. (eds) Resilience, response, and risk in water systems. Springer transactions in civil and environmental engineering. Springer, Singapore. https://doi.org/10.1007/978-981-15-4668-6_10

Sheoran, S., Kumar, S., Ramtekey, V., Kar, P., Meena, R. S., & Jangir, C. K. (2022). Current Status and Potential of Biofortification to Enhance Crop Nutritional Quality: An Overview. Sustainability, 14(6). https://doi.org/10.3390/su14063301.

Singh U, Praharaj CS, Chaturvedi SK, Bohra A (2016) Biofortification: introduction, approaches, limitations, and challenges. In: Singh U, Praharaj CS, Singh SS, Singh NP (eds) Biofortification of food crops. Springer India, pp 3–18. https://doi.org/10.1007/978-81-322-2716-8_1

Sinha K, Khare V (2017) Review on: Antinutritional factors in vegetable crops. Pharma Innov J 6(12):353–358

Siwela M, Pillay K, Govender L, Lottering S, Mudau FN, Modi AT, Mabhaudhi T (2020) Biofortified crops for combating hidden hunger in South Africa: availability, acceptability, micronutrient retention and bioavailability. Foods 9(6). https://doi.org/10.3390/foods9060815

Sunderland T, Powell B, Ickowitz A, Foli S, Pinedo-Vasquez M, Nasi R, Padoch C (2013) Food security and nutrition. Center for International Forestry Research (CIFOR), Bogor

Tian J, Bryksa BC, Yada RY (2016) Feeding the world into the future–food and nutrition security: the role of food science and technology. Front Life Sci 9(3):155–166

Ugwu FM, Oranye NA (2006) Effects of some processing methods on the toxic components of African breadfruit (Treculia africana). Afr J Biotechnol 5(22): 2329–2333

Vasconcelos IM, Oliveira JTA (2004) Antinutritional properties of plant lectins. Toxicon 44(4):385–403

Vasconcelos M, Datta K, Oliva N, Khalekuzzaman M, Torrizo L, Krishnan S, Oliveira M, Goto F, Datta SK (2003) Enhanced iron and zinc accumulation in transgenic rice with the ferritin gene. Plant Sci 164(3):371–378. https://doi.org/10.1016/S0168-9452(02)00421-1

Wang Y-H, Zou C-Q, Mirza Z, Li H, Zhang Z-Z, Li D-P, Xu C-L, Zhou X-B, Shi X-J, Xie D-T, He X-H, Zhang Y-Q (2016) Cost of agronomic biofortification of wheat with zinc in China. Agron Sustain Dev 36(3):44. https://doi.org/10.1007/s13593-016-0382-x

Waters BM, Sankaran RP (2011) Moving micronutrients from the soil to the seeds: genes and physiological processes from a biofortification perspective. Plant Sci 180(4):562–574. https://doi.org/10.1016/j.plantsci.2010.12.003

Welch RM, Graham RD (2005) Agriculture: the real nexus for enhancing bioavailable micronutrients in food crops. J Trace Elem Med Biol 18(4):299–307. https://doi.org/10.1016/j.jtemb.2005.03.001

Yahia EM, Fonseca JM, Kitinoja L (2019) Chapter 2 - Postharvest losses and waste. In: Yahia EM (ed) Postharvest technology of perishable horticultural commodities. Woodhead Publishing, pp 43–69. https://doi.org/10.1016/B978-0-12-813276-0.00002-X

Yeom WW, Kim HJ, Lee K-R, Cho HS, Kim J-Y, Jung HW, Oh S-W, Jun SE, Kim HU, Chung Y-S (2020) Increased production of α-linolenic acid in soybean seeds by overexpression of lesquerella FAD3-1. Front Plant Sci 10. https://doi.org/10.3389/fpls.2019.01812

Zhu C, Naqvi S, Gomez-Galera S, Pelacho AM, Capell T, Christou P (2007) Transgenic strategies for the nutritional enhancement of plants. Trends Plant Sci 12(12):548–555

Chapter 2
Nutritional Security Approaches for Legume Biofortification—A Major Challenge

Aisha Umar, Zubaida Yousaf, Afifa Younas, Nadia Riaz, Madiha Rashid, Arusa Aftab, and Zoya Kashif

Abstract Legumes are a primary source of protein and micronutrients. Increases in the world's population and hunger require finding the most economical and best agricultural approaches to increase the nutritional value of crops and their yields. Micronutrient-rich legumes have dual functions: reduce hunger and increase health benefits. Food legumes contain high levels of micronutrients and proteins, which can lower the chance of developing severe human disease. Biofortification is the most important method for improving legume crops. However, the biofortification potential of legumes remains unexplored. Legumes and pulses contain a variety of amino acids and micronutrients. This chapter focuses on the importance of food legumes, where soil factors influence the micronutrients of legumes, and on the biofortification programs of legumes, which facilitates sustainable plant-based food production. Different types of biofortification in pulses and legumes provide fast and effective routes to improving micronutrient concentrations. Some biofortification methods should be added to a multidisciplinary initiative to improve legume crops.

A. Umar
Institute of Botany, University of the Punjab, Lahore, Pakistan

Z. Yousaf (✉) · A. Younas · N. Riaz · A. Aftab · Z. Kashif
Department of Botany, Lahore College for Women University, Lahore, Pakistan
e-mail: zubaida.yousaf@lcwu.edu.pk

M. Rashid
Department of Botany, Lahore College for Women University, Lahore, Pakistan

Department of Botany, Division of Science and Technology, University of Education, Lahore, Pakistan

M. A. Nadeem et al. (eds.), *Legumes Biofortification*,
https://doi.org/10.1007/978-3-031-33957-8_2

1 Introduction and Background

Dramatic increases in the world's population has led to malnutrition (Bohra et al. 2014). The prevalence of micronutrient inadequacy in poor countries can be found in staple foods such as wheat and rice (Borill et al. 2014). Lentils are the fifth-healthiest food in the world, on the basis of beneficial nutritional properties, especially in Bangladesh, Sri Lanka, India, Nepal, Syria, Lebanon, Turkey, and Iran. Millions of people consume lentils in soup or dal (Raymond 2006). An increasing population creates pressure on arable land and natural resources and leads to the environmental stresses linked to climate change (Nath et al. 2017). These changes adversely affect plant growth, nutritional value, and productivity. The *overuse* of pesticides and fertilizers also damages the environment—e.g., soil and water deterioration (Kumar et al. 2017). All these problems are reflected in human health in the form of malnutrition problems in developing and developed countries (Finn 2014).

People's lifestyles (e.g., eating habits) have shifted thanks to urbanization and globalization. People have started to consume junk food on a larger scale, which has high levels of sugar, salt, and fat and low nutrient content for protein, fiber, minerals, and vitamins. Ingesting these nutrient-deficient foods has led to widespread malnutrition in Asian nations. Asian women are more vulnerable to a few forms of malnutrition, especially women of reproductive age (Kumar and Pandey 2020). Junk food consumption ultimately leads to malnutrition and syndromes (Farzana et al. 2017).

Various policies have been implemented across the world to address malnutrition, but because of natural disasters and economic priorities, malnutrition has become a significant challenge for policymakers and food scientists. Biofortification is a feasible way to reduce the malnutrition problem, and the maximum number of foods should be fortified and augmented to solve this problem. However, fortified foods are expensive and not affordable for low-income people, so an alternative solution to this issue is eating a balanced diet including pulses and legumes. These are very vital to enhance nutritional content and phytochemical content.

Recently, not much attention has been paid to the biofortification of legumes and pulses compared with that paid to biofortifying wheat, rice, and maize. In biofortification, the nutritional value of legumes is enhanced via (1) breeding, (2) transgenic techniques, or (3) agronomic practices (Bouis and Saltzman 2017). Biofortified couple lines are generated for proteins and minerals containing K, Ca, Fe, and Zn (Santos and Boiteux 2013). After biofortified varieties have been grown in country trials and local adaptation costs have been incurred, routine breeding is carried out to ensure that the characteristic remains stable (Welch et al. 2002). Over a ten-year period, the global cost of biofortification has fallen to roughly $400,000 per crop.

2 Legumes and Pulses

A large portion of human diets depends on legumes for food (Bohra et al. 2015). The United Nations declared 2016 the international year of pulses. Foods made with legumes are good targets for maximizing nutritional value. Apart from the rich nutritional value, legumes play important "ecological roles" (Soares et al. 2019). Legumes are excellent sources of amino acids, nutrients, and vitamins (Rehman et al. 2019). Their specific compounds, such soluble fibers, antioxidants, flavonoids, etc., help protect humans from developing diabetes, cancers, and cardiovascular diseases (Ferreira et al. 2020).

The word *pulse* means "thick slurry," and pulses belong to the Fabaceae, or Leguminosae, family. Pulses are part of traditional diets throughout the world and have been consumed for thousands of years. Pulses feature high protein levels, low calories, and a low glycemic index (Asif et al. 2013). Pulses like chickpeas and lentils contain ß-carotene, zeaxanthin, and lutein, which are precursors of vitamin A (Margier et al. 2018). Every pulse is a legume, but not every legume is a pulse, because only when the edible seed of a legume is utilized as a dry grain is it called a pulse. Examples of pulses include dry peas, chickpeas, lentils, and dry beans, whereas soybeans, fresh peas, and peanuts are legumes (Asif et al. 2013). Legumes have pods or fruit that contain seeds or dry grains that can fix the nitrogen content of soil. The proteins in pulses can be utilized in soups, chocolates, sauces, sausages, and confectionaries thanks to their high-quality oils and water-binding ability (Barbut 1999).

Common beans, such as other legumes, are usually combined with a starch-based food (pulse + grain) to produce a dish such as a dal (made of lentils) or a frijolada (made of beans). Beans are parts of national dishes in other places too, such as in Brazil, where feijoada is commonly eaten every day, and in the Dominican Republic, where bandera often appears in meals twice a day. In South America and Central America, people consume reheated calentao (beans) or consume mashed beans for breakfast (Blair et al. 2013).

Worldwide, 200 out of every 1000 legumes, namely *Cicer arietinum* L. (chickpea), *Cajanus cajan* L. (pigeon pea), *Vigna unguiculata* (cowpea), *Vigna mungo* L. (urdbean), *Vigna radiata* L. (mungbean), *Lens culinaris* Medik (lentil), *Phaseolus vulgaris* (French bean), *Vigna aconitifolia* (moth bean), *Macrotyloma uniflorum* (horse gram), *Pisum sativum* L. (field pea), *Glycine max* (soybean), and *Lathyrus sativus* L. (lathyrus), are cultivated for human consumption (Rac 2002).

3 Health Benefits

3.1 Cardiovascular Health

The ingestion of pulses decreases low-density lipoprotein (LDL) cholesterol, hypertension triglycerides, diabetes, and obesity. Legumes contain the highest number of hypocholesterolemic agents, which lower cholesterol levels, so eating pulses could reduce the risk of cardiovascular diseases.

3.2 Diabetes Management

Eating a combination of pulses is a sign of a healthy diet and helps to prevent healthy people from developing diabetic symptoms. People who ingest three or more servings of whole grains per day are less prone to developing type II diabetes mellitus than those who don't. Whole grains from pulses also help control glycemia. Eating a combination of healthy cereals and pulses can help people feel full for a long time, which can aid in weight loss.

3.3 Celiac Disease

Celiac is small intestine disease caused by the ingestion of gluten, which can be found in wheat, barley, and rye. Pulses can be used to produce gluten-free alternatives that help prevent vitamin B and iron deficiencies. As a result, pulse flours are great alternatives to gluten-containing cereal flours and are high in fiber and protein.

3.4 Cancer Risk

The consumption of pulses reduces the risk of developing certain cancers, such as colon, prostate, gastric, and pancreatic cancers.

4 Soil Factors Influencing Micronutrient Acquisition

The micronutrient availability of legumes depends on several factors: soil texture, organic matter, soil reaction, soil moisture, clay content, microbial activity, redox potential, nutrient interactions in soil, aeration, etc. All these factors are equally important in the micronutrient availability of plants.

The availability of micronutrients in the soil is influenced by soil texture. Coarse/sandy soils lack micronutrients, but fine-textured soils (clay) have more usable nutrients. Clays and organic soils are better at retaining water and nutrients than sandy soils are (Choudhary and Suri 2014). Sandy soils drain micronutrients, a process known as "leaching," and leached minerals are unavailable to plants. Because the source materials originally lacked certain elements, leached acidic soils will lack certain micronutrients (Kumar et al. 2014).

4.1 *Temperature and Soil Moisture*

Low moisture content and low temperatures decrease micronutrient availability because the activity of roots is reduced, leading to low rates of nutrient diffusion and nutrient dissolution. The freezing and wet soil conditions during winter months also reduces the nutrient efficiency of Zn (Deb et al. 2009). When the weather warms up and the soil becomes drier, the effects of Zn deficiency diminish. The mineralization of micronutrients from soil organic matter is also influenced by the moisture content of the soil (Choudhary and Suri 2014). The submergence of soil increases the pH and reduces the redox potential (Eh), which in turn reduces the availability of nutrients in acid soil (Karan et al. 2014). Zn and Cu increase the soil temperature while decreasing the contents of Mn and Al at high temperatures (Wallace et al. 1969).

4.2 *Soil pH*

Soil reactions measure the soils' alkalinity or acidity (pH). Soil pH represents the H^+ activity of a soil solution. The H^+ ions strongly attract negative sites and have enough power to replace other cations in the soil. Soil pH regulates the mobility, solubility, acquisition, and concentration of certain elements in soil solutions for plants (Fageria et al. 1997). In low-pH environments, protonated anions, free metal cations, and the maximum number of micronutrients are available at their peak concentrations, whereas high-pH environments favor hydroxyl complexes and carbonates. The availability of anions (SeO_4, SeO_3 and $B(OH)_4$) increases with alkalinity. Therefore, micronutrient cations are available and soluble under acidic conditions (Brady 2002). When the pH rises (from 5 to 7), the cations become more securely bound, whereas Cu, Ni, and Zn become much less soluble and exchangeable (Deb et al. 2009).

4.3 *Organic Matter*

Organic matter refers to a storehouse of important plant nutrients and heavy metals; as they decompose, they continuously provide nutrients to legume crops. For anions, this reservoir is crucial. In the presence of organic matter, the reduction in micronutrient availability is responsible for generating complexations with lignin, humic acid, and other organic compounds (Choudhary and Suri 2009). Soils that are regularly treated with manure or organic residues rarely show deficiencies in micronutrients. The excessive application of P fertilizer in highly manured soils causes Zn shortage (Das 2011; Kumar et al. 2014). Furthermore, because of natural chelation (the interaction of a micronutrient with an organic molecule), soils high in organic matter (muck or peat) lack vitamins (Deb et al. 2009).

4.4 Oxidation State

A soil's redox potential indicates the degree of oxidation or reduction in soil. Anaerobic conditions, which are caused by waterlogging, create reducing conditions in soils, while aerobic soils are well drained and create oxidizing conditions (Dass et al. 2015). At typical pH levels, the reduced forms of Fe, Cu, and Mn are more soluble than their higher-oxidation states are. High-pH (acidic) conditions and oxidation are more favorable to reduction (Brady 2002). The oxidized states of cations (micronutrients) are less soluble in neutral soil than in reduced states. Poor aeration and low pH improve the availability of micronutrient cations. The micronutrient availability of flooded soils is higher than that of aerated soils (Dass et al. 2015). Calcareous soils, high-pH soils, aerated soils, and well-drained soils are deficient in the availability of Fe, Mn, and Zn in their oxidized states, and crops grown in such soils suffer from micronutrient deficiencies (Brady 2002).

4.5 Rhizospheric Conditions

Rhizospheric conditions play important roles in the nutrient availability of soil for plants. Rhizospheric conditions are controlled by the effects of soil (Kumar et al. 2014) and by interactions between plant roots and soil microbes. The chemical composition in the rhizosphere is much different than that found in bulk soil. Under rhizospheric conditions, microorganisms and other microbes continuously produce chelating agents during the decomposition of plant and animal residues (Deb et al. 2009). Arbuscular mycorrhizal fungi (AMF) induce the many favorable changes in crops through the exudation/secretion of organic acids and chelating agents (Kumar et al. 2014), which improves the mobilization and "solubilization of nutrients" from in(organic) complexes (Suri and Choudhary 2013). Hyphal networks of AMF greatly influence the absorption of micronutrients (Harrier and Watson 2003).

5 Antinutrient Depressant

Phytic acid and mineral cations together form a mixed salt called phytate and sequester Pi phosphate inside legumes. Phytate plays a major role in stress, seed germination, and the defense against oxidative stress. In its antinutritional role, phytic acid is reduced in legumes when taking breeding and transgenic approaches (Joshi-Saha and Reddy 2015).

In common beans, *lpa1* is a low-phytic-acid line selected and used for the recognition of the *Mrp1* gene, which downregulates the phytic acid pathway (Panzeri et al. 2011). In other legumes (field peas and soybeans), *lpa* mutants are identified through EMS-based mutagenesis (Warkentin et al. 2012), whereas in chickpeas, the *CaMIPS2* gene regulates phytic acid biosynthesis (Kaur et al. 2008). Raffinose, which is synthesized by raffinose synthase, is another major antinutrient affecting a plant's nutrition potential.

Legumes contain some promoters that enhance the availability of minerals in the presence of antinutrients. Some promoters are natural plant metabolites. Inulin is a fructo-oligosaccharide found in small amounts in lentils, chickpeas, fava beans, red kidney beans, common beans, and white beans (Rastall and Gibson 2015), and it has a significant impact on the bioavailability of mineral nutrients in legumes.

6 Bioavailability of Nutrients

6.1 Protein

Legumes contain micronutrients, and many antinutrients need to be minimized to improve the bioavailability of micronutrients. The interspecific breeding (crossing of two species from the same genus) of mungbean with black gram significantly increased the quality of protein in mungbean. By taking the transgenic approach, the methionine content in legumes enhanced the sulfur in *Vicia narbonensis* (narbon beans), *Lupinus angustifolius* (lupins), and *Medicago sativa* (forage alfalfas) (Nair et al. 2013).

6.2 Miscellaneous Nutrients: Zinc (Zn), Iron (Fe), and Se

Almost two billion people in the world experience micronutrient malnutrition, known as hidden hunger (Godecke et al. 2018). In the developing world, the major contributors to hidden hunger are zinc (Zn), iron (Fe), and vitamin A (Saltzman et al. 2017). Antinutrient phytic acid in different food crops, such as common beans and field peas (*Pisum sativum*), can reduce their genetic potential, which increases the concentrations of Zn and Fe (Amarakoon et al. 2012). According to Thavarajah et al. (2009), a lentil genotype can lower the phytic acid content to lower than that in "mutated corn, wheat, soya bean, and common bean."

Deficiencies in Fe, Se, Zn, and I are common in predominantly rural countries. Deficiencies in vitamin A, I, Zn, and Fe cause nearly 20% of the deaths among five-year-old children (Prentice et al. 2008). Zn deficiency is responsible for 433,000 deaths among five-year-old children every year, whereas Se deficiency causes cardiomyopathy and osteoarthropathy (Kashin–Beck disease) (Reilly 1996). Deficiencies in vitamin B9 and folate cause several other health issues, especially in developing countries (Gupta et al. 2013).

Iron (5–15%) and zinc (18–34%) are low in bioavailability. Both are major challenges to biofortification strategies (J. Singh 2016). Zn deficiency leads to inappropriate physical growth, skeletal development, and wound healing and increases the risk of infection. Iron deficiency causes anemia and fatigue, also lowering the immune system. Se deficiency affects male fertility, impairs the human immune system, increases vulnerability to infection, impedes mental growth, and causes hypothyroidism (Khan et al. 2019).

Phytate (antinutrient compound) in legumes and pulses is responsible for the minimum bioavailability of Zn and Fe. Phytic acid forms complexes with Ca, Mg, Fe, and Cu and reduces their solubility (Biehl et al. 1995). Temperature affects the phytic acid, Fe, and zinc levels of leguminous seeds. Phytic acid is found more in legumes from areas with higher-temperature regimes compared to those with lower-temperature regimes, so phytic acid decreases in seeds exposed to low temperatures. Biofortification is a new strategy for reducing phytic acid in staple crops (Thavarajah et al. 2010). In one study, beans biofortified through breeding contained significant increases over "traditional mung bean" (Singh 2016). Zou et al. (2014) biofortified the soybean sprout by using $ZnSO_4$, which significantly enhanced the concentration of zinc, with good bioaccessibility. Selenium is a component of almost 25 human enzymes, and an intake of 55 μg/day is recommended for adults.

Iron is important in different metabolic processes, such as the electron transport chain (ETC), oxygen transport, and Deoxyribonucleic acid (DNA) synthesis. Iron deficiency leads to anemia, functional impairments that affect immunity, working capacity, and cognitive development (Abbaspour et al. 2014). Furthermore, Zn metabolism plays an important role in patients who have iron deficiency, hypogonadism, dwarfism, or hepatosplenomegaly (Prasad et al. 1963). Zn is the second-most-abundant transition metal in the human body, after iron (Read et al. 2019), and it is essential for the functioning of many important proteins, such as transcriptional factors and enzymes (Cassandri et al. 2017; Lambert et al. 2018). The COVID-19 pandemic brought Zn into the limelight because it plays a vital role in viral immunity (Wessels et al. 2020; Kumar et al. 2020).

6.3 Recommendations

Chemical dosage, in addition to adverse environmental effects, is crucial to a legume's nutritional status. Zn and P affect each other's uptake in plants (Bouain et al. 2014), so Zn fertilization reduces P uptake and causes phytates to accumulate in legume seeds, suggesting that the bioavailability of Zn is enhanced (Erdal et al. 2002). Contrariwise, the presence of maximal P can precipitate Zn and stimulate the plant's requirement for Zn (Loneragan and Webb 1993). Additionally, Zn has varied interactions with other minerals, such as Ca, B, and S (Prasad et al. 2016). Different fertifortification strategies will be used in the future to enhance the micronutrients in grains. Many efforts have been made to overcome malnutrition, taking approaches such as nutrient supplementation, biofortification, and food fortification.

We need to act in a sustainable manner to face these societal challenges, to guarantee food production with maximal nutritional content and minimal environmental impacts. A biofortification program is heavily dependent on farmers, public acceptance, and political support for its cost–benefit ratio. Biofortification is a promising strategy to lift these constraints. The World Health Organization defined *biofortification* as "the process by which the nutritional quality of food crops is improved through agronomic practices, conventional plant breeding, or modern biotechnology" (WHO 2022). Agronomical practices are improving mineral uptake via fertilizer application, nutrient solubilization and mobilization in soil (White and Broadley 2009), and improvements in the level of micronutrients (in the edible tissues of crops) and their availability to humans (Carvalho and Vasconcelos 2013).

7 Biofortification Approaches for Mineral Enrichment in Legumes

7.1 Agronomic Biofortification

In agronomic biofortification, micronutrient levels have been increased in edible tissues by using inorganic fertilizers (Prasad et al. 2014). Fe and Zn agronomically increased in legumes during crop production without the concomitant loss of yield (Shivay et al. 2016). Agronomic interventions that increase the rigidity of legislative regulations on environmental safety limit this application (Borill et al. 2014).

The following types of agronomic biofortification enhance micronutrient levels in legumes and improve the "mobilization and solubilization" of micronutrients in soil (Bouis et al. 2011):

- Seed priming
- Seed coating
- Soil or foliar fertilization
- Intercropping
- Crop rotation
- Soil microorganisms
- The pH, composition, and properties of soil

The agronomic biofortification of cereals and legumes (chickpeas, peas, and common beans) is feasible (Garg et al. 2018). Many factors have contributed to agronomic biofortification, including the application methods (soil versus foliar), the chemical nature of added nutrients, the application time, and the effects of minerals and soil components on, for example, chickpeas. The zinc–foliar application was much better than soil-comprised Zn elements (Shivay et al. 2015). Applying micronutrient fertilizers such as $FeSO_4$, $ZnSO_4$, and ZnO_2 to soil have strong potentials to improve the micronutrient quantity of crop yields in mature grains. The foliar fertilization of $FeSO_4$ and $ZnSO_4$ with etheylenediaminetetraacetic acid. (EDTA) may be a more effective treatment (Shivay et al. 2016). Soil and foliar applications improve the respective concentrations of Fe and Zn in the shoots, flowers, and grains of legumes. Fe foliar spray is effective in boosting the Fe content in peas deficient in Fe. The Fe content of grain can be significantly increased thanks to the use of foliar spray during grain filling (Kabir et al. 2016). Likewise, Zn foliar applications alone or in combination with soil Zn applications can enrich a field of pea grain (Poblaciones and Rengel 2016).

The timely application of agronomic biofortification via chemicals is the least cost-effective and most labor-intensive technique, though the labor cost can be reduced by combining the foliar Zn application with pesticide sprays (Wang et al. 2016). Chemicals leach into groundwater because of farmer's excessive use of chemical fertilizers. Soil-quality parameters such as bioavailability and mobility have contributed to the success of agronomic biofortification (Alvarez 2007). Crop physiology also affects the movement of minerals in phloem and their accumulation in seeds (White and Broadley 2011).

7.2 Genetic/Breeding Biofortification

Mere legume ingestion is not sufficient to address nutrition security. Biofortification, or the agronomical or genetic development of nutritionally rich crops, is a more feasible approach. Genetic biofortification via conventional breeding is the most acceptable, sustainable, and cost-effective strategy to yield nutritionally rich crops (Garg et al. 2018). This is a long-term remedy to the indiscriminate application of micronutrient fertilizers, the regular use of which poses huge harms to ecosystems. "Exploitable genetic variation" plays an important role in the nutritional breeding system.

The tools and technologies of modern omics are available to augment the crop biofortification program. The different omics disciplines include proteomics, ionomics, and metabolomics, which are used to understand the fundamental makeup and underlying causes of genes and their intricate network. Metabolomics provides additional value for food safety assessment programs because it uses analytical methods. Various genomic-based tools, such as quantitative trait loci (QTLs), are currently being used for legume biofortification (Bohra et al. 2015).

Legume crops could provide essential micronutrients. *Lotus japonicus* (a model legume crop) has been used for the identification of QTLs because QTLs are responsible for mineral concentrations, including those of iron and zinc. In total, 103 QTLs can be identified to determine various mineral concentrations. Likewise, forty-six QTLs are significant for minerals, such as to determine iron and zinc concentrations. In the composite interval mapping (CIM) approach, twenty-four QTLs for mineral content per seed are "mapped in Medicago RIL" (recombinant inbred line) populations to improve pulse crops (Klein and Grusak 2009).

Plant breeders are screened with the help of accessing global germplasm banks. This helps to ensure that sufficient genetic variation persists while breeding for a particular trait. This facilitates the selective nutritious breeders and cultivars of staples rich in Fe and Zn concentrations and substances that promote the bioavailability of Fe and Zn. To enhance the bioavailability of minerals in lentils, genetic improvement via biofortification has been developed as a sustainable and long-term approach. This is a low-cost method of reducing mineral deficits.

7.2.1 Plant Ionomics

This is an emerging technique for gene networks and mineral transport. Such assays require technical skills and sophisticated analytical tools to consistently scale up the meaningful inferences from multidimensional data. By examining the minerals/micronutrients in their complex networks, ionomic profiling is emerging as an attractive field to study elemental collection in living systems at the genome level (Baxter 2010). Salt et al. (2008) defined *ionome* as the "elemental composition of a living system constituting the inorganic sector." This technique, which uses nuclear properties or electronics, is currently being used to explore the elemental composition of various microorganisms (Singh et al. 2013). The ionome data sets of various organisms are available at http://www.ionomicshub.org/home/PiiMS.

Phenomics—the latest discipline—enables scientists to quickly obtain precise, reliable phenotyping data (Cobb et al. 2013). Phenomics platforms are automated, less strenuous and remarkably accurate, and it facilitates large-scale phenotypic screening. The following techniques are useful for legume biofortification:

- AAS
- Ion-beam analysis (IBA)
- X-ray fluorescence spectroscopy (XRF)
- ICP mass spectroscopy (ICP-MS)
- Laser capture microdissection (LCM) and synchrotron-based X-ray fluorescence microscopy (SXRF)
- Neutron activation analysis (NAA), among others (Djingove et al. 2013; Singh et al. 2013)

These sophisticated techniques are useful for studying the mineral dynamics of legumes and for ion profiling the mutants of model legume species (e.g., *L. japonicas*) (Chen et al. 2009).

7.3 Biofortification of Amino Acids

In humans, indispensable (essential) amino acids are not synthesized de novo; therefore, they must be consumed in the human diet. The nine essential amino acids are as follows: phenylalanine, threonine, valine, tryptophan, leucine, methionine, isoleucine, histidine, and lysine. Many crops are deficient in certain essential amino acids, such as threonine and lysine, whereas legumes (peas and beans) lack cysteine and methionine. The staple foods of most people in the world are composed of legume-and-cereal combinations, so novel varieties that maximize essential amino acid content need to be developed.

Using a recombinant storage protein is a simple approach to express the desirable profiles of amino acids—e.g., the expression of legumin (high lysine content) in peas (Molvig et al. 1997). Using an endogenous storage protein is an alternative engineering technique, like using a legume globulin, where in vitro mutagenesis is performed to mutate the codons of appropriate amino acids into essential amino acid codons (e.g., lysine and methionine). Alternatively, the insertion of additional codons into these amino acids appears promising, such as the inclusion of the ASP1 protein (synthetic protein) design to make a stable protein-like structure with whichever desired amino acids.

7.4 Nutrigenomic Biofortification Approaches

The large-scale genome sequencing data of plants has been completed thanks to deep bioinformatics analysis, to gain a deeper grasp of metabolic pathways. This is the main reason for the genesis of nutrigenomics. Nutrigenomics is a new tool to study the complex biochemical pathways of plants. The basic underlying mechanisms of the nutrients are synthesis and the accumulation of essential minerals and vitamins in plant tissues. These mechanisms increase specific micronutrient levels in the crops. After identifying the gene of interest, it is transferred into a crop species so that this species will exhibit the desired change in the nutritional content of the target tissues. This approach creates a new breeding trait that wasn't present in the germplasm.

7.4.1 Next-Generation Sequencing (NGS) Technology

NGS technology plays a central role in a breeding pipeline in that it hastens the precision of trait transfer and trait mapping. Five years ago, "second-generation technologies (SGT) like GS FLX Titanium/GS Junior, Roche/454 FLX Pyrosequencer, Solid Sequencer (Applied Biosystems), and Genome Analyzer (Solexa/Illumina) and third-generation sequencing (TGS) like Ion Torrent PGM/Proton (Life Technologies) and HiSeq/MiSeq from Illumina and Oxford Nanopore Technologies" were incorporated into classical Sanger sequencing technology (Bouis and Saltzman 2017). They have gained immense popularity because of their high throughput, low sequencing cost, and read lengths.

NGS is a de novo sequencing technology that includes "whole-genome resequencing (WGRS), whole-genome sequencing (WGS), quantitative trait mapping, Genome wide associations (GWAS), TILLING study, mutational map (MutMap), genotyping by sequencing (GBS), genomic selection (GS), whole-genome bisulfite sequencing (WGBS), reverse and fast-forward genetics analysis, Epigenetif quantitative trait loci (epiQTL) analysis, transcriptomics/differential gene expression and epigenetic (Methylated DNA immunoprecipitation (MeDIP); Chromatin immunoprecipitation (ChIP)/analysis, small Ribonucleic acid (RNA) profiling, restriction-site-associated DNA sequencing (RAD-seq), SHORE map, Exome sequencing, and QTL-seq technology (modified bulk segregant analysis)." These are widely utilized in marker-trait studies to generate "genome-wide genetic markers" and "fixed SNP genotyping arrays." Single Nucleotide Polymorphisms/haplotypes and QTL genome-resequencing approaches use DNA data from various mapping panels of crops species to identify the markers associated with agronomically important traits.

Mineral levels in legumes are affected by soil type, climatic circumstances, and weather. In the United States, a two-year study looked at a variety of micronutrients in red and green lentil genotypes. Micronutrient concentrations have been observed to differ significantly over time. Genotype–location interaction is nonsignificant (Rahman et al. 2014). For breeding, environmental factors, particularly soil type, create complications in high-Zn concentrations (Trethowan 2007). Environmental influences (soil fertility and soil type) and crop/plant characteristics (genes, seed composition, and seed characteristics) have significant impacts on mineral properties.

7.5 *Germplasm/Transgenic Biofortification*

Biofortification can be increased by using a germplasm containing cultivars, breeding lines, landraces, and wild species, which can be used for direct planting on a farmer's fields or as donors in breeding projects. This is a recent addition to breeding goals: to improve the nutritional quality of edible tissues. Germplasm screening predicts that Fe and Zn contents could be doubled in staple foods via conventional breeding. To combat hunger, developing better varieties with the highest nutritional value is a low-cost and one-time investment option.

The 240 germplasm lines panel were investigated in pigeon peas. These lines exhibited nearly four times more Zn density than the line with low Zn content (Basavarajeshwari et al. 2014). Research efforts based on "molecular marker-assisted" selection in pulse biofortification have started in Canada and the United States. In this development, high concentrations of micronutrient loci are linked with molecular markers. Mapping the gene(s)/QTL(s) and tagging them control the micronutrient status of the legumes, most of which operate in the quantitative mode of inheritance. This led to the discovery of gene(s)/QTL(s) that can explain a moderate degree of "phenotypic variation" in the micronutrient concentrations of common beans (Sompong et al. 2012) and the phytic acid content in mungbeans (Blair et al. 2010).

7.5.1 Alien Genes

Prebreeding wild species includes a large number of beneficial alien genes that are not found in the cultivated gene pool (Tanksley and McCouch 1997). Efforts to protect and collect wild relatives of several food legume crops in inter(national) gene banks are ongoing (FAO 2006). Efforts have been made to explore the genes of cultivated species and those of their wild relatives that impart resistance to multiple stresses and traits. The introgression success of alien genes in wild relatives has been limited to only a few diseases—namely those that are controlled by major/head gene(s). These major genes are accessible in the "crossable primary gene pool" (Hajjar and Hodgkin 2007).

To broaden and diversify the genetic base of cultivated germplasms, the introgression of the alien genes from wild species needs to be carried out. This will not only minimize the risk of stresses but will also advance the yield of legumes. Advancements in tissue cultures can facilitate alien gene introgression. Furthermore, evolution has changed the accessions of "wild species" toward possible cross-compatibility with "cultivated species." In this way, earlier cross-incompatible "wild species" may now cross with "cultivated species." Therefore, prebreeding efforts are immediately required, particularly for wild species, which contain useful "alien genes" that can improve the quality, yield, and stress resistance of legume species.

7.5.2 Cross Ability Group and Gene Pool Knowledge

For successful prebreeding, crossability between wild and farmed species is required. The crosses between wild species of lentils and crop species of lentils are grouped into "primary (*Lens culinaris* sp., *L. orientalis*, and *L. odemensis*), secondary (*L. nigricans* and *L. ervoides*), and tertiary gene pools (*L. tomentosus* and *L. lamottei*)" (McPhee and Hancock 2005). The traditional approach to nutrification favored the announcement of nutrient-rich "new genotypes" in breeding programs that can recover superior recombinants/segregants at higher nutritional values (Hirschi 2009). An outstanding variety of genetic tools for legumes has been uncovered thanks to advancements in genomics (Varshney et al. 2015). The platforms of next-generation sequencing (NGS) have paved the way for the rapid and cost-effective genome sequencing of different crops. HarvestPlus is a global initiative to enhance the nutritional statuses of people by using particular omics tools (Andersson et al. 2014).

Exploring the genotypes that might operate as "possible donors" in downstream breeding schemes is as simple as accessing the natural variants in a crop gene pool to determine the minerals and micronutrients of interest (Dwivedi et al. 2012). This genetic variability is tested in legume grains to determine the size of the "exploitable range" of minerals such as B, Ca, Cu, Zn, Fe, K, Mg, P, S, Mn, and Na (Beebe et al. 2000). Marked variations in Zn and Fe contents were observed in 16 mungbean genotypes (Taunk et al. 2012). The nonviable seeds of hybrids were generated from within-species crosses from the primary gene pool, using species such as *L. nigricans* and *L. ervoides*, thanks to irregular meiosis (Ladizinsky et al. 1985). The use of rescued embryos produces viable seeds in hybrids via crossing among *L. ervoides* and *L. culinaris* (Ladizinsky et al. 1985). By crossing with other "groups of species," the *L. tomentosus*—a single species group from a tertiary gene pool—does not produce viable seeds in hybrids. The identification of the primary gene pool is useful for making "genetic improvements" by producing new variability. The frequency of useful genes is higher in this pool than in the secondary and tertiary gene pools (Tullu et al. 2006).

Hybridizations between *L. nigricans* and *L. culinaris ssp. Culinaris* and between *L. ervoides* and *L. odemensis* are viable thanks to embryo rescue (Fratini and Ruiz 2006), where *L. odemensis* belong to the "secondary gene pool" and *L. ervoides* and *L. nigricans* belong to the "tertiary gene pool" (Ladizinsky and Abbo 1993). Therefore, releasing the hidden variability of wild species by determining their nutritional characteristics depends on their cross-compatibility with cultivated species (Muehlbauer et al. 2006). Hybrid fertility is determined by the wild parent's chromosomal arrangement (Ladizinsky et al. 1984). In the *L. culinaris* cross with *L. culinaris ssp. Orientalis*, each species is genetically isolated from the other species (Gupta et al. 2011).

7.6 *Marker-Assisted Breeding for Biofortification*

Marker-assisted selection (MAS) is a current alternative to genetic engineering. This increases the levels of provitamin A (beta carotene). Crop scientists have identified the genetic markers that are associated with the maximum level of provitamin A. Significant progress can be achieved through conventional breeding programs. However, transgenic approaches are potentially more advantageous than conventional breeding (Paine et al. 2005).

7.7 *Biofortification by Microbes*

Endophytic and rhizospheric microbes play important roles in crop biofortification (Ku et al. 2019). Legumes in symbiotic relationships can fix environmental N, which reduces the need for nitrogenous fertilizer, in turn decreasing agricultural CO_2 emissions and improving soil composition (Karkanis et al. 2018). The insertion of legumes into crop-rotating systems reduces greenhouse gas emissions (by up to 25%) (Ma et al. 2018). This is an ecofriendly approach to cope with the climatic effects on legumes and improve plant nutrition. Using plant growth–promoting bacteria (PGPB) is a promising green technology for the enhancement of nutrient content, from a biofortification perspective.

8 Anatomical Density of Nutrients in Legumes

Anatomical observations of seed structures indicate the micronutrient density of legumes. The seed has a thick and maternally derived testa, or seed coat, that surrounds the expanded (stable and fairly large) cotyledons and well-developed embryo. This seed structure radically differs from that of cereals, where an aleurone (thin, maternally derived) layer surrounds the specialized endosperm tissue and a less-developed embryo (in bran) (Petry et al. 2016).

Legume seeds make up 7–10% or more of the weight of beans; cotyledons make up 85% or more; and embryos make up just 2–3% of the weight of the seeds (but they are dense in nutrients) (Ariza-Nieto et al. 2007). The size of seeds varies more than their seed coat thickness does, so the legumes with smaller seeds (especially wild relatives of beans) tend to have higher percentages of seed coat compared with the cotyledons of cultivated beans. These different seed tissues of the common bean seed are targeted for biofortification (seed coat, cotyledons, and embryos). Mineral distribution, mineral inheritance, and loading minerals into each tissue are essential for making progress in breeding systems (Blair et al. 2013).

9 Mechanism of Nutrient Transportation

The acquisition, accumulation, and transport of minerals are required to maximize the bioavailability of nutrients in legume seeds. A few genes control these processes. Many genes have been identified, such as those involved in translocation to various vegetative tissues and finally to seeds (Jeong et al. 2017). However, mineral mobilization and loading minerals into different sink tissues are the tasks of many phloem-expressed genes (Braun et al. 2014). Specific transporters are required in whole-plant studies to ascertain the transporters that govern the delivery of minerals to seeds (González-Guerrero et al. 2016).

9.1 Iron Transportation

Nearly two billion people in the world experience anemia because the prolonged consumption of iron-limited diets leads to iron deficiency, or anemia (Murgia et al. 2012). Therefore, the ascorbate content in plant foods should be increase by the use of the rDNA technique, which reduces the negative impact of polyphenols and phytate in staple foods on the bioavailability of iron. This also makes food a vital source of vitamin C and other nutrients (Graham et al. 2001). Similarly, lathyrogens, saponins, proteases, and α-amylase inhibitors are in legumes, which minimize the bioavailability of micronutrients. A moderate quantity of saponin provides various health benefits, whereas large quantities of it are associated with negative symptoms such as diarrhea, abdominal pain, and vomiting (Potter et al 1980). The traditional methods, such as soaking, sprouting, and cooking, change the concentration of saponin in food.

When iron in its Fe^{3+} form is embedded in soil, it cannot be absorbed by plants, especially those in calcareous soils (Miethke and Marahiel 2007). Ascorbic acid minimizes Fe^{3+} by forming Fe(III) complexes and enhances more-soluble and more-bioavailable Fe^{2+}. Fe deficiency forces crops to grow in alkaline soil (Li et al. 2016). Plants use two strategies for Fe absorption.

Strategy I (found in nongraminaceous monocots and dicots) is characterized by the emission of protons, resulting in the acidification of the rhizosphere. This converts Fe^{3+} to Fe^{2+} by using ferric chelate reductase (FC-R)—i.e., FRO2. This transports iron (Fe^{2+}) from the root's plasma membrane to its epidermal cells by using "iron-regulated transporter 1" (IRT1), "divalent metal ion transporter 1" DMT1, or "natural resistance associated macrophage protein 1" NRAMP1 (Marschner 2012). In legumes, the Fe-uptake and Fe-transportation proteins are HA2, FRO2, and IRT1 (Santi and Schmidt 2009). The iron transport genes, which transport genes from leaves to roots, are FIT1, IRT1, bZIP23, and OPT3, which have been identified in many legumes (Xiong et al. 2012), such as *Medicago truncatula* (Tejada-Jiménez et al. 2015), soybeans (Qin et al. 2017), lentils, and chickpeas (Parveen et al. 2016). Unloading xylem to different tissues and sink cells is facilitated by the ZRT/IRT-like protein transporters (Küpper and Kochian 2010). A gene family such as NRAMP plays a significant role in iron homeostasis, while olympic protein technologies (OPTs) and yellow strip like (YSLs) are important in the loading and unloading of Fe^{2+} complexes into and out of phloem (Palmer and Guerinot 2009).

Strategy II (found in graminaceous monocots) is characterized by the formation of the phytosiderophores responsible for Fe chelation. Fe-chelating complexes are transported to the roots' epidermic cells (Gendre et al. 2006). The low 1–2 kDa weight of the organic compounds, in this case microbial siderophores, is required for the Fe uptake produced by PGPB under Fe-limiting conditions (Payne 1994). They bind Fe^{3+} at high specificity and increase Fe solubility, and this complex connects to specific receptors of bacteria. This process is an easy way to absorb Fe and convert Fe^{3+} into Fe^{2+} (Boukhalfa and Crumbliss 2002). Bacterial siderophores prepare the strategy I plants (in which legumes are included) for Fe uptake. The inoculation of chickpeas with siderophore-producer PGPB increases the Fe concentration, which increases the overall seed nutrient content (Sathya et al. 2016).

9.2 Zinc Transportation

In legumes, the uptake of Zn^{2+} and the transportation of root cells across the plasma membrane to seeds is facilitated via ZIP transporters (Palmgren et al. 2008). A ZRT/IRT-like protein (ZIP), a zinc-induced facilitator (ZIF), a heavy metal ATPase (HMA), and metal tolerance proteins (MTP) are involved in Zn transport (Hussain et al. 2004).

For the most part, Zn is transported through the symplastic pathway, but a considerable fraction of it may follow the apoplastic pathway through roots to reach the xylem (White et al. 2002). MTP1 and ZIF1 are cation diffusion facilitator (CDF) transporters that transport Zn to the vacuole, while NRAMPs mobilize the Zn from the vacuole (Haydon and Cobbett 2007) and HMAs load the Zn into the xylem (Palmgren et al. 2008). At the same time, a ZIP mediates the Zn^{2+} influx into the phloem and leaf tissues, while YSLs load Zn to the phloem and unload it onto the seeds as Zn-NA complexes (Waters and Grusak 2008).

9.3 Manganese (Mn) Transportation

Manganese serves as a cofactor in many processes, such as lipid biosynthesis and photosynthesis. The Mn^{2+} form is available in soil for plant uptake. Important transporters for Mn include NRAMP, CDF/MTP, YSL, IRT1, P-Type-ATPase, and VIT (vacuolar iron transporter) (Socha and Guerinot 2014). These transporters have broad specificity for other divalent cations (Cd, Ca, Co, Zn, Fe, Cu, and Ni). ZIP1 remobilizes the Mn from a plant's vacuoles and translocates it through the root vasculature to its shoots (Milner et al. 2013).

9.4 Phosphorus (P) Transportation

Phosphorus, in the form of phosphate (Pi) in soil, is taken from soil to a plant's root epidermal cells and impelled via H^+-ATPases. Arbuscular mycorrhizal fungi play important roles in legumes for nutrient acquisition (Bucher 2007). The numerous Pht1 genes of roots, aerial parts, and seeds imply their potential involvement in internal Pi translocation. Pht1 transporters are involved in transferring Pi into cells, while other families (Pht2, Pht3, and Pht4) are linked to transferring Pi within intercellular membranes.

9.5 Copper (Cu) Transportation

The uptake of copper and iron from soil follows strategies similar to those for nutrient transportation. The Cu in soil is available as Cu^{2+}, which is transported to root cells in a reduced form (Cu^+). The family of copper transporters (COPTs) transport the Cu through the root cells (Ryan et al. 2013). In other words, after the reduction of Cu^{2+} to Cu^+, Cu^+ is transported through the roots by COPT proteins. The COPT proteins of legumes have not yet been studied in detail (Figs. 2.1, 2.2 and Table 2.1).

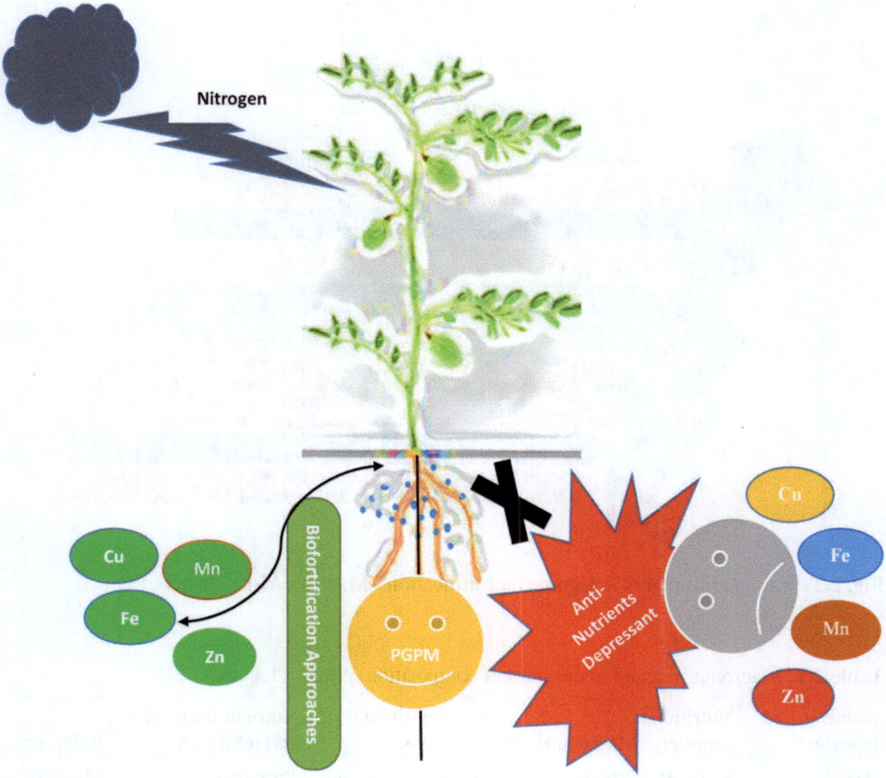

Fig. 2.1 Summary of legume biofortifications to enhance nutrients through atmospheric nitrogen fixation, PGPM, and other approaches to fortification in the presence of antinutrients, which hinder the transportation of nutrients. (Made by Aisha Umar)

10 Conclusion

Legumes are consumed as food all over the world. They contain many nutrient components, namely proteins, minerals, and vitamins. Biofortification has played a vital role in making available crucial micronutrients that can counteract malnutrition deficiencies. Malnutrition is a critical health issue in developing countries. Thus, biofortification, the most suitable method of adding nutrients, is needed to further improve legume crops. Micronutrient availability depends mainly on soil texture, soil temperature, the fauna of soil, and the availability of micronutrients (proteins, minerals, and vitamins) to legume crops.

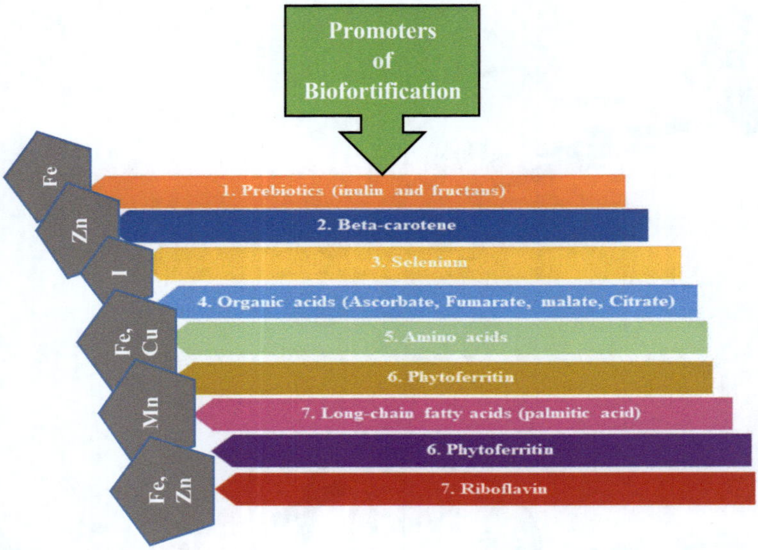

Fig. 2.2 Nutrient promoters in legume biofortification. (Made by Aisha Umar)

Table 2.1 Macronutrient and micronutrient composition of major legumes

Name (species)	Nutrition supplier	Macronutrients	Line/cultivar/ accession	Nutrient promoters and inhibitors	Reference
Pea (*P. sativum*)	S, Zn, B, Ca, Fe, K, Mg, Mn, Mo, P, S, Zn, Cu, Ni	Protein (18–31%), oil (0.6–5.5%), starch (45%), fiber (12%), sucrose (2.1%)	Aragorn, Kiflica	*Promoters* 1. Prebiotics such as inulin and fructans 2. Beta carotene 3. Selenium 4. Organic acids 5. Amino acids 6. Phytoferritin 7. Long-chain fatty acids (palmitic acid) 8. Riboflavin *Inhibitors* 1. Phytic acids 2. Fiber (cellulose, hemicellulose, lignin, cutin, suberin, etc.) 3. Phenolics 4. Haemagglutinins 5. Heavy metals 6. Tannins 7. Oxalic acids 8. Goitrogens	Ma et al. (2017)

(continued)

Table 2.1 (continued)

Name (species)	Nutrition supplier	Macronutrients	Line/cultivar/ accession	Nutrient promoters and inhibitors	Reference
Mungbean (*V. radiata*)	Fe	Protein (22–23%), oil (1.2%), starch (45%), fiber (7%), sucrose (1.1%)	Kamphaeng Saen 2, NM10–12		Prathet et al. (2012)
Common bean (*P. vulgaris* L.)	P, Fe, Zn, Fe, Ca, Cu, Ni, Mo	Protein (20–27%), oil (0.9–2.4%), starch (41.5%), fiber (10%), sucrose (5%)	G2333, G19839		Katuuramu et al. (2018)
Chickpea (*C. arietinum*)	Fe, Zn	Protein (15.5–28.2%), oil (3.1–7%), starch (44.4%), fiber (9%), sucrose (2%)	ICC 4958, ICC8261		Upadhyaya et al. (2016)
Soybean (*G. max*)	Fe,	Protein (35–42%), oil (17–21%), starch (1.5%), fiber (20%), sucrose (6.2%)	Merill		Khazaei et al. (2017)
Lentil (*L. culinaris*)	Fe, Zn, Se, Mn	Protein (23–32%), oil (0.8–2%), starch (46%), fiber (12%), sucrose (2.9%)	CDC Redberry, ILL7502		Ates et al. (2018)
Bird's-foot trefoil (*L. japonicus*)	Ca, Cu, Fe, K, Mg, Mn, Ni, P, S, Zn, Ca, Cu	–	Miyakojima MG-20, Gifu B-129		Klein et al. (2012)
Cowpea (*Vigna unguiculata*)	Fe, Zn	Protein (23.5%), oil (1.3%)	–		López-Morales et al. (2020)

In this chapter, we discussed agronomic biofortification, nutrigenomic biofortification, breeding biofortification, and microbial biofortification. These branches of biofortification offer promising enhancements to the nutrients in legumes. We also deliberated on the mechanisms of nutrient transportation. Legumes can counteract major protein demands, but a lack of experimentation limits the extent of legume biofortification and its applications.

References

Abbaspour N, Hurrell R, Kelishadi R (2014) Review on iron and its importance for human health. J Res Med Sci 19:164–174

Alvarez JM (2007) Influence of soil type on the mobility and bioavailability of chelated zinc. J Agric Food Chem 55:3568–3576. https://doi.org/10.1021/jf063236g

Amarakoon D, Thavarajah D, McPhee K, Thavarajah P (2012) Iron-, zinc-, and magnesium-rich field peas (*Pisum sativum* L.) with naturally low phytic acid: a potential food-based solution to global micronutrient malnutrition. J Food Compos Anal 27:8–13

Andersson MS, Pfeiffer WH, Tohme J (2014) Enhancing nutritional quality in crops via genomics approaches. In: Tuberosa R et al (eds) Genomics of plant genetic resources. Springer, Dordrecht, pp 417–429

Ariza-Nieto M, Blair MW, Welch RM, Glahn RP (2007) Screening of bioavailability patterns in eight bean (*Phaseolus vulgaris* L.) genotypes using the Caco-2 cell in vitro model. J Agr Food Chem 55:7950–7956

Asif M, Rooney LW, Ali R, Riaz MN (2013) Application and opportunities of pulses in food system: a review. Crit Rev Food Sci Nutr 53(11):1168–1179

Ates D, Aldemir S, Yagmur B, Kahraman A, Ozkan H, Vandenberg A, Tanyolac MB (2018) QTL mapping of genome regions controlling manganese uptake in lentil seed. G3: Genes Genomes Genet 8:1409–1416

Barbut S (1999) Determining water and fat holding. In: Methods of testing protein functionality, pp 186–225

Basavarajeshwari R, Rajashekar Reddy BH, Manoa I, Deepika M, Shankar AG (2014) Development of mapping population for grain zinc content in pigeonpea. In: 2nd International Conference on Agricultural & Horticultural Sciences, February 03–05, 2014, Hyderabad, India

Baxter I (2010) Ionomics: the functional genomics of elements. Brief Funct Genom 9:149–156

Beebe S, Gonzalez AV, Rengifo J (2000) Research on trace minerals in the common bean. Food Nutr Bull 21:387–391

Biehl RR, Baker DH, DeLuca HF (1995) 1α-hydroxylated cholecalciferol compounds act additively with microbial phytase to improve phosphorus, zinc and manganese utilization in chicks fed soy-based diets. J Nutr 125(9):2407–2416

Blair MW, Medina JI, Astudillo C, Rengifo J, Beebe SE, Machado G, Graham R (2010) QTL for seed iron and zinc concentrations in a recombinant inbred line population of Mesoamerican common beans (*Phaseolus vulgaris* L.). Theor Appl Genet 121:1059–1071

Blair MW, Izquierdo P, Astudillo C, Grusak MA (2013) A legume biofortification quandary: variability and genetic control of seed coat micronutrient accumulation in common beans. Front Plant Sci 4:275

Bohra A, Pandey MK, Jha UC, Singh B, Singh IP, Datta D, Chaturvedi SK, Nadarajan N, Varshney RK (2014) Genomics-assisted breeding in the four major pulse crops of developing countries: present status and prospects. Theor Appl Genet 127:1263–1291

Bohra A, Sahrawat KL, Kumar S, Joshi R, Parihar AK, Singh U, Singh D, Singh NP (2015) Genetics- and genomics-based interventions for nutritional enhancement of grain legume crops: status and outlook. J Appl Genet 56:151–161

Borill P, Connorton JM, Balk J, Miller AJ, Sanders D, Uauy C (2014) Biofortification of wheat grain with iron and zinc: integrating novel genomic resources and knowledge from model crops. Front Plant Sci 5:53

Bouain N, Shahzad Z, Rouached A, Khan GA, Berthomieu P, Abdelly C, Poirier Y, Rouached H (2014) Phosphate and zinc transport and signalling in plants: toward a better understanding of their homeostasis interaction. J Exp Bot 65:5725–5741

Bouis HE, Saltzman A (2017) Improving nutrition through biofortification: a review of evidence from HarvestPlus, 2003 through 2016. Glob Food Sec 12:49–58

Bouis HE, Hotz C, McClafferty B, Meenakshi JV, Pfeiffer WH (2011) Biofortification: a new tool to reduce micronutrient malnutrition. Food Nutr Bull 32:S31–S40. https://doi.org/10.1177/15648265110321S105
Boukhalfa H, Crumbliss AL (2002) Chemical aspects of siderophore mediated iron transport. Biometals 15:325–339
Brady NC (2002) The nature and properties of soils. Prentice Hall of India Private Limited, Inc, Upper Saddle River, p 621
Braun DM, Wang L, Ruan YL (2014) Understanding and manipulating sucrose phloem loading, unloading, metabolism, and signalling to enhance crop yield and food security. J Exp Bot 65:1713–1735. https://doi.org/10.1093/jxb/ert416
Bucher M (2007) Functional biology of plant phosphate uptake at root and mycorrhiza interfaces. New Phytol 173:11–26. https://doi.org/10.1111/j.1469-8137.2006.01935.x
Carvalho SMP, Vasconcelos MW (2013) Producing more with less: strategies and novel technologies for plant-based food biofortification. Food Res Int 54:961–971
Cassandri M, Smirnov A, Novelli F, Pitolli C, Agostini M, Malewicz M et al (2017) Zinc-finger proteins in health and disease. Cell Death Disco 3:17071. https://doi.org/10.1038/cddiscovery.2017.71
Chen Z, Watanabe T, Shinano T, Okazaki K, Osaki M (2009) Rapid characterization of plant nutrients with an altered ion-profile: a case study using Lotus japonicus. New Phytol 181:795–801
Choudhary AK, Suri VK (2009) Effect of organic manures and inorganic fertilizers on productivity, nutrient uptake and soil fertility in rice (Oryza sativa)–wheat (Triticum aestivum) crop sequence in western Himalayas. Curr Adv Agric Sci 1(2):65–69
Choudhary AK, Suri VK (2014) 'On farm' participatory technology development on forage cutting and nitrogen management in dual-purpose wheat (Triticum aestivum) in North–Western Himalayas. Commun Soil Sci Plant Anal 45(6):741–750
Cobb JN, DeClerck G, Greenberg A, Clark R, McCouch S (2013) Nextgeneration phenotyping: requirements and strategies for enhancing our understanding of genotype–phenotype relationships and its relevance to crop improvement. Theor Appl Genet 126:1–21
Das DK (2011) Introductory soil science. Kalyani Publishers, New Delhi, p 645
Dass A, Chandra S, Choudhary AK, Singh G, Sudhishri S (2015) Influence of field re-ponding pattern and plant spacing on rice root-shoot characteristics, yield, and water productivity of two modern cultivars under SRI management in Indian Mollisols. Paddy Water Environ. https://doi.org/10.1007/s10333-015-0477-z
Deb DL, Sakal R, Datta SP (2009) Fundamentals of soil science. Indian society of soil science. Cambridge Printing Works, New Delhi, p 728. ISBN 8190379747
Djingove R, Mihaylova V, Lyubomirova V, Tsalev D (2013) Multielement analytical spectroscopy in plant ionomics. Res Appl Spectrosc Rev 48:384–424
Dwivedi SL, Sahrawat KL, Rai KN, Blair MW, Andersson M, Pfieffer W (2012) Nutritionally enhanced staple food crops. Plant Breed Rev 34:169–262
Erdal I, Yilmaz A, Taban S, Eker S, Torun B, Cakmak I (2002) Phytic acid and phosphorus concentrations in seeds of wheat cultivars grown with and without zinc fertilization. J Plant Nutr 25:113–127
Fageria NK, Valigar VC, Wright RJ (1997) Soil environment and root growth dynamics of field crops. Recent Res Dev Agron 1:15–58
FAO (2006) Food and agriculture statistics global outlook (Produced by the Statistics Division). http://faostat.fao.org/Portals/_Faostat/documents/pdf/world.pdf. Accessed 12 Jan 2015
Farzana T, Mohajan S, Saha T (2017) Formulation and nutritional evaluation of a healthy vegetable soup powder supplemented with soy flour, mushroom, and moringa leaf. Food Sci Nutr 5:911–920
Ferreira H, Vasconcelos M, Gil AM, Pinto E (2020) Benefits of pulse consumption on metabolism and health: a systematic review of randomized controlled trials. Crit Rev Food Sci Nutr:1–12
Finn S (2014) Nutrition insecurity and malnutrition in developed countries. In: Addressing malnutrition to improve global health. Science/AAAS, Washington, DC

Fratini R, Ruiz ML (2006) Interspecific hybridization in the genus Lens applying in vitro embryo rescue. Euphytica 150:271–280

Garg M, Sharma N, Sharma S, Kapoor P, Kumar A, Chunduri V et al (2018) Biofortified crops generated by breeding, agronomy, and transgenic approaches are improving lives of millions of people around the world. Front Nutr 5:12. https://doi.org/10.3389/fnut.2018.00012

Gendre D, Czernic P, Conéjéro G, Pianelli K, Briat J-F, Lebrun M, Mari S (2006) TcYSL3, a member of the YSL gene family from the hyper-accumulator thlaspi caerulescens, encodes a nicotianamine-Ni/Fe transporter. Plant J 49:1–15

Godecke T, Stein AJ, Qaim M (2018) The global burden of chronic and hidden hunger: trends and determinants. Glob Food Sec 17:21–29. https://doi.org/10.1016/j.gfs.2018.03.004

González-Guerrero M, Escudero V, Saéz Á, Tejada-Jiménez M (2016) Transition metal transport in plants and associated endosymbionts: arbuscular mycorrhizal fungi and rhizobia. Front Plant Sci 7:1088

Graham RD, Welch RM, Bouis HE (2001) Addressing micronutrient malnutrition through enhancing the nutritional quality of staple foods: principles, perspectives and knowledge gaps. Adv Agron 70:77–142

Gupta M, Tiwari BK, Norton T (2011) Value addition and international trade. In: Pratap A, Kumar J (eds) Biology and breeding of food legumes. CABI, Oxfordshire, pp 395–404

Gupta DS, Thavarajah D, Knutson P, Thavarajah P, McGee RJ, Coyne CJ, Kumar S (2013) Lentils (*Lens culinaris* L.) a rich source of folates. J Agric Food Chem 61:7794–7799

Hajjar R, Hodgkin T (2007) The use of wild relatives in crop improvement a survey of developments over the last 20 years. Euphytica 156:1–13

Harrier LA, Watson CA (2003) The role of arbuscular mycorrhizal fungi in sustainable cropping systems. Adv Agron 42:185–225

Haydon MJ, Cobbett CS (2007) Transporters of ligands for essential metal ions in plants. New Phytol 174:499–506. https://doi.org/10.1111/j.1469-8137.2007.02051.x

Hirschi KD (2009) Nutrient biofortification of food crops. Annu Rev Nutr 29:401–421

Hussain D, Haydon MJ, Wang Y, Wong E, Sherson SM, Young J et al (2004) P-type ATPase heavy metal transporters with roles in essential zinc homeostasis in Arabidopsis. Plant Cell 16:1327–1339. https://doi.org/10.1105/tpc.020487

Jeong J, Merkovich A, Clyne M, Connolly EL (2017) Directing iron transport in dicots: regulation of iron acquisition and translocation. Curr Opin Plant Biol 39:106–113. https://doi.org/10.1016/j.pbi.2017.06.014

Joshi-Saha A, Reddy KS (2015) Repeat length variation in the 5'UTR of myo-inositol monophosphatase gene is related to phytic acid content and contributes to drought tolerance in chickpea (*Cicer arietinum* L.). J Exp Bot 66:5683–5690. https://doi.org/10.1093/jxb/erv156

Kabir AH, Paltridge N, Stangoulis J (2016) Chlorosis correction and agronomic biofortification in field peas through foliar application of iron fertilizers under Fe deficiency. J Plant Interact 11:1–4. https://doi.org/10.1080/17429145.2015.1125534

Karan AK, Kar S, Singh VK, Singh CV (2014) Effects of liming and soil moisture regimes on time changes of soil pH, redox potential, availability of native sulphur and micronutrients to rice (*Oryza sativa* L.) in acid soils. Int J Soil Sci 9(1):1–15

Karkanis A, Ntatsi G, Lepse L, Fernández JA, Vågen IM, Rewald B, Alsin AI, Kronberga A, Balliu A, Olle M et al (2018) Faba bean cultivation—revealing novel managing practices for more sustainable and competitive European cropping systems. Front Plant Sci 9:1115

Katuuramu DN, Hart JP, Porch TG, Grusak MA, Glahn RP, Cichy KA (2018) Genome-wide association analysis of nutritional composition-related traits and iron bioavailability in cooked dry beans (*Phaseolus vulgaris* L.). Mol Breed 38:44

Kaur H, Shukla RK, Yadav G, Chattopadhyay D, Majee M (2008) Two divergent genes encoding L-myo-inositol 1-phosphate synthase1 (CaMIPS1) and 2 (CaMIPS2) are differentially expressed in chickpea. Plant Cell Environ 31:1701–1716. https://doi.org/10.1111/j.1365-3040.2008.01877.x

Khan A, Singh J, Upadhayay VK, Singh AV, Shah S (2019) Microbial biofortification: a green technology through plant growth promoting microorganisms. In: Sustainable green technologies for environmental management. Springer, Singapore
Khazaei H, Podder R, Caron CT, Kundu SS, Diapari M, Vandenberg A, Bett KE (2017) Marker-trait association analysis of iron and zinc concentration in lentil (Lens culinaris Medik.) seeds. Plant Genome 10
Klein MA, Grusak MA (2009) Identification of nutrient and physical seed trait QTL in the model legume Lotus japonicus. Genome 52:677–691. https://doi.org/10.1139/g09-039
Klein MA, López-Millán A-F, Grusak MA (2012) Quantitative trait locus analysis of root ferric reductase activity and leaf chlorosis in the model legume, Lotus japonicus. Plant Soil 351:363–376
Ku YS, Rehman HM, Lam H-M (2019) Possible roles of rhizospheric and endophytic microbes to provide a safe and affordable means of crop biofortification. Agronomy 9:764
Kumar S, Pandey G (2020) Biofortification of pulses and legumes to enhance nutrition. Heliyon 6(3):e03682
Kumar A, Suri VK, Choudhary AK (2014) Influence of inorganic phosphorus, VAM fungi and irrigation regimes on crop productivity and phosphorus transformations in Okra (*Abelmoschus esculentus* L.)–Pea (*Pisum sativum* L.) cropping system in an acid Alfisol. Commun Soil Sci Plant Anal 45:953–967
Kumar A, Maurya BR, Raghuwanshi R, Meena VS, Tofazzal Islam M (2017) Co-inoculation with enterobacter and rhizobacteria on yield and nutrient uptake by wheat (*Triticum Aestivum* L.) in the alluvial soil under indo-gangetic plain of India. J Plant Growth Regul 36:608–617
Kumar A, Kubota Y, Chernov M, Kasuya H (2020) Potential role of zinc supplementation in prophylaxis and treatment of COVID-19. Med Hypotheses 144:109848. https://doi.org/10.1016/j.mehy.2020.109848
Küpper H, Kochian LV (2010) Transcriptional regulation of metal transport genes and mineral nutrition during acclimatization to cadmium and zinc in the Cd/Zn hyperaccumulator, Thlaspi caerulescens (Ganges population). New Phytol 185(1):114–129
Ladizinsky G, Abbo S (1993) Cryptic speciation in Lens culinaris. Genet Resour Crop Evol 40:1–5
Ladizinsky G, Braun D, Goshen D, Muehlbauer FJ (1984) The biological species of the genus Lens L. Bot Gaz 145:253–261
Ladizinsky G, Cohen D, Muehlbauer FJ (1985) Hybridization in the genus Lens by means of embryo culture. Theor Appl Genet 70:97–101
Lambert SA, Jolma A, Campitelli LF, Das PK, Yin Y, Albu M et al (2018) The human transcription factors. Cell 172:650–665. https://doi.org/10.1016/j.cell.2018.01.029
Li M, Ahammed GJ, Li C, Bao X, Yu J, Huang C, Yin H, Zhou J (2016) Brassinosteroid ameliorates zinc oxide nanoparticles-induced oxidative stress by improving antioxidant potential and redox homeostasis in tomato seedling. Front Plant Sci 7:615
Loneragan JF, Webb MJ (1993) Interactions between zinc and other nutrients affecting the growth of plants. In: Robson AD (ed) Zinc in soils and plants. Developments in plant and soil sciences, vol 55. Springer, Dordrecht. https://doi.org/10.1007/978-94-011-0878-2_9
López-Morales D, de la Cruz-Lázaro E, Sánchez-Chávez E, Preciado-Rangel P, Márquez-Quiroz C, Osorio-Osorio R (2020) Impact of agronomic biofortification with zinc on the nutrient content, bioactive compounds, and antioxidant capacity of cowpea bean (*Vigna unguiculata* L. Walpers). Agron 10:1460. https://doi.org/10.3390/agronomy10101460
Ma Y, Coyne CJ, Grusak MA, Mazourek M, Cheng P, Main D, McGee RJ (2017) Genome-wide SNP identification, linkage map construction and QTL mapping for seed mineral concentrations and contents in pea (*Pisum sativum* L.). BMC Plant Biol 17:43
Ma Y, Schwenke G, Sun L, Liu DL, Wang B, Yang B (2018) Modeling the impact of crop rotation with legume on nitrous oxide emissions from rain-fed agricultural systems in Australia under alternative future climate scenarios. Sci Total Environ 630:1544–1552
Margier M, Georgé S, Hafnaoui N, Remond D, Nowicki M, Du Chaffaut L, Amiot MJ, Reboul E (2018) Nutritional composition and bioactive content of legumes: characterization of pulses frequently consumed in France and effect of the cooking method. Nutrients 10:1668
Marschner P (2012) Marschner's mineral nutrition of higher plants. Elsevier, Amsterdam

Mcphee JB, Hancock RE (2005) Function and therapeutic potential of host defence peptides. Journal of peptide science: an official publication of the European Peptide Society, 11(11):677–687

Miethke M, Marahiel MA (2007) Siderophore-based iron acquisition and pathogen control. Microbiol Mol Biol Rev 71:413–451

Milner MJ, Seamon J, Craft E, Kochian LV (2013) Transport properties of members of the ZIP family in plants and their role in Zn and Mn homeostasis. J Exp Bot 64:369–381. https://doi.org/10.1093/jxb/ers315

Molvig L, Tabe LM, Eggum BO, Moore AE, Craig S, Spencer D, Higgins TJV (1997) Enhanced methionine levels and increased nutritive value of seeds of transgenic lupins (*Lupinus angustifolius* L.) expressing a sunflower seed albumin gene. Proc Natl Acad Sci U S A 94:8393–8398

Muehlbauer FJ, Cho S, Sarker A, McPhee KE, Coyne CJ, Rajesh PN, Ford R (2006) Application of biotechnology in breeding lentil for resistance to biotic and abiotic stress. Euphytica 147:149–165

Murgia I, Arosio P, Tarantino D, Soave C (2012) Crops biofortification for combating 'hidden hunger' for iron. Trends Plant Sci 17:47–55

Nair RM, Yang RY, Easdown WJ, Thavarajah D, Thavarajah P, Hughes JD, Keatinge JD (2013) Biofortification of mungbean (*Vigna radiata*) as a whole food to enhance human health. J Sci Food Agric 93(8):1805–1813

Nath D, Maurya BR, Meena VS (2017) Documentation of five potassium- and phosphorus-solubilizing bacteria for their K and P-solubilization ability from various minerals. Biocatal Agric Biotechnol 10:174–181

Paine JA, Shipton CA, Chaggar S, Howells RM, Kennedy MJ, Vernon G et al (2005) A new version of golden rice with increased pro-vitamin A content. Nat Biotechnol 23:482–487

Palmgren MG, Clemens S, Williams LE, Krämer U, Borg S, Schjørring JK et al (2008) Zinc biofortification of cereals: problems and solutions. Trends Plant Sci 13:464–473. https://doi.org/10.1016/j.tplants.2008.06.005

Palmer CM, Guerinot ML (2009) Facing the challenges of Cu, Fe and Zn homeostasis in plants. Nat Chem Biol 5(5):333–340

Panzeri D, Cassani E, Doria E, Tagliabue G, Forti L, Campion B et al (2011) A defective ABC transporter of the MRP family, responsible for the bean lpa1 mutation, affects the regulation of the phytic acid pathway, reduces seed myo−inositol and alters ABA sensitivity. New Phytol 191:70–83. https://doi.org/10.1111/j.1469-8137.2011.03666.x

Parveen F, Jaafar NI, Ainin S (2016) Social media's impact on organizational performance and entrepreneurial orientation in organizations. Management Decision

Payne SM (1994) Detection, isolation, and characterization of siderophores. Methods Enzymol 235:329–344

Petry N, Rohner F, Gahutu JB, Campion B, Boy E, Tugirimana PL et al (2016) In Rwandese women with low iron status, iron absorption from low-phytic acid beans and biofortified beans is comparable, but low-phytic acid beans cause adverse gastrointestinal symptoms. J Nutr 146:970–975. https://doi.org/10.3945/jn.115.223693

Poblaciones MJ, Rengel Z (2016) Soil and foliar zinc biofortification in field pea (*Pisum sativum* L.): grain accumulation and bioavailability in raw and cooked grains. Food Chem 212:427–433. https://doi.org/10.1016/j.foodchem.2016.05.189

Potter JD, Illman RJ, Calvert GD, Oakenfull DG, Topping DL (1980) Soya saponins, plasma lipids, lipoproteins and fecal bile acids: a double blind cross-over study. Nutr Rep Int 22:521–528

Prasad AS, Miale A, Farid Z, Sandstead HH, Schulert AR (1963) Zinc metabolism in patients with the syndrome of iron deficiency anemia, hepatosplenomegaly, dwarfism, and hypogonadism. J Lab Clin Med 61:537–549

Prasad R, Shivay YS, Kumar D (2014) Agronomic biofortification of cereal grains with iron and zinc. Adv Agro 125:55–91. https://doi.org/10.1016/B978-0-12-800137-0.00002-9

Prasad R, Shivay YS, Kumar D (2016) Interactions of zinc with other nutrients in soils and plants – a review. Indian J Fert 12:16–26

Prathet P, Somta P, Srinives P (2012) Mapping QTL conferring resistance to iron deficiency chlorosis in mungbean [*Vigna radiata* (L.) Wilczek]. Field Crop Res 137:230–236

Prentice AM, Gershwin ME, Schaible UE, Keusch GT, Victoria LG, Gordon JI (2008) New challenges in studying nutrition disease interactions in the developing world. J Clin Invest 118:1322–1329

Qin Y, Song D, Chen H, Cheng W, Jiang G, Cottrell G (2017) A dual-stage attention-based recurrent neural network for time series prediction. arXiv preprint arXiv:1704.02971

Rahman MM, Erskine W, Siddique KHM, Thavarajah P, Thavarajah D, Zaman MS, Materne MA, Mcmurray LM (2014) Selenium biofortification of lentil in Australia and Bangladesh. In: 6th international food legume research conference, TCU Place, Sakatoon, 7–11 July 2014

Rao BN (2002) Pulses and legumes as functional foods. NFI Bull 23(1):1–4

Rastall RA, Gibson GR (2015) Recent developments in prebiotics to selectively impact beneficial microbes and promote intestinal health. Curr Opin Biotechnol 32:42–46. https://doi.org/10.1016/j.copbio.2014.11.002

Raymond J (2006) World's healthiest foods: lentils (India). Health Magazine. http://www.health.com/health/article/0,23414,1149140,00.html

Read SA, Obeid S, Ahlenstiel C, Ahlenstiel G (2019) The role of zinc in antiviral immunity. Adv Nutr 10:696–710. https://doi.org/10.1093/advances/nmz013

Rehman HM, Cooper JW, Lam H-M, Yang SH (2019) Legume biofortification is an underexploited strategy for combatting hidden hunger. Plant Cell Environ 42:52–70

Reilly C (1996) Biological role of selenium. In: Selenium in food and health. Blackie, London

Ryan BM, Kirby JK, Degryse F, Harris H, McLaughlin MJ, Scheiderich K (2013) Copper speciation and isotopic fractionation in plants: uptake and translocation mechanisms. New Phytol 199:367–378. https://doi.org/10.1111/nph.12276

Salt DE, Baxter I, Lahner B (2008) Ionomics and the study of the plant ionome. Annu Rev Plant Biol 59:709–733

Saltzman A, Birol E, Oparinde A, Andersson MS, Asare-Marfo D, Diressie MT et al (2017) Availability, production, and consumption of crops biofortified by plant breeding: current evidence and future potential. Ann N Y Acad Sci 1390:104–114. https://doi.org/10.1111/nyas.13314

Santi S, Schmidt W (2009) Dissecting iron deficiency-induced proton extrusion in Arabidopsis roots. New Phytol 183(4):1072–1084

Santos CAF, Boiteux LS (2013) Breeding biofortified cowpea lines for semi-arid tropical areas by combining higher seed protein and mineral levels. Genet Mol Res 12:6782–6789

Sathya A, Vijayabharathi R, Srinivas V, Gopalakrishnan S (2016) Plant growth-promoting actinobacteria on chickpea seed mineral density: an upcoming complementary tool for sustainable biofortification strategy. 3 Biotech 6:138

Shivay YS, Prasad R, Singh RK, Pal M (2015) Relative efficiency of zinc-coated urea and soil and foliar application of zinc sulphate on yield, nitrogen, phosphorus, potassium, zinc and iron biofortification in grains and uptake by basmati rice (Oryza sativa L.). J Agric Sci 7(2):161

Shivay YS, Singh U, Prasad R, Kaur R (2016) Agronomic interventions for micronutrient biofortification of pulses. Indian J Agron 61:161–172

Singh J (2016) Biofortification of food legumes and bioavailability of nutrients. In: Biofortification of food crops. Springer, New Delhi, pp 51–60

Singh UM, Sareen P, Sengar RS, Kumar A (2013) Plant ionomics: a newer approach to study mineral transport and its regulation. Acta Physiol Plant 35:2641–2653

Soares JC, Santos CS, Carvalho SMP, Pintado MM, Vasconcelos MW (2019) Preserving the nutritional quality of crop plants under a changing climate: importance and strategies. Plant Soil 443:1–26

Socha AL, Guerinot ML (2014) Mn-euvering manganese: the role of transporter gene family members in manganese uptake and mobilization in plants. Front Plant Sci 5:106

Sompong U, Somta P, Raboy V, Srinivas P (2012) Mapping of quantitative trait loci for phytic acid and phosphorous contents in seed and seedling of mungbean (*Vigna radiata* (L.) Wilczek.). Breed Sci 62:87–92

Suri VK, Choudhary AK (2013) Effect of vesicular arbuscular mycorrhiza and applied phosphorus through targeted yield precision model on root morphology, productivity and nutrient dynamics in soybean in an acid Alfisol. Commun Soil Sci Plant Anal 44:2587–2604

Tanksley SD, McCouch SR (1997) Seed banks and molecular maps unlocking genetic potential from the wild. Science 277:1063–1066

Taunk J, Yadav NR, Yadav RC, Kumar R (2012) Genetic diversity among green gram (*Vigna radiata* L. Wilczek) genotypes varying in micronutrient content using RAPD. Indian J Biotechnol 11:48–53

Tejada-Jiménez M, Castro-Rodríguez R, Kryvoruchko I, Lucas MM, Udvardi M, Imperial J, González-Guerrero M (2015) Medicago truncatula natural resistance-associated macrophage Protein1 is required for iron uptake by rhizobia-infected nodule cells. *Plant Physiology, 168(1):258–272*

Thavarajah P, Thavarajah D, Vandenberg A (2009) Low phytic acid lentils (Lens culinaris L.): a potential solution for increased micronutrient bioavailability. J Agric Food Chem 57:9044–9049

Thavarajah P, See CT, Vandenberg A (2010) Phytic acid and Fe and Zn concentration in lentil (*Lens culinaris* L.) seeds is influenced by temperature during seed filling period. Food Chem 122(1):254–259

Trethowan RM (2007) Breeding wheat for high iron and zinc at CIMMYT: state of the art challenges and future prospects. In: Proceedings of the 7th international wheat conference, Mar del Plata

Tullu A, Buchwaldt L, Lulsdorf M, Banniza S, Barlow B, Slinkard AE, Sarker A, Tarán TD, Warkentin TD, Vandenberg A (2006) Sources of resistance to anthracnose (*Colletotrichum truncatum*) in wild Lens species. Genet Resour Crop Evol 53:111–119

Upadhyaya HD, Bajaj D, Das S, Kumar V, Gowda CL, Sharma S et al (2016) Genetic dissection of seed-iron and zinc concentrations in chickpea. Sci Rep 6:2

Varshney RK, Kudapa H, Pazhamala L, Chitikineni A, Thudi M, Bohra A, Gaur PM, Janila P, Fikre A, Kimurto P, Ellis N (2015) Translational genomics in agriculture: some examples in grain legumes. Crit Rev Plant Sci 34:169–194

Wallace A, Romney EM, Hale VQ, Hoover RM (1969) Effects of soil temperature and zinc application on yields and micronutrient content of four crop species grown together in a glasshouse. J Agron 61(4):567–568

Wang YH, Zou CQ, Mirza Z et al (2016) Cost of agronomic biofortification of wheat with zinc in China. Agron Sustain Dev 36:44. https://doi.org/10.1007/s13593-016-0382-x

Warkentin TD, Delgerjav O, Arganosa G, Rehman AU, Bett KE, Anbessa Y et al (2012) Development and characterization of low-Phytate Pea. Crop Sci 52:74–78. https://doi.org/10.2135/cropsci2011.05.0285

Waters BM, Grusak MA (2008) Whole-plant mineral partitioning throughout the life cycle in Arabidopsis thaliana ecotypes Columbia Landsberg erecta, Cape Verde Islands, and the mutant line ysl1ysl3. New Phytol 177:389–405. https://doi.org/10.1111/j.1469-8137.2007.02288.x

Welch RA, Burland V, Plunkett III G, Redford P, Roesch P, Rasko D, ... Blattner FR (2002) Extensive mosaic structure revealed by the complete genome sequence of uropathogenic Escherichia coli. Proc Natl Acad Sci 99(26):17020–17024

Wessels I, Rolles B, Rink L (2020) The potential impact of zinc supplementation on COVID-19 pathogenesis. Front Immunol 11:1712. https://doi.org/10.3389/fimmu.2020.01712

White PJ, Broadley MR (2009) Biofortification of crops with seven mineral elements often lacking in human diets – iron, zinc, copper, calcium, magnesium, selenium and iodine. New Phytol 182:49–84

White PJ, Broadley MR (2011) Physiological limits to zinc biofortification of edible crops. Front Plant Sci 2:80. https://doi.org/10.3389/fpls.2011.00080

White PJ, Whiting SN, Baker AJM, Broadley MR (2002) Does zinc move apoplastically to the xylem in roots of Thlaspi caerulescens? New Phytol 153:201–207. https://doi.org/10.1046/j.0028-646x.2001.00325.x

WHO (2022) Biofortication of staple crops. Available online: http://www.who.int/elena/titles/biofortification/en/

Xiong L, Deng Q, Tucker GJ, McDowell DL, Chen Y (2012) Coarse-grained atomistic simulations of dislocations in Al, Ni and Cu crystals. Int J Plast 38:86–101

Zou T, Xu N, Hu G, Pang J, Xu H (2014) Biofortification of soybean sprouts with zinc and bioaccessibility of zinc in the sprouts. J Sci Food Agric 94(14):3053–3060

Chapter 3
Micronutrients Enrichments in Legumes Through Agronomic and Cultural Practices

Asim Abbasi, Aiman Hina, Sehar Shahid, Safa Imtiaz, Anum Intisar, and Muhammad Tajammal Khan

Abstract Malnutrition or micronutrient deficiency is a global concern particularly in developing countries mainly due to its associated health problems. Globally, two billion people are at risk of malnutrition especially children under the age of five and pregnant women especially living in South Asia and Africa. The nutritional quality or dietary requirement is largely compromised due to lack of diversity in diet. Moreover, intake of micronutrient-deficient legume-based crops is an additional threat to global nutritional security. Hence, it is recommended to implement some cost-effective and feasible approaches in global food system to address nutritional security issues for this rapidly increasing human population. This chapter focuses on the significance of biofortification of legume-based crops as an approach to enhance crop productivity and provide viable solution to address the issues of micronutrient deficiencies. In this attempt, various innovative agronomic and cultural biofortification techniques like ferti-fortification, foliar fortification, integrated soil fertility management, seed priming, seed coating, application of different soil

A. Abbasi (✉)
Department of Environmental Sciences, Kohsar University Murree, Murree, Pakistan

A. Hina
Soybean Research Institute, Ministry of Agriculture (MOA) Key Laboratory of Biology and Genetic Improvement of Soybean (General), MOA National Centre for Soybean Improvement, State Key Laboratory for Crop Genetics and Germplasm Enhancement, Nanjing Agricultural University, Nanjing, China

S. Shahid · S. Imtiaz
Department of Botany, Kohsar University Murree, Murree, Pakistan

A. Intisar
Department of Plant Pathology, College of Agriculture, University of Sargodha, Sargodha, Pakistan

M. T. Khan
Institute of Botany, University of the Punjab, Lahore, Pakistan

Division of Science and Technology, Department of Botany, University of Education, Lahore, Pakistan

M. A. Nadeem et al. (eds.), *Legumes Biofortification*,
https://doi.org/10.1007/978-3-031-33957-8_3

amendments, suitable cropping systems, and use of green technology have been discussed in detail. These approaches significantly improved nutrient contents of targeted crops without affecting their agronomic productivity. Moreover, apart from quantitative traits, these biofortification techniques also improved the qualitative traits of crops to better alleviate hidden hunger. Thus, this chapter provides useful insights for researchers regarding the potential of these biofortification techniques to enhance crop yield and their enrichment with additional micronutrients.

1 Introduction

Micronutrient deficiencies also termed as hidden hunger not only obstruct growth and development of crop plants but also affect humans to a great extent. The nutritional quality and crop productivity are usually compromised due to soil micronutrient deficiencies which ultimately contributes to malnutrition in the human population (Sanchez and Swaminathan 2005; Nubé and Voortman 2011; Hurst et al. 2013; Kumssa et al. 2015). Due to rapid agricultural intensification, soils present in different parts of the globe usually suffer multiple nutrient deficiencies including micronutrients such as Zn, Cu, Fe Mo, Mn, and B, secondary nutrients which includes Ca, Mg, and S and certain major macronutrients such as N, K, and P (Toenniessen et al. 2008; Vanlauwe et al. 2015). Apart from these, most of the agricultural soils are also deficient in certain other nonessential micronutrients such as silicon (Si), sodium (Na), and selenium (Se) which not only enhance plant growth and development but also provide resistance against certain biotic and abiotic stresses (Datnoff et al. 2007; Marschner and Rengel 2012). Collectively, these nutrients especially the micronutrients not only dictate our mental and physical health but also play important roles in various human developmental processes. Moreover, the deficiencies of these micronutrients also hinder normal metabolic functioning of various biological compounds such as proteins and enzymes. Moreover, the micronutrient-deficient fodders also remain a major concern for existing demand and supply gaps of quality livestock produce. Moreover, micronutrient deficiency is a global phenomenon especially in areas where people consume roots, cereals, and tubers as their staple crops (Black et al. 2013; Joy et al. 2014). Thus, considering the health perspectives of livestock and humans, nutritional security has emerged as a subject of grave concern (Kennedy et al. 2020).

The term nutritional security is considered as intake of adequate quantity of food material enriched with essential nutrients. In most of the developing countries, staple crops fulfill the major portion of the daily diet (Maertens and Velde 2017). Thus, to achieve nutritional security, there is an incessant need for progression regarding development of ultra-nourishing staple crops. Various approaches such as dietary diversification, medical supplementation, and food fortification are currently being deployed in various parts of the globe to achieve nutritional security. Indeed, biofortification has emerged as a promising approach to mitigate malnutrition. It usually enhances the nutritional status of different field crops particularly their edible parts without compromising important agronomic traits such as pest and drought resistance and crop produce (Bouis et al. 2011; Saltzman et al. 2013; Klikocka and Marx 2018). Similarly, different cultural practices also aid crop plants fulfill their nutritional requirements to combat different biotic and abiotic stresses and provide humans with better nutritious food sources.

In this chapter, we reviewed and discussed various evidence regarding effectiveness of agronomic biofortification and impact of different cultural practices to enhance micronutrient availability in legume crops and potential of these approaches to alleviate hidden hunger.

2 Global Status of Micronutrients in Soil

Although most of the micronutrients are present in required quantities in different types of soils yet, their availability to crop plants is regulated by different biological and edaphic factors such as soil microbiology, soil geomorphology, soil parent material, soil pH, soil organic matter, and competing anions and cations concentrations. Upon direct contact with soil particulates, some micronutrients rapidly react with different compounds such as clay colloids, mineral complexes, carbonates, and phosphates to form chemical precipitates which hinders their availability to crop plants (Allen et al. 2001; Marschner and Rengel 2012). The deficiency of micronutrients in soils is a global phenomenon, although their concentrations vary considerably for different micronutrients (Voortman and Bindraban 2015; Monreal et al. 2016). However, the deficiency of micronutrients in different agroecosystems is also attributed to their low crop use efficiency, normally less than 10% as compared to 20–80% for other macronutrients such as N, P and K. Different analytical techniques have been used for the assessment of micronutrient deficiencies in soils (Baligar et al. 2001; Oliver and Gregory 2015).

The micronutrient deficiencies in soils mainly occurs due excessive demand of macronutrients, which most of the high yielding crop varieties demand along with lack of micronutrient supplementation. Among different micronutrients, the deficiency of B (31%) and Zn (49%) is usually greater in major areas of the globe (Dhaliwal et al. 2022). Moreover, the deficits of certain other mineral nutrients such as Fe, Mn, Cu, and Mo is believed to be 3%, 10%, 14%, and 15%, respectively. Boron is usually present in soils in the form of $BO3^{-3}$ and plays an important role in strengthening plant cell membrane and mechanical resilience. The indications for low B concentrations in soils have been reported from almost every country, particularly Thailand, Nepal, the Philippines, and India. Similarly, copper also acts as a catalyst and actively involves in synthesis of vit-A, protein, and certain enzymes needed for different plant growth and developmental processes. However, Cu deficiencies and their associated toxicities have been frequently reported from soils of Italy, Brazil, Tonga, and the Philippines. Iron is also known as an important component of not only hemoglobin but several other enzymes involved in lignin synthesis, energy transfer, and nitrogen fixation. The deficiency of bioavailable Fe forms has been reported from soils belonging to different areas of Turkey, Malta, and Mexico. Similarly, Mo, an important micronutrient and structural component of nitrogenase enzyme, performed vital functions in N fixation activities especially in rhizobium-legume symbiosis. The deficiency of this micronutrient is usually more prevalent in acidic soils as observed in New Zealand, Brazil, Nepal, and Africa. Furthermore, micronutrient Mn not only enhance the availability of Ca and P in crop plants but also facilitates the photolysis of water molecules during nitrogen metabolism and photosynthesis (Lesková et al. 2017). Although, most of the acidic soils are rich in Mn content but alkaline soils usually lack its sufficient concentrations. Hence, the concentration of Mn in soils and its bioavailability from to crop plants is highly dependent on soil pH. Significant Mn deficiencies have been reported from soils belonging to certain areas of Egypt, Pakistan, India, Syria Lebanon, and Italy. Moreover, the deficiency of Zn is usually occurring in all areas of the globe particularly, Thailand, Pakistan, Nepal, Italy, Tanzania, Syria, Iraq, India, Turkey, Mexico, and Lebanon (Dhaliwal et al. 2022).

3 Micronutrient Bioavailability

Micronutrients move from the soil to plants, their edible portion, and finally become part of human body. Various key factors regulate the efficacy of agronomic biofortification to mitigate the deficiencies of micronutrients in human body. These key factors mainly rely on bioavailability of nutrients at various stages including: presence of important nutrients in the soil and bioavailability of these nutrients for plant uptake, distribution of these nutrients in different parts of the plant, and then further translocation into their edible parts, the bioavailability of nutrients in cooked food, and physiological stage of humans which controls the process to take up and utilize these nutrients for smooth body functions (De Valença et al. 2017).

3.1 Soil to Crop

There are several factors that influence the bioavailability of micronutrients from soil to crop plants; for instance, soil aeration, organic matter content, pH of the soil, soil moistness, interaction with other mineral elements, and crop variety which defines the plant's root systems its structure and functioning (Alloway 2009). Rhizosphere of some plant can be altered by eliminating organic acids or H^+ ions that ultimately improve the micronutrient bioavailability and its uptake (Zhang et al. 2010; Marschner and Rengel 2012). The elements interact with each and consequently affect the efficiency of micronutrient uptake by roots. For example, soil phosphorus (P) can either enhance zinc (Zn) uptake or root growth (Zingore et al. 2008); however, adding P may also lead to Zn deficiency through intrusion with zinc movement from the roots and dilution effects (Singh et al. 1988). Management of soil using organic manures and lime can also change the soil properties, for example, soil pH, improvement in the bioavailability of micronutrients, and uptake by crops. Likewise, plant symbioses with arbuscular mycorrhizal fungi also help to improve the uptake of sparingly soluble nutrients in the soil such as Zn and P (Smith and Read 2010).

3.2 *Crop to Food*

There are two other important factors that influence the bioavailability of nutrients from crop to food, i.e., food processing and crop variety. Crop varieties largely determine the transfer particular micronutrients form their vegetative parts to edible portions. In the outer layer of the rice grain, protein bodies contain iron (Fe) and zinc (Zn), which is frequently detached during different food processing operations such as milling and dehusking, leaving a reduced amount of Fe and Zn in the rice (Haas et al. 2005; Zimmermann and Hurrell 2007; Hotz et al. 2015; Prakash et al. 2020). Some other crops such as wheat provide zinc even after the removal of the aleurone layer and seed coat during bread-making process (Ajiboye et al. 2015). Moreover, selenium (Se), iron (Fe), manganese (Mn), and copper (Cu) have barely vanished during the grinding of wheat grain as well as the bread-making processes (Lyons et al. 2005), which makes wheat more appropriate for agronomic biofortification. The loss of nutrients generally results from food processing; however, this processing of foods also reduces the antinutrients, consequently enhancing the micronutrient's bioavailability. For instance, when cereals are soaked in water, it reduces the phytate (an antinutrient), consequently improving the bioavailability of calcium, iron, and zinc (Hotz and Gibson 2007).

3.3 *Food to Human*

The bioavailability of micronutrients from food source to human body is also affected by several factors that are either related to the host or food itself (Gibson 2007). Dietary intake and its bioavailability to human body is an important factor and relies heavily on the nature of dietary matrix, the amount consumed, the chemical form of the diet, and the interaction between components/nutrients of the food that inhibit or improve the absorption of these nutrients in the gastrointestinal tract (Sandstroèm 2001). Inhibitors such as phytic acid or phytate and polyphenols form complexes with Zn and Fe which result in the inhibition of these essential micronutrient uptakes in the human body. However, enhancers such as ascorbic acid present in vegetables and fruits usually enhance the bioavailability of Fe (Clemens 2014). There are several other factors that also impact the bioavailability of micronutrients from food items to human body, e.g., age, genotype, gender, health, ethnicity, nutrient status, and physiological state of an individual

(Gibson 2007). Overall, the bioavailability of micronutrients at different stages from soil to crop plant and then uptake by digestion in the human body determines the full potential of agronomic biofortification. Although interactions, as well as metabolisms, diverge largely among soils, nutrients, crop varieties, and humans, potential losses of micronutrients occur at each particular stage. All these stages must be contemplated while evaluating the effectiveness of agronomic biofortification.

4 Importance of Legumes

Plants provide a score of important vitamins and minerals to humans. In certain developing countries, legume and grain crops are the major and often the sole source of human food (Myers et al. 2014). The nutritional quality of legumes is quite similar or even better than cereals as they provide ample number of vitamins, nutrients, and amino acids to their end users (Rehman et al. 2019). They possess a number of health benefits due to which their regular inclusion in human diet is highly recommended. Certain compounds such as flavonoids, antioxidants, and soluble fibers present in different legume crops are usually associated with significant reduction in occurrence of a number of human diseases including cancer, diabetes, and cardiovascular ailments (Martín-Cabrejas 2019; Ferreira et al. 2021). Moreover, the ecological benefits of legume crops usually include improved soil composition, reduction in agricultural CO_2 emissions, and reduction in nitrogenous fertilizer requirements (Karkanis et al. 2018). Furthermore, the inclusion of legume crops in cropping schemes not only enhances disease and pest resistance capability of succeeding crop but also mitigates a significant share (25%) of greenhouse gas emissions (Ma et al. 2018). In 2014, the significance of legume biofortification programs has been highlighted by the United Nations, aiming to utilize this approach up to its full potential to combat hidden hunger. However, since then little has been done in this field, and this aspect largely remained unexplored (Rehman et al. 2019).

5 Biofortification

Biofortification is a relatively new approach focused mainly on provision of enhanced micronutrients to different target crops (Huang et al. 2020). The biofortified crops usually uptake more minerals as compared to other crops and positively affect human health to fight hidden hunger (Bouis and Saltzman 2017; Praharaj et al. 2021). Biofortification of field crops is usually done with different approaches mainly includes agronomic biofortification, biofortification through conventional plant breeding, and biofortification through genetic engineering (Garg et al. 2018).

6 Agronomic Biofortification

Agronomic biofortification is an integrated approach aimed at enhancing nutritional status of crops through exogenous application of different micronutrients in field so that plants can easily absorb those nutrients which are present in relatively lesser quantity in soil. Agronomic biofortification approach is also recommended for its ability to alter the solubility and mobility of different mineral nutrients in soil profile making them available for crop plants. This particular approach involves addition of desired nutrients to soil profile either before or during growth of crop plants. This can be achieved using two different nutrient application methods. Foliar application ensures absorption of nutrient from leaf and its subsequent transportation to other plants parts, whereas soil application enhances nutrient uptake from roots to upper plant parts. However, the success of biofortification program mostly depends on the bioavailability of micronutrients largely determined by different soil properties such as pH, micronutrient application method, and application rate of micronutrients (Cakmak and Kutman 2018).

Moreover, agronomic biofortification is also an economical way to counter mineral deficiencies in humans. Many studies confirmed that crop biofortification not only alters the micronutrient status of plants but also aids in synthesis of some other compounds possessing significant nutritional values (Rizwan et al. 2019; Skrypnik et al. 2019; Adrees et al. 2021; Newman et al. 2021; Noreen et al. 2021; Puccinelli et al. 2021).

Nowadays, the primary targets of researchers for biofortification are staple crops which have global consumption. Certain cereals such as pulses, wheat, maize, rice, fodder, and oilseed crops have the potential to accumulate nutrients supplied externally and thus have been biofortified. Most of the researched biofortified crops are usually loaded with different micronutrients such as Fe, Zn, Cu, Mo etc. (Klikocka and Marx 2018). A zinc-efficient PBW 1Zn variety, possessing higher Fe (40.0 mg L^{-1}) and Zn (40.6 mg L^{-1}) contents, has been released in 2017. Up till now, several biofortified crops including sweet potatoes, squash, orange, maize, sorghum, Zn-enriched cowpeas, lentils, sorghum, wheat, rice, Fe-enriched lentils, and beans have been commercially released (De Steur et al. 2017).

Conclusively, the nutritional quality and yield of crop plants have been amended with foliar and soil application of multiple micronutrients and is considered a sustainable strategy to combat hidden hunger (Manzeke et al. 2012; Vanlauwe et al. 2015; Voortman and Bindraban 2015). Moreover, seed coating and seed priming with different micronutrients are other viable strategies for precise mineral fertilization that can stimulate plant growth and development and ultimately increase their yield (Duffner et al. 2014).

6.1 *Impact of Different Fertilization Approaches*

There are two important factors, viz., fertilizer application method and their types, which influence the efficacy of mineral fertilizer on crop performance and yield. The bioavailability of micronutrients present in different fertilizers is largely determined by their formulations. Nutrients in these formulations upon interaction with soil and other elements can cast a positive, neutral, or even a negative effect on the crop performance and yield (Rietra et al. 2015). In crops with multiple micronutrient deficiencies, the use of a sole nutrient formulation cannot bring the desired results. Moreover, use of each micronutrient formulations in separate applications is also economically not a viable option; therefore, mixtures of multi-micronutrients can be deployed to attain the required results (Singh et al. 2018).

6.1.1 Ferti-fortification

On account of manifold benefits, the application of micronutrients through the soil is the most effective and versatile method mainly to alleviate the boron (B) deficiencies (Dhaliwal and Manchanda 2009). In Zn-deficient soils, the application of $ZnSO_4{\cdot}H_2O$ (with 33% Zn) at 40 kg ha^{-1} or $ZnSO_4{\cdot}7H_2O$ (with 21% Zn) at 62.5 kg ha^{-1} has been found economical and equally effectual to alleviate the Zn deficiency. Ram et al. (Ram et al. 2015) had reported a positive impact of Zn application through soil on Zn content in rice and wheat grain as well as on yield of both crops. Moreover, Zn application through soil has also proved its positive effect on chickpea and cowpea crop regarding nutritional quality and yield of both crops (Manzeke et al. 2017). The soil zinc (Zn) application in different lentil cultivars also influences plant growth and grain production, reduces the concentration of phytate, and enhance grain Zn concentration (Rasheed et al. 2020). In soybean, adding zinc (Zn) and selenium (Se) to the soil, improved the Zn and Se concentrations in seeds, and subsequently enhanced the physiological status of crop (Dai et al. 2020).

6.1.2 Foliar Fortification

Soil fertilization can improve the level of micronutrients within the grain; nevertheless, it is not effectual for some immobile minerals and also confines the uptake of salt. Moreover, some micronutrients such as iron become futile when applied through soil as soluble Fe is readily transformed to insoluble Fe^{3+} form, consequently, limits its availability to crop plants (Dhaliwal and Manchanda 2009). In this case, foliar application of mineral fertilizers is also recommended depending upon the applied nutrient and soil condition (Dhaliwal et al. 2010). The practice is usually executed by spraying mineral fertilizers in their liquid form over plant leaves. The micronutrient either singly or in combination were applied as a solution on crop plants, where it is absorbed by the epidermis and stomata of the leaves (Dhaliwal et al. 2011). Up till now, numerous studies have proved the worth of this particular fertilization approach on different field crops (Dhaliwal et al. 2013; Kumar et al. 2016; Singh et al. 2018; Ullah et al. 2018).

In cereals and leafy vegetables, foliar application of micronutrients not only enhance nutrient uptake but also better allocate these nutrients in edible parts of crop plants (Lawson et al. 2015). In some circumstances, foliar pathways are usually more efficient in regulating nutrient uptake as mineral immobility in soil is usually avoided in this approach. Foliar application of Fe and Zn (0.5%) at different growth stages of wheat crop not only enhanced grain yield (2.5–5.1%) but also improved Fe (13.1–30.3%) and Zn (17.3–38.8%) contents in wheat grains (Dhaliwal and Manchanda 2009). Similarly, Dhaliwal et al. (2010) also reported that foliar application of Fe and Zn at different growth stages of rice cop not only improved grain Fe (22–38.2%) and Zn (30.8–44.8%) contents but also enhanced the crop yield (6–10.3%). In the same way, foliar feeding of copper (Cu) and manganese (Mn) also improved the Cu and Mn content of wheat grains (Dhaliwal et al. 2011). Maize crop also responded with increased yield and improved grain Fe and Zn contents when subjected to foliar application of both micronutrients (Dhaliwal et al. 2013). Moreover, foliar feeding of Mn and Zn also improves yield of different crops including barley, rice, and wheat (Ullah et al. 2018).

Likewise, foliar feeding of 0.2% copper (Cu) at 60 and 90 days of oat crop also enhances crop crude protein, yield, and Cu concentration (Sandhu et al. 2020). However, foliar application of fertilizers is usually not recommended in certain parts of the globe possibly due to rain obstructions and their associated operational cost (García-Bañuelos et al. 2014). Hence, the integration of foliar and soil application is often suggested as the most viable option for achieving better results (Cakmak et al. 2010; Phattarakul et al. 2012). Moreover, data regarding ferti and foliar fortification of different legume crops along with their responses has been presented in Table 3.1.

6.2 *Impact in Combination with NPK Fertilization*

The efficacy of agronomic biofortification can be influenced by the interaction between macronutrients and micronutrients. Plants having good phosphorus (P) and nitrogen (N) status showed a positive impact on the development of roots, transport of nutrients in shoots, and relocalization of these nutrients to seeds from vegetative parts of plants (Prasad et al. 2014). This results in improved concentrations of micronutrients in edible parts of plants due to their better uptake from soil, as revealed in some wheat trials where high N applications resulted in improved Fe and Zn concentrations in the endosperm of grains (Shi et al. 2010; Kutman et al. 2011). The combination of N + P fertilizer with Zn for wheat fertilization has also been prolific to enhance the yields of wheat grain (Cakmak 2004). Hence, appropriate management of N and P is crucial to enhance the efficacy of other micronutrients. Moreover, it also demonstrates the effectiveness of a more integrated approach for the soil fertility management, as explicated below.

Table 3.1 Application of micronutrients for biofortification of legume crops

Micronutrients	Crops	Responses	References
Foliar application			
Zn, Fe	Chickpea	Significant increase in yield Increase in nutrient content and nutrient uptake	Dhaliwal et al. (2021)
Fe, Mo	Lentil	Increased the grain yield, biomass yield Increase in total nitrogen uptake and protein content Increase in number of total and active nodules	Nasar and Shah (2017)
Fe, Mo	Soybean	Increased the yield Reduced the damage caused by water-deficient conditions	Thapa et al. (2021)
Soil application			
Zn	Cowpea	Increased the crop production Improved seed iron content	Mfeka et al. (2019)
Mo, Fe, B	Pigeon pea	Enhanced the productivity of pigeon pea in vertisols	Reddy et al. (2007)
Fe	Chickpea	Little amount of iron enhanced the chickpea yield and quality	Pingoliya et al. (2014)
Se	Lentil	Increased grain yield Improved biological nitrogen fixation in lentils	Ekanayake et al. (2017)

6.3 *Impact of Integrated Soil Fertility Management*

The potential of agronomic biofortification depends largely on proper soil conditions which enhance micronutrient bioavailability for plant uptake. The N and P fertilizers not only enhance the efficacy of micronutrient fertilization but also improve other certain chemical, biological, and physical characteristics of the soil that are crucial for optimization of nutrient use efficiency. Soil optimization is usually done with Integrated Soil Fertility Management (ISFM) practices. It is commonly defined as an approach of soil fertility management using different organic inputs, mineral fertilizers, and improved crop germplasm (Vanlauwe et al. 2010). Application of organic inputs along with different mineral fertilizers usually showed better results mainly because of their synergistic action. Organic resources such as animal manure and plant residues usually play important roles in sustaining soil organic matter and soil structure. Moreover, these amendments also enhance cation exchange and soil water holding capacity (Van Noordwijk et al. 1997). Moreover, where organic matter offers a more constant but slow liberation of nutrients, mineral fertilizers usually provide flexibility in their application rate, timing, and placement to synchronize the accessibility of nutrients with crop requirements (Giller 2001). The application of sole organic inputs also possesses the potential to improve the micronutrient content and their availability in soil (Traore 2006; Zingore et al. 2008; Manzeke et al. 2012; Thilakarathna and Raizada 2015).

In a study, Manzeke et al. (2014) revealed that ISFM strategies significantly improved the maize grain yield obtained from field amended with combination of Zn-enriched fertilizer, forest leaf litter, and cattle manure. Soils with long-term supplies of organic matter not only improve the total Zn content but also increases the proportion of readily accessible Zn form for crop uptake (Santos et al. 2010; Manzeke et al. 2014). Nevertheless, the application of sole organic matter is often inadequate in resource-poor farming systems for sustaining nutrient balances possibly due to inadequate provision of nutrient-rich organic matter and overall deficiency of key micronutrients in the cropping system. Hence, the application of mineral fertilizers combined with organic inputs has greater potential to improve the deficiencies of soil micronutrients. Moreover, agronomic effectiveness of a micronutrient fertilizer is frequently augmented when used in combination with different organic inputs (Vanlauwe et al. 2015). Green manures also have the potential to increase nutrient bioavailability, as reported in a study, where combined application of Zn and green manure enhanced the grain Zn contents and yield of basmati rice (Pooniya and Shivay 2013).

6.4 Seed Priming

Seed priming is defined as controlled presowing seed hydration which allows seeds to undergo necessary pregermination metabolic activities without any radical emergence from seed coat (Mcdonald 2000). Sowing of prime seeds is usually recommended for achieving uniform crop stand and better yield (Farooq et al. 2006). This approach exhibited better growth and yield response in wheat and rice crop (Farooq et al. 2012). The integrated application of Zn + Pseudomonas sp. MN12 as seed priming agents showed better results in bread wheat and proved to be highly cost-effective (Rehman et al. 2018). Similarly, seed priming of barley and wheat with Zn and Fe (2 mg L^{-1} each) increases grain yield as well as tillering of bread wheat, proving this a useful viable approach for fortification of bread wheat (Carvalho et al. 2019). Likewise, seed priming with boron using a concentration of 0.001% or 0.1% improved the crop stand establishment; however, higher concentration (0.5%) of same micronutrient usually hinders seed germination (Dhaliwal et al. 2020).

6.5 Seed Coating

Seed coating is a method of applying powdered form of micronutrients to the outer surface of seed with the help of some nonreactive sticky material. This practice influences the soil or seed at the seed–soil interface and eventually enhances the accessibility of coated as well as soil-applied nutrients. This practice is considered eco-friendly as it poses no ill effects on nontarget organisms and distributes micronutrient in uniformly in field. Several factors, i.e., soil type, soil moisture, soil fertility, nutrient and seed ratio, coated micronutrients, and coating material, can alter the potential of applied micronutrients. Seed coating with Zn has been extensively used in crop yield improvement of different field crops. In wheat, seed coating with $ZnCl_2$ and $ZnSO_4$ boosted the germination, growth, and concentration of Zn in plant tissues as compared to noncoated seeds (Rehman and Farooq 2016). Likewise, in rice crops, seed coating with B (1, 1.5, 2, 2.5, and 3 g kg^{-1}) enhanced the uniform germination and tillering on account of assimilate partitioning and water relation, ultimately resulting in the enhanced B concentration in rice grain (Rehman and Farooq 2013). Moreover, seed coating with microbial bioinoculants and B proved to be a worthy option to improve nodulation, growth, grain yield, and B concentration in chickpeas. Similarly, seed coating with Mn showed improved Mn uptake by plant, better straw and grain yield, as well as Mn concentration in wheat grains (Ullah et al. 2018). Additional data regarding effect of seed coating and seed priming on fortification of legume crops and their performances has been represented in Table 3.2.

7 Biofortification Through Different Cultural/Agronomic Soil Management Approaches

The primary objective of management approaches is to improve the state of the soil, which eventually boosts the supply of micronutrients for plant absorption and eventually enhances proportion of those nutrients in food. The efficiency with which a

Table 3.2 Seed treatments with micronutrients for biofortification of legume crops

Micronutrients	Crops	Responses	References
Seed priming			
Zn	Mung bean	Improved crop stand establishment, seedling growth, grain yield and seed Zn concentrations	Haider et al. (2020)
Zn	Green bean	Improved relative water content and mineral nutrient concentrations	Gulmezoglu et al. (2016)
B	Chickpea, lentil, cowpea	Increased seed germination, seedling length and seed nutrient contents	Kumar et al. (2020)
Seed coating			
Zn	Soybean, pigeon pea	Enhanced auxin indole-3-acetic acid (IAA) production in plant roots, which subsequently improved the overall growth Improved seed germination	Adhikari et al. (2016)
Zn, B, Fe	Chickpea	Significant increase in plant growth and yield of chickpea Higher chlorophyll content, leaf area index and stomatal conductance	Shinde et al. (2017)
Zn	Mung bean	Increased seed yield of mung bean and grain Zn concentration	Haider et al. (2020)

plant utilizes available nutrients is influenced by a number of elements, including the soil's structural, biochemical, and biological properties. Tillage, water management and integrated soil fertility management are some of the common renowned management practices (Selim 2020). Among these practices, application of organic amendments provides a number of important benefits to soil including improvement in overall soil structure, soil cation exchange capacity, and its water holding capacity. However, the particular approach usually supplies nutrients in a rate-determined manner and rarely satisfies crop's nutritional demand at the appropriate time for maximum production. Moreover, the sole use of mineral fertilizers and organic inputs is not enough to overcome the discrepancy between nutrient deficiencies and demand for better crop production. Thus, the integrated application of both approaches is more beneficial and sustainable primarily because of their complementary actions (Padbhushan et al. 2021).

With recent agricultural intensification, various soil management approaches have been utilized in different parts of the globe for maintaining soil fertility. Soil pH is one of the chemical qualities that have a significant impact on the development of plants, as well as on the distribution of nutrients and elimination of potentially hazardous poisonous compounds. Thus, neutralizing soil pH is mandatory for accomplishing qualitative and quantitative crop production. This can be achieved either by lowering the pH of calcareous soils or raising the pH of the acidic soils (Dhaliwal et al. 2022).

7.1 Application of Lime

The availability of abundant H^+ ions and a deficiency of basic cations are what give soil its acidic quality. Therefore, supplying the soil with hydroxides, carbonates, and other chemicals that operate in a basic manner from the outside may be an efficient method for reducing soil acidity. The pH of the soil regulates the solubility and mobility of micronutrients in soil; hence, even a slight change in pH has a significant impact on both the solubility and mobility of micronutrients from soil to crop plant. Frequent application of limestone in acidic soils not only improves soil fertility but also enhances crop growth and production (Barman et al. 2014). Additionally, Soltani et al. (2016) also reported existence of positive interaction between lime and Zn, which significantly increased straw and grain yield as compared to control group. The application of zinc enhanced activities of several metabollic enzymes including auxin which regulates different plant functions. This ultimately boosts the ability of rice plants to synthesize more carbohydrates and their ascending circulation to grain production and filling sites. On the other hand, the incorporation of lime into the soil also changes its structure, making it more conducive for rice production.

7.2 Application of Gypsum

The presence of a high carbonate content and a calcareous texture are typical indicators of soil has having pH > 7. Certain management practices such as deep tillage usually increase the number of carbonates in soil surface layer and increase soil pH. In addition, calcareous soils that have an abundance of calcium and magnesium might ultimately have multinutrient deficits due to the high pH of the soil, which decreases the solubility of iron and manganese. In order to lower the pH of sodic and saline-sodic soils, it has become common practice to replenish these soils using acid-producing supplements such as gypsum and elemental sulfur (Tavakkoli et al. 2022). The removal of excessive bicarbonates and exchangeable Ca and Na from the soil solution usually becomes easier with application of gypsum. The significant reduction in bicarbonates results in reduction of soil pH, which in turn improves the concentration of plant-available zinc, iron, copper, manganese, and cobalt. Moreover, application of slag-based gypsum (750 kg ha^{-1}) also cause significant improvement in nutrient uptake capacity of maize crop planted in both acidic and neutral soils (Prakash et al. 2020).

7.3 Application of Biochar

When applied to acidic soils, biochar has the potential to improve both the absorption of micronutrients and the development of plants. As a result, biochar may serve as an efficient method for the waste disposal process while simultaneously enhancing both soil fertility and crop productivity. Under iron-limited and salinized circumstances, the prospective advantages of integrated application of biochar and *Burkholderia phytofirmans* PsJN have been associated with enhanced iron bioavailability, productivity, and nutritional content of quinoa grains. Therefore, siderophore-producing microorganisms combined with organic supplements may increase the amount of iron in the grain while simultaneously lowering the antioxidant activity and the proportion of sodium to potassium in *C. quinoa*. In addition, the integrated treatment of zinc and biochar showed better outcomes in terms of increasing crop production and grain zinc proportion while simultaneously decreasing grain cadmium content in comparison to individual treatments of zinc and biochar (Farooq et al. 2020). Similarly, the integrated application of zerovalent iron (BZVI) and biochar significantly improved the Fe and Zn contents in rice crop and mitigate the concentration of toxic element Cd up to 83% as compared to untreated control. Furthermore, foliar application of ZnO NPs, either as a stand-alone treatment or in combination with biochar significantly improved the maize Zn contents, chlorophyll concentrations and plant growth parameters. In comparison to the control, the combined treatment also reduced the amount of Cd ion toxicity, MDA, and H_2O_2, while simultaneously improving the antioxidant enzyme activities. It's likely that the decreased Cd contents might improve the biomass and efficient absorption of zinc in plant through roots tips. The pH of the soil is altered by biochar, which also changes concentration of cadmium in the soil. This resulted in a decreased accessibility of cadmium and other heavy metals in soil (Khum-In et al. 2020).

7.4 Incorporation of Crop Residues

More than 1000 million tons of cereal crop residues are produced each year by developing nations. Globally, the crops that produce the relatively more residues include wheat, rice, maize, barley, soybeans, potatoes, and oilseed. Agricultural wastes are a valuable source of different nutrient elements required in proper amount for normal growth of crop plants. Crop residues also improve the accessibility of unavailable micronutrients present in soil to crop plants. Recycling of crop residues can add around 50% to 80% Zn, Cu, and Mn which were utilized by wheat and rice crop during their vegetative and reproductive phases. Under wheat–rice cropping rotation, the addition of $ZnSO_4$ along with cereal straws significantly improved the bioavailability of Zn in soil which in turn enhance the total zinc concentration of plant tissues, zinc uptake, and 1000 grain weight of crop produce (Dwivedi and Srivastva 2014). Moreover, wheat crop amended with residues of preceding clover crop also exhibited increased zinc concentration in crop straws and protein density in grains. This may be attributed to increase N supply provided by precrop residues along with formation of zinc complexes with improved concentrations of organic ligands (Khoshgoftarmanesh et al. 2017). Additionally, nutrient uptake and plant growth and development is also influenced by different abiotic stresses. Thus, understanding of interaction between crop residues with nutrients, crop production, and physical, chemical, and biological properties of soil is a prerequisite of different crop residue management practices (Dhaliwal et al. 2022).

7.5 Application of Animal Manure

The utilization of animal manure is a common component of traditional nutrient management strategies. Animal dung is a rich source of fertilizer for plants and has the potential to change the biochemical and physical characteristics of the soil. When compared to chemical fertilizers, animal manures proved their worth as a rich supply of number of plant available micronutrients including zinc, iron, and manganese. In supplement to zinc that it adds to soil, manure also stimulates various biological and biochemical processes that lead to breakdown of Zn particles that are not immediately accessible to crop plants. The use of cattle dung in conjunction with forest plant debris and Zn-based fertilizers led to significant improvements in maize grain zinc concentrations and crop yield. The addition of organic matter to the soil over an extended period of time not only raises the overall zinc content of soil but also increase the percentage of labile zinc; the form which is easily absorbed by the crop plants (Manzeke et al. 2017). The integrated application of lime and farmyard manure lower soil pH and ultimately enhanced uptake of Zn in maize crop and also increases its availability in alfisols (Behera et al. 2016). Additionally, the integrated application of lime, crop wastes and farmyard manure has considerably enhanced the nutritional status of various agricultural soils. Similarly, the exogenous application of Zn + Fe significantly increased the Fe (38.34%) and Zn (53%) contents of rice crop grown in acidic soil. Furthermore, Zn supplementation in combination with plant debris and cattle dung also increased the grain zinc content of cowpea crop grown on sandy soil (Manzeke et al. 2017). In Rabi rapeseed, fertilization with combination of sulfur, zinc, and animal manure produced superior results in terms of crop production and nutrient absorption compared to the individual treatments of nutrient fertilizers (Nayak et al. 2020). The subsequent application of iron and sulfur coupled with chicken manure has the potential to lower down polyphenol and phytate contents along with soil pH and raise grain ferritin and iron content of cereal grains grown in calcareous soil. These findings may be due to the synergistic effects since S decreases the soil pH, hence enhancing Fe solubilization and, consequently, its absorption in alkaline soil (Dhaliwal et al. 2022).

7.6 Addition of Compost

Compost is an economical and rich source of Cu and Zn fertilizers as its commercial production is much greener and cost-effective and can be easily done in any rural or urban area. Green waste compost and municipal compost enhanced grain Zn concentration (220% higher) and yield of maize crop and also limits uptake of heavy metal nickel (74% less accumulation). Application of compost gradually increases soil pH, its cation exchange capacity and C contents along with reduction in uptake of certain toxic heavy metals. Wheat plants exhibit different Zn and Cd uptake mechanism which ultimately alters Zn/Cd ratio and qualitative traits of crop plants. Rice crop amended with Zn-enriched compost + Zn-solubilizing bacteria showed better growth and yield as zinc releases gradually and steadily from ZnO over $ZnSO_4$, making it available for a longer period for necessary plant processes. The same zinc treatment also increased Zn accumulation in various parts of the crop plant particularly under Zn-deficient soils (Ulm et al. 2019).

7.7 Selection of Suitable Crops and Cropping Systems

Micronutrient availability also varies under different cropping systems which in the long run alters different soil characteristics. The quantitative and qualitative parameters of a particular crop are greatly influenced by the physical and chemical health of soil associated with crop residues or leaf litter of preceding crop. Improved soil conditions provide better opportunities for plant roots to uptake more minerals particularly immobile nutrients (Dhaliwal et al. 2022).

Moreover, besides proper cropping system, selection of suitable crop is also very important to mitigate nutrient deficiency and other related field constraints (Dragicevic et al. 2015). Thus, crop rotations must be designed using crops which complement each other and fulfill soil nutritional deficiencies. Legumes are usually complemented by cereals, e.g., in case of legumes-brassica system, the former assist in N fixation, whereas latter is useful as nutrient-rich field crop. Moreover, crops possessing high biomass should be followed by a low residue crop. Similarly, N-depleting crop should be followed by an N-fixing crop to maintain soil nutritional balance for sustainable agriculture. Moreover, it has also been reported that legume-based cropping system significantly improves soil organic C due to their higher residue contents. Hence, cultivation of leguminous crops is usually recommended in different cropping systems (Diekow et al. 2005).

7.8 Biofortification Through Green Technology

Green technology involves the utilization of different microbial agents for improving the nutrient accessibility and their uptake by crop plants. The quality of human health environmental sustainability has been largely compromised due to excessive use of different chemical fertilizers. Moreover, intrinsic plant-based strategies such as production of organic acid, phyto-siderophores, and secretions of chelators are not sufficient to fulfill nutritional deficiencies of plants. Thus, for sustaining soil fertility and plant growth, microbial fertilizers emerged as a viable option. These fertilizers usually use multiple direct and indirect mechanisms to improve plant growth which includes solubilization of P, K, and Zn, biological nitrogen fixation, and production of various hydrolytic enzymes, plant growth-promoting hormones, HCN, and siderophores. Biofortification of crop plants using microbial agents have been proved by a number of studies. Fungi and bacteria are the major microbial agents used for biofortification purposes (Kaur et al. 2020).

7.8.1 Bacterial Biofortification

Application of *Bacillus* and *Pseudomonas* sp. mitigates soil pH and converted complex Zn compounds into soluble Zn ions for crop uptake. The use of *Arthrobacter sulfonivorans* (DS-68) and *Arthrobacter* sp. DS-179 significantly improved the Fe and Zn contents of wheat crop and also enhanced crop qualitative parameters (Singh et al. 2017). Similarly, an increased average number of root tips and root surface area (1.6- to two-fold) in lesser Fe soils had been reported with inoculation of siderophore-producing *Enterococcus hirae* DS-163 and *Arthrobacter sulfonivorans* DS-68. The increase in crop root parameters directly facilitates the uptake of Fe ions. Furthermore, the micronutrient status and yield of wheat crop also improved with integrated use siderophore-producing endophytes (*Enterococcus hirae* DS-163 and *Arthrobacter sulfonivorans* DS-68) and Zn-solubilizers, *Bacillus subtilis* DS-178 and *Arthrobacter* sp. DS-179 (Singh et al. 2018).

7.8.2 Endophytic Biofortification

The availability of micronutrients in soil is also influenced by the presence of mycorrhizal fungi which act either through rhizospheric acidification, siderophore production or hyphal transport of nutrients through external mycelium to maintain soil nutritional balance. An increased grain Zn content of barley crop had been reported with colonization of *Rhizophagus irregularis* which usually shows its action through upregulation of HvZIP13 (Watts-Williams and Cavagnaro 2018). Furthermore, an enhanced Fe and Zn grain content with improved crop yield parameters had been reported with integrated application of AM fungi and PGPB. Various plant metabolic activities are greatly influenced by the presence of PGPB which ultimately stimulate plant growth and development. Similarly, AM also participates in several mechanisms including secretion of different signaling molecules which alters plant anatomy and root morphology for better and efficient nutrients uptake. Moreover, integrated application of AM fungi and PGPB makes plant more resilient against different biotic and abiotic stresses (Yadav et al. 2020). It has also been reported that wheat plants inoculated with endophytes showed significant reduction (26%) in phytic acid contents which consequently enhance grain Zn and Fe contents (Singh et al. 2018) (Table 3.3).

8 Conclusion

In this chapter, various agronomic and cultural practices regarding biofortification of legume crops have been discussed. Regular intake of micronutrients is considered essential not only for humans but also for livestock. The deficiency of certain micronutrients may not only affect plant growth but also influence human health. A significant proportion of human population in developing and underdeveloped countries rely heavily on legume-based foods and thus constantly remain prone to micronutrient deficiencies. To overcome micronutrient deficiencies in human population, agronomic biofortification emerged as a cost-effective, sustainable, and viable method that can bring fruitful results in relatively shorter time span. Moreover, different soil management approaches such as application of different inorganic and organic amendments including use of different microbial agents may either act directly as an exogenous nutrient source or may play active role in enhancing existing nutrients availability to crop plants. Moreover, the application of these agronomic/cultural inputs poses no serious ecological hazards when applied at appropriate recommended dose rates. However, still further research is needed depending on the soil type and crop for recommending a proper approach to counter

Table 3.3 Utilizing green technology for biofortification of legume crops

Microorganism	Micronutrients affected/Crops	Responses	References
Trichoderma Strains (*T. afroharzianum* T22, *T. harzianum* TH1, and *T. virens* GV41)	Fe and Zn/Lentil	Increased Fe and Zn contents in plants Improved seed germination and crop yield Reduced the risk of nutrient deficiencies	Marra et al. (2021)
Selenobacteria (Endophytes)	Se/Soybean	Improved plant growth under drought stress Increased Se content in roots, shoots, fruits and leaves	Trivedi et al. (2020)
Pseudosomonas citronellis (PC), *Pseudosomonas* sp. RA6, *Serratia* sp. S2, *Serratia marcescens* CDP13 and *Frateuria aurantia* (Symbion-K)	B, Zn and Fe/ Chickpea	Significant increase in plant B contents Significant accumulation of Zn in Shoots Increased Fe level in plants	Dogra et al. (2019)

micronutrient deficiency in our field crops. Thus, it can be anticipated that information discussed in this chapter will certainly enhance our knowledge in alleviating the micronutrient deficiencies and address nutritional security.

References

Adhikari T, Kundu S, Rao AS (2016) Zinc delivery to plants through seed coating with nano-zinc oxide particles. J Plant Nutr 39:136–146

Adrees M, Khan ZS, Hafeez M, Rizwan M, Hussain K, Asrar M, Alyemeni MN, Wijaya L, Ali S (2021) Foliar exposure of zinc oxide nanoparticles improved the growth of wheat (*Triticum aestivum* L.) and decreased cadmium concentration in grains under simultaneous Cd and water deficient stress. Ecotoxicol Environ Safe 208:111627

Ajiboye B, Cakmak I, Paterson D, De Jonge MD, Howard DL, Stacey SP, Torun AA, Aydin N, Mclaughlin MJ (2015) X-ray fluorescence microscopy of zinc localization in wheat grains biofortified through foliar zinc applications at different growth stages under field conditions. Plant Soil 392:357–370

Allen HE, Mcgrath S, Mclaughlin M, Peijnenburg W, Sauvé S, Lee C (2001) Bioavailability of metals in terrestrial ecosystems: importance of partitioning for bioavailability to invertebrates, microbes, and plants. Society of Environmental Toxicology and Chemistry

Alloway BJ (2009) Soil factors associated with zinc deficiency in crops and humans. Environ Geochem Health 31:537–548

Baligar V, Fageria N, He Z (2001) Nutrient use efficiency in plants. Commun Soil Sci Plant Anal 32:921–950

Barman M, Shukla LM, Datta SP, Rattan RK (2014) Effect of applied lime and boron on the availability of nutrients in an acid soil. J Plant Nutr 37:357–373

Behera SK, Shukla AK, Dwivedi BS, Lakaria BL (2016) Lime and zinc application influence soil zinc availability, dry matter yield and zinc uptake by maize grown on Alfisols. SOIL Discussions, 1–37

Black RE, Victora CG, Walker SP, Bhutta ZA, Christian P, De Onis M, Ezzati M, Grantham-Mcgregor S, Katz J, Martorell R (2013) Maternal and child undernutrition and overweight in low-income and middle-income countries. Lancet 382:427–451

Bouis HE, Saltzman A (2017) Improving nutrition through biofortification: a review of evidence from HarvestPlus, 2003 through 2016. Glob Food Sec 12:49–58

Bouis HE, Hotz C, Mcclafferty B, Meenakshi J, Pfeiffer WH (2011) Biofortification: a new tool to reduce micronutrient malnutrition. Food Nutr Bull 32:S31–S40

Cakmak I (2004) Identification and Correction of Widespread Zinc Deficiency in Turkey: A Success Story;(a NATO-science for Stability Project); Paper Presented to The International Fertiliser Society at a Conference in Cambridge, on 10th December 2004. International Fertiliser Society

Cakmak I, Kutman UÁ (2018) Agronomic biofortification of cereals with zinc: a review. Eur J Soil Sci 69:172–180

Cakmak I, Pfeiffer WH, Mcclafferty B (2010) Biofortification of durum wheat with zinc and iron. Cereal Chem 87:10–20

Carvalho A, Reis S, Pavia I, Lima-Brito JE (2019) Influence of seed priming with iron and/or zinc in the nucleolar activity and protein content of bread wheat. Protoplasma 256:763–775

Clemens S (2014) Zn and Fe biofortification: the right chemical environment for human bioavailability. Plant Sci 225:52–57

Dai H, Wei S, Twardowska I (2020) Biofortification of soybean (*Glycine max* L.) with Se and Zn, and enhancing its physiological functions by spiking these elements to soil during flowering phase. Sci Total Environ 740:139648

Datnoff LE, Elmer WH, Huber DM (2007) Mineral nutrition and plant disease. American Phytopathological Society (APS Press)

De Steur H, Mehta S, Gellynck X, Finkelstein JL (2017) GM biofortified crops: potential effects on targeting the micronutrient intake gap in human populations. Curr Opin Biotechnol 44:181–188

De Valença A, Bake A, Brouwer I, Giller K (2017) Agronomic biofortification of crops to fight hidden hunger in sub-Saharan Africa. Glob Food Sec 12:8–14

Dhaliwal S, Manchanda J (2009) Critical level of boron in Typic ustochrepts for predicting response of mungbean (*Phaseolus aureus* L.) to boron application. Indian J Ecol 36:22–27

Dhaliwal S, Sadana U, Khurana M, Dhadli H, Manchanda J (2010) Enrichment of rice grains with zinc and iron through ferti-fortification. Indian J Fertil 6:28–35

Dhaliwal S, Sadana U, Manchanda J (2011) Relevance and essentiality of ferti-fortification of wheat grains with manganese and copper. Indian J Fertil 7:48–55

Dhaliwal S, Sadana U, Manchanda J, Khurana M, Shukla A (2013) Differential response of maize cultivars to iron applied through ferti-fortification. Indian J Fertil 9:52–57

Dhaliwal SS, Sandhu AS, Shukla AK, Sharma V, Kumar B, Singh R (2020) Bio-fortification of oats fodder through zinc enrichment to reduce animal malnutrition. J Agric Sci Technol A 10:98–108

Dhaliwal SS, Sharma V, Shukla AK, Verma V, Behera SK, Singh P, Alotaibi SS, Gaber A, Hossain A (2021) Comparative efficiency of mineral, chelated and nano forms of zinc and iron for improvement of zinc and iron in chickpea (*Cicer arietinum* L.) through biofortification. Agronomy 11:2436

Dhaliwal SS, Sharma V, Shukla AK, Verma V, Kaur M, Shivay YS, Nisar S, Gaber A, Brestic M, Barek V (2022) Biofortification—a frontier novel approach to enrich micronutrients in field crops to encounter the nutritional security. Molecules 27:1340

Diekow J, Mielniczuk J, Knicker H, Bayer C, Dick DP, Kögel-Knabner I (2005) Soil C and N stocks as affected by cropping systems and nitrogen fertilisation in a southern Brazil Acrisol managed under no-tillage for 17 years. Soil Tillage Res 81:87–95

Dogra N, Yadav R, Kaur M, Adhikary A, Kumar S, Ramakrishna W (2019) Nutrient enhancement of chickpea grown with plant growth promoting bacteria in local soil of Bathinda, Northwestern India. Physiol Mol Biol Plants 25:1251–1259

Dragicevic V, Oljaca S, Stojiljkovic M, Simic M, Dolijanovic Z, Kravic N (2015) Effect of the maize–soybean intercropping system on the potential bioavailability of magnesium, iron and zinc. Crop Pasture Sci 66:1118–1127

Duffner A, Hoffland E, Stomph T-J, Boonstra A, Bindraban PS (2014) Eliminating zinc deficiency in rice-based systems. VFRC

Dwivedi R, Srivastva PC (2014) Effect of zinc sulphate application and the cyclic incorporation of cereal straw on yields, the tissue concentration and uptake of Zn by crops and availability of Zn in soil under rice–wheat rotation. Int J Recycl Organ Waste Agric 3:1–12

Ekanayake LJ, Thavarajah D, Mcgee R, Thavarajah P (2017) Will selenium fertilization improve biological nitrogen fixation in lentils? J Plant Nutr 40:2392–2401

Farooq M, Basra S, Tabassum R, Afzal I (2006) Enhancing the performance of direct seeded fine rice by seed priming. Plant Prod Sci 9:446–456

Farooq M, Cheema ZA, Wahid A (2012) Seed priming with boron improves growth and yield of fine grain aromatic rice. Plant Growth Regul 68:189–201

Farooq M, Ullah A, Usman M, Siddique KH (2020) Application of zinc and biochar help to mitigate cadmium stress in bread wheat raised from seeds with high intrinsic zinc. Chemosphere 260:127652

Ferreira H, Vasconcelos M, Gil AM, Pinto E (2021) Benefits of pulse consumption on metabolism and health: A systematic review of randomized controlled trials. Crit Rev Food Sci Nutr 61:85–96

García-Bañuelos ML, Sida-Arreola JP, Sánchez E (2014) Biofortification-promising approach to increasing the content of iron and zinc in staple food crops. J Elem 19

Garg M, Sharma N, Sharma S, Kapoor P, Kumar A, Chunduri V, Arora P (2018) Biofortified crops generated by breeding, agronomy, and transgenic approaches are improving lives of millions of people around the world. Front Nutr 12

Gibson RS (2007) The role of diet-and host-related factors in nutrient bioavailability and thus in nutrient-based dietary requirement estimates. Food Nutr Bull 28:S77–S100

Giller K (2001) Targeting management of organic resources and mineral fertilizers: can we match scientists' fantasies with farmers' realities? In: Integrated plant nutrient management in sub-Saharan Africa: from concept to practice. CABI Publishing, Wallingford, pp 155–171

Gulmezoglu N, Aydogan C, Turhan E (2016) Physiological, biochemical and mineral dimensions of green bean genotypes depending on Zn priming and salinity. Legume Res 39:713–721

Haas JD, Beard JL, Murray-Kolb LE, Del Mundo AM, Felix A, Gregorio GB (2005) Iron-biofortified rice improves the iron stores of nonanemic Filipino women. J Nutr 135:2823–2830

Haider MU, Hussain M, Farooq M, Nawaz A (2020) Optimizing zinc seed priming for improving the growth, yield and grain biofortification of mungbean (Vigna radiata (L.) wilczek). J Plant Nutr 43:1438–1446

Hotz C, Gibson RS (2007) Traditional food-processing and preparation practices to enhance the bioavailability of micronutrients in plant-based diets. J Nutr 137:1097–1100

Hotz C, Kabir KA, Dipti SS, Arsenault JE, Bipul M (2015) Rice fortification with zinc during parboiling may improve the adequacy of zinc intakes in Bangladesh. J Sci Food Agric 95:379–385

Huang S, Wang P, Yamaji N, Ma JF (2020) Plant nutrition for human nutrition: hints from rice research and future perspectives. Mol Plant 13:825–835

Hurst R, Siyame EW, Young SD, Chilimba AD, Joy EJ, Black CR, Ander EL, Watts MJ, Chilima B, Gondwe J (2013) Soil-type influences human selenium status and underlies widespread selenium deficiency risks in Malawi. Sci Rep 3:1–6

Joy EJ, Ander EL, Young SD, Black CR, Watts MJ, Chilimba AD, Chilima B, Siyame EW, Kalimbira AA, Hurst R (2014) Dietary mineral supplies in Africa Physiol Plant 151:208–229

Karkanis A, Ntatsi G, Lepse L, Fernández JA, Vågen IM, Rewald B, Alsiņa I, Kronberga A, Balliu A, Olle M (2018) Faba bean cultivation–revealing novel managing practices for more sustainable and competitive European cropping systems. Front Plant Sci 1115

Kaur T, Rana KL, Kour D, Sheikh I, Yadav N, Kumar V, Yadav AN, Dhaliwal HS, Saxena AK (2020) Microbe-mediated biofortification for micronutrients: present status and future chal-

lenges. In: New and future developments in microbial biotechnology and bioengineering. Elsevier, pp 1–17
Kennedy E, Jafari A, Stamoulis KG, Callens K (2020) The first Programmefood and nutrition security, impact, resilience, sustainability and transformation: Review and future directions. Glob Food Sec 26:100422
Khoshgoftarmanesh AH, Norouzi M, Afyuni M, Schulin R (2017) Zinc biofortification of wheat through preceding crop residue incorporation into the soil. Eur J Agron 89:131–139
Khum-In V, Suk-In J, In-Ai P, Piaowan K, Phaimisap Y, Supanpaiboon W, Phenrat T (2020) Combining biochar and zerovalent iron (BZVI) as a paddy field soil amendment for heavy cadmium (Cd) contamination decreases Cd but increases zinc and iron concentrations in rice grains: a field-scale evaluation. Process Saf Environ Prot 141:222–233
Klikocka H, Marx M (2018) Sulphur and nitrogen fertilization as a potential means of agronomic biofortification to improve the content and uptake of microelements in spring wheat grain DM. J Chem 2018:1–12
Kumar D, Dhaliwal S, Uppal R, Ram H (2016) Influence of nitrogen, zinc and iron fertilizer on growth parameters and yield of Parmal rice in transplanted condition. Indian J Ecol 43:115–118
Kumar V, Singhal RK, Kumar N, Bose B (2020) Micro-nutrient seed priming: a pragmatic approach towards abiotic stress management. In: New frontiers in stress management for durable agriculture. Springer, pp 231–255
Kumssa DB, Joy EJ, Ander EL, Watts MJ, Young SD, Walker S, Broadley MR (2015) Dietary calcium and zinc deficiency risks are decreasing but remain prevalent. Sci Rep 5:1–11
Kutman UB, Yildiz B, Cakmak I (2011) Improved nitrogen status enhances zinc and iron concentrations both in the whole grain and the endosperm fraction of wheat. J Cereal Sci 53:118–125
Lawson PG, Daum D, Czauderna R, Meuser H, Härtling JW (2015) Soil versus foliar iodine fertilization as a biofortification strategy for field-grown vegetables. Front Plant Sci 450
Leskowá A, Giehl RF, Hartmann A, Farga¡ Ová A, Von Wirén N (2017) Heavy metals induce iron deficiency responses at different hierarchic and regulatory levels. Plant Physiol 174:1648–1668
Lyons GH, Genc Y, Stangoulis JC, Palmer LT, Graham RD (2005) Selenium distribution in wheat grain, and the effect of postharvest processing on wheat selenium content. Biol Trace Elem Res 103:155–168
Ma Y, Schwenke G, Sun L, Li Liu D, Wang B, Yang B (2018) Modeling the impact of crop rotation with legume on nitrous oxide emissions from rain-fed agricultural systems in Australia under alternative future climate scenarios. Sci Total Environ 630:1544–1552
Maertens M, Velde KV (2017) Contract-farming in staple food chains: the case of rice in Benin. World Dev 95:73–87
Manzeke GM, Mapfumo P, Mtambanengwe F, Chikowo R, Tendayi T, Cakmak I (2012) Soil fertility management effects on maize productivity and grain zinc content in smallholder farming systems of Zimbabwe. Plant Soil 361:57–69
Manzeke GM, Mtambanengwe F, Nezomba H, Mapfumo P (2014) Zinc fertilization influence on maize productivity and grain nutritional quality under integrated soil fertility management in Zimbabwe. Field Crop Res 166:128–136
Manzeke MG, Mtambanengwe F, Nezomba H, Watts MJ, Broadley MR, Mapfumo P (2017) Zinc fertilization increases productivity and grain nutritional quality of cowpea (*Vigna unguiculata* [L.] Walp.) under integrated soil fertility management. Field Crop Res 213:231–244
Marra R, Lombardi N, Piccolo A, Bazghaleh N, Prashar P, Vandenberg A, Woo S (2021) Mineral biofortification and growth stimulation of lentil plants inoculated with trichoderma strains and metabolites. Microorganisms 10:87
Marschner P, Rengel Z (2012) Nutrient availability in soils. In: Marschner's mineral nutrition of higher plants. Elsevier, pp 315–330
Martín-Cabrejas MÁ (2019) Legumes: Nutritional quality, processing and potential health benefits. Royal Society of Chemistry
Mcdonald MB (2000) Seed priming. Seed technology and its biological basis. Sheffield Academic Press, Sheffield, pp 287–325

Mfeka N, Mulidzi RA, Lewu FB (2019) Growth and yield parameters of three cowpea (Vigna unguiculata L. Walp) lines as affected by planting date and zinc application rate. S Afr J Sci 115:1–9

Monreal C, Derosa M, Mallubhotla S, Bindraban P, Dimkpa C (2016) Nanotechnologies for increasing the crop use efficiency of fertilizer-micronutrients. Biol Fertil Soils 52:423–437

Myers SS, Zanobetti A, Kloog I, Huybers P, Leakey AD, Bloom AJ, Carlisle E, Dietterich LH, Fitzgerald G, Hasegawa T (2014) Increasing CO2 threatens human nutrition. Nature 510:139–142

Nasar J, Shah Z (2017) Effect of iron and molybdenum on yield and nodulation of lentil. ARPN J Agric Biol Sci 12:332–339

Nayak BK, Adhikary S, Pattnaik M, Pal AK (2020) Effect of Sulphur and zinc with combination of FYM on yield and uptake of nutrients in Mustard (*Brassica juncea* L.) under Alfisols of Odisha. J Pharmacog Phytochem 9:2310–2313

Newman RG, Moon Y, Sams CE, Tou JC, Waterland NL (2021) Biofortification of sodium selenate improves dietary mineral contents and antioxidant capacity of culinary herb microgreens. Front Plant Sci 12

Noreen S, Sultan M, Akhter MS, Shah KH, Ummara U, Manzoor H, Ulfat M, Alyemeni MN, Ahmad P (2021) Foliar fertigation of ascorbic acid and zinc improves growth, antioxidant enzyme activity and harvest index in barley (*Hordeum vulgare* L.) grown under salt stress. Plant Physiol Biochem 158:244–254

Nubé M, Voortman R (2011) Human micronutrient deficiencies: linkages with micronutrient deficiencies in soils, crops and animal nutrition. In: Combating micronutrient deficiencies: Food-based approaches, vol 7, p 289

Oliver MA, Gregory P (2015) Soil, food security and human health: a review. Eur J Soil Sci 66:257–276

Padbhushan R, Sharma S, Kumar U, Rana D, Kohli A, Kaviraj M, Parmar B, Kumar R, Annapurna K, Sinha AK (2021) Meta-analysis approach to measure the effect of integrated nutrient management on crop performance, microbial activity, and carbon stocks in Indian soils. Front Environ Sci 396

Phattarakul N, Rerkasem B, Li L, Wu L, Zou C, Ram H, Sohu V, Kang B, Surek H, Kalayci M (2012) Biofortification of rice grain with zinc through zinc fertilization in different countries. Plant Soil 361:131–141

Pingoliya K, Dotaniya M, Lata M (2014) Effect of iron on yield, quality and nutrient uptake of chickpea (*Cicer arietinum* L.). Afr J Agric Res 9:2841–2845

Pooniya V, Shivay YS (2013) Enrichment of basmati rice grain and straw with zinc and nitrogen through ferti-fortification and summer green manuring under Indo-Gangetic plains of India. J Plant Nutr 36:91–117

Praharaj S, Skalicky M, Maitra S, Bhadra P, Shankar T, Brestic M, Hejnak V, Vachova P, Hossain A (2021) Zinc biofortification in food crops could alleviate the zinc malnutrition in human health. Molecules 26:3509

Prakash NB, Dhumgond P, Shruthi, and Ashrit, S. (2020) Performance of slag-based gypsum on maize yield and available soil nutrients over commercial gypsum under acidic and neutral soil. Commun Soil Sci Plant Anal 51:1780–1798

Prasad R, Shivay YS, Kumar D (2014) Agronomic biofortification of cereal grains with iron and zinc. Adv Agron 125:55–91

Puccinelli M, Landi M, Maggini R, Pardossi A, Incrocci L (2021) Iodine biofortification of sweet basil and lettuce grown in two hydroponic systems. Sci Hortic 276:109783

Ram H, Sohu V, Cakmak I, Singh K, Buttar G, Sodhi G, Gill H, Bhagat I, Singh P, Dhaliwal S (2015) Agronomic fortification of rice and wheat grains with zinc for nutritional security. Curr Sci 109:1171–1176

Rasheed N, Maqsood MA, Aziz T, Jabbar A (2020) Characterizing lentil germplasm for zinc biofortification and high grain output. J Soil Sci Plant Nutr 20:1336–1349

Reddy MM, Padmaja B, Malathi S, Rao LJ (2007) Effects of micronutrients on growth and yield of pigeonpea

Rehman A-U, Farooq M (2013) Boron application through seed coating improves the water relations, panicle fertility, kernel yield, and biofortification of fine grain aromatic rice. Acta Physiol Plant 35:411–418

Rehman A, Farooq M (2016) Zinc seed coating improves the growth, grain yield and grain biofortification of bread wheat. Acta Physiol Plant 38:1–10

Rehman A, Farooq M, Naveed M, Nawaz A, Shahzad B (2018) Seed priming of Zn with endophytic bacteria improves the productivity and grain biofortification of bread wheat. Eur J Agron 94:98–107

Rehman HM, Cooper JW, Lam HM, Yang SH (2019) Legume biofortification is an underexploited strategy for combatting hidden hunger. Plant Cell Environ 42:52–70

Rietra R, Heinen M, Dimpla C, Bindraban P (2015) Effects of nutrients antagonism and synergism on fertilizer use efficiency

Rizwan M, Ali S, Ali B, Adrees M, Arshad M, Hussain A, Ur Rehman MZ, Waris AA (2019) Zinc and iron oxide nanoparticles improved the plant growth and reduced the oxidative stress and cadmium concentration in wheat. Chemosphere 214:269–277

Saltzman A, Birol E, Bouis HE, Boy E, De Moura FF, Islam Y, Pfeiffer WH (2013) Biofortification: progress toward a more nourishing future. Glob Food Sec 2:9–17

Sanchez PA, Swaminathan MS (2005) Hunger in Africa: the link between unhealthy people and unhealthy soils. Lancet 365:442–444

Sandhu A, Dhaliwal S, Shukla A, Sharma V, Singh R (2020) Fodder quality improvement and enrichment of oats with Cu through biofortification: a technique to reduce animal malnutrition. J Plant Nutr 43:1378–1389

Sandstroèm B (2001) Micronutrient interactions: effects on absorption and bioavailability. Br J Nutr 85:S181–S185

Santos S, Costa CA, Duarte AC, Scherer HW, Schneider RJ, Esteves VI, Santos EB (2010) Influence of different organic amendments on the potential availability of metals from soil: a study on metal fractionation and extraction kinetics by EDTA. Chemosphere 78:389–396

Selim MM (2020) Introduction to the integrated nutrient management strategies and their contribution to yield and soil properties. Int J Agron 2020

Shi R, Zhang Y, Chen X, Sun Q, Zhang F, Römheld V, Zou C (2010) Influence of long-term nitrogen fertilization on micronutrient density in grain of winter wheat (*Triticum aestivum* L.). J Cereal Sci 51:165–170

Shinde P, Doddagoudar S, Vasudevan S (2017) Influence of seed polymer coating with micronutrients and foliar spray on seed yield of chickpea (*Cicer arietinum* L.). Legum Res 40:704–709

Singh J, Karamanos R, Stewart J (1988) The mechanism of phosphorus-induced zinc deficiency in bean (*Phaseolus vulgaris* L.). Can J Soil Sci 68:345–358

Singh D, Rajawat MVS, Kaushik R, Prasanna R, Saxena AK (2017) Beneficial role of endophytes in biofortification of Zn in wheat genotypes varying in nutrient use efficiency grown in soils sufficient and deficient in Zn. Plant Soil 416:107–116

Singh BR, Timsina YN, Lind OC, Cagno S, Janssens K (2018) Zinc and iron concentration as affected by nitrogen fertilization and their localization in wheat grain. Front Plant Sci 9:307

Skrypnik L, Novikova A, Tokupova E (2019) Improvement of phenolic compounds, essential oil content and antioxidant properties of sweet basil (*Ocimum basilicum* L.) depending on type and concentration of selenium application. Plants 8:458

Smith SE, Read DJ (2010) Mycorrhizal symbiosis. Academic Press

Soltani SM, Hanafi MM, Samsuri AW, Muhammed SKS, Hakim MA (2016) Rice growth improvement and grains bio-fortification through lime and zinc application in zinc deficit tropical acid sulphate soils. Chem Spec Bioavailab 28:152–162

Tavakkoli E, Uddin S, Rengasamy P, Mcdonald GK (2022) Field applications of gypsum reduce pH and improve soil C in highly alkaline soils in southern Australia's dryland cropping region. Soil Use Manag 38:466–477

Thapa S, Bhandari A, Ghimire R, Xue Q, Kidwaro F, Ghatrehsamani S, Maharjan B, Goodwin M (2021) Managing micronutrients for improving soil fertility, health, and soybean yield. Sustainability 13:11766

Thilakarathna MS, Raizada MN (2015) A review of nutrient management studies involving finger millet in the semi-arid tropics of Asia and Africa. Agronomy 5:262–290

Toenniessen G, Adesina A, Devries J (2008) Building an alliance for a green revolution in Africa. Ann N Y Acad Sci 1136:233–242

Traore K (2006) Effects of soil amendments and drought on zinc husbandry and grain quality in Sahelian sorghum. Wageningen University and Research

Trivedi G, Patel P, Saraf M (2020) Synergistic effect of endophytic selenobacteria on biofortification and growth of Glycine max under drought stress. S Afr J Bot 134:27–35

Ullah A, Farooq M, Rehman A, Arshad MS, Shoukat H, Nadeem A, Nawaz A, Wakeel A, Nadeem F (2018) Manganese nutrition improves the productivity and grain biofortification of bread wheat in alkaline calcareous soil. Exp Agric 54:744–754

Ulm F, Avelar D, Hobson P, Penha-Lopes G, Dias T, Máguas C, Cruz C (2019) Sustainable urban agriculture using compost and an open-pollinated maize variety. J Clean Prod 212:622–629

Van Noordwijk M, Cerri C, Woomer PL, Nugroho K, Bernoux M (1997) Soil carbon dynamics in the humid tropical forest zone. Geoderma 79:187–225

Vanlauwe B, Bationo A, Chianu J, Giller KE, Merckx R, Mokwunye U, Ohiokpehai O, Pypers P, Tabo R, Shepherd KD (2010) Integrated soil fertility management: operational definition and consequences for implementation and dissemination. Outlook Agric 39:17–24

Vanlauwe B, Descheemaeker K, Giller KE, Huising J, Merckx R, Nziguheba G, Wendt J, Zingore S (2015) Integrated soil fertility management in sub-Saharan Africa: unravelling local adaptation. Soil 1:491–508

Voortman R, Bindraban P (2015) Beyond N and P: towards a land resource ecology perspective and impactful fertilizer interventions in Sub-Sahara Africa. VFRC Reports

Watts-Williams SJ, Cavagnaro TR (2018) Arbuscular mycorrhizal fungi increase grain zinc concentration and modify the expression of root ZIP transporter genes in a modern barley (*Hordeum vulgare*) cultivar. Plant Sci 274:163–170

Yadav R, Ror P, Rathore P, Ramakrishna W (2020) Bacteria from native soil in combination with arbuscular mycorrhizal fungi augment wheat yield and biofortification. Plant Physiol Biochem 150:222–233

Zhang F, Shen J, Zhang J, Zuo Y, Li L, Chen X (2010) Rhizosphere processes and management for improving nutrient use efficiency and crop productivity: implications for China. Adv Agron 107:1–32

Zimmermann MB, Hurrell RF (2007) Nutritional iron deficiency. Lancet 370:511–520

Zingore S, Delve RJ, Nyamangara J, Giller KE (2008) Multiple benefits of manure: the key to maintenance of soil fertility and restoration of depleted sandy soils on African smallholder farms. Nutr Cycl Agroecosyst 80:267–282

Chapter 4
General Aspects of Genetic Improvement (Traditional and Transgenic Methods) Aiming to Food Biofortification

Aiman Hina, Asim Abbasi, Safa Imtiaz, Inzamam Ul Haq, Ahtsham Ul Hassan, and Amna Chaudhry

Abstract Micronutrient deficiency has imparted an unseen burden on human population in terms of hunger, and diseases leading to higher mortality rates. Several measures such as supplements and fortified food have been taken into account to correct the nutrient deficiencies targeting upgrades in the general public health. But soon it was realized, that the best way to cure malnourishment is the use of an inclusive strategy ensuring a healthy balanced human diet. Biofortification offers significant potential for enhancing the nutritional content of important crops and has arisen as a sustainable and an economical strategy to curb nutrient deficiencies around the globe. Food biofortification is the process of increasing the nutritional content of staple foods through genetic improvement. This can be achieved through both traditional and transgenic methods. Traditional methods of genetic improvement

A. Hina (✉)
Ministry of Agriculture (MOA) National Centre for Soybean Improvement, State Key Laboratory for Crop Genetics and Germplasm Enhancement, Nanjing Agricultural University, Nanjing, China

A. Abbasi
Department of Environmental Sciences, Kohsar University Murree, Murree, Pakistan

S. Imtiaz
Department of Botany, Kohsar University Murree, Murree, Pakistan

I. Ul Haq
College of Plant Protection, Gansu Agricultural University, Lanzhou, China

A. Ul Hassan
Centre of Agricultural Biochemistry and Biotechnology (CABB), University of Agriculture Faisalabad, Faisalabad, Pakistan

A. Chaudhry
Centre of Agricultural Biochemistry and Biotechnology (CABB), University of Agriculture Faisalabad, Faisalabad, Pakistan

Centre of Agricultural Biochemistry and Biotechnology (CABB), Center for Advanced Studies in Agriculture and Food Security (CASAFS), University of Agriculture Faisalabad, Faisalabad, Pakistan

M. A. Nadeem et al. (eds.), *Legumes Biofortification*,
https://doi.org/10.1007/978-3-031-33957-8_4

include selective breeding, where plants with desirable traits are chosen for breeding to produce new varieties with improved nutritional content. This method can be time-consuming and may take many generations to produce plants with the desired traits. In addition, it is based on the availability of genetic variation within the crop species. Transgenic methods, on the other hand, involve the transfer of specific genes from one organism to another to produce plants with improved nutritional content. This can be done through various techniques such as genetic engineering and gene editing. Transgenic methods are faster and more precise than traditional methods, and can be used to introduce new traits that do not exist in the target crop species. Both traditional and transgenic methods have their own set of advantages and disadvantages. Traditional methods are generally considered to be more socially and environmentally acceptable, while transgenic methods are faster and more precise. However, the latter method still faces concerns and opposition from some sectors of society and government. In conclusion, food biofortification is a crucial strategy for addressing micronutrient malnutrition, which is a global problem that affects millions of people. Genetic improvement through both traditional and transgenic methods has the potential to improve the nutritional content of staple foods and help to combat micronutrient malnutrition.

1 Introduction

1.1 *Overpopulation, Malnutrition, and Food Security*

The constant increase in human population is spreading the wave of malnutrition worldwide which is affecting around 3 billion individuals of the total world's population (Carvalho and Vasconcelos 2013; Hoekenga 2014). In general terms, malnutrition is referred to the insufficient or surplus intake of food, causing nutrient deficiencies or improper nutrient absorption. According to the World Health Organization (WHO), malnutrition is categorized into two broad groups: undernourishment and overnourishment (WHO 2000), both conditions lead to stunted growth, obesity, impaired immunity, depression, anxiety and increase risk factors of human life (Ersado 2022; Woeltje et al. 2023). Moreover, in developing countries, where people are living under the poverty line, malnutrition is known as hidden hunger causing infectious diseases such as pneumonia, measles, diarrhea, malaria, and dengue, eventually increasing the mortality percentage to an alarming rate (WHO 2000; Gillespie et al. 2016).

Recent studies showed that almost 2 billion of the population is suffering from obesity, 140 million children are of stunted growth, and approximately 43 and 30 percent are anemic and vitamin A deficient at their early ages (McLean et al. 2009; Popkin et al. 2020; Stevens et al. 2015). Sadly, malnutrition is not only threatening existing lives but it is also risking future lives (Haseen 2005). For example, a nutrient-deficient pregnant woman would be unable to give birth to a normal and healthy child and might face the prebirth death of the fetus due to insufficient fetal growth (Mannar and Sankar 2004; UNICEF 2007).

Generally, diet trends in our societies are not diversified, i.e., absent in micronutrients (zinc, iron, iodine, and selenium) (Genc et al. 2005; Genç et al. 2009). These micronutrients are essential for human growth and in the regulation of important metabolic pathways, especially in children and women (Singh 2009). Population living in underdeveloped countries is more prone to nutrient deficiency due to the unavailability of resources and lack of awareness about a balanced diet. Consequently, low and imbalance consumption of essential elements causes nutrient deficiency which leads to multiple infectious diseases (Borrill et al. 2014; Graham et al. 2001; Kumar et al. 2015, 2019; Sinha et al. 2019; White and Broadley 2009).

To counter the problem of malnutrition, different strategies have been employed such as diversification of food, use of supplementation and food fortification to ensure nutrient-rich balanced diet. With the increasing industrialization, different manufacturers commonly recommend the use of fortified pharmaceutical supplementations (Naqvi et al. 2009). Food fortification refers to the addition of minerals and vitamins into major food-consuming items to increase the nutritional status of food products. The strategy of food fortification is marked as one of the most economical strategies to curb the state of malnourishment (Bhutta et al. 2013; Hoddinott et al. 2012; Horton et al. 2008; WHO 2000). The fast growth of commercialization and increased socio-economic status of the population has changed their preferences to use manufactured supplements instead of a naturally balanced diet (Ahn et al. 2015; Spohrer et al. 2013). However, in underdeveloped countries where the majority of people have low-income status and are illiterate, the implementation of expensive food supplement programs is difficult to achieve (Bohra et al. 2016; Kumar et al. 2015; Timmer 2003).

Therefore, it is suggested that instead of working on separate approaches, there should be a concrete, long-lasting and economical strategy to combat the problem of malnutrition (Bohra et al. 2016). Different schools of thought have agreed to the use of one comprehensive strategy of biofortification either by using nutrient-rich fertilizers, genetic manipulation of crop seeds either via inbreeding/outbreeding or via producing genetically modified crops (Cakmak 2008; Çakmak et al. 2004; Graham et al. 2007; Kumar et al. 2015; Pfeiffer and McClafferty 2007; White and Broadley 2009).

1.2 *Food Biofortification*

Food biofortification is the practice of increasing the nutritional value of crops through genetic improvement. Genetic improvement is the process of using genetic techniques to enhance the desirable traits of plants or animals. There are two main methods of genetic improvement: traditional methods and transgenic methods (Garg et al. 2018). Among other fortification strategies, biofortification is considered the most economical one without any further cost. Development and utilization of biofortified seeds could positively impact overall agriculture, as the presence of a higher amount of minerals and essential amino acids in plant seed would safeguard the plant from any biotic and abiotic attack as well as boost the crop yield (Welch and Graham 2004). For rural areas, biofortification is a viable, efficacious, simple and recommended method to eliminate the problem of malnutrition and hunger. Considering the socioeconomic status of the rural population, biofortification acts as a cheaper and sustainable solution to provide nutrient-rich food as this method does not require huge capital in the form of recurrent investments, instead initial one-time investment is applicable for the development, adaptability practices and distribution of biofortified seed among the farming community (Graham et al. 2001; Kumar et al. 2015). The fundamental objective of crop biofortification is to increase the number of trace elements, e.g., vitamin A, zinc, iron, iodine, amino acid, and proteins in major crops such as potato, wheat, rice, maize, sorghum, soybean, pearl millet, etc.

The successful implementation of biofortification is influenced by various factors. The most important is to combine the micronutrient stability feature with the increased crop yield so that the farming community could get benefited. Next, it is required to biofortified products with maximum consumption on daily basis and should have a positive impact on human health. Further, implementation of biofortification requires awareness, education and incentives to those suffering from malnutrition in rural areas to maximize the benefits of the biofortification process in terms of adaptability and smooth delivery. The benefits of this strategy, which focuses on disadvantaged families living in remote areas with limited to no access to industrially fortified food, have been identified through research on biofortification. These families may raise, consume, and market their fortified crops and frequently rely on subsistence farming. Also, biofortification can help food systems serve more nutrient-dense foods at a lower cost when properly targeted (Andersson et al. 2017). One study reported that vitamin A intake among women of reproductive age and children increased after orange sweet potatoes were made available to farmers in Africa. Multiple micronutrient deficiencies are frequently caused by nutrient-poor diets including tubers and cereals; however, biofortification can boost a crop's nutritional value by utilizing conventional breeding and agronomic biofortification procedures. However, genetic engineering-based biofortification, enabling the incorporation of various micronutrients into a single genotype could support higher levels of mineral accumulation (Van Der Straeten et al. 2020).

2 Strategies for Biofortification

Generally, three main strategies have been developed for the biofortification of crops, such as the use of agronomic practices, plant breeding schemes and biotechnological approaches (Fig. 4.1). This approach is mainly focused on the improvement of staple crops with an elevated amount of essential minerals and vitamins. One of the important influencing factors in the optimization of biofortification

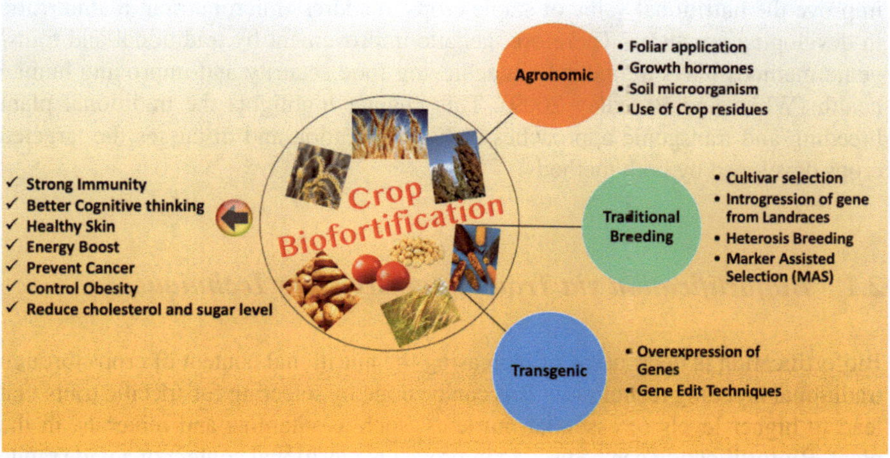

Fig. 4.1 Commonly used strategies for biofortification

strategies is neglecting postharvest losses. In the case of cereals, a huge number of trace elements, i.e., copper, zinc, and iron are stored in seed and bran and the majority of the seeds are wasted during flour milling and polishing which causes severe mineral losses (Gregorio et al. 2000; Waters and Sankaran 2011). Naturally, few crops possess antinutrients, e.g., fiber, phytate, and tannins which inhibit the biological availability of nutrients (Holme et al. 2012; Welch and Graham 2004). Moreover, with the emerging climatic changes with limited water supply, it would be difficult to maintain both crop yield and quality after fortifying crops. Additionally, a limited number of genes are reported regulating the mineral absorbance into edible parts of plants and vascular translocation of minerals in staple crops (Haas et al. 2005).

Attempts have been made to develop biofortified legumes, cereals, vegetables, and oilseed crops by using all three methods; however, the practical application of any procedure is mainly depended on the availability of genetic diversity of specific crops. The main goal of agronomical methods is to increase the solubility and mobilization of minerals in the soil, as well as the optimized use of biofertilizers, whereas traditional methods of genetic improvement include selective breeding, hybridization, and mutation breeding. These methods involve the manipulation of the genetic material through the controlled crossing of plants or animals with desirable traits, or the induction of mutations in the genetic material. These methods have been used for centuries to improve crops for increased yield, disease resistance, and nutritional content. Transgenic methods, also known as genetic engineering, involve the direct manipulation of the genetic material of an organism. This can be done by introducing new genes into the organism or by altering the expression of existing genes. Transgenic methods have been used to improve crops for increased yield, disease resistance, and nutritional content. It is a powerful tool to biofortify food crops by introducing genes that enhance their nutritional value. Food biofortification aims to improve the nutritional value of staple crops to address micronutrient malnutrition in developing countries. Therefore, genetic improvement by traditional and transgenic methods plays a vital role in achieving food security and improving human health (White and Broadley 2009). This chapter highlights the traditional plant breeding and transgenic approaches of biofortification and discusses the targeted crops developed by both methods.

2.1 Biofortification via Traditional Breeding Techniques

Biofortification is the process of increasing the nutritional content of crops through traditional breeding techniques. This can be done by selecting for specific traits that lead to higher levels of essential nutrients, such as vitamins and minerals, in the plant. Biofortification can be used to improve the nutritional content of a wide range of crops, including cereals, legumes, and vegetables.

Traditional breeding techniques have been widely used to develop biofortified crop varieties. These techniques involve selecting for specific traits in the plant, such as increased nutrient content, through repeated cycles of crossbreeding and selection. This can be done through both conventional breeding and marker-assisted selection (MAS). Conventional breeding relies on the inherent genetic variation of the plant, while MAS uses molecular markers to identify specific genes associated with the desired trait.

Irrespective of the promising progress made by traditional plant breeding in the exploitation of genetic material to develop nutrient-rich cultivars, yet, more efforts are required for practical application. The main bottlenecks in the implementation of traditional breeding schemes are the unavailability of enough variation in the genetic pool of important staple crops in addition to the lengthy and time-consuming process of gene incorporation into a single genotype to develop elite cultivars. Besides, the success of traditional breeding programs is also challenged by the negative correlation between crop yield and mineral content of grains in crops, which reduces its application (Fan et al. 2008; Garvin et al. 2006; Oury et al. 2006; Shi et al. 2013). Traditional breeding based biofortification is a widely used technique to upgrade the nutrient profile of crops. Using these strategies, a substantial number of crops have been targeted for fortification. One example of biofortification via traditional breeding techniques is the development of iron-rich pearl millet. Pearl millet is a staple food crop in many parts of Africa and India, but it is low in iron, which can lead to anemia in people who rely on it as a primary source of food. Researchers have used traditional breeding techniques to develop pearl millet varieties that are high in iron, by selecting for plants that contain higher levels of the mineral in the grain. This has been done through both conventional breeding and MAS. Another example is the biofortification of rice with vitamin A. Rice is a staple food for more than half of the world's population, but it is low in vitamin A. Vitamin A deficiency is a major public health problem, especially in developing countries. Researchers have used traditional breeding techniques to develop rice varieties that are high in vitamin A, by selecting for plants that contain higher levels of the nutrient in the grain (Garg et al. 2018; Bouis and Welch 2010).

Multiple international organizations have started initiatives to increase the nutritional value of crops using traditional breeding because this strategy is probably the fastest means of improving plants, e.g., the project of health grain was jointly carried out by fifteen countries to promote the idea of healthy and quality cereals (Bouis and Welch 2010). Many scientists reported the mechanism of nutrient absorption, utilization of genetic variation in cereals and the effect of biofortified crops on well-being of vulnerable population (Lafiandra et al. 2014; Tighe et al. 2010). According to an estimate, traditionally developed biofortified crops provide 20 percent of the extra energy for daily requirements. The estimated average requirement was reported to increase up to 25, 35 and 85 percent for zinc, iron, and provitamin A biofortified crops, respectively (Andersson et al. 2017; Bouis and Saltzman 2017).

A program named HarvestPlus was launched as a part of CGIAR based on the traditional breeding and distribution of biofortified in Africa and Asia focusing on iron, zinc and vitamin A improvement in wheat, rice, pearl millet, sweet potato and bean (Bouis and Welch 2010). One of the most well-known examples of biofortification through traditional breeding methods is the development of "Orange Fleshed Sweet Potato." The sweet potato is a staple crop in many sub-Saharan Africa countries, and vitamin A deficiency is a major public health concern in these regions. By selecting for and crossbreeding sweet potatoes with high levels of beta-carotene (a precursor to vitamin A), scientists were able to develop a variety of sweet potato that is rich in vitamin A. This variety has been successfully introduced in several African countries and it has been shown to have a significant impact on vitamin A deficiency. The statistics of HarvestPlus showed that to date, 283 biofortified varieties have been released in 30 different countries while a few are still under trials (Table 4.1). So far, HarvestPlus is facilitating 48 million individuals in rural areas with the conventionally developed biofortified crop with a future goal to target 01 billion vulnerable populations by the end of 2030 (Pfeiffer and McClafferty 2007). Multiple trials have been conducted to record the efficacy of biofortified food in the malnourished group of the population and showed improved task management with maximum critical thinking (De Moura et al. 2014; Palmer et al. 2016; Sazawal et al. 2018). The breeding method includes identifying rice lines that have a high concentration of micronutrients that are important for the human population. The discovery of rice germplasm with a greater level of micronutrients initiated numerous studies (Anandan et al. 2011; Anuradha et al. 2012; Babu et al. 2012). Traditional rice varieties appear to have higher main mineral content than contemporary high-yielding types, including Fe, Zn, Cu, and Mn (Anandan et al. 2011). Biofortification through traditional breeding methods is a sustainable, cost-effective, safe, and noninvasive approach that can help address malnutrition and micronutrient deficiencies in many parts of the world. By selecting for and crossbreeding plants that naturally have higher levels of essential nutrients, scientists are able to develop crops that can have a positive impact on public health. As the global population continues to grow,

Table 4.1 Current status of biofortified crops around the globe

Country	Targeted crop	Status Testing/Released	Fortificants
Afghanistan	Wheat	Testing	Zinc
Bangladesh	Rice	Released	Zinc
	Wheat	Released	Zinc
	Sweet potato	Released	Vitamin A
	Lentil	Released	Zinc and iron
Bhutan	Wheat	Testing	Zinc
	Potato	Testing	Zinc and iron
Cambodia	Rice	Testing	Zinc
Brazil	Bean	Released	Iron
	Rice	Testing	Zinc
	Wheat	Released	Zinc
	Cassava	Released	Vitamin A
	Maize	Released	Vitamin A
	Sweet potato	Released	Vitamin A
	Cowpea	Released	Iron and zinc
Mexico	Maize	Testing	Zinc
	Wheat	Released	Zinc
	Cassava	Testing	Vitamin A
	Maize	Testing	Vitamin A
South Africa	Wheat	Testing	Zinc
	Maize	Testing	Vitamin A
	Sweet potato	Released	Vitamin A
Zimbabwe	Bean	Released	Iron
	Pearl millet	Testing	Iron
	Wheat	Testing	Zinc
	Maize	Released	Vitamin A
	Sweet potato	Testing	Vitamin A
Pakistan	Wheat	Released	Zinc
	Maize	Testing	Vitamin A
	Lentil	Testing	Iron and zinc
India	Pearl millet	Released	Iron
	Rice	Released	Zinc
	Wheat	Released	Zinc
	Maize	Testing	Vitamin A
	Sweet potato	Released	Vitamin A
	Cowpea	Released	Iron and zinc
	Potato	Testing	Iron and zinc
	Lentil	Released	Iron and zinc
	Sorghum	Released	Zinc and iron

Source: HarvestPlus (Pfeiffer and McClafferty 2007) (https://www.harvestplus.org)

biofortification through traditional breeding methods will become an increasingly important tool in the fight against malnutrition.

2.1.1 Cultivar Selection

Selection and introduction of nutrient-rich germplasm is the prerequisite for breeding biofortified crops. Using high zinc and iron genotypes as parents in targeted breeding schemes will lead to the development of biofortified and higher yielding commercial cultivars. The crop variety having high nutriments will also increase crop production even on less fertile soil. A multiple range of germplasm has been tested to study the relationship of nutrient-rich crop with environmental and soil conditions (Vanisha et al. 2013). Therefore, germplasm screening for zinc and iron contents have been proved effective under zinc and iron deficient conditions in wheat, cassava, maize, sweet potato, rice, beans etc., (Graham et al. 2001). At ICRISAT, biofortified crops were evaluated in multi locational trials and does not find any relationship between the nutrient level of cereal grains and soil (Govindaraj et al. 2013; Kanatti et al. 2014).

In large populations, sufficient genetic diversity is required to choose potential cultivars carrying greater amounts of micronutrients which could further be used in breeding programs (Ortiz-Monasterio et al. 2007; Tong et al. 2014). For example, different studies identified nutrient-rich cultivars in oats, barely, etc. (de Bruijn et al. 2019; Doehlert et al. 2013; Loskutov and Polonskiy 2017). In pearl millet, a variety named *niadi* was identified as early maturing, large seeded, in addition to higher content of Fe and Zn, this variety have been used in multiple breeding studies (Govindaraj et al. 2013; Kanatti et al. 2014; Rai et al. 2016; Velu et al. 2011).

Additionally, studies reported that genotype x environment interaction (GEI) greatly influence the adaptation and production of any crop which makes the crop acclimatization more challenging. The GEI has reported to influence the gene expression underlying micronutrients and their percentage in biofortified crops (Govindaraj et al. 2013; Kanatti et al. 2014; Rai et al. 2016; Reynolds et al. 2005). In Pakistan, provitamin A-biofortified maize hybrids were successfully evaluated for stability analysis across multiple environments and showed higher yield (Maqbool et al. 2018).

2.1.2 Introgression of Genes from Landraces

The most widely used form of biofortification is through traditional breeding providing an economical and sustainable substitute for other methods. Though, to be successful, the trait of interest must have enough genotypic diversity. This variation could help in the elevation of the micronutrient level up to the mark. Generally, traditional breeding techniques involve the crossing of the nutrient-rich donor parent with the agronomically outstanding parent. However, the limited genetic diversity that exists in the gene pool must occasionally be taken into account in breeding methods. This could be avoided by crossing to land races, which gradually introduces the trait into the commercial varieties. Traditional breeding offers the introgression of useful genes from land races and wild relatives into modern cultivars to enhance their nutritional value. Therefore, screening of wild germplasm is crucial to explore and utilize existing genetic diversity. Studies reported that in barley, a huge number of genetic diversity rich in micronutrients remains unexplored due to the limited gene introgression experiments (Detterbeck et al. 2016; Gyawali et al. 2019; Yang et al. 2011). The few genetic studies of barley grain showed that alleles obtaining from wild ancestors are associated with the nutrient increase. Introgression of such alleles into commercial cultivar will help in developing biofortified barley cultivars (Wiegmann et al. 2019). However, this introgression of genes from ancestors is also exhausting and time-consuming in terms of a breeding complex trait such as enhancing concentrations of selenium (Se) and linolenic fatty acid in wheat and soybean, respectively (Lyons et al. 2005; Oliva et al. 2006).

Lately, participatory varietal selection (PVS) in backcross breeding program has been significantly used in biofortification of cereals. PVS, is basically performance-based assessment of any crop, which is conducted by farming communities to evaluate various important traits to choose elite variety. This PVS based backcross breeding program was conducted by collaborative efforts of CIMMYT, Punjab Agricultural University and Indian Institute of Wheat and Barley Research, resulting in the commercial release of biofortified wheat (Saini et al. 2020). The *Aegilops tasuchii*, wild relative of wheat, is enriched in Fe and Zn contents and has been exploited to improve the micronutrients percentage of modern wheat genotypes (Arora et al. 2019). So far, through introgression of genes from *Ae. Squarrosa* and *T. spelta,* four zinc biofortified wheat varieties have been developed and under cultivation in Pakistan and India with incremented zinc content (Singh and Govindan 2017; Singh et al., 2017).

2.1.3 Heterosis Breeding

The superiority of hybrid over its parents is termed as heterosis which greatly influence the success of hybrid breeding program. The genetic expression of iron and zinc is controlled by additive gene action; therefore, it is hard to achieve better-parent heterosis; instead, heterosis could be due to dominance or overdominance gene action. Therefore, for developing biofortified hybrid, it is important to screen all potential lines first and then choose the both parents with higher contents of zinc and iron to gain the mid parent heterosis. It is reported that in pearl millet, performance per se of lines positively associated with general combining ability for iron and zinc, which indicates that the selected parental lines will also act as general combiners for these trace elements (Govindaraj et al. 2013; Govindaraj et al. 2019; Kanatti et al. 2014; Kanatti et al. 2016; Velu et al. 2011). Moreover, for developing inbred lines having higher contents of iron and zinc, the working population should have maximum genetic variation. Additionally, the direction and magnitude of inbreeding also greatly influence the inbred development as it may increase or decrease the micronutrient concentrations (Rai et al. 2017). Generally, it is observed that micronutrients have high heritability which is advantageous in hybrid breeding program. So far, high yielding pearl millet hybrids with approximately 95 percent raise in iron contents have been reported (Govindaraj et al. 2019), and the *iniadi* germplasm is considered as the richest source of iron and zinc contents for developing biofortified pearl millet (Govindaraj et al. 2013; Paul et al. 2012;Velu et al. 2011). Under such situations, where source of gene introgression is same for both parents along with additive gene action, it will negatively impact the genetic diversity for other important quantitative traits by decreasing their heterosis, e.g., yield. Therefore, selective introgression of genes regulating Fe and Zn contents is recommended for breeding biofortified crops. Keeping in view, ICRISAT has identified and evaluated new germplasm sources rich in Fe and Zn in addition to *iniadi* (Rai et al. 2016). In maize, around fifteen hundred maize genotypes were screened and achieved the maximum amount of provitamin A (19 ppm) which was more than the target amount (Harjes 2000; Menkir and Torbert Rocheford 2015; Ortiz-Monasterio et al. 2007), and almost sixty-five maize provitamin A synthetic, single cross, and three-way cross hybrids were developed and released into different parts of Africa (Virk et al. 2021).

2.1.4 Marker-Assisted Breeding (MAS)

Marker-assisted breeding (MAS) is a technique used in plant breeding that utilizes molecular markers to identify specific genes or genetic regions that are associated with desirable traits in crops. This information can then be used to select plants with the desired traits for breeding. MAS can be used to biofortify crops, which is the process of increasing the nutrient content of crops through selective breeding or genetic engineering. This can be used to address micronutrient deficiencies in the diet of people in developing countries, where a lack of access to a diverse diet can lead to malnutrition. Numerous breeding and transgenic efforts are being implemented in third-world nations to develop micronutrient-rich crops (Garg et al. 2018; Nestel et al. 2006). Breeding programs also benefit from biotechnological strategies, such as a marker-assisted selection (MAS) program to increase the nutritional profile of valuable crops due to its accuracy in the targeting desirable gene (Moose and Mumm 2008).

QTL mapping has been used to study and control agronomic and quality traits in multiple crops (Elattar et al. 2021; Hina et al. 2020; Jeong et al. 2020). Through QTL, the genes regulating zinc and iron along with other important micronutrients were mapped in rice (Table 4.2). For iron and zinc concentrations in unprocessed grains of rice, five QTLs have been identified on various chromosomes (Anuradha et al. 2012). A hybrid between the Madhukar and Swarna varieties was used to map fourteen QTLs for the iron and zinc in unrefined rice seeds at the genome level (Agarwal et al. 2014). Through marker-assisted selection (MAS), different markers (SSR, SNP) were used to improve the iron and zinc concentration of rice in a double haploid mapping population (DH). QTLs for grain and micronutrients were mapped

Table 4.2 QTLs associated with nutrient enhancement in different crops

Crop	Fortificant	No. of QTLs	References
Rice	Zn, Fe	62	Anuradha et al. (2012), Calayugan et al. (2020), Dixit et al. (2019), Garcia-Oliveira et al. (2009), Ishikawa et al. (2010), Jin et al. (2013), Lee et al. (2020), Lu et al. (2008), Pradhan et al. (2020), Swamy et al. (2018a, b), Zhang et al. (2011)
	Iron (Fe)	32	Calayugan et al. (2020), Dixit et al. (2019), Garcia-Oliveira et al. (2009), Ishikawa et al. (2010), Jin et al. (2013), Lu et al. (2008), Norton et al. (2010), Pradhan et al. (2020), Stangoulis et al. (2007), Swamy et al. (2018a, b)
	Lysine	2	Peng et al. (2014), Zhong et al. (2011)
	Grain protein content (GPC)	25	Chattopadhyay et al. (2019), Kinoshita et al. (2017), Mahender et al. (2016), Qin et al. (2009), Tan et al. (2001), Yu et al. (2009), Yun et al. (2014), Zhong et al. (2011)
	Calcium	13	Garcia-Oliveira et al. (2009), Swamy et al. (2018a)
	Phosphorus (P)	12	Calayugan et al. (2020), Garcia-Oliveira et al. (2009), Swamy et al. (2018a, b)
	Potassium (K)	14	Garcia-Oliveira et al. (2009), Swamy et al. (2018a, b), Vasconcelos et al. (2003)
Barley	Zn, Fe, cd	46	Hussain et al. (2016), Reuscher et al. (2016), Sadeghzadeh et al. (2015)
	Mg, Ba, Ca, Cu, K, Mn, Si	45	Gyawali et al. (2017)
Wheat	Zn, Fe	60	Liu et al. (2019)
	Zn, Fe	10	Wang et al. (2021)
	Zn, Fe, GPC	21	Krishnappa et al. (2021)
Soybean	Zn, Fe	10	Wang et al. (2022)
	Zn, Fe	05	Kastoori Ramamurthy et al. (2014)
	Zn, Fe	07	Ning et al. (2015)
	Isoflavone content	15	Cai et al. (2018)

and candidate genes underlying zinc concentration were also identified which could incorporate further in developing zinc-rich rice varieties (Swamy et al. 2018b). Similarly, another study was performed to identify QTLs and putative genes related to both iron and zinc content in the DH population of Korean rice cultivars (Jeong et al. 2020). In other similar studies, thirteen rice grain elements were mapped to identify stable QTLs (Calayugan et al. 2020), a cross between Indica and Japonica cultivars was made and reported two stable QTLs, viz., qZn3-1 and qFe31 for zinc and iron, respectively (Lee et al. 2020), DH lines of rice were studied for iron and zinc content and found five QTLs for these trace elements (Calayugan et al. 2020).

The determination of grain protein content (GPC) is crucial, because cereal grains are the major and rich source of protein for humans (Peng et al. 2014). In milled rice, several QTLs related to GPC have been found and mapped. In a recombinant inbred lines (RILs) population, derived from a cross between the rice varieties Zhenshan97 and Nanyangzhan, all amino acids content (AAC) were characterized

(Wang et al. 2008). It was found that a candidate gene OsAAP6 is encoded by the QTL for amino acid content (Peng et al. 2014). Similarly, another QTL for GPC in rice was recently discovered on multiple chromosomes. Three environmentally stable QTLs were found, of which one qGPC1.1 regulate the amount of protein in grains while the other two (qSGPC2.1 and qSGPC7.1) control the amount of protein. Glutelin is a highly nutritive component of the human diet (Ufaz and Galili 2008; Zhang et al. 2008). One study reported a QTL linked to a gene Os01g0111900 encoding glutelin which shows upregulation during the process of seed development (Chattopadhyay et al. 2019). So far, to enhance zinc concentration of cereal grains, various QTLs have been mapped in rice (Calayugan et al. 2020; Norton et al. 2014; Swamy et al. 2018a, b), wheat (Norton et al. 2014; Velu et al. 2011) beans (Liu et al. 2019), maize (Hindu et al. 2018; Prasanna et al 2020), and pearl millet (Anuradha et al. 2017; Kumar et al. 2016, 2018; Mahendrakar et al. 2020) (Table 4.2). Further, genome-wide association studies (GWAS) were also conducted to assess the concentration of Fe and Zn in a large rice population (Pradhan et al. 2020). In barley, approximately, 3000 landraces were mapped by using single nucleotide polymorphism (SNP) markers to study the genetic variation of iron and zinc (Mamo et al. 2014). Similarly, 336 barley genotypes were mapped and identified 45 QTLs associated with eleven trace elements (Gyawali et al. 2017). In total, seventy-five QTLs were identified linked with the grain increase and high yield in wild barley by using GWAS and Nested association mapping (NAM) (Herzig et al. 2019).

2.2 *Biofortification via Transgenic Approaches*

Biofortification via transgenic approaches is a method of increasing the nutrient content of crops by introducing genes from other organisms into the plants' genome. This can be done through techniques such as genetic engineering or recombinant DNA technology. The goal of biofortification is to improve the nutritional quality of crops, especially in developing countries where people may not have access to a diverse diet and suffer from micronutrient deficiencies.

Transgenic approaches can be efficient and fast to increase the nutrient content of crops, but these methods can also raise some concerns about the safety and acceptability of these crops to consumers, farmers and regulators. One of the main concerns is the potential for unintended effects on the environment or human health. For example, transgenic crops may harm beneficial insects or other nontarget organisms, or may create new allergens or toxins. However, many studies have shown that genetically modified crops are as safe to eat as their conventional counterparts, and that the benefits to human health, the environment, and farmers outweigh the risks. Another concern is the acceptability of transgenic crops to consumers. There is a perception that genetically modified crops are unnatural or "unnatural," and this perception may be difficult to overcome. However, many people do not understand that all crops have been genetically modified over time through traditional breeding methods. Genetic engineering simply provides a more precise and efficient way to achieve the desired traits.

Selective breeding has some drawbacks, including poor heritability, linkage of undesirable traits and lack of sufficient genetic variation for micronutrients, which increases the efficacy of genetic engineering for crop improvement. The transgenic method involves incorporating genes into the crop's DNA to increase the level of micronutrients, e.g., golden rice development (Paine et al. 2005). This method uses a diverse pool of genes comprised of different organisms having phylogenetic and taxonomical differences for the transmission and expression of desirable traits among different organisms. Biofortification by metabolic engineering can provide a solution when the natural variation of any germplasm is insufficient to provide adequate levels of micronutrients in a given crop by conventional breeding techniques. For instance, transgenic approaches helped in the uptake, storage and transport of iron and zinc to develop biofortified rice (Bashir et al. 2010; Bouis and Saltzman 2017; Ishimaru et al. 2011; Wirth et al. 2009).

Similarly, the promising raise in the Zn and Fe level was achieved in transgenic rice which was higher than in conventionally bred varieties (Trijatmiko et al. 2016; Wu et al. 2019). Additionally, in the absence of naturally synthesized vitamins in any crop, transgenic approaches are the most viable and pertinent method for biofortifying crops with a particular vitamin (Pérez-Massot et al. 2013). Advancements in emerging omics such as genome editing techniques (CRISPR/Cas9, TALENs) have opened up new avenues for biofortification due to the availability of a completely sequenced genome of major staple crops (Ricroch et al. 2017). Genetically modified plants can be used to recreate specific pathways, increase the efficiency of metabolic reactions in edible tissues, redistribute micronutrients among tissues, elimination of antinutrients and use multigene transfer to increase the micronutrient bioavailability, for example, every single plant of biofortified corn have high levels of vitamin C, B9, and β-carotene (Carvalho and Vasconcelos 2013; Naqvi et al. 2009) (Table 4.3).

The transgenic biofortification practices requires long period in variety release for general cultivation, as it consists of several steps of trait identification, expression analysis and phenotypic verification. For instance, the project of Golden rice was published and shared among the scientific community after eight years of its development, but, its practical application in a rural area is still in process (Yu et al. 2009). Moreover, the slow regulatory processes are costly at the same time. In many countries, the commercialization of genetically modified crops required legal acceptance and proper regularization at the government level. For instance, *Bt* Brinjal was introduced and developed by an Indian company but remained banned for commercial cultivation for years (Inaba and Macer 2004; Watanabe et al. 2005). Yet, in order to achieve the fast growth in developing biofortified crops, joint efforts of plant breeders and molecular scientists are required. Followings are a few recent transgenic approaches for developing biofortified crops.

Table 4.3 Micronutrient enhancement in different crops through transgenic approaches

Crop	Fortificant	Incorporated gene	References
Rice	Tryptophan	Anthranilate synthase, OASA2	Saika et al. (2011), Wakasa et al. (2006)
	Lysine	Aspartate kinase and dihydrodipicolinate synthase, lysine-rich binding protein accumulation, LRP, RLRH1, and RLRH2	Liu et al. (2016), Wong et al. (2004), Yang et al. (2016)
	Methionine and cysteine	Serine acetyltransferase, Sunflower seed albumin	Nguyen et al. (2012)
	Iron (Fe) and zinc (Zn)	Iron and zinc content elevation via ferritin, NAS, IDS3, OsVIT, IRT1, OsGluB1 pro-SoyferH1 gene	Lee et al. (2009) Masuda et al. (2008), Oliva et al. (2014), Paul et al. (2012), Suzuki et al. (2008), Trijatmiko et al. (2016)
	Phytic acid	Lowering phytic acid content via lowering OsINO1 gene	Kuwano et al. (2009), Larson et al. (2000)
Maize	Total carotenoids, β-carotene	Maize PSY and CrtI, Zmpsy, PacrtI, Gllycb, Glbch, ParacrtW	Aluru et al. (2008), Zhu et al. (2008)
	Smirnoff–Wheeler pathway	Dehydroascorbate reductase (dhar)	Naqvi et al. (2009)
	Tocochromanol pathway	HGGT and HPT	Dolde and Wang (2011)
Potato	Total carotenoids β-carotene	CrtI, LCYe, CHY1, CHY2, CrtB, CrtI, CrtY Cauliflower Or (Orange) gene	Ducreux et al. (2005), Li et al. (2012), Zhu et al. (2008)
Wheat	Total carotenoids, ProVitamin A	Maize PSY and CrtI; CrtB, CrtI	Wang et al. (2014)
Tomato	Precursor b-carotene	Lycopene b-cyclase gene	Apel and Bock (2009), de la Garza et al. (2004), Welsch et al. (2007)
	Pteridine synthesis	GTP cyclohydrolase I	de la Garza et al. (2004)
	Smirnoff–Wheeler pathway	GDP-l-galactose phosphorylase	Bulley et al. (2012)
Arabidopsis	Enzymes of salvage pathway	PDX1 and PDX2	Chen and Xiong (2009)
	Tocochromanol pathway	HGGT and HPT	Cahoon et al. (2003), Yang et al. (2011)
Mustard	β-carotene	Bacterial phytoene synthase (crtB) gene	Shewmaker et al. (1999)
Strawberry	NADPH-dependent D-galacturonate reductase	GalUR GDP-l-galactose phosphorylase	Agius et al. (2003), Bulley et al. (2012)
Tobacco	Tocochromanol pathway	HGGT	Tanaka et al. (2015)
Cassava	Provitamin A	Phytoene synthase gene	Welsch et al. (2010)

2.2.1 Gene Edit Techniques

To increase the number of micronutrients in the major cereals, transgenic breeding could serve as a useful tool. The micronutrients in many staple crops can be improved due to the availability of genes for target traits. With the help of transgenic approaches, the pyramiding of genes could be possible for enhancing micronutrients, which will eventually help in combating the problem of malnutrition. There are several gene editing techniques that can be used to biofortify crops, including CRISPR-Cas9, TALENs, and ZFNs. These techniques allow for precise and targeted changes to be made to the genome of a crop, enabling the introduction of beneficial traits such as disease resistance, improved nutrient content, and increased yield. Additionally, these techniques can be used to "turn off" or remove undesirable traits without introducing foreign genes, which can be a concern with traditional genetic modification methods. Overall, gene editing techniques offer a powerful tool for developing crops that are better suited to meet the needs of farmers and consumers.

Although, biofortification via gene edit technologies is still under experimentation few studies have reported the successful application in vegetables, ornamental, and field crops. For example, knock out of lycopene ε-cyclase and non-heme β-carotene hydroxylase gene greatly increased the β-carotene content in tubers. The ability to alter germline through CRISPR/CAS system has the remarkable potential of regenerating a single cell (Sedeek et al. 2019). So far, this technique has been employed to develop biofortified wheat, rice and A*rabidopsis* by enhancing the zinc, Vit A, B1, and iron concentrations (Jaganathan et al. 2018; Ricroch et al. 2017). Recently, CRISPR/Cas9 was used to enhance β-carotene content in rice by manipulating the expression of the *Osor* gene in rice (Endo et al. 2019). Similarly, the level of carotenoids was boosted by altering the biosynthetic pathway of carotenoids via CRISPR/Cas9 system (Dong et al. 2020). In wheat, β-carotene content was improved by knocking down the metabolic pathway of provitamin A synthesis (Zeng et al. 2015). In future, these genome editing tools could be used for the swift and economic development of transgenic crops.

2.2.2 Overexpression of Genes

Overexpression of genes is a genetic engineering approach that can be used to biofortify crops. This technique involves increasing the amount of a specific gene or genes in a plant to enhance its production of beneficial traits such as increased nutrient content or improved stress tolerance. Overexpression can be achieved through various methods, including the use of strong promoters, cis-acting elements, or multicopy plasmids. The recent development of plant genetic engineering made it possible to understand the plant metabolic pathways which help in increasing the levels of trace elements in human food to alleviate the problem of malnourishment (Zimmermann and Hurrell 2002). A member of NAC transcription factors (*NAM-B1*) functions in the early maturity of wheat along with influencing the wheat grain's zinc concentration (Connorton et al. 2017). Naturally, the concentration of β-carotene is low in maize kernel and efforts have been to raise its content by using various transgenic approaches (Wong et al. 2004). A significantly greater increase was observed in the levels of β-carotene from overexpression of *crtI* and *crtB* genes in transgenic maize which could be useful in combating malnutrition (Naqvi et al. 2009; Zhu et al. 2008). In barley, the vitamin E activity has been augmented having greater levels of δ-tocopherol with inverse proportions of γ-tocopherol by expressing 2-methyl-6-phytyl benzoquinol methyltransferase genes (Van Eenennaam et al. 2003).

In higher plants, nicotianamine (NA) is synthesized by NA synthase enzyme (NAS) and is involved in metal ions transportation (Takahashi et al. 2001). In wheat, barley maize, and *Arabidopsis*, various genes encoding NAS are regulated by trace elements (Mizuno et al. 2003). With the help of recombinant DNA technology, the concentration of the NAS gene has been increased to desired levels. For example, overexpression of the barley NAS gene (HvNAS1) has upgraded the Cu, Fe, and Zn concentrations in the seeds of *Arabidopsis* and tobacco (Kim et al. 2005). In parallel, overexpression of the *HvNAS1* gene helped in improving 3 times greater iron and zinc concentrations in refined rice grains (Masuda et al. 2009). In *Arabidopsis,* a combination of *Arabidopsis* NAS gene (AtNAS1) expression along with ferritin expression resulted in the increased contents of zinc and iron. The overexpression of an endogenous NAS gene (OsNAS2) combined with Ferritin increased the iron and zinc level of wheat grains (Singh and Govindan 2017).

The three endogenous NAS genes of rice were overexpressed and resulted in a paramount increases the amounts of iron and zinc (Wirth et al. 2009). Another study reported the overexpression of OsNAS2 and OsNAS3 resulting in a substantial increase in the iron and zinc content of rice seeds (Lee et al. 2011). One study reported the overexpression of two genes (PDX1 and PDX2) resulted in two-fold raise in the vitamin B6 content of *Arabidopsis* seed (Chen and Xiong 2009). Similarly, a transgenic *Arabidopsis* was developed showing higher contents of vitamin B6, improved salinity tolerance in addition to better-sized organ growth. The folate content of tomatoes has been enhanced in transgenic tomatoes by enhancing the expression of the GTP cyclohydrolase I gene (Diaz de la Garza et al. 2004).

The vitamin B9 content of rice was enhanced in transgenic rice by overexpressing *Arabidopsis* genes (Glb-1 and GluB1) which could overcome its deficiency (Storozhenko et al. 2007). Maize crop was biofortified to improve the contents of vitamins C, A, and B9. In this transgenic corn, the *dhar* gene originating from the rice was overexpressed and showed up higher levels of ascorbic acid (Naqvi et al. 2009). Similarly, in *Nicotiana tabacum, Zea mays,* and *Arabidopsis,* the concentration of vitamin E has been enhanced by using transgenic techniques up to fifteen times greater than wild type (Dolde and Wang 2011; Tanaka et al. 2015; Yang et al. 2011). The overexpression of ferrous (Fe III) in Arabidopsis and pea mutants improve the iron uptake from soils deficient in iron (Douchkov et al. 2005; Morrissey and Guerinot 2009; Rogers and Guerinot 2002; Schröder et al. 2003).

In wheat, over-expression of the TaVIT2 gene showed higher content of wheat (Connorton et al. 2017). The overexpression of AtZIP1 (*Arabidopsis* Zn2C transporter) increased the level of iron uptake in barley roots (Ramesh et al. 2004). In rice, the concentration of cysteine has been increased by overexpression of metallothionein-like protein of rice (Lucca et al. 2001). In golden rice, iron absorption was enhanced by elevating the levels of β-carotene (Beyer et al. 2002). Biofortified pearl millet was developed with increased iron contents which increases 5–10 percent of iron uptake in a large population (Cercamondi et al. 2013; Sahu 2017).

3 Targeted Biofortified Crops

3.1 Rice (Oryza sativa)

In 2013, HarvestPlus released a zinc-rich brown rice variety for the first time in the world's history, developed by the Bangladesh Rice Research Institute containing a high amount of zinc content ranging from 20 to 22 ppm. Similarly, another conventionally bred variety Jalmagna has reported two times more levels of zinc and iron than the common variety (Gregorio et al. 2000). The development of beta-carotene-enriched Golden rice by expressing the genes regulating carotene desaturase and PSY made a significant contribution to the prevention of malnutrition (Beyer et al. 2002; Datta et al. 2003; Yu et al. 2009). Genetically modified rice high in folic acid, which plays a key role in normal pregnancy and controlling anemia, was developed by overexpression of genes that encode *Arabidopsis* aminodeoxychorismate synthase and GTP-cyclohydrolase I (Bibbins-Domingo et al. 2017; Blancquaert 2015; Storozhenko et al. 2007). By overexpression of genes encoding nicotianamine aminotransferase (Takahashi et al. 2001), OsIRT1 (Lee and An 2009; Lee et al. 2012; Trijatmiko et al. 2016; Zheng et al. 2010), nicotianamine synthase 1&2 (Lucca et al. 2002), and soybean ferritin (Trijatmiko et al. 2016; Vasconcelos et al. 2003), iron content was enhanced in rice. In addition, incorporating multiple genes regulating iron nutriment also helped in the synthesis of iron biofortification in rice (Masuda et al. 2012, 2013; Wirth et al. 2009). Also, increase in the iron bioaccumulation by reducing phytic acid concentration could enhance the iron content of rice (Hurrell and Egli 2010).

For improving zinc content in genetically modified rice, OsIRT1 (Lee and An 2009) and mugineic acid synthesis genes were overexpressed (Masuda et al. 2008). The expression of antisense RNA inhibition of starch branching enzymes and antisense waxy genes was controlled for improving resistance to amylose to control the problem of obesity (Itoh et al. 2003; Liu et al. 2003; Wei et al. 2010). Additionally, the introduction of lactoferrin in rice grain has escalated the market value of milk protein ingredients which could be used as a fundamental part of infant food (Lee et al. 2010; Nandi et al. 2002).

3.2 *Wheat (*Triticum aestivum*)*

During the past two decades, the majority of the newly released durum varieties showed a higher level of carotenoids compared to the 1970s old varieties (Digesù et al. 2009; Ficco et al. 2014). In 2006, a purple variety of wheat was released in Austria (Eticha et al. 2011), in 2014 a cultivar named PS Karkulka was released in Slovakia and in 2017 three colors (black white lines, blue and purple) of wheat varieties were released for general cultivation in India (Garg et al. 2016). For zinc and iron, wild relatives of wheat have served as the main source of variation for developing elite cultivars (Çakmak et al. 2004; Monasterio and Graham 2000; Ortiz-Monasterio et al. 2007). HarvestPlus has utilized this existing variation to develop six zinc-enriched wheat varieties in India viz., BHU 1, BHU 3, BHU 5, BHU 6, BHU 7, and BHU 18 and four in Pakistan viz., NR 419, 42, 421, and Zincol during 2014 and 2015, respectively. In a few studies, carotene desaturase and bacterial PSY genes were expressed to improve the provitamin A content in *T. aestivum* (Cong et al. 2009; Wang et al. 2014).

Further, the ferritin gene was obtained from and expressed in wheat to enhance the Fe content (Borg 2012; Xiaoyan et al. 2012). Similarly, to increase iron intake, the phytochrome gene was expressed to increase the activity of antinutrients (Brinch-Pedersen et al. 2000), whereas, for decreasing the concentration of phytic acid, ABCC13 transporter genes were silenced (Bhati et al. 2016). By increasing the expression of the *Amaranthus* albumin gene, the amount of tyrosine, cysteine, lysine and methionine was improved (Tamás et al. 2009). Moreover, the wheat crop has been experimented for improved anthocyanin production, to produce hard and resistant starch content to address the problem of obesity (Doshi et al. 2006; Sestili et al. 2010).

3.3 *Maize (*Zea mays*)*

Among the major achievements of biofortification, provitamin A maize has marked a remarkable history. Since 2013, in various African countries orange maize varieties were developed from biofortification and are grown commercially (Gannon et al. 2014). Statistics showed that the introduction and consumption of orange maize biofortified varieties in Malawi and Zimbabwe positively impact the eye vision and pupillary response of children. Maize breeders have made significant progress in evaluating multiple antioxidants in proVA maize varieties (Muzhingi et al. 2017). By expressing carotenogenic and crtB bacterial genes, levels of provitamin A were increased to develop transgenic biofortified maize (Aluru et al. 2008; Decourcelle et al. 2015). Additionally, the development of quality protein maize (QPM) having elevated levels of tryptophan and lysine is a promising achievement of maize breeders. This QPM was developed by introgressing the opaque-2 gene from landraces into modern maize cultivars. Moreover, recurrent selection has been performed to improve the carotenoids content (Palmer et al. 2016) either separately or combined with the phenolics, vitamin E (Goffman and Böhme 2001; Muzhingi et al. 2017) and antioxidants (Lago et al. 2014). The antioxidant properties of vitamin E have a key role in maintaining human health; therefore, maize biofortification is targeting the enhancement of vitamin E as a major biofortificant. By overexpressing the homogentisic acid geranylgeranyl transferase, the tocopherol and tocotrienol contents of maize were improved in concentration (Cahoon et al. 2003).

Another important vitamin for human health is vitamin C, which is considered important for preventing cardiovascular diseases, developing immunity and helping with iron intake. A study has been conducted to express the dehydroascorbate reductase to boost the levels of vitamin C in maize (Chen et al. 2003). Besides, multivitamin maize was bioengineered having a higher amount of folate, Beta carotene, and ascorbic acid by manipulating their metabolic pathways (Naqvi et al. 2009). Zeins are the major seed storage proteins in maize but have lower nutritional significance due to insufficient levels of tryptophan and lysine. Therefore, both essential amino acids of maize have been targeted for improvement by expressing the sb401 gene (obtained from potato) (Tang et al. 2013; Yu et al. 2005), and alpha-zeins via antisense dsRNA (Huang et al. 2006). A lysine-rich maize variety (Mavrea™YieldGard) was developed and released by altering the site of cis-element (Dzs10) in Mexico and Japan (Lai and Messing 2002). To achieve the balance of amino acid in maize, α-lactalbumin milk protein was highly expressed (Yang et al. 2002).

*3.4 Potato (*Solanum tuberosum*)*

Potato has a great nutritional significance with a higher caloric profile. Provitamin A level of potato tuber has been increased by incorporating three genes viz., lycopene β-cyclase, PSY, and phytoene desaturase (Ducreux et al. 2005). The beta-carotene content of tubers was enhanced by silencing the beta-carotene hydroxylase through RNA interference (RNAi) (Van Eck et al. 2007) and by expressing the lycopene β-cyclase (Song et al. 2016). Another study incorporated the *Or* gene obtained from mutants of orange cauliflower, which resulted in the improvement of carotenoids, z-carotene, and phytofluene (Lopez et al. 2008). A transgenic potato was developed with an increased level of carotenoid by overexpressing potato zeaxanthin epoxidase genes (Römer et al. 2002). The strawberry GalUR gene was overexpressed to increase the content of ascorbic acid (Upadhyaya et al. 2009). Naturally, potatoes have a lower amount of essential amino acids which was improved through inducing coexpression of two genes, viz., methionine-rich storage protein and cystathionine γ-synthase (Dancs et al. 2008). Likewise, the StMGL1 and threonine synthase genes were silenced and increased the methionine content (Huang et al. 2014; Zeh et al. 2001).

The Perilla genes encoding storage proteins and cystathionine γ-synthase were overexpressed to enhance the level of Methionine (Di et al. 2003; Goo et al. 2013).

Further, a transgenic potato was formed having higher levels of total protein content and essential amino acids by expressing the *Amaranth* albumin gene (Chakraborty et al. 2010). The cyclodextrin glycosyltransferases gene was expressed to develop high-caloric potato with a greater amount of dietary fiber. Moreover, the target of biofortification in potato was also to increase the amounts of anthocyanins and phenolic acid by overexpressing dihydroflavonol reductase and chalcone isomerase (Lukaszewicz et al. 2004).

3.5 *Tomato (*Solanum lycopersicum*)*

Among fruits, tomato is the common and cheap source of several micronutrients and vitamin C. The color of this fruit is credited to the presence of isoprenoid lycopene content. Isoprenoids comprised of numerous important compounds with an important function in carotenoids, sterols, tocopherols, ubiquinone, phytoalexins, plastoquinone, abscisic acid, cytokinin, gibberellins, and brassinosteroids. So far, successful efforts have been made to enhance the content of isoprenoid. A study reported the expression of 3-hydroxymethylglutaryl CoA for increasing the sterol content of tomato (Enfissi et al. 2005). For enhancement of lycopene, beta-carotene and phytoene contents, 1-deoxy-d-xylulose-5-phosphate synthase, lycopene beta-cyclase gene and PSY genes were expressed (Apel and Bock 2009; Enfissi et al. 2005; Fraser et al. 2007; Wurbs et al. 2007). With the help of RNAi photomorphogenesis regulatory genes were suppressed to develop fortified tomato with increased contents of both flavonoid and carotenoid (Davuluri et al. 2005). To obtain a greater amount of beta-carotene, the beta-Lcy, beta-carotene ketolase and beta-carotene hydroxylase genes were concurrently expressed (Gregorio et al. 2000; Huang et al. 2013).

Further, the tomato was also targeted to improve its vitamin content via overexpression of GDP-mannose 3′,5′-epimerase, DHAR, and coexpressing myo-inositol oxygenase 2, GDP-mannose pyrophosphorylase and arabinono1,4-lactone oxidase genes (Cronje et al. 2012; Haroldsen et al. 2011;Zhang et al. 2014). The folic acid content of tomato is a significantly important nutrient. Numerous significant attempts have been made to enhance folic acid content by overexpressing GTPCHI, and aminodeoxychorismate synthase genes in tomato (de la Garza et al. 2004; Diaz de la Garza et al. 2004). The antioxidant anthocynins of tomato, Dahlia and Rose have been increased by expression of CHI AtMYB75 (Maligeppagol et al. 2013; Muir et al. 2001; Zuluaga et al. 2008). The contents of other antioxidants such as chlorogenic acid, transresveratrol, and genistin were increased by RNAi silencing of HQT, expressing stilbene synthase, and overexpressing isoflavone synthase gene, respectively (Giovinazzo et al. 2005; Luo et al. 2008; Niggeweg et al. 2004; Shih et al. 2008).

3.6 *Soybean (*Glycine max*)*

Soybean is one of the richest sources of protein and vegetable oil. In soybean, the PSY gene (originated from bacteria) was overexpressed to enhance the level of beta carotene, seed protein, Canthaxanthin, and oleic acid (Kim et al. 2012; Pierce et al. 2015; Schmidt et al. 2015). Although, soybean consists of 40 percent proteins but at the same time lacks the number of essential amino acids containing sulfur viz., methionine and cysteine. It is reported that in soybean seed, overexpression of O-acetylserine sulfhydrylase increases the cysteine level (Kim et al. 2012). Moreover, the overexpression of zein maize protein also increases the content of cysteine and methionine (Dinkins et al. 2001). The overexpression of cystathionine γ-synthase also resulted in an increased level of methionine (Hanafy et al. 2013;Song et al. 2013).

The total oil content of soybean is 20 percent, but a heavy percentage of this oil contains unsaturated fatty acids which lower the quality of seed oil. To improve the oil quality, α-linolenic acid levels were reduced by silencing FAD3 through a siRNA-mediated gene silencing-based approach (Flores et al. 2008). In soybean, the 6-desaturase gene is involved in the conversion of linoleic acid into γ-linolenic acid (GLA) and α-linolenic acid into ω-3 fatty acids (STA). An increase in the expression of regulating genes also increases the STA and GLA content in soybean oil (Eckert et al. 2006; Sato et al. 2004). Similarly, by inhibiting the expression of the 12 oleate desaturase gene, the concentration of palmitic acid, linoleic acid, and oleic acid could be enhanced (Zhang et al. 2014). Isoflavone consumption is linked with the reduction of cardiac diseases, cancer, and menopausal symptoms (Watanabe et al. 2002). Naturally, soybean seeds have low contents of isoflavone, which has been improved by activating and suppressing the combination of R transcription factor and maize C1 genes (Yu et al. 2003).

4 Conclusion and Future Prospects

In developing world, lowering poverty, combating diseases, infrastructure development, and creating awareness are the fundamental obstacles to addressing malnutrition. These challenges are significant and ought to be handled in the same way they have been in the past. Since majority of the underdeveloped world live in rural areas, they do not have plenty of access and purchasing power to buy fortified food. To address the nutrient needs of low-income communities at a reasonable price, biofortification can be very helpful in supplying the technology through the seeds of the main staple crops. Food biofortification is becoming more widely acknowledged as a successful strategy to raise the nutritional profile of a substantial portion of the population in underdeveloped nations. Staple grain nutritional enhancement using genetic engineering and conventional breeding offers hope for food-based therapeutics for micronutrient imbalances.

To end micronutrient deficiency and significantly improve human health, biofortification can be used in conjunction with dietary diversity and nutrition education. The bioavailability of several vital vitamins and minerals could be improved through the use of recombinant DNA technology. Since these fortified food items are typically accessible to urban residents, external nutritional augmentation is of minimal value. One of the main issues is that very few commercially available biofortified transgenic crops have been developed for widespread cultivation. By combining metabolic engineering with traditional breeding, it is possible to attain hunger alleviation and malnutrition decrement substantially more quickly. In order to control population, upgraded living standards, and bring about world peace, rapid innovations in recombinant DNA technology should concentrate on long-term solutions to malnutrition as part of humanitarian intervention. Biofortification through conventional breeding or bioengineering provides developing nations another option in the broader fight against malnutrition without abandoning a mixed diet.

The future prospects of crop biofortification are promising as it has the potential to address micronutrient deficiencies in populations that rely on a limited variety of staple crops for their diet. Both traditional breeding and transgenic approaches have been used successfully to create crop varieties with higher levels of micronutrients. In the future, a combination of both traditional breeding and transgenic approaches may be used to develop biofortified crops. This will likely result in the development of new crop varieties that have improved nutrient content, disease resistance, and other desirable traits. Additionally, research and development in the field of biofortification will continue to improve the effectiveness of these methods and make it more accessible to farmers and communities in need. Overall, crop biofortification is a promising field that has the potential to improve the nutritional status of populations and provide solutions to micronutrient deficiencies.

References

Agarwal S, VGN TV, Kotla A, Mangrauthia SK, Neelamraju S (2014) Expression patterns of QTL based and other candidate genes in Madhukar× Swarna RILs with contrasting levels of iron and zinc in unpolished rice grains. Gene 546(2):430–436

Agius F, González-Lamothe R, Caballero JL, Muñoz-Blanco J, Botella MA, Valpuesta V (2003) Engineering increased vitamin C levels in plants by overexpression of a D-galacturonic acid reductase. Nat Biotechnol 21(2):177–181

Ahn R, Burke TF, McGahan AM (2015) Innovating for healthy urbanization. Springer

Aluru M, Xu Y, Guo R, Wang Z, Li S, White W et al (2008) Generation of transgenic maize with enhanced provitamin A content. J Exp Bot 59(13):3551–3562

Anandan A, Rajiv G, Eswaran R, Prakash M (2011) Genotypic variation and relationships between quality traits and trace elements in traditional and improved rice (*Oryza sativa* L.) genotypes. J Food Sci 76(4):H122–H130

Andersson MS, Saltzman A, Virk PS, Pfeiffer WH (2017) Progress update: crop development of biofortified staple food crops under HarvestPlus. Afr J Food Agric Nutr Dev 17(2):11905–11935

Anuradha K, Agarwal S, Batchu AK, Babu AP, Swamy BM, Longvah T, Sarla N (2012) Evaluating rice germplasm for iron and zinc concentration in brown rice and seed dimensions. J Phytology 4(1):19–25

Anuradha N, Satyavathi CT, Bharadwaj C, Nepolean T, Sankar SM, Singh SP et al (2017) Deciphering genomic regions for high grain iron and zinc content using association mapping in pearl millet. Front Plant Sci 8:412

Apel W, Bock R (2009) Enhancement of carotenoid biosynthesis in transplastomic tomatoes by induced lycopene-to-provitamin A conversion. Plant Physiol 151(1):59–66

Arora S, Cheema J, Poland J, Uauy C, Chhuneja P (2019) Genome-wide association mapping of grain micronutrients concentration in *Aegilops tauschii*. Front Plant Sci 10:54

Babu VR, Shreya K, Dangi KS, Usharani G, Nagesh P (2012) Genetic variability studies for qualitative and quantitative traits in popular rice (*Oryza sativa* L.) hybrids of India. Int J Sci Res Publ 2(6):1–5

Bashir K, Ishimaru Y, Nishizawa NK (2010) Iron uptake and loading into rice grains. Rice 3(2):122–130

Beyer P, Al-Babili S, Ye X, Lucca P, Schaub P, Welsch R, Potrykus I (2002) Golden rice: introducing the β-carotene biosynthesis pathway into rice endosperm by genetic engineering to defeat vitamin A deficiency. J Nutr 132(3):506S–510S

Bhati KK, Alok A, Kumar A, Kaur J, Tiwari S, Pandey AK (2016) Silencing of ABCC13 transporter in wheat reveals its involvement in grain development, phytic acid accumulation and lateral root formation. J Exp Bot 67(14):4379–4389

Bhutta ZA, Das JK, Rizvi A, Gaffey MF, Walker N, Horton S et al (2013) Evidence-based interventions for improvement of maternal and child nutrition: what can be done and at what cost? Lancet 382(9890):452–477

Bibbins-Domingo K, Grossman DC, Curry SJ, Davidson KW, Epling JW, García FA et al (2017) Folic acid supplementation for the prevention of neural tube defects: US Preventive Services Task Force recommendation statement. JAMA 317(2):183–189

Blancquaert D (2015) Van daele J, Strobbe S, Kiekens F, Storozhenko S, De Steur H, Glleynck X, Lambert W, Stove C, Van Der Straeten D (2015) Improving folate (vitamin B9) stability in biofortified rice through metabolic engineering. Nat Biotechnol 33:1076–1078

Bohra A, Jha UC, Kumar S (2016) Enriching nutrient density in staple crops using modern "-Omics" tools. In: Biofortification of food crops (pp. 85–103). Springer

Borg S (2012) Wheat ferritins: improving the iron content of the wheat grain. J Cereal Sci 56:204–213

Borrill P, Connorton JM, Balk J, Miller AJ, Sanders D, Uauy C (2014) Biofortification of wheat grain with iron and zinc: integrating novel genomic resources and knowledge from model crops. Front Plant Sci 5:53

Bouis HE, Saltzman A (2017) Improving nutrition through biofortification: a review of evidence from HarvestPlus, 2003 through 2016. Glob Food Sec 12:49–58

Bouis HE, Welch RM (2010) Biofortification—a sustainable agricultural strategy for reducing micronutrient malnutrition in the global south. Crop Sci 50:S-20–S-32

Brinch-Pedersen H, Olesen A, Rasmussen SK, Holm PB (2000) Generation of transgenic wheat (*Triticum aestivum* L.) for constitutive accumulation of an *Aspergillus phytase*. Mol Breed 6(2):195–206

Bulley S, Wright M, Rommens C, Yan H, Rassam M, Lin-Wang K et al (2012) Enhancing ascorbate in fruits and tubers through over-expression of the l-galactose pathway gene GDP-l-galactose phosphorylase. Plant Biotechnol J 10(4):390–397

Cahoon EB, Hall SE, Ripp KG, Ganzke TS, Hitz WD, Coughlan SJ (2003) Metabolic redesign of vitamin E biosynthesis in plants for tocotrienol production and increased antioxidant content. Nat Biotechnol 21(9):1082–1087

Cai Z, Cheng Y, Ma Z, Liu X, Ma Q, Xia Q et al (2018) Fine-mapping of QTLs for individual and total isoflavone content in soybean (*Glycine max* L.) using a high-density genetic map. Theor Appl Genet 131(3):555–568

Cakmak I (2008) Enrichment of cereal grains with zinc: agronomic or genetic biofortification? Plant Soil 302(1):1–17

Çakmak İ, Torun A, Millet E, Feldman M, Fahima T, Korol A et al (2004) Triticum dicoccoides: an important genetic resource for increasing zinc and iron concentration in modern cultivated wheat. Soil Sci Plant Nutr 50(7):1047–1054

Calayugan MIC, Formantes AK, Amparado A, Descalsota-Empleo GI, Nha CT, Inabangan-Asilo MA et al (2020) Genetic analysis of agronomic traits and grain iron and zinc concentrations in a doubled haploid population of rice (*Oryza sativa* L.). Sci Rep 10(1):1–14

Carvalho SM, Vasconcelos MW (2013) Producing more with less: Strategies and novel technologies for plant-based food biofortification. Food Res Int 54(1):961–971

Cercamondi CI, Egli IM, Mitchikpe E, Tossou F, Zeder C, Hounhouigan JD, Hurrell RF (2013) Total iron absorption by young women from iron-biofortified pearl millet composite meals is double that from regular millet meals but less than that from post-harvest iron-fortified millet meals. J Nutr 143(9):1376–1382

Chakraborty S, Chakraborty N, Agrawal L, Ghosh S, Narula K, Shekhar S et al (2010) Next-generation protein-rich potato expressing the seed protein gene AmA1 is a result of proteome rebalancing in transgenic tuber. Proc Natl Acad Sci 107(41):17533–17538

Chattopadhyay K, Behera L, Bagchi TB, Sardar SS, Moharana N, Patra NR et al (2019) Detection of stable QTLs for grain protein content in rice (*Oryza sativa* L.) employing high throughput phenotyping and genotyping platforms. Sci Rep 9(1):1–16

Chen H, Xiong L (2009) Enhancement of vitamin B6 levels in seeds through metabolic engineering. Plant Biotechnol J 7(7):673–681

Chen Z, Young TE, Ling J, Chang S-C, Gallie DR (2003) Increasing vitamin C content of plants through enhanced ascorbate recycling. Proc Natl Acad Sci 100(6) 3525–3530

Cong L, Wang C, Chen L, Liu H, Yang G, He G (2009) Expression of phytoene synthase1 and carotene desaturase crtI genes result in an increase in the total carotenoids content in transgenic elite wheat (*Triticum aestivum* L.). J Agric Food Chem 57(18):8652–8660

Connorton JM, Jones ER, Rodríguez-Ramiro I, Fairweather-Tait S, Uauy C, Balk J (2017) Wheat vacuolar iron transporter TaVIT2 transports Fe and Mn and is effective for biofortification. Plant Physiol 174(4):2434–2444

Cronje C, George GM, Fernie AR, Bekker J, Kossmann J, Bauer R (2012) Manipulation of L-ascorbic acid biosynthesis pathways in Solanum lycopersicum: elevated GDP-mannose pyrophosphorylase activity enhances L-ascorbate levels in red fruit. Planta 235(3):553–564

Dancs G, Kondrák M, Bánfalvi Z (2008) The effects of enhanced methionine synthesis on amino acid and anthocyanin content of potato tubers. BMC Plant Biol 8(1):1–10

Datta K, Baisakh N, Oliva N, Torrizo L, Abrigo E, Tan J et al (2003) Bioengineered 'golden' indica rice cultivars with β-carotene metabolism in the endosperm with hygromycin and mannose selection systems. Plant Biotechnol J 1(2):81–90

Davuluri G, van Tuinen A, Fraser P, Manfredonia A, Newman R, Burgess D et al (2005) Fruit-specific RNAi-mediated suppression of DET1 enhances tomato nutritional quality. Nat Biotechnol 23:890–895

de Bruijn WJ, van Dinteren S, Gruppen H, Vincken J-P (2019) Mass spectrometric characterisation of avenanthramides and enhancing their production by germination of oat (*Avena sativa*). Food Chem 277:682–690

de la Garza RD, Quinlivan EP, Klaus SM, Basset GJ, Gregory JF III, Hanson AD (2004) Folate biofortification in tomatoes by engineering the pteridine branch of folate synthesis. Proc Natl Acad Sci 101(38):13720–13725

De Moura FF, Palmer AC, Finkelstein JL, Haas JD, Murray-Kolb LE, Wenger MJ et al (2014) Are biofortified staple food crops improving vitamin A and iron status in women and children? New evidence from efficacy trials. Adv Nutr 5(5):568–570

Decourcelle M, Perez-Fons L, Baulande S, Steiger S, Couvelard L, Hem S et al (2015) Combined transcript, proteome, and metabolite analysis of transgenic maize seeds engineered for enhanced carotenoid synthesis reveals pleotropic effects in core metabolism. J Exp Bot 66(11):3141–3150

Detterbeck A, Pongrac P, Rensch S, Reuscher S, Pečovnik M, Vavpetič P et al (2016) Spatially resolved analysis of variation in barley (*Hordeum vulgare*) grain micronutrient accumulation. New Phytol 211(4):1241–1254

Di R, Kim J, Martin MN, Leustek T, Jhoo J, Ho C-T, Tumer NE (2003) Enhancement of the primary flavor compound methional in potato by increasing the level of soluble methionine. J Agric Food Chem 51(19):5695–5702

Diaz de la Garza R, Quinlivan EP, Klaus SM, Basset GJ, Gregory III JF, Hanson AD (2004) Folate biofortification in tomatoes by engineering the pteridine branch of folate synthesis. 101:13720

Digesù A, Platani C, Cattivelli L, Mangini G, Blanco A (2009) Genetic variability in yellow pigment components in cultivated and wild tetraploid wheats. J Cereal Sci 50(2):210–218

Dinkins RD, Srinivasa Reddy M, Meurer CA, Yan B, Trick H, Thibaud-Nissen F et al (2001) Increased sulfur amino acids in soybean plants overexpressing the maize 15 kDa zein protein. In Vitro Cell Dev Biol-Plant 37:742–747

Dixit S, Singh UM, Abbai R, Ram T, Singh VK, Paul A et al (2019) Identification of genomic region (s) responsible for high iron and zinc content in rice. Sci Rep 9(1):1–8

Doehlert DC, Simsek S, Thavarajah D, Thavarajah P, Ohm JB (2013) Detailed composition analyses of diverse oat genotype kernels grown in different environments in North Dakota. Cereal Chem 90(6):572–578

Dolde D, Wang T (2011) Oxidation of crude corn oil with and without elevated tocotrienols. J Am Oil Chem Soc 88(9):1367–1372

Dong OX, Yu S, Jain R, Zhang N, Duong PQ, Butler C et al (2020) Marker-free carotenoid-enriched rice generated through targeted gene insertion using CRISPR-Cas9. Nat Commun 11(1):1–10

Doshi K, Eudes F, Laroche A, Gaudet D (2006) Transient embryo-specific expression of anthocyanin in wheat. In Vitro Cell Dev Biol-Plant 42(5):432–438

Douchkov D, Gryczka C, Stephan U, Hell R, Bäumlein H (2005) Ectopic expression of nicotianamine synthase genes results in improved iron accumulation and increased nickel tolerance in transgenic tobacco. Plant Cell Environ 28(3):365–374

Ducreux LJ, Morris WL, Hedley PE, Shepherd T, Davies HV, Millam S, Taylor MA (2005) Metabolic engineering of high carotenoid potato tubers containing enhanced levels of β-carotene and lutein. J Exp Bot 56(409):81–89

Eckert H, LaVallee B, Schweiger BJ, Kinney AJ, Cahoon EB, Clemente T (2006) Co-expression of the borage Δ6 desaturase and the Arabidopsis Δ15 desaturase results in high accumulation of stearidonic acid in the seeds of transgenic soybean. Planta 224(5):1050–1057

Elattar MA, Karikari B, Li S, Song S, Cao Y, Aslam M et al (2021) Identification and validation of major QTLs, epistatic interactions, and candidate genes for soybean seed shape and weight using two related RIL populations. Front Genet 12:666440

Endo A, Saika H, Takemura M, Misawa N, Toki S (2019) A novel approach to carotenoid accumulation in rice callus by mimicking the cauliflower Orange mutation via genome editing. Rice 12(1):1–5

Enfissi EM, Fraser PD, Lois LM, Boronat A, Schuch W, Bramley PM (2005) Metabolic engineering of the mevalonate and non-mevalonate isopentenyl diphosphate-forming pathways for the production of health-promoting isoprenoids in tomato. Plant Biotechnol J 3(1):17–27

Ersado TL (2022) Causes of malnutrition. https://doi.org/10.5772/intechopen.104458

Eticha F, Grausgruber H, Siebenhandl-Ehn S, Berghofer E (2011) Some agronomic and chemical traits of blue aleurone and purple pericarp wheat (*Triticum* L.). Journal of Agricultural Science and Technology B 1:48–58

Fan M-S, Zhao F-J, Fairweather-Tait SJ, Poulton PR, Dunham SJ, McGrath SP (2008) Evidence of decreasing mineral density in wheat grain over the last 160 years. J Trace Elem Med Biol 22(4):315–324

Ficco DB, Mastrangelo AM, Trono D, Borrelli GM, De Vita P, Fares C et al (2014) The colours of durum wheat: a review. Crop Pasture Sci 65(1):1–15

Flores T, Karpova O, Su X, Zeng P, Bilyeu K, Sleper DA et al (2008) Silencing of GmFAD3 gene by siRNA leads to low α-linolenic acids (18: 3) of fad3-mutant phenotype in soybean [*Glycine max* (Merr.)]. Transgenic Res 17(5):839–850

Fraser PD, Enfissi EM, Halket JM, Truesdale MR, Yu D, Gerrish C, Bramley PM (2007) Manipulation of phytoene levels in tomato fruit: effects on isoprenoids, plastids, and intermediary metabolism. Plant Cell 19(10):3194–3211

Gannon B, Kaliwile C, Arscott SA, Schmaelzle S, Chileshe J, Kalungwana N et al (2014) Biofortified orange maize is as efficacious as a vitamin A supplement in Zambian children even in the presence of high liver reserves of vitamin A: a community-based, randomized placebo-controlled trial. Am J Clin Nutr 100(6):1541–1550

Garcia-Oliveira AL, Tan L, Fu Y, Sun C (2009) Genetic identification of quantitative trait loci for contents of mineral nutrients in rice grain. J Integr Plant Biol 51(1):84–92

Garg M, Chawla M, Chunduri V, Kumar R, Sharma S, Sharma NK et al (2016) Transfer of grain colors to elite wheat cultivars and their characterization. J Cereal Sci 71:138–144

Garg M, Sharma N, Sharma S, Kapoor P, Kumar A, Chunduri V, Arora P (2018) Biofortified crops generated by breeding, agronomy, and transgenic approaches are improving lives of millions of people around the world. Front Nutr 12

Garvin DF, Welch RM, Finley JW (2006) Historical shifts in the seed mineral micronutrient concentration of US hard red winter wheat germplasm. J Sci Food Agric 86(13):2213–2220

Genc Y, Humphries JM, Lyons GH, Graham RD (2005) Exploiting genotypic variation in plant nutrient accumulation to alleviate micronutrient deficiency in populations. J Trace Elem Med Biol 18(4):319–324

Genç Y, Verbyla A, Torun A, Cakmak I, Willsmore K, Wallwork H, McDonald G (2009) Quantitative trait loci analysis of zinc efficiency and grain zinc concentration in wheat using whole genome average interval mapping. Plant Soil 314(1):49–66

Gillespie S, Hodge J, Yosef S, Pandya-Lorch R (2016) Nourishing millions: stories of change in nutrition. Intl Food Policy Res Inst

Giovinazzo G, D'Amico L, Paradiso A, Bollini R, Sparvoli F, DeGara L (2005) Antioxidant metabolite profiles in tomato fruit constitutively expressing the grapevine stilbene synthase gene. Plant Biotechnol J 3(1):57–69

Goffman FD, Böhme T (2001) Relationship between fatty acid profile and vitamin E content in maize hybrids (*Zea mays* L.). J Agric Food Chem 49(10):4990–4994

Goo Y-M, Kim T-W, Lee M-K, Lee S-W (2013) Accumulation of PrLeg, a perilla legumin protein in potato tuber results in enhanced level of sulphur-containing amino acids. C R Biol 336(9):433–439

Govindaraj M, Rai K, Shanmugasundaram P, Dwivedi S, Sahrawat K, Muthaiah A, Rao A (2013) Combining ability and heterosis for grain iron and zinc densities in pearl millet. Crop Sci 53(2):507–517

Govindaraj M, Rai KN, Cherian B, Pfeiffer WH, Kanatti A, Shivade H (2019) Breeding biofortified pearl millet varieties and hybrids to enhance millet markets for human nutrition. Agriculture 9(5):106

Graham RD, Welch RM, Bouis HE (2001) Addressing micronutrient malnutrition through enhancing the nutritional quality of staple foods: principles, perspectives and knowledge gaps

Graham RD, Welch RM, Saunders DA, Ortiz-Monasterio I, Bouis HE, Bonierbale M et al (2007) Nutritious subsistence food systems. Adv Agron 92:1–74

Gregorio GB, Senadhira D, Htut H, Graham RD (2000) Breeding for trace mineral density in rice. Food Nutr Bull 21(4):382–386

Gyawali S, Otte ML, Chao S, Jilal A, Jacob DL, Amezrou R, Verma RPS (2017) Genome wide association studies (GWAS) of element contents in grain with a special focus on zinc and iron in a world collection of barley (*Hordeum vulgare* L.). J Cereal Sci 77:266–274

Gyawali S, Otte ML, Jacob DL, Abderrazek J, Singh Verma RP (2019) Multiple element concentration in the grain of spring barley (*Hordeum vulgare* L.) collection. J Plant Nutr 42(9):1036–1046

Haas J, Beard J, Murray-Kolb L, del Mundo A, Felix A, Gregorio G (2005) Iron-biofortified rice increases body iron in Filipino women. J Nutr 135:2823–2830

Hanafy MS, Rahman SM, Nakamoto Y, Fujiwara T, Naito S, Wakasa K, Ishimoto M (2013) Differential response of methionine metabolism in two grain legumes, soybean and azuki bean, expressing a mutated form of *Arabidopsis cystathionine* γ-synthase. J Plant Physiol 170(3):338–345

Harjes CE (2000) Lycopene Epsilon natural genetic variation in. Nat Rev Genet 1:171

Haroldsen VM, Chi-Ham CL, Kulkarni S, Lorence A, Bennett AB (2011) Constitutively expressed DHAR and MDHAR influence fruit, but not foliar ascorbate levels in tomato. Plant Physiol Biochem 49(10):1244–1249

Haseen F (2005) Malnutrition among Bangladeshi women in ultra poor households: prevalence and determinants. Retrieved September 20, 2006

Herzig P, Backhaus A, Seiffert U, von Wirén N, Pillen K, Maurer A (2019) Genetic dissection of grain elements predicted by hyperspectral imaging associated with yield-related traits in a wild barley NAM population. Plant Sci 285:151–164

Hina A, Cao Y, Song S, Li S, Sharmin RA, Elattar MA et al (2020) High-resolution mapping in two RIL populations refines major "QTL Hotspot" regions for seed size and shape in soybean (*Glycine max* L.). Int J Mol Sci 21(3):1040

Hindu V, Palacios-Rojas N, Babu R, Suwarno WB, Rashid Z, Usha R et al (2018) Identification and validation of genomic regions influencing kernel zinc and iron in maize. Theor Appl Genet 131(7):1443–1457

Hoddinott J, Rosegrant M, Torero M (2012) Investments to reduce hunger and undernutrition. Challenge Paper on Hunger and Malnutrition. Copenhagen Consensus Challenge Paper

Hoekenga OA (2014) Genomics of mineral nutrient biofortification: calcium, iron and zinc. In: Genomics of plant genetic resources (pp. 431–454). Springer

Holme IB, Dionisio G, Brinch-Pedersen H, Wendt T, Madsen CK, Vincze E, Holm PB (2012) Cisgenic barley with improved phytase activity. Plant Biotechnol J 10(2):237–247

Horton S, Mannar V, Wesley A (2008) Best practice paper: food fortification with iron and iodine. Copenhagen Consensus Center, Copenhagen Business School, Denmark

Huang S, Frizzi A, Florida CA, Kruger DE, Luethy MH (2006) High lysine and high tryptophan transgenic maize resulting from the reduction of both 19-and 22-kD α-zeins. Plant Mol Biol 61(3):525–535

Huang J-C, Zhong Y-J, Liu J, Sandmann G, Chen F (2013) Metabolic engineering of tomato for high-yield production of astaxanthin. Metab Eng 17:59–67

Huang T, Joshi V, Jander G (2014) The catabolic enzyme methionine gamma-lyase limits methionine accumulation in potato tubers. Plant Biotechnol J 12(7):883–893

Hurrell R, Egli I (2010) Iron bioavailability and dietary reference values. Am J Clin Nutr 91(5):1461S–1467S

Hussain S, Rengel Z, Mohammadi SA, Ebadi-Segherloo A, Maqsood MA (2016) Mapping QTL associated with remobilization of zinc from vegetative tissues into grains of barley (*Hordeum vulgare*). Plant Soil 399(1):193–208

Inaba M, Macer D (2004) Policy, regulation and attitudes towards agricultural biotechnology in Japan, 1

Ishikawa S, Abe T, Kuramata M, Yamaguchi M, Ando T, Yamamoto T, Yano M (2010) A major quantitative trait locus for increasing cadmium-specific concentration in rice grain is located on the short arm of chromosome 7. J Exp Bot 61(3):923–934

Ishimaru Y, Bashir K, Nishizawa NK (2011) Zn uptake and translocation in rice plants. Rice 4(1):21–27

Itoh K, Ozaki H, Okada K, Hori H, Takeda Y, Mitsui T (2003) Introduction of Wx transgene into rice wx mutants leads to both high-and low-amylose rice. Plant Cell Physiol 44(5):473–480

Jaganathan D, Ramasamy K, Sellamuthu G, Jayabalan S, Venkataraman G (2018) CRISPR for crop improvement: an update review. Front Plant Sci 9:985

Jeong O, Bombay M, Jo S, Ancheta MB, Lee J-H (2020) QTL for the iron and zinc contents of the milled grains of a doubled-haploid rice (*Oryza sativa* L.) population grown over two seasons. J Crop Sci Biotechnol 23(4):291–299

Jin T, Zhou J, Chen J, Zhu L, Zhao Y, Huang Y (2013) The genetic architecture of zinc and iron content in maize grains as revealed by QTL mapping and meta-analysis. Breed Sci 63(3):317–324

Kanatti A, Rai KN, Radhika K, Govindaraj M, Sahrawat KL, Rao AS (2014) Grain iron and zinc density in pearl millet: combining ability, heterosis and association with grain yield and grain size. Springerplus 3(1):1–12

Kanatti A, Rai KN, Radhika K, Govindaraj M (2016) Tester effect on combining ability and its relationship with line performance per se for grain iron and zinc densities in pearl millet. Crop Sci 56(2):689–696

Kastoori Ramamurthy R, Jedlicka J, Graef GL, Waters BM (2014) Identification of new QTLs for seed mineral, cysteine, and methionine concentrations in soybean [*Glycine max* (L.) Merr.]. Mol Breed 34(2):431–445

Kim S, Takahashi M, Higuchi K, Tsunoda K, Nakanishi H, Yoshimura E et al (2005) Increased nicotianamine biosynthesis confers enhanced tolerance of high levels of metals, in particular nickel, to plants. Plant Cell Physiol 46(11):1809–1818

Kim W-S, Chronis D, Juergens M, Schroeder AC, Hyun SW, Jez JM, Krishnan HB (2012) Transgenic soybean plants overexpressing O-acetylserine sulfhydrylase accumulate enhanced levels of cysteine and Bowman–Birk protease inhibitor in seeds. Planta 235(1):13–23

Kinoshita N, Kato M, Koyasaki K, Kawashima T, Nishimura T, Hirayama Y et al (2017) Identification of quantitative trait loci for rice grain quality and yield-related traits in two closely related *Oryza sativa* L subsp. japonica cultivars grown near the northernmost limit for rice paddy cultivation. Breed Sci 67:16155

Krishnappa G, Rathan ND, Sehgal D, Ahlawat AK, Singh SK, Singh SK et al (2021) Identification of novel genomic regions for biofortification traits using an SNP marker-enriched linkage map in wheat (*Triticum aestivum* L.). Front Nutr 320

Kumar S, Thirunavukkarasu N, Singh G, Sharma R, Kulkarni KS (2015) Biofortification for selecting and developing crop cultivars denser in iron and zinc. In: Nutrient use efficiency: from basics to advances. Springer, pp 237–253

Kumar S, Hash CT, Thirunavukkarasu N, Singh G, Rajaram V, Rathore A et al (2016) Mapping quantitative trait loci controlling high iron and zinc content in self and open pollinated grains of pearl millet [*Pennisetum glaucum* (L.) R. Br.]. Front Plant Sci 7:1636

Kumar S, Hash CT, Nepolean T, Mahendrakar MD, Satyavathi CT, Singh G et al (2018) Mapping grain iron and zinc content quantitative trait loci in an iniadi-derived immortal population of pearl millet. Genes 9(5):248

Kumar S, Palve A, Joshi C, Srivastava RK (2019) Crop biofortification for iron (Fe), zinc (Zn) and vitamin A with transgenic approaches. Heliyon 5(6):e01914

Kuwano M, Mimura T, Takaiwa F, Yoshida KT (2009) Generation of stable 'low phytic acid' transgenic rice through antisense repression of the 1d-myo-inositol 3-phosphate synthase gene (RINO1) using the 18-kDa oleosin promoter. Plant Biotechnol J 7(1):96–105

Lafiandra D, Riccardi G, Shewry PR (2014) Improving cereal grain carbohydrates for diet and health. J Cereal Sci 59(3):312–326

Lago C, Cassani E, Zanzi C, Landoni M, Trovato R, Pilu R (2014) Development and study of a maize cultivar rich in anthocyanins: coloured polenta, a new functional food. Plant Breed 133(2):210–217

Lai J, Messing J (2002) Increasing maize seed methionine by mRNA stability. Plant J 30(4):395–402

Larson SR, Rutger JN, Young KA, Raboy V (2000) Isolation and genetic mapping of a non-lethal rice (*Oryza sativa* L.) low phytic acid 1 mutation. Crop Sci 40(5):1397–1405

Lee S, An G (2009) Over-expression of OsIRT1 leads to increased iron and zinc accumulations in rice. Plant Cell Environ 32(4):408–416

Lee S, Jeon US, Lee SJ, Kim Y-K, Persson DP, Husted S et al (2009) Iron fortification of rice seeds through activation of the nicotianamine synthase gene. Proc Natl Acad Sci 106(51):22014–22019

Lee JH, Kim HS, Suh SC, Rhim SL (2010) Development of transgenic rice lines expressing the human lactoferrin gene. J Plant Biotechnol 37(4):556

Lee S, Persson DP, Hansen TH, Husted S, Schjoerring JK, Kim YS et al (2011) Bio-available zinc in rice seeds is increased by activation tagging of nicotianamine synthase. Plant Biotechnol J 9(8):865–873

Lee S, Kim Y-S, Jeon US, Kim Y-K, Schjoerring JK, An G (2012) Activation of rice nicotianamine synthase 2 (OsNAS2) enhances iron availability for biofortification. Mol Cells 33(3):269–275

Lee S-M, Kang J-W, Lee J-Y, Seo J, Shin D, Cho J-H et al (2020) QTL analysis for Fe and Zn concentrations in rice grains using a doubled haploid population derived from a cross between rice (*Oryza sativa*) cultivar 93-11 and milyang 352. Plant Breed Biotechnol 8(1):69–76

Li L, Yang Y, Xu Q, Owsiany K, Welsch R, Chitchumroonchokchai C et al (2012) The Or gene enhances carotenoid accumulation and stability during post-harvest storage of potato tubers. Mol Plant 5(2):339–352

Liu Q, Wang Z, Chen X, Cai X, Tang S, Yu H et al (2003) Stable inheritance of the antisense Waxy gene in transgenic rice with reduced amylose level and improved quality. Transgenic Res 12(1):71–82

Liu X, Zhang C, Wang X, Liu Q, Yuan D, Pan G et al (2016) Development of high-lysine rice via endosperm-specific expression of a foreign LYSINE RICH PROTEIN gene. BMC Plant Biol 16(1):1–13

Liu J, Wu B, Singh RP, Velu G (2019) QTL mapping for micronutrients concentration and yield component traits in a hexaploid wheat mapping population. J Cereal Sci 88:57–64

Lopez A, Van Eck J, Conlin B, Paolillo D, O'Neill J, Li L (2008) Effect of the cauliflower Or transgene on carotenoid accumulation and chromoplast formation in transgenic potato tubers. J Exp Bot 59(2):213–223

Loskutov I, Polonskiy V (2017) Content of β-glucans in oat grain as a perspective direction of breeding for health products and fodder. Agric Biol 52(4):646–657

Lu K, Li L, Zheng X, Zhang Z, Mou T, Hu Z (2008) Quantitative trait loci controlling Cu, Ca, Zn, Mn and Fe content in rice grains. J Genet 87(3):305

Lucca P, Hurrell R, Potrykus I (2001) Approaches to improving the bioavailability and level of iron in rice seeds. J Sci Food Agric 81(9):828–834

Lucca P, Hurrell R, Potrykus I (2002) Fighting iron deficiency anemia with iron-rich rice. J Am Coll Nutr 21(sup3):184S–190S

Lukaszewicz M, Matysiak-Kata I, Skala J, Fecka I, Cisowski W, Szopa J (2004) Antioxidant capacity manipulation in transgenic potato tuber by changes in phenolic compounds content. J Agric Food Chem 52(6):1526–1533

Luo J, Butelli E, Hill L, Parr A, Niggeweg R, Bailey P et al (2008) AtMYB12 regulates caffeoyl quinic acid and flavonol synthesis in tomato: expression in fruit results in very high levels of both types of polyphenol. Plant J 56(2):316–326

Lyons G, Ortiz-Monasterio I, Stangoulis J, Graham R (2005) Selenium concentration in wheat grain: is there sufficient genotypic variation to use in breeding? Plant Soil 269(1):369–380

Mahender A, Anandan A, Pradhan SK, Pandit E (2016) Rice grain nutritional traits and their enhancement using relevant genes and QTLs through advanced approaches. Springerplus 5(1):1–18

Mahendrakar MD, Parveda M, Kishor P, Srivastava RK (2020) Discovery and validation of candidate genes for grain iron and zinc metabolism in pearl millet [*Pennisetum glaucum* (L.) R. Br.]. Sci Rep 10(1):1–16

Maligeppagol M, Chandra GS, Navale PM, Deepa H, Rajeev P, Asokan R et al (2013) Anthocyanin enrichment of tomato (*Solanum lycopersicum* L.) fruit by metabolic engineering. Curr Sci:72–80

Mamo BE, Barber BL, Steffenson BJ (2014) Genome-wide association mapping of zinc and iron concentration in barley landraces from Ethiopia and Eritrea. J Cereal Sci 60(3):497–506

Mannar M, Sankar R (2004) Micronutrient fortification of foods—rationale, application and impact. Indian J Pediatr 71(11):997–1002

Maqbool MA, Aslam M, Beshir A, Khan MS (2018) Breeding for provitamin A biofortification of maize (*Zea mays* L.). Plant Breed 137(4):451–469

Masuda H, Suzuki M, Morikawa KC, Kobayashi T, Nakanishi H, Takahashi M et al (2008) Increase in iron and zinc concentrations in rice grains via the introduction of barley genes involved in phytosiderophore synthesis. Rice 1(1):100–108

Masuda H, Usuda K, Kobayashi T, Ishimaru Y, Kakei Y, Takahashi M et al (2009) Overexpression of the barley nicotianamine synthase gene HvNAS1 increases iron and zinc concentrations in rice grains. Rice 2(4):155–166

Masuda H, Ishimaru Y, Aung MS, Kobayashi T, Kakei Y, Takahashi M et al (2012) Iron biofortification in rice by the introduction of multiple genes involved in iron nutrition. Sci Rep 2(1):1–7

Masuda H, Kobayashi T, Ishimaru Y, Takahashi M, Aung MS, Nakanishi H et al (2013) Iron-biofortification in rice by the introduction of three barley genes participated in mugineic acid biosynthesis with soybean ferritin gene. Front Plant Sci 4:132

McLean E, Cogswell M, Egli I, Wojdyla D, De Benoist B (2009) Worldwide prevalence of anaemia, WHO vitamin and mineral nutrition information system, 1993–2005. Public Health Nutr 12(4):444–454

Menkir A, Torbert Rocheford B (2015) Maziya-Dixon & Sherry Tanumihardjo. Euphytica 205:203–217

Mizuno D, Higuchi K, Sakamoto T, Nakanishi H, Mori S, Nishizawa NK (2003) Three nicotianamine synthase genes isolated from maize are differentially regulated by iron nutritional status. Plant Physiol 132(4):1989–1997

Monasterio I, Graham RD (2000) Breeding for trace minerals in wheat. Food Nutr Bull 21(4):392–396

Moose SP, Mumm RH (2008) Molecular plant breeding as the foundation for 21st century crop improvement. Plant Physiol 147(3):969–977

Morrissey J, Guerinot ML (2009) Iron uptake and transport in plants: the good, the bad, and the ionome. Chem Rev 109(10):4553–4567

Muir SR, Collins GJ, Robinson S, Hughes S, Bovy A, Ric De Vos CHV et al (2001) Overexpression of petunia chalcone isomerase in tomato results in fruit containing increased levels of flavonols. Nat Biotechnol 19(5):470–474

Muzhingi T, Palacios-Rojas N, Miranda A, Cabrera ML, Yeum KJ, Tang G (2017) Genetic variation of carotenoids, vitamin E and phenolic compounds in Provitamin A biofortified maize. J Sci Food Agric 97(3):793–801

Nandi S, Suzuki YA, Huang J, Yalda D, Pham P, Wu L et al (2002) Expression of human lactoferrin in transgenic rice grains for the application in infant formula. Plant Sci 163(4):713–722

Naqvi S, Zhu C, Farre G, Ramessar K, Bassie L, Breitenbach J et al (2009) Transgenic multivitamin corn through biofortification of endosperm with three vitamins representing three distinct metabolic pathways. Proc Natl Acad Sci 106(19):7762–7767

Nestel P, Bouis HE, Meenakshi JV, Pfeiffer W (2006) Biofortification of staple food crops. J Nutr 136(4):1064–1067

Nguyen HC, Hoefgen R, Hesse H (2012) Improving the nutritive value of rice seeds: elevation of cysteine and methionine contents in rice plants by ectopic expression of a bacterial serine acetyltransferase. J Exp Bot 63(16):5991–6001

Niggeweg R, Michael AJ, Martin C (2004) Engineering plants with increased levels of the antioxidant chlorogenic acid. Nat Biotechnol 22(6):746–754

Ning L, Sun P, Wang Q, Ma D, Hu Z, Zhang D et al (2015) Genetic architecture of biofortification traits in soybean (*Glycine max* L. Merr.) revealed through association analysis and linkage mapping. Euphytica 204(2):353–369

Norton GJ, Deacon CM, Xiong L, Huang S, Meharg AA, Price AH (2010) Genetic mapping of the rice ionome in leaves and grain: identification of QTLs for 17 elements including arsenic, cadmium, iron and selenium. Plant Soil 329(1):139–153

Norton GJ, Douglas A, Lahner B, Yakubova E, Guerinot ML, Pinson SR et al (2014) Genome wide association mapping of grain arsenic, copper, molybdenum and zinc in rice (*Oryza sativa* L.) grown at four international field sites. PLoS One 9(2):e89685

Oliva M, Shannon J, Sleper D, Ellersieck M, Cardinal A, Paris RL, Lee J (2006) Stability of fatty acid profile in soybean genotypes with modified seed oil composition. Crop Sci 46(5):2069–2075

Oliva N, Chadha-Mohanty P, Poletti S, Abrigo E, Atienza G, Torrizo L et al (2014) Large-scale production and evaluation of marker-free indica rice IR64 expressing phytoferritin genes. Mol Breed 33(1):23–37

Ortiz-Monasterio JI, Palacios-Rojas N, Meng E, Pixley K, Trethowan R, Pena R (2007) Enhancing the mineral and vitamin content of wheat and maize through plant breeding. J Cereal Sci 46(3):293–307

Oury F-X, Leenhardt F, Remesy C, Chanliaud E, Duperrier B, Balfourier F, Charmet G (2006) Genetic variability and stability of grain magnesium, zinc and iron concentrations in bread wheat. Eur J Agron 25(2):177–185

Paine JA, Shipton CA, Chaggar S, Howells RM, Kennedy MJ, Vernon G et al (2005) Improving the nutritional value of Golden Rice through increased pro-vitamin A content. Nat Biotechnol 23(4):482–487

Palmer AC, Healy K, Barffour MA, Siamusantu W, Chileshe J, Schulze KJ et al (2016) Provitamin A carotenoid–biofortified maize consumption increases pupillary responsiveness among Zambian children in a randomized controlled trial. J Nutr 146(12):2551–2558

Paul S, Ali N, Gayen D, Datta S, Datta K (2012) Molecular breeding of Osfer 2 gene to increase iron nutrition in rice grain. GM Crops Food 3(4):310–316

Peng B, Kong H, Li Y, Wang L, Zhong M, Sun L et al (2014) OsAAP6 functions as an important regulator of grain protein content and nutritional quality in rice. Nat Commun 5(1):1–12

Pérez-Massot E, Banakar R, Gómez-Galera S, Zorrilla-López U, Sanahuja G, Arjó G et al (2013) The contribution of transgenic plants to better health through improved nutrition: opportunities and constraints. Genes Nutr 8(1):29–41

Pfeiffer WH, McClafferty B (2007) HarvestPlus: breeding crops for better nutrition. Crop Sci 47:S-88–S-105

Pierce EC, LaFayette PR, Ortega MA, Joyce BL, Kopsell DA, Parrott WA (2015) Ketocarotenoid production in soybean seeds through metabolic engineering. PLoS One 10(9):e0138196

Popkin BM, Corvalan C, Grummer-Strawn LM (2020) Dynamics of the double burden of malnutrition and the changing nutrition reality. Lancet 395(10217):65–74

Pradhan S, Pandit E, Pawar S, Naveenkumar R, Barik S, Mohanty S et al (2020) Linkage disequilibrium mapping for grain Fe and Zn enhancing QTLs useful for nutrient dense rice breeding. BMC Plant Biol 20(1):1–24

Prasanna BM, Palacios-Rojas N, Hossain F, Muthusamy V, Menkir A, Dhliwayo T et al (2020) Molecular breeding for nutritionally enriched maize: status and prospects. Front Genet 10:1392

Qin Y, Kim S, Sohn J (2009) QTL analysis of protein content in double-haploid lines of rice. Korean J Crop Sci/Hanguk Jakmul Hakhoe Chi 54(2):165–171

Rai KN, Yadav O, Govindaraj M, Pfeiffer WH, Yadav H, Rajpurohit B et al (2016) Grain iron and zinc densities in released and commercial cultivars of pearl millet (*Pennisetum glaucum*). Indian J Agric Sci 86(03):11–16

Rai KN, Govindaraj M, Kanatti A, Rao AS, Shivade H (2017) Inbreeding effects on grain iron and zinc concentrations in pearl millet. Crop Sci 57(5):2699–2706

Ramesh SA, Choimes S, Schachtman DP (2004) Over-expression of an Arabidopsis zinc transporter in Hordeum vulgare increases short-term zinc uptake after zinc deprivation and seed zinc content. Plant Mol Biol 54(3):373–385

Reuscher S, Kolter A, Hoffmann A, Pillen K, Krämer U (2016) Quantitative trait loci and inter-organ partitioning for essential metal and toxic analogue accumulation in barley. PLoS One 11(4):e0153392

Reynolds M, Pellegrineschi A, Skovmand B (2005) Sink-limitation to yield and biomass: a summary of some investigations in spring wheat. Ann Appl Biol 146(1):39–49

Ricroch A, Clairand P, Harwood W (2017) Use of CRISPR systems in plant genome editing: toward new opportunities in agriculture. Emerg Topics Life Sci 1(2):169–182
Rogers EE, Guerinot ML (2002) FRD3, a member of the multidrug and toxin efflux family, controls iron deficiency responses in Arabidopsis. Plant Cell 14(8):1787–1799
Römer S, Lübeck J, Kauder F, Steiger S, Adomat C, Sandmann G (2002) Genetic engineering of a zeaxanthin-rich potato by antisense inactivation and co-suppression of carotenoid epoxidation. Metab Eng 4(4):263–272
Sadeghzadeh B, Rengel Z, Li C (2015) Quantitative trait loci (QTL) of seed Zn accumulation in barley population clipper X Sahara. J Plant Nutr 38(11):1672–1684
Sahu A (2017) Acquisition Related Genes Involved in Early stage iron deficiency response in foxtail millet (Setaria italica). Ph. D. dissertation/master's thesis). College of Agriculture. Indira Gandhi…
Saika H, Oikawa A, Matsuda F, Onodera H, Saito K, Toki S (2011) Application of gene targeting to designed mutation breeding of high-tryptophan rice. Plant Physiol 156(3):1269–1277
Saini DK, Devi P, Kaushik P (2020) Advances in genomic interventions for wheat biofortification: a review. Agronomy 10(1):62
Sato S, Xing A, Ye X, Schweiger B, Kinney A, Graef G, Clemente T (2004) Production of γ-linolenic acid and stearidonic acid in seeds of marker-free transgenic soybean 1. Crop Sci 44(2):646–652
Sazawal S, Dhingra U, Dhingra P, Dutta A, Deb S, Kumar J et al (2018) Efficacy of high zinc biofortified wheat in improvement of micronutrient status, and prevention of morbidity among preschool children and women-a double masked, randomized, controlled trial. Nutr J 17(1):1–10
Schmidt MA, Parrott WA, Hildebrand DF, Berg RH, Cooksey A, Pendarvis K et al (2015) Transgenic soya bean seeds accumulating β-carotene exhibit the collateral enhancements of oleate and protein content traits. Plant Biotechnol J 13(4):590–600
Schröder I, Johnson E, De Vries S (2003) Microbial ferric iron reductases. FEMS Microbiol Rev 27(2–3):427–447
Sedeek KE, Mahas A, Mahfouz M (2019) Plant genome engineering for targeted improvement of crop traits. Front Plant Sci 10:114
Sestili F, Janni M, Doherty A, Botticella E, D'Ovidio R, Masci S et al (2010) Increasing the amylose content of durum wheat through silencing of the SBEIIagenes. BMC Plant Biol 10(1):1–12
Shewmaker CK, Sheehy JA, Daley M, Colburn S, Ke DY (1999) Seed-specific overexpression of phytoene synthase: increase in carotenoids and other metabolic effects. Plant J 20(4):401–412
Shi R-L, Tong Y-P, Jing R-L, Zhang F-S, Zou C-Q (2013) Characterization of quantitative trait loci for grain minerals in hexaploid wheat (*Triticum aestivum* L.). J Integr Agric 12(9):1512–1521
Shih C-H, Chen Y, Wang M, Chu IK, Lo C (2008) Accumulation of isoflavone genistin in transgenic tomato plants overexpressing a soybean isoflavone synthase gene. J Agric Food Chem 56(14):5655–5661
Singh M (2009) Effect of trace element deficiencies in soil on human and animal health. Bull Indian Soc Soil Sci 27:75–101
Singh R, Govindan V (2017) Zinc-biofortified wheat: harnessing genetic diversity for improved nutritional quality. Science Brief: Biofortification No. 1 (May 2017). CIMMYT, HarvestPlus, and the Global Crop Diversity Trust. Bonn, Germany.(https://croptrust.org/our-work/science-briefs/)
Singh SP, Keller B, Gruissem W, Bhullar NK (2017) Rice NICOTIANAMINE SYNTHASE 2 expression improves dietary iron and zinc levels in wheat. Theor Appl Genet 130(2):283–292
Sinha P, Davis J, Saag L, Wanke C, Salgame P, Mesick J et al (2019) Undernutrition and tuberculosis: public health implications. J Infect Dis 219(9):1356–1363
Song S, Hou W, Godo I, Wu C, Yu Y, Matityahu I et al (2013) Soybean seeds expressing feedback-insensitive cystathionine γ-synthase exhibit a higher content of methionine. J Exp Bot 64(7):1917–1926
Song X-Y, Zhu W-J, Tang R-M, Cai J-H, Chen M, Yang Q (2016) Over-expression of StLCYb increases β-carotene accumulation in potato tubers. Plant Biotechnol Reports 10(2):95–104

Spohrer R, Larson M, Maurin C, Laillou A, Capanzana M, Garrett GS (2013) The growing importance of staple foods and condiments used as ingredients in the food industry and implications for large-scale food fortification programs in Southeast Asia. Food Nutr Bull 34(2_suppl1):S50–S61

Stangoulis JC, Huynh B-L, Welch RM, Choi E-Y, Graham RD (2007) Quantitative trait loci for phytate in rice grain and their relationship with grain micronutrient content. Euphytica 154(3):289–294

Stevens GA, Bennett JE, Hennocq Q, Lu Y, De-Regil LM, Rogers L et al (2015) Trends and mortality effects of vitamin A deficiency in children in 138 low-income and middle-income countries between 1991 and 2013: a pooled analysis of population-based surveys. Lancet Glob Health 3(9):e528–e536

Storozhenko S, De Brouwer V, Volckaert M, Navarrete O, Blancquaert D, Zhang G-F et al (2007) Folate fortification of rice by metabolic engineering. Nat Biotechnol 25(11):1277–1279

Suzuki M, Morikawa KC, Nakanishi H, Takahashi M, Saigusa M, Mori S, Nishizawa NK (2008) Transgenic rice lines that include barley genes have increased tolerance to low iron availability in a calcareous paddy soil. Soil Sci Plant Nutr 54(1):77–85

Swamy B, Kaladhar K, Anuradha K, Batchu AK, Longvah T, Sarla N (2018a) QTL analysis for grain iron and zinc concentrations in two *O. nivara* derived backcross populations. Rice Sci 25(4):197–207

Swamy BM, Descalsota GIL, Nha CT, Amparado A, Inabangan-Asilo MA, Manito C et al (2018b) Identification of genomic regions associated with agronomic and biofortification traits in DH populations of rice. PLoS One 13(8):e0201756

Takahashi M, Nakanishi H, Kawasaki S, Nishizawa NK, Mori S (2001) Enhanced tolerance of rice to low iron availability in alkaline soils using barley nicotianamine aminotransferase genes. Nat Biotechnol 19(5):466–469

Tamás C, Kisgyörgy BN, Rakszegi M, Wilkinson MD, Yang M-S, Láng L et al (2009) Transgenic approach to improve wheat (*Triticum aestivum* L.) nutritional quality. Plant Cell Rep 28(7):1085–1094

Tan Y, Sun M, Xing Y, Hua J, Sun X, Zhang Q, Corke H (2001) Mapping quantitative trait loci for milling quality, protein content and color characteristics of rice using a recombinant inbred line population derived from an elite rice hybrid. Theor Appl Genet 103(6):1037–1045

Tanaka H, Yabuta Y, Tamoi M, Tanabe N, Shigeoka S (2015) Generation of transgenic tobacco plants with enhanced tocotrienol levels through the ectopic expression of rice homogentisate geranylgeranyl transferase. Plant Biotechnol:15.0702 a

Tang M, He X, Luo Y, Ma L, Tang X, Huang K (2013) Nutritional assessment of transgenic lysine-rich maize compared with conventional quality protein maize. J Sci Food Agric 93(5):1049–1054

Tighe P, Duthie G, Vaughan N, Brittenden J, Simpson WG, Duthie S et al (2010) Effect of increased consumption of whole-grain foods on blood pressure and other cardiovascular risk markers in healthy middle-aged persons: a randomized controlled trial. Am J Clin Nutr 92(4):733–740

Timmer CP (2003) Biotechnology and food systems in developing countries. J Nutr 133(11):3319–3322

Tong L-T, Liu L-Y, Zhong K, Yan W, Guo L-N, Zhou S-M (2014) Effects of cultivar on phenolic content and antioxidant activity of naked oat in China. J Integr Agric 13(8):1809–1816

Trijatmiko KR, Dueñas C, Tsakirpaloglou N, Torrizo L, Arines FM, Adeva C et al (2016) Biofortified indica rice attains iron and zinc nutrition dietary targets in the field. Sci Rep 6(1):1–13

Ufaz S, Galili G (2008) Improving the content of essential amino acids in crop plants: goals and opportunities. Plant Physiol 147(3):954–961

UNICEF B (2007) Child and Mother Nutrition Survey 2005. Bangladesh Bureau of Statistics and UNICEF, Dhaka, 1207

Upadhyaya CP, Young KE, Akula N, Soon Kim H, Heung JJ, Oh OM et al (2009) Over-expression of strawberry D-galacturonic acid reductase in potato leads to accumulation of vitamin C with enhanced abiotic stress tolerance. Plant Sci 177(6):659–667

Van Der Straeten D, Bhullar NK, De Steur H, Gruissem W, MacKenzie D, Pfeiffer W et al (2020) Multiplying the efficiency and impact of biofortification through metabolic engineering. Nat Commun 11(1):1–10

Van Eck J, Conlin B, Garvin D, Mason H, Navarre D, Brown C (2007) Enhancing beta-carotene content in potato by RNAi-mediated silencing of the beta-carotene hydroxylase gene. Am J Potato Res 84(4):331–342

Van Eenennaam AL, Lincoln K, Durrett TP, Valentin HE, Shewmaker CK, Thorne GM et al (2003) Engineering vitamin E content: from Arabidopsis mutant to soy oil. Plant Cell 15(12):3007–3019

Vanisha K, Atwal A, Dhaliwal S, Banga SK (2013) Assessment of diverse sesame (*Sesamum indicum* L.) germplasm for mineral composition. J Plant Sci Res 29(1):29

Vasconcelos M, Datta K, Oliva N, Khalekuzzaman M, Torrizo L, Krishnan S et al (2003) Enhanced iron and zinc accumulation in transgenic rice with the ferritin gene. Plant Sci 164(3):371–378

Velu G, Rai K, Muralidharan V, Longvah T, Crossa J (2011) Gene effects and heterosis for grain iron and zinc density in pearl millet (Pennisetum glaucum (L.) R. Br). Euphytica 180(2):251–259

Virk P, Andersson M, Arcos J, Cherian B, Velu G, Chatrath R (2021) Potential of biofortified wheat to alleviate hidden hunger. In: Sareen S, Sharma P, Singh C, Jasrotia P, Singh GP, Sarial AK (eds) Improving cereal productivity through climate smart practices. Elsevier WP Woodhead Publishing, Cambridge, pp 91–96

Wakasa K, Hasegawa H, Nemoto H, Matsuda F, Miyazawa H, Tozawa Y et al (2006) High-level tryptophan accumulation in seeds of transgenic rice and its limited effects on agronomic traits and seed metabolite profile. J Exp Bot 57(12):3069–3078

Wang L, Zhong M, Li X, Yuan D, Xu Y, Liu H et al (2008) The QTL controlling amino acid content in grains of rice (*Oryza sativa*) are co-localized with the regions involved in the amino acid metabolism pathway. Mol Breed 21(1):127–137

Wang C, Zeng J, Li Y, Hu W, Chen L, Miao Y et al (2014) Enrichment of provitamin A content in wheat (*Triticum aestivum* L.) by introduction of the bacterial carotenoid biosynthetic genes CrtB and CrtI. J Exp Bot 65(9):2545–2556

Wang Y, Xu X, Hao Y, Zhang Y, Liu Y, Pu Z et al (2021) QTL mapping for grain zinc and iron concentrations in bread wheat. Front Nutr 8:680391

Wang H, Jia J, Cai Z, Duan M, Jiang Z, Xia Q et al (2022) Identification of quantitative trait loci (QTLs) and candidate genes of seed Iron and zinc content in soybean [*Glycine max* (L.) Merr.]. BMC Genomics 23(1):1–14

Watanabe S, Uesugi S, Kikuchi Y (2002) Isoflavones for prevention of cancer, cardiovascular diseases, gynecological problems and possible immune potentiation. Biomed Pharmacother 56(6):302–312

Watanabe KN, Sassa Y, Suda E, Chen C-H, Inaba M, Kikuchi A (2005) Global political, economic, social and technological issues on trasngenic crops. Plant Biotechnol 22(5):515–522

Waters BM, Sankaran RP (2011) Moving micronutrients from the soil to the seeds: genes and physiological processes from a biofortification perspective. Plant Sci 180(4):562–574

Wei C, Zhang J, Chen Y, Zhou W, Xu B, Wang Y, Chen J (2010) Physicochemical properties and development of wheat large and small starch granules during endosperm development. Acta Physiol Plant 32(5):905–916

Welch RM, Graham RD (2004) Breeding for micronutrients in staple food crops from a human nutrition perspective. J Exp Bot 55(396):353–364

Welsch R, Maass D, Voegel T, DellaPenna D, Beyer P (2007) Transcription factor RAP2. 2 and its interacting partner SINAT2: stable elements in the carotenogenesis of Arabidopsis leaves. Plant Physiol 145(3):1073–1085

Welsch R, Arango J, Bär C, Salazar B, Al-Babili S, Beltrán J et al (2010) Provitamin A accumulation in cassava (*Manihot esculenta*) roots driven by a single nucleotide polymorphism in a phytoene synthase gene. Plant Cell 22(10):3348–3356

White PJ, Broadley MR (2009) Biofortification of crops with seven mineral elements often lacking in human diets–iron, zinc, copper, calcium, magnesium, selenium and iodine. New Phytol 182(1):49–84

WHO G (2000) Turning the tide of malnutrition: responding to the challenge of the 21st century

Wiegmann M, Thomas WT, Bull HJ, Flavell AJ, Zeyner A, Peiter E et al (2019) Wild barley serves as a source for biofortification of barley grains. Plant Sci 283:83–94

Wirth J, Poletti S, Aeschlimann B, Yakandawala N, Drosse B, Osorio S et al (2009) Rice endosperm iron biofortification by targeted and synergistic action of nicotianamine synthase and ferritin. Plant Biotechnol J 7(7):631–644

Woeltje MM, Evanoff AB, Helmink BA, Culbertson DL, Maleta KM, Manary MJ, Trehan I (2023) Community-Based management of acute malnutrition for infants under 6 months of age is safe and effective: analysis of operational data. Public Health Nutr 26(1):246–255

Wong J, Lambert R, Wurtzel E, Rocheford T (2004) QTL and candidate genes phytoene synthase and ζ-carotene desaturase associated with the accumulation of carotenoids in maize. Theor Appl Genet 108(2):349–359

Wu TY, Gruissem W, Bhullar NK (2019) Targeting intracellular transport combined with efficient uptake and storage significantly increases grain iron and zinc levels in rice. Plant Biotechnol J 17(1):9–20

Wurbs D, Ruf S, Bock R (2007) Contained metabolic engineering in tomatoes by expression of carotenoid biosynthesis genes from the plastid genome. Plant J 49(2):276–288

Xiaoyan S, Yan Z, Shubin W (2012) Improvement Fe content of wheat (*Triticum aestivum*) grain by soybean ferritin expression cassette without vector backbone sequence. J Agric Biotechnol

Yang S-H, Moran DL, Jia H-W, Bicar EH, Lee M, Scott MP (2002) Expression of a synthetic porcine α-lactalbumin gene in the kernels of transgenic maize. Transgenic Res 11(1):11–20

Yang W, Cahoon RE, Hunter SC, Zhang C, Han J, Borgschulte T, Cahoon EB (2011) Vitamin E biosynthesis: functional characterization of the monocot homogentisate geranylgeranyl transferase. Plant J 65(2):206–217

Yang Q-Q, Zhang C-Q, Chan M-L, Zhao D-S, Chen J-Z, Wang Q et al (2016) Biofortification of rice with the essential amino acid lysine: molecular characterization, nutritional evaluation, and field performance. J Exp Bot 67(14):4285–4296

Yu O, Shi J, Hession AO, Maxwell CA, McGonigle B, Odell JT (2003) Metabolic engineering to increase isoflavone biosynthesis in soybean seed. Phytochemistry 63(7):753–763

Yu J, Peng P, Zhang X, Zhao Q, Zhu D, Sun X et al (2005) Seed-specific expression of the lysine-rich protein gene sb401 significantly increases both lysine and total protein content in maize seeds. Food Nutr Bull 26(4_suppl3):S312–S316

Yu Y-H, Li G, Fan Y-Y, Zhang K-Q, Min J, Zhu Z-W, Zhuang J-Y (2009) Genetic relationship between grain yield and the contents of protein and fat in a recombinant inbred population of rice. J Cereal Sci 50(1):121–125

Yun B-W, Kim M-G, Handoyo T, Kim K-M (2014) Analysis of rice grain quality-associated quantitative trait loci by using genetic mapping. Am J Plant Sci 2014:05

Zeh M, Casazza AP, Kreft O, Roessner U, Bieberich K, Willmitzer L et al (2001) Antisense inhibition of threonine synthase leads to high methionine content in transgenic potato plants. Plant Physiol 127(3):792–802

Zeng J, Wang X, Miao Y, Wang C, Zang M, Chen X et al (2015) Metabolic engineering of wheat provitamin A by simultaneously overexpressing CrtB and silencing carotenoid hydroxylase (TaHYD). J Agric Food Chem 63(41):9083–9092

Zhang W, Bi J, Chen L, Zheng L, Ji S, Xia Y et al (2008) QTL mapping for crude protein and protein fraction contents in rice (*Oryza sativa* L.). J Cereal Sci 48(2):539–547

Zhang Y, Li X, Chen W, Yi B, Wen J, Shen J et al (2011) Identification of two major QTL for yellow seed color in two crosses of resynthesized *Brassica napus*. Mol Breed 28(3):335–342

Zhang L, Yang X-D, Zhang Y-Y, Yang J, Qi G-X, Guo D-Q et al (2014) Changes in oleic acid content of transgenic soybeans by antisense RNA mediated posttranscriptional gene silencing. Int J Genom 2014

Zheng L, Cheng Z, Ai C, Jiang X, Bei X, Zheng Y et al (2010) Nicotianamine, a novel enhancer of rice iron bioavailability to humans. PLoS One 5(4):e10190

Zhong M, Wang L-Q, Yuan D-J, Luo L-J, Xu C-G, He Y-Q (2011) Identification of QTL affecting protein and amino acid contents in rice. Rice Sci 18(3):187–195

Zhu C, Naqvi S, Breitenbach J, Sandmann G, Christou P, Capell T (2008) Combinatorial genetic transformation generates a library of metabolic phenotypes for the carotenoid pathway in maize. Proc Natl Acad Sci 105(47):18232–18237

Zimmermann MB, Hurrell RF (2002) Improving iron, zinc and vitamin A nutrition through plant biotechnology. Curr Opin Biotechnol 13(2):142–145

Zuluaga DL, Gonzali S, Loreti E, Pucciariello C, Degl'Innocenti E, Guidi L et al (2008) *Arabidopsis thaliana* MYB75/PAP1 transcription factor induces anthocyanin production in transgenic tomato plants. Funct Plant Biol 35(7):606–618

Chapter 5
Contribution of Conventional Breeding Approaches in Legumes Biofortification

Hafiz Muhammad Ahmad, Sadaf Zahra, Sadaf Oranab, Shazia Arif, Shehnaz Zakia, Gul-E-Zahra, Aamir Raina, Muhammad Zubair Khan, Flavien Shimira, Ghassan Zahid, Saira Bano, and Mahmood-Ur-Rahman

Abstract Deficiency of important micronutrients in human diet is usually known as hidden hunger. Globally, malnutrition affects the life of about 2 billion people. Especially, the life of pregnant women and children of developing countries is affected very badly. In the past, plant breeders majorly focused to increase the crop productivity by improving resistance against biotic and abiotic stresses. A very less attention was given to improve the nutritional accumulations of crops. Recently, biofortification of crop plants has been considered an objective of major breeding

H. M. Ahmad (✉) · S. Zahra · Mahmood-Ur-Rahman
Department of Bioinformatics and Biotechnology, Government College University Faisalabad, Faisalabad, Pakistan

S. Oranab
Department of Biochemistry, Government College University Faisalabad, Faisalabad, Pakistan

S. Arif · M. Z. Khan
Department of Plant Breeding and Molecular Genetics, University of Poonch, Rawalakot, Pakistan

S. Zakia
Botanical Science Division, Pakistan Museum of Natural History, Islamabad, Pakistan

Gul-E-Zahra
Institute of Botany, University of the Punjab, Lahore, Pakistan

A. Raina
Mutation Breeding Laboratory, Department of Botany, Aligarh Muslim University, Aligarh, India

F. Shimira
Department of Horticulture, Institute of Natural and Applied Sciences, Çukurova University, Adana, Turkey

G. Zahid
Department of Biotechnology, Institute of Natural and Applied sciences, Çukurova University, Adana, Turkey

S. Bano
Department of Plant Breeding and Genetics, University of Sargodha, Sargodha, Pakistan

M. A. Nadeem et al. (eds.), *Legumes Biofortification*,
https://doi.org/10.1007/978-3-031-33957-8_5

programmes by combining conventional breeding and genetic engineering tools. This chapter focuses on various approaches adopted by conventional breeding programmes to fortify the food legumes for bioavailability, translocation and uptake of micronutrients. We also highlighted the strategies of legume breeders to improve the vitamins and diminish the anti-nutrients. In the end, we shed light on the challenges and limitations of conventional breeding approaches to fortify legumes.

1 Introduction

Legumes are considered the powerhouse of energy, standing among the important staple food crops after cereals grown by human beings in various civilizations worldwide (Mousavi-Derazmahalleh et al. 2019). Legume seeds are believed to be the most economical meat substitute supplying valuable proteins with essential amino acids profile, carbohydrates, vitamins and important minerals required for the proper functioning of the human body (Didinger and Thompson 2021; Roorkiwal et al. 2021). Food legumes such as beans, mung bean, broad bean, chickpea, lentils, lupins, peanuts and other podded plants are widely used in human diet (Ahmad et al. 2012; Didinger and Thompson 2021). Legumes seeds are rich source of good quality proteins and dietary fibres that generally contain essential amino acid lysine (Didinger and Thompson 2021). Protein contents of peas and beans are about 17–20%, whereas lupin and soybeans contain 38–45% proteins that is important for normal body growth and development (Mahto et al. 2022). The daily dietary reference value of nutrient components for adults is 8 to 18 mg for Fe, 8 to 11 mg for Zn and 750 mg of Ca depending on gender, which is usually not possible to fulfil and hence results in micronutrient deficiency (O'Neill et al. 2020). Besides their potential roles in the sustainability of healthy food systems, legumes also contribute to human nutrition, food security and are also associated with reduced risk factors for chronic disease (Kurek et al. 2022). A study on human health proved that use of lentil in daily diet may increase the blood selenium concentration (Thavarajah et al. 2010). According to FAO stat 2018, about 92 million tons pulses are produced globally out of which 42 million tons are produced in Asia (Kumar and Pandey 2020).

Malnutrition of micronutrients usually caused due to supply of insufficient or poor quality of nutrients in daily diet (Mahto et al. 2022). The scarcity of micronutrients in the diet is also termed hidden hunger that adversely affects the body's normal development and physical functions like immunodeficiency, retarded physical and mental growth (Shahzad et al. 2021). Malnutrition also results in different infectious diseases like malaria, diarrhoea, measles in developing countries (Shahzad et al. 2021). Deficiencies of β-carotene, folic acid, Fe and Zn are global issues and affect more than two billion people in Asia, Africa and Latin America (Shahzad et al. 2021). Globally, there are approximately 32.8% of pregnant ladies, 32.5% of non-pregnant ladies and 41.7% of youngsters under the age of 5 are suffering from iron deficiency and potentially leading to restriction in intrauterine development, low birth weight, protein malnutrition and persistent energy deficiency (Kumar and Pandey 2020). The industrialized solution for this micronutrient deficiency is food fortification by enhancing nutrient content and biofortification.

Biofortification is an approach to enhancing the dietary value of crops with the assistance of transgenic techniques like breeding and agronomic practices. The approaches used for biofortification aimed to target and regulate the metabolic pathways such as transportation, root uptake mechanism, remobilization, storage and enhanced minerals concentration (Roorkiwal et al. 2021). Biofortification acts as an advanced technology involving the transfer of the genes directly in selected genotype to reduce mineral deficiency. Efficient biofortification could be achieved by increasing the bioavailability of micronutrients and their concentration, improving the retention of minerals and decreasing the concentration of nutrients that minimize their absorption (Roorkiwal et al. 2021). Vitamins C, D, and E, choline, niacin and provitamin A act as promoters and increase the absorption of Se, P, Zn, Ca, Fe, methionine and tryptophan, while certain polyphenols and phytate decrease the micronutrient bioavailability to crops (Rehman et al. 2019; Shahzad et al. 2021). Transgenic approaches to protein, vitamins and mineral contents have been significantly observed in pigeon pea, common beans, field pea and chicken pea (Rehman et al. 2019). Biofortified crops proved better to withstand adverse environmental conditions with a higher concentration of micronutrients. Biofortification by using techniques of plant breeding is one of the most effective and economic approaches among different strategies to reduce the micronutrient deficiencies, and it could easily be available to people living in rural and remote areas where the access to fortified food is limited (Jha and Warkentin 2020). Biofortification has improved legumes' nutritional quality and contents in the last decade, although numerous demanding situations should be addressed to maximize the successful use of biofortified foods. Mass selection technique was successfully adopted to identify the dry pea genotypes with improved yield and nutritional contents and the results indicated that there was huge diversity in minerals and phytic acid contents among genotypes (Thavarajah et al. 2022). Biofortification refers to crops with increased nutrient density developed using different modern and conventional breeding approaches. Nowadays access to nutritionally balanced food is important for the overall growth and development of organisms including human beings. However, most food crops lack adequate amounts of critical micronutrients; increasing nutrient density is important to mitigate the adverse effects of malnutrition. The number of undernourished people increased from 636.8 million to 811 million over the last 10 years (FAO 2022) (Fig. 5.1). An estimated 22% of children (149.2 million) under the age of five suffered from stunting due to severe acute malnutrition and is expected to worsen under the shadow of the COVID-19 pandemic. Malnutrition is an underlying cause of death of more than 2.6 million children each year, which accounts for one-third of total child deaths worldwide. It is also a leading cause of physical and mental developmental disorders, diseases and premature deaths (Development Initiatives 2018). With such a gloomy scenario it is important to have major changes in agriculture and its allied sectors. The first and most important change would be a shift from producing more food quantities to producing nutritionally rich food in

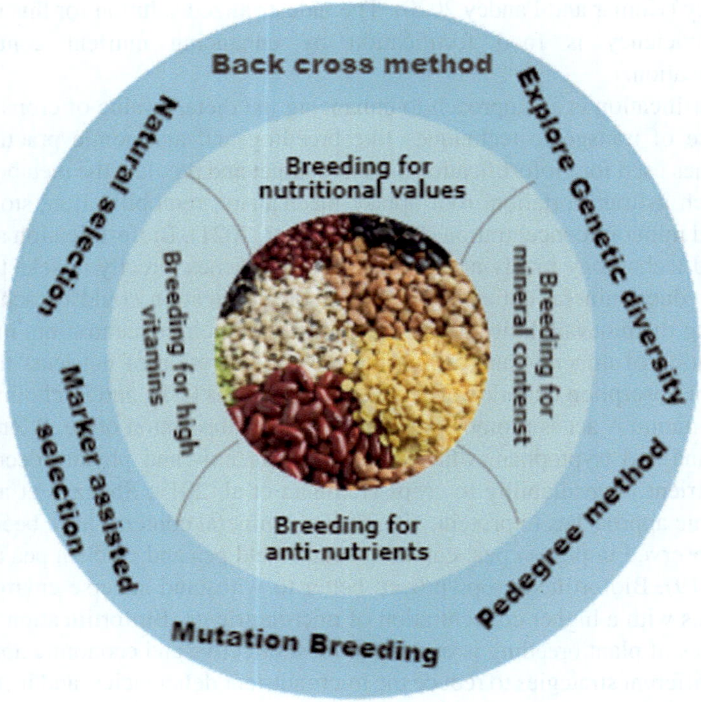

Fig. 5.1 Conventional breeding techniques to improve the food legumes

adequate amounts. Here, the biofortification of crops would play a pivotal role in providing adequate quantities of nutrients to the poor population. The biofortified cereals, legumes, vegetables and fruits are providing adequate levels of micronutrients to targeted populations. Crops are the major source of nutrients in the developing world and their biofortification of crops would play a vital role in making the nutritionally available food for the masses (Table 5.1).

2 Exploitation of Genetic Variations for Micronutrients Improvement

Genetic variation can be used to improve the targeted nutritional traits by selecting favourable alleles (Kumar et al. 2016). The exploitation of genetic variation is a tool to start new research strategies for the biofortification of legumes using the wild relatives to fight hidden hunger and provide nutritious food to a major portion of the global population. Genetic variation uses conventional breeding methods along with modern genomic approaches to exploit the genetic diversity to boost up the micronutrient contents in legume seeds (Roorkiwal et al. 2021; Mahto et al. 2022). Existing genetic diversity, trait inheritance, gene activity, trait linkage, accessible

Table 5.1 Nutritional contents of various food legumes

Name of legumes	Protein contents	Lipid contents (%)	Vitamins	CHOs (%)	Dietary fibres	References
Glycine max	37–42	17–19	A, D, E and K	35	20	Kumar and Pandey (2020)
Cicer arietinum	17–22	4–8	Folic acid, tocopherol, vitamin B complex	40–60	18–22	Madurapperumage et al. (2021)
Vigna mongo	20–25	2.1–2.7	A, B6 and E	55–65	7	Kumar and Pandey (2020)
Vicia faba	26.1–38	0.7–3.2	Folates, vitamin C	55.1–71.4	6.4–34.9	Martineau-Côté et al. (2022)
Cajanus cajan	19.5–22.9	1.49	Folates, vitamin B complex, thiamin	62.78	10	Talari and Shakappa (2018)
Pisum sativum	22.24–31.59	1.66–2.22	Thiamin, riboflavin, niacin, vitamin B6	27.80 to 34.78	16.81–40.63	Brigide et al. (2014), Kumar and Pandey (2020)
Lens culinaris	23.3–25.88	1.93–2.15	β-Carotene	54.08–55.81	6.99–8.14	Kumar et al. (2016)
Arachis hypogaea	25.80	49.24	Folates, niacin, pyridoxine	16.13	8.5	Arya et al. (2016)

screening procedures and diagnostic tools are all utilized to assess prospective genetic gains (Baker et al. 2019). Lot of genetic diversity is available between current legumes and their wild relatives (Rehman et al. 2019). However, wild species contain a high number of beneficial foreign genes that are no longer found in the farmed gene pool. Efforts to gather and protect wild relatives of diverse food legume crops in national and international gene banks have been underway. Many species have previously shown cross-compatibility with cultivated varieties in a number of studies (Kumar et al. 2016). Because of this compatibility, foreign genes from wild species have been effectively introgressed which are regulated by significant genes. Different breeding techniques may be used to exploit this genetic diversity to boost the micronutrients content of food legumes. A good genetic biofortification technique requires both high micronutrients concentration and high yield under different environmental circumstances. Plant breeding techniques are used to screen the large genetic variation available for selective traits. However, advanced genetic approaches such as quantitative trait loci (QTL) mapping and marker-assisted selection are used to identify the genes controlling micronutrients concentration in a particular crop. The exploitation of genetic variation can be a sustainable solution for malnutrition by linking agriculture to nutrition and health. Micronutrients comprise very small portion of the total weight; therefore, successful breeding programme with precise estimation is prerequisite (Brigide et al. 2014; Shahzad et al.

2021). So malnutrition could be mitigated by a precise breeding method, including identifying and utilizing the factors affecting the nutritional traits (Roorkiwal et al. 2021). An alternative is the selection of plant varieties with superior ability to accumulate certain nutrients in the grain. Variations in seed coat colour also affect the nutrients contents in food legumes. Research reported that the white-coloured seed coat of the Andean bean possessed lower contents of phenolic compounds and antioxidant activities than genotypes with mottled or red seed coat genotypes (Rehman et al. 2019). In food legumes several fortified varieties have been released throughout the world through conventional breeding (Rehman et al. 2019).

3 Breeding for Micronutrients Improvement

Breeding of high nutrients cultivars, rich in Zn, Fe, Ca, K and the substances make the bioavailability of these nutrients is an effective way to minimize the nutrient deficiency (Marques et al. 2021; Shahzad et al. 2021). Different strategies are being used to develop the biofortified legume crops. The biofortification strategy strives to enrich legumes with minerals such as Fe, Zn, Sr, I and others so that individuals who eat such grains get more of them (Madurapperumage et al. 2021; Mahto et al. 2022). Even though legumes are rich in mineral contents, however their bioavailability is very low, consequently dwindling the use of legumes as a mineral source (Praharaj et al. 2021). Because micronutrient deficiency or hidden hunger is more common in low-income nations, where customers have limited purchasing power, they can't afford vitamin supplements or a micronutrient-rich diet (Rehman et al. 2019). Improved nutritional status of regularly consumed food grains is the best sustainable method for treating micronutrient deficiency in such circumstances (Rehman et al. 2019; Kumar and Pandey 2020). A successful biofortification method should guarantee that grain yield is enhanced or maintained, as well as the grain micronutrients contents are increased for considerable beneficial health effects, and grain performance is consistent across settings (Praharaj et al. 2021). By creating genotypes with high levels of Zn in edible plant portions, plant breeding and/or transgenic techniques give a promising and long-term strategy to alleviate micronutrient deficiencies (Xia et al. 2020). Though establishing a genotype is expensive and complex, it provides a long-term advantage as there are no recurrent costs. Because considerable genetic variety exists in the germplasms of key cereal crops, breeding for high micronutrients concentration is conceivable (Jha and Warkentin 2020). Identifying appropriate genetic variation and selecting parents, long-term crossing and backcrossing, stabilization of target characteristics across multiple climates conditions and acclimatization of biofortified genotypes to regional agricultural management practices are the minimum stages required in breeding (Maqbool et al. 2020). Improved nutritional quantity was reported in mung bean when this crop was crossed with mash bean through interspecific hybridization (Abbas et al. 2019).

4 Breeding for Vitamin-Enriched Legumes

Vitamins are organic compounds vital for energy production, but inadequacy of vitamins is directly linked with various human health losses (Madurapperumage et al. 2021). Such losses can be effectively overcome by delivering vitamin-rich food to an impoverished population. Biofortification through conventional breeding is the most effective approach which enables various food plants to be enriched with such vitamins (O'Neill et al. 2020). Using new breeding techniques, including modification of metabolic pathways of vitamins, can enhance the biofortification progress in a variety of plant species like legumes. The diversity of wild relatives and cross compatible species-rich in essential vitamins can be used by selective breeding to increase such nutrients in legume seeds (Garg et al. 2018). The planned identifications of specific factors associated with nutritional traits and their proper utilization in a selective breeding program could help in mitigating malnutrition. The important vitamins found in legumes belong to the vitamins B-group like folate, thiamin, niacin and riboflavin. Folic acid is also present in abundant amount; however, the availability of vitamin C and ascorbic acid is very lower in legumes (Rehman et al. 2019). It is suggested that if legumes are used in combination with foods high in vitamin C, iron absorption can be increased. The biochemical composition of the legumes differs among various forms of seeds exhibiting high variability in case of their vitamin content. The contents of lipo-soluble vitamins are also very low in legumes, except α-tocopherol (vitamin E) in soybean and peanut. The γ-tocopherol form in a few legumes is most abundant, with the highest level being described in peas, pigeon peas and lentils (Amarovicz 2009). Various strategies comprising metabolic engineering, classical breeding and mutation breeding have been initiated to enhance the vitamin contents of legumes (Mène-Saffrané and Pellaud 2017). Furthermore, enrichment in α-tocopherol contents can be obtained by traditional breeding utilizing naturally high α-tocopherol level alleles detected in QTL studies of legume germplasms and then by introgression into commercial varieties to increase vitamin E (Mène-Saffrané and Pellaud 2017). Some legumes seeds are a good source of carotenoids, most commonly β-carotene, cryptoxanthin, lutein and zeaxanthin; for example, the concentration of β-carotene in the case of chickpea was more in comparison with some crops genetically modified for β-carotene. The traits for carotenoid contents have high heritability that is not much affected by the environment (Abbo et al. 2005). Therefore, identifying such barriers more likely connected with the carotenoids biosynthesis pathway will improve the modifying strategies for producing legumes enriched with carotenoids (Schmidt et al. 2015). This trait is highly heritable, and variations found in legume germplasm for this trait can be used in breeding programs. In chickpea cross of wild relatives with Israeli cultivar showed more β-carotene and lupine contents as compared to their parent lines (Abbo et al. 2005). Seed coat colour is important trait that influence the of β-carotene and lupin contents in soybean so this trait may be keep in mind while breeding the legumes for higher carotene contents (Gebregziabher et al. 2022).

5 Breeding for Anti-nutrients

Anti-nutrients generate indigestible complexes with nutrients and proteins and impair the bioavailability of micronutrients to human body. Abundant intake of anti-nutrients through diet may become toxic to the body (Samtiya et al. 2020; Martineau-Côté et al. 2022). Anti-nutrients can have a large negative impact on food nutrition; hence, lowering these contents in meal is an important objective to boost human nutrition. Most legumes contains anti-nutritional factors, such as tannins, phytic acid, digestive enzyme inhibitors, oxalate and lectins that can decrease the bioavailability and uptake of proteins and minerals during digestion and induce toxic effects (Martineau-Côté et al. 2022). Anti-nutrient compounds restrict the bioavailability of essential micronutrients, ultimately resulting in malnutrition and various diseases like anaemia, beriberi, night blindness, rickets and scurvy more prevalent among the population. Plant molecular biology and genetic modification techniques currently allow for the reduction or elimination of anti-nutrients in staple plant foods and a large boost of promoter substance levels in these foods. When trying to develop food crops as sources of micronutrients for humans, plant breeders and molecular scientists should attentively examine the approach of boosting promoter chemicals in food crops. Biofortification is a balanced method to combat mineral shortages. It entails improving the nutritional content of food crops using either classic plant breeding or current biotechnology (Jha and Warkentin 2020). In the last decade, biofortification via plant breeding has increased the nutrient quality of pulse crops and has gained traction. Several studies on pulse crops have found genetic diversity for critical micronutrients in accessible gene pools, with successful breeding lines employed in breeding and related genotypic markers for marker-assisted breeding selection (A.M. Pérez-de-Castro et al. 2012). For accurate marker–trait association, gene discovery, functional marker creation and their deployment in routine breeding programmes, next-generation sequencing (NGS) and genotyping technologies must be applied (Scheben et al. 2018). Anti-nutritional agents such as aponins, tannins, phytic acid, gossypol, lectins, protease inhibitors, amylase inhibitors, raffinose and goitrogens are present in edible crops (Samtiya et al. 2020). Anti-nutritional substances mix with nutrients and create lower nutrient bioavailability, a big problem. Other variables, such as trypsin inhibitors and phytates, found mostly in legumes and grains, limit the digestibility of nutrients and mineral assimilation. Wild lima beans (*Phaseolus lunatus*) contains a toxic compound cyanogenic glycoside its high quantity consumption may cause respiratory distress in human body (Shlichta et al. 2014). Another legume grass pea (*Lathyrus sativus*) possess a non-protein amino acid β-N-oxalyl-l-α,β-diaminopropionic acid which causes lower limb paralysis (Yigzaw et al. 2001). Consumption of alkaloid glycoside present in faba bean (*Vicia faba*) causes hemolysis (Crépon et al. 2010). An anti-nutrient compound trypsin present in mung bean (*Vigna radiata*) is the cause of indigestibility and reduce the bioavailability of minerals (Ullah et al. 2014). A comparative study on lentil crop showed that phytic acid concentration is influenced by temperature. Higher the temperature lower will be the phytic acid concentration whereas in cool temperature phytic concentration reduced in same genotypes (Thavarajah et al. 2010).

6 Identification of QTL/Genes for Nutrients/Vitamins Accumulation

Traditionally, screening for beneficial alleles influencing improved concentration for specific nutritional properties was applied in legumes to determine existing natural variation. This assisted in identifying genetic variability that may be used as a donor for transferring helpful genes into the background of cultivated genotypes, as well as for usage as a biofortified variety if the detected variant is already a high producing cultivar. Legume cultivars were tested over many years and in several places and resulted in the generation of maps that illustrated variability for certain micronutrients using geostatistics (AbdelRahman et al. 2016). Advances in tissue culture techniques, on the other hand, may now be ready to aid in embryo rescue, which encouraged alien gene expression from the secondary gene pool. Furthermore, evolutionary pressures cause modifications in wild species germplasm in the direction of probable cross-compatibility with farmed species. As a result, formerly incompatible wild and domesticated species can now cross (Singh et al. 2021). It's time to revisit the crossability links between modern legume cultivars and wild species. As a result, pre-breeding activities are critical, especially among wild species that carry important foreign genes for biofortification features. Moreover, wild relatives are major sources of novel variety generated by recombination breeding techniques (Lyzenga et al. 2021). Furthermore, the utilization of multi-parent populations has aided in the discovery of quantitative trait loci (QTL). As a result, a number of QTL with high mapping resolution have been found for breeding programs. The markers for this QTL are a valuable tool for enhancing selection efficiency in breeding projects employing marker-assisted selection (Phan and Sim 2017). For QTL detection in wild populations or germplasm collections, association mapping is a powerful technique that utilizes past recombination events. In comparison to linkage analysis, this mapping technique offers various advantages, including improved mapping accuracy, reduced time and a larger number of alleles to mine (Hu et al. 2016). For nutritional qualities, there is genetic heterogeneity among cultivated germplasm that may be used in breeding operations. Several nutritional properties in lentil were strongly influenced by environmental factors and the genetic composition of a given genotype (Kahraman et al. 2004). As a result, efforts should be focused on location-specific nutritional characteristic breeding. Several crops have identified wild relatives as possible providers of nutritional characteristics. However, wild relatives have been discovered as important genetic resources for other agronomical qualities, but such variants have not yet been investigated in lentils (Kumar et al. 2016). Hence, wild relatives will be prioritized in the future for providing appropriate pre-breeding materials for biofortification-related features. Genomic analysis is becoming an important aspect of breeding efforts. In the case of legumes, significant progress has already been done in terms of developing genetic resources. However, in the genetic biofortification of legume crops, these genomic resources have not been completely used. Through clustering, molecular markers may be utilized to identify and map genomic areas that affect the expression of nutritional characteristics. These advancements may make it easier to dissect the complicated genetics that regulate dietary characteristics. The markers linked with beneficial genes/QTL impacting biofortification features can be employed in marker-assisted breeding to generate biofortified legume cultivars promptly and cost-effectively in the near future (Kahraman et al. 2004). QTL mapping to explain the genetic determinants for seed mineral deposition in soybean seeds, explicitly for phosphorus (P), calcium (Ca), magnesium (Mg), iron (Fe) and zinc (Zn) minerals. The study made extensive use of recombinant inbred lines and cultivated soybean accessions (Wang et al. 2022). Considering several genetic loci are engaged in the metabolic pathways of mineral deposition in seeds, the molecular mechanisms regulating mineral element deposition are genetically sophisticated. These discoveries will lay the groundwork for mineral biofortification, mostly through MAS breeding (Wang et al. 2022).

7 Marker-Assisted Selection for Dietary Improvement of Legumes

Marker-assisted selection (MAS) enables rapid identification of elite breeding material based on DNA, RNA and/or protein markers, making this method genotype-based selection (Wang et al. 2022). These markers can be found throughout the gene of interest or linked to a gene that determines a trait of interest (Boopathi 2013). Further people more rely on conventionally biofortified crops as compared to genetically modified crops (Marques et al. 2021). MAS assists breeders in increasing selection efficiency, precision and intensity, as well as the selection of satisfactory gene combinations in early generations, leading to increased genetic gain. Nevertheless, before employing a marker for the selection of promising crops in a segregating generation, the magnitude of the target QTL's effects and precise chromosomal position are critical for exploring the benefits of MAS. Markers linked to a trait must be validated across multiple genetic backgrounds. To increase the accessibility of genomic resources and data in important species, the use of extremely polymorphic and reproducible markers like SNP and SSR in marker–trait association assessment is a key problem for MAS utilization in breeding programs. Furthermore, the cooperation between biotechnologists and plant breeders should be improved to allow for a more systematic and efficient use of MAS. Countering these obstacles may enable the adoption of challenging genomic data sets and the production of enhanced cultivars involving MAS by breeders. In cereal crops, marker-assisted pyramiding has been widely used to combine multiple genes and QTLs (Ahmad et al. 2015). This combination is seen in legume crops, such as soybean, whereas successful gene and QTL pyramiding may be dependent on a close marker–trait association. A well-established and close relationship between markers and target traits has already been described in pulses. As a result, breeders are working actively to employ them for marker-assisted pyramiding in pulse crops and legumes (Ahmad et al. 2018).

MAS is often used in cultivar development and genetic enhancement. For instance, in faba bean (*Vicia faba L.*), the primer focus was on the development of molecular markers for selecting resistance to diseases. MAS is currently being utilized to improve seed quality in faba beans by targeting zero tannins. Thus, these polyphenolic substances had an inhibition effect on the activity of digestive enzymes in humans. To develop the tannin-free cultivars, breeders are looking for the presence of two recessive genes, *zt-1* and *zt-2*, responsible for tannin absence. A reliable SCAR marker such as SCC5551, releasing a high accuracy prediction of the *zt-1* gene, was approved in a sizable population of faba beans, demonstrating its potential as a cost-effective tool for MAS. Similarly, a practical SCAR marker associated to the *zt-2* gene was identified, which led to the elaboration of tannin-free faba bean cultivars. The abovementioned discoveries will enable gene pyramiding and intensify the development of new faba bean cultivars with optimized nutritional value for immediate consumption (Sallam et al. 2016). Furthermore, recent advancements in whole-genome sequencing and comparative genomic approaches, particularly for lentil (*Lens esculenta*), have greatly assisted in the mapping of genes and QTLs for important agronomic traits in the lentil, as well as the elaboration of functional markers for MAS. Using QTL analysis, genes controlling Fe uptake in lentil populations have been identified. Several molecular markers, comprising SSRs, AFLPs and SNPs, have been used in population genotyping to develop biofortified lentil cultivars. Polymorphic SSR markers were employed in the same fashion to characterize a dozen lentil genotypes with stable Fe and Zn concentrations in their grain (Kumar et al. 2016). Accuracy rate of MAS and genomic selection (GS) for the seed weight (SW) trait using a genome-wide association study in a large population of soybean (Glycine max) based on a high number of single nucleotide polymorphisms (SNPs). In the mixed model that included these loci, they discovered two minor-effect loci associated with SW that explained 83.4% of the phenotypic variation. They also reported high prediction precisions for GS and MAS using cross-validation, with 0.75–0.87 and 0.62–0.75, respectively. Their findings could contribute to the discovery of genes controlling the SW trait in soybean. Likewise, moreover, Yang et al. (2015) used next-generation sequencing (NGS) to develop a significant number of reliable SSR markers for MAS in pea breeding, retaining approximately 841 stable amplifications of perceptible polymorphisms within 24 genotypes of cultivated pea (*Pisum sativum* L.) and wild relatives (*P. fulvum* Sm.). Alongside that, 33 polymorphic SSR markers were highlighted as being compelling in F2 generation. This discovery may be useful for future research into pea quality enhancement and the development of biofortified pea cultivars. Several reports have highlighted advancements in chickpea (*Cicer arietinum*) research, and a few SNP markers controlling iron and zinc concentrations have also been discovered (Kumar et al. 2016). The common bean is one of the legume crops being used in the large strategy to combat zinc deficiency in developing countries. This strategy is built on a combination of conventional breeding and MAS. Thus, zinc content in the seeds of legume crops is regulated by several genes engaged in mineral uptake, transport and deposition from soils to seeds. Numerous DNA markers and quantitative trait loci (QTLs) linked with seed zinc accumulation have already been identified and validated in common beans. There are approximately 28 different QTLs associated with grain zinc content and their associated SSR/SNP markers that have already been observed in common bean. Studies focusing on the zinc content of common bean seeds are now incorporating other molecular techniques, such as the Genome Wide Association Study (GWAS). As a result, this technique employs diverse crop germplasm to scan the entire genome to provide a clear perception of the candidate genes responsible for the expression of the trait of interest (Abbas et al. 2019). Other QTL-based studies in common bean revealed moderate phenotypic variation for Fe and Zn concentrations in seeds (Kumar et al. 2016).

8 Mutation Breeding for Nutrient-Enhanced Legumes

Different approaches have been designed and implemented from time to time to achieve the goals of biofortification. Among various approaches mutation breeding has played a vital role in developing biofortified crops. More than 1150 mutant varieties with improved nutritional quality have been developed and officially released (MVD-2022). Mutation breeding has proven a suitable technique for improving various attributes such as yielding potential, adaptability, stress tolerance and nutrient quality of economically important crops such as cowpea (Rasik et al. 2022; Raina et al. 2022a, b), faba bean (Khursheed et al. 2018a, 2019), urdbean (Goyal et al. 2021a, b), mung bean (Wani et al. 2017), chickpea (Laskar et al. 2015; Raina et al. 2017), lentil (Laskar et al. 2018; Raina et al. 2022a) and black cumin (Tantray et al. 2017; Amin et al. 2019). However, one of the most important attributes is the biofortification of crops. The success of mutation breeding aimed at biofortification is determined by several factors such as selection of appropriate mutagen, mutagen dose and duration of the treatment. The dose employed must be optimum that could induce maximum frequency of mutation with least biological damage (Goyal et al. 2019, 2020a, b). In addition to mutagen attributes, selection of plant material is also important. Depending on the degree of utilization, the crop can be selected, and it is always preferred to improve the grain quality and nutritional status of staple crops (Raina et al. 2016, 2018). Developing crops with increased nutrient density may prove a sustainable approach to mitigate the devastating impact of malnutrition. Several workers have employed mutation breeding technique to improve the nutritional status of several crops. For instance, Raina et al. (2020) treated two varieties of cowpea with different doses of gamma rays and sodium azide employed individually and in combination. They were successful in developing 11 mutant lines with increased micronutrient concentration. Similarly, Laskar et al. (2018) developed lentil mutant lines that showed increased micronutrient density. They treated traditional lentil cultivars with different single and combined doses of gamma rays and hydrazine hydrates. Khursheed et al. (2018b) while working with mutagenized faba bean reported few biofortified lines and emphasized the role of mutation breeding in developing biofortified crops. Wang et al., 2022 attempted to incorporate zinc biofortification in rice through mutation

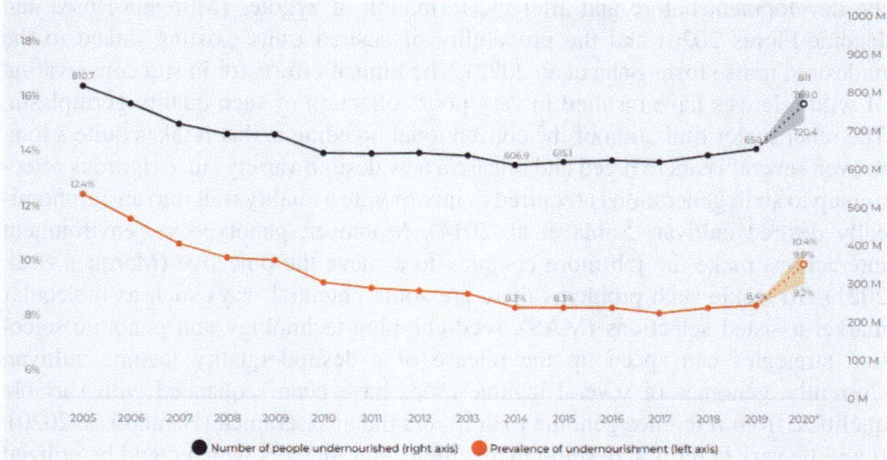

Fig. 5.2 The number of undernourished people in the world continued to rise in 2020. Between 720 and 811 million people in the world faced hunger in 2020. NOTES: * Projected values for 2020 in the figure are illustrated by dotted lines. Shaded areas show lower and upper bounds of the estimated range that considers statistical uncertainty. (Source: FAO)

breeding by treating nine popular rice varieties with ethyl methanesulphonate. They were successful in isolating few mutant lines with increased zinc content. On the basis of the literature, it is evident that mutation breeding has played a vital role in accomplishing the goals of biofortification of crops; however, the technique has been employed in limited crops. Therefore, more crops should be mutagenized to observe the mutagen-induced alterations in the contents of micro and macronutrients. This way the devastating effects of malnutrition can be mitigated to a greater extent (Fig. 5.2).

9 Challenges and Limitations of Legumes Biofortification Through Conventional Breeding

Although there are several benefits of breeding for legumes biofortification through conventional breeding methods, numerous limitations exist, the most prominent of which is the dependence on availability of the diverse gene pool of the crop under focus (Jha and Warkentin 2020; Marques et al. 2021). Using conventional breeding methods, no crop can be biofortified in the absence of enough genetic diversity and only transgenic approaches remains helpful (Ahmad et al. 2012). Low levels of heritability and linkage drag also make conventional methods almost ineffective. It has also been observed that some of modern-day crops show lower nutritive quality in comparison with their wild relatives (Didinger and Thompson 2021). But, there are limitations of using wild relatives of several crops due to their immense underrepresentation in the global gene bank (Pandey et al. 2008), reproductive hindrances in the development before and after the formation of zygotes (Munguía-Rosas and Jácome-Flores 2020) and the probability of desired traits existing linked to the undesired traits (Joshi-Saha et al. 2022). The limited efforts for in situ conservation of wild relatives have resulted in very poor collection of such quality germplasm. The other major limitation of the conventional breeding is that it takes quite a long to even several years to breed and release a new desired variety since rigorous selection up to sixth generation is required to incorporate a quality trait into an agronomically desired cultivar (Sobia et al. 2014). Moreover, genotype vs. environment interactions make the job more complex to achieve the objective (Marques et al. 2021). To tackle such problems there are some potential ways such as molecular marker-assisted selections (MAS), seed-chipping-technology and genomic selection strategies can speed up the release of a desired quality legume cultivar. Currently, genomes of several legume crops have been sequenced with variable qualities, from reference genome to draft and the unassembled (Ha and Lee 2020). It will be very helpful in developing the molecular markers which could be utilized in high-throughput arrangements to identify the linkages/marker–trait associations of such markers for high-yielding legumes. Moreover, the whole-genome sequence data offers its use in genome-wide association studies (GWAS) to identify and incorporate the micronutrient genes in legumes in addition to the use of these markers in marker-assisted recurrent selection (MARS) and the marker-assisted backcross breeding (MABC). At present, molecular markers which are associated or linked with QTLs for zinc (Zn) and/or iron (Fe) have been identified in very few legume species (Joshi-Saha et al. 2022).

However, to achieve the success at large, a potential collaboration among plant breeders, genetic engineers and nutrition experts is crucial, and the biofortification of legumes has huge potential to overcome the hidden hunger among the poor and the developing states (Garg et al. 2018). In either case, to breed for a legume rich in nutritional quality, it is also central to understand the biochemical, physiological as well as molecular mechanisms of entire processes which are essential for their biofortification. Additionally, in order to break the negative linkage between yield or yield enhancing traits and the nutrient contents, there is an endless need to supplement the conventional breeding with the targeted and random mutagenesis. It has been reported that polyphenols and phytates are among such anti-nutrient substances which diminish the biological availability of the mineral nutrients. For example, polyphenols normally bind themselves with non-heme iron and minimize its absorption. Hence, to enhance the biological availability of the mineral nutrients, the lower phytic acid contents are considered very important. It has also been reported that the phytic acid found in numerous legumes is involved in controlling and regulation of different abiotic and biotic stress resistance system.

The ever-increasing demand of nutritionally rich legume crops is a challenge to be met. This challenge can be solved by incorporating the legume breeding projects into the national level breeding networks and programs. Moreover, policy reforms regarding markets and investments opportunities in legume are bitterly required which should be incentive and subsidies based on the farmers growing the biofortified legumes (Joshi-Saha et al. 2022).

10 Conclusion

Conventional breeding techniques have been intensively used to develop the legume varieties with improved nutritional contents. Some common bean varieties such as MAC42, MAC44, CAB2 and PVA1438 having high zinc and iron contents were developed through conventional breeding approaches in Bangladesh. Similarly, plant breeders released high zinc contents lentil varieties such as Alemaya, Shital, Sisir and Simal. Through molecular breeding approaches, several QTL identified for both nutrient and vitamins contents. These QTLs contain several genes related to nutrients and vitamins and can be used as candidates for future studies aimed at developing biofortified legume crops.

References

Abbas G, Ahsan M, Asghar MJ et al (2019) Genetic improvement in nutritional quality of mungbean through interspecific hybridization with mashbean. Iran J Sci Technol Trans A Sci 43:2173–2184. https://doi.org/10.1007/s40995-019-00746-w

Abbo S, Molina C, Jungmann R et al (2005) Quantitative trait loci governing carotenoid concentration and weight in seeds of chickpea (Cicer arietinum L.). Theor Appl Genet 111:185–195. https://doi.org/10.1007/s00122-005-1930-y

AbdelRahman MAE, Natarajan A, Hegde R (2016) Assessment of land suitability and capability by integrating remote sensing and GIS for agriculture in Chamarajanagar district, Karnataka, India. Egypt J Remote Sens Sp Sci 19:125–141. https://doi.org/10.1016/j.ejrs.2016.02.001

Ahmad HM, Ahsan M, Ali Q, Javed I (2012) Genetic variability, heritability and correlation studies of various quantitative traits of mungbean (Vigna radiate L.) at different radiation levels. Int Res J Microbiol 3:352–362

Ahmad HM, Ur-Rahman M, Azeem F, Ali Q (2015) QTL mapping for the improvement of drought tolerance in cereal crops: an overview. Life Sci J 12:102–108. https://doi.org/10.7537/marslsj1204s15.14

Ahmad HM, Mahmood-Ur-Rahman AF et al (2018) QTL mapping for crop improvement against abiotic stresses in cereals. J Anim Plant Sci 28:1558–1573

Amarovicz R (2009) Natural antioxidants as a subject of research. Eur J Lipid Sci Technol 111:1053–1055. https://doi.org/10.1002/ejlt.200900216

Amin R, Wani MR, Raina A et al (2019) Induced morphological and chromosomal diversity in the mutagenized population of black cumin (Nigella sativa L.) using single and combination treatments of gamma rays and ethyl methane sulfonate. Jordan J Biol Sci 12(1):23–33

Arya SS, Salve AR, Chauhan S (2016) Peanuts as functional food: a review. J Food Sci Technol 53:31–41. https://doi.org/10.1007/s13197-015-2007-9

Baker L, Muir P, Sample SJ (2019) Genome-wide association studies and genetic testing: understanding the science, success, and future of a rapidly developing field. J Am Vet Med Assoc 255:1126–1136. https://doi.org/10.2460/javma.255.10.1126

Boopathi NM (2013) Genetic mapping and marker assisted selection. Springer India, India

Brigide P, Canniatt-Brazaca SG, Silva MO (2014) Nutritional characteristics of biofortified common beans. Food Sci Technol 34:493–500. https://doi.org/10.1590/1678-457x.6245

Crépon K, Marget P, Peyronnet C et al (2010) Nutritional value of faba bean (Vicia faba L.) seeds for feed and food. F Crop Res 115:329–339. https://doi.org/10.1016/j.fcr.2009.09.016

Didinger C, Thompson HJ (2021) Defining nutritional and functional niches of legumes: a call for clarity to distinguish a future role for pulses in the dietary guidelines for Americans. Nutrients 13:1100. https://doi.org/10.3390/nu13041100

FAO (2022) The State of Food Security and Nutrition in the World 2022. Repurposing food and agricultural policies to make healthy diets more affordable. Rome, FAO. https://doi.org/10.4060/cc0639en

Garg M, Sharma N, Sharma S et al (2018) Biofortified crops generated by breeding, agronomy, and transgenic approaches are improving lives of millions of people around the world. Front Nutr 5. https://doi.org/10.3389/fnut.2018.00012

Gebregziabher BS, Zhang S, Ghosh S et al (2022) Origin, maturity group and seed coat color influence carotenoid and chlorophyll concentrations in soybean seeds. Plan Theory 11:848. https://doi.org/10.3390/plants11070848

Goyal S, Wani MR, Laskar RA et al (2019) Induction of Morphological Mutations and Mutant Phenotyping in Black gram [Vigna mungo (L.) Hepper] using Gamma Rays and EMS. Vegetos 32(4):464–472

Goyal S, Wani MR, Laskar RA et al (2020a) Mutagenic Effectiveness and Efficiency of Individual and Combination Treatments of Gamma Rays and Ethyl Methanesulfonate in Black Gram [Vigna mungo (L.) Hepper]. Adv Zool Bot 8(3):163–168

Goyal S, Wani MR, Laskar RA et al (2020b) Performance Evaluation of Induced Mutant Lines of Black gram (Vigna mungo (L.) Hepper). Acta Fytotech Zootech 23(2):70–77

Goyal S, Wani MR, Raina A, Laskar RA, Khan S (2021a) Phenotypic diversity in mutagenized population of urdbean (Vigna mungo (L.) Hepper). Heliyon 7(5):p.e06356 IF 3.775

Goyal S, Wani MR, Laskar RA, Raina A et al (2021b) Quantitative assessments on induced high yielding mutant lines in urdbean [Vigna mungo (L.) hepper]. Legume Science. https://doi.org/10.1002/leg3.125

Ha J, Lee S-H (2020) Updates on legume genome sequencing. pp 1–18

Hu S, Lübberstedt T, Zhao G, Lee M (2016) QTL mapping of low-temperature germination ability in the maize IBM Syn4 RIL population. PLoS One 11:e0152795. https://doi.org/10.1371/journal.pone.0152795

Jha AB, Warkentin TD (2020) Biofortification of pulse crops: status and future perspectives. Plan Theory 9:73. https://doi.org/10.3390/plants9010073

Joshi-Saha A, Sethy SK, Misra G et al (2022) Biofortified legumes: present scenario, possibilities and challenges. F Crop Res 279:108467. https://doi.org/10.1016/j.fcr.2022.108467

Kahraman A, Kusmenoglu I, N. Aydin A et al (2004) QTL mapping of winter hardiness genes in lentil. Crop Sci 44:13–22

Khursheed S, Raina A, Laskar RA, Khan S (2018a) Effect of gamma radiation and EMS on mutation rate: their effectiveness and efficiency in faba bean (Vicia faba L.) Caryologia: International Journal of Cytology, Cytosystematics and Cytogenetics, 71(4):397–404. https://doi.org/10.1080/00087114.2018.1485430IF0.791

Khursheed S, Raina A, Wani MR et al (2018b) Quantitative analysis of genetic parameters in the mutagenized population of faba bean (Vicia faba L.). Res Crop 19(2):276–284

Khursheed S, Raina A, Parveen K, Khan S (2019) Induced phenotypic diversity in the mutagenized populations of faba bean using physical and chemical mutagenesis. J Saudi Soc Agric Sci 18:113–119

Kumar S, Pandey G (2020) Biofortification of pulses and legumes to enhance nutrition. Heliyon 6:e03682. https://doi.org/10.1016/j.heliyon.2020.e03682

Kumar J, Gupta DS, Kumar S et al (2016) Current knowledge on genetic biofortification in lentil. J Agric Food Chem 64:6383–6396. https://doi.org/10.1021/acs.jafc.6b02171

Kurek MA, Onopiuk A, Pogorzelska-Nowicka E et al (2022) Novel protein sources for applications in meat-alternative products—insight and challenges. Foods 11:957. https://doi.org/10.3390/foods11070957

Laskar RA, Khan S, Khursheed S et al (2015) Quantitative analysis of induced phenotypic diversity in chickpea using physical and chemical mutagenesis. J Agron 14(3):102–112

Laskar RA, Wani MR, Raina A, Amin R, Khan S (2018) Morphological characterization of gamma rays induced multipodding mutant (mp) in lentil cultivar Pant L 406. Int J Radiat Biol 94(11):1049–1053 IF 3.647

Lyzenga WJ, Pozniak CJ, Kagale S (2021) Advanced domestication: harnessing the precision of gene editing in crop breeding. Plant Biotechnol J 19:660–670. https://doi.org/10.1111/pbi.13576

Madurapperumage A, Tang L, Thavarajah P et al (2021) Chickpea (Cicer arietinum L.) as a source of essential fatty acids – a biofortification approach. Front Plant Sci 12. https://doi.org/10.3389/fpls.2021.734980

Mahto RK, Ambika SC et al (2022) Chickpea biofortification for cytokinin dehydrogenase via genome editing to enhance abiotic-biotic stress tolerance and food security. Front Genet 13. https://doi.org/10.3389/fgene.2022.900324

Maqbool A, Abrar M, Bakhsh A et al (2020) Biofortification under climate change: the fight between quality and quantity. In: Environment, climate, plant and vegetation growth. Springer International Publishing, Cham, pp 173–227

Marques E, Darby HM, Kraft J (2021) Benefits and limitations of non-transgenic micronutrient biofortification approaches. Agronomy 11:464. https://doi.org/10.3390/agronomy11030464

Martineau-Côté D, Achouri A, Karboune S, L'Hocine L (2022) Faba bean: an untapped source of quality plant proteins and bioactives. Nutrients 14:1541. https://doi.org/10.3390/nu14081541

Mène-Saffrané L, Pellaud S (2017) Current strategies for vitamin E biofortification of crops. Curr Opin Biotechnol 44:189–197. https://doi.org/10.1016/j.copbio.2017.01.007

Mousavi-Derazmahalleh M, Bayer PE, Hane JK et al (2019) Adapting legume crops to climate change using genomic approaches. Plant Cell Environ 42:6–19. https://doi.org/10.1111/pce.13203

Munguía-Rosas MA, Jácome-Flores ME (2020) Reproductive isolation between wild and domesticated chaya (Cnidoscolus aconitifolius) in sympatry. Plant Biol 22:932–938. https://doi.org/10.1111/plb.13140

O'Neill LM, Dwyer JT, Bailey RL et al (2020) Harmonizing micronutrient intake reference ranges for dietary guidance and menu planning in complementary feeding. Curr Dev Nutr 4. https://doi.org/10.1093/cdn/nzaa017

Pandey A, Tomer AK, Bhandari DC, Pareek SK (2008) Towards collection of wild relatives of crop plants in India. Genet Resour Crop Evol 55:187–202. https://doi.org/10.1007/s10722-007-9227-4

Pérez-de-Castro AM, Vilanova S, Cañizares J et al (2012) Application of genomic tools in plant breeding. Curr Genomics 13:179–195. https://doi.org/10.1109/cdc.1996.577319

Phan NT, Sim SC (2017) Genomic tools and their implications for vegetable breeding. Hortic Sci Technol 35(2):149–164. https://doi.org/10.12972/kjhst.20170018

Praharaj S, Skalicky M, Maitra S et al (2021) Zinc biofortification in food crops could alleviate the zinc malnutrition in human health. Molecules 26:3509. https://doi.org/10.3390/molecules26123509

Raina A, Laskar RA, Khursheed S, Khan S et al (2017) Induce physical and chemical mutagenesis for improvement of yield attributing traits and their correlation analysis in chickpea. Int Lett Nat Sci. 61:14–22

Raina A, Laskar RA, Wani MR, Khan S (2022a) Chemical mutagenesis: Role in Breeding and Biofortification of lentil (Lens culinaris Medik) mutant lines. Molecular Biology Reports. https://doi.org/10.1007/s11033-022-07678-6

Raina A, Laskar RA, Wani MR et al (2022b) Comparative mutagenic effectiveness and efficiency of gamma rays and sodium azide in inducing chlorophyll and morphological mutants of cowpea. Plants 11:1322. https://doi.org/10.3390/plants11101322

Raina A, Laskar RA, Khursheed S et al (2016) Role of mutation breeding in crop improvement-past, present and future. Asian J Agric Res 2(2):1–13

Raina A, Laskar RA, Tantray YR et al (2020) Characterization of induced high yielding Cowpea Mutant lines using physiological, biochemical and molecular markers. Sci Rep 10:3687. IF 5.012

Rasik S, Raina A, Laskar RA, Wani MR, Reshi Z, Khan S (2022) Lower doses of Sodium azide and Methyl methanesulphonate improved yield and pigment contents in vegetable cowpea [Vigna unguiculata (L.) Walp.]. S Afr J Bot 148:727–736. https://doi.org/10.1016/j.sajb.2022.04.034

Rehman HM, Cooper JW, Lam H-M, Yang SH (2019) Legume biofortification is an underexploited strategy for combatting hidden hunger. Plant Cell Environ 42:52–70. https://doi.org/10.1111/pce.13368

Roorkiwal M, Pandey S, Thavarajah D et al (2021) Molecular mechanisms and biochemical pathways for micronutrient acquisition and storage in legumes to support biofortification for nutritional security. Front Plant Sci 12. https://doi.org/10.3389/fpls.2021.682842

Sallam A, Arbaoui M, El-Esawi M et al (2016) Identification and verification of QTL associated with frost tolerance using linkage mapping and GWAS in winter faba bean. Front Plant Sci 7. https://doi.org/10.3389/fpls.2016.01098

Samtiya M, Aluko RE, Dhewa T (2020) Plant food anti-nutritional factors and their reduction strategies: an overview. Food Prod Process Nutr 2:6. https://doi.org/10.1186/s43014-020-0020-5

Scheben A, Batley J, Edwards D (2018) Revolution in genotyping platforms for crop improvement. Adv Biochem Eng Biotechnol 164:37–52. https://doi.org/10.1007/10_2017_47

Schmidt MA, Parrott WA, Hildebrand DF et al (2015) Transgenic soya bean seeds accumulating β-carotene exhibit the collateral enhancements of oleate and protein content traits. Plant Biotechnol J 13:590–600. https://doi.org/10.1111/pbi.12286

Shahzad R, Jamil S, Ahmad S et al (2021) Biofortification of cereals and pulses using new breeding techniques: current and future perspectives. Front Nutr 8. https://doi.org/10.3389/fnut.2021.721728

Shlichta JG, Glauser G, Benrey B (2014) Variation in cyanogenic glycosides across populations of wild lima beans (Phaseolus lunatus) has no apparent effect on bruchid beetle performance. J Chem Ecol 40:468–475. https://doi.org/10.1007/s10886-014-0434-0

Singh M, Malhotra N, Singh K (2021) Broadening the genetic base of cultivated chickpea following introgression of wild Cicer species-progress, constraints and prospects. Genet Resour Crop Evol 68:2181–2205. https://doi.org/10.1007/s10722-021-01173-w

Sobia A, Ahmad HM, Awan SI et al (2014) Estimation of genetic variability, heritability and correlation for some morphological traits in spring wheat. J Biol Agric Healthc 4:10–16

Talari A, Shakappa D (2018) Role of pigeon pea (Cajanus cajan L.) in human nutrition and health: a review. Asian J Dairy Food Res. https://doi.org/10.18805/ajdfr.DR-1379

Tantray A, Raina A, Khursheed S, Amin R, Khan S (2017) Chemical Mutagen affects Pollination and Locule Formation in Capsules of Black Cumin (Nigella sativa L.) . Int J Agric Sci 8(1):108–118

Thavarajah D, Thavarajah P, See C-T, Vandenberg A (2010) Phytic acid and Fe and Zn concentration in lentil (Lens culinaris L.) seeds is influenced by temperature during seed filling period. Food Chem 122:254–259. https://doi.org/10.1016/j.foodchem.2010.02.073

Thavarajah D, Lawrence TJ, Powers SE et al (2022) Organic dry pea (Pisum sativum L.) biofortification for better human health. PLoS One 17:e0261109. https://doi.org/10.1371/journal.pone.0261109

Ullah R, Ullah Z, Al-Deyab SS et al (2014) Nutritional assessment and antioxidant activities of different varieties of Vigna radiata. Sci World J 2014:1–5. https://doi.org/10.1155/2014/871753

Wani MR, Dar AR, Tak A et al (2017) Chemo-induced pod and seed mutants in mungbean (Vigna radiata L. Wilczek). SAARC J Agric 15(2):57–67

Wang H, Jia J, Cai Z et al (2022) Identification of quantitative trait loci (QTLs) and candidate genes of seed iron and zinc content in soybean [Glycine max (L.) Merr.]. BMC Genomics 23:146. https://doi.org/10.1186/s12864-022-08313-1

Xia H, Wang L, Qiao Y et al (2020) Elucidating the source–sink relationships of zinc biofortification in wheat grains: a review. Food Energy Secur 9. https://doi.org/10.1002/fes3.243

Yang T, Fang L, Zhang X, Hu J, Bao S, Hao J et al. (2015) High-throughput development of SSR markers from Pea (Pisum sativum L.) based on next generation sequencing of a purified Chinese commercial variety. PLoS ONE 10(10):e0139775. https://doi.org/10.1371/journal.pone.0139775

Yigzaw Y, Larsson N, Gorton L et al (2001) Liquid chromatographic determination of total and β-N-oxalyl-l-α,β-diaminopropionic acid in Lathyrus sativus seeds using both refractive index and bioelectrochemical detection. J Chromatogr A 929:13–21. https://doi.org/10.1016/S0021-9673(01)01183-9

Chapter 6
Contribution of Biotechnological Approaches to Micronutrient Improvements in Legumes

Afifa Younas, Zubaida Yousaf, Madiha Rashid, Nadia Riaz, Arusa Aftab, Munazzam Jawad Shahid, Hamna Yasin, Zainab Shahzadi, Zainab Maqbool, Ayesha Feroz, Sana Sarwar Khan, and Tahira Zaheer

Abstract Legumes are an important group food. The productivity and dietary value of legumes have declined because of climate change. For legumes, global warming and abrupt meteorological conditions are the main problems stemming from climatic change. As result of these changes, more than 2.5 billion people experience hidden hunger, which is hunger caused by dietary deficiencies from a lack of micronutrients in food. The major micronutrients, namely zinc (Zn), iron (Fe), and selenium (Se), heavily influence human health. The symbiotic relationship between plants' rhizospheres and microbes has conferred several beneficial effects to plant growth, development, and nourishment and has protected plants from biotic and abiotic stresses. In this chapter, we discuss biotechnological and genetic modified techniques that could be useful in improving legume crops.

There is a dearth of research on fortifying legumes to improve their quality. Nutritional and biologically beneficial food crops can also be used to manage certain diseases. Biotechnological techniques have been used to improve specific functional properties of food plants by adding bioactive compounds. The increased availability of nutrients would significantly improve human health, especially in developing countries. This chapter will lay out recent biotechnological approaches to improving the availability of micronutrients in legumes and to conferring resistance to the abiotic and biotic stresses that adversely affect legume crops.

A. Younas · Z. Yousaf (✉) · N. Riaz · A. Aftab · H. Yasin · Z. Shahzadi · Z. Maqbool · A. Feroz · S. S. Khan · T. Zaheer
Department of Botany, Lahore College for Women University, Lahore, Pakistan
e-mail: zubaida.yousaf@lcwu.edu.pk

M. Rashid
Department of Botany, Lahore College for Women University, Lahore, Pakistan

Department of Botany, Division of Science and Technology, University of Education, Lahore, Pakistan

M. J. Shahid
Department of Environmental Sciences, University of Jhang, Jhang, Pakistan

M. A. Nadeem et al. (eds.), *Legumes Biofortification*,
https://doi.org/10.1007/978-3-031-33957-8_6

1 History of Beans

The wild common bean can be found all across Central and South America (Gepts and Debouck 1991). All the cultivated varieties grown in the world today descend from two pre-Columbian domestication events of wild populations in western Mexico (Kaplan and Lynch 1999; Kwak et al. 2009) and in central Peru (Chacón et al. 2005). In each region, human selection has produced dozens of landraces. After 1492, the common bean was introduced to Southwestern Europe, the Mediterranean region (Angioi et al. 2010), Africa (Westphal 1974), and parts of Asia, and finally, it returned to the Americas.

The common bean was domesticated in central Mexico and southern America approximately 800 years ago. This domestication resulted in two major genetic pools: the Mesoamerican, which extends from northern Mexico to Colombia, and the Andean, which stretches from Peru to Argentina (Rodriguez et al. 2016). Furthermore, dried beans were domesticated and brought from the Americas to the Iberian Peninsula approximately 500 years ago, quickly spreading into Europe, Africa, and Asia (Santalla et al. 2002; Angioi et al. 2010).

Archaeological studies in Mexico have dated the origin of the common bean (*Phaseolus vulgaris* L.) back ≥7000 years (Piperno and Smith 2012). There is no evidence of wild beans in archaeological remains, implying that it was already an established cultivated and widespread crop on the American continents (Schmutz et al. 2014). The American continents are thought to be the origin of the wild common bean and the source of its spread. Nevertheless, its center of origin remains unknown. It is now thought that this plant was domesticated in Mexico and the Southern Andes approximately 8000 years ago (Bitocchi et al. 2012). During the pre-Columbian era, common beans spread throughout North, Central, and South America. The "three sisters" (winter squash, maize, and bean) were at the heart of many Native American agricultural peoples' cropping systems (Ngapo et al. 2021).

A significant amount of evidence has shown that the common bean arrived in France in 1508, most likely unfit for human consumption at that time. Fuchs (1542–1543), in his first review, reported that the European common bean is a climbing herb with red or white flowers, along with red, white, or yellow spotless seeds. Roesslin described it further in 1550, Oellinger in 1553, and Dodonaeus in 1554 (Zeven 1997; Piergiovanni and Lioi 2010). However, Fuchs and Dodonaeus only mentioned that this bean plant has a propensity for climbing. However, Krell and Hammer (2008) recently mentioned *P. vulgaris* or its synonyms in their selection of old manuscripts (1493–1774). In Europe, Mesoamerican biotypes are not common, so McClean et al. (1993) hypothesized that the dispersed germplasm in Europe was native to the Andes.

The term *bean* has been used haphazardly without a specific botanical denotation and remains in use. For example, in the epic poem "The Iliad" (Homer 1998), the world's oldest work of Greek literature, "dark fava beans" have at times been translated as "dark (common) beans." Beans have been referred to as dolichos by Teophrastus, the founder of botany, and fasiolos by Dioscorides in ancient Greek. The bean was also thought to have been brought to Rome from the Greek–Roman city of Phaselis in ancient Lycia—hence the genus name (Comes 1909). It has also been proposed that the term *phaselus* derives from the ancient Greek "phaselusa," a small sailing vessel with a bean-shaped hull (Quattrocchi 2012). In Latin, the word *phaselus* is also used to describe a yacht (Corrado 2022). The species name is derived from the fact that beans have always been considered a common and plebeian food: *vulgaris* means "of the masses" (Alcock 2004; Merrow 2020).

Previous studies have shown that most *Phaseolus* species are in Mesoamerica, where most of the diversification has occurred within the past 4–6 Ma (Delgado-Salinas et al. 2006). This diversification occurred because of tectonic events (Delgado-Salinas et al. 2006) in the Late Miocene period (Nieto-Samaniego et al. 1999; Alva-Valdivia et al. 2000).

2 Beans as Sources of Micronutrients

2.1 *Mineral and Protein Content of Fava Beans (*Vicia faba* L.) Seed*

According to Khazaei and Vandenberg (2020), more than half of the global population is at risk of negative outcomes from lacking more than one essential nutrient. The elevated number of essential minerals in pulse seeds has been found to be harmful to human health. Although previously this abnormal concentration of tannin was reported in twenty-five accessions of fava beans, with low tannin in twelve and normal tannin in thirteen accessions. Inductively coupled plasma mass spectrometry (ICP-MS) was used to examine the seed mineral concentrations, and near-infrared (NIR) spectroscopy was used to determine the protein content. It was observed that minerals were positively affected by location and year, particularly for calcium and protein. Genotypic and environmental interactions were found to slightly affect magnesium, calcium, cobalt, zinc, and protein. Higher concentrations of calcium, magnesium, manganese, and cadmium were found in low-tannin accessions compared with normal tannin accessions. It was also observed that low-tannin accessions had 1.9% higher protein content compared to normal tannin-containing accessions. The higher estimated heritability of seeds in terms of mineral and protein content suggested that these species were genetically capable of modifying their mineral composition (Khazaei and Vandenberg 2020).

2.2 *FAO Report*

The most important pulse crop for global production is common beans (*Phaseolus vulgaris* L.), followed by other pulse crop, such as peas (*Pisum sativum* L.), cowpeas (*Vigna unguiculata* L. Walp.), chickpeas (*Cicer arietinum* L.), and lentils (*Lens culinaris* Medik.). China is the largest producer of fava beans, followed by Ethiopia and Australia, in that order.

2.3 Protein Content in Beans

Fava bean protein content varies depending on the seed dry matter (24–35%) when trying to make it the most abundant protein-enriched crop (Crépon et al. 2010; Robinson et al. 2019). In addition to using fava beans as protein supplements for human health, animals also consume them. They can be consumed in many ways, such as in a broth or as a paste for humans and as unadulterated seeds for animals (Khazaei and Vandenberg 2020).

2.4 Micronutrients in Beans

The common bean is also an important crop because of its high nutrient composition, with carbohydrates, vitamins, micronutrients, and protein. This plant also has higher iron content than barley, corn, rice, or wheat crops. Beans are high in a variety of micronutrients, many of which are often deficient in human diets. Bean seeds contain more minerals—i.e., iron, magnesium, zinc, and calcium—than other cereal crops (White and Broadley 2009). They are cultivated worldwide, more specifically in Asia, Africa, Europe, and North, Central, and South America. There are two globally cultivated varieties: Andean beans (large seeds) and Mesoamerican beans (small seeds) (Broughton et al. 2003). Several studies have been conducted to investigate the nutrient diversity of pulse crop species.

2.4.1 Iron and Zinc Content in Beans

Beebe et al. (2000) studied the enhanced micronutrient content, specifically iron and zinc, of common beans. The authors evaluated more than a thousand cultivated accessions and found a mean iron concentration of 55 mg/kg. The calculated zinc ranged from 21 to 54 mg/kg, with a mean value of 35 mg/kg. The above-measured observations show that this crop has genetic variations to enhance iron content and zinc content by up to 80% and 50%, respectively.

2.4.2 Beans as Sources of Micronutrients and Other Beneficial Nutrients

According to the literature, beans are the cheapest sources of proximate and mineral nutrition to meet the dietary requirements of many countries (Guzman-Maldonado et al. 2003; Hillocks et al. 2006). When supplemented with other cereals and carbohydrate-enriched food, beans help to provide basic nutrition and reduce cholesterol and cancer risk (Singh 1999). However, their concentrations can fluctuate depending on genetic and environmental factors (Grusak 2002).

2.4.3 Low-Tannin Beans Rich in Micronutrients

Low-tannin white-flowered fava beans have high contents of calcium, magnesium, iron, zinc, protein, and carbohydrates. These minerals have been found to be in low quantities in human diets, and white flour made from low-tannin beans can be used to counteract these deficiencies (Khazaei and Vandenberg 2020).

2.4.4 Deficiency of Micronutrients

Fe deficiency is the most prevalent micronutrient deficiency in the world, impacting over two billion people, the majority of whom depend on beans in their main meals, according to recent reports (Welch and Graham 1999). Children with diets high in starchy foods had Zn shortages (Ranum 1999). Further, 40% of the iron consumed in underdeveloped countries comes from cereals and legumes (Rosado et al. 2007). Food legumes typically have high levels of iron and other mineral elements (Beebe et al. 2000; Grusak 2002). The fact that nonstaple foods, especially those derived from animals, are the richest sources of bioavailable micronutrients yet are out of the price range of the majority of people in underdeveloped nations is a root cause of micronutrient deficiency and a key barrier to resolving this deficiency. In order to create cultivars with higher performance or cultivars that are tailored to specific environments and/or farming systems, bean breeders have also extensively selected plants on the basis of their yield and other agronomic traits, such as resistance to biotic and abiotic stresses, upstanding plant layout, growth habits, lodging resistance, and maturity (Taràn et al. 2002). Although there is genetic heterogeneity for this feature in germplasm collections, seed nutrient composition is not currently a precise selection criterion for plant breeding (Vreugdenhil et al. 2004).

According to the available research, the mineral concentration ranges in most leguminous seeds are comparable (Wang et al. 2003). Therefore, identifying cultivars with high amounts of Fe and Zn could significantly enhance the micronutrients of people whose diets heavily rely on common beans. Deficits in iron and zinc have been linked to anemia, stunted growth and development in children, and low performance in adults (George and Susan 2010). To maintain appropriate iron and zinc delivery with a diet predominantly constituted of staple foods such as common beans, certain plant-breeding techniques are needed (*Phaseolus vulgaris* L.).

In this study, the variation in Fe and Zn concentrations among genotypes of common beans cultivated in Tanzania's four main bean-growing regions was evaluated for breeding purposes. Ninety genotypes were tested at the Sokoine University of Agriculture in a screen house. An atomic absorption spectrophotometer (AAS) was used to check for the presence of iron and zinc in collected, dried, and ground seeds and leaves. The best genotypes were found to have the highest iron and zinc levels, which differed between genotypes in both seeds and leaves. The results showed a strong and favorable association.

2.5 Cause of Reduction in Bioavailability of Micronutrients

Fava beans contain mineral-rich proteins. The bioavailability of these micronutrients may be hindered by antinutritional elements, such as phytates and tannins. Most of the phytic acids (PAs) or phytates are stored in the seeds of plants (Raboy 1997). A high percentage of PA in food can cause micronutrient deficiencies in Ca, Mg, Zn, Mn, K, Fe, and/or Ba (Shunmugam et al. 2015). Phytic acids negatively affect the edibility of proteins, starches, minerals, and other dietary nutrients. The concentration of PA in the seeds of fava beans is 1% (Afinah et al. 2010). Using processing methods such as soaking, cooking, baking, and fermentation has led to a significant decrease in PA level (Multari et al. 2015; Gupta et al. 2015).

3 Factors that Limit Productivity

The following are some of the factors that limit the production of legumes.

3.1 Optimum Land Preparation

The best land preparation for seed germination, seedling emergence, and subsequent legume growth is urgently needed. The yields of legumes can be significantly increased by using simple management approaches. Zero tillage is the most frequently used method for growing mungbeans, black grams, cowpeas, khesaris (lathyrus), and lentils in rice fallows.

Although zero tillage lessens the risk of crop failure due to early season drought, lowers field preparation costs, and allows for the timely seeding of numerous legumes, yield levels are often poor. To ensure optimal growth conditions and realize the full potential of yields for grain legume crops, limited tillage should be encouraged in both lowland and highland rice fields of the Indo-Gangetic Plains (IGP), especially in medium- to heavy-textured soils. Legume crops can be sown in strips to enhance deeper root development and subsurface water access (Ramakrishna et al. 2000).

3.2 Time of Sowing

The date to start legume sowing is determined by factors such as the length of the rice/wheat crop, the cropping sequence, and the time between harvesting the rice/wheat crop and preparing the field for legume cultivation. Adjustments can be made within the available window to improve sowing timeliness. To synchronize sowing with the proper soil moisture for good crop establishment several changes in field preparation and sowing procedures may be required. In rare circumstances, it may be necessary to adjust the variety or sowing date of the preceding crop in order to advance the sowing date of the following legumes to avoid drought stress during the later phases of growth (Pooniya et al. 2015).

3.3 Water Management

Because legumes are prone to waterlogging, they require appropriate drainage. In semiarid and arid locations, legumes are produced mostly as rainfed crops. Lentils produced in the summer, on the other hand, demand a lot of water, especially during the critical period when soil moisture becomes a limiting factor. In pulse physiology, the vegetative, blooming, and fruiting stages are critical, so closely following an irrigation schedule throughout these stages is critical. Alternate-furrow irrigation (AFI) and water-conservation strategies are viable approaches in water-scarce places (Sharma et al. 2005).

3.4 Tillage

To achieve perfect conditions for seed germination, seed establishment, and plant growth in arable land, agricultural tools must be used to treat the soil. The main purpose of tillage is to provide good soil and soil conditions for crop establishment and early root and shoot growth. Tillage is required for the kharif legume, which includes a spinning plow, two cross wrinkles, and a plank.

To maintain appropriate moisture in rabi legumes, soil plows and irrigation should be started before planting. The use of rotary tillers and cropping systems to recycle summer mungbean waste boosts the system's productivity, profitability, and soil health. Crop residue integration in organic fields using tillage techniques is also a useful way to regulate crop fertilizer usage while improving the soil's organic matter composition (Pooniya et al. 2014).

3.5 Nutrient Management

For legumes to achieve and maintain higher productivity, they need an adequate and balanced supply of plant nutrients. Nutrient management for legumes in rice-wheat cropping systems is complicated and has previously received little attention. Because the majority of studies on fertilizer use for legumes are based on particular crops, the results are applicable only to a restricted number of situations.

Although legumes obtain a lot of their nitrogen from biological nitrogen fixation (BNF), a beginning dose of 10–15 kg N ha has been advised. Although microbiologists contend that this lowers nodulation and nitrogen fixation, it is frequently recommended. Late-sown legumes, on the other hand, react to treatments of up to 40 kg N ha in fields with low rhizobial populations. The application of nitrogen may benefit not only the legumes but also subsequent cereal crops, possibly because the legumes improve the physical conditions of soil (Singh and Chahal 2020).

3.6 Symmetrical Planting

The amount of space required is determined by the type of crop, variety, planting season, and cropping system. Most short-lived legumes require a small amount of land, whereas long-term types thrive in larger areas. Appropriate planting densities in fields and vegetables result in the more efficient use of solar radiation, which translates into higher yields. Grain legumes that are planted in the first week of June will have the largest percentage of pods or seeds. Grain yield decrease because of the tighter and broader gap performance values in different kinds and when seeding after this date.

Growing green peas at a distance of 10–20 cm is more than enough space to produce good yields. Because of the relatively mild temperatures, long vegetative growth time, and extensive branching, kharif farming requires more spacing and a smaller plant population than summer crops do.

3.7 Weed Management

The most frequent fusarium wilt disease, which is part of the root rot complex, causes severe yield losses in mungbeans. On green and urban peas, fusarium wilt, sterile mosaic, phytophthora blight, yellow mosaic, *Cercospora* spp., and white rust cause substantial damage. In India, 250 bug species attack legumes. Nearly a dozen of them cause considerable crop damage. Each year, pest damage results in the loss of roughly 2–2.4 million tonnes of legumes, worth around 6000 rupees (INR 6000) (Singh et al. 2022).

3.8 Abnormal Soil Condition

Legumes prefer neutral soil responses and are sensitive to acidic, salty, and alkaline conditions, and most legumes are phosphorus deficient. As a result, P must be given special consideration in legume production systems. The soil in the northwest of India, in particular, has a high pH, in contrast to the respective soils of the east and the northeast, which are characterized by acidic conditions. These soil conditions lead to micronutrient deficiencies, which result in severe nutrient shortages. Acute shortages of zinc, iron, boron, molybdenum, and secondary elements like sulfur, particularly in legumes, reduce the production quality and quantity of legumes (Kumar et al. 2014) (Fig. 6.1).

Fig. 6.1 Factors that limit the productivity of legumes

4 Biotechnology Approaches and Beans

The following are some of the biotechnological tools employed to improve beans (Obembe 2019):

1. Molecular markers for the assessment of genetic diversity and marker-assisted breeding
2. Plant tissue culture for mass propagation
3. Genetic modification for novel trait integration
4. Omics

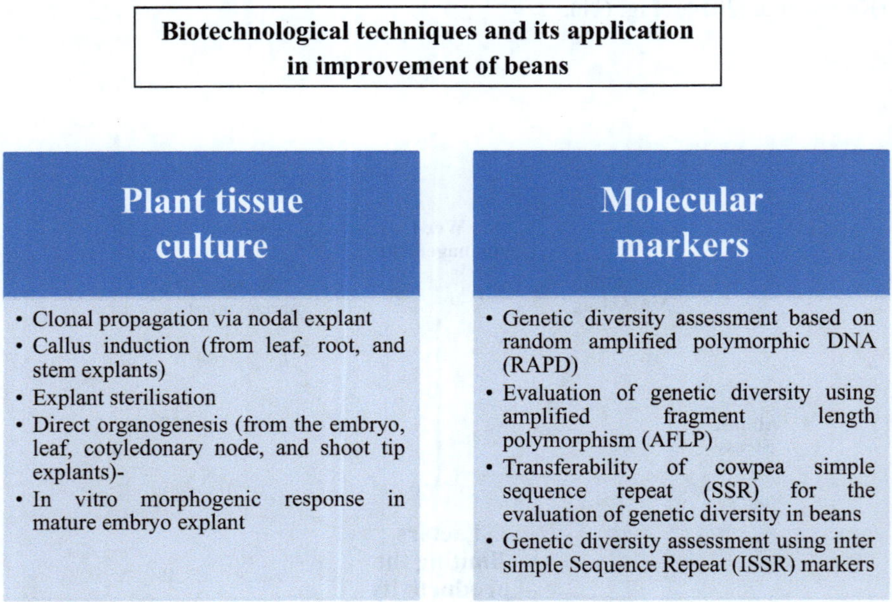

5 Micropropagation Techniques for Improving Beans

For the sterilization of bean explants, Aliyu and Adesoye (2007) showed that using 0.1% mercuric chloride was the best solution. For landraces of beans, Otsoseng (2005) created a nodal explant-based clonal propagation technique. N-phenyl-N′-1,2,3 thidiazol-5-ylurea [TDZ], 6-(γ,γ-dimethylallylamino) purine [2iP], and 6-benzylaminopurine [BAP] cytokinins were added to an Murashige and Skoog (MS) basal medium by the author to retain the nodal segments of stems. In both the stage of culture establishment and the stage of shoot proliferation, BAP produced a more favorable effect than the other cytokinins.

Applying growth regulators known to decrease callogenesis, such as gibberellic acid (GA3) and 2,3,5-triiodobenzoic acid (TIBA), had no beneficial effects in the same study. Some shoots rooted in the presence of the auxins—namely naphthalene acetic acid (NAA) and indole-3-butyric acid (IBA)—but irregular adventitious root development was seen in vitro. IBA generated more roots than NAA did, which resulted in a more satisfying impact. According to the author, cuttings of beans developed adventitious roots whether auxin was present or not. The callus induction of two AYB accessions, SSSWN56 (brown seed) and SSSWN75 (gray seed) from the root, stem, and leaf explants was studied by Akande et al. (2009). According to the authors, the maximal callus percentage (100%) was seen in stem explants from both accessions grown on media supplemented with 1.5 mg/L of kinetin (KIN) and NAA, compared to the minimal callusing seen in root explants. In plant growth regulator (PGR)-free and indole acetic acid (IAA)-fortified media, callus induction was not seen.

5.1 Analysis of Genetic Diversity

Morphological characterization, which is influenced by environmental influences, has been the subject of numerous studies. To confirm that variance is hereditary and not caused by the environment, molecular studies are crucial. Evaluating genetic diversity is important when breeding African beans. For the preservation and use of germplasm resources, obtaining accurate information on genetic variability is essential.

Random amplified polymorphic DNA (RAPD) markers were used in the genetic analysis of AYB by Moyib et al. (2008) and Popoola et al. (2017), separately, whose similarity indices ranged from 0.42 to 0.96 and from 0.72 to 0.93, respectively. According to Moyib et al. (2008), the eight clusters formed by the cluster analysis for the twenty-four accessions had a similarity index of 0.80. According to a principal component analysis (PCA), the first three main components accounted for 30.0, 22.0, and 8.60—or 60.98%—of the overall variation seen among the twenty-four AYB accessions. The phylogenetic tree and the structure were supported by the PCAs. According to their phenotypic differences, primarily tuber formation, tuber flesh color, and seed shape (demonstrating a predicative relationship between genotype and phenotype), these accessions were divided into three categories.

6 Techniques for Improving Beans Breeding

The common bean is the most important cultivated crop among bean species. Thanks to the many classical methods used to improve bean breeding, the regeneration ability of beans has become limited. Now, many biotechnological processes and methods are needed to improve common beans, and many biotechnological techniques have been used.

6.1 Micropropagation and Morphogenesis

Many meristems cultures are used for mass propagation, the removal of seedborne infections, and the preservation of germplasms, because they are essential for germplasm cryopreservation. Some achievements have been made from investigating the in vitro meristem cultures of some grain legumes, supplying them with different concentrations of benzyl adenine (BA) (1–10 μM) for bean regeneration. Some amino acids are also included, such as L- methionine and L- cysteine HCl, in regeneration media to reduce the volume of growing explants. The conditions of cultivation, such as temperature variability and the content used for media (e.g., hormonal combinations and osmotic drugs), are important. Apical meristem plant regeneration needs a BA concentration of (1–10 μM) and an in vitro optimal temperature of 26 °C.

Whole plants developed from an apical meristem culture supplied with coconut water, honey water, and light, without any growth regulators. Phaseolus vulgaris regeneration through meristem culture bud initiation and elongation process depends on the concentration of BA, isopentenyl adenine (2ip), GA3, and cytokinin. Micropropagation developed fifty-one meristem and shoot-tip cultures and axillary buds used for multiplication. Through meristems, cultures that have hormones produced multiple buds that were comparable to the regenerated buds, thanks to the breakdown of the cytokinin–auxin ratio necessary for plant regeneration. Several reports have provided a protocol for seed production without disease and for shoot production from explants and from callus cells.

For plant regeneration, screening the genetic lines, age, and physiological state of parent plants is helpful. The first indirect organogenesis protocol was published by Muhammad et al. in 2007. Genetic factors are important for in vitro responses. By developing a system for organogenesis in *Phaseolus vulgaris* and *Phaseolus coccineus*, more shoots are developed from *Phaseolus coccineus*. The limitations on regeneration are due to low frequency and the source of the explant. Some seed legumes have a high tendency to produce roots rather than shoots, and the frequency of root initiation must be high. For example, forage legumes form roots at low frequencies without being affected by auxin and cytokinin concentrations. Cytokinins and auxins heavily influence the regeneration ability of beans, and many legumes' species require higher cytokinin concentrations. Through direct organogenesis, the regenerated bean plants showed that the buds that developed from the subepidermal parenchyma cells of the node have adventitious structures. The bean regeneration process depends on the genotype, type of explants, concentration of cytokinins, duration of subcultures, and inhibition of phenols released into a medium. Explants that are used for the regeneration of beans also vary in type depending on the cotyledonary and primal nodal cultures of common bean cultivars.

The three methods for multiple-shoot formation and plantlet production in beans use BAP and NAA combinations: (1) The biolistic apical meristem method achieved bean regeneration through direct organogenesis; (2) the BA-treated donor seedlings of petiole and juvenile leaf explants underwent shoot organogenesis; and (3) the organogenic responses of early-germinated beans grown in dark conditions were also determined. In vitro bean embryogenesis and the organogenesis of BA, TDZ, and CPPU play essential roles in determination, and the molecular and physiological roles of cytokinin are under investigation. A plant's ability to regenerate in cell culture media is related to growth regulators and some other factors. Other additives also influence growth media. For example, jasmonic acid helps in the shoot induction of *Phaseolus vulgaris* and develops a system for shoot organogenesis to occur in beans when cotyledonary explants are used as the initiating materials and when the nitrogen source is glutamine. These are some of the methods for the indirect organogenesis of somaclonal genetic variations.

6.2 Synthetic Seeds

Synthetic seeds are similar to zygotic seeds, or somatic embryos enclosed in a coating. Synthetic seeds come in many types: somatic embryos enclosed in water gel; coated, uncoated, or dried somatic embryos; fluid carriers containing suspended somatic embryos; and shoot buds enclosed in water gel. In Japan, only at a smaller scale and with moderate-size seeds where each contained an enclosed somatic embryo in a water gel were synthetic seeds studied with F by using hybrids of celery and lettuce. Synthetic seeds have many applications, and they save time and money and last a long time compared with vegetatively propagated crops. They also avoid the time-consuming process of transferring plants from in vitro conditions to field conditions.

Other applications include the development of male sterile lines, parental lines for the hybrid production of crops, and the elite genotypes of woody plants that have long juvenile phases, which are preserved and multiplied. Development of male sterile lines is not used as extensively, but before its widespread usage, somaclonal variation was minimized. The quality of the embryos must be high for larger-scale productions of interested species, and the technique should be more cost-effective than current micropropagation techniques (Brown and Thorpe 1995).

6.3 Pathogen Eradication

Pathogen removal is another bean-improvement technique, and it plays an important role in the eradication of pathogens. Vegetatively propagated crops are more infected with pathogens. For example strawberries are vulnerable to over sixty viruses and mycoplasmas, and these vulnerabilities require that mother plants be replaced every year. Most of the time, the presence of viruses and pathogens is not obvious, but when they infect crops, they reduce the yield and quality of the crops. In China, virus-free potatoes are produced through in vitro culturing, increasing yields by up to 150%. Care should be taken while propagating seeds because this can remove viruses from plants as only 10% of viruses are transmitted through seeds. The distribution of viruses in plants is not equal, and the apical meristems are less susceptible to viruses. This is why an apical meristem is called the virus-free area of a plant. The excision and culturing of apical meristems connected with thermotherapy and chemotherapy have been successfully used to produce pathogen-free plants for micropropagation (Brown and Thorpe 1995).

6.4 Germplasm Preservation

Conserving germplasm is an alternate to using seed banks, filed collections, or in vitro storage because it provides slow growth conditions, such as low temperature and growth-controlling factors, to media; This is also called cryopreservation or desiccated synthetic seed. All these technologies have been developed for plants and to directly control or stop their growth and metabolic activities. Some of the limitations of the common methods used for all species and genotypes include their high costs, possible somaclonal variation, and the selection of an unintentional cell type in stored material. For example, aneuploidy occurs at low temperatures and under nonoptimal conditions, conferring a selective growth advantage to a cell (Brown and Thorpe 1995).

6.4.1 Advanced Genomic Tools in Beans

Many genomic technologies can be used to improve beans. A high-throughput genetic investigation into several African yam beans characteristics was conducted using diversity array technology sequencing (DArTseq). The DArTseq method relies on a set of restriction enzymes to simplify the genome. Genomic representations use microarray hybridization to determine whether certain pieces are present or absent. This technology is increasingly being utilized to analyze and study the diversity of various crops to improve gene bank conservation. By using this affordable technology, a few orphan crops and some tropical crops have been examined (Huttner et al. 2005).

One technique for determining the association between single-nucleotide polymorphisms (SNPs) and variance in a certain phenotypic characteristic is the genome-wide association study (GWAS). This method has been used on numerous crops to identify the genes influencing certain features, and it eliminates several of the problems in conventional breeding (Table 6.2). When utilizing GWAS, the degree of linkage disequilibrium (the degree of nonrandom association between alleles at distinct loci) and potentially misleading associations resulting from the population structure and genetic relatedness are taken into account (Luo et al. 2020).

SNPs and phenotypic variation based on linkage disequilibrium are used in GWAS to pinpoint the loci in charge of a given characteristic. At the Genetic Resources Center (GRC) of the International Institute of Tropical Agriculture (IITA), in Ibadan, Nigeria, GWASs on beans are now being conducted. Oluwole

Table 6.2 Collection of legume transcriptome data and web sources

Web resource	Legume	Type of information	Reference
LegumeIP	*Medicago* sp., *Lotus* sp., soybean	RNA-sequence data, microarray data	http://plantgrn.noble.org/LegumeIP/
MtGEA	*Medicago* sp.	Microarray data	http://mtgea.noble.org/v3/
LjGEA	*Lotus* sp.	Microarray data	http://ljgea.noble.org/v2/
CTDB	Chickpea	RNA-sequence data	http://www.nipgr.res.in/ctdb.html
SoyPLEX	Soybean	Microarray data	http://www.plexdb.org/plex.php?database=Soybean
SoySeq	Soybean	RNA-sequence data	http://www.soybase.org/soyseq/

et al. (2020) used the DArTseq method to produce 3.6 K of SNPs from a GWAS on the nutritional characteristics of beans. About fifty potential quantitative trait loci (QTLs) that are connected to a seed's starch, protein, and oil concentrations were discovered by these scientists.

The third-largest family of higher plants, with more than 20,000 species, legumes have had major effects on agriculture and on human and livestock nutrition, which consist of legumes. But mostly, legume grains produce low yields. The reduction in the adaptability and productivity of legumes is due mainly to major biotic and abiotic stresses from fungal and viral diseases, drought, frost, chilling, insect pests, salinity, mineral toxicities, and water logging. Several restrictions limit crop quality and productivity, which are being managed by enhanced management strategies and by conventional plant-breeding methods. But sometimes the required traits are missing from existing genetic resources.

The development of a model system for investigating genetically important characteristics initially helped to advance legume biology. For *Arabidopsis thaliana*, two legumes, namely *Lotus japonicus* and *Medicago truncatula*, have emerged as model legume plant systems thanks to their autogamous natures, prolific seed production levels, short generation times, and small and diploid genomes. Powerful genetic and genomic tools have been developed, including genome sequencing, the construction of genetic and physical maps for each model species, and the isolation of expressed sequence tags (ESTs). The genetic models of these two valuable species help in studies on the molecular genetics of various agronomic characteristic related to the increased productivity of legumes. A legume's genome contains an increasing wealth of genetic and genomic data and a high degree of synteny. Advances in plant genomics have recently shifted from model systems to economically important varieties of plant species. The release of the genomic sequence of plants in the past and that of many legumes, such as *Lotus japonicus*, *Glycine max*, *Cajanus cajan*, *Cicer arietinum*, and *Medicago truncatula*, in the present have led to the development and use of many comprehensive tools for sequence assembly, functional annotation, high-throughput expression via microarray platforms, large cDNA and gDNA libraries, and transformation systems. These tools have been developed for a wide range of species, including many important legumes. Recently, a new challenge has emerged: the incorporation of these various tools to better evaluate the genome structures and functions of legumes. To solve these problems, more-comprehensive approaches for various qualitative and quantitative analyses of gene-expression products are needed. The development of these approaches will provide better options to understand functional genomes and their regulatory mechanisms by combining computational approaches with translational genomics (Gupta et al. 2014) (Table 6.1).

6.4.2 Bioinformatics

Bioinformatics is a technique that refers to the organization of biological data for logical evaluation that involves biology, computer science, and information technology. This field combines data from different fields, like omics, to produce novel results (Kumar and Chordia 2017). As a result of the development of high-throughput omic technologies and the creation of a sizable data set, bioinformatics has expanded. There are many online databases from which information can be downloaded, some of which include the legume information system (LegumeInfo.org) and the chickpea transcriptome database (CTDB). Bean analysis has made use of bioinformatic technologies. For instance, the GWAS of the AYB nutritional trait was analyzed through trait analysis by association, evolution, and linkage (TASSEL) (Oluwole et al. 2020).

6.4.3 Genome Editing

It is possible to analyze gene regulation, genomic variations, genome evolution, and genome sequence data thanks to genomics, the study of genomes. Not all facets of genetics have yet been used to enhance beans. The crop's entire genome needs to be sequenced in order to develop a successful breeding program. As a result, the Whole-Genome Sequencing Project is currently being carried out for beans by the Alliance for Accelerated Crop Improvement in Africa (ACACIA) (ACACIA 2020). The precise detection of multiple QTL sites in the bean's genome and the discovery

Table 6.1 Modern biotechnological tools applied to legumes

Sr. No.	Biotechnological tools	Function
1	Bioinformatics	Verification and identification of DNA sequences and DNA motifs
2	Proteomic technology	Identification of protein stresses and responses
3	Genetic modification	Modification of genes or traits—e.g., against biotic and abiotic stresses
4	Genome-wide associated study (GWAS)	Identification of gene loci for different plant traits
5	Genome editing—CRISPR/Cas9	Improvements to genes or gene expressions in plants

of markers would both be greatly enhanced by the availability of a comprehensive whole-genome sequence (Zheng et al. 2021).

6.4.4 Genome Sequencing

The nuclear genomes of legumes vary widely: from 370 Mbp of *Lablab niger* to 13,000 Mbp of *Vicia faba*. Completing the sequence information for some important legumes is a major task these days. Whole-genome sequencing has been completed for legume genera such as *Medicago*, *Glycine*, and *Lotus*. Drafts of genome sequences for the *Cajanus* and *Cicer* genera are also available. In some other legumes, the genome sequencing is in various stages of development. Those expected to be completed soon include mungbeans, alfalfas, peanuts, peas, cowpeas, and some other common beans. The whole-genome sequencing of these crops can be completed with the help of next-generation sequencing. Thanks to short read lengths, method-specific sequencing errors, and the absence of physical clones, next-generation sequencing has more advantages than capillary-based sequencing. It provides assembly and sequence accuracy to challenges related to sequencing. It also allows plant genomes to be assessed at their functional levels (Gupta et al. 2014).

6.4.5 Genetic Maps

Different types of populations and molecular markers have been used for the development of genetic linkage maps. Genetic maps were first developed in the 1990s with the help of morphological markers, isozymes, seed protein genes, and random amplified polymorphic DNA markers (RAPDs). Later, microsatellites, simple sequence repeats (SSRs), expressed sequence tags (ESTs), EST-SSRs, and single-nucleotide polymorphism (SNP) development were used in genetic studies and the breeding of many legumes. The genetic map of fava bean was identified as having nineteen genomic factors forming four linkages groups. Translocation lines can be used to match the different loci of morphological observations, such as flower color and seed coat color, to their respective chromosomes.

6.4.6 Sequencing-Based Trait Mapping

Sequencing-based trait mapping has provided a thorough understanding of trait mapping at the sequence level and also facilitated plant selection at nucleotide levels in legumes. The two mostly commonly used techniques to conduct high-resolution trait mapping to attain large-scale genome-wide SNPs are genotyping by sequencing (GBS) and genome-wide association studies (GWASs). A large number of SNPs for trait mapping can be identified through GBS, while GWAS is a promising approach to sequencing many crops. Both GBS and GWAS provide beneficial advantages over other sequencing techniques insofar as they don't require prior genome information, have low genotyping costs, and provide high-density genotyping data. The construction of high-density linkage maps, analyses of genetic diversity, and GBS applications to improve horticulture crops in legumes are performed by GBS and GWAS. Some other trait mapping approaches do not require the process of whole-genome sequencing to map populations. Instead, the whole analysis is conducted by comparing crops, using QTL sequence (Seq), bulked segregant analysis (BSA)-Seq, and bulk segregant RNA (BSR)-Seq. These approaches have been used in legume breeding to construct a high-density trait-specific map (Rai et al. 2018).

6.4.7 Molecular Marker

Several problems related to conventional breeding have been solved by the identification of various molecular markers and genes related to agronomic traits. Several physical and genetic maps of legumes have been constructed by using molecular markers. Isozyme markers, restriction fragment length polymorphism (RFLP), amplified fragment length polymorphism (AFLP), random amplification of polymorphic DNA (RAPD), and various simple sequence repeat (SSR) markers have been used to study genetic mapping and diversity. Genomic DNA libraries are used to generate SSR markers. Gene-based molecular markers with transcriptome sequences that confer resistance to environmental stresses have been developed from EST-SSR markers. The genome of different species and the accession of legumes could be sequenced and resequenced to identify the coding and noncoding regions of the markers (Ghangal et al. 2020).

6.4.8 Transcriptome Sequence

Transcriptome sequencing based on expressed sequence tags (ESTs) has used legume tissues to generate many single-pass sequences for legumes under different stress conditions. High-throughput, deep-sequencing technologies have been used for the transcriptomic sequencing of wild and cultivated legumes. This can help reveal the evolutionary relationships among legumes. The high transcriptome sequence of wild legume varieties that are more commonly consumed in human diets also illuminates the various platforms of next-generation sequencing (NGS) and several of their hidden biological pathways and processes (Ghangal et al. 2020).

The cheapest methods for sequencing, such as Sanger FLX/454 and Illumina, have developed EST libraries for legumes—e.g. hyacinth beans resist stresses. This practice has been followed for other beans, such as chickpeas and pigeon peas. The hybrid transcript assembly congtigs (TACs) is formed by using transcriptomic data and comparative analysis. Many transcriptomes have been developed for different legume crops at the National Center for Genome Resource (NCGR), which is associated with United State Department of Agriculture (USDA). Expressed sequence tags (ESTs), SNPs, and insertions/deletions (InDels) are useful transcriptomic resources for breeding programs that aim both to produce legumes that can resist biotic and abiotic stresses and to improve the yield and quality of legumes (Rai et al. 2018).

6.4.8.1 Improving Micronutrients in Legumes through Transcriptomics

Improvements to biological variety are beneficial for molecular breeding and for progress in crops. Identifying the transpiration of a genome and producing more markers for particular traits will improve legume breeding systems. The sequencing and resequencing of legume varieties have made physical mutations and functional genomics possible. Next-generation sequencing (NGS) and the emergence of different techniques, such as molecular markers, ESTs, quantitative trait locus (QTL) identification, genotyping, high-density linkage maps, single-nucleotide polymorphisms (SNPs), and transcription factors added to current breeding techniques, have accelerated the provision of accurate data to scientists. EST generation delivers a fast and simple method of identifying the novel gene responsible for sustainable agricultural production. The NGS provides a quick and economical way of sequencing transcriptomes (Morozova et al. 2009). Four legume gene-sequencing species, gene discovery, and the generation of ESTs are represented in Table 6.1.

Transcriptomics can be used to diagnose and profile diseases in beans. Scientists can employ RNA-sequencing (RNA-Seq) to find disease-associated SNPs. It can also be used to research how plants and pathogens interact in order to formulate effective control strategies. Dual RNA-Seq technology can be utilized to outline the identifications of pathogens and crops. Thanks to the development of transcriptomics, it is now possible to discover the genetic factors and traits that can resist biotic and abiotic stresses and the genes that are responsible for particular physical expressions (Lowe et al. 2017).

A study was performed on the basis of the genes in chickpeas to identify drought-responsive genes and molecular markers. The study observed 435,018 reads and 21,491 ESTs. The gene sequence for fava beans and the gene sequence for *Medicago* were estimated to have undergone 41,141 ranged tentative unique sequences (TUSs) for the putative gene assembly produced. The transcranial ultrasound stimulation (TUS) technique recognized the markers, including inter simple sequence regeats (ISSR) (2088), conserved ortholog sets (COS) (387), single nucleotid polymorphism (SNP) (495), and simple sequence repeats (SSR) (728) (Hiremath et al. 2011). Pyrosequencing technology was used, and about two million arrangements with a mean length of 372 base pairs (bps) were produced in chickpeas. The new transcript's long and short reads, which predicted the best results for about 34,760 transcripts' 1020 bps, were represented. The results mapped 4.8% of the total chickpea genome (Garg et al. 2011).

Another study documented 20,162 ESTs and 48,796 bacterial artificial chromosomes (BACs) (Varshney et al. 2009). The 2000 SSR markers were reported in another study. The whole-genome sequencing for chickpeas generated 80,329 sequence tags (Molina et al. 2008). The transcriptomes of Sri Lankan wing beans were analyzed in a study by Vatanparast et al. (2016). The study recorded almost 804,757 new transcripts, which generated 16,115 contigs and revealed substantial genome information. It was observed that 97,241 singleton transcripts and 12,956 SSRs were also generated from the data set. The wing bean genome documented 2594 repeat primers and 5190 SNPs. In one study, the seed transcriptomics of Pongamia (*Millettia pinnata*) was observed by using Illumina's sequencing technology. The outcome showed that eighty-three billion reads cover 53,586 assembled unigenes with a mean length of 787 bp. The information about unigenes contains 73.90% and 44.93% similar proteins to those in the NCBI and Swiss protein databases, respectively.

In total, 364 unigenes were involved in oil biosynthesis and accumulation. These unigenes had an ability desirable in upcoming functional genomics research. Almost 5710 ESTs and SSRs have documented 7.39 kb of density arrangements (Huang et al. 2016). One study revealed a PLAC8 transcription that is symbiotically associated with N_2 and is particularly expressed in the roots of legumes. It consists of a cysteine-rich region and a region that regulates cell numbers (Libault and Stacey 2010). The cytochrome P450 showed significant expression. One study on P450 revealed that these cytochromes also participate in the synthesis and production of bioproducts. *Lotus japonicus* L. contains remorin one gene in its root nodule, and the overexpression of remorin one gene causes an increase in the nodule figure (Tóth et al. 2012).

In a study by Joshi et al. (2013), two soybean species (*Glycine max* and *Glycine soja*) were resequenced. The results identified 425 genes; these genes were not found in *Glycine soja*. Twelve genes associated with seed development and two genes interlaced with lipid metabolism were also identified. Sequencing and phylogenetically studying these soybean species confirmed that these two cultivars belonged to the same ancestor. The resequencing of ninety chickpea genotypes was performed. The study identified 122 essential genes; these genes were used in the latest breeding experiments. Six genomic regions (50–200 kb) were also identified. The resistant genetic traits are beneficial against an assortment of diseases (Varshney et al. 2013). The resequencing of stable mutation genotypes was performed to locate the mutation and express it in physical appearance with linking transformation, offer the ground as a genomic tool, and to define gene tasks (Afzal et al. 2020). A similar study was performed in Arabidopsis to identify the fast neutron mutagenesis that induced the deletion and replacement of single base pairs, determined by using the old chip method (Belfield et al. 2012). Moreover, 2700 coding regions were identified by Leshchiner et al. (2012) by using old chip methods that influence mutant phenotypes and using 17,000 SNPs. Also, 0.1 kb deletions were identified in three genetically mutated lines. The efficiency of old chip methods could improve other tasks and allow mutated expressions to be removed (Tables 6.2 and 6.3).

6.4.9 Proteomics

Another excellent omic tool for examining the functional components of proteins in a particular circumstance is proteomics. It can be used to keep track of potential genes or proteins involved in biological processes at a specific stage in a particular tissue. Multiple genes' functions can be determined by using proteomics in a single experiment, and new genes can be found that can be employed to combat biotic and abiotic stresses. Breeding initiatives for legumes, especially beans, may be made more sustainable by combining the omic approach with these efforts (Afzal et al. 2019).

Global interest has been shifting toward legumes because they provide 27% of the world's principal crops (Duc et al. 2015). They have become the diet of choice for low-income people in developing nations because of their comparatively higher protein, fiber, carbohydrate, mineral, and vitamin contents and relatively low cost compared to nonvegetarian meals. Cowpeas, peas, lupins, and *Medicago* sp. are the main legume crops undergoing proteome research. Through these investigations, the proteins involved in stress tolerance, seed physiology, and plant growth and development have been discovered (Fecht-Christoffers et al. 2003).

To reduce hidden hunger, the nutritional quality of food crops must be improved. Deficits in micronutrients are for the most part responsible for that hidden hunger. Mental illness, poor health, and low productivity have also resulted in this problem. Globally, nearly two billion people, or one in four people, experience such hidden hunger (Assembly 2014; Tubiello et al. 2013). Additionally, this hidden hunger can hinder socioeconomic progress, particularly in low- and middle-income nations. By supplying wholesome food to large populations, this issue might be solved. Moreover, according to recent studies, 70% more food will be needed to feed the global human population by the time it reaches nine billion people, in 2050 (Zargar et al. 2011).

Proteins that are differently expressed, their subcellular localization, post translational modification (PTMs), and protein–protein interactions could all be studied by taking a proteomic approach. Studying post-translational modification, however, continues to be a significant barrier for proteomic research. The model legume plant *Medicago truncatula* (commonly called barrel medic) has been used for the majority of proteome research in legumes. But this is not considered a common crop that is consumed by humans as food. Therefore, proteomic studies of agricultural legumes are urgently needed to understand the unique molecular and genetic pathways of legumes (de Bruijn 2020). To solve the problems from food insecurity and nutritional deficiencies, comibining genomes and proteomes with other omic methods is crucial. These integrated omic techniques may also be used to find other trait-specific biomarkers.

Proteogenomics is a rapidly developing field of biological study that combines genomics and proteomics to better understand the mechanisms driving alterations in plants that have been caused by various stressors. In 2004, Jaffe coined the term *proteogenomics* (Jaffe et al. 2004). Several genes and proteins that are crucial to several biological processes have recently been found by using proteogenomics.

Table 6.3 Workbenches and bioinformatic software packages available for transcriptome data analysis

Process	Package	Description	Reference
Design of RNA-Seq experiment	Scotty	Calculate different gene expressions or products	http://scotty.genetics.utah.edu/
	sizeRNA	Measure the sample size	https://cran.rproject.org/web/packages/ssizeRNA/index.html
	RNAtor	Measure the optimal factors for tools	https://rdrr.io/bioc/PROPER/
Quality control	FastQP	Evaluate the features	https://github.com/mdshw5/fastqp
	FastQC	Control the quality	http://www.bioinformatics.babraham.ac.uk/projects/fastqc/
	AfterQC	Remove the errors, perform read trimming	https://github.com/OpenGene/AfterQC
	QC-Chain	Control the quality	
	FASTX-Toolkit	Perform read trimming	http://hannonlab.cshl.edu/fastx_toolkit/
	NGS QC Toolkit	Control the quality	http://www.nipgr.res.in/ngsqctoolkit.html
	PRINSEQ	Generate a summary of the statistics of a sequence	http://prinseq.sourceforge.net/manual.html
	SolexaQA	Analyze sequence quality statistics and generate visual illustration	http://solexaqa.sourceforge.net/
	mRIN	Measure mRNA integrity	http://zhanglab.c2b2.columbia.edu/index.php/MRIN
	clean_reads	Cleanse NGS reads	http://bioinf.comav.upv.es/clean_reads/
	cutadapt	Remove adapter sequences	https://cutadapt.readthedocs.org/en/stable/
	Deconseq	Eliminate contamination from sequence data	http://deconseq.sourceforge.net/
	htSeqTools	Control the quality, remove excess artifacts	http://www.bioconductor.org/packages/2.9/bioc/html/htSeqTools.html
	SEECER	Precisely sequence errors	http://sb.cs.cmu.edu/
	UCHIME	Observe chimeric sequences	http://drive5.com/uchime

Alignment tools	Bowtie	Align the small reads	http://bowtie-bio.sourceforge.net/index.shtml
	BWA	Map low-divergent sequence against a large genome	http://bio-bwa.sourceforge.net/
	Mosaik	Generate a short gap containing a sequence aligner	http://bioinformatics.bc.edu/marthlab/Mosaik
	GNUMAP	Use the Needleman–Wunsch algorithm-based aligner	http://dna.cs.byu.edu/gnumap/
	WHAM	Align the HTS	http://research.cs.wisc.edu/wham/
	BBMap	Align reads directly to the transcriptome	http://sourceforge.net/projects/bbmap/
	STAR	Sense splice junctions	https://github.com/alexdobin/STAR
Quantitative analysis/transcriptome reconstruction	Cufflinks	Calculate new transcript isoform appearance	http://cufflinks.cbcb.umd.edu/
	Scripture	Reconstruct the transcriptome	http://www.broadinstitute.org/software/scripture/
	RNA eXpress	Extract and annotate the transcripts	http://www.rnaexpress.org/
	SLIDE	Discover isoforms	https://sites.google.com/site/jingyijli/
	StringTie	Assemble the RNA-sequence alignment into possible transcripts	http://www.ccb.jhu.edu/software/stringtie/
	Alexa-Seq	Analyze gene expression	http://www.alexaplatform.org/alexa_seq/downloads.htm
	ERANGE	Standardize and quantify the genes	http://woldlab.caltech.edu/rnaseq
	GFOLD	Express genes differentially	http://compbio.tongji.edu.cn/~fengjx/GFOLD/gfold.html
	NEUMA	Estimate RNA abundance on the basis of mRNA isoforms	http://neuma.kobic.re.kr/
	SpliceSEQ	Investigate alternative mRNA splicing patterns	http://bioinformatics.mdanderson.org/main/SpliceSeq:Overview

(continued)

Table 6.3 (continued)

Process	Package	Description	Reference
miRNA prediction and analysis	miRDeep-P	Inquire into miRNA	http://faculty.virginia.edu/lilab/miRDP/
	miRPlant	Analyze miRNA	https://sourceforge.net/projects/mirplant/
Freely available workbenches	BioJupies	Investigate	http://biojupies.cloud/
	Galaxy	Investigate	http://galaxyproject.org/
	RobiNA	Examine	http://mapman.gabipd.org/web/guest/robin
	NGSUtils	Explore	http://ngsutils.org/
	MeV	Examine	

However, little is known about the mechanisms through which these genes and proteins function in different pathways (Alexandersson et al. 2014). By using a bioinformatic method, integrating proteomics into current genomic data shed light on the cellular processes in agricultural plants.

In order to identify prospective genomic loci, such as transporters and other candidate genes important for nutrition transport and accumulation, it is essential to use this information to establish a relationship between differential protein expression and stress-derived mRNAs. Additionally, when using a nutrigenomics method, the correct and accurate enrichment of food crops may result from the discovery of nutrition-contributing biomarkers and associated QTLs, and proteomics can be used to further analyze the translated products for stability.

Proteomics contains a high-throughput analysis of all cellular proteomes present and expressed in a specific cell, tissue, or organ of a species at specific developmental stages. The protein structure and the modification of the stress-induced regulatory functions of proteins encoded by specific genes can be predicted by determining a proteome's composition, thanks to the large proteome coverage from improved quantitative analysis. Proteomics has become indispensable in crop research because it studies the response mechanisms of different stress-specific traits. The information obtained from such analyses can be used to construct proteome mapping that will help in comparing the protein profiles of different genotypes. Proteomic analysis has been carried out in many model legumes, including *Medicago* sp., *Lotus* sp., and soybean, which provide sufficient proteomic data on and reference maps of these legumes, which in turn provide information on the abiotic and biotic stress on plants. However, many legumes still lack proteomic data, which can be increased by increasing the availability of transcriptome data and legume genome sequences. Both differential expression and comparative proteome studies have substantially contributed to interpretations of stress-specific protein responses in legume crops, such as responses to dehydrin proteins in chickpeas, drought stress in common beans, and drought/salt stress in groundnuts (Rai et al. 2018).

6.4.10 Metabolomics

Specialized metabolites that determine the biochemical phenotype of a cell or tissue are produced in plants and are often thought to be the results of gene expression (Sumner et al. 2003). Measurements of cellular metabolites, both quantitative and qualitative, offer comprehensive perspectives on an organism's metabolic condition, which may be used to track and evaluate gene activity (Fiehn et al. 2000). Additionally, metabolomics has made a substantial contribution to the study of stress biology by discovering a variety of substances, including waste products from stress metabolism, molecules involved in stress-signal transduction, and plant acclimatization (Larrainzar et al. 2009; Weckwerth 2003).

Currently, metabolite-profiling and metabolic fingerprinting methods are employed in the rapidly evolving field of metabolomics. Several analytical approaches, including separation and identification, have been used to cover the large spectrum of metabolites (Doerfler et al. 2013, 2014; Scherling et al. 2010). Gas chromatography (GC) has been used for the separation of volatile and primary metabolites such as sugars and amino acids; liquid chromatography (LC) for the separation of secondary metabolites; capillary electrophoresis (CE) for ionic metabolite separation, and ultraperformance liquid chromatography (UPLC) for metabolite separation (Scherling et al. 2010; Soga 2007; Soga et al. 2003; Weckwerth 2011). UPLC is a potent technology that outperforms traditional high-performance liquid chromatography (HPLC) on resolution, sensitivity, and throughput.

New and improved metabolomic data, industrial platforms, and plant-engineering methods can be used to refine and increase the production of plant-derivative and nonplant compounds and to search for new genes. The metabolomics behind the stress-tolerance mechanism in legumes can be induced. However, metabolomic studies face many challenges when evaluating large numbers of target metabolites under stress conditions. Until now, the profiling studies on other legume crops have provided very limited data on metabolomics. For example, *Lotus* sp. and peas were subjected to abiotic stress, common beans to heavy-metal stress, and chickpeas to biotic stress. Legume crops are sources of flavonoids, which have medicinal and therapeutic activities, and metabolomic studies can facilitate isoflavonoid profiling in response to abiotic and biotic stresses. When recombinant DNA technology combines with metabolic approaches, the combination can unveil the mechanism behind the basic resistance in legume crops and reaffirm the use of biotechnology in breeding management and crop improvement. Extensive metabolic investigations into coincidences with genomics-assisted breeding strategies is, therefore, crucial for improving crops and for analyzing the regulatory networks that govern plant growth and reproduction against climate extremes (Rai et al. 2018).

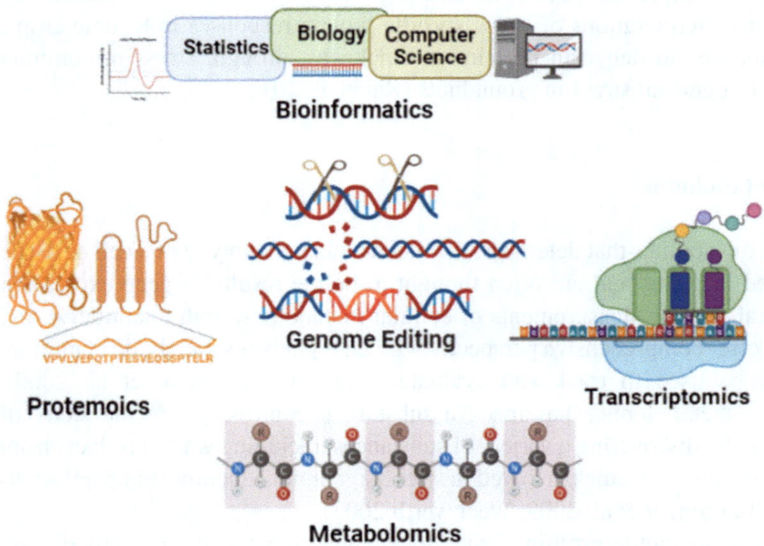

6.4.11 Linkage Maps

Linkage maps of legumes can be created by using different molecular markers, such as isozyme markers. It can also be carried out by using interspecific crosses among species of the same genus. An integrated genetic linkage map of legumes is consequently a sequence of different morphological and molecular markers, such as RFLP, RAPD, and isozyme markers. Recombinant inbred line (RIL) population mapping can be taken as a reference for the construction of genetic maps of high-density legumes. Further, the reference mapping can be used for the development of comprehensive maps of all cultivars enriched with markers. Locus-based genetic maps can also be constructed, which may help in the identification of novel loci in legume cultivars. Molecular marker screenings from transcriptomic sequences have been used to map loci in the parental and offspring lines of legumes such as chickpeas. Interspecific mapping is used to identify the genes of crucial morphological characteristics. The sequence tagged microsatellite site (STMS) markers used for genetic map construction, as well as inter simple sequence repeats (ISSRs) and resistance-gene analogs, can be used to map legume populations. The mapping of specific populations instead of interspecific populations can lead to the development of linkage maps that contain diverse agronomical characteristics and molecular markers. The inheritance of various features, such as seed size, could be used in linkage mapping for either interspecific populations or intraspecific populations of legumes (Ghangal et al. 2020).

6.4.12 QTLs for Agronomically Important Traits

Many studies have been conducted to join genomic regions and traits that are agronomically important, such as resistance to biotic stresses (diseases and fungal infections) and/or abiotic stresses. Identifying the QTLs associated with specific responses can help elevate the agronomic features in legumes. Various assumed genes can be identified in major genomic regions. Genome-wide association studies (GWASs) and QTLs related to disease resistance can be finely mapped. The attempts at resequencing the whole genome revealed many genomic regions that were under positive selection for various agronomic traits in different legumes, traits such as yield, nitrogen fixation, and resistance to abiotic stress. A pool-based DNA analysis can be conducted to identify the QTLs related to different features in legumes, such as seed weight, size, color, yield, filling rate, etc. The environmental influences on agronomic traits are significantly less impactful than the genetic influences (Ghangal et al. 2020).

For instance, a study carried out by Hossain et al. (2010) found two major QTLs accounting 20% of seed size variation in chickpeas. Both loci also showed genetic relationships between seed size and seed weight, which accounted for 37% of this variation. The roles of environmental factors such as rainfall, soil type, and temperature were also assessed and were determined to be limited.

6.4.13 Gene-Expression Data Sets

Genes across various stages in the development of tissues play potential roles in plants' adaptability to stress conditions, which is expressed differently depending on whether the stress is biotic or abiotic. This affects the development of organs and other tissues. Transcriptomic sequencing has helped in the detection of various genes or ESTs related to diverse stress responses. However, other methods, such as EST sequencing, microarray, the super serial analysis of gene expression (SuperSAGE), and suppression subtractive hybridization, have also been employed for the detection of genes responsive to abiotic stress. A transcriptome analysis and a deep SuperSAGE analysis can be performed for the detection of numerous genes responsive to stresses such as salt, drought, desiccation, low temperature, etc. Stress-tolerant and stress-susceptible genes can be identified, and their upregulation or downregulation mechanisms can be explained. The genes expressed under stress conditions may involve transcription regulation, photosynthesis, secondary metabolite biosynthesis, hormonal responses, energy metabolism, and osmopectorant metabolism. Some genes are upregulated under specific stress conditions. For example, in chickpeas, 4954 genes are specifically upregulated in drought-tolerant genotypes and 5545 genes are upregulated in salinity-tolerant genotypes under drought stress and salinity stress, respectively. A microarray analysis can be carried out to profile the disease and immune state of legumes, and those genotypes can be compared and contrasted to shed light on gene expressions. Gene-expression switches during host-specific resistance. a global network and a subcellular analysis have been used for the identification or detection of core genes regulated by and core genes related to disease or immunity in legumes. Gene-expression studies profile the different tissue types and developmental stages of legumes. Transcript abundance could be measured by high-throughput sequencing. A gene-expression resource for legumes that consists of tissue samples obtained from different organs and different developmental stages can help identify the differentially expressed genes involved in various mechanisms, such as nodulation, flowering, root and seed development, and stress-responsive genes. This can also help generate a QTL hotspot for that legume. Differential gene expression can be observed in the same or different cultivars, according to a specific study that carried out a transcriptomic analysis (Ghangal et al. 2020).

6.4.14 Noncoding RNAs

The expression of protein coding genes is regulated by noncoding RNAs because they do not code for functional proteins. In legumes, many micro-RNAs that directly target mRNA are involved in processes such as plant development and biotic and abiotic stress responses. The roles of these micro-RNAs (miRNA) analyzed in different tissues of plant suggested their involvement in the regulation of transcriptome-factor-encoding genes. Moreover, they can be responsible for the gene expression of tissues, organs, or specific developmental stages. Long intergenic noncoding RNAs (lincRNAs) are usually represented by transcripts that are >200 bp long, but they do not have the ability to encode functional proteins. For instance, 2248 lincRNAs have been reported in chickpeas as functionally annotated by a network propagation algorithm. Comprehensive expression profiles for legumes can reveal the presence or expression of different lincRNAs in various developmental stages and in actively dividing cells. These lincRNAs can be associated with stress-responsive genes or any function of legumes during developmental stages (Ghangal et al. 2020).

6.4.15 DNA Methylation

There has been a rapid increase in epigenetic studies for the regulation of gene expression. Eukaryotic gene expression is widely studied because many epigenetic factors, such as DNA methylation and histone modifications, are mostly responsible for their regulation. The methylation-sensitive amplified polymorphism (MSAP) method helps profile the DNA methylation changes in legume genotypes under any stress. Single-base resolution methylome analyses of different tissues and organs reveal many differentially methylated regions (DMRs) between the cultivated and wild varieties of legumes on the basis of a large number of CG contents (the number of G's and C's in the primer as a percentage of the total bases) or Crustacean Hyperglycemic Hormone (CHH) contents abundant in organs or tissues. In the future, many more studies are expected to scrutinize epigenetic regulation in different biological contexts (Ghangal et al. 2020).

6.4.16 Genome-Wide Mapping of Epigenetic Markers

Epigenomic-assisted breeding (EAB) is defined as the mapping of epigenetic factors by manipulating DNA methylation or the demethylation markers that are linked with the heritable epialleles of a specific trait. It involves two techniques: chromatin-immunoprecipitated sequencing (Chlp-Seq) and whole-genome bisulfite sequencing (WGBS), used mainly for the mapping of epigenetic factors. WGBS, accompanied by the NGS technique, rapidly and vigorously scans the whole genome for the occurrence of 5-methlycytosine as a symmetrical cytosine methylation. It is used as a major epigenetic marker that changes the gene-expression regulatory mechanism and cellular metabolism under difficult environmental conditions. This NGS-based WGBS approach is very useful in understanding the molecular mechanism of epigenetic modification in plants and the methylation pattern in segregated offspring generations. Recently, epigenetic recombinant inbred lines (epiRIL) have been generated by crossing the homozygous recessive plant with its wild type to qualify the heritable phenotypic effect variations on several important traits. Now, this technique is extensively used to shed light on the de novo DNA methylation mechanism containing basic heterosis lines of parental and hybrid populations of legumes through siRNA profiling. This can be further prolonged for other legume crops so that epigenetic markers can be mapped under stress conditions (Rai et al. 2018).

6.4.16.1 Challenges of Legume Transformations

DNA must be transferred into a plant to ensure the success of plant transformation. Legumes show resistance to transformation. It has been documented that some tissues of legume crops have the potential to transfer and some have the ability to regenerate; these two possibilities are not always in the same tissue. A stable and efficient transformation protocol is necessary to combat transformation resistance (Bhowmik et al. 2021). An agrobacterium-mediated gene transformation protocol is used in most of the legume crops, whereas a biolistic protocol is followed in a few cases. The CRISPR/Cas9 protocol has been widely applied to soybeans (Zhang et al. 2020). Until recently, forty-one cultivars of soybeans have been released by the CRISPR/Cas9 system (Bhowmik et al. 2021). The main issue of using the CRISPR/Cas9 protocol for large seeded legumes is rooting in vitro, in classes, and in editing systems. In legumes, method specification and some obstructions prevent large-scale transformations. For an effective genetic transformation, conventional and classical transformation approaches cannot be used. CRISPR/Cas9 is an effective technique to reduce these concerns and obstructions. The commercial fruitful production of some legume species can be realized only by generating stable and reproducible methods. Until now, owing to poor in vitro renewal regeneration methods, the attempted transformations of many legumes have not been successful. These issues cannot be solved by taking old genetic transformation approaches (Dewir et al. 2016; Negi et al. 2021).

It has been observed that some countries that produce large quantities of legumes have been resistant to the production of GMO (genetically modified organism) crops because of a lack of acceptance in their local markets. The advancement of

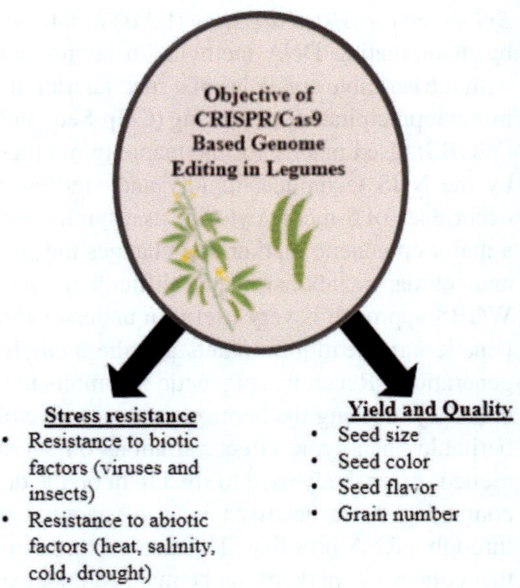

Fig. 6.2 Key objectives of CRISPR/Cas9-based genome editing in legumes

effective transformation techniques is vital not only to reduce hunger but also to validate the role of genes in targeted crops (Bhowmik et al. 2021). Various methods, such as sonication-assisted agrobacterium transformation, have earned serious attention because they hasten the process of genetic transformation in legume crops. Significant strategies to enhance the rate of genetic transformation in legume include optimizing explants, increasing the affinity in host–plant relationships, and refining culture media additives (Sehaole 2022). Further, obstructions to legume transformation and the methods to overcome these obstructions must be investigated (Fig. 6.2).

6.4.16.2 CRISPR/Cas9 Classes and Editing Systems

The CRISPR/Cas9 method can be divided into two classes: I and II. This classification was created on the basis of the function and structure of Cas9 and has been further classified into six more types (Makarova et al. 2015). Types I, III, and IV fall under class I, whereas types II, V, and VI fall under class II. Types I, II, and V can identify and cut DNA, type VI mutates RNA, and type III can improve DNA and RNA. Previous studies have explained the systems that are derived from similar forefathers. During the evolution from class II, class I is coded by a particular function as a gene and loses a section of the added Cas9 gene. Types II and V are easy to handle and are useful in agriculture.

6.4.16.3 Base Editing

Two systems—namely, an adenine-base editor (ABE) and a cytosine-base editor (CBE)—have been extensively used in modern genome-editing systems. The coordinated transfer of DNA into another DNA without damaging the DNA is possible thanks to ABE and CBE. ABE converts adenine to guanine, and CBE converts cytosine to thymine (Zhu et al. 2020). The excision results can be yielded by following sequential steps. The initial step is started by the deamination of cytosines to uridines in nontarget components. The targeted strand designed by Cas9 starts the different repair pathway of DNA, converts the U–G gap into a U–A base arrangement, consequently generating the favored T–A base-pairing exchange and subsequently the DNA duplication. The preferred editing competence is increased by using uracil glycosylase inhibitor (UGI). UGI can stop Uracil-DNA glycosylase (UDG) action to deaminate the cytosines to apyrimidinic places (Komor et al. 2016). Recently, a twin base-pairing scheme acted as an additional base corrector to a cytosine-base editor, and an adenine-base editor has been introduced, implementing the conversion of A to T and G to C (Lin et al. 2020; Grunewald et al. 2020). ABE and CBE are experimental tools and have been primarily exercised in numerous plants. Some defects, such as low editing efficiency and the formation of byproducts, need to be explained in upcoming research.

6.4.16.4 Prime Editing (PE)

The Maloney murine leukemia virus has linked an enzyme (reverse transcriptase) with the C terminal of CRISPR/Cas9 H840A to a supple connector in the prime editors (Anzalone et al. 2019). This combination merges pegRNA, a major base editor, and an enzyme (reverse transcriptase) through an editing array. The equilibrium between three flap entertaining edits without editing five flaps is created when the Cas9 nickase enzyme produces an incision in the selected site; the three ends of the double-stranded DNA and template initiate the reverse transcription (Anzalone et al. 2019). Various experimental studies have shown that variations can produce alterations in plant protoplasts, long lasting genotypes with less editing efficiency (Buttet and Dolar 2020). Many variables can affect plant-editing capacity, such as the nature of the reverse transcriptase, heat factors, template length, prime binding site (PBS) length, and second nick (Tang et al. 2020). PE has been completed in maize. It can be concluded that PE can be a noteworthy genome-editing tool for harvests, especially for legumes.

6.4.16.5 CRISPR/Cas9 Applications to Improve Major Legume Crops

A CRISPR-based genome editing tool was applied to essential legumes such as soybeans, chickpeas, lentils, *Medicago truncatula*, etc. Legumes are grown all over the globe. It was documented that the total production of soybeans yielded 306 million tonnes in 2016 (FAOSTAT 2017); that of lentils was 6.3 million tonnes in 2018 (FAOSTAT 2017); cowpeas yielded 7.4 million tonnes in 2016 (Nkomo et al. 2021); and chickpea production yielded 11.5 million tonnes in 2017 (FAOSTAT 2019; Merga and Haji 2019). CRISPR/Cas9 has been documented to improve harvest volume, quality, and resistance against biotic and abiotic stress (Fig. 6.2). The CRISPR/Cas9 tool is used to improve various traits of crops, and it will continue to improve the traits of commercially important crops (Rasheed et al. 2021). CRISPR/Cas9 has gained much attention in legume trait improvement. Several plant transformation mechanisms are available, such as agrobacteria, plant protoplasts, and transforming *Arabidopsis thaliana* (Narusaka et al. 2012). Cas12a, a new kind of Cas nuclease, offers free genetic editing. Cpf1 brings various site-specific nucleotide deletions. Cpf1 protein synthesizes crRNAs into protoplasts (Table 6.4).

6.4.17 Applications in Crop Improvement

The aforementioned methods and techniques have been widely used to efficiently analyze and improve genotypes. Different plant genotypes—namely transcripts, proteins, and metabolites—were identified and collected, and these biomarkers for plant metabolism were subjected to different environmental conditions, such as day–night rhythms. Incorporating metabolites and transcript data demonstrations assisted in revealing the connection between mRNA operation and the subtleties of

Table 6.4 Key Applications of CRISPR/Cas9 to improve traits of major legume crops

Crop	Genes	Traits	Tool	Delivery system	Reference
Soybean	*GmIPK1*	Hairy roots	Cas9 (SpCas9)	*A. rhizogenes*	Carrijo et al. (2021)
	GmFEI2 and *GmSHR*	Hairy roots	Cas9	*Agrobacterium rhizogenes K599*	Cai et al. (2015)
	DD20, DD43	Resistance to herbicides	Cas9	Particle bombardment	Li et al. (2015)
	Glyma06g14180, Glyma08g02290	Hairy roots and nodules	Cas9	*A. rhizogenes*	Sun et al. (2015)
	Glyma03g36470, Glyma14g04180	Hairy roots	Cas9	*Agrobacterium rhizogenes* K599	Di et al. (2019)
	Green forescent protein gene	Hairy roots	Cas9	*Agrobacterium rhizogenes* K599	Zheng et al. (2021)
	GmFT2a and GmFT5a	Delayed flowering	*Cas9*	*Agrobacterium tumefa ciens* strain EHA105	Cai et al. (2020)
	GmFAD2–1 A and *GmFAD2–1B*	Seed oil profile	Cas9	*Agrobacterium rhizogenes* K599	Do et al. (2019)
	GmPDS11	Biosynthesis of carotenoids	Cas9	*Agrobacterium rhizogenes* K599	Du et al. (2016)
	GmF3H1, GmF3H2	Resistance to soybean mosaic virus	Cas9	*Agrobacterium rhizogenes* K599	Zhang et al. (2020)
	GmLox1, GmLox2	Soybean seed favor	Cas9	*A. tumefaciens* GV3101	Wang et al. (2020)
	GmSPL9	Altered pod number	Cas9	*Agrobacterium tumefaciens U3*	Bao et al. (2019)
	FAD2-A	Fatty acid content	CRISPR/Cas12a	*Agrobacterium rhizogenes* strain K599	Duan et al. (2021)
	GmHSP17.9	Seed yield	Cas9	*Agrobacterium rhizogene* K599	Yang et al. (2022)
Chickpea	*Cry1Aabc*	Resistance to pod borer	Cas9	*Agrobacterium tumefa ciens* strain EHA105	Das et al. (2017)
	OsNAS2 and *CaNAS2*	Fe biofortification	Cas9	*Agrobacterium tumefa ciens* strain AGL1	Das et al. (2019)
	4CL, (RVE7)	Drought tolerance	Cas9	Protoplast transformation using polyethylene glycol (PEG)	Badhan et al. (2021)

(continued)

Table 6.4 (continued)

Crop	Genes	Traits	Tool	Delivery system	Reference
Peanut	*AhNFR1*	Improved rhizobia inoculation	Cas9	*Agrobacterium rhizogenes* strain K599	Shu et al. (2020)
	ahFAD28	Increased linolic acid	Cas9	*Agrobacterium rhizogenes* strain K599	Yuan et al. (2019)
	Ara h 2	Improved nutritional properties	Cas9	Polyethylene glycol	Biswas et al. (2022)
	FAD2B	Oleic acid content	Cas9	*Agrobacterium tumefaciens* strain GV3101	Wei et al. (2022)
Cowpea	*VuSPO11–1*	Improved regeneration efficiency	Cas9	*Agrobacterium auxotrophic* strain LBA4404	Che et al. (2021)
Alfalfa	*NOD26*	Enhanced protein content for nutritious values	Cas9	*A. tumefaciens* LBA4404	Bottero et al. (2021)
	MsSGR	Green plants	Cas9	*A. tumefaciens* strain EHA105	Wolabu et al. (2020)
Lotus japonicus	*LjLB3*	Nodule development for nitrogen fixation	Cas9	*A. rhizogenes* LBA1334	Wang et al. (2016)

Table 6.5 Legume-genome-sequencing results, the discovery of genes, and ESTs generated

Sr. #	Legume species	No. of genes	No. of ESTs	Reference
1	*Cajanus cajan*	48,680	25,640	Varshney et al. (2012)
2	*Cicer arientum* (chickpea)	28,269	46,064	Varshney et al. (2013)
3	*Lotus japonicus* (trefoil)	46,430	1,530,030	Schmutz et al. (2010)
4	*Medicago truncatula* (barrel medic)	47,845	286,175	Young et al. (2011)

secondary metabolisms. Manipulating these approaches in QTL-based marker-assisted breeding methods produces a noticeable expansion.

Several studies on legumes have focused on DNA markers, proteomics, and metabolomics. Such efforts need to be made for crops. Marker-assisted selection can hasten the procedure and the recognition of essential traits in the early years of the assortment process. However, the effectiveness of these markers should first be analyzed with numerous samples and by using strong statistical powers (Gupta et al. 2014) (Table 6.5).

6.4.17.1 RNA Sequencing in Legumes

Transcriptome RNA sequencing is an advanced technique for profiling genes. This RNA-sequencing technique provides information about gene characteristics, gene-expression analysis, and functional genomic studies, even when little information is known about the studied genome. Next-generation sequencing—genome sequencing, RNA sequencing (transcriptome sequencing), and resequencing (DNA sequencing)—offers information on legume growth and expansion systems by using genome evolution and structural design studies (Orourke et al. 2014). RNA sequencing is the cheapest method for quantification. It obtains efficient, precise, and wide-ranging results. Chip-based methods have been replaced by NGS (modern plant omic techniques) for genomic studies and need precision on bioinformatics and data analysis. The complete transcript profiling of the seeds and similar phenotypes can offer genes of interest and their functions, which can regulate the desired traits and provide a beneficial platform from which a gene map of the species could be assembled. Regarding genome-wide transcriptome profiling, RNA-based sequencing techniques were used for different crops, such as rice (Davidson et al. 2012), maize, chickpeas, (Mahdavi Mashaki et al. 2018), and *Raphanus sativas* (Arun-Chinnappa and McCurdy 2015). RNA sequencing has been applied to study the development, nitrogen fixation, composition, and stress adaptation of the following legumes: *M. truncatula* (Boscari et al. 2013), soybeans (Atwood et al. 2014), *Lupin* sp. (O'Rourke et al. 2013), and alfalfas (Yang et al. 2011). RNA sequencing has been used to analyze micro-RNA in *M. truncatula*, common beans, and soybeans (Turner et al. 2012; Simon et al. 2009). The results of the soybean genome investigation revealed that 90.4% of gene models are transcriptionally active, that there are 46,430 high-confidence genes, and that there are 19,780 lower-confidence genes (Schmutz et al. 2010).

Three upper-group tissues, ground tissues, and seed tissues were predicted by a hierarchical clustering analysis using tissue and development transcripts (Severin et al. 2010a,b). In a transcriptomic investigation, extraordinarily high-throughput next-generation RNA-sequencing established a precedent (Garg and Jain 2013). Transcriptomic research benefits from the availability of software that assembles genome-sequencing data (O'Rourke et al. 2013). The greatest of the legume crops' RNA-sequencing data have been gathered, although research on *Medicago sativa*, chickpeas (Garg et al. 2011), and peanuts has received the most attention (Zhang et al. 2012), though lentils (Kaur et al. 2011) and common beans (Kalavacharia et al. 2011) have also undergone remarkable development.

By using the genotype of the fava bean, RNA-sequencing techniques were employed to identify the root transcriptome of differentially expressed genes (DEGs) under drought stress (Hassawi 2). Samples were taken at the vegetative and blooming stages. The results of the data analysis suggested that there were 198,155 unigenes with an average length of 738 bp and roughly 624.8 M quality reads. The majority of the unigenes were downregulated at both stages (vegetative and reproductive), according to the results. Furthermore, 14,097 upregulated genes were found at the vegetative stage, compared to 15,366 upregulated genes at the blooming stage. At the flowering stage, 20,144 genes were downregulated, whereas at the vegetative stage, 22,737 genes were downregulated (Afzal et al. 2020).

Different regulatory proteins, including kinases and phosphatases, transcription factors (TFs), plant hormones, and functional proteins (including enzymes for osmoprotection, detoxification, and transportation). The expression and upregulation of drought stress–responsive genes was induced by these genes. At the expression level, significant change was seen in the genes under drought stress. The qT-PCR results are consistent with the sequencing data, which is helpful for functional genomics and understanding plant tolerance mechanisms (Alghamdi et al. 2018).

6.4.17.2 Computational Resources for RNA-Seq Transcriptome Analysis

Several studies have obtained transcriptome data for a select few legumes (*Medicago* sp., soybeans, chickpeas, and *Lotus*, sp.) (Garg and Jain 2013). The ability to handle and make these data available to researchers, thanks to transcriptome analysis, will aid them in providing and analyzing the specific transcription activity of certain genes at particular developmental stages.

These numerous experiments are further aided by the ability to describe and annotate the functions of these genes in legumes. Characterizations and measurements of the transcriptome are now possible because of next-generation sequencing. The development of advanced bioinformatic software has been aided by the characterization and quantification of the transcriptome. Performing a transcriptome analysis is important for any research on organisms (Yang and Kim 2015).

The software used routinely for the RNA-sequencing workflow is shown in Table 6.5. It starts with designing the RNA-sequence experiments and progresses to performing quality control on the sequence reads, quantitative expression analysis, annotation, and alignment.

7 Conclusion

Currently, more than half of the world's population is facing the same problem: malnutrition. Improvements to the micronutrients in legumes can themselves improve agriculture systems so that they can addresses food security.

This chapter focused on biotechnological approaches to improving the micronutrients in legumes. The most common micronutrient deficiency problems are affecting human health all over the world. The latest biotechnological techniques need to be applied to produce micronutrient-enriched crop yields. These modern biotechnological techniques have been used in molecular markers, genetics, transcriptomics, metabolomics, and proteomics. The application of biotechnology techniques can help to unlock the genetic potential of legumes, which can lead to further research opportunities.

References

ACACIA (2020) Full-genome sequencing and annotation of the African yam bean (Sphenostylis stenocarpa). Available at https://acaciaafrica.org/bioinformatics-community-practice/full-genome-sequencing-and-annotation-of-the-african-yam-bean/. Accessed 30th Apr 2020

Afinah S, Yazid AM, Anis Shobirin MH, Shuhaimi M (2010) Phytase: application in food industry. Int Food Res J 17:13–21

Afzal M, Alghamdi SS, Migdadi HH, Khan MA, Nurmansyah MSB, El-Harty E (2019) Legume genomics and transcriptomics: from classic breeding to modern technologies. Saudi J Biol Sci 27:543

Afzal M, Alghamdi S, Migdadi H, Khan M, Nurmansyah M, S. and El-Harty, E. (2020) Legume genomics and transcriptomics: from classic breeding to modern technologies. Saudi J Biol Sci 27(1):543–555. https://doi.org/10.1016/j.sjbs.2019.11.018

Akande SR, Balogun MO, Ogunbodede BA (2009) Effects of plant growth regulators and explant types on callus formation in African Yam Bean (Sphenostylis stenocarpa (Hochst. Ex A. Rich) Harms) Kasetsart. J Nat Sci 43:442–448

Alcock JP (2004) Around the table of the Romans: food and feasting in Ancient Rome. Gastronomica 4(1):103

Alexandersson E, Jacobson D, Vivier MA, Weckwerth W, Andreasson E (2014) Field-omics—understanding large-scale molecular data from field crops. Front Plant Sci 5:286

Alghamdi SS, Khan MA, Ammar MH, Sun Q, Huang L, Migdadi HM, El-Harty EH, Al-Faifi SA (2018) Characterization of drought stress-responsive root transcriptome of faba bean (*Vicia faba* L.) using RNA sequencing. Biotechnology 8(12):502

Aliyu RE, Adesoye A (2007) Effect of sterilising agents on *in vitro* germination of three explant types of African Yam Bean (*Sphenostylis stenocarpa* (Hochst Ex. A. Rich) Harms). J Biol Sci Res 2:36–39

Alva-Valdivia LM, Goguitchaichvili A, Ferrari L, Rosas-Elguera J, Urrutia-Fucugauchi J, Zamorano-Orozco JJ (2000) Paleomagnetic data from the trans-mexican volcanic belt: implications for tectonics and volcanic stratigraphy. Earth Planets Space 52:467–478

Angioi SA et al (2010) Beans in Europe: origin and structure of the European landraces of Phaseolus vulgaris L. Theor Appl Genet 121(5):829–843

Anzalone AV, Randolph PB, Davis JR, Sousa AA, Koblan LW, Levy JM et al (2019) Search-and-replace genome editing without double-strand breaks or donor DNA. Nature 576:149–157

Arun-Chinnappa KS, McCurdy DW (2015) De novo assembly of a genome-wide transcriptome map of *Vicia faba* (L.) for transfer cell research. Front Plant Sci 6:217

Assembly UG (2014) Report of the Special Rapporteur on the right to food, Olivier de Schutter, final report: the transformative potential of the Right to Food

Atwood SE, O'Rourke JA, Peiffer GA, Yin T, Majumder M, Zhang C, Cianzio SR, Hill JH, Cook D, Whitham SA, Shoemaker RC (2014) Replication protein A subunit 3 and the iron efficiency response in soybean. Plant Cell Environ 37(1):213–234

Badhan S, Ball AS, Mantri N (2021) First report of CRISPR/Cas9 mediated DNA-free editing of 4CL and RVE7 genes in chickpea protoplasts. Int J Mol Sci 22(1):396

Bao A, Chen H, Chen L, Chen S, Hao Q, Guo W, Qiu D, Shan Z, Yang Z, Yuan S, Zhang C, Zhang X, Liu B, Kong F, Li X, Zhou X, Tran LP, Cao D (2019) CRISPR/Cas9-mediated targeted mutagenesis of GmSPL9 genes alters plant architecture in soybean. BMC Plant Biol 19:131

Beebe S, Gonzalez AV, Rengifo J (2000) Research on trace minerals in the common bean. Food Nutr Bull 21(4):387–391

Belfield EJ, Gan X, Mithani A, Brown C, Jiang C, Franklin K, Alvey E, Wibowo A, Jung M, Bailey K, Kalwani S (2012) Genome-wide analysis of mutations in mutant lineages selected following fast-neutron irradiation mutagenesis of Arabidopsis thaliana. Genome Res 22(7):1306–1315

Bhowmik P, Konkin D, Polowick P, Hodgins CL, Subedi M, Xiang D, Kagale S (2021) CRISPR/Cas9 gene editing in legume crops: Opportunities and challenges. Legume Science 3(3):e96

Biswas S, Bridgeland A, Irum S, Thomson MJ, Septiningsih EM (2022) Optimization of prime editing in rice, peanut, chickpea, and cowpea protoplasts by restoration of GFP activity. Int J Mol Sci 23:9809

Bitocchi E, Nanni L, Bellucci E, Rossi M, Giardini A, Zeuli PS, Logozzo G, Stougaard J, McClean P, Attene G (2012) Mesoamerican origin of the common bean (Phaseolus vulgaris L.) is revealed by sequence data. Proc Natl Acad Sci 109(14):E788–E796

Boscari A, Del-Giudice J, Ferrarini A, Venturini L, Zaffini AL, Delledonne M, Puppo A (2013) Expression dynamics of the Medicago truncatula transcriptome during the symbiotic interaction with Sinorhizobium meliloti: which role for nitric oxide? Plant Physiol 161(1):425–439

Bottero E, Massa G, González M, Stritzler M, Tajima H, Gómez C, ... Soto G (2021) Efficient CRISPR/Cas9 genome editing in alfalfa using a public germplasm. Front Agron 3:661526

Buttet S, Dolar V (2020) How much do food prices matter for men's and women's body weight?. Int J Food Agri Econom (IJFAEC) 8(1128-2020-990):125–142

Broughton WJ, Hernandez G, Blair MW, Beebe SE, Gepts P, Vanderleyden J (2003) Beans (*Phaseolus* spp.) – *Model Food Legumes*. Plant Soil 252:55–128

Brown DCW, Thorpe TA (1995) Crop improvement through tissue culture. World J Microbiol Biotechnol 11:409–415

Cai Y, Chen L, Liu X, Sun S, Wu C, Jiang B et al (2015) CRISPR/Cas9-mediated genome editing in soybean hairy roots. PLoS One, 10, e0136064

Carrijo DRA, Priebe DDAM, Stringhini MLF (2021) AVALIAÇÃO NUTRICIONAL E CONSUMO ALIMENTAR DE GESTANTES ATENDIDAS EM PRÉ-NATAL DE ALTO RISCO: REGIÃO METROPOLITANA DE GOIÂNIA E INTERIOR DE GOIÁS. Revista Destaques Acadêmicos 13(3)

Chacón SMI, Pickersgill SB, Debouck DG (2005) Domestication patterns in common bean (*Phaseolus vulgaris* L.) and the origin of the Mesoamerican and Andean cultivated races. Theor Appl Genet 110(3):432–444

Che P, Chang S, Simon MK, Zhang Z, Shaharyar A, Ourada J, ... Jones TJ (2021) Developing a rapid and highly efficient cowpea regeneration, transformation and genome editing system using embryonic axis explants. The Plant J 106(3):817–830

Comes O (1909) Del fagiuolo comune (Phaseolus vulgaris, L.)

Corrado G (2022) Food history and gastronomic traditions of beans in Italy. J Ethn Foods 9(1):1–9

Crépon K, Marget P, Peyronnet C, Carrouée B, Arese P, Duc G (2010) Nutritional value of fabEa bean (*Vicia faba* L.) seeds for food and feed. Field Crops Res 115:329–339

Das A, Datta S, Thakur S, Shukla A, Ansari J, Sujayanand G, Chaturvedi SK, Kumar P, Singh N (2017) Expression of a chimeric gene encoding insecticidal crystal protein Cry1Aabc of Bacillus thuringiensis in chickpea (Cicer arietinum L.) confers resistance to gram pod borer (Helicoverpa armigera Hubner.). Front Plant Sci 8:1423

Das SK, Shethi KJ, Hoque M, Sarker R (2019) Agrobacterium-mediated genetic transformation of lentil (Lens culinaris Medik.) with chitinase gene followed by in vitro flower and pod formation. Plant Tissue Cult Biotechnol 29:99–109

Davidson RM, Gowda M, Moghe G, Lin H, Vaillancourt B, Shiu SH, Buell CR (2012) Comparative transcriptomics of three Poaceae species reveals patterns of gene expression evolution. Plant J 71(3):492–502

de Bruijn FJ (ed) (2020) The model legume Medicago truncatula, 2 volume set. Wiley

Delgado-Salinas A, Bibler R, Lavin M (2006) Phylogeny of the genus *Phaseolus* (Leguminosae): a recent diversification in an ancient landscape. Syst Bot 31:779–791

Dewir YH, Murthy HN, Ammar MH, Alghamdi SS, Al-Suhaibani NA, Alsadon AA, Paek KY (2016) In vitro rooting of leguminous plants: difficulties, alternatives, and strategies for improvement. Hortic Environ Biotechnol 57:311–322

Di Y-H, Sun X-J, Hu Z, Jiang Q-Y, Song G-H, Zhang B et al (2019) Enhancing the CRISPR/Cas9 system based on multiple GmU6 promoters in soybean. Biochem Biophys Res Commun 519:519–823

Doerfler H, Lyon D, Nägele T, Sun X, Fragner L, Hadacek F et al (2013) Granger causality in integrated GC–MS and LC–MS metabolomics data reveals the interface of primary and secondary metabolism. Metabolomics 9:564–574. https://doi.org/10.1007/s11306-012-0470-0

Doerfler H, Sun X, Wang L, Engelmeier D, Lyon D, Weckwerth W (2014) mzGroupAnalyzer-Predicting pathways and novel chemical structures from untargeted high-throughput metabolomics data. PLoS ONE 9:e96188. https://doi.org/10.1371/journal.pone.0096188

Do PT, Nguyen CX, Bui HT, Tran LT, Stacey G, Gillman JD et al (2019) Demonstration of highly efficient dual gRNA CRISPR/Cas9 editing of the homeologous GmFAD2–1A and GmFAD2–1B genes to yield a high oleic, low linoleic and α-linolenic acid phenotype in soybean. BMC Plant Biol 19:311

Du H, Zeng X, Zhao M, Cui X, Wang Q, Yang H, Cheng H, Yu D (2016) Efficient targeted mutagenesis in soybean by TALENs and CRISPR/Cas9. J Biotechnol 217:90–97

Duan L, Li SJ, Su C, Sirichamorn Y, Han LN, Ye W, Chen HF (2021) Phylogenomic framework of the IRLC legumes (Leguminosae subfamily Papilionoideae) and intercontinental biogeography of tribe Wisterieae. Mol Phylogenet Evol 163:107235

Duc G, Agrama H, Bao S, Berger J, Bourion V, De Ron AM et al (2015) Breeding annual grain legumes for sustainable agriculture: new methods to approach complex traits and target new cultivar ideotypes. Crit Rev Plant Sci 34(1–3):381–411

FAOSTAT (2017) http://www.fao.org/faostat/en/#home

FAOSTAT (2019) http://www.fao.org/faostat/en/#home

Fecht-Christoffers MM, Braun H-P, Lemaitre-Guillier C, VanDorsselaer A, Horst WJ (2003) Effect of manganese toxicity on the proteome of the leaf apoplast in cowpea. Plant Physiol 133(4):1935–1946. https://doi.org/10.1104/pp.103.029215

Fiehn O, Kopaka J, Trethewey RN, Willmtizer L (2000) Identification of uncommon plant metabolites based on calculation of elemental compositions using gas chromatography and quadrupole mass spectrometry

Garg R, Jain M (2013) Transcriptome analyses in legumes: a resource for functional genomics. https://doi.org/10.3835/plantgenome2013.04.0011

Garg R, Patel RK, Jhanwar S, Priya P, Bhattacharjee A, Yadav G, Bhatia S, Chattopadhyay D, Tyagi AK, Jain M (2011) Gene discovery and tissue-specific transcriptome analysis in chickpea with massively parallel pyrosequencing and web resource development. Plant Physiol 156:1661

George MT, Susan NM (2010) Diversity of common bean (Phaseolus vulgaris L.) genotypes in iron and zinc contents under screenhouse conditions. Afr J Agric Res 5(8):738–747

Gepts P, Debouck DG (1991) Origin, domestication, and evolution of the common bean (Phaseolus vulgaris L.). In: van Schoonhoven A, Voysest O (eds) Common bean: research for crop improvement. CABI, Wallingford

Ghangal R, Singh VK, Khemka NK, Rajkumar MS, Garg R, Jain M (2020) Updates on genomic resources in chickpea for crop improvement. In: Legume genomics. Humana, New York, NY, pp 19–33

Grünewald J, Zhou R, Lareau CA, Garcia SP, Iyer S, Miller BR, ... Joung JK (2020) A dualdeaminase CRISPR base editor enables concurrent adenine and cytosine editing. Nat Biotechnol 38(7):861–864

Grusak MA (2002) Enhancing mineral content in plant food products. J Am Col Nutr 21(90003):178–183

Gupta S, Nadarajan N, Gupta DS (2014) Legumes in omic era: retrospects and prospects. In: Legumes in the omic era. Springer, New York, NY, pp 1–14

Gupta RK, Gangoliya SS, Singh NK (2015) Reduction of phytic acid and enhancement of bioavailable micronutrients in food grains. J Food Sci Technol 52:676–684

Guzman-Maldonado SV, Martinez O, Acosta-Gallegos JA, GuevaraLara F, Paredes Lopez O (2003) Putative quantitative trait loci for physical and chemical components of common bean. Crop Sci 43:1029–1035

Hillocks RJ, Madata CS, Chirwa R, Minja EM, Msolla S (2006) Phaseolus bean improvement in Tanzania, 1959–2005. Euphytica 150:215–231

Hiremath PJ, Farmer A, Cannon SB, Woodward J, Kudapa H, Tuteja R, ... Varshney RK (2011) Large-scale transcriptome analysis in chickpea (Cicer arietinum L.), an orphan legume crop of the semi-arid tropics of Asia and Africa. Plant Biotechnol J 9(8):922–931

Homer (1998) The Illiad. Penguin

Hossain S, Ford R, McNeil D, Pittock C, Panozzo JF (2010) Inheritance of seed size in chickpea (Cicer arietinum L.) and identification of QTL based on 100-seed weight and seed size index. Aust J Crop Sci 4

Huang J, Guo X, Hao X, Zhang W, Chen S, Huang R, Gresshoff PM, Zheng Y (2016) De novo sequencing and characterization of seed transcriptome of the tree legume *Millettia pinnata* for gene discovery and SSR marker development. Mol Breed 36(6):75

Huttner E, Wenzl P, Akbari M, Caig V, Carling J, Cayla C, Evers M, Jaccoud D, Peng K, Patarapuwadol S, Uszynski G, Xia LW, Yang S, Kilian A (2005. Diversity arrays technology: a novel tool for harnessing the genetic potential of orphan crops) Proceedings of the discovery to delivery: BioVision Alexandria 2004. In: Proceedings of the 2004 Conference of the World Biological Forum. CABI Publishing, pp 145–155

Jaffe JD, Berg HC, Church GM (2004) Proteogenomic mapping as a complementary method to perform genome annotation. Proteomics 4(1):59–77

Joshi T, Valliyodan B, Wu J-H, Lee S-H, Xu D, Nguyen HT (2013) Genomic differences bee: Paper Presented at the BMC Genomics

Kalavacharia V, Liu Z, Meyers BC, Thimmapuram J, Melmaiee K (2011) Identification and analysis of common bean (*Phaseolus vulgaris* L.) transcriptomes by massively parallel pyrosequencing. BMC Plant Biol 11:135–1107

Kaplan L, Lynch TF (1999) Phaseolus (Fabaceae) in archaeology: AMS radiocarbon dates and their significance for pre-Colombian agriculture. Econ Bot 53(3):261–272

Kaur S, Cogan NO, Pembleton LW, Shinozuka M, Savin KW, Materne M, Forster JW (2011) Transcriptome sequencing of lentil based on second-generation technology permits large-scale unigene assembly and SSR marker discovery. BMC Genomics 12(1):265

Khazaei H, Vandenberg A (2020) Seed mineral composition and protein content of faba beans (*Vicia faba L.*) with contrasting tannin contents. Agronomy 10(4):511

Komor AC, Kim YB, Packer MS, Zuris JA, Liu DR (2016) Programmable editing of a target base in genomic DNA without double-stranded DNA cleavage. Nature 533:420–424

Krell K, Hammer K (2008) 500 jahre gartenbohne (Phaseolus vulgaris L.) in Europa. In: Botanik, einführungsgeschichte und genetische ressourcen. Schriften VEN 7, Cremlingen-Schandelah, p 100

Kumar A, Chordia N (2017) Role of bioinformatics in biotechnology. Res Rev Biosci 12(1):116

Kumar A, Suri VK, Choudhary AK (2014) Influence of inorganic phosphorus, VAM fungi, and irrigation regimes on crop productivity and phosphorus transformations in okra (Abelmoschus esculentus L.)–pea (Pisum sativum L) cropping system in an acid Alfisol. Commun Soil Sci Plant Anal 45(7):953–967

Kwak M, Kami JA, Gepts P (2009) The putative Mesoamerican domestication center of Phaseolus vulgaris is located in the Lerma-Santiago basin of Mexico. Crop Sci 49(2):554–563

Larrainzar E, Wienkoop S, Scherling C, Kempa S, Ladrera R, Arrese-Igor C, ... González EM (2009) Carbon metabolism and bacteroid functioning are involved in the regulation of nitrogen fixation in Medicago truncatula under drought and recovery. Mol Plant Microbe Interact 22(12):1565–1576

Leshchiner I, Alexa K, Kelsey P, Adzhubei I, Austin-Tse CA, Cooney JD, Anderson H, King MJ, Stottmann RW, Garnaas MK, Ha S (2012) Mutation mapping and identification by whole-genome sequencing. Genome Res 22(8):1541–1548

Li X, Zhao J, Tan Z, Zeng R, Liao H (2015) GmEXPB2, a cell wall b-expansin, affects soybean nodulation through modifying root architecture and promoting nodule formation and development. Plant Physiol 169:2640–2653

Libault M, Stacey G (2010) Evolution of FW2.2-like (FWL) and PLAC8 genes in eukaryotes. Plant Signal Behav 5:1226–1228

Lin Q, Zong Y, Xue C, Wang S, Jin S, Zhu Z et al (2020) Prime genome editing in rice and wheat. Nature

Lowe R, Shirley N, Bleackley M, Dolan S, Shafee T (2017) Transcriptomics technologies. PLoS Comput Biol 13(5):e1005457

Luo Z, Szczepanek A, Abdel-Haleem H (2020) Genome-wide association study (GWAS) analysis of camelina seedling germination under salt stress condition. Agronomy 10:1444

Mahdavi Mashaki K, Garg V, Nasrollahnezhad Ghomi AA, Kudapa H, Chitikineni A, Zaynali Nezhad K, ... Thudi M (2018) RNA-Seq analysis revealed genes associated with drought stress response in kabuli chickpea (Cicer arietinum L.). PLoS One 13(6):e0199774

Makarova LE, Dudareva LV, Petrova IG (2015) The content of phenolic compounds in the pea seedling root exudates depends on the size of their roots and inoculation of bacteria mutualistic and antagonistic type of interactions. J Stress Physiol Biochem 11(3):94–103

McClean PE, Myers J, Hammond JJ (1993) Coefficient of parentage and cluster analysis of North American dry bean cultivars. Crop Sci 33:190–197

Merga B, Haji J (2019) Economic importance of chickpea: Production, value, and world trade. Cogent Food Agric 5(1):1615718

Merrow A (2020) Caesar's great success: sustaining the Roman army on campaign. Frontline Books

Molina C, Rotter B, Horres R, Udupa SM, Besser B, Bellarmino L, Baum M, Matsumura H, Terauchi R, Kahl G, Winter P (2008) Super SAGE: the drought stress-responsive transcriptome of chickpea roots. BMC Genomics 9:553

Morozova O, Hirst M, Marra MA (2009) Applications of new sequencing technologies for transcriptome analysis. Annu Rev Genomics Hum Genet 10:135–151

Moyib OK, Gbadegesin MA, Aina OO, Odunola OA (2008) Genetic variation within a collection of Nigerian accessions of African yam bean (*Sphenostylis stenocarpa*) revealed by RAPD primers. Afr J Biotechnol 7:1839–1846

Muhammad AH, Tariq M, Ulhaque MI, Muhammad ZK (2007) Mycoflora associated with lentil (Lens esculenta Moench) seeds from five localities of Punjab, Pakistan. Pak J Bot 39(3):903–906

Multari S, Stewart D, Russell WR (2015) Potential of fava bean as future protein supply to partially replace meat intake in the human diet. Compr Rev Food Sci Food Saf 14:511–522

Narusaka Y, Narusaka M, Yamasaki S, Iwabuchi M (2012) Methods to transfer foreign genes to plants. Transgenic plants-advances and limitations

Negi VS, Pathak R, Thakur S, Joshi RK, Bhatt ID, Rawal RS (2021) Scoping the need of mainstreaming indigenous knowledge for sustainable use of bioresources in the Indian Himalayan region. Environ Manage 1–12

Ngapo TM, Bilodeau P, Arcand Y, Charles MT, Diederichsen A, Germain I, Liu Q, MacKinnon S, Messiga AJ, Mondor M (2021) Historical indigenous food preparation using produce of the three sisters intercropping system. Foods 10(3):524

Nieto-Samaniego A, Ferrari L, Alaniz-Alvarez S, Labarthe-Hernandez G, Rosas-Elguera J (1999) Variation of Cenozoic extension and volcanism across the southern Sierra Madre Occidental volcanic province, Mexico. Geol Soc Am Bull 111:347–363

Nkomo GV, Sedibe MM, Mofokeng MA (2021) Production constraints and improvement strategies of cowpea (Vigna unguiculata L. Walp.) genotypes for drought tolerance. Int J Agron 2021:1–9

O'Rourke JA, Yang SS, Miller SS, Bucciarelli B, Liu J, Rydeen A et al (2013) An RNA-Seq transcriptome analysis of orthophosphate-deficient white lupin reveals novel insights into phosphorus acclimation in plants. Plant Physiol 161(2):705–724

Obembe O (2019) Subdue and dominate the Earth: plant biotechnology for sustainable development, Inaugural lecture series, vol 9. Covenant Utniversity, Ota

Oluwole OO, Olomitutu OE, Paliwal R, Oyatomi OA, Abberton MT, Obembe OO (2020) Preliminary assessment of the association between DArT-Seq SNP and some nutritional traits in african yam bean. Trop J Nat Prod Res 4:877–879

Orourke JA, Bolon YT, Bucciarelli B, Vance CP (2014) Legume genomics: understanding biology through DNA and RNA sequencing. Ann Bot 113(7):1107–1120

Otsoseng O (2005) African yam bean: morphology, clonal propagation and nitrogen fixation (Doctoral dissertation). http://eprints.nottingham.ac.uk/12946/1/417103.pdf

Piergiovanni AR, Lioi L (2010) Italian common bean landraces: history, genetic diversity and seed quality. Diversity 2(6):837–862

Piperno DR, Smith BD (2012) The origins of food production in Mesoamerica. The Oxford Handbook of Mesoamerican Archaeology Oxford University Press, Oxford, pp 151–164

Pooniya V, Choudhary AK, Sharma SN, Bana RS, Rana DS, Rana KS (2014) Mungbean (Vigna radiata) residue recycling and varietal diversification for enhanced system productivity and profitability in basmati rice (Oryza sativa)–wheat (Triticum aestivum)–mungbean cropping system. In: Proceedings of National Symposium on Crop Diversification for Sustainable Livelihood and Environmental Security, held during 18–20 November 2014 at PAU, Ludhiana, pp. 629–630

Pooniya V, Choudhary AK, Dass A, Bana RS, Rana KS, Rana DS et al (2015) Improved crop management practices for sustainable pulse production: an Indian perspective. Indian J Agric Sci 85(6):747–758

Popoola J, Adebayo BM, Adegbite A, Omonhinmin C, Adewale D (2017) Fruit morphometric and RAPD evaluation of intraspecific variability in some accessions of african yam bean (*Sphenostylis stenocarpa* hochst. Ex. A. Rich. Harms). Annu Res Rev Biol 14:1–10

Quattrocchi U (2012) CRC world dictionary of medicinal and poisonous plants: common names, scientific names, eponyms, synonyms, and etymology (5 Volume Set). CRC press

Raboy V (1997) Accumulation and storage of phosphate and minerals. In: Larkins BA, Vasil IK (eds) Cellular and molecular biology of plant seed development. Kluwer Academic Publishers, Dordrecht, pp 441–447

Rai KK, Rai N, Rai SP (2018) Recent advancement in modern genomic tools for adaptation of *Lablab purpureus* L, to biotic and abiotic stresses: present mechanisms and future adaptations. Acta Physiol Plant 40(9):1–29

Ramakrishna A, Gowda CLL, Johansen C (2000) Management factors affecting legumes production in the Indo-Gangetic plains. In: Johansen C, Duxbury JM, Virmani SM, Gowda CLL, Pande S, Joshi PK (eds) Legumes in rice and wheat cropping systems of the Indo-Gangetic plain: constraints and opportunities, pp 156–165. ICRISAT, Patancheru, India; and Cornell University, New York, USA

Ranum P (1999) Zinc enrichment of cereal staples. Cereal Foods World 44:604–605

Robinson GHJ, Balk J, Domoney C (2019) Improving pulse crops as a source of protein, starch and micronutrients. Nutr Bull 44:202–215

Rodriguez M, Rau D, Bitocchi E, Bellucci E, Biagetti E, Carboni A et al (2016) Landscape genetics, adaptive diversity, and population structure in *Phaseolus vulgaris*. New Phytol 209:1781–1794. https://doi.org/10.1111/nph.13713

Rosado JL, Lopez P, Morales M, Munoz E, Allen LH (2007) Bioavailability of energy, nitrogen fat, zinc, iron and calcium from rural and urban Mexican diets. Br J Nutr 68:45–58

Santalla M, Rodiño AP, De Ron AM (2002) Allozyme evidence supporting southwestern Europe as a secondary center of genetic diversity for the common bean. Theor Appl Genet 104:934–944. https://doi.org/10.1007/s00122-001-0844-6

Scherling C, Roscher, C, Giavalisco P, Schulze ED, Weckwerth W (2010) Metabolomics unravel contrasting effects of biodiversity on the performance of individual plant species. PLoS ONE 5:e12569. https://doi.org/10.1371/journal.pone.0012569

Schmutz J, Cannon SB, Schlueter J, Ma J, Mitros T, Nelson W, Hyten DL, Song Q, Thelen JJ, Cheng J, Xu D (2010) Genome sequence of the palaeopolyploid soybean. Nature 463(7278):178

Schmutz J, McClean PE, Mamidi S, Wu GA, Cannon SB, Grimwood J, Jenkins J, Shu S, Song Q, Chavarro C (2014) A reference genome for common bean and genome-wide analysis of dual domestications. Nat Genet 46(7):707–713

Sehaole EKM (2022) Genetic transformation in agro-economically important legumes

Severin AJ, Peiffer GA, Xu WW, Hyten DL, Bucciarelli B, O'Rourke JA, Bolon YT, Grant D, Farmer AD, May GD, Vance CP (2010a) An integrative approach to genomic introgression mapping. Plant Physiol 154(1):3–12

Severin AJ, Woody JL, Bolon YT, Joseph B, Diers BW, Farmer AD, Muehlbauer GJ, Nelson RT, Grant D, Specht JE, Graham MA (2010b) RNA-Seq Atlas of Glycine max: a guide to the soybean transcriptome. BMC Plant Biol 10(1):160

Sharma DP, Singh MP, Gupta SK, Sharma NL (2005) The response of pigeon pea to short term water stagnation in a moderately sodic soil under field conditions. J Indian Soc Soil Sci 53(2):243–248

Shu R, Meng Q, Miao L, Liang H, Chen J, Xu Y, ... Zhang H (2020) Genome analysis of a novel clade b betabaculovirus isolated from the legume pest Matsumuraeses phaseoli (Lepidoptera: Tortricidae). Viruses 12(10):1068

Shunmugam ASK, Bock C, Arganosa GC, Georges F, Gray GR, Warkentin TD (2015) Accumulation of phosphorus-containing compounds in developing seeds of low-phytate pea (*Pisum sativum* L.) mutants. Plan Theory 4:1–26

Simon SA, Meyers BC, Sherrier DJ (2009) MicroRNAs in the rhizobia legume symbiosis. Plant Physiol 151:1002–1008

Singh PS (ed) (1999) Common bean improvement in the twenty-first century. Kluwer Academic Publishers, Netherlands, p 405

Singh A, Chahal HS (2020) Organic grain legumes in India: potential production strategies, perspective, and relevance. In: Legume crops – prospects, production and uses. IntechOpen. https://doi.org/10.5772/intechopen.93077

Singh R, Pandey A, Verma PK (2022) Comparative genomic analysis of GARP transcription factor family in legumes and identification of stress-responsive candidate genes. J Plant Growth Regul 1–16

Soga T (2007) Capillary electrophoresis-mass spectrometry for metabolomics. Methods Mol Biol 358:129–137. https://doi.org/10.1007/978-1-59745-244-1_8

Soga T, Ohashi Y, Ueno Y, Naraoka H, Tomita M, Nishioka T (2003) Quantitative metabolome analysis using capillary electrophoresis mass spectrometry. J Proteome Res 2:488–494. https://doi.org/10.1021/pr034020m

Sumner LW, Mendes P, Dixon RA (2003) Plant metabolomics: large-scale phytochemistry in the functional genomics era. Phytochemistry 62(6):817–836

Sun J, Miller JB, Granqvist E, Wiley-Kalil A, Gobbato E, Maillet F ... Oldroyd GE (2015) Activation of symbiosis signaling by arbuscular mycorrhizal fungi in legumes and rice. The Plant Cell 27(3):823–838

Tang J, Wan Y, Zhao M, Zhong H, Zheng JS, Feng F (2020) Legume and soy intake and risk of type 2 diabetes: a systematic review and meta-analysis of prospective cohort studies. Am J Clin Nutr 111(3):677–688

Taràn B, Michaels TE, Pauls KP (2002) Genetic mapping of agronomic traits in common bean. Crop Sci 42:544–556

Tóth K, Stratil TF, Madsen EB, Ye J, Popp C, Antolín-Llovera M, Grossmann C, Jensen ON, Schüßler A, Parniske M, Ott T (2012) Functional domain analysis of the remorin protein LjSYMREM1 in *Lotus japonicus*. PLoS One 7(1):e30817

Tubiello FN, Salvatore M, Rossi S, Ferrara A, Fitton N, Smith P (2013) The FAOSTAT database of greenhouse gas emissions from agriculture. Environ Res Lett 8(1):015009

Turner M, Yu O, Subramanian S (2012) Genome organization and characteristics of soybean microRNAs. BMC Genomics 13:169

Varshney RK, Hiremath PJ, Lekha P, Kashiwagi J, Balaji J, Deokar AA, Vadez V et al (2009) A comprehensive resource of drought- and salinity-responsive ESTs for gene discovery and marker development in chickpea (Cicer arietinum L.). BMC Genom 10:523

Varshney RK, Hiremath PJ, Lekha P, Kashiwagi J, Balaji J, Deokar AA, Vadez V, Xiao Y, Varshney RK, Chen W, Li Y, Bharti AK, Saxena RK, Schlueter JA, Donoghue MT, Azam S, Fan G, Whaley AM, Farmer AD (2012) Draft genome sequence of pigeonpea (*Cajanus cajan*), an orphan legume crop of resource-poor farmers. Nat Biotechnol 30(1):83

Varshney RK, Mohan SM, Gaur PM, Gangarao N, Pandey MK, Bohra A, Janila P (2013) Achievements and prospects of genomics-assisted breeding in three legume crops of the semi-arid tropics. Biotechnol Adv 31(8):1120–1134

Vatanparast M, Shetty P, Chopra R, Doyle JJ, Sathyanarayana N, Egan AN (2016) Transcriptome sequencing and marker development in winged bean (*Psophocarpus tetragonolobus*; Leguminosae). Sci Rep 6:29070

Vreugdenhil D, Aarts MG, Koornneef M, Nelissen M, Ernst WHO (2004) Natural variation and QTL analysis for cation mineral contents in seeds of *Arabidopsis thaliana*. Plant Cell Environ 27:828–839

Wang TL, Domoney C, Hedley CL, Casey R, Grusak MA (2003) Update on seed quality traits. Plant Physiol 131:886–891

Wang L, Wang L, Tan Q, Fan Q, Zhu H, Hong Z, Zhang Z, Duanmu D (2016) Efficient inactivation of symbiotic nitrogen fixation related genes in Lotus japonicus using CRISPR-Cas9. Front Plant Sci 7:1333

Wang L, Sun S, Wu T, Liu L, Sun X, Cai Y, Li J, Jia H, Yuan S, Chen L (2020) Natural variation and CRISPR/Cas9-mediated mutation in GmPRR37 affect photoperiodic flowering and contribute to regional adaptation of soybean. Plant Biotechnol J 18:1869–1881

Weckwerth W (2003) Metabolomics in systems biology. Annu Rev Plant Biol 54:669–689

Weckwerth W (2011) Unpredictability of metabolism- the key role of metabolomics science in combination with next-generation genome sequencing. Anal Bioanal Chem 400:1967–1978. https://doi.org/10.1007/s00216-011-4948-9

Wei Z, Maxwell TM, Robinson B, Dickinson N (2022) Legume nutrition is improved by neighbouring grasses. Plant and Soil, 475(1–2), 443–455

Welch RM, Graham RD (1999) A new paradigm for world agriculture: meeting human needs, productive, sustainable, and nutritious. Field Crops Res 60:1–10

Westphal, E. (1974). Pulses in Ethiopia, their Taxonomy and Agricultural Significance. Wageningen: Centre for Agricultural Publishing and Documentation

White PJ, Broadley MR (2009) Biofortification of crops with seven mineral elements often lacking in human diets–iron, zinc, copper, calcium, magnesium, selenium and iodine. New Phytol 182:49–84

Wolabu TW, Park JJ, Chen M, Cong L, Ge Y, Jiang Q, ... Wang Z (2020) Improving the genome editing efficiency of CRISPR/Cas9 in Arabidopsis and Medicago truncatula. Planta 252:1–14

Yang IS, Kim S (2015) Analysis of whole transcriptome sequencing data: workflow and software. Genome Inform 13(4):119

Yang SS, Tu ZJ, Cheung F, Xu WW, Lamb JF, Jung HJG, Carroll P, Gronwald JW (2011) Using RNA-Seq for gene identification, polymorphism detection and transcript profiling in two alfalfa genotypes with divergent cell wall composition in stems. BMC Genomics 12(1):1–19

Yang J, Lan L, Jin Y, Yu N, Wang D, Wang E (2022) Mechanisms underlying legume–rhizobium symbioses. J Integr Plant Biol 64(2):244–267

Young ND, Debellé F, Oldroyd GE, Geurts R, Cannon SB, Udvardi MK, Benedito VA, Mayer KF, Gouzy J, Schoof H, Van de Peer Y (2011) The Medicago genome provides insight into the evolution of rhizobial symbioses. Nature 480(7378):520

Yuan M, Zhu J, Gong L, He L, Lee C, Han S, et al (2019) Mutagenesis of FAD2 Genes in Peanut with CRISPR/Cas9 Based Gene Editing. BMC Biotechnol 19:24. https://doi.org/10.1186/s12896-019-0516-8

Zargar SM, Nazir M, Agarwal GK, Rakwal R (2011) Omics based interventions for climate proof crops. Genomics Appl Biol 2

Zeven AC (1997) The introduction of the common bean (*Phaseolus vulgaris* L.) into Western Europe and the phenotypic variation of dry beans collected in The Netherlands in 1946. Euphytica 94:319–328

Zhang J, Liang S, Duan J, Wang J, Chen S, Cheng Z, Li Y (2012) De novo assembly and characterization of the transcriptome during seed development, and generation of genic-SSR markers in peanut (*Arachis hypogaea* L.). BMC Genomics 13(1):90

Zhang Y, Meenu M, Yu H, Xu B (2020) An investigation on phenolic and antioxidant capacity of underutilized food legumes consumed in China. Foods 9(4):438
Zheng T, Li P, Li L, Zhang Q (2021) Research advances in and prospects of ornamental plant genomics. Hortic Res 8(1)
Zhu L, Li W, Deng Z, Li H, Zhang B (2020) The composition and antioxidant activity of bound phenolics in three legumes, and their metabolism and bioaccessibility of gastrointestinal tract. Foods 9(12):1816

Web Sources

http://plantgrn.noble.org/LegumeIP/
http://mtgea.noble.org/v3/
http://ljgea.noble.org/v2/
http://www.nipgr.res.in/ctdb.html
http://www.plexdb.org/plex.php?database=Soybean
http://www.soybase.org/soyseq/
http://scotty.genetics.utah.edu/
https://cran.r-project.org/web/packages/ssizeRNA/index.html
https://rdrr.io/bioc/PROPER/
https://github.com/mdshw5/fastqp
http://www.bioinformatics.babraham.ac.uk/projects/fastqc/
https://github.com/OpenGene/AfterQC
http://hannonlab.cshl.edu/fastx_toolkit/
http://www.nipgr.res.in/ngsqctoolkit.html
http://prinseq.sourceforge.net/manual.html
http://solexaqa.sourceforge.net/
http://zhanglab.c2b2.columbia.edu/index.php/MRIN
http://bioinf.comav.upv.es/clean_reads/
https://cutadapt.readthedocs.org/en/stable/
http://deconseq.sourceforge.net/
http://www.bioconductor.org/packages/2.9/bioc/html/htSeqTools.html
http://sb.cs.cmu.edu/
http://drive5.com/uchime
http://bowtie-bio.sourceforge.net/index.shtml
http://bio-bwa.sourceforge.net/
http://bioinformatics.bc.edu/marthlab/Mosaik
http://dna.cs.byu.edu/gnumap/
http://research.cs.wisc.edu/wham/
http://sourceforge.net/projects/bbmap/
https://github.com/alexdobin/STAR
http://cufflinks.cbcb.umd.edu/
http://www.broadinstitute.org/software/scripture/
http://www.rnaexpress.org/
https://sites.google.com/site/jingyijli/
http://www.ccb.jhu.edu/software/stringtie/
http://www.alexaplatform.org/alexa_seq/downloads.htm
http://woldlab.caltech.edu/rnaseq
http://compbio.tongji.edu.cn/~fengjx/GFOLD/gfold.html
http://neuma.kobic.re.kr/
http://bioinformatics.mdanderson.org/main/SpliceSeq:Overview

http://faculty.virginia.edu/lilab/miRDP/
https://sourceforge.net/projects/mirplant/
http://biojupies.cloud/
http://galaxyproject.org/
http://mapman.gabipd.org/web/guest/robin
http://ngsutils.org/

Chapter 7
Nutritional Enrichment in Legumes Through Omics Approaches

Afifa Younas, Zubaida Yousaf, Nadia Riaz, Madiha Rashid, Munazzam Jawad Shahid, Arusa Aftab, Zainab Maqbool, Hamna Yasin, Zainab Shehzadi, Ayesha Feroz, Tahira Zaheer, and Sana Sarwar Khan

Abstract Legumes belong to the Fabaceae family are nutritionally rich in protein, especially black-eyed peas, mungbeans, soybeans, and peas. Legumes are important sources of plant protein in many diets. They are also an excellent reservoir of dietary fiber and complex carbohydrates, resulting in a low glycemic index (GI). Legumes are considered to be valuable sources of potentially functional ingredients, and a remarkable shift toward the increased consumption of legume proteins has been noticed in the past decade. Legumes are pods or fruits containing seeds or dry grains and have a tendency to fix nitrogen in soil. In the past twenty years, the scientific community have made great strides in different fields thanks to the development of high-throughput omic technologies. Four major types of omics have emerged: genomics, transcriptomics, proteomics, and metabolomics.

Functional foods are natural products of plants that have health benefits beyond necessary nutrition. Efforts to identify functional foods in our diet and their beneficial aspects have been limited to a few crops. Advances in sequencing and the availability of different omic technologies have given opportunities to utilize these tools to enhance the functional components of certain foods to ensure nutritional security.

This chapter provides insights into omic studies that have been carried out to determine the active components of crops and improve them by enhancing the functional compounds in legume plants. The functional foods that are being used to improve staple foods need to be characterized in order to more effectively tackle malnutrition and hunger.

A. Younas · Z. Yousaf (✉) · N. Riaz · A. Aftab · Z. Maqbool · H. Yasin · Z. Shehzadi · A. Feroz · T. Zaheer · S. S. Khan
Department of Botany, Lahore College for Women University, Lahore, Pakistan
e-mail: zubaida.yousaf@lcwu.edu.pk

M. Rashid
Department of Botany, Lahore College for Women University, Lahore, Pakistan

Department of Botany, Division of Science and Technology, University of Education, Lahore, Pakistan

M. J. Shahid
Department of Environmental Sciences, University of Jhang, Jhang, Pakistan

M. A. Nadeem et al. (eds.), *Legumes Biofortification*,
https://doi.org/10.1007/978-3-031-33957-8_7

1 Nutrients in Legumes

The term *legume* is a derivative of a botanical word—Leguminosae, which are the dry seeds of the family Papilionaceae, which is the third largest family of higher plants. Legumes are also called pulses. Soybeans are the most cultivated legumes in the world, followed by fava beans and peas. They have made positive contribution to soil fertility by promising nitrogen availability to plants thanks to their symbiotic relationship with rhizobia. The daily consumption of legumes is highly commended. In the Middle Ages, legumes were sources of high energy in human diets thanks to their containing starch and protein. Nowadays, the inclusion of legumes in daily diets has been replaced by other foods, like cereals, potatoes, etc. Many factors are responsible for their current low consumption: the expanding effect of legumes, their simple and rustic image, their being perceived as animal fodder, their low cultivation level, etc. (Erbersdobler et al. 2017).

Nearly a thousand varieties of legumes are now known, among which only twenty are cultivated for human consumption. Some common legumes used as food are chickpeas (*Cicer arietinum* L.), pigeon peas (*Cajanus cajan* L.), mungbeans (*Vigna radiata* L.), cowpeas (*Vigna unvuiculata*), urdbeans (*Vigna mungo* L.), French beans (*Phaseolus vulgaris*), field peas (*Pisum sativum* L.), lentils (*Lens culinaris* Medik), horse grams (*Macrotyloma uniflorum*), soybeans (*Glycine max*), moth beans (*Vigna aconitifolia*), grass peas (*Lathyrus sativus* L.), and a few others. In combination with other cereals, pulses can be used as tools in processing and developing new food products, such as breads, pastas, snacks, soups, cereal bar fillings, meats, and bakery products (Kumar and Pandey 2020).

Legumes are used as staple foods for a large portion of the world's population, especially for low-income people. Legumes' seeds contain valuable numbers of carbohydrates, fibers, proteins, and important structures of essential amino acids. Sulfur-containing amino acids are limiting. Legumes are reported to have some non-nutritional compounds as well. These compounds have toxic effects when the absorption of nutrients is reduced. However, these non-nutritional compounds also have many bioactive compounds, such as antioxidants and hypolipidemic, hypoglycemic, and anticarcinogenic properties, depending on the number of legumes consumed. These beneficial properties of non-nutritional compounds have been scientifically proven.

People in countries that consume a lot of legumes have lower risks of colorectal cancer. Studies have suggested that legumes are substitutes for preventive chemotherapy against various cancers, such as colon cancer. Legume grains are important for human health, more so for people in low-income countries (Sánchez-Chino et al. 2015). Percentages of some of the principal constituents of commonly used legume seeds and their nutrient compositions, in terms of g/100 g, are in Tables 7.1 and 7.2 below, respectively;

Before the use of potatoes, legumes and cereals were the important sources of starch in human diets because starch is one of their major components. They contain relatively high levels of carbohydrates, fats, dietary fibers, proteins, etc. Peas and

Table 7.1 Principal constituents of some major legume seeds

Legume	Protein (%)	Oil (%)	Starch (%)	Fiber (%)	Sucrose (%)
Soybean	35.1–42	17.7–21.0	1.5	20	6.2
Common bean	20.9–27.8	0.9–2.4	41.5	10	5
Pea	18.3–31	0.6–5.5	45	12	2.1
Fava bean	26.1–38	1.1–2.5	37.5–45.6	7.5–13.1	0.4–2.3
Lentil	23–32	0.8–2	46	12	2.9
Chick pea	15.5–28.2	3.1–7	44.6	9	2
Cow pea	23.5	1.3	–	–	–
Mungbean	22.9–23.6	1.2	45	7.0	1.1
Pigeon pea	19.5–22.9	1.3–3.8	44.3	10	2.5

Reference: Kumar and Pandey (2020)

Table 7.2 Nutrient composition of the most-consumed legumes (g/100 g)

Components	Chickpea	Pea	Lentil	Broad bean	Bean	Soybean	Peanut
Proteins	23.6	21.9	20.6	26.6	21.6	36.9	24.8
Carbohydrates	62.3	52.5	56.4	35.4	47.8	6.1	19.0
Fibers	3.8	10.4	6.83	31.3	18.4	20.9	3.1
Lipids	6.4	2.3	2.15	1.8	1.6	18.1	49.7
Ashes	3.7	3.0	2.8	4.1	4	4.7	2

Reference: Sánchez-Chino et al. (2015)

fava beans are rich in starch content. Soybeans contain considerable numbers of fats and proteins but have a low level of functional carbohydrates Lupines are high in dietary fibers and proteins. Therefore, soybeans and lupines can be used to produce isolates and concentrates of proteins for technological and nutritional purposes. White lupines contain more proteins and fats but have low fiber content than blue and yellow lupines. Therefore, white lupines are more useful in daily nutrition (Erbersdobler et al. 2017).

A series of treatments were conducted before the consumption of legume seeds, consisting of mostly soaking the legumes in water before cooking them and before germination. These treatments help promote the deliciousness of legumes and also accelerate bionutrient availability. The number of bionutrients is also enhanced thanks to the inactivation of various enzymes, such as proteases, inhibitors, and hemagglutinins, and that of many other compounds. A mature legume seed is composed of a seed coat (testa), a cotyledon, and an embryonic axis. The energy reservoirs that form the anatomical structure of seeds are present within the cotyledon in the form of protein bodies and starch granules. Although the chemical composition of legumes varies from species to species, the high concentration of proteins, fibers, and carbohydrates and the low level of lipids are significant in each species (Sánchez-Chino et al. 2015). Detailed studies on each nutrient in legumes are presented below.

2 Carbohydrates

Legumes contain a special proportion of carbohydrates, which is their distinguishing characteristic. Carbohydrates make legumes a desirable food, and their slow digestion rate makes them a food with a low GI (Sánchez-Chino et al. 2015). This low glycemic index has made them preferable for people with diabetes as it keeps blood glucose levels under control. Additionally, legumes are gluten-free and can be used by people with celiac disease (Maphosa and Jideani 2017).

About 6–62% of legume seeds contain carbohydrates, mainly starch. The starch in a legume may change the legume's structure for the starch's functional activity. It is transformed into a resistant starch during processing and acts as fiber. Many other carbohydrates are also in legumes. Monosaccharides (ribose, galactose, glucose, and fructose), disaccharides (maltose and sucrose), and α-galactosides (stachyose, raffinose, and verbascose) are also present. After reaching the colon, α-galactosides are fermented by bacteria. The fermentation results in the production of oxygen, carbon dioxide, and methane. These gases result in flatulence because the human digestive system lacks the enzyme used to degrade polysaccharides—α-galactosidase (Sánchez-Chino et al. 2015) (Table 7.3).

3 Proteins

Plant-derived proteins are healthy and sustainable substitutes for animal-based proteins and will reduce the latter's immense consumption in human diets. One unit study found a certain correlation between uptakes of animal-based food and an increase in early cardiovascular death. Food from plants is associated with lower early cardiovascular mortality and lower overall mortality. As a result of overpopulation, more plant-derived proteins need to be consumed in human diets. Legumes are rich in protein content at 20–40% of their dry matter, which is relatively higher than cereals, whose protein content amounts to only 10–15% of their dry matter. The isolates and concentrates of proteins are most often used in the nutrition-rich special diets of athletes. However, high-quality protein foods are needed for the growing tendency toward vegetarian and vegan diets (Erbersdobler et al. 2017). Pulses rich in proteins can be used as additives in various food substances, such as

Table 7.3 Concentrations of α-galactosides in various legumes (% on dry matter basis)

Carbohydrates	Bean (%)	Lentil (%)	Chickpea (%)	Pea (%)	Broad bean (%)	Lupin (%)	Soybean (%)
Raffinose	<0.05–0.93	0.3–1.0	0.4–1.2	0.3–1.6	0.1–0.3	0.5–1.1	0.5–1.3
Stachyose	0.5–4.1	1.7–3.1	2.0–3.6	1.3–5.5	0.7–3.1	0.9–7.4	2.2–4.3
Verbascose	0.06–4.00	0.6–3.1	0.6–4.2	1.6–4.2	1.7–3.1	0.6–3.44	0.0–0.3
Total	2.6–6.6	3.0–7.1	7.4–7.5	5.1–8.7	3.1–4.2	7.4–9.5	2–6

Reference: Sánchez-Chino et al. (2015)

sauces, chocolates, soups, confectionaries, sausages, etc. because of their water- and oil-binding aptitudes (Kumar and Pandey 2020).

The protein concentration in legumes is affected by various factors, including genetic and environmental circumstances. A seed assembles proteins in its cotyledon during its growth period, which provides nourishment through freely available amino acids, ammonia, and carbon. The carbon in amino acids acts as a skeleton for the seeds during germination. The large fractions of storage proteins are found mainly in legume seeds, classified on the basis of their solubility: globulins, albumins, prolamins, and glutelins. About 70% of the total protein amount is occupied by the globulins, which are soluble in saline solution and have a high quantity of legumin and vicilin. Albumins, which are water soluble, make up 10–12% of the total protein content. Glutelins account for 10–20% of the total protein content and are easily soluble in diluted acids and in alkaline solutions, whereas prolamins make up very little to none of the protein content in seeds. Prolamins are soluble in a 50–80% ethanol solution. Sulfur-containing amino acids appear in low concentrations in legumes. Tryptophan has received more attention than lysine amino acids. Combining legumes with different cereals in a dietary regime provides amino acids that are essential for balanced nutrition (Sánchez-Chino et al. 2015).

The global consumption of soybeans is rather important. Its production is above 330 million tonnes. Of its contents, 37–42% are proteins, mainly storage proteins such as globulin, β-conglycinin, and glycinin. The soybean proteins lack sulfur-containing amino acids but have all the functional essential amino acid necessary for the human body—e.g., lysine, leucine, isoleucine, methionine, cysteine, tyrosine, threonine, tryptophan, valine, histidine, and phenylalanine.

Mungbeans are widely cultivated in Asia, which produces approximately 90% of the global supply. The proteins in mungbeans, which are also storage proteins, account for 20–25% of mungbean dry weight. They are composed of 60% globulin and 25% albumin, along with many other essential amino acids, including leucine (1.847%), isoleucine (1.008%), phenylalanine (1.443%), valine (1.237%), arginine (1.672%), lysine (1.664%), methionine (0.286%), tryptophan (0.26%), histidine (0.695%), and threonine (0.782%). The digestibility of mungbean protein is 70%, which is higher than that of soybean protein, at 65%. Protein concentrations in peas range from 21.2% to 32.9%, among which the major storage proteins are globulins. The nutritional value of globulins varies depending on their amino-acid profiles. Arginine, valine and methionine are present in higher concentrations than the concentrations of glutamic acid and cysteine (Kumar and Pandey 2020) (Table 7.4).

Table 7.4 Protein concentrations in legumes, according to Osborne's classification

Legume	Globulins (%)	Albumins (%)	Glutelins (%)	Prolamins (%)
Bean	35–39.3	27.6–36.6	0.2–0.4	0.1–0.2
Lupin	71.8	11.2	5.5	1.0
Pea	66	21	12	Not determined
Soybean	90	10	0	0

Reference: Sánchez-Chino et al. (2015)

4 Fibers

The amount of fiber material in legumes is another reason for humans to consume legume seeds. The fiber content varies depending on the species, variety, and processing of seeds. In legumes, concentrations of fiber content range from 8% to 27.5%, and around 3.3% to 13.8% of this is soluble fiber. The cotyledon's cell wall contains various polysaccharides: 55% are pectic substances, 9% are cellulose, and 6–12% or 16–18% are noncellulosic glucans with no starch. However, inside the seed coat, the cellulose content is higher: 35–57% (Sánchez-Chino et al. 2015).

5 Lipids

Legumes have either low fat content or high fat concentrations, so they can be categorized into two classes on the basis of their various lipid contents. Legumes with low fat content are chickpeas, beans, lentils, and broad beans, which have about 1–6% lipid content. Peanuts and soybeans are classified into the category of legumes with high fat content because 50% and 18% of their contents are lipids, respectively. High concentrations of triglycerides, monounsaturated, and fatty acids are also prominent in legumes (Maphosa and Jideani 2017). Lastly, 52.5% of soybean content is linolenic and 7.5% is linolenic acids (Sánchez-Chino et al. 2015).

6 Minerals

Mineral accessibility is defined by mineral content, mineral–mineral interactions, and the level of tannic and phytic acids, which differ among legumes. The mineral content depends in part on the conditions at the time of cultivation but mainly on the level of phytic acid. Soybeans have the highest phytic acid content—i.e., >2%—followed by fava beans and peas (Erbersdobler et al. 2017). These minerals are essential for proper physiological responses in humans. For example, calcium regulates bone health, copper is required for enzyme activity and proper iron metabolism, and chromium and zinc are responsible for the metabolism of carbohydrates and lipids. Zinc is also necessary for plasma membrane stability and protein synthesis, and iron is used for hemoglobin synthesis and antioxidant activities. Sodium content is comparatively low in legumes. On the other hand, although legumes are rich in iron, little of this iron is available to human, which decreases the bioavailability of iron in legumes. To increase iron absorption from legumes, they can be combined with foods that are rich in vitamin C, which can also prevent anemia, especially in women (Maphosa and Jideani 2017).

Legumes contain important amounts of different minerals required by humans. For instance, chickpeas contain 5 mg 100^{-g} of iron (Fe), 4.1 mg 100^{-g} of zinc (Zn), and 138 mg 100^{-g} or 160 mg 100^{-g} of calcium (Ca). Half the daily required intake

Table 7.5 Most important minerals in legumes

Minerals	Peas	Fava beans	Lupines (blue)	Soybeans
Potassium (g)	1.0	1.2	1.1	1.8
Calcium (g)	0.05	0.14	0.24	0.21
Magnesium (g)	0.12	0.15	0.13	0.22
Iron (mg)	5.2	6.7	5.4	8.0
Copper (mg)	0.66	1.1	0.6	1.2
Zinc (mg)	3.2	4.1	5.1	4.2
Selenium (μg)	1.6	2.0	4.7	19

Reference: Erbersdobler et al. (2017)

of Fe—i.e., 1.05 mg/day for men and 1.46 mg/day for women—and Zn—i.e., 4.2 mg/day for men and 3.0 mg/day for women—can be obtained from 100 g of chickpea seeds. Also, 200 g of seeds can be consumed for their magnesium (Mg), where 260 mg/day is needed by men and 220 mg/day is needed for women. However, chickpeas contain lower iron amounts—5.45 mg 100^{-g}—than other pulses. Lentils and beans contain 8.60 mg 100^{-g} and 7.48 mg 100^{-g} of iron, respectively. Beans can be sources of about 10–20% of daily nutrients in terms of the minerals required by adult humans. They contain high proportions of Fe, Mg, and Mn and low proportions of Zn, Cu, and Ca. These values of Fe, Zn, and Ca are lower than those in animal-based foods. Legumes also contain trace elements of selenium (Se), an essential trace element in human nutrition. Chickpeas, for example, contain 8.2 mg of Se and contain many other trace elements, such as 10.2 mg of aluminum (Al), 0.12 mg of chromium (Cr), 0.26 mg of nickel (Ni), 0.48 mg of lead (Pb), and 0.01 mg of cadmium (Cd) (Sánchez-Chino et al. 2015) (Table 7.5).

7 Vitamins

Legumes are rich in vitamins, mainly the B group of vitamins, such as thiamin, riboflavin, and folate. But many fat-soluble vitamins and vitamin C are not present in them. Folate as an important nutrient and has been reported for its therapeutical significance. It is used to diminish the threat of neural tube imperfections such as spina bifida in newborn babies (Maphosa and Jideani 2017) (Table 7.6).

8 Bioactive and Non-nutrient Compounds

Many bioactive compounds, such as phytochemicals and antioxidants, are in legumes. Lignin, saponins, alkaloids, isoflavones, phytoestrogens, phytates, trypsin, and chymotrypsin inhibitors are nontoxic chemicals but are known as antinutrients. These non-nutrient compounds have physiological effects on the digestibility of proteins and the availability of some bioactive compounds, such as minerals.

Table 7.6 Most important vitamins in legumes

Vitamins	Peas	Fava beans	Lupins (blue)	Soybeans
α-tocopherol (mg) γ-tocopherol mg	0.11 5.0	0.08 Not determined	1.1 15.3	6.5 23.0
Thiamine (mg) (vitamin B_1)	0.7	0.255	0.32	1.0
Riboflavin (mg) (vitamin B_2)	0.27	0.29	0.59	0.46
Pyridoxine (mg) (vitamin B_6)	0.12	0.37	0.4	1.1
Folates (µg)	274	423	40	250

Reference: Erbersdobler et al. (2017)

Table 7.7 Non-nutrients in common legumes (% on dry matter basis)

Legume	Polyphenols	Phytic acids	Tannins	α-galactosides
Common bean	0.3	1.0	0	3.1
Brown bean	1.0	1.1	0.5	3.0
Pea	0.2	0.9	0.1	5.9
Lentil	0.8	0.6	0.1	3.5
Fava bean	0.8	1.0	0.5	2.9
Chickpea	0.5	0.5	0	3.8
Soybean	0.4	1.0	0.1	4.0
Pigeon pea	0.2	0.1	0	0

Reference: Maphosa and Jideani (2017)

Antinutrients are heat sensitive, so when legumes are cooked for consumption, they pose no threats to human health. The decontamination of legumes before processing can be carried out by dehulling, soaking, boiling, steaming, sprouting, roasting, and fermenting them. Most of the antinutrient compounds have antioxidant activities, which are used to manage some cancers, heart diseases, bone diseases, and other chronic diseases. Some of the non-nutrients in legumes are listed in Table 7.7.

Although legumes contain large proportions of nutritional compounds, many people have rejected the consumption of legumes in their daily diets because of the presence of these non-nutritional compounds, which may have positive or negative effects on their health. The effect of antinutrient compounds completely depends on the how many legumes a person consumes. These non-nutrient compounds accumulate in seeds as defense mechanisms against attacks from parasites, fungi, and herbivores and to maintain the growth of their plants during unfavorable conditions. Non-nutrient compounds in legumes are classified into two major groups. The first group contains all the protein compounds, such as lectins, agglutinins, protease inhibitors (e.g., trypsin and chymotrypsin inhibitors), and bioactive peptides. The second group contains all the nonprotein compounds, such as alkaloids, phytic acids, phenolic compounds (tannins), and saponins (Sánchez-Chino et al. 2015).

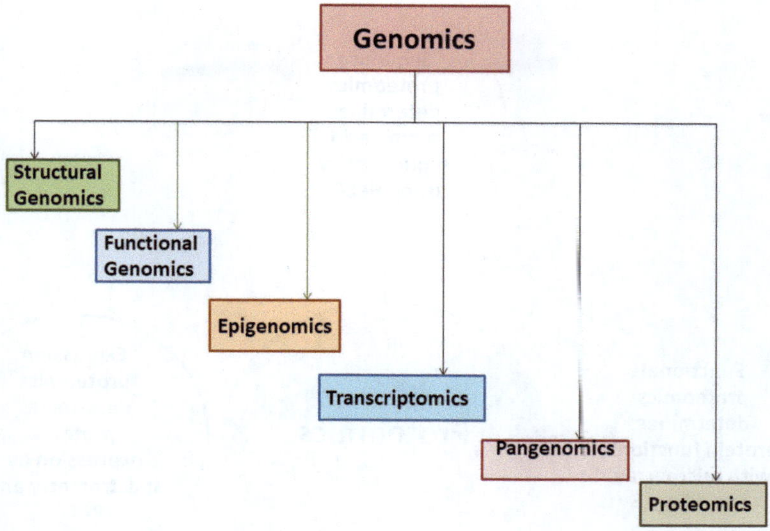

Fig. 7.1 Technologies developed from genomics

Some of the other bioactive compounds in legumes include polyphenols and their derivatives, including flavanols, flavans, tocopherols, and anthocyanins. Legumes contain 321–2404 μg of glutathione and tocopherol per 100 g. Tannins inhibit protein digestion thanks to their reducing nature and are thus considered undesirable. The color of a legume's seed coat is now being associated with antioxidant properties. The denser the color of the seed coat, the higher the value of antioxidant material in a legume. For example, black beans, red kidney beans, black grams, and Bambara groundnuts possess high antioxidant and anticarcinogenic potentials (Maphosa and Jideani 2017) (Figs. 7.1 and 7.2).

9 Multiomic Techniques for Legumes

9.1 Multiomics

Omic technologies have revolutionized an emerging scientific era. These techniques are based on the comprehensive study or detection of genes (genomics), proteins (proteomics), Ribonucleic acid (RNAs) (transcriptomics), metabolites (metabolomics), and the quantitative aspects of medical imaging (radiomics). Multiomic techniques are widely used in the basic research on and clinical treatment of many diseases. Genomics and transcriptomics, based on next-generation sequencing (NGS), shed light on genomes, especially diseased genomes, and can also find differently expressed genes, which are responsible for the establishment of different molecular subtypes and stratifications (Lu and Zhan 2018). Integrating mutiomic

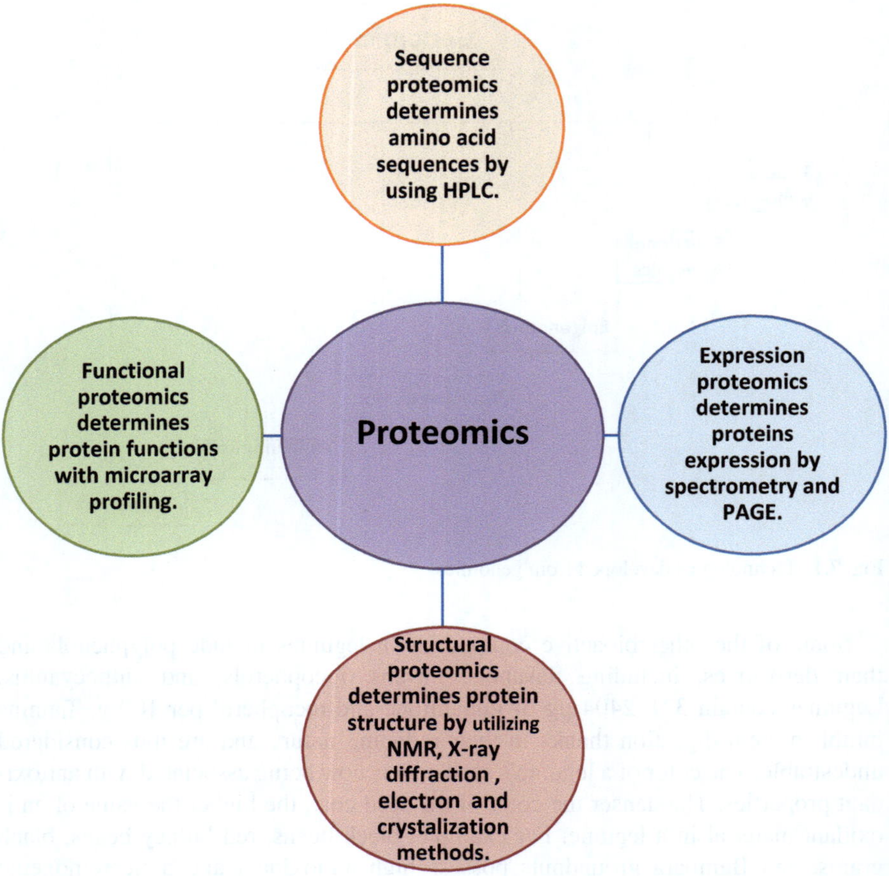

Fig. 7.2 Proteomics categories

data is also being used to design a single model for trait extrapolation. Transcriptomic data and genomic data can be combined to understand the agronomic traits of plants. Transcriptomic, genomic, and metabolomic data can be used to study phenotypic hybrids (Xu et al. 2021).

9.2 Genomics

In recent decades, omic technologies have gained the interest of researchers and emerged as a promising new technology. Such approaches gained importance and became valuable for discovering the molecular and genetic bases of crop development thanks to their various levels of modification for mineral nutrients, metabolites, proteins, transcripts, and DNA in response to physiological and environmental stresses (Muthamilarasan et al. 2019). Plant systems corresponding to molecular compositions have been revealed by multiple omic approaches, including genomics, phenomics, transcriptomics, ionomics, proteomics, mutagenomics, and metabolomics. Legume crops are important for the global food supply and for food security. The instability and low yields of legume grains, compared with cereal crops, is a massive constraint to expanding legume crop cultivation.

The past few decades have witnessed remarkable growth in genomics-derived legume crops. Approaches based on genomics have fast-tracked the growth and development of different varieties of crops with agriculturally superior traits. Genomics helps researchers optimally use resources to improve the predictability of field performance. Legume crop improvements based on genomics have reduced a prolonged generation time to a short generation time thanks to rapid-generation technologies (RGT). Genomics supports hybrid breeding in legumes to obtain good-quality crop yields (Bohra et al. 2021). Such robust approaches are also used in the identification of senescence, yields, and stress responses—essential components to making legume crops more economical (Deshmukh et al. 2014).

Genomics is the study of genes, gene products, and genomes, focusing mainly on the evolution, function, structure, mapping, and editing of genomes. Several technologies have been developed from genomics, including structural and functional genomics, epigenomics, transcriptomics, pangenomics, and proteomics (Pennie et al. 2001). Genomics illustrates the genetic variation in legumes and enhances the breeding efficiency of crops, thus producing improved legume crop species (Bohra et al. 2021).

9.3 *Structural Genomics*

Structural genomics incorporates sequence polymorphism and chromosomal organization. It enables biologists to use DNA markers to create genetic and physical maps in order to identify certain traits, and it provides proteins with 3D structures by encoding genes in legumes (Sato et al. 2010; Skolnick et al. 2000; Tettelin et al. 2005). In legumes, the most used DNA marker technologies can be divided into two categories: non-PCR-based and PCR-based approaches. Non-polymerase chain-reaction-based approaches include restriction fragment-length polymorphism (RFLP). In RFLP, restriction fragments of DNA are formed through digestion with restriction endonuclease enzyme and separated by using gel electrophoresis. Such fragments help in the detection of polymorphism by undergoing southern blotting using chemically labeled probes of DNA. Polymerase chain-reaction-based approaches include random amplified polymorphic DNAs (RAPDs), amplified fragment-length polymorphisms (AFLPs), and single-nucleotide polymorphisms (SNPs) (Govindaraj et al. 2015; Sato et al. 2010).

9.4 Functional Genomics

Functional genomics sheds light on the functions of particular genes in regulating specific traits, also called traits of interest. A variety of information provided by structural genomics is used in functional genomics in the experimental evaluation of gene functions. Numerous biotechnological tools have been developed for isolating, colonizing, and characterizing specified genes. Functional genomics requires following arduous procedures—expressed sequence tag (EST), complementary DNA and amplified fragment-length polymorphism (cDNA-AFLP) sequencing, and suppression subtractive hybridization (SSH)—for gene detection. It facilitates assessing genetic variation in legume species by improving traits in legumes such that they become resistant to several biotic and abiotic stresses (Muthamilarasan et al. 2019; Mishra et al. 2014). After the publication of the reference genome sequences of different legumes, many wild and cultivated genomes of legumes, their genetic diversities, and their selection signatures have been identified. An accurate reference genome can enhance functional genomics research and the de novo assembly of genomes on the basis of using advanced sequencing technologies.

Genomic approaches can be used for the discovery of the genes associated with productivity as well. Hybrid analysis, fine mapping, and expression analysis can also be performed. Various agronomic traits related to the genes responsible for flowering and yield can be analyzed, and a more-productive cultivation of legumes can be performed. The genes identified for better productivity can be used to make hybrids that are ecologically adaptable and produce higher yields. The bionutrients in legumes are immensely important in human diets. Sequencing and the sequencing of the genomes of legumes can help in identifying the gene that can increase the availability of these nutrients to organisms. It can also result in an increase in seed size, oil content, etc. Genomic analysis has helped in the identification of only a few genes that regulate legume productivity. The flowering-time-related genes could be used to plant legumes in new ecological locations (Benlloch et al. 2015). Genes responsible for nutrient contents, such as for oil and protein, could be used to improve the seed composition of legumes. Scientists need a platform that integrates global data on reference genome series because the number of wild and cultivated legume germplasms are increasing day by day. Such a platform would expand the range of available germplasm resources from legumes (Wang et al. 2003).

9.5 Pangenomics

Pangenomics deals with pangenomes and is the study of the core genome or genomic sequences of all individuals of a species. It allows for the characterization and mining of the genes responsible for genetic variation among legume species (Hu et al. 2020). Pangenomics consists of a set of accessory genes (present in some individuals) and core genes (present in all). Accessory genes, also known as dispensable genes, act as disease-resistant genes, are quite valuable, and play significant roles in maintaining the diversity and quality of legume crops (Zhou et al. 2015).

Pangenomes can facilitate the construction of core genomes and enable genomic comparisons in order to identify lineage-specific genes, genes with copy number variations, and genes with large-effect mutations. They are based on long read sequencing, which can be used for the detection of many genetic variations that cannot be identified by short sequence reads. They provide a platform for evaluating and discovering the evolution and functional genomics of legumes.

9.6 *Epigenomics*

Epigenomics refers to epigenetic changes at the genomic level, such as the methylations of small RNAs or DNAs and histone modifications (Tettelin et al. 2005). Epigenomics in legumes contributes to growth at the cellular level and the development of environmental stress responses by revealing genetic regulations, thus improving the adaptability and productivity of legume crops (Windels et al. 2021).

Over the past few decades, knowledge related to approaches to epigenomics has been rapidly increasing. When combined, genomics and epigenetics are termed *epigenomics*. In the 1940s, Conard Waddington was the first to introduce the term *epigenetics*, which deals with heritable changes without altering DNA sequences or genetic code, leading to gene expression modification and changes in phenotype. It does not change DNA sequences but rather leads to changes through two mechanisms, namely histone modification and DNA methylation. These changes generate stress-responsive genes, which increase tolerance in response to stress. Epigenomics, unlike genomics, is influenced by different biotic and abiotic stresses. Epigenetic events can be analyzed at several stages of legume development to assess abnormalities related to plant diseases and stress conditions. Legumes are exposed different abiotic stresses, so understanding legume responses to such stresses helps in developing resistant crops, which in turn helps to fulfill increasing worldwide legume demand (Deans and Maggert 2015; Muthamilarasan et al. 2019; Chang et al. 2020; Yung et al. 2021).

9.6.1 DNA methylation

Cytosine DNA methylation is the most-used modification in epigenetics. It affects DNA by placing the S-adenosyl-L-methionine methyl group into the carbon number 5 position on the cytosine ring, resulting in 5mC-5-methylcytosine. In plants such as legumes, the methylation of DNA passes from one generation to the next generation and acts as an epigenetic marker in memory (Gupta and Garg 2020). DNA methylation can be identified through bisulfite sequencing (Cokus et al. 2014). It empowers in retaining genome integrity, gene expression regulation and suppresses transposable elements.

DNA methylation plays a significant role in legume responses to a variety of abiotic stresses, such as cold, heat, drought, and salt stresses (Akhter et al. 2021). Drought stress is a serious problem in agriculture worldwide and induces different morphophysiological and biochemical changes, along with epigenetic modifications—i.e., DNA methylation. Under distinct conditions in the genomic region, DNA methylation refers to regions of differential methylation (Anjum et al. 2011; Trenberth et al. 2014; Kim et al. 2015; Wu et al. 2015). Salt stresses alter the process of DNA methylation and regulate tolerance to salt in various legume species (Karan et al. 2012). Methylation-sensitive amplified polymorphism (MSAP) has been used under salt stress for DNA methylation quantification (Zhong et al. 2009; Pandey et al. 2016). In general, plants can tolerate temperatures below optimal levels. A 10–15 °C increase in temperature is considered as heat stress. The determination of heat-stress tolerance in general is acquired by thermotolerance (acquired and basal) and regulated by DNA methylation. Temperature stress causes changes in the level of DNA methylation (Lim et al. 2013; Clarke et al. 2004). Cold stress from chilling temperatures (0–15 °C) and freezing temperatures (<0 °C) affects legume productivity. To acclimate against cold stress, legumes have developed some sophisticated mechanisms, including epigenetic regulations (Ding et al. 2020; Shi et al. 2018; Garg et al. 2014; Rakei et al. 2016).

9.7 Transcriptomics

Transcriptomics is the study of transcriptomes, which deal with the production of RNA transcripts (Raza et al. 2021). Transcriptomics is a dynamic technique for analyzing the expression of genes in response to or against any stimuli during a particular period. Such an approach helps researchers to assess the transcriptional activities of genes at distinct stages of legume development and legume reproduction. It also defines the characteristics and functions of genes in legumes (Yang and Kim 2015). This strategy uses microarray analysis, digital gene or RNA profiling, next-generation genotyping, serial analyses of gene expressions (SAGEs), and RNA-Seq for sustainable agriculture and breeding programs (Morozova et al. 2009; Kawahara et al. 2012).

Transcriptomics is becoming a foundation of many plant studies thanks to the development of sequencing technology. RNA sequencing, also known as transcriptome analysis, and NGS are used for the balanced, high-output detection of all expressed transcripts. It provides new perceptions into molecular profiles and signaling pathways at a systemic biological level and can also recognize helpful gene markers for the efficient breeding of legumes (Garg and Jain 2013). RNA-sequencing data cover the overall model of the metabolic activities of storage compounds during seed development in legumes. For instance, it has enabled the gene network modeling of seed lipids and protein deposition in soybeans. Applying transcriptomics to the different tissues and developmental stages of legumes could construct RNA-sequencing graphics of clustered gene expression profiles on the basis of hierarchy, highly expressed genes, and legume-specific genes. Transcriptome sequencing has revealed that gene-encoding storage proteins, lipids, and starch enzymes are highly expressed at the embryonic stage, suggesting their immediate deposit into seeds before desiccation. Some seeds in the dry stages have expressed other genes marked with water-shortage-related hydrophilic proteins. These proteins facilitate nutrient preservation and cellular structure in legumes during seed desiccation. Many transcription factors have been identified as seed development regulators, such as APETALA2 (AP2), VIVIPAROUS1/ABI3-LIKE (VAL), fertilization independent endosperm (FIE), GLABRA2 (GL2), PICKLE (PKL), and DNA binding with one finger (DOF4) factors. PKL is an embryonic development activator, while FIE inhibits premature endosperm development. The interactions between transcription factors are quite complex, especially during the embryonic development and germination of legumes, because they involve several factors at every stage (Afzal et al. 2020).

Transcriptomic analysis provides a deep understanding of molecular and genetic responses that trigger adaptation to environmental stresses in legumes. The data can be exceptionally valuable for determining differences among the gene expression of stress-tolerant and stress-sensitive genotypes. These transcriptomic data can be used to facilitate the development of the stress-tolerant genotypes essential for legume breeding. Comparative transcriptome analyses on legumes have uncovered different stresses, which have led to the identification of functional regulator genes acting individually or mutually in stresses. These can be used in stress-tolerant breeding. Protein kinase, phosphatases, and many transcriptomic factors, such as ethylene response factors, myeloblastoses, the absence of apical meristems, and WRKY families, are included in these genes. Transcriptomic studies allow for the comparative genomic analysis of wild and cultivated crop relatives, the identification of target genes crucial for breed improvement, etc. (Perez-de-Castro et al. 2012). For instance, RNA sequencing has been used to study the expression profiles of legumes under alkaline stress, which provide insights into the functions of genes responsive against alkaline stress. Transcriptomic research is a high-output technology used to study all the transcript levels of multiple genes with biological functions from a large data set. It has become the most common approach for solving biological problems thanks to the discovery of new transcripts, detailed transcripts, graphs, and precise metabolic pathways for identification. Because it has large-scale data-processing capabilities, it reduces the resequencing cost. Current analysis methods and basic assumptions need re-evaluation and adjustment to cope with large numbers of omic data owing to the continuous development of sequencing technology. Specific biological applications, such as plant transformation, can be obtained from cell- and tissue-specific transcriptomic techniques. Single-cell transcriptomics provides accurate spatial and temporal data on biological responses. The integration of transcriptomics into genomic techniques and other omic techniques could optimize legumes' regulatory network and refine core gene aspirants for breeding (Verdier et al. 2008).

9.8 Proteomics

Proteomics is a type of analysis focused on proteomes, which deal with profiling protein expressions. In the science of legumes, proteomic analysis has been deemed a new hope. It enables the interpretation of specific proteins adaptive to biotic and abiotic stress conditions (Rathi et al. 2016). Proteomics can be further divided into four categories: sequence proteomics, expression proteomics, structural proteomics, and functional proteomics (Mosa et al. 2017; Aizat and Hassan 2018).

Proteins govern biological activities and are the executors of different biological processes, including physiological functions. Proteomics is the study of the expressions and functions of proteins. Proteomics has developed very fast since its inception, rapidly increasing the amount of proteomic information and the number of plant genome resources and EST sequence libraries available. It helps in the identification of proteins and provides basic information about protein expression in legumes. It can be widely used in research on different topics of legume biology, such as growth, development, stress, root–nodule interactions, and changes in protein expression during the life cycle of legumes (Rose et al. 2004). Before proteomics, the study of proteins relied solely on gel electrophoresis, but advances in liquid chromatography mass spectrometry (LC-MS) have increased output and made proteomics possible while improving the efficiency of proteome research. High-resolution proteome maps of legumes can be constructed through proteomics; however, 2D gel electrophoresis integration and multidimensional protein identification technology, such as LC-MS, have improved the metabolic profiling of processes that occur during seed filling in legumes (Ramalingam et al. 2015). Tissue-specific proteins can be detected from the proteome profiling of the leaves, roots, and hypocotyls of young legume seedlings. Proteomics is being widely used for analyzing enzyme expression and enzyme regulatory functions that accumulate in seeds during storage. Isobaric tags of relative and absolute quantitation (iTRAQ) have been analyzed to detect the proteins related to root hair and seed development. For instance, the seed storage and seed oil proteins detected by using this method are believed to be the main seed reserves in soybeans (Komatsu and Ahsan 2009).

Proteomics can be used to analyze the differences in protein expression profiles and oil synthesis between the cultivars and parental variety of legumes. It can be widely used in stress biology for the identification of key proteins. For example, GmDGAT1–2 transgenic soybeans that have high oil contents were analyzed through quantitative proteomics and lipidomics. The overexpression of GmDGAT1–2 stimulates the downregulation of lipoxygenase and the upregulation of oleosin and significantly changes the composition of fatty acids. Flood- and drought-response mechanisms in legumes can also be analyzed by proteomics (Hossain et al. 2013). Stress-sensitive tissues can be identified in legumes at different developmental stages on the basis of protein profile. These stress responses could be documented in a tissue-specific manner. Quantitative proteomics has revealed the agitation in the proteins of legumes that are related to the metabolic pathways of fatty acids, the abundance of proteins associated with initiation, and elongation and desaturation processes. A study combining proteomics and physiological data would help in identifying the effects of environmental factors such as temperature stress and humidity stress on embryos, cotyledons, leaves, and pods.

The scope of proteome research on legumes remains quite narrow compared with that of other crops. On the other hand, it provides a starting point for the functional genomics studies of biotic and abiotic stresses and the natural product biosynthesis mechanism of legumes.

The information acquired from the proteomic data can help in the identification of novel proteins, the determination of protein expression patterns for corresponding genes, and the improvement of molecular cloning. Proteome reference maps of legumes can advance research on legumes. When combined with genomics and transcriptomics, proteomics could facilitate the screening of elite alleles responsible for the end development of different molecular markers while advancing the molecular breeding of legumes (Cao et al. 2022). Single-cell transcriptomic studies on plants are trending thanks to recent advancements in protoplasts and sequencing technologies, but single-cell proteomic studies need more development to achieve equivalent output. Single-cell proteomics has naturally lowered the volume of samples, which challenges the conventional sample preparation methods and LC-MS sensitivity, which is the major difficulty of adopting this technique. Nevertheless, the protein data more heavily suggest closely controlling many cellular processes than the transcriptome data do, and the protein data are highly efficient thanks to an output technique for analyzing factors such as protein expression, interaction, and modification, which are essential to understanding the molecular mechanisms that underly plant phenotypes.

9.9 Metabolomics

Metabolomics is defined as the qualitative and quantitative analysis of small metabolites with relative molecular weights less than 1000 in a tissue or cell. Metabolomics is an important approach in system biology, and its development will have future research implications for legumes. These analyses can reveal metabolic signaling pathways, gene-discovery resources, metabolic engineering, and explanations of regulatory pathways (Hall 2011). Quantitative metabolomic techniques, such as liquid chromatography electrochemistry mass spectrometry (LC-EC-MS), thin-layer chromatography (TLC), gas–liquid chromatography mass spectrometry (GC/LC-MS), NMR, Fourier transfer infrared spectroscopy (FTIR), capillary electrophoresis LC-MS, and direct infusion mass spectrometry (DIMS), are used for plant metabolite detection. The most commonly used plant metabolomic techniques are LC-MS, GCMS, NMR, and capillary electrophoresis LC-MS. The results from plant metabolomics are directly related to plant phenotypes, unlike those of transcriptomics and proteomics. A dual effect of manipulating and controlling gene transcription and protein expression has been observed in metabolites. Metabolomics can reveal delicate changes in regulatory mechanisms. One metabolomic study included plant sample collection, metabolite isolation, and metabolite detection (type, content, and assay) techniques to build metabolic fingerprints. Metabolomics can be used to gain a comprehensive understanding of plant metabolic processes, and changes can be analyzed by combining bioinformatics with integrated metabolic pathways (Patel et al. 2021).

The seed composition of legumes can be modified after their metabolic regulations have been characterized. The metabolites involved in the key pathways of seed development can be studied. Variations in metabolites, metabolite–metabolite correlations, and bioactive compound profiles for different cultivars can be observed. The effect of seed dry weight, seed coat, seed color, maturity from the abundance of metabolites, the responses of several metabolites on seed maturation, and nutrition enhancement in legumes can also be observed. The use of metabolomics could be used to detect the effects of abiotic stress on legume metabolites. The exposure of some legumes to abiotic stress enhances the biosynthesis of secondary metabolites. The levels of polyphenols, alkaloids, phytochemicals, and phenylpropanoids increase in response to drought stress. Metabolomic research can be used to discover the metabolomic markers required for crop improvement (Borisjuk et al. 2003).

In the past decade, metabolomics has provided new perspectives on clarifying plant metabolic pathways and genetic buildup. Metabolomics has determined the isoflavone profile in soybeans and the metabolites related to seed dry weight, maturity, and differentiating disease-resistant varieties from disease-susceptible varieties under stress. Because the plants have complex metabolic pathways and broad mechanisms of product synthesis, there are some early-stage gaps between metabolomic findings on and practical applications for legume breeding. No single metabolomic analysis alone can cover all the aspects. The methods used in metabolomics also differ according to the research purpose and objectives, and their techniques have different requirements for processing samples. It also fails to assimilate well with other approaches, narrowing its area of research. Obtaining comprehensive information on plant metabolites is still difficult. Future research on metabolomics should aim to formulate strategies to solve these methodological issues (Lin et al. 2014).

9.10 Phenomics

Given the rapid development in sequencing technology and the absolute number of plant materials to be tested, gathering phenotypic data will be crucial for plant breeding in the future. The plant phenotypes are rather complex and dynamic in that they are easily affected by environmental factors. Manually investigated plant phenotypes are less efficient and come with many errors. Plant phenomics is receiving meaningful attention as an approach to solve the plant phenotype problem (Yang et al. 2020). Many articles have been published on high-throughput phenotyping in legumes, mostly over the past decade. The breeders are becoming aware of how to accelerate legume breeding thanks to phenotypic studies. One defined growth method is followed for plant growth and development, but the harvested plants different in their cotyledon size, stress resistance, and metabolic functions. Minute differences in development can cause striking changes in the physical and anatomical characteristics of plants. Plant phenomics controls high-throughput, high-resolution phenotyping techniques and platforms to attain phenotype data before and after plant production. Phenomics is characterized by a large number of trait data and an ability to divide a trait into many smaller units of traits for testing purposes (Falk et al. 2020).

For instance, the accurate legume yield estimation and classification of efficient mature pods by reconstructing time-course multispectral high-throughput image data could be achieved by using professional uncrewed aerial vehicles (UAVs), each equipped with a high-definition dual camera and multispectral equipment. The estimation accuracy of legumes' physiological, biochemical (like chlorophyll content and nitrogen concentration), and biophysical parameters, such as leaf-area index and biomass (aboveground fresh and dry), has been improved thanks to the fusion of high-spatial resolution (e.g., a red, green, and blue (RGB) image) with multispectral and thermal data from UAV systems. The color and texture characteristics of early-season RGB images of canopies have been used for the prediction of legume yield, maturity, and seed size. The high-precision sampling of canopy traits such as height, leaf wilting, area, and temperature are being estimated by the increasing use of professional UAVs with expensive hyperspectral, multispectral, and thermal-imaging equipment. Two-dimensional (2D) images cannot help identify higher-dimensional phenotypic traits, and in order to estimate certain morphological traits, calibration is still needed. For this purpose, researchers have used three-dimensional (3D) reconstructions of plant morphologies from 2D image sequences by using the fully open-source structure-from-motion (SfM) and multiview stereo (MVS) methods. For instance, high-density 3D painted clouds from plant image data can be established to obtain plant height and growth phenotype data. In 2020, 3D reconstruction technology was used by researchers to analyze soybean growth patterns and soybean phenotypic fingerprints throughout the growth period. Moreover, 3D reconstructions are cost effective, and this method can replace many expensive laser scanners and can automate certain procedures (Zhu et al. 2020).

Plant phenomic research has entered a stage of rapid growth with the development of robotics, remote sensing, visualization, and artificial intelligence (AI). Unparalleled challenges are present, however, owing to the large number of data and pictures. To solve this problem, artificial intelligence and machine learning (e.g., algorithms) make predictions and learn from data without using any obvious programming techniques. A new field within machine learning has emerged since 2006, termed *deep learning*. It contains various approaches, such as convolutional neural networks (CNNs), recurrent neural networks (RNNs), and multilayer perceptrons (MLPs). CNNs are widely used for plant phenotyping and commonly include the following deep-learning frameworks: TensorFlow, Caffe, and PyTorch. Advancements in cloud computing and graphic processing unit (GPU) parallel computing have supported the development of deep-learning programs. Today, modern, multisite, high-frequency, and standardized phenotypes and standardized storage for phenomics data are still needed. Nevertheless, a large number of image-based tasks have come with remarkable results, which have been obtained from low-cost sensors and deep learning. These methods enable the efficient, accurate, and timely predictions of the legume phenotypes in different regions and at different scales. They also reveal the regional variations in and evolution of the phenotypical traits of legumes and assist in their breeding and cultivation (Jiang and Li 2020).

In spite of the current advancements in plant phenomics, the identification of plant phenotypes has been limited, owing to descriptions of external physical traits and a failure to address their internal characteristics; therefore, they are not applicable in practical breeding.

Additionally, other issues still need to be researched. Phenotyping apparatuses are expensive and require personnel with specific biological and technical backgrounds to follow the standardized process. Some equipment operates under specific environment and weathers conditions. Thus, the data selection can be subject to huge deviations because of these weather conditions. AI techniques and computed tomography (CT) imaging have been used in legumes breeding. It will be beneficial to breeders to filter useful data out of massive data sets and integrate them with other biological data to perform deep data mining for the selection of new varieties (Burridge et al. 2020).

9.11 Ionomics

All the minerals, nutrients, and trace elements in an organism are referred to as the ionome (Huang and Salt 2016). The meaning of the word *metallome*has been expanded from its original meaning to now encompass physiologically important nonmetals (Ziegler et al. 2013). Additionally, the ionome contains both essential and nonessential components. This description emphasizes and encompasses a number of important ideas in the study of the ionome. First of all, the study of the ionome is based on the idea that it should give an overview of the functional status of a complex biological organism. This information is stored in the quantitative and qualitative patterns of macro- and micronutrients in the different tissues and cells of the organism. Such an approach is based primarily on the early research of Salt et al. (2008), who came up with the idea that a quantitative metabolite composition may be indicative of a certain physiological or pathological state. First, the accurate and simultaneous assessment of as many ionome components as feasible is required to fully extract this information from the ionome. Second, the value of ionomics comes from its capacity to precisely record data on an organism's functional status under various circumstances. These problems might be caused by genetic and developmental variations or by biotic and/or abiotic influences.

9.12 Microbiome

Plants have a diverse microbiome that aids in their development, defense, and absorption of nutrients. In legumes, nitrogen-fixing microbial associations are productive and well understood, but they are few in cereals, including maize. Plants provide significant contributions to the soil's organic carbon pool in the form of lysed cells, mucilage, and root exudates (Jones et al. 2009). A complex combination of simple and complex sugars, organic and amino acids, fatty acids, and vitamins can be found in root exudates. The amount and forms of exudates might vary depending on the plant genotype, growth stage, soil texture, soil nutrients, water-holding capacity of soil, and—most critically—the rhizosphere's microbial populations. In addition, the rhizosphere's root exudate composition could affect the soil's microbial population and the availability of mineral elements, particularly nitrogen and phosphorus (Turner et al. 2013).

Additionally, the composition of the root exudate acts as a complex for the attraction of certain bacterial and eukaryotic populations (Turner et al. 2013). More notably, some microbiotas develop by secreting certain carbon sources. For example, dicarboxylates in tomato root exudates promote the establishment of biocontrol strains of *Pseudomonas*, whereas pea plants exude homoserine to favor the growth of *Rhizobium leguminosarum* (Baetz and Martinoia 2014). Plants may protect themselves by secreting phytochemicals that can stop the development of certain microorganisms. The capacity to withstand these toxins might be a key factor in the colonization of the plant. For instance, the primary antibacterial benzoxazinoid generated by maize (a nonlegume plant) is both tolerable and attractive to the plant growth-promoting rhizobacteria (PGPR) *Pseudomonas putida*. However, compared with wild-type plants, transgenic plants that express the genes for opine biosynthesis have been demonstrated to remodel existing rhizosphere populations to enhance the populations of opine-catabolizing microbes.

Legume nodule symbiotic nitrogen fixers contribute a considerable quantity of nitrogen to both wild and cultivated environments (Brewin 2004). Perennial forage legumes can fix as much nitrogen as traditional farming methods can through the application of nitrogen fertilizers. In addition, the rhizosphere receives a significant amount of nitrogen and other vital plant nutrients through rhizodeposition from legumes. Thanks to the direct impact of British National Formulary (BNF) and the indirect impact of rhizodeposition, there was a reported 28% increase in nitrogen availability in temperate woods (Lai et al. 2022) In North America, certain leguminous plants have primarily been grown for livestock and poultry feed, including dry beans (*Phaseolus vulgaris* L.), fava beans (*Vicia faba* subsp. *minor* L.), and peas (*Pisum sativum* L.) (Ashworth et al. 2015).

Most of the common legumes for human consumption include dried beans (*Phaseolus vulgaris* L.), cowpeas (*Vigna unguiculata* L.), chickpeas (*Cicer arietinum* L.), lentils (*Lens esculenta* L.), peanuts (*Arachis hypogea* L.), and pigeon peas (*Cajanus cajan* L.). All these legumes can fix nitrogen, and intercropping or crop rotation are common reasons for growing them. The growth, development, and maturation phases of soybeans have been significantly influenced by nitrogen fixation via the symbiotic interaction of the root system with rhizobacteria. It is possible to transform an increase in plant components, such as soybean pods, into an increase in nitrogen-fixation capability. The levels of nitrogen derived from fixation (Ndfs), soil, and 15 N-labeled fertilizer (Ndff) in field-grown soybeans (*Glycine max* cv. *chippewa*) were 47%, 50%, and 3%, respectively, upon physiological maturity.

In comparison to soil nitrogen (43%), 55% more nitrogen was contributed to soybean pods and seeds through nitrogen fixation (Thilakarathna et al. 2016). Forage legumes might produce larger yields and vital nitrogen for the soil so that they can be cultivated in a variety of climatic conditions. The majority of the hot and dry regions of Earth's grasslands are covered by the four primary forage legumes: alfalfas (*Medicago sativa* L.), red clovers (*Trifolium pratense* L.), subterranean clovers (*T. subterraneum* L.), and white clovers (*T. repens* L.). The cool-season perennial forage legume alfalfa (*Medicago saliva* L.) symbiotically associates with soil bacteria in its root nodules to receive nitrogen from the soil and the BNF. Alfalfas are becoming more and more desired as animal feed in grass pastures because of its ability to supply fixed nitrogen (Thilakarathna et al. 2016; Ledgard and Steele 1992) (Fig. 7.3).

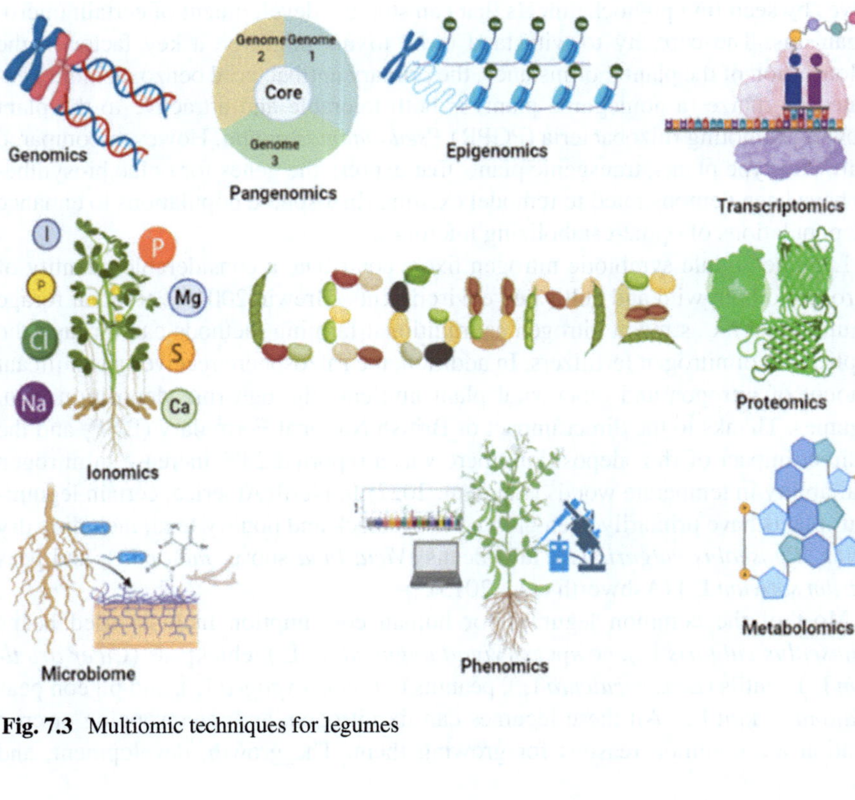

Fig. 7.3 Multiomic techniques for legumes

10 Multiomic Approaches to Enhancing the Nutritional Quality and Productivity of Legumes

To identify the genes and genomic regions prevailing in specific plant traits, a comprehensive whole-genome-sequence data analysis is crucial. A gold standard genome sequence has become significant for biological purposes: It occurs when the genes are being described with their functions and the genome is sequenced. The key to understanding the different traits of legumes at the genetic level is sequencing their genomes. However, there are many concerns related to different legumes; for example, in the case of pigeon peas, the main concern about breeding and improving its trait is that it is naturally an outcross with the help of insects, which can cause inadequacy problems in maintaining the breeding programs and a pure line. In legumes, because pollination is uncontrolled, it can result in the loss of many key traits, such as disease resistance, productivity, and seed control. There are also some unwanted traits of seeds that are not required for variety adoption and breeding programs, such as the cleistogamous flowers, which can self-pollinate.

The use of modern genomic tools can help lower these hurdles in the way legume breeding. The rapid advancement of genome-sequencing technologies has led to an explosion of data. The data can be collected from gene sequencing, transcriptomics, proteomics, and metabolomics. These sequencing tools help in the determination of cellular network variability and systemic functions. Molecular determinants can be integrated with this information. Multiomic approaches also offer opportunities to increase crop yields by enhancing crop-quality traits. These are also essential for producing biofortified crops, which reduce the risk of malnourishment. To eliminate hunger and reduce malnutrition, the qualitative and quantitative traits of legumes that underly complex genetic structural designs need to be identified.

The quantitative trait locus (QTLs) of many legumes have been mapped over the past several years. For instance, in pigeon peas, the QTLs related to plant type, earliness (including days of flowering), growth habit, and disease resistance have been developed in recent years, but studies on seed quality and nutritional and nutraceutical traits have not yet been conducted. The development of legumes that have high-quality nutritive traits and enhanced yields is urgently needed. To enhance breeding, trait-improvement programs, and the functional validity of legumes, different gene-sequencing techniques—including marker-assisted selection (MAS), genome-wide association studies (GWASs), phenome-wide association studies (PheWASs), genome-phenome-wide association studies (GPWASs), transcriptome-wide association studies (TWASs), proteome-wide association studies (PWASs), and metabolite-based genome-wide association studies (mGWASs)—can be employed. The resequencing of legumes will elevate researchers' understanding of the current level of genetic diversity, population structure, and phenotypically associated genes and will improve genetic diversity in order to develop superior breeds.

Currently, seed protein content (SPC) and its connection with four agronomic traits, including seed weight, seed yield, seed growth habits, and the number of days to the first flowering, have been analyzed by conducting a genotyping-by-sequencing (GBS) analysis. This analysis was carried out on five populations of pigeon peas, and its data were recorded. In total, 192 main QTLs (M-QTLs) and 573 epistatic QTLs (E-QTLs) were found in the four traits. Among the main QTLs, fourteen were associated with seed protein content, sixteen with seed weight, seventeen with seed yield, nineteen with seed growth habits, and twenty-four with the number of days to the first flowering, all of which had major effects (PVE $\geq$ 10%). The construction of comprehensive high-density genetic maps is necessary to map the genomic regions associated with these traits. Another important achievement of these multiomic techniques is that they not only acquire molecular data but also enable powerful data, mining tools, and machine-learning algorithms to predict and explain them. Deep-learning models can help in accurately evaluating the flow of information from DNA genomic sequences to molecular phenotypes.

The whole-genome-sequencing information has aided in the development of new possibilities in legume genomics. However, the development of a high-quality reference genome to promote effective gene discovery and molecular breeding is still needed. For the development of superior, nutritionally rich legumes, their corresponding genes need to be mapped and tagged and the origins of allelic differences at specific loci need to be determined (Singh et al. 2020).

10.1 Current Challenges and Future Directions

In today's increasing population growth, one of the major global challenges is the increasing demand of food—more importantly, nutritious food. Efforts have been made to meet the demands of food, which has decreased the diversity of food. The green revolution started in the 1940s, helping to increase the food production of high-yield varieties, but no steps were taken to provide appropriate micronutrients, including vitamins and minerals. Because of this, malnutrition has become a serious problem in many human societies, one that needs immediate solutions.

Malnutrition is a complex condition of inadequate and inappropriate nutrition, and it is a problem in both developed and developing countries. Micronutrient deficiencies cause chronic and clinical problems in more than two billion people (Jain et al. 2019). Rapidly increasing population growth has led to deficiencies in macro and micro, or trace, elements—i.e., macronutrients and micronutrients. The macronutrients that are essential for human health are potassium, calcium, magnesium, phosphorous, sodium, sulfur, and chlorine, and the essential micronutrients, or trace elements, are zinc, copper, boron, fluorine, cobalt, selenium, etc. The terms *micronutrient deficiency* and *malnutrition* are used interchangeably.

The current health statuses of more than half of the world's population, in most cases infants, kindergarteners, and pregnant people, meet the critical criteria of micronutrient malnutrition (MNM). Currently, Southeast Asia and sub-Saharan Africa are the regions with the most undernourished populations. The average uptake of Ca is higher in people living in developed countries compared to those in developing countries. According to a world health report (2000), these people have notable health deficiencies in vitamin A, L, Fe, and Zn, but other deficiencies, such as that in folate, have also been reported. Globally, over 30% of people are Fe deficient and two billion are iodine deficient, which makes them vulnerable to malaria and infectious diseases (Bohra et al. 2015). There are many causes of malnutrition, but the leading cause is globalization, in which people are more dependent on readymade food items that contain high amounts of sugar, salt, and fat but lower amounts of nutrients such as protein, minerals, and vitamins, which is the main cause of malnutrition. Fortification techniques can increase nutrition levels, but they are very expensive and are not affordable by people in rural areas, who are the main casualties of malnutrition. To overcome this problem, the already-existing nutrients of certain crops can be used, and legumes are first on the list (Banerjee et al. 2021).

Omic technologies comprise four disciplines: genomics, transcriptomics, proteomics, and metabolomics. Genome editing uses these approaches, specifically computational and bioinformatic tools, for data science applications such as artificial intelligence and machine learning, which have helped in studying the biological processes and crop-prediction models related to genomic selection. These approaches have been used to obtain nutrient-rich, balanced human diets, have enhanced the key component of nutrients, and have removed wasteful compounds so that people can consume more food. Biofortification is a process that increases the micronutrient content of crops such as rice, maize, and wheat by using conventional breeding and transgenic methods to solve nutrient deficiencies. Proteomics, which is the total protein complement for crop improvement and includes an array of tools to improve food quality and composition, has recently been revolutionized. There have been many advancements in the field of crop improvement, such as phenomics, machine learning, nanotechnology, and artificial intelligence. Assessing these traits for breeding plants and developing food security is now a global issue insofar as food insecurity and food waste can be reduced by enhancing the nutritional value of crops. Many initiatives have been taken to estimate and prevent food waste, but the current methodologies still pose challenges. These technologies alleviate malnutrition by reducing food waste, stabilizing the bioeconomy, and converting food waste to bioproducts (Nayak et al. 2021). Innovative technologies in metabolomics for extraction and microencapsulation can be used to enhance the functional food of plants, and epidemiological studies on omic technologies should be conducted. With the help of omic techniques, in the future, advances in food omics and nutrigenomics can be used to increase food security and explore more crops, such as fruits, vegetables, and medicinal plants. Now, multiomic technologies need to be used to clarify functional food research and enhance the nutrient components of plants. Nutrigenomics has shown the interactions among functional foods in human health, and it can suggest scientifically support, personized diets (Nayak et al. 2021).

Nutritional breeding is considered one of the main techniques for improving food security in legume crops thanks to its high protein and mineral contents. Further, more research is required to develop a metabolic system responsible for nutrient synthesis, their accumulation, and their transportation to different parts of plants. Enhancing the nutrient profiles of legumes through genetic engineering needs international approval as a multidisciplinary approach to crop improvement. Today, legumes are entering a new field of genomic-aided marker-assisted breeding and biotechnological approaches to increase their global production (Banerjee et al. 2021).

New technologies are emerging in the fields of ionomics and metabolomics—providing data on minerals, metabolite dynamics, and relevant gene networks—and are increasing access to sequencing and genotyping systems so that the novel genes in the crop gene pool or QTLs that are responsible for the production of improved mineral contents in plants can be identified (Bohra et al. 2015).

11 Conclusion

Legumes are great sources of food and nutrition because they contain large amounts of proteins, minerals, and micronutrients. Obtaining proper nutrition is a major problem for half of the world's population. Modern omic techniques, namely genomics, proteomics, metabolomics, phenomics, and ionomics, can improve legumes' genetic composition, molecular networks, and physiological and biochemical compositions. These omic approaches are involved in the nutrient enrichment of and mineral accumulation in legumes' edible parts. Through genetic modifications, molecular and metabolic pathways can advance nutrient-dense legume crops.

This chapter focused mainly on the improvement strategies that depend on variations available in gene pools. Omic approaches should be adopted because they have shown promising results. Recently, nutritional enrichment has been the primary aim of improving legume crops, especially their seed quality, quantity, genetic composition, protein expressions, and ion exchanges. We expect that omic approaches will become the most effective strategies for the nutritional enrichment of legumes and will improve food security all over the world.

References

Afzal M, Alghamdi SS, Migdadi HH, Khan MA, Mirza SB, El-Harty E (2020) Legume genomics and transcriptomics: from classic breeding to modern technologies. Saudi J Biol Sci 27(1):543–555

Aizat WM, Hassan M (2018) Proteomics in systems biology. In: Omics applications for systems biology, pp 31–49

Akhter Z, Bi ZZ, Ali K, Sun C, Fiaz S, Haider FU, Bai JP (2021) In response to abiotic stress, DNA methylation confers epigenetic changes in plants. Plan Theory 10:520

Anjum, SA, Wang LC, Farooq M, Hussain M, Xue LL, Zou CM (2011) Brassinolide application improves the drought tolerance in maize through modulation of enzymatic antioxidants and leaf gas exchange. J Agron Crop Sci 197:177–185. https://doi.org/10.1111/j.1439-037X.2010.00459

Ashworth AJ, West CP, Allen FL, Keyser PD, Weiss SA, Tyler DD. Taylor AM, Warwick KL, Beamer KP (2015) Biologically fixed nitrogen in legume intercropped systems: comparison of nitrogen-difference and nitrogen-15 enrichment techniques. Agron J 107:2419–2430

Banerjee J, Sharma A, Singh Y, Singh SP (2021) Nutritional enhancement in legumes using recent plant breeding and biotechnological. Front Crop Improv 9(2):85–90

Baetz U, Martinoia E (2014) Root exudates: the hidden part of plant defense. Trends Plant sci 19(2):90–98

Benlloch R, Berbel A, Ali L, Gohari G, Millán T, Madueño F (2015) Genetic control of inflorescence architecture in legumes. Front Plant Sci 6:543

Bohra A, Sahrawat KL, Kumar S, Joshi R, Parihar AK, Singh U, Singh NP (2015) Genetics and genomics-based interventions for nutritional enhancement of grain legume crops: status and outlook. J Appl Genet 56(2):151–161

Bohra A, Jha UC, Satheesh Naik SJ, Mehta S, Tiwari A, Maurya AK, Varshney RK (2021) Genomics: shaping legume improvement. In: Genetic enhancement in major food legumes, pp 49–89

Borisjuk L, Rolletschek H, Wobus U, Weber H (2003) Differentiation of legume cotyledons as related to metabolic gradients and assimilate transport into seeds. J Exp Bot 54(382):503–512

Brewin NJ (2004) Plant cell wall remodelling in the rhizobium-legume symbiosis. Crit Rev Plant Sci 23:293–316

Burridge JD, Rangarajan H, Lynch JP (2020) Comparative phenomics of annual grain legume root architecture. Crop Sci 60(5):2574–2593

Cao P, Zhao Y, Wu F, Xin D, Liu C, Wu X et al (2022) Multi-omics techniques for soybean molecular breeding. Int J Mol Sci 23(9):4994

Chang YN, Zhu C, Jiang J, Zhang HM, Zhu JK, Duan CG (2020) Epigenetic regulation in plant abiotic stress responses. J Integr Plant Biol 62:563–580

Clarke SM, Mur LA, Wood JE, Scott IM (2004) Salicylic acid dependent signaling promotes basal thermotolerance but is not essential for acquired thermotolerance in Arabidopsis thaliana. Plant J 38:432–447

Cokus SJ, Feng S, Zhang X, Chen Z, Merriman B, Haudenschild CD (2014) Shotgun bisulphite sequencing of the Arabidopsis genome reveals DNA methylation patterning. Nature 452:215–219

Deans C, Maggert KA (2015) What do you mean, "Epigenetic"? Genetics 19:887–896

Deshmukh R, Sonah H, Patil G, Chen W, Prince S, Mutava R, Nguyen HT (2014) Integrating omic approaches for abiotic stress tolerance in soybean. Front Plant Sci 25:244

Ding Y, Shi Y, Yang S (2020) Molecular regulation of plant responses to environmental temperatures. Mol Plant 13:544–564

Erbersdobler HF, Barth CA, Jah-reis G (2017) Legumes in human nutrition. Nutrient content and protein quality of pulses. Ernahr Umsch 64(9):134–139

Falk KG, Jubery TZ, Mirnezami SV, Parmley KA, Sarkar S, Singh A et al (2020) Computer vision and machine learning enabled soybean root phenotyping pipeline. Plant Methods 16(1):1–19

Garg R, Jain M (2013) Transcriptome analyses in legumes: a resource for functional genomics. Plant Genome 6(3):plantgenome2013-04

Garg R, Kumari R, Tiwari S, Goyal S (2014) Genomic survey, gene expression analysis and structural modeling suggest diverse roles of DNA methyltransferases in legumes. PLoS One 9:88947

Govindaraj M, Vetriventhan M, Srinivasan M (2015) Importance of genetic diversity assessment in crop plants and its recent advances: an overview of its analytical perspectives. Genetics Research International. Hindawi Publishing Corporation, pp 1–12

Gupta K, Garg R (2020) Method for bisulfite sequencing data analysis for whole-genome level DNA methylation detection in legumes. Legume Genomics:127–145

Hall RD (2011) Plant metabolomics in a nutshell: potential and future challenges. In: Biology of plant metabolomics, Annual plant reviews, vol 43, pp 1–24

Hossain Z, Khatoon A, Komatsu S (2013) Soybean proteomics for unraveling abiotic stress response mechanism. J Proteome Res 12(11):4670–4684

Hu H, Yuan Y, Bayer PE, Fernandez CT, Scheben A, Golicz AA, Edwards D (2020) Legume pangenome construction using an iterative mapping and assembly approach. In: Legume genomics, pp 35–47

Huang XY, Salt DE (2016) Plant ionomics: from elemental profiling to environmental adaptation. Mol Plant 9:787–797

Jain S, Rustagi A, Kumar D, Yusuf MA, Shekhar S, Sarin NB (2019) Meeting the challenge of developing food crops with improved nutritional quality and food safety leveraging proteomics and related omic techniques. Biotechnol Lett 41(4):471–481

Jiang Y, Li C (2020) Convolutional neural networks for image-based high-throughput plant phenotyping: a review. Plant Phenomics 2020:2020

Jones CK, Bergstrom JR, Tokach MD, DeRouchey JM, Goodband R, Nelssen JL, Dritz SS (2009) Efficacy of commercial enzymes in diets containing various concentrations and sources of dried distillers grains with soluble for nursery pigs. J Anim Sci 88(6):2084–2091

Karan R, DeLeon T, Biradar H, Subudhi PK (2012) Salt stress induced variation in DNA methylation pattern and its influence on gene expression in contrasting rice genotypes. PLoS One 7(6):40203

Kawahara Y, Oono Y, Kanamori H, Matsumoto T, Itoh T, Minami E (2012) Simultaneous RNA-Seq analysis of a mixed transcriptome of rice and blast fungus interaction. PLoS One 7(11):49

Kim JM, Sasaki T, Ueda M, Sako K, Seki M (2015) Chromatin changes in response to drought, salinity, heat, and cold stresses in plants. Front Plant Sci 6:114

Komatsu S, Ahsan N (2009) Soybean proteomics and its application to functional analysis. J Proteome 72(3):325–336

Kumar S, Pandey G (2020) Biofortification of pulses and legumes to enhance nutrition. Heliyon 6(3):e03682

Lai H, Gao F, Su H, Zheng P, Li Y, Yao H (2022) Nitrogen distribution and soil microbial community characteristics in a legume–cereal intercropping system: a review. Agronomy 12(8):1900. https://doi.org/10.3390/agronomy12081900

Ledgard SF, Steele KW (1992) Biological nitrogen fixation in mixed legume/grass pastures. Plant Soil 141(1):137–153

Lim SD, Cho HY, Park YC, Ham DJ, Lee JK, Jang CS (2013) The rice RING finger E3 ligase, OsHCI1, drives nuclear export of multiple substrate proteins and its heterogeneous overexpression enhances acquired thermotolerance. J Exp Bot 64:2899–2914

Lin H, Rao J, Shi J, Hu C, Cheng F, Wilson ZA et al (2014) Seed metabolomic study reveals significant metabolite variations and correlations among different soybean cultivars. J Integr Plant Biol 56(9):826–836

Lu M, Zhan X (2018) The crucial role of multiomic approach in cancer research and clinically relevant outcomes. EPMA J 9(1):77–102

Maphosa Y, Jideani VA (2017) The role of legumes in human nutrition. In: Functional food-improve health through adequate food, vol 1, p 13

Mishra S, Panda SK, Sahoo L (2014) Transgenic Asiatic grain legumes for salt tolerance and functional genomics. Rev Agric Sci 2:21–36

Morozova O, Hirst M, Marra MA (2009) Applications of new sequencing technologies for transcriptome analysis. Annu Rev Genomics Hum Genet 10(1):135–151

Mosa KA, Ismail A, Helmy M (2017) Omics and system biology approaches in plant stress research. In: Plant stress tolerance: an integrated omics approach, pp 21–34

Muthamilarasan M, Singh NK, Prasad M (2019) Multi-omics approaches for strategic improvement of stress tolerance in underutilized crop species: a climate change perspective. Adv Genet 103:1–38

Nayak SN, Aravind B, Malavalli SS, Sukanth BS, Poornima R, Bharati P, Puppala N (2021) Omic technologies enhance plant-based functional foods. An overview. Front Genet 12

Pandey MK, Roorkiwal M, Singh VK, Ramalingam A, Kudapa H, Thudi M et al (2016) Emerging genomic tools for legume breeding: current status and future perspectives. Front Plant Sci 7:455

Patel MK, Pandey S, Kumar M, Haque MI, Pal S, Yadav NS (2021) Plants metabolome study: emerging tools and techniques. Plan Theory 10(11):2409

Pennie WD, Woodyatt NJ, Aldridge TC, Orphanides G (2001) Application of genomics to the definition of the molecular basis for toxicity. Toxicol Lett 120(1–3):353–358

Perez-de-Castro AM, Vilanova S, Cañizares J, Pascual L, Blanca JM, Diez MJ, Picó B (2012) Application of genomic tools in plant breeding. Curr Genomics 13(3):179–195

Rakei A, Maali-Amiri R, Zeinali H, Ranjbar M (2016) DNA methylation and physio-biochemical analysis of chickpea in response to cold stress. Protoplasma 253:61–76

Ramalingam A, Kudapa H, Pazhamala LT, Weckwerth W, Varshney RK (2015) Proteomics and metabolomics: two emerging areas for legume improvement. Front Plant Sci 6:1116

Rathi D, Gayen D, Gayali S, Chakraborty S, Chakraborty N (2016) Legume proteomics: progress, prospects, and challenges. Proteomics 16(2):310–327

Raza A, Tabassum J, Kudapa H, Varshney RK (2021) Can omics deliver temperature resilient ready-to-grow crops? Crit Rev Biotechnol 41:1–24

Rose JK, Bashir S, Giovannoni JJ, Jahn MM, Saravanan RS (2004) Tackling the plant proteome: practical approaches, hurdles and experimental tools. Plant J 39(5):715–733

Salt DE, Baxter I, Lahner B (2008) Ionomics and the study of the plant ionome. Annu Rev Plant Biol 59:709–733

Sánchez-Chino X, Jiménez-Martínez C, Dávila-Ortiz G, Álvarez-González I, Madrigal-Bujaidar E (2015) Nutrient and nonnutrient components of legumes, and its chemopreventive activity: a review. Nutr Cancer 67(3):401–410

Sato S, Isobe S, Tabata S (2010) Structural analyses of the genomes in legumes. Curr Opin Plant Biol 13(2):146–152

Shi Y, Ding Y, Yang S (2018) Molecular regulation of CBF signaling in cold acclimation. Trends Plant Sci 23:623–637

Singh N, Rai V, Singh NK (2020) Multi-omics strategies and prospects to enhance seed quality and nutritional traits in pigeonpea. Nucleus 63(3):249–256

Skolnick J, Fetrow JS, Kolinski A (2000) Structural genomics and its importance for gene function analysis. Nat Biotechnol 18(3):283–287

Tettelin H, Masignani V, Cieslewicz MJ, Donati C, Medini D, Ward NL, Fraser CM (2005) Genome analysis of multiple pathogenic isolates of Streptococcus agalactiae: implications for the microbial "pan-genome". Proc Natl Acad Sci 102(39):13950–13955

Thilakarathna MS, McElroy MS, Chapagain T, Papadopoulos YA, Raizada MN (2016) Belowground nitrogen transfer from legumes to non-legumes under managed herbaceous cropping systems. A review. Agron Sustain Dev 36:58

Trenberth KE, Dai A, Van Der Schrier G, Jones PD, Barichivich J, Briffa KR, Sheffield J (2014) Global warming and changes in drought. Nat Clim Chang 4(1):17–22

Turner TR, James EK, Poole PS (2013) The plant microbiome. Genome Biol 14:209. https://doi.org/10.1186/gb-2013-14-6-209

Verdier J, Kakar K, Gallardo K, Le Signor C, Aubert G, Schlereth A et al (2008) Gene expression profiling of M. truncatula transcription factors identifies putative regulators of grain legume seed filling. Plant Mol Biol 67(6):567–580

Wang TL, Domoney C, Hedley CL, Casey R, Grusak MA (2003) Can we improve the nutritional quality of legume seeds? Plant Physiol 131(3):886–891

Windels D, Dang TT, Chen Z, Verdier J (2021) Snapshot of epigenetic regulation in legumes. Legume Sci 3(3):60

Wu H, Xu TL, Feng H, Chen L, Li B, Yao B, Qin ZH, Jin P, Conneely KN (2015) Detection of differentially methylated regions from whole-genome bisulfite sequencing data without replicates. Nucleic Acids Res 43:141

Xu Y, Zhao Y, Wang X, Ma Y, Li P, Yang Z et al (2021) Incorporation of parental phenotypic data into multi-omic models improves prediction of yield-related traits in hybrid rice. Plant Biotechnol J 19(2):261–272

Yang IS, Kim S (2015) Analysis of whole transcriptome sequencing data: workflow and software. Genomics Inform 13(4):119

Yang W, Feng H, Zhang X, Zhang J, Doonan JH, Batchelor WD et al (2020) Crop phenomics and high-throughput phenotyping: past decades, current challenges, and future perspectives. Mol Plant 13(2):187–214

Yung WS, Wang Q, Huang M, Wong FL, Liu A, Ng MS, Li KP, Sze CC, Li MW, Lam HM (2021) Priminginduced alterations in histone modifications modulate transcriptional responses in soybean under salt stress. Plant J 1

Zhong L, Xu YH, Wang JB (2009) DNA-methylation changes induced by salt stress in wheat Triticum aestivum. Afr J Biotechnol 8:6201–6207

Zhou Z, Jiang Y, Wang Z, Gou Z, Lyu J, Li W, Tian Z (2015) Resequencing 302 wild and cultivated accessions identifies genes related to domestication and improvement in soybean. Nat Biotechnol 33(4):408–414

Zhu R, Sun K, Yan Z, Yan X, Yu J, Shi J et al (2020) Analysing the phenotype development of soybean plants using low-cost 3D reconstruction. Sci Rep 10(1):1–17

Ziegler G, Terauchi A, Becker A, Armstrong P, Hudson K, Baxter I (2013) Ionomic screening of field-grown soybean identifies mutants with altered seed elemental composition. Plant Genome 6:1–9

Chapter 8
How Genome Editing Can Be Helpful in the Biofortification of Legumes

Zakir Ullah, Javed Iqbal, Banzeer Ahsan Abbasi, Shumaila Ijaz, Javad Sharifi-Rad, Tabassum Yaseen, Siraj Uddin, and Tariq Mahmood

Abstract Legumes are consumed as a staple food in low-income countries to meet growing nutritional demands and are used as rich sources of protein and micronutrients. However, they are unable to deliver enough calories, because they are deficient in several essential nutrients and are rich in antinutrient compounds that prevent the absorption of micronutrients. Billions of people in the world are being malnourished as a result, leading to disease in the population. Researchers have worked predominantly to identify new and improved genetic features in legumes, such as high yield, stress tolerance, and nutritional quality, in consideration of the significant roles that they play in human nutrition and agricultural productivity. Genome-editing (GE) technologies have provided new opportunities for the development of biofortified legumes. We evaluate the experiences from the past and various strategies to demonstrate the necessity of biofortifying pulses to reduce malnutrition and hidden hunger and provide the general population with enriched diets. Grain legumes could be biofortified with micronutrients as a potential method to improve the nutritional value of food. Biofortified grain legumes have been produced using a variety of technologies, including classical techniques, genomics, transcriptomics, proteomics, and metabolomics. Moreover, NBTs are utilized to biofortify grain legumes by

---

Zakir Ullah and Javed Iqbal contributed equally with all other contributors.

Z. Ullah · S. Uddin · T. Mahmood
Department of Plant Sciences, Faculty of Biological Sciences, Quaid-i-Azam University Islamabad, Islamabad, Pakistan

J. Iqbal (✉) · T. Yaseen
Department of Botany, Bacha Khan University, Charsadda, Khyber Pakhtunkhwa, Pakistan

B. A. Abbasi
Department of Botany, Rawalpindi Women University, Rawalpindi, Pakistan

S. Ijaz
Department of Botany, Division of Science and Technology, University of Education, Lahore, Pakistan

J. Sharifi-Rad
Facultad de Medicina, Universidad del Azuay, Cuenca, Ecuador, Cuenca, Ecuador

M. A. Nadeem et al. (eds.), *Legumes Biofortification*,
https://doi.org/10.1007/978-3-031-33957-8_8

silencing, overexpressing, and transferring genes from other species. Developing nutrient-dense and antinutrient-free grain legumes holds considerable promise for transgenic, small single-stranded, 21–23 nucleotide, non-coding RNA molecules (micro-RNA), and genome-editing technologies. Using genomics and genome-editing-based methods, we highlight recent attempts to manipulate the genes/QTLs controlling biofortification and antinutrient buildup in legumes. Technologies for genome editing have opened new opportunities for breeding legumes to improve important agronomic traits. With the help of these innovative technologies and methods, we intend to accelerate the development of varieties of legume crops high in micronutrients and low in antinutritional elements, which will help in the biofortification of legumes and address issues such as malnutrition and hidden hunger.

1 Introduction

One of the most important concerns in the world is to provide a balanced diet to a fast-growing global population. The improvement of crops in terms of tolerating extreme climatic challenges, such as drought and heat stress, and can provide a balanced diet for people, but the changing environment is exerting significant pressures on agricultural output (Lewis 2005). In these conditions, pulses have become vital components in the food chain that can appeal to a sustainable source of vitamins, lipids, and proteins (Lewis 2005; Bauchet et al. 2019). Leguminous crops also play a significant part in sustainable farming by improving soil quality via symbiotic nitrogen fixation and the discharge of high-quality organic compounds into soils (Varshney et al. 2019). Even though legumes are ecologically significant and have positive health effects, antinutrients such as phytates, lectins, tannins, protease inhibitors, and calcium oxalates interfere with the absorption of essential minerals and reduce them in individuals, and these negative properties have effects on the cultivation of legumes (Afzal et al. 2020).

Biofortification refers to the process of enhancing the nutritional value of dietary legumes by using fertilizer, selective breeding, or genetic manipulation. Legumes that have been biofortified can increase production and generate more revenue, which benefits the economy and decreases both hunger and infectious diseases (Jain et al. 2017).

Currently, the main objective of biofortification is to accelerate genetic advancements in yield, stress tolerance, and nutritional quality. Over the past fifty years, pedigree- and performance-based selection have been the basic techniques for genetically improving legumes. However, enhanced bean varieties with significant agronomic features have been produced because of the widespread application of innovative genomics techniques and high-throughput phenomics, which have accelerated genetic improvements in legumes (Mousavi-Derazmahalleh et al. 2019). Though using genetic engineering implies the insertion of foreign genetic material, the objective of targeted crop improvement has changed because of the development of new molecular approaches. As part of improving pulses by using recently developed genome-editing (GE) techniques, the agricultural output must rise to meet the needs of the expanding population (Varshney et al. 2015, 2019). The goals of numerous studies on legume are to alter their genomes to fulfill global food requirements and to improve the legumes themselves, and they must meet the following requirements:

- A genome-edited legume must be efficient and successful in eliminating micronutrient deficiencies.
- Edited legumes must produce a considerable amount of food for the people and be profitable.
- Biofortification via genome editing must overcome the shortcomings and limitations of supplementation.
- The secondary metabolites involved in directly transferring the genes in elite genotypes include carbohydrates, lipids, proteins, minerals, vitamins, and other nutrients. This modern technique is also used to improve mineral shortage.
- The focus must be on farmers, and consumers must learn to accept consuming gene-edited legumes.

Over the past ten years, massive advancements have been made in the structural analysis and sequencing of legume genomes. The identification and use of molecular markers in the selection of complex traits and the production of improved cultivars of grain legume crops have been made possible by these fundamental resources (Varshney et al. 2019; Bibikova et al. 2002). The mentioned strategies can both reduce antinutrients and enhance promoters, resulting in increased bioavailability of micronutrients as well as their concentration (Li et al. 2011; Zhang et al. 2011).

This chapter focuses on current developments in genome-editing technology and how they might be useful in improving legume crops through precision breeding (Fig. 8.1). Gene editing involves making precise, targeted alterations to an organism's genome by using synthetic nucleases and cellular DNA repair pathways.

2 Concerns over Food Security Administration

In 2015, the United Nations (UN) established seventeen sustainable development goals (SDGs). A unified roadmap for peace and prosperity of people and the planet, both now and in the future, is provided by the 2030 Agenda for Sustainable Development, which was accepted by all United Nations Member States in 2015. SDG3 aims to "ensure healthy lives and promote well-being for all at all ages," according to its mission statement. "Nutrition is the foundation of good health, and without it, people cannot survive, learn, ward off illnesses, or lead fulfilling lives." A study by the European Food Safety Authority found that our daily diets fall short of the daily dietary reference values for several nutrients, including 8–11 mg of zinc, 8–18 mg of iron, and 750 mg of calcium, depending on gender (Cong et al. 2013). The UN states that having access to a sufficient amount of nutritious, safe food that meets one's dietary needs and food preferences is essential for living an

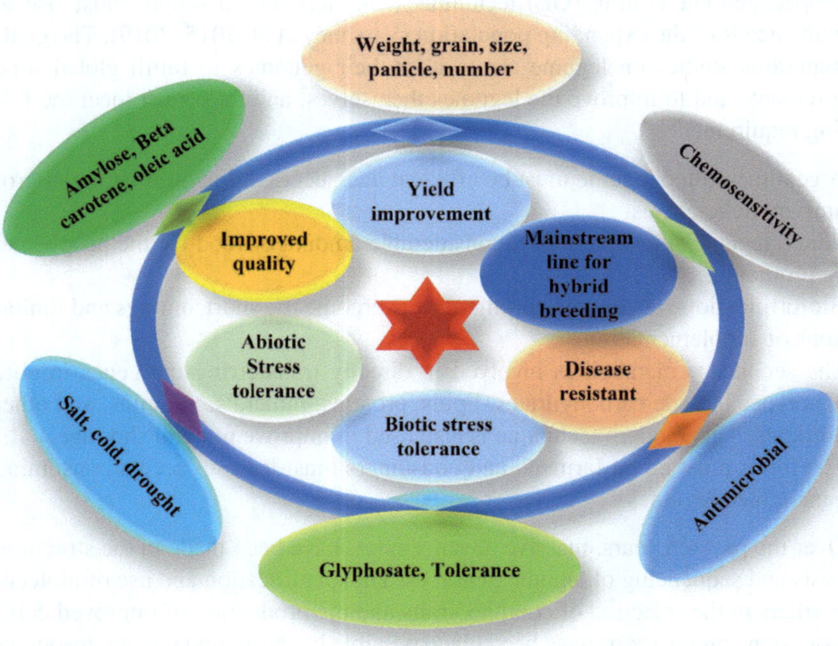

Fig. 8.1 Current developments in genome-editing expertise and how they might be useful in improving legume crops

active and healthy life. Moreover, Food and Agriculture Organization of the United Nations (FAO) research claims, 2.3 billion people worldwide experienced moderate to severe food insecurity in 2021, while 11.7% of the world's population had severe food insecurity, a rising percentage that reflects a worsening situation. In 2020, more than three billion people worldwide could not afford nutritious food.

In a quiet epidemic, micronutrient insufficiency slowly compromises the immune system, stunts physical and intellectual development, and can even result in death (Cong et al. 2013; Mali et al. 2013). With more than 2 billion people affected, micronutrient insufficiency, commonly known as hidden hunger, is exceedingly widespread (Lloyd and Carroll 2005). In many low-income countries, this shortage increases the risk of infectious illnesses and fatalities from pneumonia, measles, malaria, and diarrhea (Fig. 8.2).

3 Initiatives Taken to Combat Poor Nutrition and Food Insecurity

- The FAO claims that now is the ideal time for governments to show their support for the agricultural industry.

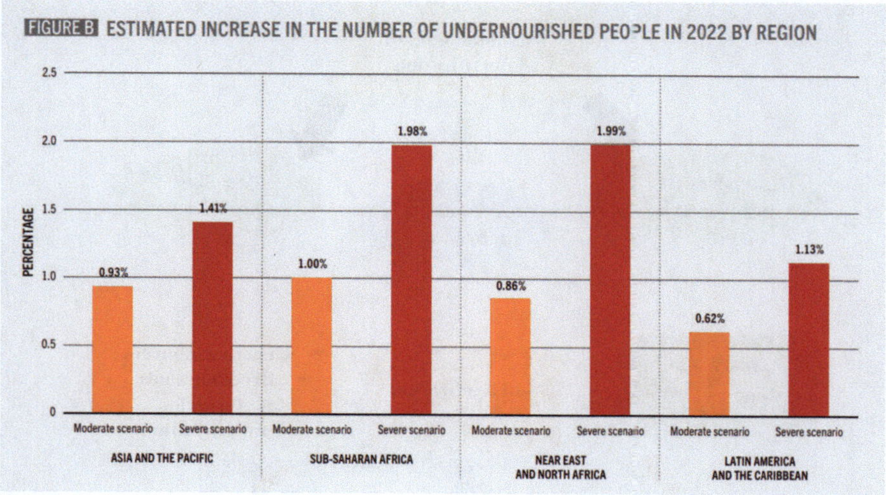

Fig. 8.2 In 2021, 2.3 billion people worldwide experienced moderate to severe malnutrition. (Source: FAO)

- From 2013 to 2018, a worldwide average of over USD 630 billion was spent on agriculture. However, a lot of this funding not only distorts the market but also does not reach many farmers, harms the environment, and does not encourage the development of wholesome foods.
- The FAO is urging the community to refocus its support to make it easier for people to purchase more reasonably priced nutritional goods.
- By 2050, the World Economic Forum predicts that 60% more food will be required to feed everyone on the planet.
- The global agriculture industry, however, is ill-equipped to meet this demand.
- To biofortify food for the global population, innovative technologies should be implemented.

3.1 *Innovative Technologies to Combat Poor Nutrition and Food Insecurity*

- The biofortification method is used for the dietary value of legumes
- Breeding, transgenic techniques, agronomic practices, and microbial approaches have been implemented to mitigate malnutrition (Fig. 8.3).

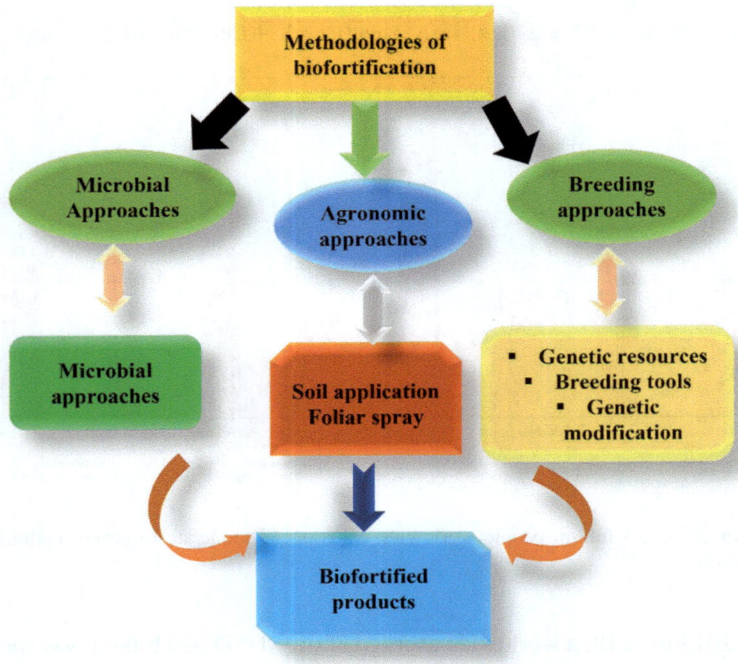

Fig. 8.3 Various approaches to legume biofortification

3.2 A Recent Approach to Legume Biofortification

3.2.1 Genome-Editing Technology

Genome-editing approaches, including clustered regularly interspaced palindromic repeat (CRISPR)/Cas9, transcription activator-like effector nucleases (TALENs), zinc-finger nucleases (ZFNs), and mega-nucleases (MNs), appear to be faster, cheaper, and more accurate than alternative technologies, such as transgenic methods and conventional practices (Adli 2018). According to a study conducted by the University of Illinois, TALENs-based gene editing (Table 8.1) is five times better than CRISPR/Cas9 in heterochromatin, a highly dense type of DNA (Malzahn et al. 2017). MNs are endodeoxyribonucleases that identify the targeted location and adopt changes. They are combined with a 12–45 bp DNA sequence. They can cause mutations and are important in the advancement of legume crops (Ashokkumar et al. 2020). ZFNs and TALENs have advantages and limitations, but CRISPR/Cas9 is more effective and dependable. Although genome editing is the main application of these approaches, they take time and require specialized expertise (Fig. 8.4).

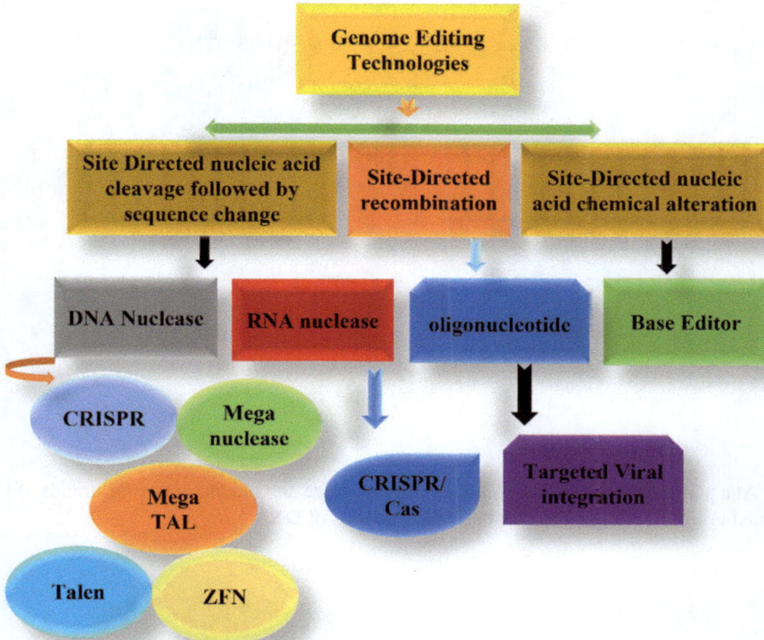

Fig. 8.4 A recent approach to legume biofortification

3.2.1.1 Transcription Activator-Like Effector Nucleases (TALENs) and Zinc-Finger Nucleases (ZFNs)

The development of nucleases (e.g., TALENs and ZFNs) that cut only the target DNA was the true breakthrough of genome editing. The earliest nucleases were an isolated 5′-GGATG-3′-recognized dimeric-type IIS restriction enzyme from Flavobacterium okeanokoites (FokI1)-based endonucleases, called ZNFs, which were based on transcription factor zinc-finger proteins. Long DNA sequences are identified by linked series of ZFNs on a highly selective target, and ZFNs recognize trinucleotide DNA sequences. The specificity of nearby ZFs is affected and the processing time is increased in an array with linked series of ZFN units. Furthermore, the final configuration is difficult to predict because the FokI dimer domain permits dsDNA synthesis only at locations where two ZFNs are attached (Fig. 8.5). To reduce the off-target effect, two ZNFs domains that concurrently detect and bind distinct but nearby nucleotide sequences located in the target have been developed (Clasen et al. 2016; Bo et al. 2019).

A bacterial transcription activator-like effector (TALE) protein and the endonuclease2 FokI combine to form TALENs. Like ZFNs, TALENs also achieve target detection and specificity through DNA–protein interactions. A single TALE unit in a TALEN identifies a single nucleotide, and several TALEs eventually bind to a

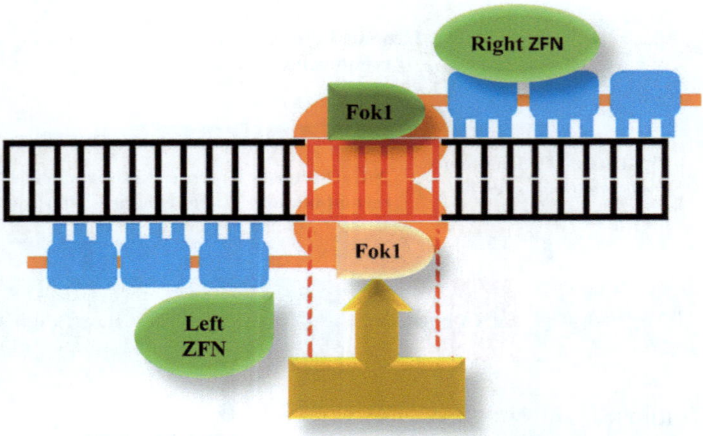

Fig. 8.5 At a particular position, two ZFN motifs recognize and bind to opposing strands of DNA, and coupled FokI dimer units preferentially bind strands of DNA

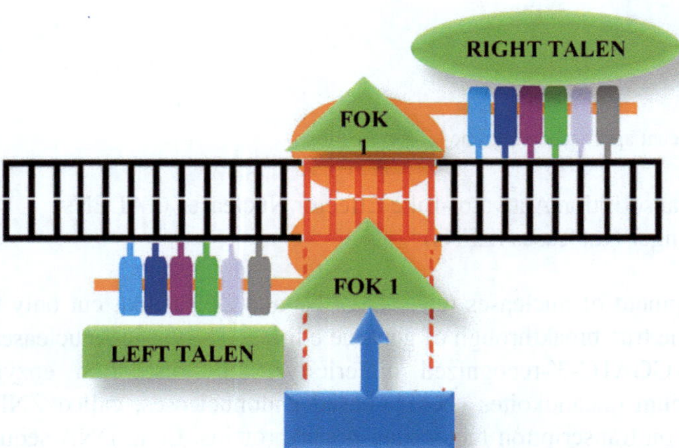

Fig. 8.6 TALENs' two motifs recognize and bind to a DNA's opposing strands at a precise location, and connected FokI dimer units specifically prevent DNA

longer sequence (Fig. 8.6). Engineering TALENs is easier than engineering ZFNs because each TALE unit functions on a single nucleotide without changing the adjacent TALE's binding specificity. Two TALENs attach in proximity at opposing strands of the target DNA and are connected by ZFNs that resemble the FokI endonuclease, which requires dimerization to cleave a target DNA (Bo et al. 2019; Bortesi and Fischer 2015).

3.2.1.2 Mechanism of Clustered Regularly Interspaced Palindromic Repeat (CRISPR)/Cas9

In 2013, CRISPR/Cas9 gene-editing technology, a third-generation gene-editing approach, was employed for the first time in legumes. Since then, it has emerged as the most innovative genome-editing method (Cantos et al. 2014). Using sophisticated computational models and biotechnological methods, miRNA fine-tuning is being used to improve crops. The three steps of the Cas9 system are as follows: the acquisition of a spacer, the processing of the CRISPR ribonucleic acid (crRNA), and the specific editing of targets. Figure 8.7 describes GE via the CRISPR/Ca9 method. The Cas9 protein and single-guided RNA (sgRNA) complex provide the basis of the initial step. To match the Cas9 protein with the gene, guided RNA must first recognize the area of DNA that is being targeted.

Cas9 starts to act as a biological scissor to cut DNA, which causes double-stranded breaks, much like the guided RNA was unwinding the DNA. The repairing of double-stranded DNA is usually initiated by two naturally occurring repairing processes: homology direct repair (HDR) and nonhomologous end joining (NHEJ). These are nucleic acids' repair mechanisms. These repair mechanisms prevent the genome's random insertion of broken template parts (Čermák et al. 2011). Microhomology-mediated end joining (MMEJ), the third repairing process, is an error-prone restoration technique that involves placing microhomologous arrangements into the damaged ends before linking them (Cheng et al. 2021; Johnson et al. 2015).

3.2.1.3 Comparisons Between CRISPR, TALEN, and ZFNs

Table 8.1 An in-depth comparison of CRISPR, ZFNs, TALENs, and MNs and other genome-editing techniques

Functions	MNs	ZFNs	TALENs	CRISPR/Cas9
Target recognition efficacy	High	High	High	Very high
Cost	High	High	High	Small/low
PAM	Absent	Absent	Absent	Present
Off-target effects	Low	Low	Low	Small
Multiplexing	Challenging	Challenging	Challenging	Easy
Editing efficacy	Low	Low	Low	High
RNA/protein	Protein based	Protein based	Protein based	RNA/protein component
Restoring pathways	MMEJ, HDR, NHEJ	MMEJ, HDR, NHEJ	MMEJ, HDR, NHEJ	MMEJ, HDR, NHEJ
Targeted DNA chain size	12–45 bp	9–18 bp	30–40 bp	20 bp
Protein type	Restriction endonuclease Fok 1	Restriction endonuclease Fok 1	Restriction endonuclease Fok 1	Cas9 protein, Cas10, Cas12

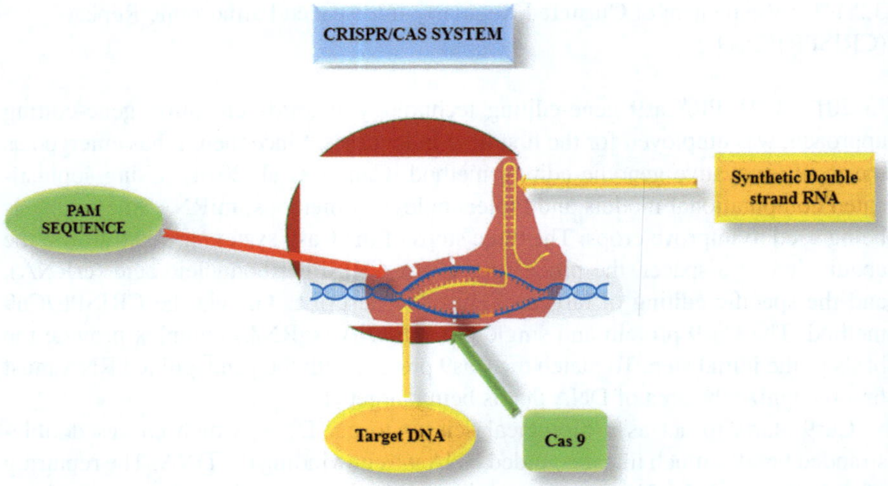

Fig. 8.7 A brief schematic of CRISPR/Cas9 operation showing how it preferentially nicks the target double-stranded DNA

4 Challenges of Legume Modifications and Solutions

A successful plant transformation requires the ability to incorporate DNA into the plant. Legumes are repellant as well as resistant to transformation (Christian et al. 2013). This is given additional complexity by the understanding that not all tissues in legume crops are capable of both changing and regenerating. Only for this reason is a reliable and effective transformation protocol needed, one that doesn't only produce a renewal mechanism but combines a vector from the beginning (Christian et al. 2010).

- Most legume crops employ agrobacterium-mediated genomic transformation
- The biolistic approach has been employed in a few instances
- Plant regeneration from seed tissues is one example that has been effective

One of the legumes that have been widely examined is the soybean, in which the genetic change is complete. The single crop for which the use of CRISPR/Cas9 has been extensively described is soybean (Clasen et al. 2016). Rooting large-seeded legumes in vitro is a significant problem as well. Method specificity and a few additional obstacles inhibit large-scale transformation in legumes. Effective genetic transformation cannot be achieved using conventional or outdated transformation approaches, but CRISPR/Cas9 can ease these worries. Some species of legumes are produced successfully for commercialization because of dependable and predictable regeneration techniques, however, inadequate in vitro rooting during renewal has hindered the success of regeneration techniques for many other legumes. Using outdated breeding methods will not be able to solve these problems (Cong et al. 2013).

Culturing genetically modified (GM) legumes has been met with resistance in many cultures. Even if there is no demand for transgenic cultivars, the progress of effective transformation procedures is significant in that it confirms the use of specific genes in legumes (Cong et al. 2013). In speeding up the genetic transformation of legumes, various approaches have gained popularity:

- Transforming sonication-assisted agrobacteria
- Boosting the rate of genetic transformation in legumes by making the explant more effective
- Increasing the affinity of the host–plant relationship
- Refining culture media additives

Investigations into the obstacles to the transformation of legumes and potential solutions still need to be conducted. New concepts will emerge as molecular science advances, shedding light on the rapidly increasing rate of legume change.

4.1 Genome Editing Legumes via CRISPR/Cas9

The CRISPR-based gene-editing approach has been developed for several model legume crops, including chickpeas, soybeans, lentils, *Medicago truncatula*, and others. Large areas of the world are used to raise legumes. In total, 306 million tonnes of soybeans were produced in 2016 (FAO 2017), 6.3 million tonnes of lentils in 2018, 11.5 million tonnes of chickpeas in 2017, and 7.4 million tonnes of cowpeas in 2016 (Haun et al. 2014). Crop quality, productivity, and tolerance to biotic and abiotic stresses were all improved by using CRISPR/Cas9. Recent improvements to legumes have been due to the calibration of CRISPR/Cas9. We have covered how genome-editing technologies have enhanced critical traits in legumes. We now focus on how genome editing via CRISPR, and other techniques, is helpful in biofortification.

4.1.1 Soybeans

One of the most significant crops of legumes is soybeans. A major aspect of soybeans is delayed flowering, which results in an increase in nodes, a change in the shape of the flowers, an increase in internode length, and an increase in plant height. In one experiment, the Cas9-based mutagenesis method was used to create mutants of GmFT2a, which led to a delay in soybean flowering. For gene transfer, *A. tumefaciens* was utilized (Martínez-Fortún et al. 2017). By crossing GmFT5a with the ft2a mutant, the soybean research team continued to mutagenize the ft2aft5a mutants. Compared with the natural type, double mutants' flowering dates were reported to be thirty-one days later, and they produced more pods and seeds when the days were short. The modified gene was transferred into soybean by using the *A. tumefaciens* strain EHA105. Therefore, by focusing on the flowering and blooming time gene, we can boost production and thereby improve food safety (Kim et al. 2021).

Four GmSPL9 genes were edited by using a sophisticated gene-editing method based on Cas9, and transformation was conducted by using *A. tumefaciens*. The mutants displayed a different phenotype. This method produced many soybean mutants with various combinations of changed loci and various pod and branch counts. The increase in pods would result in a rise in soybean yield and grain production. The CRISPR/Ca9 gene-editing technology simultaneously enhanced the critical seed-related properties of an oil, isoflavone, and viral disease resistance (Komor et al. 2016; DeFrancesco 2011). With the use of CRISPR, a sizable gene library made up of a hundred soybean genes has been generated.

Similarly, several mutants have recently been produced (Li et al. 2012). Without external growth regulators, the hairy root can grow swiftly. Hairy roots are frequently utilized as transgenic instruments for the generation of metabolites, and in investigations into gene activities in soybean plants, hairy roots are frequently utilized as transgenic instruments (Li et al. 2013). Following the deletion of GmIPK1, the gene in the hairy root of soybeans was edited by using Cas9, demonstrating the effectiveness of the Cas9 tool. A phytic acid synthase enzyme was encoded by this gene. *A. rhizogenes* was used for the transformation (Li et al. 2011).

Moreover, the Cpf1 nuclease, also known as Cas12a, is another strong and effective Cas nuclease. For transformation, the *A. rhizogenes* strain K599 was employed. The outcomes demonstrated the usefulness of this method in enhancing the four-gene cluster in soybeans. Similarly, GmHSP17.9 was effectively transformed by *A. rhizogene* K599 to produce more seeds by using the CRISPR/Cas9 tool (Li et al. 2012). The most important breeding goal is to enhance soybean output, and more seeds result in higher yields.

4.1.2 Chickpeas

Chickpeas, which are grown all over the world, are the second-most significant and valued of the legume crops (Shu et al. 2020). The genomic sequence of the Desi chickpea (*C. arietinum*) was published in 2013 (Liang et al. 2014), but sadly, attempts to create a unique genetic variation by using reliable genome-editing techniques are infrequent because there is no reliable plant-regeneration system. In addition, chickpeas now contain a transgenic plant-generating system (Liu et al. 2019). The transgenic plants of chickpeas bearing chimeric genes encoded the CryAabc protein of *Bacillus thuringiensis* to thwart pod borers. Being resistant to pod borers is essential because they are responsible for the drastic reduction in chickpea yields that is occurring globally. The EHA105 strain of *A. tumefaciens* conducted the transformation. In a comparable way, this recently developed, reliable transformation technology was used to create stable transgenic lines of chickpeas. For the transformation of the gene, strain AGL1 of *A. tumefaciens* was employed. The stress-tolerance genes AtBAG4 and TlBAG and the genes for Fe biofortification (OsNAS2 and CaNAS2) were all successfully expressed in the stable transgenic lines (Liu et al. 2020). One of the essential plant nutrients is iron (Fe). Among the ecological solutions to combat Fe insufficiency, which is decreasing chickpea output, is the Fe biofortification of chickpeas (Lor et al. 2014). The use of modern gene-editing practices (e.g., CRISPR) for the development of agricultural features in chickpeas is made possible by whole-genome sequencing, the reference genome, and the successful history of whole-plant transformation. Therefore, the most promising method to combat drought stress is to create drought-tolerant chickpeas by using CRISPR/Cas9 (Lou et al. 2017).

Moreover, many chickpea cultivars are herbicide sensitive. An essential enzyme involved in the synthesis of lignin is encoded by the gene. Chickpeas' ability to withstand drought was boosted through effective gene deletion. This work involved protoplast-mediated transformation. Plant hormones and photosynthetic pigments both depend on the carotenoid production process. The basis for future gene knockouts was established by the discovery and characterization of many genes involved in the synthesis of carotenoids in chickpeas (Lyzenga et al. 2019; Ma et al. 2015).

These recent experiments have proven that gene editing is feasible in chickpeas, providing a solid foundation for the future discovery and development of traits. We are hopeful that CRISPR/Ca9-mediated gene editing may open new avenues for gene editing in chickpeas for a number of essential characteristics.

4.1.3 Mungbeans

A large crop, the mungbean (*Vigna radiata*), is grown and eaten predominantly in Southeast Asia, including China, Pakistan, and India. Of the total area used by legumes in these nations, 85% is taken up by mungbeans (Miao et al. 2013). The mungbean is high in proteins, fiber, and vital amino acids (Ntui et al. 2020). Owing to its nonsynchronal development attitude and the lack of superior cultivars available, mungbean production is quite low (Pramanik et al. 2021). Multiple cycles of domestication and breeding have led to the successful release of a large diversity of mungbeans. Thus, crop enhancement using contemporary genome-editing techniques can also be conducted more quickly, effectively, and reliably. To improve mungbeans throughout the initial stages of crop development, various markers, including single-sequence repeats (SSR) and single-nucleotide polymorphism (SNP), have been designed. Early mungbean features include yield, nutrient content, and resistance to abiotic stress (Ramalingam et al. 2011).

Cas9 gene editing in cowpeas provides evidence of the effectiveness of using Cas9 in mungbeans. By knocking down the kinase gene, cowpeas were able to fix nitrogen in a symbiotic manner (Rashid et al. 2017). Because there has not been much CRISPR/Cas9 use in mungbeans, there is a lot of room and opportunity to employ the method to promote self-sufficiency and accomplish desired goals (Shan et al. 2013a).

For future investigations, ensuring that all information is up to date would be useful. Future gene-engineering efforts in mungbeans could start with the goal of improving quality traits and disease resistance. The global cultivation of mungbeans could be increased with the development of climate-resilient mungbean variants.

4.1.4 Lentils

Among the legumes, lentils (*Lens culinaris*) account for 6% of the total production of dry pulses. Numerous abiotic stressors, including drought, salinity, cold, and elevated temperatures, significantly limit its yield. Lentils' limited genetic diversity inhibits the production of stress-tolerant cultivars. Germplasm screening is being carried out to find plants with excellent qualities, including high yield, nutrient efficacy, water-use efficiency, disease-fighting ability, and high mineral uptake (Shan et al. 2013b). The lentil-breeding program's ongoing goal is to create cultivars with better climatic stability and desirable meal qualities. New features in lentils are mostly the results of the genetic diversity that already exists in legumes. Important developments in achieving high yield and meal quality in lentils call for the generation of novel allelic sources within legume germplasm pools.

Targeted genome editing, such as CRISPR/Cas9, can successfully increase the yield and quality of cultivars and combine conventional approaches into an approach that could help advance new steps toward the successful production of new cultivars in order to address all these questions and current issues. *A. tumefaciens*-based transformation techniques have been extensively researched (Shu et al. 2020); however, their success rate has often been less than 1% (Szczepek et al. 2007). The regeneration of explants from different plant sources, such as shoots, roots, and embryo apices, has successfully achieved genetic change. One of the finest techniques to successfully regenerate explants was to use in vitro plant regeneration, but breeders concluded that extensive review was necessary to better understand the difficulties and to boost productivity (Cermák et al. 2015). Soon, method optimization with a proper blend of inorganic media and hormones will be necessary because the number of shoots produced per explant directly impacts agrobacterium and Cas9 transformation efficiency. Random mutagenesis's innovative techniques have advanced our understanding of the genes responsible for seed size (Gaj et al. 2013). Other than these features, novel gene-editing approaches can also be used to target disease resistance, herbicide tolerance, and pod shattering.

Lentils from the current wild-type gene pool have long-lasting resistance to abiotic stressors such as cold and drought. The inability to isolate reproductive components, hybrid infertility, and imprecise genetic variety are barriers to employing these resources more extensively in lentil breeding. However, there is still much work to be carried out in the genome transformation of lentils to move genes around genomic restrictions (Kelliher et al. 2017). A new age of gene editing for lentil herbicide tolerance has started. The uses of gene editing have expanded thanks to the

Ala251Thr mutation in the psbA chloroplast gene that causes herbicide resistance (Tilman et al. 2011). A more convenient, less expensive, and more precise method of altering genes to enhance lentils can be provided by using CRISPR/Cas9 genome editing.

4.1.5 Peanuts

One of the major legume crops grown worldwide is the peanut (*Arachis hypogaea*). The top two producers of peanuts are India and China, with the United States coming in at number five (Tsai et al. 2014). Because of its polyploidy and sterile DNA, peanut breeding has always been difficult (Shukla et al. 2009). The successful application of various molecular approaches for its genetic enhancement has been made clear by a current study on the peanut genome. Future research on the novel alleles of desirable features that can be pursued by the CRISPR/Cas9 approach will result from the integration of DNA recombinant technology and operational genomics (Ainley et al. 2013). The generation of transgenic peanuts via agrobacterium-mediated transformation is more interesting and less competent than that of Arabidopsis and rice (Wada et al. 2020). In this instance, *A. rhizogenes* has been frequently sprayed to change the hairy roots of peanuts, but there is no proof that the transformed roots have produced any plants (Wang et al. 2014).

A successful and repeatable approach for producing and transforming the Cas9/gRNA complex is required. The creation and the transformation of the Cas9/gRNA complex are essential to assess the potential role of CRISPR/Cas9 in peanuts. One of the key breeding goals for peanuts is higher oil content. Oil is widely used in industry and provides benefits such as high shelf life and antioxidant properties. In one study, the gene responsible for converting oleic acid into linoleic acid, ahFAD28, was edited by using the CRISPR/Cas9 gene-editing technology. Using *A. rhizogenes* strain K599 to convert the gene, CRISPR/Cas9 was able to modify the gene to introduce the desired features. The specific mutation of this gene was carried out in the peanut protoplast and culture (Naik et al. 2022; Wang et al. 2014). CRISPR/Cas9 technology was used in a different experiment, where it was used to modify the allergen gene (Ara h 2) in peanuts. For individuals who are allergic to peanuts, a mutation in this gene increases the nutritional value of peanuts. Protoplast was successfully isolated, and the polyethylene glycol (PEG) approach was used to convert it. The results demonstrated the utility of the protoplast transformation method as a quick and precise tool for temporary expression (Zegeye et al. 2022; Wu et al. 2020) and investigated the role of the gene AhNFRI in the growth of root nodules using the CRISPR/Cas9 technology. Strain K599 of *A. rhizogenes* was employed for the transformation.

The findings suggest that by using the peanut hairy root transformation technology, CRISPR-Cas9 could be employed to specifically investigate the function of the genes in roots for functional genomic research (Wu et al. 2014). Peanut plants produce root nodules, which fixate nitrogen and promote growth and development. We can boost root nodule growth and boost peanut production. This research collectively established the immense potential of the CRISPR/Cas9 tool to precisely alter any desired gene in peanuts. Recently, Wei et al. (Ran et al. 2017) effectively used the CRISPR/Cas9 technology to delete the FAD2B gene in peanuts. The resultant mutants contained 80% or more oleic acid. For plant transformation, they used the *A. tumefaciens* strain GV3101. In the future, additional genes will be able to be modified by using the CRISPR/Cas9 method.

4.1.6 Cowpeas

The cowpea (*Vigna unguiculata*), first domesticated in Africa, is now grown all over the world. It goes by the name black-eyed pea as well. Larger amounts of fats, minerals, and carbohydrates are in them (Ran et al. 2017). According to estimates, 200 million people in Africa consume cowpea (Sun et al. 2013). The cowpea is one of the world's most significant and valuable legume crops. It contains 25% more protein than average and important amino acids such as lysine. Cowpeas are recalcitrant to genetic transformation, which limits the application of CRISPR/Cas9 in this plant. This resistance is a result of inadequate agrobacterium-based DNA transfer to the targeted cell, ineffective transgenic assortment techniques for workable transgenic plant recovery, and the lack of a flexible in vitro shoot regeneration system (Zafar et al. 2020). Iqbal et al. (2020) reported that cowpeas recently had an effective transformation ratio of 37%. To produce mutant plants, these authors modified the GmEF1A2 gene by using the CRISPR/Cas9 gene-editing system. Gene transfer was carried out by using agrobacterium strain LBA4404. The VuSPO11-1 was altered to create hybrids (Zeng et al. 2020). For this reason, a second rapid CRISPR/Cas9 technique was created in 48 hours by using leaflet invasion by agrobacterium.

By expressing the Cas9 protein by using AtPS5A, researchers have explored a variety of designs. It has been used to make mutant plants after the deletion of VuSPO11-1, and the frequency of mutation in the transient array was found to be 3.9% (Zhang et al. 2016). Future targeted gene editing in cowpeas may give rise to the extensive use of CRISPR/Cas9 gene manipulation.

Other legumes could benefit from the increased transformation efficiencies provided by the principles described here. But it's important to find a reliable technique for genetic modification. The effective application of CRISPR/Cas9 in cowpeas could be made possible by conducting more genetic investigations (Zhang et al. 2013).

4.1.7 Alfalfas

One of the most important crops in the legume family and the queen of forages is alfalfa (*Medicago sativa* L.). The CRISPR/Cas9 tool's overall effectiveness in the polyploid alfalfa genome is limited. For successful gene editing in alfalfas, an improved CRISPR/Cas9 system will be required.

Editing the MsSGR gene in alfalfas by using CRISPR/Cas9 proved effective. The findings revealed a sizable color diversity in mutants. To draw in insects and birds for effective pollination, color variation is important. The mutants had a greenish color and demonstrated that knocking off alfalfa genes by using CRISPR/Cas9 might have important long-term effects (Zhang et al. 2010).

In a similar manner, Zhang et al. (2014) used CRISPR/Cas9 to modify the NOD26 gene in alfalfas to boost their protein levels. Mutants displayed transgenic alfalfas with strong beta-glucuronidase (GUS) activity after genes were transferred by using *Agrobacterium tumefaciens*. Tetraploid alfalfas are one of the most important sources of vegetable protein used in the production of milk and meat in temperate zones. These outcomes showed prospective ways for CRISPR/Cas9 in alfalfas to be used in future research to produce high-yielding, nutrient-rich, and disease-resistant varieties.

Furthermore, the primary goals of pulse biofortification through transgenic breeding are to enrich important amino acids, to fortify them with iron and zinc, and to decrease the number of antinutrient chemicals (Fig. 8.8). The overexpression of

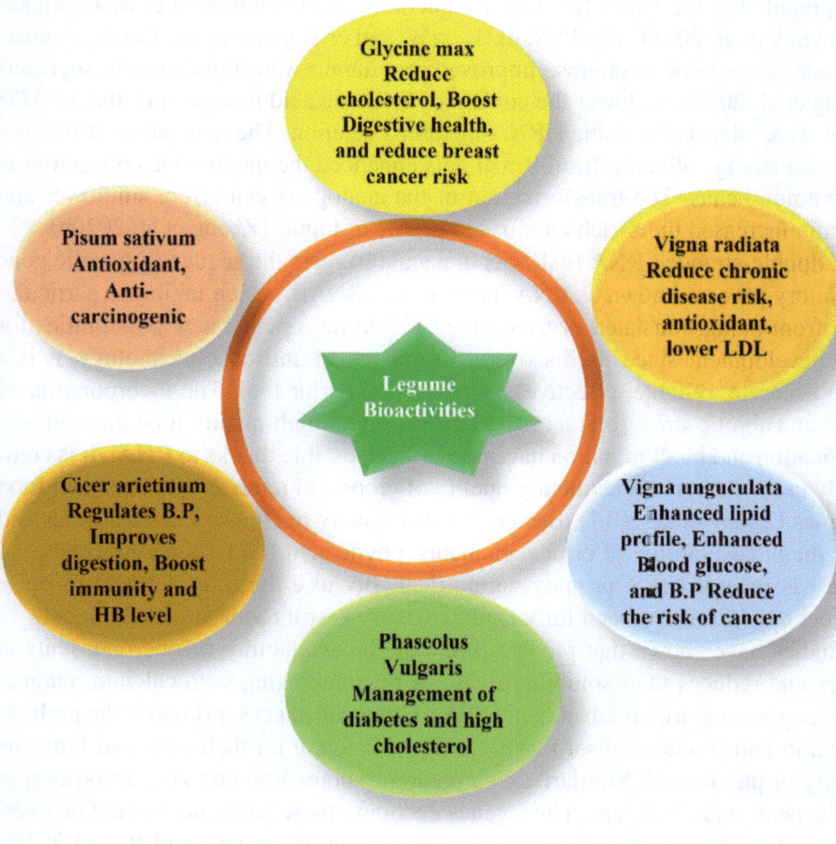

Fig. 8.8 Genome editing and its application in some important legume crops

heterologous proteins rich in these amino acids helped to overcome the lack of sulfur-rich amino acids. A cysteine-rich protein derived from maize, called 27 kDa-zein, was introduced and overexpressed in several pulse crops to improve the nutritional value of cysteine amino acids (Zhang et al. 2020a). Similarly, methionine content was increased in carbon beans and lupins by overexpressing S-rich proteins. Upregulating aspartate kinase and the 2S albumin storage protein with a seed-specific promoter boosted methionine content in Brazil nuts, and fourfold higher methionine accumulation was accomplished in Brazil nuts seeds (Zhang et al. 2019). Transformed adzuki beans accumulated more free tryptophan thanks to the rice OASA1D transgene (Zheng et al. 2021). Because chickpeas have higher iron bioavailability, the genes for glycine max ferritin and chickpea NAS2 were inserted and upregulated (Zhou et al. 2014).

Moreover, the y-synthase gene also increased the content of methionine in soybeans that overexpressed cystathionine. Increased cysteine and methionine contents were achieved by overexpressing the maize zein protein in soybeans. Increased cysteine concentration in seeds was another benefit of O-acetyl serine sulfhydrylase overexpression. The lysine content of seeds was increased in transgenic soybeans by upregulating the genes for dihydrodipicolinic acid synthase and aspartokinase (Bhowmik et al. 2018). The PSY, bkt1, crtW, and crtB genes as well as the overexpression of carotene desaturase improved provitamin A accumulation in soybeans (Wang et al. 2020). To lower the content of linolenic acid in soybeans, the 3-FAD3 genes were silenced by using siRNA-mediated deletion. The renovation of methionine-rich storage albumin from Brazil nuts enhanced the methionine concentration in common beans. The transformation of the analogous gene from sunflower and albumin increased the S-rich amino acid profile of lupins (Zhang et al. 2020b).

A double-stranded RNA (dsRNA) molecule triggers the sequence-specific gene regulatory process known as RNA interference (RNAi), which inhibits a particular gene from being translated or transcribed. RNAi has opened new possibilities for crop development since its discovery. Compared to antisense technology, it is a more accurate, reliable, effective, and overall superior tool. The incorporation of biotic and abiotic stress tolerances and the supply of high-quality food through biofortification and bioelimination have been made possible thanks to RNAi. It is commonly used to improve the dietary quality of crops and remove pollutants and food allergies (Yadav et al. 2017). Because of its capacity to chelate micronutrients and limit the bioavailability of critical elements, phytic acid (PA), found in cereals and pulses, is regarded as a primary antinutrient. Because the role of RNAi in pulse nutrient uptake has not been fully explored, there is still room for development.

Studies have shown that phytate limits the bioavailability of micronutrients in pulses and reduces their solubility qualities by complexing with calcium, magnesium, copper, and iron (Ochatt et al. 2018). Other substances, primarily the prebiotics inulin and fructans, absorb iron, zinc, and calcium on their own and limit the activity of phytic acid. Similarly, β-carotene promotes iron and zinc absorption in lentils, peas, and chickpeas. Thus, genes encoding these substances could be overexpressed to increase the bioavailability of micronutrients (Ali and Borrill 2020). Furthermore, it has been suggested that selenium increases iodine's bioavailability in lentils, peas, and chickpeas. But some inhibitors that impede bioavailability require more research. To develop nutritionally enriched pulse crops, the biochemical processes involved in the generation of antinutrients should be investigated, and genes that play a crucial role should be silenced or knocked out by using RNAi (Carvalho and Vasconcelos 2013).

Saponins are antinutrient substances that are beneficial at lower concentrations but can act as antinutrients when consumed in larger amounts (Simkin 2019). Furthermore, 13 OSC genes, 246 P450 genes, and 112 uridine diphosphate glycosyltransferases (UGTs) were found to be involved in the manufacture of saponins in an *A. thaliana*-based study. To minimize saponin production, it is necessary to identify and remove important regulatory genes in the saponin biogenesis pathway (Shoab and Hefferon 2022). Similar steps must be taken to identify and modify the genes that make other antinutrients, such as lathyrogens, protease inhibitors, and amylase, to decrease the number of these molecules (Lal et al. 2020). Increased mineral concentrations were seen in transgenic chickpea lines with the cytokinin oxidase (CaCKX6) gene expressed specifically in the roots, including Cu (26–62%), Mg (13–21%), Zn (27–62%), Fe (22–48%), K (11–27%), and P (5–19%) (Nogué et al. 2016). The rice (OsNas2) and chickpea (CaNas2) Nicotianamine Synthase (NAS) genes were identified, and both are elaborate in Fe absorption and transportation in plants. They both catalyze the manufacture of nicotinamide (NA). It has been successful at creating firm transgenic chickpea lines carrying the GUS (uidA) and Fe-biofortification genes (OsNAS2 and CaNAS2) (Koç and Karayiğit 2021).

Up to 94% more methionine was present in lupin thanks to increased levels of gene expression for sunflower seed albumin (Wu et al. 2015). It has been demonstrated that expressing the bacterial PSY (phytoene synthase) gene can increase the provitamin A, oleic acid, and seed protein content of soybeans (Goredema-Matongera et al. 2021). The overexpression of PSY and carotene desaturase in soybeans has also been observed to enhance beta-carotene levels (Kawakami and Bhullar 2018). Cysteine and methionine, two necessary amino acids that contain sulfur, are insufficient in soybean proteins. O-acetyl serine sulfhydrylase gene overexpression was used by Johns and Eyzaguirre (2007) to demonstrate enhanced cysteine content in soybeans. It has been found that overexpressing cystathionine-synthase increases the amount of methionine in soybeans (Ricroch et al. 2017). The isoflavone content of soybeans was increased by manipulating the maize C1 and R transcription factor-driven gene (Wilson 2021). CRISPR/Cas9 gene editing has unlocked a new capacity for functional genomics in legume grain crops. The ability to supply the DNA/RNA components with the regeneration of a complete plant is the fundamental goal of genetic transformation, including CRISPR/Cas9 gene editing. Most dietary legumes are resistant to absorbing and integrating foreign DNA, and many of them are also recalcitrant (Saltzman et al. 2017).

The availability of efficient plant transformation and whole-plant regeneration technologies, a supportive regulatory framework, and consumer acceptance of gene-edited crops are all required for the achievement of genome editing and for legume improvement.

5 Output and Expected Results Following Genome Editing

Both governmental and private breeders consider gene editing to be a promising breeding strategy for creating new crop varieties. Most nations experienced initial confusion over the regulation of gene editing, which has been addressed over the past three years. With the development of the technical (i.e., regeneration and transformation) potential of legumes, this new generation of genome-editing technology will significantly advance research on breeding legumes to increase production and enhance tolerance to biotic and abiotic stresses. The choices for precisely and effectively altering genes through the addition or deletion of genetic material have expanded because of recent developments in genome-editing technology (Fig. 8.9).

In this chapter, we examined several methods and materials that can be used for both the creation and delivery of genes and the identification of genetic changes. New opportunities for functional genomics and the enhancement of numerous features in legume grain crops have been made possible thanks to genome-editing technology. However, the successful application of genome editing for legume enhancement depends on the availability of efficient techniques for plant transformation and regeneration. Figure 8.9 depicts the output of legumes after biofortification.

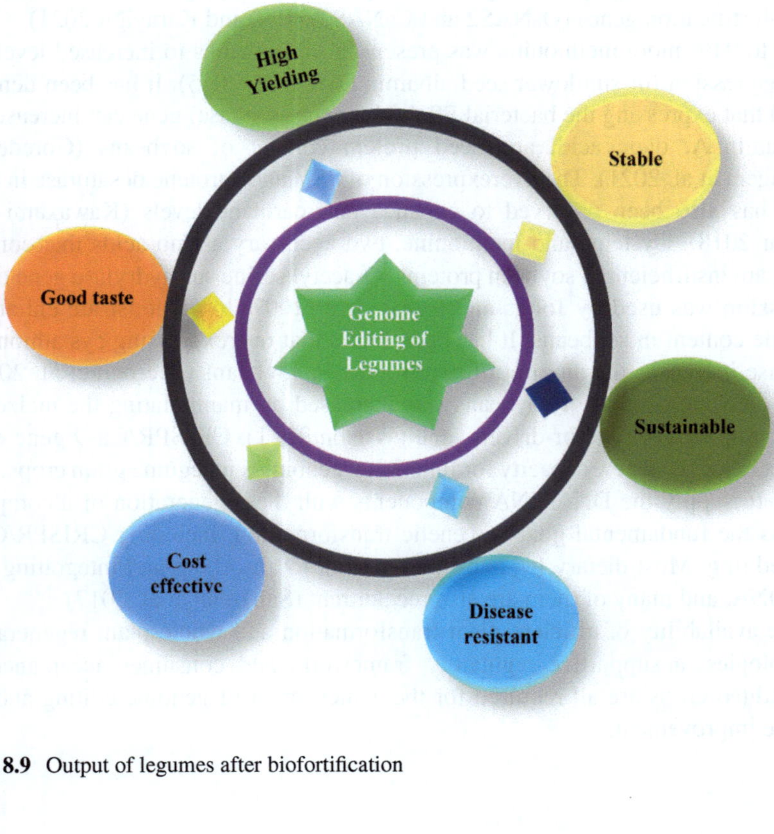

Fig. 8.9 Output of legumes after biofortification

6 Conclusion and Future Points of View

Genome editing is the most popular and adaptable tool for improving legumes. The efficient, multiplexing, integrity, simplicity, and highly precise character of the genome-editing technologies discussed here, as well as their attractive survival landscapes, point to how legume breeding is conducted and paves the way for future generations. This chapter encompasses the initiatives being carried out to improve the nutritional value of major pulses in future key regions. To raise the level of micronutrients and their bioavailability, it is possible to take advantage of the variety in Fe and Zn contents and other micronutrients in legume crops. The bioavailability and intake of legumes can be further improved by reducing antinutritional factors, such as phytate, trypsin, and chymotrypsin inhibitors. A study of the literature on transcriptomics, biotechnology, genomics, and phonemics has shown that this novel approach to improving legumes is effective. To assist the quick development of this technology and make transgenic crops suitable for consumption, the regulation of these crops has also been made systematically simpler. To achieve nutritional security, one significant goal of legume improvement programs is to find genetic constitutions with low antinutrient content in the germplasm, through genome editing. Because the wild-type relatives of many legume crops lack these antinutrients, they could be employed in prebreeding initiatives. Recent research should be recognized as opening the door for the future use of transgenic and gene-editing techniques in legumes to improve dietary quality. The widespread application of genome editing for improving legumes is already a reality. However, the process of genome editing presents ethical issues that society and researchers must address on a large scale.

References

Adli M (2018) The CRISPR tool kit for genome editing and beyond. Nat Commun 9:1911. https://doi.org/10.1038/s41467-018-04252-2

Afzal M, Alghamdi SS, Migdadi HH, Khan MA, Nurmansyah M, S. B., & El-Harty, E. (2020) Legume genomics and transcriptomics: from classic breeding to modern technologies. Saudi J Biol Sci 27:543–555

Ainley WM, Sastry-Dent L, Welter ME, Murray MG, Zeitler B, Amora R, Corbin DR, Miles RR, Arnold NL, Strange TL, Simpson MA (2013) Trait stacking via targeted genome editing. Plant Biotechnol J 11(9):1126–1134

Ali MW, Borrill P (2020) Applying genomic resources to accelerate wheat biofortification. Heredity 125(6):386–395

Ashokkumar S, Jaganathan D, Ramanathan V, Rahman H, Palaniswamy R, Kambale R, Muthurajan R (2020) Creation of novel alleles of fragrance gene OsBADH2 in rice through CRISPR/Cas9 mediated gene editing. PLoS One 15(8):e0237018

Bauchet GJ, Bett KE, Cameron CT, Campbell JD, Cannon EKS, Cannon SB, Carlson JW, Chan A, Cleary A, Close TJ, Cook DR, Cooksey AM, Coyne C, Dash S, Dickstein R, Farmer AD, Fernandez-Baca D, Hokin S, Jones ES et al (2019) The future of legume genetic data resources: challenges, opportunities, and priorities. Legume Sci 1:e16

Bhowmik P, Ellison E, Polley B, Bollina V, Kulkarni M, Ghanbarnia K, Song H, Gao C, Voytas DF, Kagale S (2018) Targeted mutagenesis in wheat microspores using CRISPR/Cas9. Sci Rep 8:6502

Bibikova M, Golic M, Golic KG, Carroll D (2002) Targeted chromosomal cleavage and mutagenesis in Drosophila using zinc-finger nucleases. Genetics 161:1169–1175

Bo W, Zhaohui Z, Huanhuan ZH, Xia WA, Binglin LI, Lijia Y, Xiangyan HA, Deshui Y, Xuelian Z, Chunguo W, Wenqin S (2019) Targeted mutagenesis of NAC transcription factor gene, OsNAC041, leading to salt sensitivity in rice. Rice Sci 26(2):98–108

Bortesi L, Fischer R (2015) The CRISPR/Cas9 system for plant genome editing and beyond. Biotechnol Adv 33(1):41–52

Cantos C, Francisco P, Trijatmiko KR, Slamet-Loedin I, Chadha-Mohanty PK (2014) Identification of "safe harbor" loci in indica rice genome by harnessing the property of zinc-finger nucleases to induce DNA damage and repair. Front Plant Sci 5:302

Carvalho SM, Vasconcelos MW (2013) Producing more with less: strategies and novel technologies for plant-based food biofortification. Food Res Int 54(1):961–971

Cermak T, Doyle EL, Christian M, Wang L, Zhang Y, Schmidt C, Baller JA, Somia NV, Bogdanove AJ, Voytas DF (2011) Efficient design and assembly of custom TALEN and other TAL effector-based constructs for DNA targeting. Nucleic Acids Res 39(12):e82

Čermák T, Baltes NJ, Čegan R, Zhang Y, Voytas DF (2015) High-frequency, precise modification of the tomato genome. Genome Biol 16(1):1–5

Cheng Z, Sun Y, Yang S, Zhi H, Yin T, Ma X, Zhang H, Diao X, Guo Y, Li X, Wu C (2021) Establishing in planta haploid inducer line by edited SiMTL in foxtail millet (Setaria italica). Plant Biotechnol J 19(6):1089

Christian M, Cermak T, Doyle EL, Schmidt C, Zhang F, Hummel A, Bogdanove AJ, Voytas DF (2010) Targeting DNA double-strand breaks with TAL effector nucleases. Genetics 186(2):757–761

Christian M, Qi Y, Zhang Y, Voytas DF (2013) Targeted mutagenesis of Arabidopsis thaliana using engineered TAL effector nucleases. G3: Genes Genomes Genet 3(10):1697–1705

Clasen BM, Stoddard TJ, Luo S, Demorest ZL, Li J, Cedrone F, Tibebu R, Davison S, Ray EE, Daulhac A, Coffman A (2016) Improving cold storage and processing traits in potato through targeted gene knockout. Plant Biotechnol J 14(1):169–176

Cong L, Ran FA, Cox D, Lin S, Barretto R, Habib N, Hsu PD, Wu X, Jiang W, Marraffini LA, Zhang F (2013) Multiplex genome engineering using CRISPR/Cas systems. Science 339(6121):819–823

DeFrancesco L (2011) Move over ZFNs: a new technology for genome editing may put the zinc finger nuclease franchise out of business, some believe. Not so fast, say the finger people. Nat Biotechnol 29(8):681–685

FAO (2017) The future of food and agriculture – trends and challenges. FAO, Rome

Gaj T, Gersbach CA, Barbas CF (2013) ZFN, TALEN, and CRISPR/Cas based methods for genome engineering. Trends Biotechnol 31(7):397–405

Goredema-Matongera N, Ndhlela T, Magorokosho C, Kamutando CN, van Biljon A, Labuschagne M (2021) Multinutrient biofortification of maize (Zea mays L.) in Africa: current status, opportunities and limitations. Nutrients 13(3):1039

Haun W, Coffman A, Clasen BM, Demorest ZL, Lowy A, Ray E, Retterath A, Stoddard T, Juillerat A, Cedrone F, Mathis L (2014) Improved soybean oil quality by targeted mutagenesis of the fatty acid desaturase 2 gene family. Plant Biotechnol J 12(7):934–940

Iqbal Z, Iqbal MS, Ahmad A, Memon AG, Ansari MI (2020) New prospects on the horizon: genome editing to engineer plants for desirable traits. Curr Plant Biol 1(24):100171

Jain A, Roorkiwal M, Pandey MK, Varshney RK (2017) Status and prospects of genomic selection in legumes. In: Varshney RK, Roorkiwal M, Sorrells ME (eds) Genomic selection for crop improvement: new molecular breeding strategies for crop improvement. Springer, Cham, pp 131–147

Johns T, Eyzaguirre PB (2007) Biofortification, biodiversity, and diet: a search for complementary applications against poverty and malnutrition. Food Policy 32(1):1–24

Johnson RA, Gurevich V, Filler S, Samach A, Levy AA (2015) Comparative assessments of CRISPR-Cas nucleases' cleavage efficiency in planta. Plant Mol Biol 87(1):143–156

Kawakami Y, Bhullar NK (2018) Molecular processes in iron and zinc homeostasis and their modulation for biofortification in rice. J Integr Plant Biol 60(12):1181–1198

Kelliher T, Starr D, Richbourg L, Chintamanani S, Delzer B, Nuccio ML, Green J, Chen Z, McCuiston J, Wang W, Liebler T (2017) MATRILINEAL, a sperm-specific phospholipase, triggers maize haploid induction. Nature 542(7639):105–109

Kim D, Hager M, Brant E, Budak H (2021) Efficient genome editing in wheat using Cas9 and Cpf1 (AsCpf1 and LbCpf1) nucleases. Funct Integr Genomics 21(3):355–366

Koç E, Karayiğit B (2021) Assessment of biofortification approaches used to improve micronutrient-dense plants that are a sustainable solution to combat hidden hunger. J Soil Sci Plant Nutr:1–26

Komor AC, Kim YB, Packer MS, Zuris JA, Liu DR (2016) Programmable editing of a target base in genomic DNA without double-stranded DNA cleavage. Nature 533(7603):420–424

Lal MK, Kumar A, Kardile HB, Raigond P, Changan SS, Thakur N et al (2020) Biofortification of vegetables. In: Advances in agri-food biotechnology. Springer, Singapore, pp 105–129

Lewis GP (2005) Legumes of the world. Royal Botanic Gardens, Kew, Richmond

Li T, Huang S, Zhao X, Wright DA, Carpenter S, Spalding MH, Weeks DP, Yang B (2011) Modularly assembled designer TAL effector nucleases for targeted gene knockout and gene replacement in eukaryotes. Nucleic Acids Res 39(14):6315–6325

Li T, Liu B, Spalding MH, Weeks DP, Yang B (2012) High-efficiency TALEN-based gene editing produces disease-resistant rice. Nat Biotechnol 30(5):390–392

Li JF, Norville JE, Aach J, McCormack M, Zhang D, Bush J, Church GM, Sheen J (2013) Multiplex and homologous recombination–mediated genome editing in Arabidopsis and Nicotiana benthamiana using guide RNA and Cas9. Nat Biotechnol 31(8):688–691

Liang Z, Zhang K, Chen K, Gao C (2014) Targeted mutagenesis in Zea mays using TALENs and the CRISPR/Cas system. J Genet Genomics 41(2):63–68

Liu G, Li J, Godwin ID (2019) Genome editing by CRISPR/Cas9 in sorghum through biolistic bombardment. Methods Mol Biol 1931:169–183. https://doi.org/10.1007/978-1-4939-9039-9_12

Liu H, Wang K, Jia Z, Gong Q, Lin Z, Du L, Pei X, Ye X (2020) Efficient induction of haploid plants in wheat by editing of TaMTL using an optimized Agrobacterium-mediated CRISPR system. J Exp Bot 71(4):1337–1349

Lloyd A, Plaisier D Carroll GN (2005) Targeted mutagenesis using zinc-finger nucleases in Arabidopsis. Proc Natl Acad Sci U S A, 102 (6) 2232–2237. doi: https://doi.org/10.1073/pnas.0409339102

Lor VS, Starker CG, Voytas DF, Weiss D, Olszewski NE (2014) Targeted mutagenesis of the tomato PROCERA gene using transcription activator-like effector nucleases. Plant Physiol 166(3):1288–1291

Lou D, Wang H, Liang G, Yu D (2017) OsSAPK2 confers abscisic acid sensitivity and tolerance to drought stress in rice. Front Plant Sci 8:993

Lyzenga WJ, Harrington M, Bekkaoui D, Wigness M, Hegedus DD, Rozwadowski KL (2019) CRISPR/Cas9 editing of three CRUCIFERIN C homoeologues alters the seed protein profile in Camelina sativa. BMC Plant Biol 19(1):1–6

Ma L, Zhu F, Li Z, Zhang J, Li X, Dong J, Wang T (2015) TALEN-based mutagenesis of lipoxygenase LOX3 enhances the storage tolerance of rice (Oryza sativa) seeds. PLoS One 10(12):e0143877

Mali P, Yang L, Esvelt KM, Aach J, Guell M, DiCarlo JE, Norville JE, Church GM (2013) RNA-guided human genome engineering via Cas9. Science 339:823–826

Malzahn A, Lowder L, Qi Y (2017) Plant genome editing with TALEN and CRISPR. Cell Biosci 7:21. https://doi.org/10.1186/s13578-017-0148-4

Martínez-Fortún J, Phillips DW, Jones HD (2017) Potential impact of genome editing in world agriculture. Emerg Top Life Sci 1(2):117–133

Miao J, Guo D, Zhang J, Huang Q, Qin G, Zhang X, Wan J, Gu H, Qu LJ (2013) Targeted mutagenesis in rice using CRISPR-Cas system. Cell Res 23(10):1233–1236

Mousavi-Derazmahalleh M, Bayer PE, Hane JK, Valliyodan B, Nguyen HT, Nelson MN, Erskine W, Varshney RK, Papa R, Edwards D (2019) Adapting legume crops to climate change using genomic approaches. Plant Cell Environ 42:6–19

Naik BJ, Shimoga G, Kim SC, Manjulatha M, Subramanyam C, Palem RR, Kumar M, Kim SY, Lee SH (2022) CRISPR/Cas9 and nanotechnology pertinence in agricultural crop refinement. Front Plant Sci 461

Nogué F, Mara K, Collonnier C, Casacuberta JM (2016) Genome engineering and plant breeding: impact on trait discovery and development. Plant Cell Rep 35(7):1475–1486

Ntui VO, Tripathi JN, Tripathi L (2020) Robust CRISPR/Cas9 mediated genome editing tool for banana and plantain (Musa spp.). Curr Plant Biol 21:100128

Ochatt S, Conreux C, Moussa Mcolo R, Despierre G, MagninRobert J-B, Raffiot B (2018) Phytosulfokine-alpha, an enhancer of in vitro regeneration competence in recalcitrant legumes. Plant Cell Tissue Organ Cult 135:189–201

Pramanik D, Shelake RM, Park J, Kim MJ, Hwang I, Park Y, Kim JY (2021, 1878) CRISPR/Cas9-mediated generation of pathogen-resistant tomato against tomato yellow leaf curl virus and powdery mildew. Int J Mol Sci 22(4)

Ramalingam S, Kandavelou K, Rajenderan R, Chandrasegaran S (2011) Creating designed zinc-finger nucleases with minimal cytotoxicity. J Mol Biol 405(3):630–641

Ran Y, Liang Z, Gao C (2017) Current and future editing reagent delivery systems for plant genome editing. Sci China Life Sci 60(5):490–505

Rashid B, Tariq M, Khalid A, Shams F, Ali Q, Ashraf F, Ghaffar I, Khan MI, Rehman R, Husnain T (2017) Crop improvement: new approaches and modern techniques. Plant Gene Trait 8(3):18–30

Ricroch A, Clairand P, Harwood W (2017) Use of CRISPR systems in plant genome editing: toward new opportunities in agriculture. Emerg Top Life Sci 1(2):169–182

Saltzman A, Birol E, Oparinde A, Andersson MS, Asare-Marfo D, Diressie MT et al (2017) Availability, production, and consumption of crops biofortified by plant breeding: current evidence and future potential. Ann N Y Acad Sci 1390(1):104–114

Shan Q, Wang Y, Chen K, Liang Z, Li J, Zhang Y, Zhang K, Liu J, Voytas DF, Zheng X, Zhang Y (2013a) Rapid and efficient gene modification in rice and Brachypodium using TALENs. Mol Plant 6(4):1365–1368

Shan Q, Wang Y, Li J, Zhang Y, Chen K, Liang Z, Zhang K, Liu J, Xi JJ, Qiu JL, Gao C (2013b) Targeted genome modification of crop plants using a CRISPR-Cas system. Nat Biotechnol 31(8):686–688

Shoab E, Hefferon K (2022) Crop biofortification and food security. In: Plant nutrition and food security in the era of climate change. Academic Press, pp 423–436

Shu H, Luo Z, Peng Z, Wang J (2020) The application of CRISPR/Cas9 in hairy roots to explore the functions of AhNFR1 and AhNFR5 genes during peanut nodulation. BMC Plant Biol 20(1):1–5

Shukla VK, Doyon Y, Miller JC, DeKelver RC, Moehle EA, Worden SE, Mitchell JC, Arnold NL, Gopalan S, Meng X, Choi VM (2009) Precise genome modification in the crop species Zea mays using zinc-finger nucleases. Nature 459(7245):437–441

Simkin AJ (2019) Genetic engineering for global food security: photosynthesis and biofortification. Plan Theory 8(12):586

Sun Z, Li N, Huang G, Xu J, Pan Y, Wang Z, Tang Q, Song M, Wang X (2013) Site-S pecific gene targeting using transcription activator-like effector (TALE)-B ased nuclease in Brassica oleracea. J Integr Plant Biol 55(11):1092–1103

Szczepek M, Brondani V, Büchel J, Serrano L, Segal DJ, Cathomen T (2007) Structure-based redesign of the dimerization interface reduces the toxicity of zinc-finger nucleases. Nat Biotechnol 25(7):786–793

Tilman D, Balzer C, Hill J, Befort BL (2011) Global food demand and the sustainable intensification of agriculture. Proc Natl Acad Sci 108(50):20260–20264

Tsai SQ, Wyvekens N, Khayter C, Foden JA, Thapar V, Reyon D, Goodwin MJ, Aryee MJ, Joung JK (2014) Dimeric CRISPR RNA-guided FokI nucleases for highly specific genome editing. Nat Biotechnol 32(6):569–576

Varshney RK, Kudapa H, Pazhamala L, Chitikineni A, Thudi M, Bohra A, Gaur PM, Janila P, Fikre A, Kimurto P, Ellis N (2015) Translational genomics in agriculture: some examples in grain legumes. Crit Rev Plant Sci 34:169–194

Varshney RK, Pandey MK, Bohra A, Singh VK, Thudi M, Saxena RK (2019) Toward the sequence-based breeding in legumes in the post-genome sequencing era. Theor Appl Genet 132:797–816

Wada N, Ueta R, Osakabe Y, Osakabe K (2020) Precision genome editing in plants: state-of-the-art in CRISPR/Cas9-based genome engineering. BMC Plant Biol 20(1):1–2

Wang Y, Cheng X, Shan Q, Zhang Y, Liu J, Gao C, Qiu JL (2014) Simultaneous editing of three homoeoalleles in hexaploid bread wheat confers heritable resistance to powdery mildew. Nat Biotechnol 32(9):947–951

Wang J, Kuang H, Zhang Z, Yang Y, Yan L, Zhang M, Song S, Guan Y (2020) Generation of seed lipoxygenase-free soybean using CRISPR-Cas9. Crop J 8:432–439

Wilson AK (2021) Will gene-edited and other GM crops fail sustainable food systems? In: Rethinking food and agriculture. Woodhead Publishing, pp 247–284

Wu X, Scott DA, Kriz AJ, Chiu AC, Hsu PD, Dadon DB, Cheng AW, Trevino AE, Konermann S, Chen S, Jaenisch R (2014) Genome-wide binding of the CRISPR endonuclease Cas9 in mammalian cells. Nat Biotechnol 32(7):670–676

Wu Z, Banuelos GS, Lin ZQ, Liu Y, Yuan L, Yin X, Li M (2015) Biofortification and phytoremediation of selenium in China. Front Plant Sci 6:136

Wu M, Liu H, Lin Y, Chen J, Fu Y, Luo J, Zhang Z, Liang K, Chen S, Wang F (2020) In-frame and frame-shift editing of the Ehd1 gene to develop Japonica rice with prolonged basic vegetative growth periods. Front Plant Sci 19(11):307

Yadav R, Mehrotra M, Singh AK, Niranjan A, Singh R, Sanyal I, Lehri A, Pande V, Amla D (2017) Improvement in Agrobacteriummediated transformation of chickpea (Cicer arietinum L.) by the inhibition of polyphenolics released during wounding of cotyledonary node explants. Protoplasma 254:253–269

Zafar K, Khan MZ, Amin I, Mukhtar Z, Yasmin S, Arif M, Ejaz K, Mansoor S (2020) Precise CRISPR-Cas9 mediated genome editing in super basmati rice for resistance against bacterial blight by targeting the major susceptibility gene. Front Plant Sci 11:575

Zegeye WA, Tsegaw M, Zhang Y, Cao L (2022) CRISPR-based genome editing: advancements and opportunities for rice improvement. Int J Mol Sci 23(8):4454

Zeng Z, Han N, Liu C, Buerte B, Zhou C, Chen J, Wang M, Zhang Y, Tang Y, Zhu M, Wang J (2020) Functional dissection of HGGT and HPT in barley vitamin E biosynthesis via CRISPR/Cas9-enabled genome editing. Ann Bot 126(5):929–942

Zhang F, Maeder ML, Unger-Wallace E, Hoshaw JP, Reyon D, Christian M, Li X, Pierick CJ, Dobbs D, Peterson T, Joung JK (2010) High frequency targeted mutagenesis in Arabidopsis thaliana using zinc finger nucleases. Proc Natl Acad Sci 107(26):12028–12033

Zhang F, Cong L, Lodato S, Kosuri S, Church GM, Arlotta P (2011) Efficient construction of sequence-specific TAL effectors for modulating mammalian transcription. Nat Biotechnol 29:149–153

Zhang Y, Zhang F, Li X, Baller JA, Qi Y, Starker CG, Bogdanove AJ, Voytas DF (2013) Transcription activator-like effector nucleases enable efficient plant genome engineering. Plant Physiol 161(1):20–27

Zhang H, Zhang J, Wei P, Zhang B, Gou F, Feng Z, Mao Y, Yang L, Zhang H, Xu N, Zhu JK (2014) The CRISPR/C as9 system produces specific and homozygous targeted gene editing in rice in one generation. Plant Biotechnol J 12(6):797–807

Zhang H, Gou F, Zhang J, Liu W, Li Q, Mao Y, Botella JR, Zhu JK (2016) TALEN-mediated targeted mutagenesis produces a large variety of heritable mutations in rice. Plant Biotechnol J 14(1):186–194

Zhang Z, Hua L, Gupta A, Tricoli D, Edwards KJ, Yang B, Li W (2019) Development of an Agrobacterium-delivered CRISPR/Cas9 system for wheat genome editing. Plant Biotechnol J 17(8):1623–1635

Zhang M, Liu Q, Yang X, Xu J, Liu G, Yao X, Ren R, Xu J, Lou L (2020a) CRISPR/Cas9-mediated mutagenesis of Clpsk1 in watermelon to confer resistance to Fusarium oxysporum f. sp. niveum. Plant Cell Rep 39(5):589–595

Zhang P, Du H, Wang J, Pu Y, Yang C, Yan R, Yang H, Cheng H, Yu D (2020b) Multiplex CRISPR/Cas9-mediated metabolic engineering increases soya bean isoflavone content and resistance to soya bean mosaic virus. Plant Biotechnol J 18:1384–1395

Zheng Z, Ye G, Zhou Y, Pu X, Su W, Wang J (2021) Editing sterol side chain reductase 2 gene (StSSR2) via CRISPR/Cas9 reduces the total steroidal glycoalkaloids in potato. All Life 14(1):401–413

Zhou H, Liu B, Weeks DP, Spalding MH, Yang B (2014) Large chromosomal deletions and heritable small genetic changes induced by CRISPR/Cas9 in rice. Nucleic Acids Res 42(17):10903–10914

Chapter 9
Transgenic Strategies and Genome Editing for Nutritional Enrichment

Faizan Ali, Akhtar Hameed, Abdul Rehman, Sohaib Sarfraz, and Nasir Ahmed Rajput

Abstract The agriculture sector is facing severe challenges in producing high-quality food materials in significant quantities because of climate change and food insecurity. Therefore, efforts are needed to ensure global food security with millions of people experiencing malnutrition, and this challenge is expected to further intensify in the near future. Conventional approaches have been used for trait enhancement in crops, but these approaches are often laborious and time-consuming and have failed to keep up with rising food demands in recent times. In this regard, molecular approaches for various crops have shown some promising results with the recent introduction of site-specific genome-editing technologies such as transcription activator-like effector nucleases (TALENS) and CRISPR (clustered regularly interspaced short palindromic repeat)/CRISPR-assisted 9 (Cas9) systems. CRISPR technology has gained fame over the past few years because the technology generates DNA-free target mutations in plants, avoiding any possibility of introducing foreign DNA into host cells. This technology is also more cost-effective and less time-consuming than other gene-editing technologies. CRISPR/Cas-based gene-editing approaches also provide an escape from many GMO restrictions, increasing societal approval and paving the way to meeting the global agricultural demands of the future. In this chapter, we will discuss some basic concepts on gene editing and its application for nutrition enrichment and global food security.

F. Ali (✉) · A. Rehman · S. Sarfraz · N. A. Rajput
Department of Plant Pathology, University of Agriculture Faisalabad, Faisalabad, Punjab, Pakistan
e-mail: faizan.ali95@uaf.edu.pk; rehman.abdul@uaf.edu.pk; sarfraz.sohaib@uaf.edu.pk; nasir.ahmed@uaf.edu.pk

A. Hameed
Department of Plant Pathology, Institute of Plant Protection, MNS- University of Agriculture Multan, Multan, Punjab, Pakistan
e-mail: akhtar.hameed@mnsuam.edu.pk

M. A. Nadeem et al. (eds.), *Legumes Biofortification*,
https://doi.org/10.1007/978-3-031-33957-8_9

1 Introduction

Plants are key components of maintaining a sustainable life in that they technically serve as key sources for all four Fs; food, feed, fuel, and fiber (Hendrickson et al. 2008). Earlier human life primarily focused on meeting these demands by hunting animals, but as time passed and civilization progressed, humankind became more aware of the domestication of plant crops and how to use these plants to meet basic necessities. This change of events led to numerous spells of rapid food production over time, leading to increases in global population. During the eighteenth and nineteenth centuries, new discoveries, ranging from the steam engine to electricity and digital devices, created a fusion between surplus food and innovation, which ultimately resulted in an explosive growth in birthrates and increasing populations (Mohajan 2019; Galloway and Cowling 2002; Romer 1990).

A growing population has been a major concern since the early twentieth century. Many of the famines (i.e., the Irish famine) and events of economic turmoil (i.e., the Great Depression) exposed the vulnerability of global efforts to satisfy the basic needs of human beings and showed how easily the inability to meet global food demands can become a complete disaster (Romer 1990; Fotheringham et al. 2013). Lessons learned from the tragic events of the past prompt new thinking about the world and have forced humankind to make efforts in certain directions to avoid such catastrophic events and so that enough food supplies can be secured for the global population. In response to these threats, the Green Revolution of the 1970s witnessed an immense increase in crop yields across the globe, especially among a large number of middle- and low-income countries, and was instrumental in improving socioeconomic outlooks for many countries. This revolution was powered mainly by the development of high-yield cereal crop varieties, along with use of modern weed killing, pest-controlling synthetic chemicals, fertilizers, and other agrochemicals (Tilman et al. 2002). This revolution introduced an era of stable food supplies and reduced hunger around the globe for the better half of the past century and the early years of the twenty-first century; however, the overall impact of these approaches was not properly assessed, which resulted in issues such as the runoff of harmful chemicals, soil degradation, and unintended climatic crises (Pingali 2012). Now, this plus resource scarcity, reduced availability of fertile and arable land, and the increasing negative impacts of climate change are grave threats to global food security.

Climate change and human activities have created some major issues for sustainable agricultural production: Higher levels of carbon dioxide, increasing global temperatures, heat waves, droughts, and resource depletions will lower yields across the world (Nelson et al. 2009; Gliessman 2015). Meanwhile, the UN's estimates suggest that the global population will reach more than ten billion by the middle of the twenty-first century, and to feed this immense population, an increase of about 70% in production output is required (Prosekov and Ivanova 2018). Securing food for such a population is a complex task as agricultural production systems need to be adjusted in accordance with different agro-climatic zones and soil conditions to attain maximum growth (Tubiello et al. 2007).

Breeding practices have been used for a long time in plants with the specific aim of enhancing key agricultural traits. Over time, four key approaches have been used for this purpose: crossbreeding practices, mutation breeding, transgenic approaches, and (most recently) genome-editing approaches. Crossbreeding is among the earliest approaches focusing on trait enhancement by targeting the sexual crossbreeding of plants with certain desired characters, and it was the most common approach during the Green Revolution era. However, a key limiting factor for the wide-scale adoption of this approach was that it could be used only for traits that were part of parental genome, and under the low availability of genetic variability, the method tended to lose its efficiency. Mutation breeding focuses on the use of mutagens (chemical or radioactive) to introduce mutations into genomes, creating new genetic combinations and increasing the genetic variability of crop plants (Holme et al. 2019). Although this approach was useful for the introduction of new allelic combinations, a key challenge was screening for desirable traits and the cultivar type's possessing those traits, making the whole process laborious while undermining its efficiency. Transgenic breeding was the most common technique for trait enhancement during late 1990s and early 2000s; this technique was based on introducing foreign genes with desired traits into other plants. Although the results were promising, with increased yields and better nutrition profiles evident among treated plants, a major challenge to obtaining regulatory approval for GMCs is still restricting the potential benefits of this approach (Raman 2017). Recent advancements in molecular biology have now led to a fourth key approach to plant trait enhancement, based on breeding crops with site-specific genome editing using molecular scissors and gene-editing tools. These techniques tend to introduce a site-specific mutation into a plant genome to achieve the desired status without introducing any foreign DNA into the host cell (Chen et al. 2019). The most common of these approaches, with wide-scale applications, is the CRISPR (clustered regularly interspaced short palindromic repeat)/Cas (CRISPR-associated) approach, and its use has been rapidly growing for trait enhancement in several crops (e.g., wheat, rice, and potatoes) over the past few years (Zhu et al. 2020).

The central idea of food security is access to healthy, nutritious food as part of people's daily diets, regardless of socioeconomic boundaries. Currently, over 800 million people around the globe are malnourished, and this challenge is more prevalent in underdeveloped countries (Jensen 2002). Not taking a sustainable approach to maintaining steady increases in food production would spell disaster because the food security researcher community predicts stagnant growth in food supplies in the face of resource scarcity and threats from global warming (Eitelberg et al. 2015). Modern genetic-based breeding approaches, though, provide a key lifeline for the effective manipulation of plant genomes and can introduce new traits in order to meet the food demands of the future (Cook et al. 2014).

In order to solve the massive challenge of sustainable food production and global agriculture, genomic approaches can ensure global food security while maintaining balance among changing climate conditions, socioeconomic impacts, sustainable production, and global food production (Fig. 9.1).

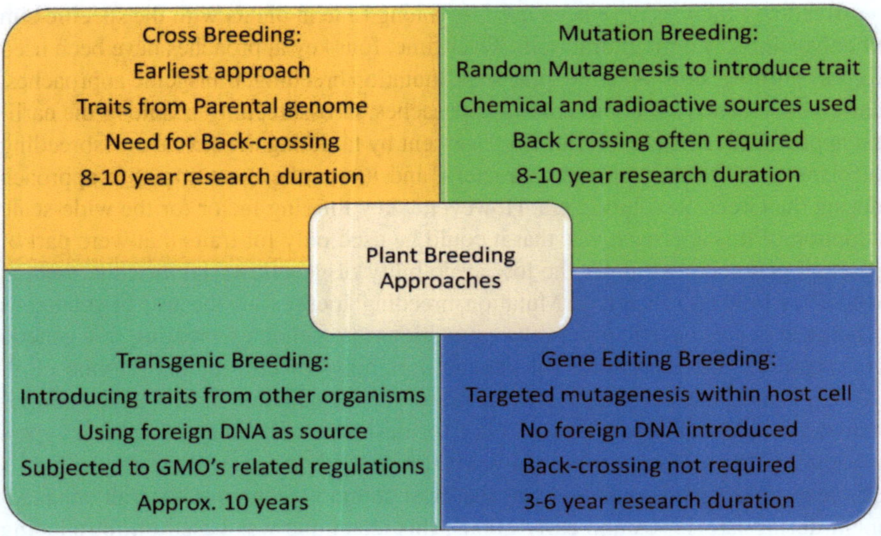

Fig. 9.1 Common plant-breeding approaches

2 Transgenic Approaches

Guaranteeing food availability for everyone is among the basic human norms and is a part of one of the UN's sustainable development goals: "zero hunger" worldwide. This includes not only the quantity of food but also its quality, as over 800 million people in the world experience malnutrition (Von Grebmer et al. 2008). The situation is worse in developing countries, where accessing numerous mineral- and vitamin-rich foods (e.g., fruit, meat, and fish) is not easy for everyone. People are more dependent on cereal crops to meet their daily food demands, and these do not pack enough vitamins and minerals in them to satisfy human needs. Therefore, this lack of vitamins and minerals has given rise to several malnourishment-related diseases (i.e. scurvy, night blindness, and rickets), posing grave risks to public health (Semba 2012). Although different nutrient supplements have often been used as solutions to this issue, a long-term solution remains a priority. For this purpose, efforts are being made to improve the nutritional traits of less-nutrient-rich crops (e.g., cereal crops), through a process called biofortification. In this transgenic approach, genetic encoding for nutrient enrichment is directly introduced into host cells. These cells then act as green production units that produce minerals and vitamins that generate genetic information introduced into the plant cells. Some of the reported works related to transgenic-based biofortification are discussed in this chapter.

The transgenic approach has been used for the vitamin enrichment of crop plants to make up for vitamin deficiencies. Research has been conducted on enhancing vitamin A and vitamin E contents in numerous crops, with positive results. Vitamin A, an essential nutrient, is required for the normal functioning of the human body, specifically key biological process such as normal growth, cell integrity, immune responses, and reproduction activity (Aslam et al. 2017). The human body can produce additional quantities of vitamin A from beta carotene (provitamin) precursor molecules, but cereal crops normally have lower contents of it. Therefore, efforts have been made to metabolically engineer biochemical pathways in crops to trigger beta-carotene biosynthesis (Ye et al. 2000). The development of golden rice was a key milestone on this journey, for which biosynthesis pathways in rice endosperms were engineered to increase the production of beta carotene. Its initial development included introducing a phytoene synthase from *Narcissus pseudonarcissus* (daffodil), causing the accumulation of vitamin A precursor molecules, followed by introducing a combination of phytoene synthase (psy) and lycopene β-cyclase (β-lcy) coding sequences from *N. pseudonarcissus*, under the influence of a glutelin promoter and together with bacterial phytoene desaturase from *E. uredovora*. The engineering of metabolic pathways resulted in the accumulation of beta carotene in rice grains, which emitted a golden color, from which the name golden rice derived for this newly developed variety (Beyer et al. 2002).

Vitamin E is another major essential compound needed for the normal functioning of the human body, playing a key role in preventing molecular oxidation and unsaturated fatty acid polymerization. Initial efforts involved the conversion of ϒ-tocopherol to α/β-tocopherol by using a combination of genes from Arabidopsis and Synechocystis PCC6803, encoding for ϒ-tocopherol methyltransferase (ϒ-TMT), indicating the possibility of enhancing vitamin E content in plants (Shintani and Dellapenna 1998). A similar approach was later used in soybean crops, where a combination of genes encoding for 2-methyl-6-phytylbenzoquinol (MPBQ) methyltransferase and ϒ-TMT were introduced to increase vitamin E content; an eightfold-plus increase in concentration has been observed (Van Eenennaam et al. 2003). In addition to this, vitamin biofortification has also been carried out in other crops (e.g., maize and lettuce) (Nunes et al. 2009; Naqvi et al. 2009).

Amino acids are other important components for the human diet, especially the essential amino acids that are not synthesized by the human body under normal conditions. Nine amino acids are known to be essential amino acids and are continuous required by the human body. Cereal and legume crops are the most essential components of the human diet, and these crops are often deficient in some of these essential amino acids, which is why researchers are interested in developing new varieties with higher nutritious values. The sunflower seed's albumin-coding gene, AmA1, has been introduced in chickpeas to increase methionine content, and the same gene has also increased protein content in potato crops (Chakraborty et al. 2000; Chiaiese et al. 2004). In most of the higher plants, the syntheses of lysine, threonine, and methionine take place from aspartic acid through a biochemical pathway with a feedback control mechanism. Two enzymes—namely aspartate kinase (AK) and dihydrodipicolinate synthase (DHPS)—play important roles in these pathways. Bacterial homologs for these enzymes expressed in Arabidopsis have improved lysine content in seeds. The introduction of DHPS into corn stimulated higher levels of lysine in soybean and canola crops, and increased concentrations of lysine and threonine were evident in corn (Falco et al. 1995).

Enhancing mineral concentration in plants is a rather difficult task, because unlike organic compounds, they are not synthesized by plants but rather are taken up from their surrounding environment. The most common approach used for mineral biofortification increases the uptake of minerals in the harvestable part of crops—i.e., fruits or vegetables. A second approach increases the content of digestible minerals available during digestion. Research has been carried out on enhancing the uptake of iron and zinc minerals by crop plants, by increasing the activity of certain chelating biomolecules, namely pyhtosiderophores. Success has been reported by using the naat-A and naat-B genes from barley when expressed in rice, which improved the synthesis of nicotianamine aminotransferases, which plays a vital role in siderophore synthesis (Takahashi et al. 2001). Soybean-based ferritin has also been used for mineral biofortification in crop plants (Goto et al. 1999).

In order to fight global food insecurity, both the quantity and the quality of food crops need to be improved. Transgenic approaches have provided early successes in the biofortification of crops, enabling humankind to combat malnutrition. Although this approach has its own set of challenges, such as wider adoptability, regulatory challenges, and public perception, dealing with these challenges can make transgenic approaches valuable assets in integrated efforts to combat global hunger.

3 Genome-Editing Concept and Technologies

Genome editing, or genome engineering, is an approach in molecular biology that allows for the insertion, alteration, and/or deletion of genetic information in DNA. A key difference that separates genome editing from previous mutagenesis approaches is that it allows the modification of genomes to be specific rather than produce random mutations (Bak et al. 2018). This approach has been used to alter traits in various life-forms, including plants, animals, and bacteria. The most common type of gene-editing technology is the CRISPR/Cas technology; others include zinc-finger nucleases (ZFNs) and transcription activator-like effector nucleases (TALENs) (Table 9.1).

Table 9.1 Comparison of the ZFN, TALEN, and CRISPR technologies

Technology	ZFN	TALEN	CRISPR
Target detection	Protein–DNA interaction	Protein–DNA interaction	DNA–RNA interaction
Construction	Target specific protein engineering	Target specific protein engineering	Single-guide RNA (sgRNA) used
Delivery	Two ZFNs required for the target	Two TALENs required for the target	sgRNA with Cas protein required
Multiplexing	Challenging	Challenging	Feasible
Affordability	Resource- and time-consuming approach	Affordable but time-consuming approach	Affordable approach

Recent advances in biological sciences have generated a better and more-efficient gene-editing technology that is based on programmable nucleases that enable site-specific nuclease (SSN) activity. These technologies have been reported to produce double-strand breakage at the point of their programmed target sequence. The natural process for DNA repair results in either knockout from a nonhomologous end join (NHEJ) or knockin for a new gene of interest thanks to homology-directed repair (HDR) (Symington and Gautier 2011). The following subsections detail some of the most important gene-editing technologies.

3.1 Zinc-Finger Nuclease (ZFN)

ZFN is among the earliest developed gene-editing technologies and is based on a combination of a zinc-finger-mediated DNA-binding domain for target identification and a FokI nuclease domain for cleavage activity, and it acts on a protein–DNA recognition system (Kim et al. 1996). The nuclease domain for the ZFN has been engineered for enhanced catalytic activity. The technology has over time been used on multiple crop plants, with positive results (Petolino 2015; Townsend et al. 2009; Zhang et al. 2010). Although there has been some initial success, wide-scale commercial applications are absent because of its costly, laborious procedure and the complex technical challenges impeding its application (Ramirez et al. 2008).

3.2 Transcription Activator-Like Effector Nucleases (TALENs)

Plant pathogenic bacteria such as *Xanthomonas* spp. secrete a specific kind of molecule—transcription activator-like effector (TALE)—which has a DNA-binding domain of 33–35 amino acids in a series (Boch et al. 2009). These DNA-binding domains are re-engineered according to a specific target sequence and later on are combined with a sequence-specific, independent FokI nuclease domain. A re-engineered TALEN molecule is then used for the site-specific gene modification of different life-forms (Li et al. 2011). A major challenge to the wide-scale application of TALEN is the high similarity of the TALE recognition sequence and the complexity of the procedure involved in increasing the specificity of the recognition domain for specific target sites in DNA.

3.3 CRISPR/Cas Proteins

First observed in bacterial species *Streptococcus pyogenes*, the CRISPR system was originally discovered as a type of bacterial defense mechanism against phage viruses and was later on harnessed and re-engineered for targeted genome modification

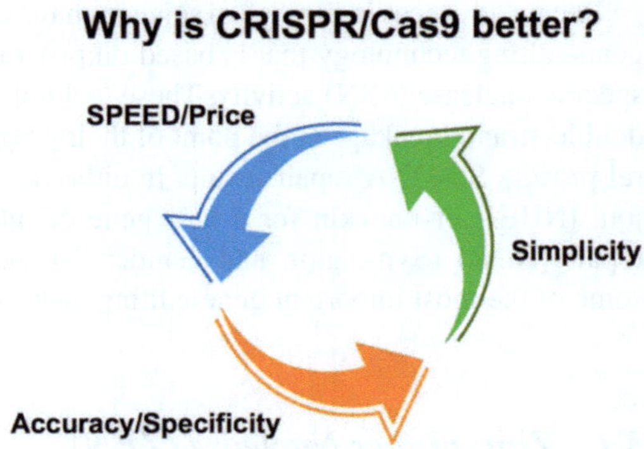

Fig. 9.2 CRISPR/Cas—a better technology

purposes. The most common of these systems is the CRISPR/Cas9 system, which uses a Cas9 protein molecule for genome engineering. The process combines a Cas9 molecule and a single-guide RNA (sgRNA) for the targeted modification of genes; it introduces a double-strand break at the target site, introducing targeted mutations into living cells. The CRISPR system uses an engineered version of sgRNA to enable identification on the basis of base pairings between sgRNA and the DNA target, making it much easier to manipulate and work with compared with ZFNs and TALENs (Jinek et al. 2012). These systems have exhibited relatively higher degrees of efficient modification in comparison to their predecessors, with positive results observed in all cellular life-forms, including plants and animals (Doudna and Charpentier 2014) (Fig. 9.2).

3.3.1 History of CRISPR Systems

CRISPR is a form of bacterial immunity against various phage viruses. This system was discovered in bacteria for the first time in 1987 by a Japanese scientist who discovered a long continuously repetitive sequence of DNA separated by a small segment of spacer DNA with an unknown function (Ishino et al. 1987). Later, in the 1990s, closely resembling long repetitive sequences were also observed in the archaea *H. mediterrane*, and further, similar sequences were discovered in a few other archaea. On the basis of these discoveries in archaea and bacteria, Francisco Mojica deduced that both archaea and bacteria possessed these segments of DNA and that the presence of these segments of DNA indicate that they play important roles in these microbes (Mojica et al. 1995). By the end of 2002, all the key elements of CRISPR loci had been identified, including the Cas (CRISPR-associated) genes, but the true function of the CRISPR loci was still unknown (Jansen et al. 2002). In 2005, after a massive bioinformatics analysis, it was deduced that many of the spacer sequences were a perfect match for various phages and conjugative plasmid DNA segments. Later, it was proposed that CRISPR is an adaptive immunity system in bacteria and that multiple-spacer DNAs are the memories of past invasions. In 2007, a group of scientists, including Rodolphe Barrangou and Philippe

Horvath, while working on lactic acid bacteria *S. thermophilus* for Danisco DuPont, provided the first experimental evidence for CRISPR's being an adaptive immunity system. Dr. Barrangou and his team focused on isolating phage-resistant bacteria and found that bacterial strains that have phage sequences in their CRISPR loci are resistant to specific phage viruses. Also, a larger number of spacer DNA enhances bacterial resistant to various phage viruses, and slight mutations in the corresponding phage DNA can result in the loss of resistance (Barrangou et al. 2007). This also showed a direct correlation between the numbers of spacer sequences acquired and an increased range of phage resistance. During 2011–2012, Virginijus Siksnys, a Lithuanian biochemist, created in vitro CRISPR/Cas9, crRNA, and tracrRNA complex, showing that a combination of Cas9 and artificially programmed crRNA and tracrRNA can be used in gene-editing experiments. In the same year, the team of Jennifer Doudna and Emmanuelle Charpentier performed a similar experiment, where they used Cas9 from *Streptococcus pyogenes* in combination with artificially synthesized crRNA and tracrRNA; in addition to this, they showed that the two RNAs can be fused to form a single-guide RNA (sgRNA) (Jinek et al. 2012). The creation of sgRNA was a huge step forward in the nascent field of CRISPR genome editing because it resolved many issues related to the creation, handling, and implementation of two separate RNAs. Following the discoveries of the artificial synthesis of crRNA and the creation of sgRNA, CRISPR soon became the headline of numerous major studies around the world as scientists began to use it heavily and started to produce positive results in a short period of time. The creation of artificial sgRNA laid the foundation of a wide range genome-editing experiments on various species. Numerous scientists, including the likes of Feng Zhang, George Church, and Keith Joung, conducted successful gene-editing experiments on various types of mammalian cells (Cong et al. 2013; Mali et al. 2013). What was initially observed as an immune system of bacteria has now become an efficient tool for genetic manipulation. The use of Cas9 for genetic editing in mammalian cells has shown the ease of using prokaryotic Cas9 RNA–protein complexes for introducing gene into various eukaryotic cellular forms.

In a similar manner to how CRISPR/Cas9 has been used in mammalian cells, CRISPR/Cas9 genome editing is also being widely used in plants for their genetic enhancement. Creating new, effective genetic combinations and producing beneficial genetic variations are essential for ensuring global food security. Conventional plant-breeding practices have been used for centuries to improve plants, followed by the creation of genetically modified plants (GMPs), which ensures optimal increases in crop yields. But the success of these approaches has been concealed by issues such as long duration, losing genetic variations among plants, compatibility issues among various species, and political controversies over GMPs. In recent years, the field of plant science has entered into a new phase, where site-specific genome editing is emerging as a worthy means for improving plants. Among various site-specific genome-editing techniques, CRISPR/Cas9-based gene editing has proven to be the most effective. The CRISPR/Cas9 has several advantages over other site-specific genome-editing techniques: It is cost-effective, easy to handle, and easy to implement, and it enables effective multiplex gene editing. The use of

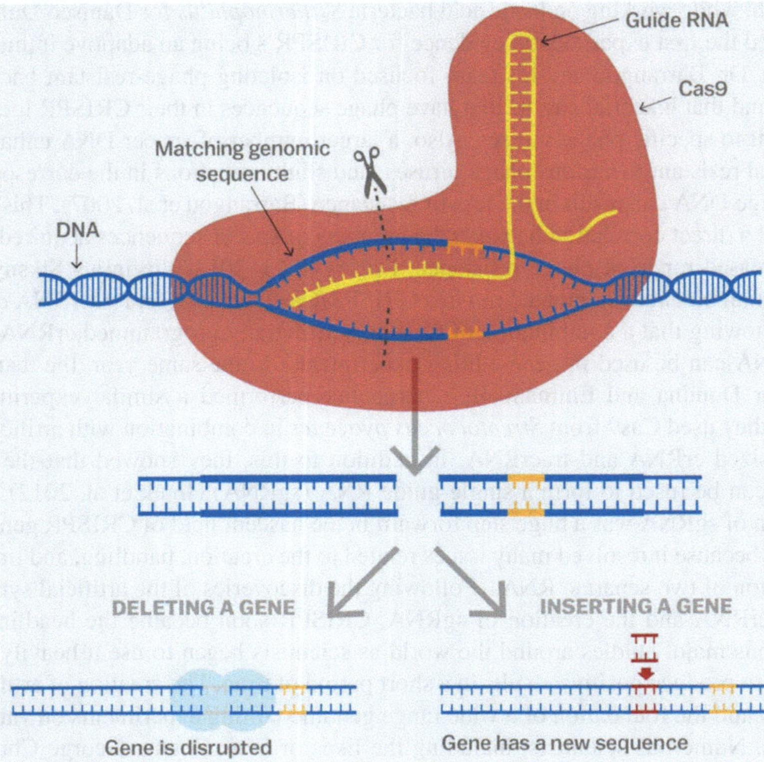

Fig. 9.3 Molecular mechanism of CRISPR/Cas9

CRISPR/Cas9 in plants was first reported in 2013, when it was used to introduce various sorts of mutations into plants (Li et al. 2013). The system is now being used in plants of greater importance, including wheat, rice, and maize (Fig. 9.3).

4 Applications of CRISPR in Food and Agriculture

Improving crop traits is a major focus of today's agricultural research. Over time, the focus on increasing yields and quality trait enhancement has greatly increased. Plants are key sources for numerous essential minerals, vitamins, and fiber, which make them important for global food security. For this purpose, numerous technologies have been used, and with the advent of modern genome-editing technologies, precise trait enhancement has gained serious attention in recent times (Gaj et al. 2013). A lack of nutrients in plants can lead to several health complications, and efforts are being made to address this challenge. One of the approaches involves engineering the root systems in plants because they carry out most of their vital functions related to growth and nutritional uptake. Therefore, re-engineering these roots can help us to overcome these challenges by improving root systems to promote better nutrient uptake and alleviate water stress (Comas et al. 2013; White et al. 2013).

The root system of a plant is complex trait influenced by multiple genes, and thus, improving it, especially via the traditional breeding approach, poses a great challenge (Ramireddy et al. 2018). Numerous factors, including the availability of nutrients, distribution, mobility, soil texture, etc., tend to interact with and influence root genetics. In order to absorb all these different nutrients, roots must deeply penetrate into soil to increase their absorption rates, which can further help deal with drought conditions (White et al. 2013). Studies have shown that various quantitative trait loci tend to regulate the root system. Also, cytokinins tend to have a negative impact on root growth. However, the degradation of cytokinin can lead to better root system development and the accumulation of more nutrients and better survival under drought conditions (Werner et al. 2010). Research has been carried out on multiple crops in order to stimulate the growth of root hair systems, enabling deeper penetration into soil and improving the accumulation of nutrients. Early efforts include the induction of root hair development via *Agrobacterium tumefaciens*–mediated transformation in soybean crops. The CRISPR/Cas9-based genome-engineering approach was used mainly to target the specific genes involved in the hairy growth of roots, including GmFEI2 and GmSHR. The final results of this study have shown mutated hairy root systems in gene-edited plants, indicating a higher degree of root hair growth (Cai et al. 2015).

In addition to this, gene-editing approaches can also be used in the metabolic engineering of crops to improve biochemical profiles and obtain a higher degree of metabolite content. Recent studies have used CRISPR-based knockin approaches for this task: One of the early studies focused on the insertion of carotenoid biosynthesis sequences into rice plants. The experiment was based on the insertion of two carotenoid synthesis genes at two safe harbor sites in the genome without damaging the genome (Dong et al. 2020). Recently, the multiplex gene-editing approach has gained great interest among the research community, and studies have yielded effective results from this approach in various crops. Multigene manipulation was carried out in tomato crops; for this purpose, four genes involved in lycopene metabolism were targeted, and the resultant mutant showed a 2.5-fold increase in lycopene content in tomatoes (Li et al. 2018). A similar approach was used in soybean crops, in which three key genes responsible for isoflavonoid synthesis were targeted (i.e., GmF3H1, GmF3H2, and GmFNSII), and a corresponding triple-mutated specimen, which exhibited a twofold increase in isoflavone content, was obtained (Zhang et al. 2020). Similarly, two genes in soybean crops were targeted to create mutants with improved oil contents. For this purpose, two genes (i.e., GmFAD2-1A and GmFAD2-2A) were targeted in soybean plants, and the corresponding mutants produced showed a nearly threefold increase in oleic content and an increase in protein content (Wu et al. 2020).

The gene-editing approach has been attracting serious attention in recent years for the nutritional trait enhancement of crops, especially with the ongoing long-term screening for transgenic plants. CRISPR-based approaches are being used in the genetic manipulation of several crops, where the technology focuses on manipulating the traits responsible for the efficient removal of antinutritional metabolites. These results point to a potential application of this technology in improving nutritional quality in and eliminating allergens from crop plants. These efforts are made possible mainly by re-engineering metabolic pathways in a manner that enhances the synthesis of useful metabolites via genetic manipulation. This precise editing at target sites with minimal off-target effects guarantees that the changes made to generating nutrition-rich crops remain sustainable in the long run.

5 Conclusion

The nutritional enhancement of food products is urgently needed, and with a booming global population, the challenge will only be further exacerbated. In this regard, a combination of approaches, including the conventional approach, the transgenic approach, and the gene-editing approach, can provide a viable way of overcoming this challenge. Many pre-existing (transgenic-based) products are waiting to enter the market, owing to legal challenges and public approval. Efforts can be made to improve public understanding and build confidence, thus paving the way to getting these products on the market. In the long run, CRISPR-based gene-editing technology is a valuable addition in to research tool box. This technology has emerged as a breakthrough approach in modern agricultural research, and with continuous upgrades to pre-existing CRISPR/Cas9 technology and the addition of new CRISPR-based systems, the potential for more-efficient genome editing is increasing. It has become a fast, efficient, easy-to-manage tool for introducing enhanced agronomic traits into crop plants. Therefore, the technology provides a new and superior way of developing new crop varieties with better traits in shorter amounts of time. At this rapid pace and with the efficient application of this technology, the world will soon benefit from the better traits in crops on the market thanks to the short amount of time needed for their development. This will help fight against large-scale malnutrition and hunger and ensure global food security for the current and future generations.

References

Aslam MF, Majeed S, Aslam S, Irfan JA (2017) Vitamins: key role players in boosting up immune response-a mini review. Vitam Miner 6:1

Bak RO, Gomez-Ospina N, Porteus MH (2018) Gene editing on center stage. Trends Genet 34:600–611

Barrangou R, Fremaux C, Deveau H, Richards M, Boyaval P, Moineau S, Romero DA, Horvath P (2007) CRISPR provides acquired resistance against viruses in prokaryotes. Science 315:1709–1712

Beyer P, Al-Babili S, Ye X, Lucca P, Schaub P, Welsch R, Potrykus IJ (2002) Golden rice: introducing the β-carotene biosynthesis pathway into rice endosperm by genetic engineering to defeat vitamin A deficiency. J Nutr 132:506S–510S

Boch J, Scholze H, Schornack S, Landgraf A, Hahn S, Kay S, Lahaye T, Nickstadt A, Bonas U (2009) Breaking the code of DNA binding specificity of TAL-type III effectors. Science 326:1509–1512

Cai Y, Chen L, Liu X, Sun S, Wu C, Jiang B, Han T, Hou W (2015) CRISPR/Cas9-mediated genome editing in soybean hairy roots. PLoS One 10:e0136064

Chakraborty S, Chakraborty N, Datta A (2000) Increased nutritive value of transgenic potato by expressing a nonallergenic seed albumin gene from Amaranthus hypochondriacus. Proc Natl Acad Sci U S A 97:3724–3729

Chen K, Wang Y, Zhang R, Zhang H, Gao C (2019) CRISPR/Cas genome editing and precision plant breeding in agriculture. Annu Rev Plant Biol 70:667–697

Chiaiese P, Ohkama-Ohtsu N, Molvig L, Godfree R, Dove H, Hocart C, Fujiwara T, Higgins T, Tabe LM (2004) Sulphur and nitrogen nutrition influence the response of chickpea seeds to an added, transgenic sink for organic sulphur. J Exp Bot 55:1889–1901

Comas LH, Becker SR, Cruz VMV, Byrne PF, Dierig DA (2013) Root traits contributing to plant productivity under drought. Front Plant Sci 4:442

Cong L, Ran FA, Cox D, Lin S, Barretto R, Habib N, Hsu PD, Wu X, Jiang W, Marraffini LA, Zhang F (2013) Multiplex genome engineering using CRISPR/Cas systems. Science 339:819–823

Cook C, Martin L, Bastow R (2014) Opportunities in plant synthetic biology. J Exp Bot 65:1921–1926

Dong OX, Yu S, Jain R, Zhang N, Duong PQ, Butler C, Li Y, Lipzen A, Martin JA, Barry KW, Schmutz J, Tian L, Ronald PC (2020) Marker-free carotenoid-enriched rice generated through targeted gene insertion using CRISPR-Cas9. Nat Commun 11:1179

Doudna JA, Charpentier E (2014) The new frontier of genome engineering with CRISPR-Cas9. Science 346:1258096

Eitelberg DA, Van Vliet J, Verburg PH (2015) A review of global potentially available cropland estimates and their consequences for model-based assessments. Glob Chang Biol 21:1236–1248

Falco S, Guida T, Locke M, Mauvais J, Sanders C, Ward R, Webber P (1995) Transgenic canola and soybean seeds with increased lysine. Biotechnology (N Y) 13:577–582

Fotheringham AS, Kelly MH, Charlton M (2013) The demographic impacts of the Irish famine: towards a greater geographical understanding. Trans Inst Br Geogr 38:221–237

Gaj T, Gersbach CA, Barbas CF III (2013) ZFN, TALEN, and CRISPR/Cas-based methods for genome engineering. Trends Biotechnol 31:397–405

Galloway JN, Cowling EB (2002) Reactive nitrogen and the world: 200 years of change. Ambio 31:64–71

Gliessman S (2015) Agroecology: the ecology of sustainable food systems, 3rd edn. Taylor & Francis, Boca Raton

Goto F, Yoshihara T, Shigemoto N, Toki S, Takaiwa F (1999) Iron fortification of rice seed by the soybean ferritin gene. Nat Biotechnol 17:282–286

Hendrickson J, Sassenrath G, Archer D, Hanson J, Halloran JJ (2008) Interactions in integrated US agricultural systems: the past, present and future. In: Renewable agriculture and food systems, vol 23, pp 314–324

Holme IB, Gregersen PL, Brinch-Pedersen H (2019) Induced genetic variation in crop plants by random or targeted mutagenesis: convergence and differences. Front Plant Sci 10:1468

Ishino Y, Shinagawa H, Makino K, Amemura M, Nakata A (1987) Nucleotide sequence of the iap gene, responsible for alkaline phosphatase isozyme conversion in Escherichia coli, and identification of the gene product. J Bacteriol 169:5429–5433

Jansen R, Embden JDV, Gaastra W, Schouls LM (2002) Identification of genes that are associated with DNA repeats in prokaryotes. Mol Microbiol 43:1565–1575

Jensen HH (2002) Food insecurity and the food stamp program. Am J Agric Econ 84:1215–1228

Jinek M, Chylinski K, Fonfara I, Hauer M, Doudna JA, Charpentier E (2012) A programmable dual-RNA–guided DNA endonuclease in adaptive bacterial immunity. Science 337:816–821
Kim Y-G, Cha J, Chandrasegaran S (1996) Hybrid restriction enzymes: zinc finger fusions to Fok I cleavage domain. Proc Natl Acad Sci U S A 93:1156–1160
Li J-F, Norville JE, Aach J, Mccormack M, Zhang D, Bush J, Church GM, Sheen J (2013) Multiplex and homologous recombination–mediated genome editing in Arabidopsis and Nicotiana benthamiana using guide RNA and Cas9. Nat Biotechnol 31:688–691
Li T, Huang S, Jiang WZ, Wright D, Spalding MH, Weeks DP, Yang B (2011) TAL nucleases (TALNs): hybrid proteins composed of TAL effectors and FokI DNA-cleavage domain. Nucleic Acids Res 39:359–372
Li X, Wang Y, Chen S, Tian H, Fu D, Zhu B, Luo Y, Zhu H (2018) Lycopene is enriched in tomato fruit by CRISPR/Cas9-mediated multiplex genome editing. Front Plant Sci 9:559
Mali P, Yang L, Esvelt KM, Aach J, Guell M, Dicarlo JE, Norville JE, Church GM (2013) RNA-guided human genome engineering via Cas9. Science 339:823–826
Mohajan H (2019) The first industrial revolution: Creation of a new global human era. J Soc Sci Humanit 5(4):377–387
Mojica F, Ferrer C, Juez G, Rodríguez-Valera F (1995) Long stretches of short tandem repeats are present in the largest replicons of the Archaea Haloferax mediterranei and Haloferax volcanii and could be involved in replicon partitioning. Mol Microbiol 17:85–93
Naqvi S, Zhu C, Farre G, Ramessar K, Bassie L, Breitenbach J, Perez Conesa D, Ros G, Sandmann G, Capell T, Christou P (2009) Transgenic multivitamin corn through biofortification of endosperm with three vitamins representing three distinct metabolic pathways. Proc Natl Acad Sci U S A 106:7762–7767
Nelson GC, Rosegrant MW, Koo J, Robertson R, Sulser T, Zhu T, Ringler C, Msangi S, Palazzo A, Batka M (2009) Climate change: Impact on agriculture and costs of adaptation. Intl Food Policy Res Inst
Nunes A, Kalkmann DC, Aragao FJ (2009) Folate biofortification of lettuce by expression of a codon optimized chicken GTP cyclohydrolase I gene. Transgenic Res 18:661–667
Petolino JF (2015) Genome editing in plants via designed zinc finger nucleases. In Vitro Cell Dev Biol Plant 51:1–8
Pingali PL (2012) Green revolution: impacts, limits, and the path ahead. Proc Natl Acad Sci U S A 109:12302–12308
Prosekov AY, Ivanova SA (2018) Food security: the challenge of the present. Geoforum 91:73–77
Raman R (2017) The impact of Genetically Modified (GM) crops in modern agriculture: a review. GM Crops Food 8:195–208
Ramireddy E, Hosseini SA, Eggert K, Gillandt S, Gnad H, Von Wirén N, Schmülling T (2018) Root engineering in barley: increasing cytokinin degradation produces a larger root system, mineral enrichment in the shoot and improved drought tolerance. Plant Physiol 177:1078–1095
Ramirez CL, Foley JE, Wright DA, Müller-Lerch F, Rahman SH, Cornu TI, Winfrey RJ, Sander JD, Fu F, Townsend JA (2008) Unexpected failure rates for modular assembly of engineered zinc fingers. Nat Methods 5:374–375
Romer CD (1990) The great crash and the onset of the great depression. Q J Econ 105:597–624
Semba RD (2012) The historical evolution of thought regarding multiple micronutrient nutrition. J Nutr 142:143S–156S
Shintani D, Dellapenna D (1998) Elevating the vitamin E content of plants through metabolic engineering. Science 282:2098–2100
Symington LS, Gautier J (2011) Double-strand break end resection and repair pathway choice. Annu Rev Genet 45:247–271
Takahashi M, Nakanishi H, Kawasaki S, Nishizawa NK, Mori S (2001) Enhanced tolerance of rice to low iron availability in alkaline soils using barley nicotianamine aminotransferase genes. Nat Biotechnol 19:466–469
Tilman D, Cassman KG, Matson PA, Naylor R, Polasky S (2002) Agricultural sustainability and intensive production practices. Nature 418:671–677

Townsend JA, Wright DA, Winfrey RJ, Fu F, Maeder ML, Joung JK, Voytas DF (2009) High-frequency modification of plant genes using engineered zinc-finger nucleases. Nature 459:442–445

Tubiello FN, Soussana J-F, Howden SM (2007) Crop and pasture response to climate change. Proc Natl Acad Sci U S A 104:19686–19690

Van Eenennaam AL, Lincoln K, Durrett TP, Valentin HE, Shewmaker CK, Thorne GM, Jiang J, Baszis SR, Levering CK, Aasen ED (2003) Engineering vitamin E content: from Arabidopsis mutant to soy oil. Plant Cell 15:3007–3019

Von Grebmer K, Fritschel H, Nestorova B, Olofinbiyi T, Pandya-Lorch R, Yohannes Y (2008) Global hunger index: the challenge of hunger 2008

Werner T, Nehnevajova E, Köllmer I, Novák O, Strnad M, Krämer U, Schmülling T (2010) Root-specific reduction of cytokinin causes enhanced root growth, drought tolerance, and leaf mineral enrichment in Arabidopsis and tobacco. Plant Cell 22:3905–3920

White PJ, George TS, Gregory PJ, Bengough AG, Hallett PD, Mckenzie BM (2013) Matching roots to their environment. Ann Bot 112:207–222

Wu N, Lu Q, Wang P, Zhang Q, Zhang J, Qu J, Wang N (2020) Construction and analysis of GmFAD2-1A and GmFAD2-2A soybean fatty acid desaturase mutants based on CRISPR/Cas9 technology. Int J Mol Sci 21:1104

Ye X, Al-Babili S, Kloti A, Zhang J, Lucca P, Beyer P, Potrykus I (2000) Engineering the provitamin A (β-carotene) biosynthetic pathway into (carotenoid-free) rice endosperm. Science 287:303–305

Zhang F, Maeder ML, Unger-Wallace E, Hoshaw JP, Reyon D, Christian M, Li X, Pierick CJ, Dobbs D, Peterson T (2010) High frequency targeted mutagenesis in Arabidopsis thaliana using zinc finger nucleases. Proc Natl Acad Sci U S A 107:12028–12033

Zhang P, Du H, Wang J, Pu Y, Yang C, Yan R, Yang H, Cheng H, Yu D (2020) Multiplex CRISPR/Cas9-mediated metabolic engineering increases soya bean isoflavone content and resistance to soya bean mosaic virus. Plant Biotechnol J 18:1384–1395

Zhu H, Li C, Gao C (2020) Applications of CRISPR–Cas in agriculture and plant biotechnology. Nat Rev Mol Cell Biol 21:661–677

Chapter 10
Biofortification of Legume Hybrids Obtained Through Intergeneric Hybridization

Aslı Küçükrecep, Şükran Yıldız, and Dilek Tekdal

Abstract Agricultural biotechnology has allowed scientists to introduce novel features into commonly consumed staple crops to enhance their productivity. One of the most critical applications for obtaining high-yielding cultivars is hybridization. Intergeneric hybrids are progenies obtained from parents who belong to distinct genera. The fertilization of distant relatives could pose a problem. Because of a lack of genetic information in one parent, incompatibility between parents can develop, resulting in pre- and postpollination occurrences. Pollen germination failure, poor pollen penetration through stigma, poor pollen tube growth, and pollen arresting in gynoecium are prefertilization barriers. Inadequate endosperm growth resulting in embryo abortion owing to a lack of nutrition, hybrid sterility, and lethality induced by chromosomal or genetic variations are instances of postfertilization barriers. Biofortification can be an alternative method to prevent these problems. This review focuses on the following topics: (i) the importance of hybridization for crop development; (ii) problems encountered in the hybridization of crop plants, including legumes; and (iii) the importance of mineral nutrition in legume hybrids.

1 The Importance of Hybridization in Crop Development

Somatic hybridization is also called protoplast fusion or somatic cell fusion and refers to the fusion of plant protoplasts from the somatic cells of different species or the same species. Regeneration is required to obtain hybrid plants from fused protoplasts (Hoffmann-Tsay et al. 1994). It has become essential to produce hybrids from sexually incompatible species with the techniques developed by protoplast

A. Küçükrecep · D. Tekdal (✉)
Department of Biotechnology, Faculty of Science, Mersin University, Mersin, Turkey
e-mail: dilektekdal@mersin.edu.tr

Ş. Yıldız
Department of Agricultural Biotechnology, Faculty of Agriculture, Akdeniz University, Antalya, Turkey

M. A. Nadeem et al. (eds.), *Legumes Biofortification*,
https://doi.org/10.1007/978-3-031-33957-8_10

fusion. In recent years, the rapid development of somatic cell genetics has made possible intra- and interspecies gene transfer via somatic hybridization. The somatic hybridization process includes (1) protoplast isolation, (2) protoplast fusion, (3) the selection of somatic hybrids, and (4) culturing somatic hybrids for whole-plant regeneration (Tomar and Dantu 2010).

Protoplast fusion requires the approach, adhesion, and union of two protoplast cells. The approach of protoplasts to each other is determined by many electrostatic forces arising from the potential at the cell surface (Bhojwani and Razdan 1986). The selection of heterokaryon cells resulting from fusion is crucial for somatic hybridization (Veltcheva et al. 2005).

Before proceeding to somatic hybridization, we must address the genetic variations among existing crops. Fusing protoplasts isolated from their somatic tissues by trying to carry genetic information from one species to another provides essential opportunities for plant improvement studies (Belete 2018).

Crop production can be affected by biotic stresses such as bacterial, fungal, and viral diseases or by abiotic stress such as adverse environmental conditions. The most crucial key to increasing crop yields is the genetic improvement of cultivated species to withstand biotic and abiotic stresses and meet the dietary needs of an increasing population. For crops to be more resistant to the environment and biotic stress and to increase their yields, it is imperative to cut and use the gene of interest from related or distant culture plant species. Because of some limiting factors, such as sexual incompatibility, it is challenging and time-consuming to transfer the desired traits to the cultured species through traditional breeding methods (Grosser and Chandler 2000; Johnson and Veilleux 2001; Orczyk et al. 2003). Protoplast fusion, developed and successfully used by Melchers and Labib (1974), is an unconventional method that shortens the crop improvement process (Melchers and Labib 1974). Interspecific and intergeneric crosses, which are difficult to improve with conventional breeding methods, can be easily achieved. The obtained hybrid plants can be obtained by transferring, via fusion, a resistant characteristic against any biotic or abiotic stress to another plant that may be susceptible to diseases (Shuro 2018).

1.1 Interspecific Hybridization

Interspecific hybridization is a vital process in plant species' ecological adaptation and evolution. Hybridization leads to gene flow, and new combinations of genes not found in a species can be created (Sikka and Joshi 1960). Thanks to many studies, it is becoming increasingly clear that hybridization between species sometimes results in the formation of entirely new species (Chapman and Burke 2007; Rieseberg et al. 2003; Rieseberg 1995; Soltis and Soltis 2009).

In a study conducted on alliums, interspecific hybridization was reviewed for the cross-compatibility and transfer of a desired gene. Because alliums have high medicinal values, it is vital to develop these species. However, biennial crop cycles make it difficult to rehabilitate these species. For this reason, interspecific hybridization, which can provide gene transfer, was part of a study on improving alliums. Interspecific hybridization enables the transfer of genes that can provide resistance to pests and to abiotic and biotic stresses, which can minimize or eliminate the use of plant protection agents. This method helps increase genetic diversity by providing new gene sources and is helpful in the genetic improvement of alliums. With the information obtained in that study, the breeder was given an idea about using certain species to improve the genetic basis of alliums (Benke et al. 2021).

1.2 Intergeneric Hybridization

In order to feed the increasing world population, the productivity of crops must be increased and their genomes enriched. Gene transfers are required to enrich the genome, and diversity must be exploited. For example, rice varieties, one of the most used crops, have become very sensitive to biotic and abiotic stresses because of environmental effects and pests, and their yield has decreased (Shakiba and Eizenga 2014). It is very important to improve rice varieties and develop new breeding strategies to increase the germplasm available. It can be used for breeding species closely related to the crops of interest and distantly related genera. Studies have been carried out to use distantly related species in improvement programs to increase the studied product's resistance properties and obtain high yields (Zeigler et al. 2013). Hybridization is a very strategic method to increase plants' genetic diversity and obtain crops with increased yields. Successful crosses can be made between two species of the same genus, called interspecific hybridization, or between two different genera of the same family, called intergeneric hybridization (Ballesfin et al. 2018).

Intergeneric hybridization is an essential tool to increase genetic diversity and improve species in many similar crop species in yellow mustard and rapeseeds (Brown et al. 1997) and in turnips and radishes (Bang et al. 2007; D'Hont et al. 1995; Hu et al. 2002). Produced by using the tissue culture technique of intergeneric hybrids, it is also a tool to preserve the rare characteristics of species in danger of extinction. It is very important that the hybrids obtained by intergeneric hybridization, which is one of the techniques used to increase crop yields via genetic transfer, successfully turn into a whole plant. For this, intergeneric hybrids can be obtained via tissue culture techniques. A protoplast fusion study to transfer a trait of interest used rice and barley to produce an intergeneric hybrid (Kisaka et al. 1998). Ballesfin et al. (2018) successfully obtained intergeneric hybrids between *Oryza sativa* L. and *Leersia perrieri* (*A. camus*) by using the embryo-recovery technique (Ballesfin et al. 2018).

Hylocereus is a genus of high interest in horticulture, while *Selenicereus* and *Epiphyllum* are species grown for ornamental purposes only. The intergeneric hybridization method was used to develop and improve these species (Tel-Zur et al. 2012). An intergeneric cross between *Hylocereus* and *Selenicereus* and one between

Epiphyllum and *Selenicereus* were developed. Genetic compatibility between parent breeds indicates successful interspecies hybridization.

As can be understood from the aforementioned studies, the protoplasts used in hybridization are excellent resources for improving existing species and increasing genetic diversity. However, there are some problems with this technology. Different levels of somatic incompatibility can be achieved in somatic hybrids from fusion combinations containing distant species, which can lead to the undesirable performance of hybrids (Fahleson et al. 1994). At the same time, hybrids resulting from aneuploidy, growth retardation and late development, loss of growth vigor, and abnormal morphology may be challenging to use in hybrids resulting from symmetrical or asymmetric fusion (Leino et al. 2003). Despite these disadvantages, successful results have been obtained from intergeneric somatic hybridization in some studies, and the hybrids obtained will probably be used for cultivar development (Grosser and Chandler 2002; Xia et al. 2001).

2 Problems Encountered in Hybridization in Crop Plants

Traditional breeding methods aim to develop new varieties by using existing variations in the target population. This purpose is to develop new varieties through selection from hybrid progeny obtained by crossbreeding between parents with desired characteristics. It aims to transfer the desired gene to the cultivars cultivated in intraspecies and interspecies hybridizations (Uysal et al. 2007). Interhybridization studies in plant breeding have been carried out to transfer genes that provide resistance against various biotic and abiotic stress factors from wild species to cultured forms or any economically important trait from one species to another related species (Christie 1987).

The first culture hybrid in plants was made between *Nicotiana rustica* and *N. paniculata* in a study conducted by Koelreuter in 1760. The F1 generation obtained from interspecies hybrids is generally sterile (Demir 1990).

Various problems are encountered in obtaining successful hybrids, and these problems are examined under two headings: prefertilization problems and postfertilization problems. Reproductive isolation falls into two basic categories. Among these mechanisms, those that play a role in preventing mating and fertilization are called prezygotic. In contrast, mechanisms that prevent the survival or reproduction of the hybrid zygote or hybrid offspring are called postzygotic (Rieseberg and Carney 1998).

Abnormal pollen germination or failure to germinate on the stigma, loss of the pollen tube before it reaches the egg or ovary, or a lack of fertilization are prefertilization problems. Hybrid necrosis is classified as a postfertilization problem (Bajaj 1990). In addition, cell and tissue necrosis and hybrid necrosis—which are characterized by low growth rates and, in some cases, lethality—are among the problems encountered (Bomblies 2009). Hybrid necrosis is molecularly well defined and genetically explained by the Bateson–Dobzhansky–Muller (BDM) model (Chen et al. 2016). According to this model, as ancestral species diverge, each lineage develops nondestructive mutations in its natural genomes, but when two genomes recombine, they negatively interact in a hybrid (Bomblies 2006).

Although abortion of the zygote or young embryo after fertilization is the most common problem in interspecies hybridization, this problem can be solved by the in vitro culture of the developing embryo in the ovary or the ovule containing the ovule of the zygote at any stage of development (Pierik 1997). The biggest obstacles to using wild gene resources in classical breeding studies are incompatibility and sterility. The commitment that expresses the transfer of desired genes and unwanted genes to hybrids poses a significant problem (Koca 2015).

Although the success rate of interbreeding is less than that of intergeneric hybrids, new species can be obtained with increased genetic variability thanks to the combination of different genomes in the crossing of distant relatives (Saxena et al. 2013). Because successfully obtaining hybrids varies depending on the degree of kinship and genetic relationships between the hybridized species, attention has been paid to the fact that the parents used in the crossing studies carried out until recently were of the same species (Uysal et al. 2007). It is vital to use varieties with different ploidy levels to obtain successful results in interspecific and intergeneric hybridizations (Deniz and Özer 1990).

Although various problems are encountered, approximately 30%–35% of flowering plants in nature today have emerged as a result of interspecies hybridization and the chromosome folding that follows this process (Stebbins 1971). Plant research aims to understand and overcome the hybridization barriers and thereby increase and improve the gene pools of plants. The studies carried out by transferring the genes related to desired characteristics to the cotton plant in order to increase its use in the food industry had the following features:

1. The cessation of endosperm development
2. Differences in DNA sequences in genomes
3. Differences in ploidy levels
4. The death of the hybrids
5. Sterility in the hybrids

Successful crosses could not be realized because of such problems, and the embryo culture technique was used to overcome this problem (Basal 2002).

2.1 Problems Encountered in Hybridization in the Brassicaceae

Raphanobrassica is an intergeneric hybrid obtained from *R. sativus* and *B. oleracea* plants and was reported by Karpechenko in 1928 (Karpechenko 1928). Cross-incompatibility has adversely affected the development of hybrids in extensive intergeneric hybridization studies conducted to develop strains resistant to biotic and abiotic stress factors (Tsunoda et al. 1980). When the pistillate parent used in Brassicaceae species is selected from a self-compatible line, it has been observed that the pollen tubes grow well and penetrate the ovule. It has been reported that the barriers to prefertilization observed in intergeneric hybridization are similar to the self-incompatibility observed within the species (Kaneko and Bang 2014).

In 1968, Wilmar and Hellendoorn evaluated postfertilization barriers to the growth and development of embryos by using *B. oleracea* in their work (Wilmar and Hellendoorn 1968). Various researchers determined that low hybrid embryo development was due to abnormally developing endosperm in intergeneric hybridization studies and reported their hypotheses on what could have caused this result. These hypotheses are listed as follows:

1. Endosperm equilibrium number (Johnston et al. 1980)
2. The activation of polar nuclei (Nishiyama and Yabuno 1978)
3. Genomic imprinting in the endosperm (Kinoshita 2007)

Tonosaki et al. (2013) conducted intergeneric hybridization studies between *B. rapa* and *R. sativus*. Studies have revealed that hybrid embryos degenerate at an early stage of their development, and they conducted development studies of in vitro procedures to overcome this problem (Bang et al. 2009; Tonosaki et al. 2013).

2.2 Problems Encountered in Hybridization in the Solanaceae

In order to develop varieties with high resistance to biotic and abiotic stresses, extensive hybridization studies have been carried out on tomato plants because wild tomato taxa are resistant to stress factors and have rich qualities of quality characteristics. The survival of tomato cultivars depends on the successful transfer of resistance genes from wild strains to cultivars to be successfully produced. Transferring the desired economic characteristics from wild species takes more time than transferring within the same variety. The problem of transferring undesired characteristics to the varieties to be obtained arises (Kalloo 1991).

Because *Solanum lycopersicoides* Dun. is a plant not found in *Lycopersicon* but has essential features such as resistance to cold stress, extreme drought, and many diseases, crossing the two genera provides access to the relevant genes and desired characterization features (Ji et al. 2004). By crossing *S. lycopersicoides* and *S. sitiens*, F1 hybrids are readily available and show high fertility and normal meiotic behavior (Rick 1979). Despite this high success achieved in interspecies crossing, in intergeneric hybridization studies using *L. esculentum* and *S. lycopersicoides*, styles reject *L. esculentum* pollen, and incompatibility problems arise from chromosomal genetic effects and male sterility (Chetelat et al. 1997).

2.3 Problems Encountered in Hybridization in the Poaceae

It has been reported that the low frequency of various alleles in modern wheat cultivars had reduced the existing diversity, and many of these alleles are in danger of being genetically lost (Alvarez and Guzmán 2018). These emerging problems have paved the way for variant searches where studies use various wheat subspecies and old varieties (Alvarez and Guzmán 2018). In addition, intergeneric hybridization studies have started to be carried out by using *Triticum* and *Aegilops* genera (Hajjar and Hodgkin 2007; Feuillet et al. 2008; Schneider et al. 2008).

Although extensive studies and applications have been carried out for many years to transfer the beneficial properties of various plants to wheat through crossing studies, it has not always been reported that the grain quality has improved. For example, the presence of secalin, which is transferred together with the relevant genes, as a result of crossing studies carried out to improve resistance and yield against various diseases and adaptation to stress factors, brings with it adverse effects such as relatively lower grain quality, increased stickiness and low volume in the dough obtained (Smartt 1979).

In the intergeneric hybridization studies carried out between *Hordeum jubatum* and *Secale cereale* in the same family, it was observed that the obtained hybrid seeds were destroyed after fertilization. Although the hybrid embryos grew thanks to endosperm incompatibility, it was determined that their development stopped ahead of time (de Fernandes et al. 2000).

2.4 Problems Encountered in Hybridization in the Fabaceae

Legumes are the third-most-prominent family in the world, and they contribute to the enrichment of the soil they are in by binding the free nitrogen of the air to the soil as a result of the symbiotic relationship that they establish with the *Rhizobium* bacteria in their roots (Vural et al. 2015). Thanks to their nutritional value, low fat content, and high protein content, they are essential in managing diseases that require special diets, such as cancer, diabetes, and cardiovascular diseases (Arslan 2019).

Plant breeders have conducted various studies to increase the diversity and size of the gene pool of this important plant group. Although many intergeneric hybridization studies have been attempted to obtain functional variants, most attempts have reportedly failed (McComb 1975).

In Knobloch's review in 1972, he reported eight possible intergeneric hybridizations in the Fabaceae family. All of the genera that he reported include crosses between *Phaeoleae* members of the species in the Papilionoideae subfamily (Knobloch 1972).

Phaseolus vulgaris, an essential member of the legume family, is a plant that is not resistant to root rot and bacterial diseases, and *P. acutifollius* carries genes that provide resistance against these disease agents. Although transferring these resistant genes to *P. vulgaris* may contribute to developing varieties resistant to these disease factors, incompatibility problems have been observed in intergeneric and interspecific hybridization (Uysal et al. 2007).

Peanut (*Arachis hypogea*) is a protein-rich plant that ranks third among edible oils globally in terms of production (Uysal et al. 2007). The first hybridization among species in the genus Arachis was reported in 1952 by Krapovickas and Rigoni, in a study with *A. hypogaea* and *A. correntina* (Krapovickas and Rigoni 1957). Breeding studies have aimed to enrich the protein content of peanuts, increase the yield, and develop varieties resistant to diseases and pests.

Studies conducted by Varisai (1973) and Raman (1976) have shown that the similarity to the parent of hybrids assumed to be obtained between species is open to debate insofar of being productive and lacking the vital characteristics of the pollinator parent (Raman 1959; Varisai Muhammad 1973). In crossbreeding, problems such as incompatibility or the slow development of hybrids have been encountered (Uysal et al. 2007).

Again, no success was reported in the crossing studies using *C. microphyllum*, *C. pinnatifidum*, and *C. arietinum* plants, the lattermost of which are also in the legume family, and as a result of crosses with more wild species, cultivars suitable for culture were obtained (Smartt 1979). In studies with *C. arietinum* and *C. echinospermum*, it has been reported that they are sterile, although viable hybrids were obtained (Ladizinsky and Adler 1976).

Although Sacinadze (1961) reported that hybridization could be obtained in his study with *Phaseolus vulgaris* and *Glycine hispida* in his study in 1961, he later reported, in 1967, that the morphological and biochemical changes aimed to be achieved by hybridization disappeared from subsequent generations (McComb 1975; Saginadze 1961).

Thanks to its high protein content and easy digestibility, *Vigna radiata*, preferred in nutrition, is a sensitive plant against pod cracking and yellow mosaic virus, and *V. munga* is resistant to the diseases mentioned above (Uysal et al. 2007). Although hybridizations were efficiently carried out in the studies conducted, reciprocal hybridizations resulted in unsuccessful results (Bajaj 1990).

In Tsitsin's study in 1946, he conducted intergeneric hybridization between *Caragana arborescens* and *Pisum sativum* and reported that although peas showed changes, they were not hereditary. However, crosses were carried out in both directions, and seeds could not be obtained (Tsitsin 1946).

Although a spontaneous hybridization between *Lens esculenta* and *Vicia sativa* was thought to have taken place, crosspollination as a result of the karyological analyses showed that the hybrid formed by eliminating the genomes of the *Lens esculenta* plant had the same genome as *Vicia sativa* (Zadrail 1960).

3 The Importance of Biofortifying Crops

Grain production may decrease owing to various abiotic stress factors in legumes, and plants tend to be vulnerable to stress factors. Plants need mineral nutrients at every stage of their development, and obtaining the necessary minerals is an essential variable in obtaining maximum yields from plants. Potassium (K), phosphorus (P), zinc (Zn), iron (Fe), and selenium (Se) are among the essential nutrients involved in various morphological and biochemical processes (Hossain et al. 2020). Biofortification is a method that can increase the nutritional value of crops with agronomic applications and various transgenic techniques (Bouis and Saltzman 2017).

Approximately half of the world's population is exposed to some form of mineral deficiency, and mineral deficiency is one of the biggest health problems experienced today (Pfeiffer and McClafferty 2007). With its cost-effective and sustainable features, biofortification is an application that can provide a solution to the problem of malnutrition, especially in developed countries, and it aims to enrich the nutrient content of products (Dwivedi et al. 2012). In addition to eliminating mineral deficiencies in human nutrition, crop viability, seedling formation, and durability can be increased in plants enriched with minerals such as zinc and iron (Blair 2013).

In addition, various studies have shown that the low concentrations of minerals such as zinc in seeds may be sensitive to various stress factors of the plants obtained from these seeds (Obata et al. 1999). The difficulties experienced in obtaining hybrid individuals from hybrid seeds obtained in breeding studies may also be caused by various mineral deficiencies. Various enrichment studies on minerals may provide an advantage in obtaining hybrid individuals.

In various studies, it has been reported that zinc-enriched seeds perform better in germination, obtaining products from these seeds, and various yield characteristics, such as seedling health (Cakmak et al. 1996).

Various studies have shown that the inheritance of micronutrient traits is controlled by various genes (Blair and Izquierdo 2012). The possibility that this trait shows multigenic inheritance, that unwanted genes are transferred to hybrid individuals in individuals obtained by the hybridization method, or that these traits cannot be transferred to hybrid individuals should be considered one of the reasons that make it challenging to obtain hybrid individuals.

In addition to obtaining the desired properties in breeding studies for the improvement of beans, high zinc and high iron contents are among the producers' goals. On the bean plant, a vital legume, its rate of iron (Fe), its placement in the embryo and cotyledon, its seed, and its seed coat varies between genotypes, and in particular, 4%–22% of the total iron is located in the seed coat. In addition, the embryonic axis is also rich in iron and contains 2%–3% of the existing iron (Ariza-Nieto et al. 2007). In hybridization studies, it has been reported that iron levels can reach 90+ ppm levels. Interspecies crosses with *Phaseolus dumosus* (*Phaseolus polyanthus*) and *Phaseolus coccineus* have reported hybrids containing up to 127 ppm of iron (Blair 2013). Although promising developments have been reported, it is known that there have been significant regressions in interspecies hybridization thanks to studies conducted with various species (Blair 2013).

A different study reported that the chance of success in interspecies crossing between methionine-rich (1.8–2.0 g/kg) black lentils and mungbeans was significantly increased (Kumar and Pandey 2020). Various studies have been reported on producing sulfur-rich *Vicia narbonensis*, *Medicago sativa*, and *Lupinus angustifolius* to obtain legumes rich in methionine (Nair et al. 2013).

4 Conclusions

The ever-increasing world population and the day-by-day decreases in agricultural areas appear as factors that threaten the food security of a large part of the global population. Given the problems above, high-quality and high-yield products need to be obtained. Although there have been various advances in obtaining crops with desirable characteristics thanks to the breeding studies that have been conducted for many years, faster, practical, cost-effective, and target-oriented studies are needed today. Biofortification is considered an application that can provide advantages both in achieving success in breeding studies and in transferring the traits aimed to be gained by hybridization to crops with different applications. The lack of access to food and the consumption of foods that lack essential minerals (such as iron and zinc), called hidden hunger, pose threats to the food security of the world population. Enriching the mineral content of easily accessible foods with high protein content, such as legumes, can be a new option to break down the crossing barriers experienced by similar plant groups.

References

Alvarez JB, Guzmán C (2018) Interspecific and intergeneric hybridization as a source of variation for wheat grain quality improvement. Theor Appl Genet 131(2):225–251. https://doi.org/10.1007/s00122-017-3042-x

Ariza-Nieto M, Blair MW, Welch RM, Glahn RP (2007) Screening of iron bioavailability patterns in eight bean (*Phaseolus* vulgaris L.) genotypes using the Caco-2 cell in vitro model. J Agric Food Chem 55(19):7950–7956

Arslan M (2019) Küresel İklim Değişikliğinin Olumsuz Etkilerine Karşı Ümitvar Baklagiller olarak Mürdümük (*Lathyrus sativus* L.) ve Burçak (Vicia ervilia L.)'ın Önemi. Adnan Menderes Üniversitesi Ziraat Fakültesi Dergisi 16(1):97–104

Bajaj Y (1990) In vitro production of haploids and their use in cell genetics and plant breeding. İçinde Haploids in crop improvement I (ss. 3–44). Springer

Ballesfin MLE et al (2018) Development of an intergeneric hybrid between *Oryza sativa* L. and *Leersia perrieri* (A. Camus) Launert, p. 18045

Bang S et al (2007) Production and characterization of intergeneric hybrids between *Brassica oleracea* and a wild relative *Moricandia arvensis*. Plant Breed 126(1):101–103

Bang SW, Ueno O, Wada Y, Hong SK, Kaneko Y, Matsuzawa Y (2009) Production of *Raphanus* sativus (C3)-Moricandia arvensis (C3-C4 intermediate) monosomic and disomic addition lines with each parental cytoplasmic background and their photorespiratory characteristics. Plant Prod Sci 12(1):70–79

Basal H (2002) Gossypollu bitki gossypolsuz tohum özelliğinin kültürü yapılan pamuk (*Gossypium hirsutum* L.) türlerine aktarılması. Mustafa Kemal Üniversitesi Ziraat Fakültesi Dergisi 7(1–2):45–50

Belete T (2018) A review on somatic hybridization and its utilization in crop improvement. Int J African Asian Stud 43:24–34

Benke AP et al (2021) Interspecific hybridization in Allium crops: status and prospectus. Genet Resour Crop Evol:1–9

Bhojwani SS, Razdan MK (1986) Plant tissue culture: theory and practice. Elsevier

Blair MW (2013) Mineral biofortification strategies for food staples: the example of common bean. J Agric Food Chem 61(35):8287–8294. https://doi.org/10.1021/jf400774y

Blair MW, Izquierdo P (2012) Use of the advanced backcross-QTL method to transfer seed mineral accumulation nutrition traits from wild to Andean cultivated common beans. Theor Appl Genet 125(5):1015–1031

Bomblies K (2006) Hybrid incompatibility: when opposites attract with a fatal outcome. Curr Biol 16(14):R542–R544

Bomblies K (2009) Too much of a good thing? Hybrid necrosis as a by-product of plant immune system diversification. Botany 87(11):1013–1022

Bouis HE, Saltzman A (2017) Improving nutrition through biofortification: a review of evidence from HarvestPlus, 2003 through 2016. Glob Food Sec 12:49–58

Brown J et al (1997) Intergeneric hybridization between *Sinapis* alba and *Brassica* napus. Euphytica 93(2):163–168

Cakmak I, Yilmaz A, Kalayci M, Ekiz H, Torun B, Braun H (1996) Zinc deficiency as a critical problem in wheat production in Central Anatolia. Plant Soil 180(2):165–172

Chapman MA, Burke JM (2007) Genetic divergence and hybrid speciation. Evolution 61(7):1773–1780

Chen C, E Z, Lin H-X (2016) Evolution and molecular control of hybrid incompatibility in plants. Front Plant Sci 7:1208

Chetelat RT, Cisneros P, Stamova L, Rick CM (1997) A male-fertile *Lycopersicon esculentum* × *Solanum lycopersicoides* hybrid enables direct backcrossing to tomato at the diploid level. Euphytica 95(1):99–108

Christie BR (1987) CRC handbook of plant science in agriculture. CRC series in agriculture (USA)

D'Hont A et al (1995) Identification and characterisation of sugarcane intergeneric hybrids, *Saccharum officinarum* x Erianthus *arundinaceus*, with molecular markers and DNA in situ hybridisation. Theor Appl Genet 91(2):320–326

de Fernandes MI, Zanatta AC, Prestes AM, Caetano VD, Barcellos AL, Angra DC, Pandolfi V (2000) Cytogenetics and immature embryo culture at Embrapa Trigo breeding program: transfer of disease resistance from related species by artificial resynthesis of hexaploid wheat (*Triticum aestivum* L. em. Thell). Genet Mol Biol 23(4):1051–1062

Demir İ (1990) Genel Bitki Islahı Ege Üniversitesi Ziraat Fakültesi Yayınları. Ofset Basımevi, 367

Deniz B, Özer İ (1990) Yumak ile çim cinslerinde türler ve cinsler arası melezlemeler. Atatürk Üniversitesi Ziraat Fakültesi Dergisi 21(1)

Dwivedi SL, Sahrawat KL, Rai KN, Blair MW, Andersson MS, Pfeiffer W (2012) Nutritionally enhanced staple food crops

Fahleson J, Eriksson I, Glimelius K (1994) Intertribal somatic hybrids between *Brassica napus* and *Barbarea vulgaris*—production of in vitro plantlets. Plant Cell Reports 13(7):411–416

Feuillet C, Langridge P, Waugh R (2008) Cereal breeding takes a walk on the wild side. Trends Genet 24(1):24–32

Grosser J, Chandler JL (2000) Somatic hybridization of high yield, cold-hardy and disease resistant parents for citrus rootstock improvement. J Horticult Sci Biotechnol 75(6):641–644

Grosser J, Chandler J (2002) New Citrus rootstocks via protoplast fusion in XXVI International Horticultural Congress: Genetics and Breeding of Tree Fruits and Nuts 622

Hajjar R, Hodgkin T (2007) The use of wild relatives in crop improvement: a survey of developments over the last 20 years. Euphytica 156(1):1–13

Hoffmann-Tsay S-S et al (1994) Design, synthesis and application of surface-active chemicals for the promotion of electrofusion of plant protoplasts. Bioelectrochem Bioenerget 34(2):115–122

Hossain A, EL Sabagh A, Erman M, Fahad S, Islam T, Bhatt R, Hasanuzzaman M (2020) Nutrient management for improving abiotic stress tolerance in legumes of the family Fabaceae. In: Hasanuzzaman IM, Araújo S, Gill SS (eds) The plant family Fabaceae: biology and physiological responses to environmental stresses. Springer, pp 393–415. https://doi.org/10.1007/978-981-15-4752-2_15

Hu Q et al (2002) Intergeneric hybrids between *Brassica napus* and *Orychophragmus violaceus* containing traits of agronomic importance for oilseed rape breeding. Theor Appl Genet 105(6):834–840

Ji Y, Pertuzé R, Chetelat RT (2004) Genome differentiation by GISH in interspecific and intergeneric hybrids of tomato and related nightshades. Chromosom Res 12(2):107–116

Johnson AAT, Veilleux RE (2001) Somatic hybridization and application in plant breeding. John Wiley & Sons

Johnston S, Den Nijs T, Peloquin S, Hanneman R (1980) The significance of genic balance to endosperm development in interspecific crosses. Theor Appl Genet 57(1):5–9

Kalloo G (1991) Interspecific and intergeneric hybridization in tomato. In: Genetic improvement of tomato. Springer, pp 73–82

Kaneko Y, Bang SW (2014) Interspecific and intergeneric hybridization and chromosomal engineering of Brassicaceae crops. Breed Sci 64(1):14–22. https://doi.org/10.1270/jsbbs.64.14

Karpechenko GD (1928) Polyploid hybrids of Raphanus *sativus* L. x *Brassica oleracea* L. Zeitschrift für induktive Abstammungs-und Vererbungslehre 48(1):1–85

Kinoshita T (2007) Reproductive barrier and genomic imprinting in the endosperm of flowering plants. Genes Genet Syst 82(3):177–186

Kisaka H et al (1998) Intergeneric somatic hybridization of rice (*Oryza sativa* L) and barley (*Hordeum vulgare* L) by protoplast fusion. Plant Cell Reports 17(5):362–367

Knobloch IW (1972) Intergeneric hybridization in flowering plants. Taxon 21(1):97–103

Koca AS (2015) Çeltikte (*Oryza sativa* L.) endosperm destekli olgun embriyo kültüründe kallus oluşumu ve bitki rejenerasyonuna tohum iriliğinin etkisi

Krapovickas A, Rigoni VA (1957) Nuevas especies de Arachis: Vinculadas al problema del origen del maní. Darwin 11(3):431–455

Kumar S, Pandey G (2020) Biofortification of pulses and legumes to enhance nutrition. Heliyon 6(3):e03682. https://doi.org/10.1016/j.heliyon.2020.e03682

Ladizinsky G, Adler A (1976) The origin of chickpea *Cicer arietinum* L. Euphytica 25(1):211–217

Leino M et al (2003) Brassica napus lines with rearranged Arabidopsis mitochondria display CMS and a range of developmental aberrations. Theor Appl Genet 106(7):1156–1163

McComb JA (1975) Is intergeneric hybridization in the Leguminosae possible? Euphytica 24(2):497–502. https://doi.org/10.1007/BF00028219

Melchers G, Labib G (1974) Somatic hybridisation of plants by fusion of protoplasts. Mol Gener Genet 135(4):277–294

Nair RM, Yang R, Easdown WJ, Thavarajah D, Thavarajah P, de Hughes JA, Keatinge J (2013) Biofortification of mungbean (*Vigna radiata*) as a whole food to enhance human health. J Sci Food Agric 93(8):1805–1813

Nishiyama I, Yabuno T (1978) Causal relationships between the polar nuclei in double fertilization and interspecific cross-incompatibility in Avena. Cytologia 43(2):453–466

Obata H, Kawamura S, Senoo K, Tanaka A (1999) Changes in the level of protein and activity of Cu/Zn-superoxide dismutase in zinc deficient rice plant, *Oryza sativa* L. Soil Sci Plant Nutr 45(4):891–896

Orczyk W et al (2003) Somatic hybrids of Solanum tuberosum–application to genetics and breeding. Plant Cell Tissue Organ Cult 74(1):1–13

Pfeiffer WH, McClafferty B (2007) HarvestPlus: breeding crops for better nutrition. Crop Sci 47:S-88
Pierik RLM (1997) In vitro culture of higher plants. Springer Science & Business media
Raman V (1959) Studies in the genus Arachis. VI. Investigation on 30-chromosomed interspecific hybrids. Indian Oilseeds J 3:157–161
Rick C (1979) Biosystematic studies in Lycopersicon and closely related species for Solanum
Rieseberg LH (1995) The role of hybridization in evolution: old wine in new skins. Am J Bot 82(7):944–953
Rieseberg LH, Carney SE (1998) Plant hybridization. New Phytol 140(4):599–624. https://doi.org/10.1046/j.1469-8137.1998.00315.x
Rieseberg LH et al (2003) Major ecological transitions in wild sunflowers facilitated by hybridization. Science 301(5637):1211–1216
Saginadze G (1961) Hybridization of Phaseolus *vulgaris* (L.) Savi x *Glycine hispida* (Moench.) Maz. Preliminary communication. Tbilisskii Bot Inst Trudy 21:301–308
Saxena KB, Kumar RV, Tikle AN, Saxena MK, Gautam VS, Rao SK, Khare DK, Chauhan YS, Saxena RK, Reddy BVS (2013) ICPH 2671–the world's first commercial food legume hybrid. Plant Breed 132(5):479–485
Schneider A, Molnár I, Molnár-Láng M (2008) Utilisation of *Aegilops* (goatgrass) species to widen the genetic diversity of cultivated wheat. Euphytica 163(1):1–19
Shakiba E, Eizenga GC (2014) Genetics, and U. Improvement. InTech, Unraveling the secrets of rice wild species, pp 1–58
Shuro AR (2018) Review paper on the role of somatic hybridization in crop improvement
Sikka S, Joshi AB (1960) Cotton breeding in India, A Monograph, vol 1, New Delhi, pp 336–340
Smartt J (1979) Interspecific hybridization in the grain legumes—a review. Econ Bot 33(3):329–337
Soltis PS, Soltis DE (2009) The role of hybridization in plant speciation. Annu Rev Plant Biol 60:561–588
Stebbins G (1971) Chromosomal evolution in higher plants. Addison-Wesley, Reading
Tel-Zur N et al (2012) Intergeneric hybridization within the tribe Hylocereeae, subfamily Cactoideae (Cactaceae). Israel J Plant Sci 60(3):325–334
Tomar UK, Dantu PK (2010) Protoplast culture and somatic hybridization. In: Cellular and biochemical science. IK International House Pvt. Ltd, New Delhi, pp 876–891
Tonosaki K, Michiba K, Bang S, Kitashiba H, Kaneko Y, Nishio T (2013) Genetic analysis of hybrid seed formation ability of *Brassica rapa* in intergeneric crossings with Raphanus sativus. Theor Appl Genet 126(3):837–846
Tsitsin (1946) Studies on vegetative and sexual hybridization between herbaceous and woody plants
Tsunoda S, Hinata K, Gómez-Campo C (1980) *Brassica* crops and wild allies
Uysal H, Seyis F, Orhan K (2007) Tarla bitkilerinde melezleme bariyerlerinin aşilmasinda alternatif bir yöntem: Embriyo kültürü. Anadolu Tarım Bilimleri Dergisi 22(1):116–122
Varisai Muhammad S (1973) Cytogenetic investigations in the genus *Arachis* L. α. Triploid hybrids and their derivatives. Madras Agr J 60:1414–1427
Veltcheva M et al (2005) In vitro regeneration and genetic transformation of common bean (*Phaseolus vulgaris* L.)—Problems and progress. Sci Horticult 107(1):2–10
Vural M, Erdem O, Ergin E, Erkol I (2015) Baklagillerin Kraliçesi Eber Sarısı-Piyan (*Thermopsis turcica*) Tür Koruma Eylem Planı. Doğa Koruma Milli Parklar Genel Müdürlüğü, Afyonkarahisar Şube Müdürlüğü, Afyon
Wilmar J, Hellendoorn M (1968) Embryo culture of Brussels sprouts for breeding. Euphytica 17(1):28–37
Xia G et al (2001) Study on salt tolerance of F3–F5 hybrids originated from somatic hybridization between wheat and Agropyron elongatum. Shandong Agric Sci 39(6) 12–14
Zadrail K (1960) Cocak'—A vetch with lentil-like seeds as an edible legume and as a forage plant
Zeigler RS et al (2013) Food security, climate change and genetic resources. In: Plant genetic resources and climate change. CABI, Wallingford, pp 1–15

Chapter 11
The Importance of Plant Growth-Promoting Rhizobacteria in the Biofortification of Legumes

Dilek Tekdal

Abstract Sudden climate changes have made their impact being felt in Turkey and all over the world in recent years, suggesting that humanity may face severe problems in the future. Among the most pressing of these problems is the possibility of drought due to increasing global temperatures and the possibility of associated high yield reductions in plants used as human food. Since new agricultural lands cannot be created worldwide, the most acceptable rational approach is to make maximum use of existing agricultural lands. Increasing the concentration of bioavailable micronutrients in consumable crop tissues named biofortification is one of the strategies that should be developed against the problem of hunger that may occur in the future. Existing biofortification methods, which include some agronomic approaches, conventional plant breeding, and genetic engineering, have not always been effective in biofortification. Plant growth-promoting rhizobacteria (PGPR) have been recognized as a potential strategy for the biofortification of essential crops. This review will pay attention to the significance of legumes in human food and identify the importance of PGPR in plant nutrition and mineral nutrition of legume crops with PGPR to cope with hidden hunger.

1 Introduction

After the industrial revolution, the use of fossil fuels and the increase in greenhouse gas emissions in the atmosphere are the main reasons behind climate change (Huxel 2019). It is stated that the sudden changes in the climate will increase with the effect of the human factor, and if it is not prevented, it will cause serious problems. In addition to climate change, severe decreases in agricultural areas show that the risk of serious conditions such as hunger will be high. The temperature changes affect the precipitation regimes and time and, therefore, the amounts of underground water

D. Tekdal (✉)
Department of Biotechnology, Faculty of Science, Mersin University, Mersin, Turkey
e-mail: dilektekdal@mersin.edu.tr

M. A. Nadeem et al. (eds.), *Legumes Biofortification*,
https://doi.org/10.1007/978-3-031-33957-8_11

resources and surface waters. The evaporation that occurs due to the increase in temperature further increases and causes a decrease in the irrigation water capacity. It is thought that with the increase of evaporation, more humid air will prevail on the earth, the precipitation regime will change in some regions, an increase in heavy precipitation will be observed, and the risk of flooding will increase due to these precipitations. It is predicted that the expected increase in evaporation will increase the risk of drought (Bayraç 2016). With the increase in seawater temperature, many species living in the sea and oceans, especially fishing, are threatened (Johansen et al. 2014). Water stress reduces the water or turgor potential in the plant tissues to levels that negatively affect optimum development (Kudoyarova et al. 2013). Acute water shortage occurs in plants with a sudden increase in air temperature or a rapid decrease in humidity (Kiliçaslan et al. 2020). Drought stress affects plants' different structures, such as vegetative growth, yield, water relations, and photosynthesis (Farooq et al. 2009).

Globally, productive agricultural areas have shrunk in recent decades. Desertification, salinization, and soil erosion are human-caused declines linked to unsustainable land management. However, urbanization is the primary cause of this loss of agricultural land. People from rural areas migrate to cities, searching for better economic and social possibilities (Cohen 2006). In addition to the loss of farming land, the world's population is increasing exponentially; the global population is expected to exceed 9 billion by 2050, implying that 70–100% more food production will be required to ensure food security (da Dias 2020). For these reasons, obtaining maximum efficiency from the limited area has been included in the priorities of agricultural strategies. Since the arable land in the world has reached the limit, the process that needs to be done is not to expand the cultivation areas but to increase the amount of product taken from the unit area. As a result of attempts such as agriculture on steep slopes that are not suitable for agricultural production, deliberate drying of lakes to make agricultural areas, and artificial lakes for water and electricity needs, the risk of deterioration of the ecological balance and reduction of biodiversity has emerged as a result of attempts such as disturbing the balance of the areas in the valley (Bullock et al. 2001).

In order to reduce the water used in the agricultural sector, it is essential to use new techniques that reduce water losses and reduce the use of excess fertilizers and pesticides that cause soil and water pollution. For this reason, the use of microorganisms in agricultural production, which provides the nutrients needed for the plants to carry out their vital activities related to their growth and development, and plays a role in the intake of these nutrients, comes to the fore.

2 Plant Growth-Promoting Rhizobacteria

In recent years, the importance of microorganisms that naturally exist in the soil and interact beneficially with plant roots has increased day by day. Free-living soil bacteria of the genus *Azotobacter*, *Agrobacterium*, *Bacillus*, *Beijerinckia*, *Burkholderia*,

Clostridium, *Enterobacter*, *Erwinia*, *Flavobacterium*, *Klebsiella*, *Micrococcus*, *Pseudomonas*, *Rhizobium*, *Rhodobacter*, *Rhodospirrilum*, *Serratia*, *Xanthomonas*, *Alspirtazobacter* of *Arspircaillium*, and *Azobacter* have a plant growth-promoting effect by colonizing the plant root zone (Nazir et al. 2018). Among these microorganisms, plant growth-promoting root bacteria (plant growth-promoting rhizobacteria, PGPR) have an important position due to their antagonistic effects and increased plant growth and yield (Beneduzi et al. 2012). PGPR positively affect parameters such as germination rate, yield, leaf area, nitrogen ratio, protein ratio, drought tolerance, root and stem weight, and provide positive features such as delaying leaf aging and developing resistance to some diseases (Çakmakçi 2005; Roriz et al. 2020).

The most studied group of PGPR are rhizobacteria, which can colonize near the soil surface, root surface, and in the rhizosphere, promoting plant growth (Kloepper and Schroth 1978). Inoculating plants with these bacteria directly affects the development of roots and shoots and helps to increase product quality by increasing biomass production (Küçük and Almaca 2020). The following steps are involved in the colonization of PGPR: (1) seed exudates are used to replicate in the zone surrounding the seed (spermosphere), (2) adhesion to root surfaces, (3) inoculation onto the seed, and (4) colonization of expanding root systems (Vacheron et al. 2013; Zhang et al. 2019).

Plant growth-promoting rhizobacteria are a viable alternative to the overuse of chemical fertilizers due to their ability to solubilize a wide range of nonsoluble minerals such as phosphorus and the synthesis of essential hormones that aid plant growth and biological control. Rhizobacteria promote plant growth through nitrogen fixation, the bioavailability of phosphorus, uptake of iron by plants with the help of siderophores, production of plant hormones such as auxin, cytokinin, and gibberellin, and reduction of plant ethylene level (Bal et al. 2013). PGPR positively affect parameters such as germination rate, yield, leaf area, nitrogen ratio, protein ratio, drought tolerance, root and stem weight, and provide positive features such as delaying leaf aging and developing resistance to some diseases (Yadegari et al. 2010).

Bacteria living in the rhizosphere constitute 7–15% of the plant root area. These bacteria, which use amino acids and sugars secreted in this region as a rich energy and nutrient source, benefit from carbon and nitrogen sources leaking from the rhizosphere (Pinton et al. 2000).

Islam et al. (2013) investigated the effects of N-fixing bacteria such as *Novosphingobium* and *Pseudomonas* on plant nutrient uptake in tomatoes and peppers. As a result of the study, the researchers reported that *Pseudomonas* sp. RFNB3 bacterial strain increased the amount of P by 67% according to the control application, *Novosphingobium* sp. RFNB21 bacterial strain increased the amount of N by 66% and the amount of K by 61% (Islam et al. 2013). As a result of PGPR applications, phosphate dissolves, and its uptake by the plant increases. As a result of the application of seven different PGPR isolates to determine the effect on the development, water, and nutrient uptake of the tomato plant, the fresh and dry weights of the tomato plants treated with PGPR were higher than those of the control plant, and the N, P, K, Ca, and Mg amount increased in the PGPR-applied groups (Fan et al. 2017).

3 Legumes

Legume crops are vital in terms of food security. Legumes, which are a joint family worldwide from tropical to temperate and cold belts, are annual or perennial herbaceous plants, some of which are woody or in the form of shrubs. Although intense cultivation of essential crops such as maize, rice, and wheat may offer enough calories for people to thrive, their protein level is frequently inadequate in several critical amino acids (Tharanathan and Mahadevamma 2003).

Leguminous plants belonging to the Fabaceae family are essential sources of energy and protein; important complex carbohydrates (dietary fibers), vitamin C, riboflavin, and niacin, and their seeds contain copper, iron, zinc, calcium, and omega-3 fatty acids. In addition to its nutritional values, it contains phenolic compounds with antioxidant activity and prebiotics such as tannins and oligosaccharides, and some bioactive components such as phytate, lectin, and enzyme inhibitors that are thought to play a role in satiety. Legumes (Fabaceae), which have economic value and are the third largest family globally, contribute to the enrichment of soils with nitrogen. They generally contain bacteria species belonging to the *Rhizobium* genus and show symbiotic characteristics in their roots. *Rhizobium* bacteria convert the free nitrogen of the air into nitrite or nitrate form that plants can use (Öğütçü et al. 2008; Uyanık et al. 2011). Fabaceae, placed in the Fabales order according to the APG IV system, includes six subfamilies: (1) Cercidoideae LPWG, (2) Detarioideae Burmeist., (3) Duparquetioideae LPWG, (4) Dialioideae LPWG, (5) Papilionoideae DC., and (6) Caesalpinioideae DC (Azani et al. 2017).

Legumes are considered a good source of phenolic compounds found in plants that play a role in their antioxidant, antimicrobial, and cytotoxic effects. Legumes, alternative food for diseases like diabetes and obesity, which require special diets, low-fat, and high-fiber content, are plant groups with various economic advantages, especially in developing countries. In developing countries, protein availability for low-income people is less than one-third of the recommended requirements. Legumes have played an essential role in human nutrition for many years due to their high-protein content. As a result of research, it has been reported that legumes have great importance in the better management of chronic diseases such as cardiovascular diseases, diabetes, and cancer and their nutritional protective role and preferable fatty acid content, reducing cholesterol and glycemic index (Arslan 2019). Legumes, which have a high export potential economically, and provide various contributions to employment, can quickly enter crop rotation and play an essential role in reducing fallow areas (Bolat et al. 2017).

4 The Role of PGPR in Legume Biofortification

Balanced nutrition is the adequate and correct intake of the required compounds and minerals from the consumed plant and animal products. Insufficient intake of vitamins and minerals required for body metabolism from consumed foods is malnutrition. It is known that millions of people globally are undernourished in terms of energy and protein needs, and nearly 2 billion people suffer from deficiencies in microelements such as Fe, B, Zn, and Se (Welch 2002; Khalid et al. 2015). For this reason, it has become essential to increase the nutritional value of the products consumed, especially for people living in undernourished, underdeveloped, and developing countries.

Biofortification eliminates deficiency in humans by increasing the concentration of vitamins and minerals, which are commonly deficient in society, in the products most consumed by society (Qaim et al. 2007). Unlike traditional fortification, biofortification aims to increase the micronutrients in the structure of the plant during the development period instead of adding minerals during food production. Plant growth-promoting rhizobacteria have been shown to improve soil fertility and crop productivity by producing siderophore, in addition to fortifying the iron content of food crops (Rana et al. 2012). Siderophores are chemical compounds with low molecular weight and a strong affinity for iron. PGPR create siderophores, release them into the environment, displace iron by forming an iron chelate complex, and then translocate toward the plant by sprouting roots. The plant absorbs these iron siderophores complexes via transporter proteins found on the plasma membrane of the root. According to Sharma et al. (2013), the use of PGPR enhanced the iron content of rice grains (Sharma et al. 2013). The use of PGPR to boost plant nutrient content is one example of a biofortification method that appears to be very promising. Some research has shown that PGPR can be used to biofortify legumes, with the majority being chickpeas, mungbeans, and soybeans. In their study, Khalid et al. (2015) aimed rhizobacterial inoculation to boost iron intake and overall growth and output of chickpeas and reported that if the soil is treated with more Fe, PGPR can assist plants in uptaking the extra Fe (Khalid et al. 2015). Co-inoculation of fodder galega plants with *Rhizobium galegae* bv. *orientalis* HAMBI 540 and *Pseudomonas trivialis* 3Re27 boosted nodule numbers and nitrogen content (Egamberdieva et al. 2010). In common beans, co-inoculation with *Pseudomonas* sp. LG and *Rhizobium phaseoli* strain 123 increased plant growth and N and phosphorus content (P) (Stajković et al. 2011).

PGPR increased chickpea plant development by increasing P-solubilization and indole acetic acid synthesis (Khalid et al. 2015). Inoculating soybean and chickpea seeds with fluorescent *Pseudomonas* that produce siderophores boosted plant growth and yield (Praveen Kumar et al. 2015). In beans, siderophore-producing strains increased growth variables such as shoot and root dry weight (Omidvari et al. 2010). Furthermore, it has been observed that siderophore synthesis by PGPR boosted soybean growth in nonsterilized soil conditions (Cattelan et al. 1999).

5 Conclusion and Prospects

The world is seriously affected by environmental damage caused by human intervention in the workings of nature, misuse, and a massive population of natural resources pressure. These adverse situations carry the risk of causing undesirable consequences, such as the world's global food not being enough to feed the whole world in the coming years. In order to feed the increasing population, agricultural production must increase significantly in the coming decades without harming nature. In order to minimize these risks, sustainable and environmentally friendly agricultural practices such as biofortification come to the fore. Increasing the quality of food with biofortification contributes to the development of healthy individuals worldwide. It is thought that biofortification studies in legumes, which constitute an important part of human nutrition, will intensify daily with increasing climatic problems. Soil-friendly PGPR have been successfully applied to provide legume biofortification but have not been sufficiently developed. However, it is thought that these applications will increase in the future due to their increasing importance.

References

Arslan M (2019) Küresel İklim Değişikliğinin Olumsuz Etkilerine Karşı Ümitvar Baklagiller olarak Mürdümük (*Lathyrus sativus* L.) ve Burçak (*Vicia ervilia* L.)'ın Önemi. Adnan Menderes Üniversitesi Ziraat Fakültesi Dergisi 16(1):97–104. https://doi.org/10.25308/aduziraat.452077

Azani N, Babineau M, Bailey CD, Banks H, Barbosa AR, Pinto RB, Boatwright JS, Borges LM, Brown GK, Bruneau A, Candido E, Cardoso D, Chung KF, Clark RP, Conceição ADS, Crisp M, Cubas P, Delgado-Salinas A, Dexter KG, Zimmerman E (2017) A new subfamily classification of the leguminosae based on a taxonomically comprehensive phylogeny. Taxon 66(1):44–77. https://doi.org/10.12705/661.3

Bal HB, Das S, Dangar TK, Adhya TK (2013) ACC deaminase and IAA producing growth promoting bacteria from the rhizosphere soil of tropical rice plants. J Basic Microbiol 53(12):972–984. https://doi.org/10.1002/jobm.201200445

Bayraç HN (2016) Türkiye' de İklim Değişikliğinin Tarım Sektörü Üzerine Etkileri. Eskişehir Osmangazi Üniversitesi İİBF Dergisi 11(1):23–48

Beneduzi A, Ambrosini A, Passaglia LMP (2012) Plant growth-promoting rhizobacteria (PGPR): Their potential as antagonists and biocontrol agents. Genet Mol Biol 4:1044–1051. http://www.scielo.br/pdf/gmb/v35n4s1/20.pdf

Bolat M, Unuvar FI, Dellal I (2017) Türkiye' de Yemeklik Baklagillerin Gelecek Eğilimlerinin Belirlenmesi. Tarım Ekonomisi Araştırmaları Dergisi 3(2):7–18

Bullock JM, Pywell RF, Burke MJW, Walker KJ (2001) Restoration of biodiversity enhances agricultural production. Ecol Lett 4(3):185–189. https://doi.org/10.1046/j.1461-0248.2001.00215.x

Çakmakçi R (2005) Bitki Gelişimini Teşvik Eden Rizobakterilerin Tarımda Kullanımı.Pdf. Atatürk Üniversitesi Ziraat Fakültesi Dergisi 36(1):97–107

Cattelan AJ, Hartel PG, Fuhrmann JJ (1999) Screening for plant growth-promoting rhizobacteria to promote early soybean growth. Soil Sci Soc Am J 63(6):1670–1680. https://doi.org/10.2136/sssaj1999.6361670x

Cohen B (2006) Urbanization in developing countries: current trends, future projections, and key challenges for sustainability. Technol Soc 28(1–2):63–80. https://doi.org/10.1016/j.techsoc.2005.10.005

da Dias JCS (2020) Integrated plant breeding strategies for harmony between modern agriculture production and the environment. Cutting-Edge Res Agric Sci 3(January):1–37. https://doi.org/10.9734/bpi/cras/v3/5393d

Egamberdieva D, Berg G, Lindström K, Räsänen LA (2010) Co-inoculation of *Pseudomonas* spp. with Rhizobium improves growth and symbiotic performance of fodder galega (*Galega orientalis* Lam.). Eur J Soil Biol 46(3–4):269–272. https://doi.org/10.1016/j.ejsobi.2010.01.005

Fan X, Zhang S, Mo X, Li Y, Fu Y, Liu Z (2017) Effects of plant growth-promoting rhizobacteria and N source on plant growth and N and P uptake by tomato grown on calcareous soils. Pedosphere 27(6):1027–1036. https://doi.org/10.1016/S1002-0160(17)60379-5

Farooq M, Wahid A, Kobayashi N, Fujita D, Basra SMA (2009) Review article Plant drought stress: effects, mechanisms and management. Agron Sustain Dev 29:185–212

Huxel M (2019) Climate Change Policy: A Case Study Analysis of Three American Cities. In: ProQuest Dissertations and Theses. https://search.proquest.com/dissertations-theses/climate-change-policy-case-study-analysis-three/docview/2240037152/se-2?accountid=41849

Islam MR, Sultana T, Joe MM, Yim W, Cho JC, Sa T (2013) Nitrogen-fixing bacteria with multiple plant growth-promoting activities enhance growth of tomato and red pepper. J Basic Microbiol 53(12):1004–1015. https://doi.org/10.1002/jobm.201200141

Johansen JL, Messmer V, Coker DJ, Hoey AS, Pratchett MS (2014) Increasing ocean temperatures reduce activity patterns of a large commercially important coral reef fish. Glob Chang Biol 20(4):1067–1074. https://doi.org/10.1111/gcb.12452

Khalid S, Asghar HN, Akhtar MJ, Aslam A, Zahir ZA (2015) Biofortification of iron in chickpea by plant growth promoting rhizobacteria. Pak J Bot 47(3):1191–1194

Kiliçaslan SC, Yildirim E, Ekİncİ M, Kul R (2020) Kuraklık Stresinin Fasulyede Bitki Gelişimi, Bazı Fizyolojik ve Biyokimyasal Özellikler Üzerine Etkisi (The effect of drought stress on plant growth, some physiological and biochemical properties of bean). Erciyes Univ J Inst Sci Technol 36(2):264–273

Kloepper J, Schroth M (1978) Plant growth-promoting rhizobacteria on radishes. In: Fourth International Conference on Plant Pathogen Bacteria 2(1):879–882

Küçük Ç, Almaca A (2020) Bitki gelişimini Teşvik Eden Rizobakteriler tarafından Üretilen Metabolitler ve Bitki Gelişimine Etkileri. Int J Life Sci Biotechnol *3*(1):81–94. https://doi.org/10.38001/ijlsb.624435

Kudoyarova GR, Kholodova VP, Veselov DS (2013) Current state of the problem of water relations in plants under water deficit. Russ J Plant Physiol 60(2):165–175. https://doi.org/10.1134/S1021443713020143

Nazir N, Kamili AN, Shah D (2018) Mechanism of Plant Growth Promoting Rhizobacteria (PGPR) in enhancing plant growth – a review. Int J Manage Technol Eng 8(7):709–721

Öğütçü H, Algur ÖF, Elkoca E, Kantar F (2008) The determination of symbiotic effectiveness of Rhizobium strains isolated from wild chickpeas collected from high altitudes in Erzurum. Turk J Agric For 32(4):241–248. https://doi.org/10.3906/tar-0801-35

Omidvari M, Sharifi RA, Ahmadzadeh M, Dahaji PA (2010) Role of fluorescent pseudomonads siderophore to increase bean growth factors. J Agric Sci 2(3):242–247. https://doi.org/10.5539/jas.v2n3p242

Pinton R, Varanini Z, Nannipieri P (2000) The rhizosphere as a site of biochemical interactions among soil components, plants, and microorganisms. In: The Rhizosphere (p. 18)

Praveen Kumar G, Desai S, Reddy G, Leo Daniel Amalraj E, Rasul A, Mir Hassan Ahmed SK (2015) Seed Bacterization with Fluorescent *Pseudomonas* spp. Enhances Nutrient Uptake and Growth of *Cajanus cajan* L. Commun Soil Sci Plant Anal 46(5):652–665. https://doi.org/10.1080/00103624.2015.1005219

Qaim M, Stein AJ, Meenakshi JV (2007) Economics of biofortification. Agric Econ 37(S1):119–133. https://doi.org/10.1111/j.1574-0862.2007.00239.x

Rana A, Joshi M, Prasanna R, Shivay YS, Nain L (2012) Biofortification of wheat through inoculation of plant growth promoting rhizobacteria and cyanobacteria. Eur J Soil Biol 50:118–126. https://doi.org/10.1016/j.ejsobi.2012.01.005

Roriz M, Carvalho SMP, Castro PML, Vasconcelos MW (2020) Legume biofortification and the role of plant growth-promoting bacteria in a sustainable agricultural era. Agronomy 10(3). https://doi.org/10.3390/agronomy10030435

Sharma A, Shankhdhar D, Shankhdhar SC (2013) Enhancing grain iron content of rice by the application of plant growth promoting rhizobacteria. Plant Soil Environ 59(2):89–94. https://doi.org/10.17221/683/2012-pse

Stajković O, Delić D, Jošić D, Kuzmanović D, Rasulić N, Knežević-Vukčević J (2011) Improvement of common bean growth by co-inoculation with Rhizobium and plant growth-promoting bacteria. Roman Biotechnol Lett 16(1):5919–5926

Tharanathan RN, Mahadevamma S (2003) Grain legumes - a boon to human nutrition. Trends Food Sci Technol 14(12):507–518. https://doi.org/10.1016/j.tifs.2003.07.002

Uyanık M, Afshar K, Rezaeieh P, Delen Y, Gürbüz B (2011) Baklagillerde Bakteri Aşılaması ve Azot Fiksasyonu. Ziraat Mühendisliği 357:8–12. https://dergipark.org.tr/en/pub/zm/issue/52109/680918

Vacheron J, Desbrosses G, Bouffaud ML, Touraine B, Moënne-Loccoz Y, Muller D, Legendre L, Wisniewski-Dyé F, Prigent-Combaret C (2013) Plant growth-promoting rhizobacteria and root system functioning. Front Plant Sci 4(SEP):1–19. https://doi.org/10.3389/fpls.2013.00356

Welch RM (2002) The impact of mineral nutrients in food crops on global human health. Plant Soil 247(1):83–90. https://doi.org/10.1023/A:1021140122921

Yadegari M, Asadi Rahmani H, Noormohammadi G, Ayneband A (2010) Plant growth promoting rhizobacteria increase growth, yield and nitrogen fixation in *Phaseolus vulgaris*. J Plant Nutr 33(12):1733–1743. https://doi.org/10.1080/01904167.2010.503776

Zhang Y, Gao X, Shen Z, Zhu C, Jiao Z, Li R, Shen Q (2019) Pre-colonization of PGPR triggers rhizosphere microbiota succession associated with crop yield enhancement. Plant Soil 439(1–2):553–567. https://doi.org/10.1007/s11104-019-04055-4

Chapter 12
Plant–Microbe Interaction for Legume Biofortification: Present Status and Future Challenges

Muhammad Waseem, Mehtab Muhammad Aslam, Aisha Lawan Idris, Oswald Nkurikiyimfura, and Zhang Di

Abstract Climate change and the rapidly growing global population, coupled with the problem of hidden hunger, necessitates the implementation of environmentally friendly agriculture practices to boost crop nutritional value and productivity. An effective solution for this is the use of plant growth–promoting bacteria (PGPB) in legume biofortification, which offers numerous health benefits and decreases the risk of various diseases. Legumes, being a significant source of plant proteins, can engage in symbiotic nitrogen (N) fixation, solubilize phosphorus (P), reduce CO_2 emissions, improve plant resistance to pathogens, and enhance soil exploration, ultimately leading to improved plant growth and soil preservation. However, the potential of microbe-mediated legume biofortification has not yet been fully explored. This chapter focuses on the significance of microbe-mediated legume biofortification in improving plant nutritional value, agronomic traits, and yields. It also emphasizes the need for the integration of genetic, biochemical, physiological, and environmental data to achieve this. Hence, the use of beneficial rhizobacteria as biofertilizers constitutes a cost-effective and promising approach for sustainable agriculture and the resolution of food security issues around the world.

M. Waseem (✉)
Department of Botany, University of Narowal, Narowal, Punjab, Pakistan

College of Horticulture, Hainan University, Haikou, Hainan, China
e-mail: m.waseem@uon.edu.pk

M. M. Aslam (✉) · Z. Di
State Key Laboratory of Agrobiotechnology, the Chinese University of Hong Kong, Shatin, Hong Kong

A. L. Idris
College of Life Sciences, Fuzhou, China

O. Nkurikiyimfura
Fujian Provincial Key Laboratory of Agroecological Processing and Safety Monitoring, Fujian Agriculture and Forestry University, Fuzhou, China

M. A. Nadeem et al. (eds.), *Legumes Biofortification*,
https://doi.org/10.1007/978-3-031-33957-8_12

1 Introduction

By 2050, the global population is expected to reach 9.7 billion, requiring 333.67 million tons of food and water (Alexandratos and Bruinsma 2012; Poveda 2021). The intensive application of synthetic fertilizers and intercropping have resulted in the depletion of soil nutrients, the degradation of soil structure, and environmental pollution. The production of food with negligible residual chemicals and environment impact could lead way to developing ecologically sustainable agriculture (Suman et al. 2022). Plants require an adequate amount of macro- and micronutrients, sufficient water, light, and specific temperatures for their optimal growth and maximal yield (Dhuldhaj and Pandya 2017). Therefore, plant–microbe interaction at the root–soil interface is the most fruitful way to reduce environmental pollution, nutrient deficiencies, and food security issues.

Legumes are excellent food crops and rich in proteins, nutrients, dietary fiber, complex carbohydrates, and several other bioactive molecules (Sathya et al. 2017). The symbiotic interaction between leguminous crops and *Rhizobium* represents 65% of the N required for agriculture. Rhizobia have also shown other desirable traits, including the synthesis of growth-promoting hormones such as indole-3-acetic acids (IAA), cytokinins, and gibberellins (GA) (Dakora and Phillips 2002, Dakora et al. 2015). These phytohormones play diverse roles in stimulating plant growth and productivity. The better adaptability of legumes with cereals crops supports livelihood resilience for smallholder farmers.

Legume biofortification offers a way to achieve naturally enriched nutritive crops through conventional breeding strategies, agronomic practices, or genetic engineering (Malik and Maqbool 2020). These strategies of biofortification are not well suited to conventional breeding and agronomic practices for the long term and could cause deficiencies in a few essential nutrients, such as the iron (Fe) required for wheat. On the contrary, the biofortification of legumes using microbial associations is a promising technique for developing sustainable crops with enhanced micronutrients, improved yields, and better soil fertility. Plants can make the most of beneficial soil bacteria because they facilitate mineral solubilization, more-efficient nutrient uptake, and phytohormone biosynthesis (i.e. cytokinin, auxin) (Bowen and Rovira 1999; Gopalakrishnan et al. 2015a). Beneficial microbes are key engineers that can restore the biogeochemical cycle and protect the agroecosystem (Kaur et al. 2020). Additionally, plant growth–promoting microorganisms (PGPMs) significantly improve plant growth status through phosphorus solubilization, phytohormones synthesis, metabolites, nitrogen fixation, and ethylene mitigation. In this context, PGPMs strengthen soil structures and increase soil fertility, which returns a supply of nutrients and phytohormones to plants, ultimately enhancing plant tolerance to salinity, drought, and metal toxicity (Singh and Singh 2017; Sun et al. 2021). In microbe-assisted biofortification, the inoculation of plants with beneficial microbes affects crop production and crops' nutritional values (Sura-de Jong et al. 2015). It has been reported that endophytic microbes are more effective relative to rhizosphere microbes because they more closely interact with plants (Reiter et al.

2002). Wheat and rice, which are used as daily consumed food crops in several regions of the world, are also facing nutrient deficiencies. Several studies have shown that the fungi and bacteria belonging to either *Acinetobacter*, *Bacillus*, *Klebsiella*, *Piriformospora indica*, or *Rhizophagus intraradices* are important for zinc (Zn) and selenium (Se) biofortification in wheat (Durán et al. 2014, 2015; Padash et al. 2016). Cereal crops fortified with minerals and other nutrients can minimize the risk of deficiencies (Bouis et al. 2011). Current agricultural practices are concentrating on increasing crop productivity and grain yields but also focus on increasing the production of nutritionally enriched staple food crops, which would help fight against global food security (Khush et al. 2012).

1.1 Importance of Biofortification

The human body requires twenty-one essential nutrients that can be supplied only through an appropriate diet. Around 60% of people in the world are facing Fe deficiency, 30% facing Zn and iodine (I) deficiencies, and 15% facing Se deficiency, while calcium (Ca), magnesium (Mg), and cooper (Cu) deficiencies are commonly found in people living in developing countries (Thacher et al. 2006). These nutrient deficiencies may cause diseases, some of them fatal, including cancer, birth defects, cardiovascular diseases, stunted growth, osteoporosis, and many others (Calton 2010). The reduction of micronutrients in the staple food crops has affected more than two million people, led to hidden hunger, and become a global threat (White and Broadley 2005, 2009). Therefore, the biofortification of legume crops, in coordination with selective breeding, to achieve nutritional demand is encouraged as part of an important strategy to improve crop quality and yields under infertile soil conditions.

The excessive and disproportionate application of chemical fertilizers to improve crop nutrients may over time accumulate in soil and cause soil contamination and thus become ineffective. Furthermore, biofortification is an effective technique for increasing the nutritional quality of staple food crops (Khush et al. 2012) and for increasing crop yields (Pfeiffer and McClafferty 2007) without adversely affecting the environment. It is now being used for raising the nutritional value of legume crops and their yields. For instance, biofortification can increase the nutritional value of a few important stable food crops, such as maize, rice, pearl millet, and wheat, for which biofortification was being achieved through conventional and molecular breeding strategies for several years. To date, the bioavailability of nutrients to crops is achieved through breeding strategies, which is cost-effective. Unlike conventional or molecular breeding, biofortification is a cost-effective, long-term process that can reach underserved areas. Finally, microbe-assisted biofortification is another promising strategy for increasing microbiome-mediated micronutrient availability to plants.

The evolution of microbial symbiosis plays a key role in the various biological and ecological processes that plants carry out and assists in various nutrient cycles through the mineralization and decomposition of organic matter or litter and by transforming inorganic nutrients to plant-usable forms. Plant growth–promoting rhizobacteria (PGPR) can influence plant development, growth, and nutrient supply through solubilization, chelation, and oxidation or reduction in soil (Pfeiffer and McClafferty 2007). In addition, plant root–microbe interactions initiate root exudations, which may adversely impact certain properties of soil physiology, such as pH and nutrient solubility (Dakora and Phillips 2002). Therefore, the use of beneficial microbes as biofertilizers is a cost-effective technique offering a promising alternative to using chemical fertilizer.

1.2 Strategies for Legume Biofortification

Micronutrient deficiencies such as those in Fe, I, Zn, and vitamin A are major threats to human health, particularly during pregnancy and early childhood (Stewart et al. 2010). To mitigate these micronutrient deficiencies, the biofortification of the stable food crops to increase nutrient contents offers a sustainable solution to enhance the availability of micronutrients (Yadav et al. 2020). Therefore, biofortification is a low-cost technique that contributes to the development of sustainable agriculture. The main strategies that have been successfully adopted to increase the nutritional quality of plant-based foods are agronomic biofortification, microbe-assisted biofortification, conventional plant breeding, and genetic engineering (Kaur et al. 2020). Agronomic biofortification is the use of mineral fertilizer to improve the mineralization and solubilization of nutrients in soil (White and Broadley 2009). Microbe-assisted biofortification is the application of microbial inoculants as biofertilizers to solubilize micronutrients in soil, to make it accessible for plant uptake. One strategy of using beneficial microbial inoculum is to reduce environmental toxicity and ensure that they are accessible to farmers because they are cost-effective and easy to cultivate, thereby maintaining the biogeochemical cycle of the environment (Prasanna et al. 2016). Conventional breeding programs tend to improve nutritional quality by enhancing the availability of edible plant tissues via genetic engineering (Singh et al. 2022).

2 Microbe-Mediated Biofortification of Different Legumes

2.1 White Lupin Biofortification

White lupin is favored among legumes thanks to its potential to improve nitrogen fixation and nutrient deficiency and to provide significant human food in marginal soils with low water supply (Atnaf et al. 2020). Because of the nitrogen fixation capacity of lupin, it may be used to increase soil fertility and rehabilitate damaged areas. When lupins (white and Andean) are intercropped with other seasonal annual legumes, the result may be increased grain yields and high forage/fodder with greater protein content (Mikić et al. 2013). A cool-season, environmentally friendly protein crop, white lupin does not require nitrogen or phosphorus fertilizer, and it prefers well-drained soils with low lime content (3%). In fine clay and silt fractions, lime inhibits lupin's ability to take the soil Fe, which is needed by nodules for N fixation (Arncken et al. 2020). Precipitation during the crucial time of vegetation is one of the most essential elements influencing crop yields (Mazur et al. 2019). The lupin family is better adapted to a wide range of abiotic stresses than other legumes are, and they can regenerate soil in poor and polluted soils. Because of its ability to absorb Cd, Zn, and other heavy metals through root nodulation, it might serve as a pioneer plant to prevent soil erosion and as a prospective phytoremediator (Fernández-Pascual et al. 2007).

White lupin is one of the most protein-rich legume seeds (Kohajdova et al. 2011), and they have been utilized for more than 3000 years. They are used not just as food but also for therapeutic purposes, although their use as food is not regarded as safe, owing to their high alkaloid content (Janusz 2017). Researchers have recently focused on breeding and producing lupin cultivars with low alkaloid contents, high protein contents, and short vegetative durations (Sujak et al. 2006). However, the domestication and the breeding of this legume have resulted in a reduction in alkaloid content, cultivar classes, and breeding lines (Kroc et al. 2017), so this crop could be a valuable raw material. To treat illnesses resulting partly from dietary habits—i.e., type II diabetes, obesity, and cardiovascular diseases—healthy nutrition is vital. White lupin seeds also provide several health advantages to combat the problems associated with cholesterol levels, high blood pressure, and insulin resistance. Given the estimates of an ever-increasing number of deaths due to these illnesses, functional foods developed with the utilization of lupin seeds will become increasingly important (Martins and Bento 2007; Martirosyan and Singh 2015). These researchers believe that lupin could be a nutraceutical and functional food.

White lupin is among the few crops capable of creating stunning structures, known as cluster roots, and have a specific physiology dedicated to the efficient accumulation of inorganic phosphate (P_i) (Lambers et al. 2013; Aslam et al. 2020, 2021a). Despite being a pivotal micronutrient, plants have developed various strategies to improve P_i remobilization and uptake from soil to plant (Kurlovich et al. 2002; Lamont and Pérez-Fernández 2016). In addition, mycorrhizal symbioses increase soil exploration in most terrestrial ecosystems, but white lupins lack such associations. Instead, soil microorganisms can encourage the growth of cluster roots (Lamont and Pérez-Fernández 2016). White lupins' ability to absorb approximately five times more P_i per root length unit than legumes with mycorrhizal associations, such as soybean (Martínez-Villaluenga et al. 2006), indicate their high potential for improving the nutrient uptake efficiency of crops (Lambers et al. 2013; Aslam et al. 2021b).

Soil bacteria such as *Bradyrhizobium* sp. are those that most commonly infect Lupinus plants and cause the development of root nodules (González-Sama et al. 2004). Field studies in two soil types showed that inoculating white lupin seed exclusively with *bradyrhizobia* or *rhizobacteria* substantially improved root dry matter, nodulation status, and the total N contents of plants growing in clay loam or sandy soils. Furthermore, lupin yield and its qualities showed a similar trend thanks to improvements to nodulation, plant vigor, and nitrogen fixation ability. For example, the largest percentage increase in white lupin yield (approximately 47.46%) was observed when grown in sandy soil and inoculated only with *bradyrhizobia*, and this value increased by 75.36% to 99.33% as a result of coinoculation with other *rhizobacteria* (Abdel-Wahab et al. 2008).

PGPB were used to analyze nutrient enrichment in white lupins. PGPB increase the effectiveness of biological nitrogen fixation, promoting the growth and increasing the yields of white lupins. The experiment included different inoculation variants, such as seeds inoculated with nitroflora, nitragine, *Bacillus subtillis* strain, and *Pseudomonas fluorescens* strain, and the few that were coinoculated with *Rhizobium* from nitroflora or nitragine + *Bacillus subtillis* or + *P. fluorescens* showed that all seed vaccinations resulted in positive responses in white lupins. The highest seed production was obtained following seed inoculation with *Rhizobium* (derived from nitragine) and the coinoculation of *P. fluorescens* with *Rhizobium* (Sulewska et al. 2019). Another study found that two bacteria (*P. brenneri* LJ215 and *Paenibacillus glycanilyticus* LJ121) boost plant growth in sterilized and seminatural environments (Ferchichi et al. 2019). Therefore, microbial biofortification is a promising technique to make nutritious, safe food containing increased levels of vitamins and macronutrients.

2.2 Soybean Biofortification

Soybean (*Glycine max*) crops are considered primary sources of oil worldwide and are essential thanks to their commercial importance and nutritional value (Singh et al.; Valliyodan et al. 2016), as they are good sources of vegetable oil and proteins (de Santos Silva et al. 2017; Singh and Kumar 2019). Soybeans are extremely important for animal feed and also have many other industrial applications (Singh and Kumar 2019). From a nutraceutical perspective, soybeans contain polyunsaturated fatty acids and good-quality fats (Tidke et al. 2015; Kamshybayeva et al. 2017). Being an excellent source of Ca, Zn, and Fe elements, which are marginally ingested, soy seeds have a higher percentage of minerals (5%) than cereal seeds (1%) (Board 2013; Kahraman 2017). As a source of bioactive peptides, soybean proteins also offer unique health benefits in that they help prevent chronic illnesses associated with aging, such as obesity, cardiovascular disease, decreased immune function, and cancer. Among the major sources of plant protein, soybeans and their products are considered important because they contain large amounts of vital amino acids and are beneficial for human health. Therefore, the demand for its grain will likely increase in the future (de Santos Silva et al. 2017).

Micro- and macronutrients are present in variable amounts in various parts of the soybean and usually are absorbed from soil. Improvements to soil's micro- and macronutrients status through microbial biofortification can enhance the nutritional status of crops and can contribute to decreasing nutrient deficiencies in humans. Microbial fertilizers and pesticides can be made from beneficial microorganisms associated with plants (Vryzas 2016), and new biotechnology tools develop stress tolerance and ameliorate nutrient efficiency in crops for a sustainable agroecosystem. Soybeans form symbiotic relationships with nitrogen-fixing rhizobia—a key bacterium for N nutrition and for agricultural production because soybean root residues supply N to cultivated crops during crop rotations (Herridge et al. 2008; Peoples et al. 2009).

Most cultivated plants, including soybeans, are colonized by arbuscular mycorrhizal fungi, which transport nutrients (P, N, and K) from the soil and assist in conferring tolerance to biotic and abiotic stresses, including salinity, heavy metals, drought, and pathogens (Smith and Read 2010). For instance, soybean-associated bacterial or fungal endophytes offer a range of plant growth–stimulating qualities that inhibit dangerous microorganism growth. These include phytohormone synthesis, biological nitrogen fixation (diazotrophic endophytes), 1-aminocyclopropane-1-carboxylate (ACC) deaminase biosynthesis, phosphate solubilization, antimicrobial metabolites, and siderophore biosynthesis and release (Santoyo et al. 2016). In soybeans, *B. aryabhattai* was investigated for its Zn solubilization capability. In soybean rhizosphere soils, *B. aryabhattai* encouraged auxin accumulation and glucosidase, dehydrogenase, microbial biomass-*C*, and microbial respiration while decreasing pH, calcium carbonate complex with Zn, and organic complexes (Ramesh et al. 2014a, b). Furthermore, only twenty (20) of the one hundred fifteen (115) isolates from soybeans were effective in solubilizing the insoluble Zn compounds with various Bacillus species, including *B. anthracis*, *B. cereus*, *B. inaquosorum*, *B. tequilensis*, *B. thuringiensis*, and *B. subtilis*. Among these, *B. anthracis* and *B. cereus* were the most efficient in increasing Zn concentrations in soybeans and seeds (Khande et al. 2017).

In addition, the rhizospheric soil of soybeans (genotype = PK 1024) contained *Enterobacter cloacae* subsp. dissolvent. This strain is capable of producing IAA, siderophores, ammonia, mineralized phytate, and solubilized K, Zn, and P, among other plant growth–promoting features, and it boosts soybean shoot weight (13.77%) and soybean seed weight (16.09%). When soybeans were inoculated with *E. cloacae*, they proved to have increased concentrations of Fe, Cu, Mn, P, and N in the shoots of the soybeans, which ultimately improved overall plant growth (Ramesh et al. 2014a, b). Another study showed that the application of *B. japonicum* intensifies the root nodulation, grain production, plant dry mass, and total seed protein content of soybeans in Uzbekistan's salinized soils (Egamberdiyeva et al. 2004). Furthermore, coinoculating PGPB such as *Bradyrhizobium* increases soybean seed production by up to 44% per hectare compared to using a single inoculant (Prakamhang et al. 2015). Therefore, the microbe-mediated biofortification of soybeans significantly improves their nutritional value and reduces the risk of malnutrition in humans who consume soybeans.

2.3 Common Bean Biofortification

The common bean (*Phaseolus vulgaris*) is the grain legume cultivated for its edible seeds as dry beans and its fresh vegetables as green beans and pods. Globally, in 2018, 24.7 million tons of fresh beans and pods and 30.4 million tons of grain were produced from 34.5 million hectares dedicated agricultural lands. In general, the common bean is a rich source of proteins, vitamins, complex carbohydrates, dietary fiber, and minerals (Broughton et al. 2003; Hayat et al. 2014). Beans are good sources of dietary protein, vital in human nutrition (Broughton et al. 2003). The most prevalent storage proteins in common bean seeds include phaseolins, globulins, legumins, albumins, lectins, and lectin-related proteins (Sparvoli et al. 2015). Among these proteins, globulins account for up to 50% of the common bean's total protein (Vitale and Bollini 1995).

The common bean is nutritionally useful because of its low lipid content and high vitamin, mineral, and protein contents (Messina 1999; Broughton et al. 2003; Paredes et al. 2009; Hayat et al. 2014; Rebello et al. 2014). Among the total lipid contents (which account for only about ~2% of total content) are beneficial exogenic polyunsaturated fatty acids such as oleic acid, linoleic acid, and palmitic acid (Hayat et al. 2014). Among the others, phospholipids and triacylglycerols are the most abundant lipid components in beans, with minor components such as diacylglycerols present in trace amounts. In contrast, beans have the greatest mineral concentration among agricultural legumes (Campos-Vega et al. 2011) and are essential sources of inorganic minerals, including Zn, Fe, P, Al, and Cu (Broughton et al. 2003; Shimelis and Rakshit 2005). Furthermore, the biological functions of common beans are associated with various health-promoting effects, such as the prevention and/or regulation of chronic illnesses such as cancers, coronary heart disease, diabetes, and obesity (Messina 2014).

A sustainable approach for increasing common bean output in developing countries is to inoculate them with symbiotic and associative bacteria to enhance plant growth (de Souza and de Brito Ferreira 2017). Rhizobium inoculation significantly increases N, P, K, Mg, and Ca absorption levels in common bean plant tissues, which subsequently enhance nodulation, plant dry mass, and final productivity (Goettsch et al. 2017). Some studies have shown that soil microorganisms such as rhizobacteria have various effects on soil quality and pH, mineral solubilization, and the nutrient absorption of plants (Saharan and Nehra 2011).

Consistent with the coinoculation of common beans with *Rhizobium* and *P. fluorescens* by Yadegari and Rahmani (2010), Khaitov et al. (2020), the coinoculation of *Rhizobium phaseoli* R9 and *Mesorhizobium ciceri* R6 significantly improved plant height, grain production, root/shoot biomass, the number of nodules/plants, nodule dry mass, and root length. Similarity, the coinoculation of the rhizobial strains was effective insofar as it increased seed yield by 35.1% and 37.9%. The bacterial coinoculation of common beans with *Azospirillum* sp., *Bacillus*, and *Rhizobium* sp. improved plant growth, productivity, and nodulation capacity (Massoud et al. 2009). Meanwhile, a combination of *Pseudomonas*, *Rhizobium*, and *Bacillus* significantly increased plant growth and N and P contents in the common bean. It was found that phosphate solubilization, IAA, siderophores, and ammonia ultimately contribute to plant growth (Knezevic-vukcevic 2011). Hungria et al. discovered that inoculating the common bean with *Rhizobium tropici* enhanced its dry nodule weight, root weight, and shoot weight by 33%, 32%, and 26%, respectively (Hungria et al. 2003). These findings might be explained by the efficacy of symbiosis between various rhizobial strains and the common bean, which subsequently increased nutrient uptake in low quantities of accessible soil N (Goettsch et al. 2017). Therefore, the microbe-mediated biofortification of the common bean is a significant technique for reducing mineral deficiencies and human health issues.

2.4 Chickpea Biofortification

Chickpeas (*Cicer arietinum*), which are self-pollinating pulse crops (Varshney et al. 2013), contain forty-four perennial and annual species (Zohary and Hopf 2000; Kerem et al. 2007). Among these, kabuli and desi (Knights and Hobson 2016) have gained popularity in Western and Southeast Asia and are consumed canned, in hummus, in salads/soups (as raw-seeds), and in channa dal. Chickpeas are excellent nutritional sources comprising carbohydrates (50%–58%), fats (3.8%–10.20%), proteins (15–22%), micronutrients (<1%), and moisture (7%–8%) (Jukanti et al. 2012; USDA, 2021). Chickpea carbohydrates such as sugar alcohols, fructo-oligosaccharides, raffinose oligosaccharides, inulin, and resistant starches are highly prebiotic (Peterbauer and Richter 2001; Johnson et al. 2020), regulating gut microbiota and promoting human health (Roberfroid et al. 2010). In comparison with lentils and field peas, chickpeas are rich in proteins, 18% of which is kabuli and 18.2% is desi (Upadhyaya et al. 2016). Likewise, chickpeas are high in amino acid contents, such as arginine, lysine, methionine, and cysteine (Jukanti et al. 2012). Furthermore, several minerals, including iron (Fe), zinc (Zn), and selenium (Se), have also been found in chickpeas.

The aim of one of the United Nations' sustainable development goals (SDGs) is to reduce malnutrition and minimize global hunger in the forthcoming years. The biofortification of staple food crops increases micronutrient concentrations, which has proved critical in combating global hunger and malnutrition. Microbes and chickpeas are involved in a specific method that makes nutrients available for plant absorption to supply plants with nutrients. Several studies have shown that beneficial microorganisms have effects on growth, nodulation, and final yield (El-Mokadem et al. 1989; Saini et al. 2015; Martínez-Hidalgo and Hirsch 2017; Rahman and Monira 2018; Kumari et al. 2019; Zaheer et al. 2019), on grain protein content (Pellegrino and Bedini 2014; Saini et al. 2015; Rahman and Monira 2018), and on biofortification (Pellegrino and Bedini 2014). Microorganisms can enhance plant growth by promoting the absorption of essential nutrients such as nitrogen (N), phosphorus (P), and potassium (K). They also boost the activities of antioxidant enzymes such as peroxidase dismutase (POD) and superoxide dismutase (SOD).

Additionally, the accumulation of root organic acids reduces the pH of the rhizosphere. These effects are attributed to the microorganisms' abilities to produce siderophores (Verma and Yadav 2012), solubilize minerals such as phosphorus (Ansari et al. 2015; Aslam et al. 2022), increase root exudation (Akrami et al. 2012; Israr et al. 2016), chelate iron (Gopalakrishnan et al. 2015b), fix atmospheric nitrogen (Verma and Yadav 2012; Gopalakrishnan et al. 2015b; Ortega García et al. 2016), and synthesize phytohormones (Verma and Yadav 2012; Yadav and Verma 2014; Gopalakrishnan et al. 2015b).

PGPB have emerged as a desirable option for enhancing the sustainability of agricultural systems worldwide, thanks to their environmental friendliness, low production costs, and reduced consumption of nonrenewable resources (Gopalakrishnan et al. 2015b), and they are widely used in chickpea cultivation. The use of PGPB has been found to be effective in improving the growth and yield of chickpea plants. One example is the use of bacteria from the genus *Pseudomonas*, which have been found to stimulate growth and increase yields in chickpeas (Joshi et al. 2019). Additionally, PGPB have been found to reduce stem-rot disease and root dryness, which are caused by pathogens such as *Sclerotinias clerotiorum* and *Rhizoctonia bataticola* (Patel et al. 2011). Another study found that in chickpeas inoculated with *Pseudomonas* sp. strain AZ5 and *Bacillus* sp. strain AZ17, P- and Zn-solubilizing bacterial strains increased Zn and P uptake, along with grain yield, nodule number, and nodule dry mass. *Pseudomonas* sp. strain AZ5 was found to produce better results (Zaheer et al. 2019). Other studies have found that *Serratia marcescens* isolates increased crop grain production in fertile irrigated soils and nutrient-deficient rainfed soils (Zaheer et al. 2016). Moreover, *Streptomyces* sp. strains had higher root and sprout mass at 30 days after sowing (DAS) with an aggressive mass/number of nodules. In addition to this, yield-related traits such as leaf area, leaf, and stem masses at 60 DAS, pod number, and pod mass were also increased to gain better grain yields at harvest (Gopalakrishnan et al. 2015b). Furthermore, in chickpea plants, *Azospirillum lipoferum* (FK1) has been found to increase salinity tolerance by increasing biomass, nutrient uptake, chlorophyll accumulation, gas exchange, enzymatic/nonenzymatic antioxidant levels, and phenolic/flavonoid contents (Mazumdar et al. 2020). Coinoculation with different types of PGPB has been found to have a positive impact on crop production. For example, coinoculation with *Bacillus halotolerans* FSZ 47 and *Mesorhizobium* sp. FCAP 26 resulted in increased plant growth, development, and seed production (Mohammadi et al. 2010). Lastly, coinoculation with *B. lentus*, *Trichoderma harzianum*, and *P. putida* resulted in more grain production and to higher contents of K_2O, P_2O_5, Fe, N, and Mg in both grains and leaves (Jat and Ahlawat 2006).

Microorganisms such as arbuscular mycorrhizal fungi (AMF) have positive impacts on chickpea crop productivity and nutrient uptake. Studies have found that the inoculation of chickpea plants with AMF, such as *Rhizophagus irregularis* and *Funneliformis mosseae*, leads to increases in plant biomass and seed yields and improve the nutritional contents of grains and the concentrations of proteins, Fe, and Zn in grains (Solaiman et al. 2012; Pellegrino and Bedini 2014; Rahman and Monira 2018). Additionally, these effects are further enhanced when local inoculum is used, as opposed to foreign inoculum (Pellegrino and Bedini 2014). Research has also shown that the combination of AMF inoculation and *Rhizobium etli* —a PO_4^{3-}-solubilizing bacterium—can significantly increase plant growth and yield indicators (Pramanik and Bera 2012). These results suggest that the use of plant growth–promoting rhizobacteria and AMF, either alone or in combination, can significantly enhance yields and improve the nutritional quality of chickpea grains.

3 Biofortification of Legumes Using Plant Growth–Promoting Microbes

Research has shown that the main barrier to absorbing micronutrients occurs at the root–soil interface (Welch 2001). One potential solution to this issue is to use microbes, known as "invisible engineers" of soil health, which play crucial roles in various biogeochemical cycles (Gadd 2010). PGPB can be found in the rhizosphere or as endophytes and can impact plant growth and development through multifaced mechanisms, such as N fixation, growth hormone production, and essential nutrient solubilization for Zn, P, and K (Bhattacharyya and Jha 2012). The use of PGPB is considered a promising approach for the biofortification of crops such as legumes (Roriz et al. 2020). Legumes are key symbiotic N fixating plants: They improve soil composition, reduce the amount N fertilizer and CO_2 emissions, and offer a broad spectrum of resistance to diseases and pests (Roriz et al. 2020). Biofortification has been used to enhance the nutritional content of pulse crops, with a focus on micronutrients such as Fe, Zn, Se, iodine, folates, and carotenoids—essential components of the various metabolic processes required for normal human development and growth (Jha and Warkentin 2020). Given that plant-based foods that are often low in key micronutrients are consumed by most of the world's population (Jha and Warkentin 2020), PGPB need to be applied to enhance the nutrient content of legumes for healthy human consumption. Table 12.1 lists the studies that used plant growth–promoting rhizobacteria to carry out the biofortification of legumes crops.

3.1 Biofortification of Legumes Using Rhizobacteria

Using plant growth–promoting bacteria (PGPB) for biofortification programs is a promising technique for enhancing the nutritional value of food crops. Studies have shown that inoculation with PGPB, including free-living bacteria, those in a symbiotic relationship with a plant's rhizosphere, and those that developed via endophytic colonization, can have positive effects on various crops, particularly legumes such as soybeans, chickpeas, and mungbeans (Roriz et al. 2020). PGPB can impact plant growth thanks to many of their traits, including nitrogen fixation, the production of growth hormones, the secretion of siderophores, and the solubilization of essential

Table 12.1 Biofortification of legume crops with plant growth–promoting rhizobacteria

Legumes	Rhizobacteria	Fortified Mineral	References
Chickpeas	Siderophore producer	Iron	Khalid (2015)
Galga plants	*Pseudomonas trivialis* 3Re27 and *Orientalis* HAMBI 540	Nitrogen	Egamberdieva et al. (2010)
Chickpeas and pigeon peas	*Brevibacterium antiquum* SRI-158, *Pseudomonas plecoglossicida* SRI-156, *P. monteilii* SRI-360, *Enterobacter ludwigii* SRI-211, *E. ludwigii* SRI-229, *Acinetobacter tandoii* SRI-305, and *Bacillus altitudinis* SRI-178	Iron, zinc, copper, manganese, and calcium	Gopalakrishnan et al. (2016)
Lentils and peas	*Pseudomonas* sp. PGERs17, *Pseudomonas* sp. NARs1, and *Rhizobium leguminosarum*-PR1	Nitrogen, phosphorus, and iron	Mishra et al. (2011, 2012)
Common beans Soybeans	*Pseudomonas* sp. LG and *Rhizobium phaseoli*	Nitrogen and phosphorus	Kne (2011)
	Bacillus aryabhattai strains (MDSR7 and MDSR14)	Zinc	Ramesh et al. (2014a, b)
Mungbeans	*Pseudomonas putida* MPJ6 and *Pantoea dispersa* MPJ9	Iron	Patel et al. (2018)
Soybeans	*Streptomyces griseoflavus* P4 and *Bradyrhizobium japonicum* SAY3-7	Nitrogen, phosphorus, potassium, calcium, and magnesium	Htwe et al. (2018)
Mungbeans	*Bacillus aryabhattai* S10 and *B. subtilis* ZM63	Nitrogen, phosphorus, and potassium	Dogra et al. (2019)
Chickpeas	Symbion-K (*Frauteria aurantia*), *Pseudomonas* sp. RA6, *P. citronellis* (PC), *Serratia* sp. S2, and *Serratia marcescens* CDP-13)	Increased macro- and micronutrient concentrations	Dogra et al. (2019)
Soybeans	*Bacillus*	Zinc	Sharma et al. (2012)
Chickpeas	*Acinetobacter* sp.	Iron	Sathya et al. (2016)
Chickpeas	*Enterobacter* sp. MN17	Zinc	Ullah et al. (2020)
Soybeans	*Paraburkholderia megapolitana*, *Alcaligenes faecalis*, and *Stenotrophomonas maltophilia*	Selenium	Trivedi et al. (2020)
Chickpeas	*Streptomyces*	Iron, Zinc, Calcium, Copper, Manganese, and Magnesium	Sathya et al. (2016)

micronutrients such as zinc, phosphorus, and potassium (Bhattacharyya and Jha 2012). PGPB offer promising crop biofortification alternatives in that they can increase nutrient content and improve overall plant health (Roriz et al. 2020).

Iron is an abundant element in the Earth's crust is one of the critical parts of plant content. Its assimilation is crucial for human health, and iron deficiencies pose significant challenges (Lurthy et al. 2021). Zinc is also an essential nutrient for plants, humans, and microorganisms to carry out the entire arrays of their physiological functions. Biofortification is an approach aimed at increasing the bioavailability of micronutrients such as zinc and iron in crops, including legumes (Hafeez et al. 2013). Studies have shown that coinoculation with plant growth–promoting bacteria (PGPB) can improve the nutritional content of crops. For example, the nodule numbers and nitrogen contents of galega were significantly improved via coinoculation with *P. trivialis* 3Re27 and *R. galegae* bv. *orientalis* HAMBI 540 (Egamberdieva et al. 2010). In pulses such as peas and lentils, improved chlorophyll accumulation, iron content, and nitrogen and phosphorus uptake, as well as nodulation and leghemoglobin, were improved thanks to coinoculation with *Pseudomonas* sp. NARs1, R. *leguminosarum*-PR1, and *Pseudomonas* sp. PGERs17 (Mishra et al. 2011, 2012). In common beans, nitrogen and phosphorus contents were improved upon *Pseudomonas* sp. LG and *Rhizobium phaseoli* coinoculation, subsequently resulting in improved plant growth (Kne 2011). Similarly, in soybeans and wheat, inoculation with *B. aryabhattai* strains (MDSR7 and MDSR14) enhanced Zn uptake in Zn-deficient soils (Ramesh et al. 2014a, b). Additionally, chickpeas and pigeon peas inoculated with *Acinetobacter tandoii* SRI-305 and *E. ludwigii* SRI-229 displayed significant increases in Zn, Fe, Cu, Ca, and Mn uptake levels. Furthermore, these bacteria are able to promote root-shoot growth and development, nodulation, crop production and yield, and the nutritional factors in soil (Gopalakrishnan et al. 2016). Siderophore-producing bacteria such as *P. putida* MPJ6 and *Pantoea dispersa* MPJ9 could increase Fe content under Fe-deficient conditions, by increasing the Fe content of mungbeans 3.4-fold, protein content 2.0-fold, and carbohydrate content 1.5-fold following *P. dispersa* inoculation (Patel et al. 2018). However, in soybeans, the absorption of other nutrients, including N, P, K, Ca, and Mn, were improved thanks to coinoculation with *Streptomyces griseoflavus* P4 and *Bradyrhizobium japonicum* SAY3–7 (Htwe et al. 2018). Coinoculation of mungbeans with *B. aryabhattai* S10 and *B. subtilis* ZM63 improved their nutritional composition in terms of N, P, and K. The inoculation of two varieties of chickpeas with five PGPB, namely *Pseudomonas* sp. RA6, *Serratia marcescens* CDP-13, *Serratia* sp. S2, *Frauteria aurantia* (Symbion-K), and *P. citronellis* (PC), increased macronutrient and micronutrient accumulation (Dogra et al. 2019). It has been found that the presence of Zn-solubilizing bacteria in soil can enhance the absorption of Zn in soybean seeds (Sharma et al. 2012). Additionally, research has shown that the Fe concentration in chickpeas can be significantly increased through inoculation with siderophore-producing bacteria, with observed increases of 81% in roots and 75% in shoots (Khalid 2015). Similarly, a study found that chickpea seed coinoculation with nineteen *Acinetobacter* species encouraged an increase in iron content by up to 38% (Sathya et al. 2016).

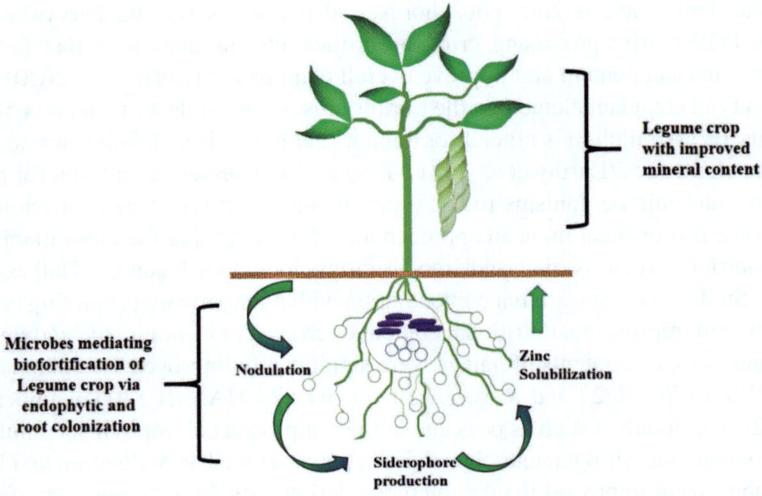

Fig. 12.1 Plant–microbe interaction for legume biofortification

Selenium is an essential trace element that plays a crucial role in animal health (Ligowe et al. 2020). It serves multiple physiological functions and is a basic component of a variety of selenium-containing proteins, making it vital for overall well-being (Ye et al. 2020). The importance of functional food production that contains high levels of essential elements such as selenium and iodine has been well established in that their deficiency can have significant impacts on human health (Golubkina et al. 2021). Research has revealed that soil microbial processes have significant impacts on the availability of inorganic selenium in soil. Microbes can alter soil properties and the redox chemistry of selenium to make selenium more bioavailable to plants. They can also affect root morphology and stimulate plant growth by secreting certain compounds, which aids in the uptake of selenium. Additionally, the expression of certain genes and proteins associated with selenium metabolism are increased, and the inoculation of certain microorganisms leads to the accumulation of certain metabolites, further contributing to the absorption of selenium (Yang et al. 2021). In leguminous plants, the phenomenon of nodulation is critical for nitrogen fixation, and iron-containing proteins play crucial roles in this process (Terpolilli et al. 2012).

The use of nodulation as a strategy for acquiring iron in leguminous plants has been well established, with research showing that rhizobia, which are bacteria that live in symbiosis with legumes, can enhance iron acquisition through the secretion of siderophores and the synthesis of Fe-binding proteins (Ku et al. 2019). This has been demonstrated in various legume species, such as *Phaseolus vulgaris*, *Pisum sativum*, and *Lens culinaris* (Mishra et al. 2011, 2012; Slatni et al. 2012). Microbial processes within roots are crucial for legume biofortification, and further research on microbial ecology is needed to fully understand these interactions and key regulatory processes (Fig. 12.1). Microbes adopt different means, such as nodulation, zinc solubilization, and siderophore production, to improve the mineral content and nutritional quality of legume crops.

3.2 *Legume Biofortification Using Endophytic Microbes*

The use of microbiomic and endophytic components as inoculants can improve plant productivity and micronutrient concentrations in plants (Singh et al. 2018a, b). For example, in chickpeas, the combination of Zn-coated seeds and a Zn-solubilizing endophyte, *Enterobacter* sp. MN17, has been shown to improve grain yields and Zn availability (Ullah et al. 2020). Similarly, endophytic seleno-bacteria from *Ricinus communis* plants, identified as *Paraburkholderia megapolitana*, *Alcaligenes faecalis*, and *Stenotrophomonas maltophilia*, have been found to improve the growth of soybean crops 7.4-fold compared to the controls (Trivedi et al. 2020). Plant growth–promoting actinobacteria have also been found to significantly improve the seed mineral density of chickpeas, including minerals such as Fe (10%–38%), Zn (13%–30%), Ca (14%–26%), Cu (11%–54%), Mn (18%–35%), and Mg (14%–21%), when grown under field conditions (Sathya et al. 2016). While research on bacterial and fungal endophytes is limited, some studies have demonstrated endophytes' potential for Fe/Zn biofortification in wheat and rice. Mechanisms that enhance Fe/Zn uptake through the inoculation of endophytes include the chelation of Fe by siderophore-producing microorganisms, the excretions of root organic exudates (for altering soil pH), rhizospheric microbe proton extrusions, and phytohormone synthesis for gibberellic acid, ethylene, and auxin (Singh et al. 2018a, b). Further research on the endophytes that can mediate biofortification in legumes, and their morphological and physiological traits, is needed to gain a better understanding of their potential for improving the nutrient content of these important food crops.

4 Conclusion and Prospects

Undoubtedly, legumes are highly important food crops because they can meet human nutrition needs and develop sustainable agriculture. Plant–microbe associations are part of an environmentally friendly promising approach to not only improving the nutrition quality of crops but also reducing environmental pollution by minimizing the input of chemicals. Currently, climate change and the rapid increase in the global population are fundamental concerns, and the microbe-assisted biofortification of legumes is probably the most influential and cost-effective way to deal with hidden hunger. Meanwhile, understanding microbes, identifying the desirable traits of legumes, and learning how they interact under different environmental conditions and in different soil compositions would be meaningful. The primary purpose of legumes, for human health, is to provide nutrients and minerals, and insufficient investigations have restricted their application in cereal crops. Nonetheless, the microbe-assisted biofortification of legumes can support the Green Revolution and achieve sustainable environment.

Author Contributions MMA and MW conceived the study design. MMA, OW, AI, and MW drafted the manuscript and designed the figures. MW, ZD, and MAA critically reviewed the manuscript. All the authors have read and agreed to the published version of the manuscript.

Conflict of Interest The authors declare no conflicts of interest.

References

Abdel-Wahab A, Mekhemar G, Badawi FSF, Shehata HS (2008) Enhancement of nitrogen fixation, growth, and productivity of bradyrhizobium-lupin symbiosis via co-inoculation with rhizobacteria in different soil types. J Agric Chem Biotechnol 33(1):469–484

Akrami M, Khiavi HK, Shikhlinski H, Khoshvaghtei H (2012) Bio controlling two pathogens of chickpea Fusarium solani and Fusarium oxysporum by different combinations of Trichoderma harzianum, Trichoderma asperellum and Trichoderma virens under field condition. Int J Agric Sci Res 1(3):41–45

Alexandratos N, Bruinsma J (2012) World agriculture towards 2030/2050: the 2012 revision

Ansari MF, Tipre DR, Dave SR (2015) Efficiency evaluation of commercial liquid biofertilizers for growth of Cicer aeritinum (chickpea) in pot and field study. Biocatal Agric Biotechnol 4(1):17–24

Arncken C, Klaiss M, Wendling M, Messmer M (2020) Cultivation of white lupin-A cool-season and environmentally friendly protein crop

Aslam MM, Akhtar K, Karanja JK, Haider FU (2020) Understanding the adaptive mechanisms of plant in low phosphorous soil. Plant stress physiology. IntechOpen, London

Aslam MM, Karanja JK, Yuan W, Zhang Q, Zhang J, Xu W (2021a) Phosphorus uptake is associated with the rhizosheath formation of mature cluster roots in white lupin under soil drying and phosphorus deficiency. Plant Physiol Biochem 166:531–539

Aslam MM, Waseem M, Zhang Q, Ke W, Zhang J, Xu W (2021b) Identification of ABC transporter G subfamily in white lupin and functional characterization of L. albABGC29 in phosphorus use. BMC Genom **22**(1):1–14

Aslam MM, Waseem M, Weifeng X, Qamar MTU. Identification and expression analysis of phosphate transporter (PHT) gene family in Lupinus albus cluster root under phosphorus stress. Int J Biol Macromol. 2022 Apr 30;205:772–781. https://doi.org/10.1016/j.ijbiomac.2022.03.085. Epub 2022 Mar 21. PMID: 35331794.

Atnaf M, Wegary D, Tesfaye K, Dagne K, Mazengia Y, Ayalew B, Melak A, Jaleta M (2020) Exploring forgotten opportunity: White Lupin development for food, feed, cash, health, and soil fertility management in Ethiopia. Cogent Environ Sci 6(1):1813451

Bhattacharyya PN, Jha DK (2012) Plant growth–promoting rhizobacteria (PGPR): emergence in agriculture. World J Microbiol Biotechnol 28(4):1327–1350

Board J (2013) A comprehensive survey of international soybean research: genetics, physiology, agronomy and nitrogen relationships

Bouis HE, Eozenou P, Rahman A (2011) Food prices, household income, and resource allocation: socioeconomic perspectives on their effects on dietary quality and nutritional status. Food Nutr Bull 32(1_suppl1):S14–S23

Bowen G, Rovira A (1999) The rhizosphere and its management to improve plant growth. Adv Agron 66:1–102

Broughton WJ, Hernandez G, Blair M, Beebe S, Gepts P, Vanderleyden J (2003) Beans (Phaseolus spp.)–model food legumes. Plant Soil 252(1):55–128

Calton JB (2010) Prevalence of micronutrient deficiency in popular diet plans. J Int Soc Sports Nutr 7:24

Campos-Vega R, Vergara-Castañeda H, Oomah B (2011) Functional food sources: beans in sight. Nova Science Publishers, Inc., Hauppauge, pp 1–56

Dakora, F.D., & Phillips, D.A. (2002). Root exudates as mediators of mineral acquisition in low-nutrient environments. Plant and Soil, 245, 35–47.

Dakora F, Matiru V, Kanu A (2015) Rhizosphere ecology of lumichrome and riboflavin, two bacterial signal molecules eliciting developmental changes in plants. Front Plant Sci 6

de Santos Silva, F. C., T. Sediyama, R. de Cássia Teixeira Oliveira, A. Borém, F. L. de Silva, A. R. G. Bezerra and A. F. de Silva (2017). Economic importance and evolution of breeding. Soybean Breeding. Springer: 1–16.

de Souza JEB, de Brito Ferreira EP (2017) Improving sustainability of common bean production systems by co-inoculating rhizobia and azospirilla. Agric Ecosyst Environ 237:250–257

Dhuldhaj U, Pandya U (2017) Implementation of biofortification technology by using PGPR for sustainable agricultural production. Agriculturally important microbes for sustainable agriculture. Springer, pp 63–79

Dogra N, Yadav R, Kaur M, Adhikary A, Kumar S, Ramakrishna W (2019) Nutrient enhancement of chickpea grown with plant growth promoting bacteria in local soil of Bathinda, Northwestern India. Physiol Mol Biol Plants 25(5):1251–1259

Durán P, Acuña JJ, Jorquera MA, Azcón R, Paredes C, Rengel Z, de la Luz Mora M (2014) Endophytic bacteria from selenium-supplemented wheat plants could be useful for plant-growth promotion, biofortification and Gaeumannomyces graminis biocontrol in wheat production. Biol Fertil Soils 50(6):983–990

Durán P, Acuña JJ, Gianfreda L, Azcón R, Funes-Collado V, Mora ML (2015) Endophytic selenobacteria as new inocula for selenium biofortification. Appl Soil Ecol 96:319–326

Egamberdieva D, Berg G, Lindström K, Räsänen LA (2010) Co-inoculation of Pseudomonas spp. with Rhizobium improves growth and symbiotic performance of fodder galega (Galega orientalis Lam.). Eur J Soil Biol 46(3):269–272

Egamberdiyeva D, Qarshieva D, Davranov K (2004) Growth and yield of soybean varieties inoculated with Bradyrhizobium spp in N-deficient calcareous soils. Biol Fertil Soils 40(2):144–146

El-Mokadem M, Helemish F, Abou-Bakr Z, Sheteawi S (1989) Associative effect of Azospirillum lipoferum and Azotobacter chroococcum with Rhizobium spp. on mineral composition and growth of chickpea (Cicer arietinum) on sandy soils. Zentralblatt für Mikrobiologie 144(4):255–265

Ferchichi N, Toukabri W, Boularess M, Smaoui A, Mhamdi R, Trabelsi D (2019) Isolation, identification and plant growth promotion ability of endophytic bacteria associated with lupine root nodule grown in Tunisian soil. Arch Microbiol 201(10):1333–1349

Fernández-Pascual M, Pueyo JJ, Felipe M, Golvano MP, Lucas MM (2007) Singular features of the Bradyrhizobium-Lupinus symbiosis

Gadd GM (2010) Metals, minerals and microbes: geomicrobiology and bioremediation. Microbiology 156(3):609–643

Goettsch LH, Lenssen AW, Yost RS, Luvaga ES, Semalulu O, Tenywa M, Miiro R, Mazur RE (2017) Improved production systems for common bean on Ferralsol soil in south-central Uganda. Afr J Agric Res 12(23):1959–1969

Golubkina N, Moldovan A, Kekina H, Kharchenko V, Sekara A, Vasileva V, Skrypnik L, Tallarita A, Caruso G (2021) Joint biofortification of plants with selenium and iodine: new field of discoveries. Plants 10(7):1352

González-Sama A, Lucas MM, De Felipe MR, Pueyo JJ (2004) An unusual infection mechanism and nodule morphogenesis in white lupin (Lupinus albus). New Phytologist:371–380

Gopalakrishnan S, Sathya A, Vijayabharathi R, Varshney RK, Gowda C, Krishnamurthy L (2015a) Plant growth promoting rhizobia: challenges and opportunities. 3 Biotech 5(4):355–377
Gopalakrishnan S, Srinivas V, Alekhya G, Prakash B, Kudapa H, Rathore A, Varshney RK (2015b) The extent of grain yield and plant growth enhancement by plant growth–promoting broad-spectrum Streptomyces sp. in chickpea. SpringerPlus 4(1):1–10
Gopalakrishnan S, Vadlamudi S, Samineni S, Sameer Kumar CV (2016) Plant growth promotion and biofortification of chickpea and pigeonpea through inoculation of biocontrol potential bacteria, isolated from organic soils. SpringerPlus 5(1):1882
Hafeez FY, Abaid-Ullah M, Hassan MN (2013) Plant growth–promoting rhizobacteria as zinc mobilizers: a promising approach for cereals biofortification. In: Maheshwari DK, Saraf M, Aeron A (eds) Bacteria in agrobiology: crop productivity. Springer Berlin Heidelberg, Berlin, Heidelberg, pp 217–235
Hayat I, Ahmad A, Masud T, Ahmed A, Bashir S (2014) Nutritional and health perspectives of beans (Phaseolus vulgaris L.): an overview. Crit Rev Food Sci Nutr 54(5):580–592
Herridge DF, Peoples MB, Boddey RM (2008) Global inputs of biological nitrogen fixation in agricultural systems. Plant Soil 311(1):1–18
Htwe AZ, Moh SM, Moe K, Yamakawa T (2018) Effects of co-inoculation of Bradyrhizobium japonicum SAY3-7 and Streptomyces griseoflavus P4 on plant growth, nodulation, nitrogen fixation, nutrient uptake, and yield of soybean in a field condition. Soil Sci Plant Nutr 64(2):222–229
Hungria M, Campo RJ, Mendes IC (2003) Benefits of inoculation of the common bean (Phaseolus vulgaris) crop with efficient and competitive Rhizobium tropici strains. Biol Fertil Soils 39(2):88–93
Israr D, Mustafa G, Khan KS, Shahzad M, Ahmad N, Masood S (2016) Interactive effects of phosphorus and Pseudomonas putida on chickpea (Cicer arietinum L.) growth, nutrient uptake, antioxidant enzymes and organic acids exudation. Plant Physiol Biochem 108:304–312
Janusz P (2017) White lupin (Lupinus albus L.)–nutritional and health values in human nutrition–a review. Czech J Food Sci 35(2):95–105
Jat R, Ahlawat I (2006) Direct and residual effect of vermicompost, biofertilizers and phosphorus on soil nutrient dynamics and productivity of chickpea-fodder maize sequence. J Sustain Agric 28(1):41–54
Jha AB, Warkentin TD (2020) Biofortification of pulse crops: status and future perspectives. Plants 9(1):73
Johnson N, Johnson CR, Thavarajah P, Kumar S, Thavarajah D (2020) The roles and potential of lentil prebiotic carbohydrates in human and plant health. Plants People Planet 2(4):310–319
Joshi D, Chandra R, Suyal DC, Kumar S (2019) Impacts of bioinoculants Pseudomonas jesenii MP1 and Rhodococcus qingshengii S10107 on chickpea (Cicer arietinum L.) yield and soil nitrogen status. Pedosphere 29(3):388–399
Jukanti AK, Gaur PM, Gowda C, Chibbar RN (2012) Nutritional quality and health benefits of chickpea (Cicer arietinum L.): a review. Br J Nutr 108(S1):S11–S26
Kahraman A (2017) Nutritional value and foliar fertilization in soybean. J Elementol 22(1)
Kamshybayeva G, Atabayeva S, Kenzhebayeva S, Domakbayeva A, Utesheva S, Nurmahanova A, Asrandina S, Zhuniszhan A, Turpanova R (2017) The importance of soybean (Glycine max) as a source of biologically valuable substances. Int J Biol Chem 10(2):23–27
Kaur T, Rana KL, Kour D, Sheikh I, Yadav N, Kumar V, Yadav AN, Dhaliwal HS, Saxena AK (2020) Microbe-mediated biofortification for micronutrients: present status and future challenges. In: New and future developments in microbial biotechnology and bioengineering, vol Elsevier, pp 1–17
Kerem Z, Lev-Yadun S, Gopher A, Weinberg P, Abbo S (2007) Chickpea domestication in the Neolithic Levant through the nutritional perspective. J Archaeol Sci 34(8):1289–1293
Khaitov B, Vollmann J, Yeong Pyon J, Park K (2020) Improvement of salt tolerance and growth in common bean (Phaseolus vulgaris L.) by co-inoculation with native rhizobial strains. J Agric Sci Technol 22(1):209–220

Khalid S (2015) Biofortification of iron in chickpea by plant growth promoting rhizobacteria. Pak J Bot 47(3):1191–1194
Khande R, Sharma SK, Ramesh A, Sharma MP (2017) Zinc solubilizing Bacillus strains that modulate growth, yield and zinc biofortification of soybean and wheat. Rhizosphere 4:126–138
Khush GS, Lee S, Cho J-I, Jeon J-S (2012) Biofortification of crops for reducing malnutrition. Plant Biotechnol Reports 6:195–202
Kne J (2011) Improvement of common bean growth by co-inoculation with Rhizobium and plant growth–promoting bacteria
Knezevic-vukcevic J (2011) Improvement of common bean growth by co-inoculation with Rhizobium and plant growth–promoting bacteria. Rom Biotechnol Lett 16:5919–5926
Knights E, Hobson K (2016) Chickpea overview
Kohajdova Z, KaroVičoVá J, Schmidt Š (2011) Lupin composition and possible use in bakery-a review. Czech J Food Sci 29(3):203–211
Kroc M, Rybiński W, Wilczura P, Kamel K, Kaczmarek Z, Barzyk P, Święcicki W (2017) Quantitative and qualitative analysis of alkaloids composition in the seeds of a white lupin (Lupinus albus L.) collection. Genet Resour Crop Evol 64(8):1853–1860
Ku Y-S, Rehman HM, Lam H-M (2019) Possible roles of rhizospheric and endophytic microbes to provide a safe and affordable means of crop biofortification. Agronomy 9(11):764
Kumari N, Mondal S, Mahapatra P, Meetei T (2019) Effect of biofertilizer and micronutrients on yield of chickpea
Kurlovich B, Kartuzova L, Heinanen J, Benken I, Chmeleva Z, Bernatskaya M (2002) Biochemical composition. Chapter 9. In: Lupins (Geography, classification, genetic resources and breeding). OY International Express, St. Petersburg, pp 241–268
Lambers H, Clements JC, Nelson MN (2013) How a phosphorus-acquisition strategy based on carboxylate exudation powers the success and agronomic potential of lupines (Lupinus, Fabaceae). Am J Bot 100(2):263–288
Lamont BB, Pérez-Fernández M (2016) Total growth and root-cluster production by legumes and proteas depends on rhizobacterial strain, host species and nitrogen level. Ann Bot 118(4):725–732
Ligowe IS, Young SD, Ander EL, Kabambe V, Chilimba ADC, Bailey EH, Lark RM, Nalivata PC (2020) Selenium biofortification of crops on a Malawi Alfisol under conservation agriculture. Geoderma 369:114315
Lurthy T, Pivato B, Lemanceau P, Mazurier S (2021) Importance of the rhizosphere microbiota in iron biofortification of plants. Front Plant Sci 12
Malik KA, Maqbool A (2020) Transgenic crops for biofortification. Front Sustain Food Syst 4
Martínez-Hidalgo P, Hirsch AM (2017) The nodule microbiome: N_2-fixing rhizobia do not live alone. Phytobiomes 1(2):70–82
Martínez-Villaluenga C, Sironi E, Vidal-Valverde C, Duranti M (2006) Effects of oligosaccharide removing procedure on the protein profiles of lupin seeds. Eur Food Res Technol 223(5):691–696
Martins J, Bento O (2007) As leguminosas como alimentos funcionais: o caso das dislipidémias e das doenças cardiovasculares. Revista de Ciências Agrárias 30(1):385–399
Martirosyan DM, Singh J (2015) A new definition of functional food by FFC: what makes a new definition unique? Func Foods Health Dis 5(6):209–223
Massoud O, Morsy EM, El-Batanony NH (2009) Field response of snap bean (Phaseolus vulgaris L.) to N_2 fixers Bacillus circulans and arbuscular mycorrhizal fungi inoculation through accelerating rock phosphate and feldspar weathering. Aust J Basic Appl Sci 3:844–852
Mazumdar D, Saha SP, Ghosh S (2020) Isolation, screening and application of a potent PGPR for enhancing growth of Chickpea as affected by nitrogen level. Int J Veg Sci 26(4):333–350
Mazur V, Mazur K, Pantsyreva H (2019) Influence of the technological aspects growing on quality composition of seed white lupine (Lupinus albus L.) in the Forest Steppe of Ukraine. Ukrain J Ecol 9(1):50–55

Messina MJ (1999) Legumes and soybeans: overview of their nutritional profiles and health effects. Am J Clin Nutr 70(3):439 s–450 s

Messina V (2014) Nutritional and health benefits of dried beans. Am J Clin Nutr 100(suppl_1):437S–442S

Mikić A, Ćupina B, Mihailović V, Krstić Đ, Antanasović S, Zorić L, Đorđević V, Perić V, Srebrić M (2013) Intercropping white (Lupinus albus) and Andean (Lupinus mutabilis) lupins with other annual cool season legumes for forage production. South Afr J Bot 89:296–300

Mishra PK, Bisht SC, Ruwari P, Joshi GK, Singh G, Bisht JK, Bhatt J (2011) Bioassociative effect of cold tolerant Pseudomonas spp. and Rhizobium leguminosarum-PR1 on iron acquisition, nutrient uptake and growth of lentil (Lens culinaris L.). Eur J Soil Biol 47(1):35–43

Mishra PK, Bisht SC, Mishra S, Selvakumar G, Bisht J, Gupta H (2012) Coinoculation of Rhizobium leguminosarum-PR1 with a cold tolerant Pseudomonas sp. improves iron acquisition, nutrient uptake and growth of field pea (Pisum sativum L.). J Plant Nutr 35(2):243–256

Mohammadi K, Ghalavand A, Aghaalikhani M (2010) Effect of organic matter and biofertilizers on chickpea quality and biological nitrogen fixation. World Acad Sci Eng Technol 4:966–971

Ortega García M, Shagarodsky Scull T, Dibut Álvarez BL, Ríos Rocafull Y, Tejeda González G, Gómez Jorrin LA (2016) Influencia de la interacción entre el cultivo del garbanzo (Cicer arietinum L.) y la inoculación con cepas seleccionadas de Mesorhizobium spp. Cultivos Tropicales 37:20–27

Padash A, Shahabivand S, Behtash F, Aghaee A (2016) A practicable method for zinc enrichment in lettuce leaves by the endophyte fungus Piriformospora indica under increasing zinc supply. Scientia Horticulturae 213:367–372

Paredes M, Becerra V, Tay J (2009) Inorganic nutritional composition of common bean (Phaseolus vulgaris L.) genotypes race Chile. Chilean J Agric Res 69(4):486–495

Patel D, Singh H, Shroff S, Sahu J (2011) Antagonistic efficiency of Pseudomonas strains against soil borne disease of chickpea crop under in vitro and in vivo. Elixir Agric 30:1774–1777

Patel P, Trivedi G, Saraf M (2018) Iron biofortification in mungbean using siderophore producing plant growth promoting bacteria. Environ Sustain 1(4):357–365

Pellegrino E, Bedini S (2014) Enhancing ecosystem services in sustainable agriculture: biofertilization and biofortification of chickpea (Cicer arietinum L.) by arbuscular mycorrhizal fungi. Soil Biol Biochem 68:429–439

Peoples M, Brockwell J, Herridge D, Rochester I, Alves B, Urquiaga S, Boddey R, Dakora F, Bhattarai S, Maskey S (2009) The contributions of nitrogen-fixing crop legumes to the productivity of agricultural systems. Symbiosis 48(1):1–17

Peterbauer T, Richter A (2001) Biochemistry and physiology of raffinose family oligosaccharides and galactosyl cyclitols in seeds. Seed Sci Res 11(3):185–197

Pfeiffer WH, McClafferty B (2007) HarvestPlus: breeding crops for better nutrition. Crop Sci 47(S3):S-88–S-105

Poveda J (2021) Insect frass in the development of sustainable agriculture. A review. Agron Sustain Dev 41(1):1–10

Prakamhang J, Tittabutr P, Boonkerd N, Teamtisong K, Uchiumi T, Abe M, Teaumroong N (2015) Proposed some interactions at molecular level of PGPR coinoculated with Bradyrhizobium diazoefficiens USDA110 and B. japonicum THA6 on soybean symbiosis and its potential of field application. Appl Soil Ecol 85:38–49

Pramanik K, Bera A (2012) Response of biofertilizers and phytohormone on growth and yield of chickpea (Cicer aritinum L.). J Crop Weed 8(2):45–49

Prasanna R, Nain L, Rana A, Shivay YS (2016) Biofortification with microorganisms: present status and future challenges. In: Biofortification of food crops. Springer, pp 249–262

Rahman GM, Monira S (2018) Performance of biofertilizers on growth and yield of chickpea

Ramesh A, Sharma SK, Sharma MP, Yadav N, Joshi OP (2014a) Inoculation of zinc solubilizing Bacillus aryabhattai strains for improved growth, mobilization and biofortification of zinc in soybean and wheat cultivated in Vertisols of central India. Appl Soil Ecol 73:87–96

Ramesh A, Sharma SK, Sharma MP, Yadav N, Joshi OP (2014b) Plant growth–promoting traits in Enterobacter cloacae subsp. dissolvens MDSR9 isolated from soybean rhizosphere and its impact on growth and nutrition of soybean and wheat upon inoculation. Agric Res 3(1):53–66

Rebello CJ, Greenway FL, Finley JW (2014) Whole grains and pulses: a comparison of the nutritional and health benefits. J Agric Food Chem 62(29):7029–7049

Reiter B, Pfeifer U, Schwab H, Sessitsch A (2002) Response of endophytic bacterial communities in potato plants to infection with Erwinia carotovora subsp. atroseptica. Appl Environ Microbiol 68(5):2261–2268

Roberfroid M, Gibson GR, Hoyles L, McCartney AL, Rastall R, Rowland I, Wolvers D, Watzl B, Szajewska H, Stahl B (2010) Prebiotic effects: metabolic and health benefits. Br J Nutr 104(S2):S1–S63

Roriz M, Carvalho SMP, Castro PML, Vasconcelos MW (2020) Legume biofortification and the role of plant growth–promoting bacteria in a sustainable agricultural era. Agronomy 10(3):435

Saharan B, Nehra V (2011) Plant growth promoting rhizobacteria: a critical review. Life Sci Med Res 21(1):30

Saini R, Dudeja SS, Giri R, Kumar V (2015) Isolation, characterization, and evaluation of bacterial root and nodule endophytes from chickpea cultivated in Northern India. J Basic Microbiol 55(1):74–81

Santoyo G, Moreno-Hagelsieb G, del Carmen Orozco-Mosqueda M, Glick BR (2016) Plant growth–promoting bacterial endophytes. Microbiol Res 183:92–99

Sathya A, Vijayabharathi R, Srinivas V, Gopalakrishnan S (2016) Plant growth–promoting actinobacteria on chickpea seed mineral density: an upcoming complementary tool for sustainable biofortification strategy. 3 Biotech 6(2):138

Sathya A, Vijayabharathi R, Gopalakrishnan S (2017) Plant growth–promoting actinobacteria: a new strategy for enhancing sustainable production and protection of grain legumes. 3 Biotech 7(2):102–102

Sharma SK, Sharma MP, Ramesh A, Joshi OP (2012) Characterization of zinc-solubilizing Bacillus isolates and their potential to influence zinc assimilation in soybean seeds. J Microbiol Biotechnol 22(3):352–359

Shimelis EA, Rakshit SK (2005) Proximate composition and physico-chemical properties of improved dry bean (Phaseolus vulgaris L.) varieties grown in Ethiopia. LWT-Food Sci Technol 38(4):331–338

Singh G, Kumar A (2019) Synteny analysis of Glycine max and Phaseolus vulgaris revealing conserved regions of NBS-LRR coding genes. Biosci Biotechnol Res Commun 12(1):124–133

Singh J, Singh AV (2017) Microbial strategies for enhanced phytoremediation of heavy metal-contaminated soils. Environmental pollutants and their bioremediation approaches. CRC Press, pp 257–272

Singh D, Geat N, Rajawat M, Mahajan M, Prasanna R, Singh S, Kaushik R, Singh R, Kumar K, Saxena A (2018a) Deciphering the mechanisms of endophyte-mediated biofortification of Fe and Zn in wheat. J Plant Growth Regul 37(1):174–182

Singh D, Geat N, Rajawat MVS, Prasanna R, Kar A, Singh AM, Saxena AK (2018b) Prospecting endophytes from different Fe or Zn accumulating wheat genotypes for their influence as inoculants on plant growth, yield, and micronutrient content. Ann Microbiol 68(12):815–833

Singh B, Goutam U, Kukreja S, Siddappa S, Sood S, Sharma J, Bhardwaj V (2022) Biofortification strategies to improve iron concentrations in potato tubers: lessons and future opportunities. Potato Res 65(1):51–64

Singh G, Ratnaparkhe M, Kumar A. Comparative analysis of transposable elements from

Slatni T, Dell'Orto M, Salah IB, Vigani G, Smaoui A, Gouia H, Zocchi G, Abdelly C (2012) Immunolocalization of H+-ATPase and IRT1 enzymes in N_2-fixing common bean nodules subjected to iron deficiency. J Plant Physiol 169(3):242–248

Smith SE, Read DJ (2010) Mycorrhizal symbiosis. Academic Press

Solaiman A, Rabbani M, Hossain D, Hossain G, Alam M (2012) Influence of phosphorus and inoculation with Rhizobium and AM fungi on growth and dry matter yield of chickpea. Bangladesh J Sci Res 25(1):23–32

Sparvoli F, Bollini R, Cominelli E (2015) Nutritional value. In: Grain legumes. Springer, pp 291–325

Stewart CP, Dewey KG, Ashorn P (2010) The undernutrition epidemic: an urgent health priority. Lancet 375(9711):282

Sujak A, Kotlarz A, Strobel W (2006) Compositional and nutritional evaluation of several lupin seeds. Food Chem 98(4):711–719

Sulewska H, Ratajczak K, Niewiadomska A, Panasiewicz K (2019) The use of microorganisms as bio-fertilizers in the cultivation of white lupine. Open Chem 17(1):813–822

Suman J, Rakshit A, Ogireddy SD, Singh S, Gupta C, Chandrakala J (2022) Microbiome as a key player in sustainable agriculture and human health. Front Soil Sci 2:12

Sun Z, Yue Z, Liu H, Ma K, Li C (2021) Microbial-assisted wheat iron biofortification using endophytic Bacillus altitudinis WR10. Front Nutr 8:704030

Sura-de Jong M, Reynolds RJB, Richterova K, Musilova L, Staicu LC, Chocholata I, Cappa JJ, Taghavi S, van der Lelie D, Frantik T (2015) Selenium hyperaccumulators harbor a diverse endophytic bacterial community characterized by high selenium resistance and plant growth promoting properties. Front Plant Sci 6:113

Terpolilli JJ, Hood GA, Poole PS (2012) What determines the efficiency of N_2-fixing Rhizobium-legume symbioses? Adv Microb Physiol 60:325–389

Thacher TD, Fischer PR, Strand MA, Pettifor JM (2006) Nutritional rickets around the world: causes and future directions. Ann Trop Paediatr 26(1):1–16

Tidke SA, Ramakrishna D, Kiran S, Kosturkova G, Ravishankar G (2015) Nutraceutical potential of soybean: review. Asian J Clin Nutr **7**(2):22–32

Trivedi G, Patel P, Saraf M (2020) Synergistic effect of endophytic selenobacteria on biofortification and growth of Glycine max under drought stress. South Afr J Bot 134:27–35

Ullah A, Farooq M, Hussain M (2020) Improving the productivity, profitability and grain quality of kabuli chickpea with co-application of zinc and endophyte bacteria Enterobacter sp. MN17. Arch Agron Soil Sci 66(7):897–912

Upadhyaya HD, Bajaj D, Narnoliya L, Das S, Kumar V, Gowda C, Sharma S, Tyagi AK, Parida SK (2016) Genome-wide scans for delineation of candidate genes regulating seed-protein content in chickpea. Front Plant Sci **7**:302

Valliyodan B, Qiu D, Patil G, Zeng P, Huang J, Dai L, Chen C, Li Y, Joshi T, Song L (2016) Landscape of genomic diversity and trait discovery in soybean. Sci Reports 6(1):1–10

Varshney RK, Song C, Saxena RK, Azam S, Yu S, Sharpe AG, Cannon S, Baek J, Rosen BD, Tar'an B (2013) Draft genome sequence of chickpea (Cicer arietinum) provides a resource for trait improvement. Nat Biotechnol 31(3):240–246

Verma JP, Yadav J (2012) Evaluation of plant growth promoting rhizobacteria and their effect on plant growth and grain yield of chickpea (Cicer arietinum L.) under sustainable agriculture Production. Agric Sci Eng (ICASE2012):127

Vitale A, Bollini R (1995) Legume storage proteins. In: Seed development and germination, pp 73–102

Vryzas Z (2016) The plant as metaorganism and research on next-generation systemic pesticides–prospects and challenges. Front Microbiol **7**:1968

Welch R (2001) Perspectives on the micronutrient nutrition of crops

White PJ, Broadley MR (2005) Biofortifying crops with essential mineral elements. Trends Plant Sci 10(12):586–593

White PJ, Broadley MR (2009) Biofortification of crops with seven mineral elements often lacking in human diets--iron, zinc, copper, calcium, magnesium, selenium and iodine. New Phytol 182(1):49–84

Yadav J, Verma JP (2014) Effect of seed inoculation with indigenous Rhizobium and plant growth promoting rhizobacteria on nutrients uptake and yields of chickpea (Cicer arietinum L.). Eur J Soil Biol 63:70–77

Yadav AN, Rastegari AA, Yadav N, Kour D (2020) Advances in plant microbiome and sustainable agriculture. Springer

Yadegari M, Rahmani HA (2010) Evaluation of bean (Phaseolus vulgaris) seeds inoculation with Rhizobium phaseoli and plant growth promoting Rhizobacteria (PGPR) on yield and yield components. Afr J Agric Res 5(9):792–799

Yang D, Hu C, Wang X, Shi G, Li Y, Fei Y, Song Y, Zhao X (2021) Microbes: a potential tool for selenium biofortification. Metallomics 13(10)

Ye Y, Qu J, Pu Y, Rao S, Xu F, Wu C (2020) Selenium Biofortification of Crop Food by Beneficial Microorganisms. J fungi (Basel, Switzerland) 6(2):59

Zaheer A, Mirza BS, Mclean JE, Yasmin S, Shah TM, Malik KA, Mirza MS (2016) Association of plant growth–promoting Serratia spp. with the root nodules of chickpea. Res Microbiol 167(6):510–520

Zaheer A, Malik A, Sher A, Qaisrani MM, Mehmood A, Khan SU, Ashraf M, Mirza Z, Karim S, Rasool M (2019) Isolation, characterization, and effect of phosphate-zinc-solubilizing bacterial strains on chickpea (Cicer arietinum L.) growth. Saudi J Biol Sci 26(5):1061–1067

Zohary D, Hopf M (2000) Domestication of plants in the Old World: The origin and spread of cultivated plants in West Asia, Europe and the Nile Valley. Oxford University Press

Chapter 13
Improving Iron Nutrition in Legumes to Overcome Hidden Hunger

Umed Ali, Tahmina Shar, Mahpara Khatoon Bhutto, and Ghulam Hussain Jatoi

Abstract Iron (Fe) is a trace element and an essential component in human nutrition. Cytochromes and myoglobin are the forms of Fe found in different types of cells and that act as major carriers of oxygen in the heme group of hemoglobin. Fe deficiency is a worldwide problem in underdeveloped and industrialized countries. Nearly two billion people have been affected by Fe deficiency, causing a serious health problem, namely iron-deficiency anemia (IDA). Food crops are major sources of Fe, and among them, legume crops are considered vital sources of iron intake in humans. Legumes can play key roles in improving Fe nutrition to overcome hidden hunger. By monitoring this scenario, the World Health Organization (WHO) has provided a strategy to improve nutritional qualities of cereal and legume crops, namely biofortification, which uses techniques such as agronomic methods, crop genetics and breeding technologies, the application of modern biotechnology, and the use of various microorganisms. Legume crops can play an important part in the biofortification process by improving Fe content and producing an Fe-rich and cost-effective dietary source for human health. In this chapter, we discuss the Fe status in pulses and focus on agronomic practices, breeding methods, and biotechnological approaches, along with the role of soil microbes in improving levels of iron in legume plants and overcoming hidden hunger problems in the global population.

U. Ali (✉)
Department of Botany, Mir Chakar Khan Rind University, Sibi, Balochistan, Pakistan

T. Shar
Department of Plant Breeding and Genetics, Lasbela University of Agriculture, Water & Marine Sciences, Uthal, Balochistan, Pakistan

M. K. Bhutto
Department of Agronomy, Sindh Agriculture University, Tandojam, Sindh. Pakistan

G. H. Jatoi
Department of Plant Pathology, Faculty of Crop Protection, Sindh Agriculture University, Tandojam, Sindh, Pakistan

M. A. Nadeem et al. (eds.), *Legumes Biofortification*,
https://doi.org/10.1007/978-3-031-33957-8_13

1 Introduction

Mineral nutrients play key roles in the metabolism and homeostasis of all living organisms, especially humans. Every essential mineral nutrient has various functions in the human body and is equally important for the human diet. The absence of or a deficiency in these minerals develops various diseases and metabolic disorders in our bodies. To overcome these diseases and metabolic disorders, the status of these minerals in the balanced diets of the global human population should be improved. Mineral sources, their quantities, their bioavailability, and their use in the biofortification of food grains need to be identified to overcome malnourishment (Stein 2010). Therefore, the nutritional quality of cereal and legume crops can be improved by understanding the mechanisms of mineral biofortification and bioavailability.

Humans require twenty-one mineral elements in our diets to maintain the physiological and biochemical functions of our bodies (Martínez-Ballesta et al. 2010). Deficiencies in iron (Fe), zinc (Zn), and iodine (I) are the most serious in the world because they have the most significant negative effects on human health (Ezzati et al. 2004; Horton et al. 2009). Because of the redox properties of iron, it is considered the third-most-essential element for the survival of human beings, after two other elements (Connorton et al. 2017). Fe is the fourth-most-abundant nutrient in soil, but unfortunately, it is not in a readily available form for uptake and absorption by the roots of many plants (Eng et al. 1998). The low accessibility of this element remains high in calcareous soils, which comprise 30% of the soils in the world. This is why the crops grown in that soil show severe Fe deficiency symptoms and reduced productivity (Kobayashi et al. 2016). The plants grown in such soils face iron-deficiency chlorosis (IDC), a condition in which a plant is deficient in iron.

Sustainably producing foods with their maximal nutrient values has become a recently demand, but our modern agriculture is more focused on obtaining higher yields while ignoring the nutritional status of the crop plants (Shaikh and Saraf 2017). Humans can take up Fe from both animal and plant sources, but the crop plants are considered the main sources of Fe in the ecosphere (Gibson et al. 2010). Various grain legumes—i.e., field peas (*Pisum sativum*), chickpeas (*Cicer arietinum*), lentils (*Lens culinaris*), pigeon peas (*Cajanus cajan*), common beans (*Phaseolus vulgaris*), and soybeans (*Glycine max*)—are considered Fe rich and contribute most of the nutrients that we require, including Fe. Iron-deficiency anemia (IDA) is a serious problem faced by countries that have low food availability and poor diet diversity. IDA has affected two billion people all around the world. Pregnant women most commonly experience the symptoms of IDA, and they may give birth to a child that is already deficient in Fe. Small children and nonpregnant women also face IDA problems.

Nowadays, experts are trying to overcome microelement deficiencies in humans, also called hidden hunger. One possible mechanism to fight hidden hunger is biofortification. Biofortification is a strategy for enhancing the content and bioavailability of microelements in edible seed crops through various methods, such as agronomic practices and plant breeding (Allen et al. 2006; Vasconcelos et al. 2017). The WHO has taken the initiative on overcoming these problems by following biofortification methods. The WHO is running various biofortification projects, such as the biofortification of Fe in the Fabaceae family crops, *Oryza sativa* (rice), *Ipomoea batatas* (sweet potatoes), and *Manihot esculenta* (cassavas) (Allen et al. 2006). The biofortification of Fe is made possible via agronomic methods, crop genetics and breeding technologies, the applications of modern biotechnology, and the use of various microorganisms (Vasconcelos et al. 2017). Modern fields of research, such as molecular biology and genetic engineering, have been used instead of conventional breeding. But successful Fe biofortification via these techniques requires a vast molecular knowledge about Fe uptake; Fe accumulation; Fe transportation, storage, and remobilization; and Fe enhancement in crop plants (Vasconcelos et al. 2017).

In the current scenario, grain legumes can be considered usable for Fe biofortification because they contain high nutritional levels and can be cost-effective sources of Fe. Biofortification in legumes can benefit the populations of low-income countries (Suttle 2010). Legume crops are considered a staple food in parts of Africa and Asia and can act as basic nourishment in poverty scenarios (Allen et al. 2006). In this chapter, we discuss iron, its nutritional value and functions in human health, the current status of iron in crops (particularly legumes), biofortification strategies e.g., agronomic practices, conventional and molecular breeding, biotechnological tools, and the use of plant-associated microorganisms to improve iron content in plants to overcome hidden hunger.

2 Functions of Fe in the Human Body

Enormous quantities of minerals and vitamins are essentially required for the proper functioning of the human body. Deficiencies in such minerals are called hidden hunger (Trijatmiko et al. 2016; Daly et al. 2017; Sharma et al. 2017).

Fe is the most important essential micronutrient required by all living organisms and an important element required in human nourishment and the constituents of the human body (Eng et al. 1998). Its biological significance relies on its reactivity, which allows for the emergence of reversible one-electron oxidation–reduction processes to transition between ferrous (Fe^{2+}) and ferric (Fe^{3+}) forms of Fe (Mann and Truswell 2017). Different age groups need different amounts of iron, depending on their age factors. Also, the amount of Fe required for women is different from that required for men. The prescribed intake of Fe in our daily diets is 8–18 mg per day, but this can be extended to 30 mg day^{-1} for pregnant women (Vasconcelos et al. 2017). Depending on the type of iron consumed, the human body can absorb only between 5% to 35% of the ingested amount (Abbaspour et al. 2014).

A deficiency in iron or iron-deficiency anemia (IDA) causes unavoidable physiological problems—especially in children because Fe deficiency affects the cognitive development and growth of children (Ouf and Jan 2015). The amount of oxygen that is transported to the muscles of the human body is decreased because of Fe deficiency, which affects the overall physical performance and working efficiency of human beings. It weakens the immune system, which ultimately increases the risk of infection. Iron-deficiency anemia increases the risk factors during perinatal period, which affects both mother and child, and it leads to infant mortality. It has been observed that the breakdown of a few neurotransmitters, the level of thyroid hormones, and the actions of some enzymes that depend on Fe are all adversely affected by a shortage of Fe in human body. Because of these high-risk problems occurring in human body due to iron inefficiency, the WHO has started a chain of projects to mitigate iron-deficiency issues, particularly in the developing nations of the world (Hassan et al. 2016).

Iron-deficiency problems in developed countries are most commonly caused by absorption disorders, blood loss, or not consuming a diet rich in Fe from plants. Different diets contain different types of iron in their bioavailability, depending on the source of that food. Hemic iron, just like hemoglobin and myoglobin, provides advanced bioavailability, and it is found mainly in meat, fish, and shellfish (WHO 2015). On the other hand, various organic and inorganic sources contain nonhemic Fe in different chemical forms (Fig. 13.1). The various sources of nonhemic Fe in different foods come in the form of ferric citrates, ferric phosphates, phytates, oxalates, or hydroxides, which are all molecules of low molecular weight, whereas lactoferrin, leghemoglobin, and ferritin are all considered compounds of high molecular weight. Different parts of the plant are used to fulfill the human body's requirements for nonhemic Fe, through the consumption of seeds from the Fabaceae family plants, grains from rice, dry nuts, and the green leaves of vegetables (Geissler and Singh 2011; WHO 2015).

Similarly, iron absorption is influenced by the amount of iron in the human body and by enhancers such as ascorbic acid and specific muscle tissue proteins. Maintaining an iron-rich diet is one method to combat Fe deficiency. However, the WHO supports food biofortification techniques to boost Fe nutrition in the human population in order to overcome hidden hunger (Abbaspour et al. 2014).

3 Current Status of Fe in Legumes and Other Crops

The nonhemic form of Fe is found in pulses, cereals, and vegetables, in both organic and inorganic forms. Hence, each iron type has a different capacity for absorption and a different level of digestion efficiency (Theil and Briat 2004).

Legumes are considered important sources of nutrients in the human diet; they account for 20% of daily human protein intake. Legume crops are also healthy and affordable source of digestible fiber and nutrients such as Fe (Table 13.1.) (Garcia et al. 2008; Guillamón et al. 2010). Fe is found in two forms in foods: hemic iron types, which support the arrangement and structure of various proteins, and non-hemic iron types—i.e. ferric citrates, phosphates, phytates, hydroxides, and oxalates—which are compounds with low molecular weights and which are most commonly found in the foods that humans consume (Zielińska-Dawidziak 2015).

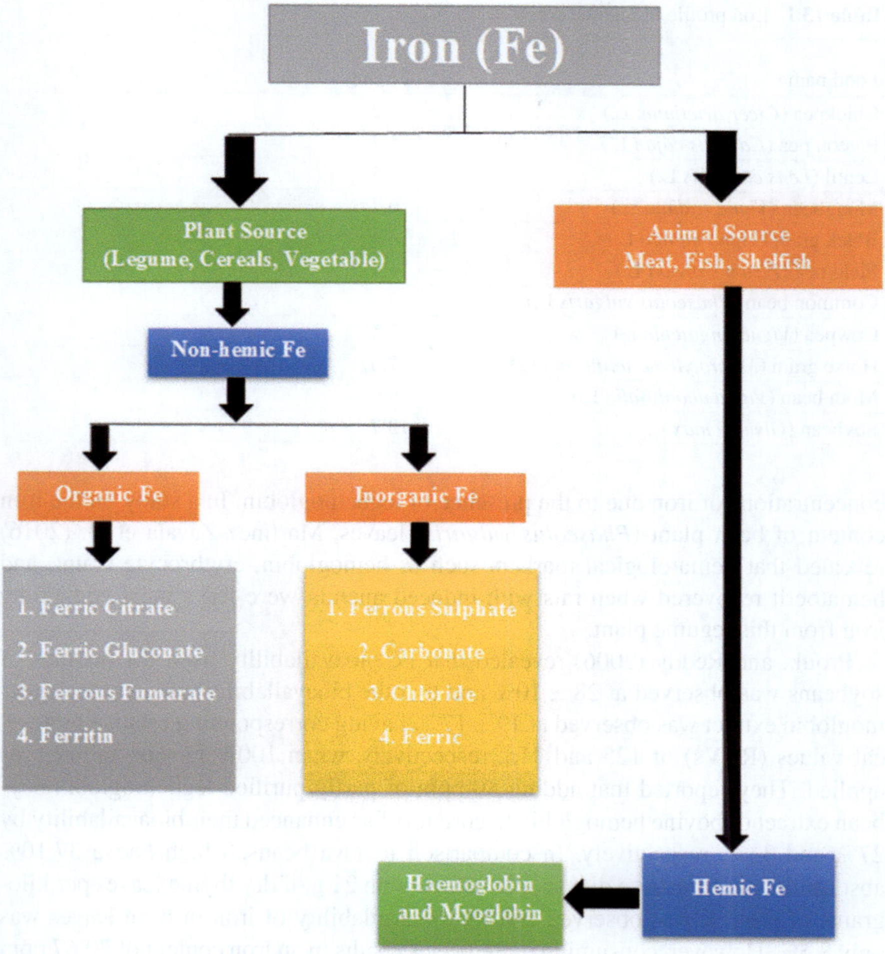

Fig. 13.1 Sources and forms of iron in food

Ferritin and leghemoglobin are the proteins found in vegetables that contain the hemic form of Fe. Edible parts of plants vary in ferritin distribution. The ferritin concentration in seeds can be between 50 and 70 mg kg^{-1}, while the portion of iron content in those amounts of ferritin are 10 mg kg^{-1} (Valdes-Miramontes et al. 2015). The root nodules of soybeans and lupines contain both the ferritin and leghemoglobin forms of Fe, and these nodules have more iron content than soybean leaves do (Chungopast et al. 2017). According to a study conducted by Valdes-Miramontes et al. (2015), the Fe content in the roots was higher (70 mg per 100 g) than that in the seeds (6.12 mg per 100 g) of *L. rotundiflorus*. This higher concentration of iron in roots is due to the presence of nodules, which are rich sources of leghemoglobin protein and contain high amounts of Fe. Neudorf (2015) concluded a study of root nodules from the Fabaceae family of plants and found that they have high

Table 13.1 Iron profile of legume crops

Food name	Fe (mg/100 g)
Chickpea (*Cicer arietinum* L.)	6.2
Pigeon pea (*Cajanus cajan* L.)	5.2
Lentil (*Lens culinaris* L.)	7.5
Mungbean (*Vigna radiata* L.)	6.7
Black gram (*Vigna mungo* L.)	8.4
Field pea (*Pisum sativum* L.)	4.4
Common bean (*Phaseolus vulgaris* L.)	3.4
Cowpea (*Vigna unguiculata* L.)	7.5
Horse gram (*Macrotyloma uniflorum* L.)	7.0
Moth bean (*Vigna aconitifolia* L.)	9.6
Soybean (*Glycine max* L.)	13.7

concentrations of iron due to the presence of leghemoglobin. In a study on the iron content of bean plant (*Phaseolus vulgaris*) leaves, Martínez-Zavala et al. (2016) revealed that hematological markers such as hemoglobin, erythrocyte count, and hematocrit recovered when rats with induced anemia were fed a meal containing iron from this legume plant.

Proulx and Reddy (2006) revealed that Fe bioavailability from the nodules of soybeans was observed at 28 ± 10% and that the bioavailability of soybean leghemoglobin extract was observed at 19 ± 17%, having corresponding relative biological values (RBVs) of 125 and 113, respectively, when 100% ferrous sulfate was applied. They reported that adding 50 ppm of partly purified leghemoglobin soybean extract or bovine hemoglobin to corn tortillas enhanced their bioavailability by 27% and 33%, respectively. In comparison to fava beans, which had a 37.10% absorption rate for a base diet supplemented with 21 g of dry thyme leaves per kilograms of food, it was observed that the bioavailability of iron in bean leaves was only 8.5%. However, consuming these leaves results in an iron content of 70.67 ppm (Rosado et al. 2005; Yossef 2010). Valdes-Miramontes et al. (2015) imparted that the bioavailability of Fe was 13.80% in root nodules and 13.70% in the cooked seeds of *L. rotundiflorus*.

4 Legume Crops Are Best Choice for Fe Biofortification

Soybean, field pea, chickpea, bean, lentil, and peanut crops belong to the Leguminosae (Fabaceae) family, and their seeds serve as edible sources of nutrients for human being (Mann and Truswell 2017). Legume crops possess low fat contents, high quantities of carbohydrates, vitamins, proteins, minerals, and dietary fiber. Legume or pulse crops are considered the most effective source of nutrients to satisfy the nutritional requirements of human beings (Erbersdobler et al. 2017). The consumption of fresh fruits, vegetables, and legumes has increase in recent years because the consumption of plants has been widely recommended over that of animals (Fabbri and Crosby 2016). Given the increasing population of the world, plant-derived protein is the best option for the future (Erbersdobler et al. 2017). The information provided by the European food information for consumer regulations claimed that 100 g of soybeans, lupines, field peas, and common beans contained 40% to 60% of the Fe needed in our daily diets (Erbersdobler et al. 2017). This shows that legume plants are key sources of dietary Fe (Balk and Schaedler 2014).

Numerous studies have focused on examining the various traits of legume crops related to the accumulation of iron (Blair et al. 2013; Tan et al. 2017). Blair et al. (2013) revealed an important method for the biofortification of iron: He found that some of the genes associated with the seed coat and cotyledonary iron are located on the linkage groups of B04 and B11 in common beans. Pulse legumes are rich in proteins compared with other crops, and these legumes have high values of starch, which is what makes pulse seeds a complete nutritional complement for the human diet (Tan et al. 2017). Field peas (*Pisum sativum* L.) alone can deliver 28% to 68% of the daily suggested intake of iron (Amarakoon et al. 2012). The consumption of microminerals in gram, another legume crop, was evaluated for Fe by using ten commercially cultivated gram genotypes. The authors concluded that these genotypes can provide 4.6–6.7 mg per 100 g of Fe content (Thavarajah 2012). Another study, conducted on lentils (*Lens culinaris* L.), confirmed that lentils are strong dietary sources of iron with high bioavailability (DellaValle et al. 2013).

With high nutritional values, grain legumes possess proteases, amylases, lectins, saponins, some phenolic compounds, and PA in their makeup. These compounds are considered as antinutrient compounds (Magalhães et al. 2017). The color of seeds and the sensory characteristics of plants are associated with phenolic compounds, which have antiallergenic, antiatherogenic, anti-inflammatory, antibacterial, and antioxidant properties. In addition to these properties, phenolic compounds have the potential property of cardiac protection (Balasundram et al. 2006). Phenolic acids, flavonoids, and condensed tannins are the most abundant phenolic compounds. These compounds occur mostly as attachments or as efficient byproducts (Balasundram et al. 2006; Magalhães et al. 2017). Generally, phytates fix elements such as iron and zinc and reduce absorption ratios in plants (Mann and Truswell 2017). In legumes, phosphorus is stored in the shape of PA, which is known as an inhibitor of food. This is because PA chelates with iron and makes it unavailable for humans (Gupta et al. 2015). Gupta et al. (2015) showed that when iron content is increased in plants, PA content also increases; these increased values of PA decrease the absorption ratio of iron.

Cooking and processing legume grains is the best way to reduce these toxins and antinutrient compounds and thus improve the absorption of Fe (Fabbri and Crosby 2016). Ascorbic acid is also present in legumes, which combats and helps overcome the negative effects of these inhibitors, containing phytate and polyphenols, in the Fe absorption process of plants (Abbaspour et al. 2014).

The crops of the Fabaceae family play key roles in ecological sustainability because legumes can fix nitrogen (N) through the nodules in their roots (Denton et al. 2017). The formation of root nodules occurs in the symbiosis of *Rhizobium* bacteria and plant roots (Popp and Ott 2011). The closeness of *Rhizobium* bacteria to the root hair cells of legumes results in legume infections, and a symbiosome membrane develops in the cortical cells of legume roots (Popp and Ott 2011; Brear et al. 2013). Nitrogen content affects the transportation of other elements from the roots to the shoots in a plant body. The symbiosis between N-fixing bacteria and legumes produces higher amounts of Fe. Here, iron is utilized in the manufacture of a protein called leghemoglobin and of the nitrogenase enzymes and cytochromes of the electron transport chain in the roots of host plants (Brear et al. 2013).

Another benefit of legume crops is that they are grown in the fallow period of other crops to increase fertility of soils, they can also be used as cover crops (Dapaah and Vyn 1998). They can be cultivated to control weeds, to recycle nutrients and manage pests, to increase soil N content, and to enhance the productivity of crops (Tonitto et al. 2006; Lundgren and Fergen 2011; O'Reilly et al. 2011). Alfalfas, red clovers, and crimson clovers, are the legume crops mostly cultivated for the above purposes. But the constraining problem is the life cycle of these crops (Coombs et al. 2017). Furthermore, some input is required to establish these kinds of systems, but the benefits of these systems far outweigh their costs (Snapp et al. 2005).

Green manuring is a technique in which legumes are used as cover crops and are incorporated in soil to enhance the fertility and nutritional status of soils (Toda and Uchida 2017). Symbiosis between a legume crop and a bacteroid can also be considered a good practice in green manuring because the N-fixing rhizobacteria in soil are reusable (Brear et al. 2013). The intercropping of nodule bearing crops close to weeds is another environment friendly practice (Wang et al. 2017). The concentration of iron in plants can also be enhanced when peanuts are intercropped with maize, wheat with chickpea, and guava with sorghum (Zuo and Zhang 2009), but further study is required to understand the mechanism behind this positive effect.

5 Raising Iron Content in Plants via Biofortification

Biofortification is defined as a procedure of improving the nutritious content of food grains. This procedure follows a series of techniques, including agronomic methods, crop genetics and breeding technologies, modern applications of biotechnology, and the use of various microorganisms. The objectives of this procedure are to improve the micronutrient quantities of foods and to reduce the antinutrients that negatively affect nutrients' capacity to be absorbed by humans. The availability of certain phenolic compounds and toxins and the state of oxidation and that of all food matrices influence the bioavailability and solubility of minerals (Etcheverry et al. 2012). Food crops are the main sources of minerals available for human consumption. The appropriate amounts of mineral elements and vitamins in plant-based diets can support healthy human lives (Welch and Graham 2004). Mineral deficiency issues have been resolved in recent decades with the help of fortification methods for food crops and taking dietary supplements rich in these elements (Tan et al. 2017). The production of staple food crops that are rich in essential mineral elements and that can be grown and distributed by using current agricultural methods can be sustainably provided by following biofortification techniques (Díaz-Gómez et al. 2017).

The Fe biofortification of crops can be accomplished in a variety of ways (Fig. 13.2). Agronomic practices and conventional and molecular plant-breeding techniques are some of the available agricultural methods, whereas the genetic-engineering methods include modifications to the genetic makeup of an organism, fertilization procedures, modern biotechnology tools, and the use of microorganisms (Van Der Straeten et al. 2017).

5.1 Agronomic Practices to Improve Iron Nutrition in Legumes

An adequate and balanced diet for humans is one that provides all the energy, essential amino acids (i.e. lysine and methionine), essential vitamins (i.e. A, B, C, D and E), and essential minerals (i.e., iron, zinc, iodine and selenium and folic acids) that they need. Deficiencies in micronutrients, such as iron, zinc, and iodine, and a deficiency in vitamin A in plants and soil has led the human population to the major problem of malnutrition, which is among the primary factors of the human diseases in developing countries. Micronutrient deficiencies have serious negative impacts on human health and development characteristics, such as physical development, immune system maturation, cognitive development, maternal mortality, etc. Higher yields of agricultural crops that contain these nutrients are needed to feed the rapidly increasing population of the world and to sustain human well-being.

The biofortification of legumes through agronomic practices is the strategy to enhance the nutrients, vitamins, and minerals in crops. Management practices can also be adopted to improve the iron content in legume crops and to thus combat hidden hunger.

5.1.1 Cultivation Practices and Fertilizer Application

Using legume genotypes/varieties can accumulate large amounts of iron from soil is the best option to avoid Fe deficiency and bioavailability problems (Hansen et al. 2003). Numerous fertilizers and cultural management practices can be applied in combination or alone to prevent or remedy the deficiency of iron in different crops. Various agronomic strategies—including lowering the pH values of soil, spraying iron through Fe liquid fertilizers, fertilizing the soil by using chelated/complexed Fe fertilizers, applying Fe fertilizers in band placements near the root zone of crops, using high-iron-accumulation legume crops as companion crops, changing the

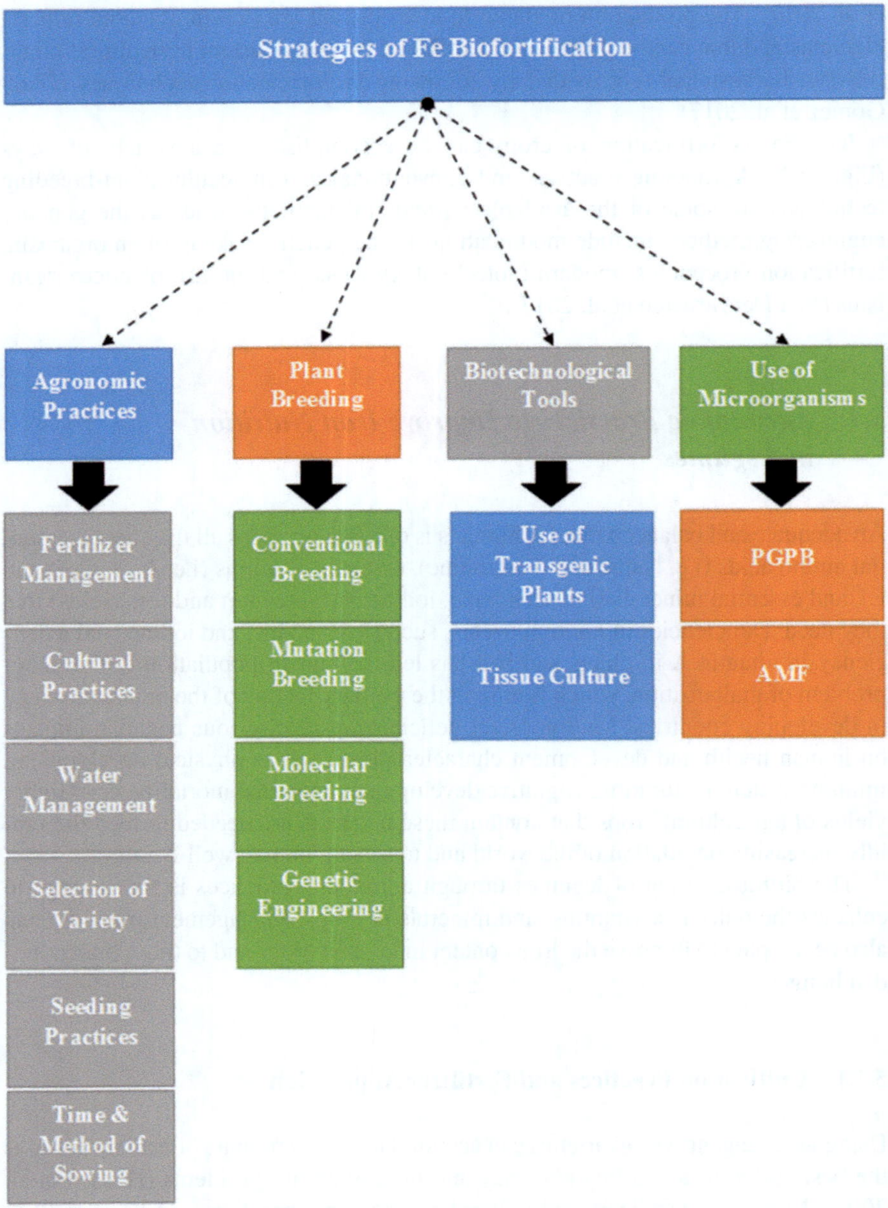

Fig. 13.2 Different strategies to improve Fe content in legume plants

schedule of irrigation and drainage, increasing soil fertility, and trying various sowing methods—can improve the Fe status of legume plants (Hopkins et al. 2005).

5.1.1.1 Low Soil pH Can Enhance Fe Content

Artificially lowering pH values can effectively overcome Fe deficiency (Olson 1950; Lucena 2003). For example, a tenfold increase in the activity of hydrogen ion and a thousandfold increase in Fe solubility can be achieved when the pH of soil drops from 7.5 to 6.5. Although the cost of the material that lowers the pH of soil and makes it more acidic in calcareous soils (which have high buffering capacities) is not very high, before applying that material or lowing the pH, the value of crops and their market prices should be kept in mind. A poorly buffered sandy soil may require 0.5 tonnes of elemental S per hectare to lower the soil pH from alkaline to neutral, whereas several tonnes of S is required for the soils to have high buffering capacities to obtain same results (Tisdale and Nelson 1966). Furthermore, the soil pH values can over time increase up into the alkaline range if the irrigation water has a high lime content. Because of its low cost, its wide availability, and its ease of use, S is the most commonly used acidifying agent (Tisdale and Nelson 1966). Ali et al. (2020) conducted experiments on a hydroponic culture with different pH values and revealed that Fe content in the roots was higher at the pH value of 6.0 than at 7.0 or 8.0. Several chemical compounds, either solid or liquid—i.e., ammonium polysulfide, aluminum sulfate, phosphoric acid, and sulfuric acid—are used to lower the pH values of soils (Horneck et al. 2005). Adding ammoniacal fertilizers to soils that have low buffering capacities and relatively low excess lime could be beneficial in lowering the pH of soils and increasing iron bioavailability (Whitney et al. 1991).

5.1.1.2 Foliar Method of Applying Fe Fertilizers

Liquid fertilizers containing Fe in either chelated or nonchelated form can temporarily rectify Fe deficiency (Godsey et al. 2003). The absorbed Fe is used by the plant if these materials are applied such that the leaf surfaces adhere to the plants. The main advantages of foliar applications of Fe are that they avoid the reactions occurring in alkaline/calcareous soils during iron fixation (Mengel 1995).

The foliar sprays require applying an adjuvant substance, which acts in the absorption and distribution of Fe on the leaf surface. After the leaves absorb the Fe, it is internally distributed. The foliar application of Fe immediately supplies iron to leaves, but this is a temporary remedy to combat Fe deficiency (Godsey et al. 2003). However, the foliar application of Fe inhibits the natural ability of plants to combat iron-deficient conditions (Römheld 1986). As compared to soil iron availability, foliar sprays have few lasting advantages. Because of Fe's immobility, once it becomes part of a cellular structure, it shows its appearance on the tissues where it was supplied (Vose 1982). Plant leaves that emerge after foliar applications of Fe always face Fe deficiency until the plants take up soil Fe with their roots or until the environment changes and makes the Fe available to plants (Anderson 1982).

Foliar applications of Fe immediately overcome the visual Fe deficiency symptoms of leaves. In some reports, it is stated that the foliar application of Fe may enhance the yields of crops but in most cases, the effect of foliar applications of Fe is minimum. Randall (1981) conducted experiments and treated his soybean crop with foliar applications of FeDDHA (ethydiaminedhephen acetic-Fe). He reveled that the Fe applied to the leaves of soybeans corrected the Fe deficiency of those parts of the plants but did not play any role in increasing crop yields. Another experiment, by Goos and Johnson in 2000, which was conducted at different location, reported that foliar applications of FeEDTA (Ferric Ethylenediaminetetraacetic Acid) corrected the chlorosis in soybean varieties that were deficient in Fe and increased the seed yield in only one location, but the effect of Fe on seed yield was not observed in most of the locations. In this study, the authors concluded that it would be better to select Fe-efficient varieties to uptake more Fe from their soils rather than to use foliar applications of Fe (Goos and Johnson 2000).

Cautionary measures should be taken while spraying foliar Fe because foliar applications of Fe in excess amounts will result in tissue necrosis and because injuries from foliar applications of fertilizer are prevalent. Ferrous sulfate is the most commonly applied nonchelated source of foliar Fe. Citrates, lignosulfates, gluconates, fulvates, etc. are only a few examples of the various complexed and chelated (EDHA, EDDHA (ethylenediamine-N,N′-bis(2-hydroxyphenylacetic acid)), DTPA, HEDTA, etc.) Fe sprays that are available in the market. Furthermore, only authentic materials that have detailed instructions printed on their labels and that are obtained from reputable companies should be used.

Field crops most often require multiple sprays of foliar applications of iron to enhance the yields of crops to some extent, but it makes this strategy too expensive if it is used for low-value crops or if the market prices are low (Anderson 1982; Godsey et al. 2003). It would be more expensive if the foliar Fe fertilizers are mixed with any foliar applications of pesticides (Mallarino et al. 2001). In this case, caution must be exercised because mixing Fe with any other substance may result in the inactivation of other compounds.

5.1.1.3 Organic and Inorganic Fertilizers Containing Fe

Inorganic Fe fertilizers are often useless when applied to soil under alkaline pH conditions because iron in soil reacts with other anions and because that reaction makes very insoluble complexes, such as ferric hydroxide. The combination of chelated iron and compounds that are organic in nature can decrease the formation rate of insoluble Fe (Lucena 2003). Fe is essentially grasped by an efficient chelation that maintains the level of Fe in the soil until it is used by plants and microorganisms. The exceptions are that any chemical or microbial decomposition can degrade the Fe hold by soil solution; it also makes a strong bonding with organic matter (OM) or mineral elements present in soil, or it is inactivated when iron is exchanged with any other cation (Álvarez-Fernández et al. 2002; Lucena 2003). The efficiency of a compound with chelated iron could also be hampered by leaching because there is a negative charge on synthetic chelates, so it rejects soils that have negative charges (Abadía et al. 2004). The rain or irrigation water leaches the chelated Fe from the root zone of plants. If it does not rain or if there's no delays in irrigation, the rain or irrigation water can provide Fe to the plants for a long time. Fe must be available to plants in the early growth stages because that is when the Fe deficiency in plants is most severe (Álvarez-Fernández et al. 2002). The effectiveness of synthetically chelated iron is also constrained by the ratio of salts in soil and the type of soil (Siebner-Freibach et al. 2004).

Alkaline/calcareous soils are notorious for causing iron chlorosis, though most Fe chelates are likewise ineffective in these kinds of soils. The EDDHA only and its related types of Fe chelates are effective against soils deficient in Fe (Abadía et al. 2004). Other forms of chelated Fe, such as EDHA, HEDTA and DTPA and Fe complex fertilizers such as citrates, fulvates, lignosulfates, and gluconates, are not very beneficial or effective compared to EDDHA or similar chelated Fe against the conditions where calcareous soils are present. It is because the other chelates can temporarily sustain iron in soils (Lucena 2003). On the other hand, other chelated forms of iron, except EDDHA, are used to treat only moderate Fe deficiencies when applied on leaves in some noncalcareous soils. It has been reported that chelated Fe fertilizers can slow the release of iron. This is beneficial because this system reduces the chances of leaching and helps alleviate iron-deficiency chlorosis (Yehuda et al. 2003; Goos et al. 2004). Goos et al. (2004) revealed that the efficacy of chelated forms of iron could not be improved, regardless of whether they are applied with polymer coatings. However, it has also been reported that polymer coatings reduced the ability of FeEDDHA in a greenhouse experiment. Fe fertilizers such as ferrous sulfate can be rather beneficial for slow-releasing methods of Fe to plants (Morikawa 2004; Singh et al. 2004).

In noncalcareous soils, several organic chemicals may act as activators or enhancers for Fe when fixed with any form of iron, and in calcareous soils, high amounts of organic chemicals are required. Although organic compounds may not be more effective than FeEDDHA in calcareous soils, complexing Fe components, such as Fe complexes and nonspecific humic compounds, are found in biosolid substances and the organic matter fabric of soils (Siebner-Freibach et al. 2004). These organic compounds, like chelates, have carbon in their composition, so they are beneficial for chemical and microbial degradation. The ratio for the degradation of organic materials completely depends on the soil moisture, soil temperature, and soil oxidation and is affected by the types and species of soil microorganisms (Tisdale and Nelson 1966).

Organic materials (animal dung, human excreta, etc.), which contain large amounts of Fe, applied in large quantities in either raw form or composted form can alleviate Fe deficiency in plants (Anderson 1982). The inorganic wastes of industry are also sources of Fe, but their availability and effects are not equal to those of inorganic fertilizers (Wallace et al. 1976). Although organic compounds are less efficient than chemical fertilizers in supplying nutrients, they can be beneficial in adding Fe to the soil if they are applied in large quantities. The commercial Fe fertilizers contain a consistent amount of Fe, but organic biosolids do not. Their Fe quantity is irregular, and it depends mostly on microbial activities to degrade the material. In temperate regions, iron deficiency in soil occurs mostly in the early growing seasons, because the ability of microorganisms to degrade soil compounds is very low in such zones. This causes the availability of Fe to be inadequate in early seasons. It is also crucial to appropriately manage organic materials, to minimize nitrogen immobilization, salt toxicities, and particular concentrations of cations and anions and to correctly solve the problems of pathogen and weed infestations.

Organic matter plays a complex and unpredictable function in the prevention or treatment of Fe deficiency. Organic matter not only acts as a source of Fe for soil but also results in loosening the soil. Soils that have low bulk density contain more oxygen and less carbon dioxide, which has less bicarbonate effects on Fe bioavailability (Lucena 2000). When easily degradable organic matter is added to soil, a mineralization process occurs, in which microbial respiration causes the amount of carbon dioxide to rise (Lucena 2000). According to study results, the rate of available Fe was decreased because bicarbonates formed in the soil.

The carbon added by the application of organic matter increases the activities of microorganisms. Several microorganisms, called siderophores, release chelates that can mobilize Fe. Iron uptake by plants is enhanced by these siderophores. Also, they are less likely to leach or degrade than synthetic forms of chelated Fe (Siebner-Freibach et al. 2004).

5.1.1.4 Concentrated Fertilizers and Their Placement Methods

Iron-deficiency chlorosis in plants can be removed by applying inorganic iron fertilizers via the band-placement method. These fertilizers should by applied near or on the sides of seeds (Morikawa 2004; Singh et al. 2004). This is because the growing roots of plants are in direct contact with fertilizer grains, suggesting that the rapid accumulation of Fe by plant roots can prevent Fe losses. Goos et al. (2004) reported that the ferrous sulfate applied to Fe-deficient soybean crops did not increase the growth of the crop, but its growth was significantly improved when the ferrous sulfate was applied along with ammonium sulfate and citric acid. The controlled release of Fe fertilizers has a major contribution to overcoming iron deficiency in a variety of crops, but owing to their high costs, their frequency of use remains relatively low compared with other Fe fertilizers.

Soaking the seeds in chelated fertilizer (FeEDDHA) has demonstrated some benefits in preventing Fe deficiency (Wiersma 2005), but the same results have not been consistently observed (Goos and Johnson 2000). Goos and Johnson (2000) conducted an experiment and reported that plowing the seed after soaking it in Fe fertilizer decreased their iron deficiency and increased the yield of soybean crops when the seed was sown in wide rows (76 cm), while the seeds sown in narrow rows did not show these kinds of results. They concluded that selecting Fe-deficiency-resistant cultivars is a better option than soaking seeds in Fe fertilizers.

Low levels of fertilizer salts and less attention are required when fertilizers are in direct contact with seeds. Such levels occur especially when salt-intolerant species are cultivated or when these plants are grown in saline soils or with saline water (Ayers and Westcot 1985). The strategy of soaking seeds with iron fertilizer has been studied on some crops, but further experiments on their rate and their placement methods are needed to examine the efficiency levels and the toxicity levels in different cropping patterns.

5.1.2 Seeding Methods and Water Management to Improve Fe Status

Agronomic practices are effective measures to combat Fe-deficiency symptoms in plants, and some manipulations of these techniques can eliminate Fe deficiency. The techniques include managing poor-drainage problems, compacting soil layers, minimizing salinity, increasing soil fertility, scheduling irrigation, arranging seed placement, and applying herbicides.

The problem of iron-deficiency chlorosis is often observed in poorly drained fields (Hansen et al. 2003). Chemical and biological processes that take place in saturated conditions are complicated and difficult to anticipate (Marschner 1986). Poor root activity brought on by insufficient oxygen, cold soil temperatures, and high bicarbonate levels contributes to the issue of rising carbon dioxide in saturated soils (El-Shatnawi and Makhadmeh 2001). Root tissues and the microorganisms that solubilize Fe do not have ability to respond well to Fe-stress conditions under a shortage of oxygen and low temperatures (Marschner 1986). In the soils that feature poor drainage and a cold temperature, the chemical reaction capability or mineralization of Fe in organic material is decreased (Marschner 1986). Poorly drained soil often becomes compacted, which creates problems for root growth. In general, crops grown in these kinds of soils exhibit symptoms of Fe deficiency. Improving the drainage of lower areas of fields can minimize the compaction issue of soils and also enhance the availability of Fe because of improvements in oxygen/moisture ratio. Good root growth is exceptional in this system. Tillage practices in areas where soil compaction is a problem can improve the soil's structure and the bioavailability of Fe, although the effect of tillage is not long lasting (Tisdale and Nelson 1966).

Areas that have good water drainage and good soil structures face fewer problems from iron-deficiency chlorosis (Hansen et al. 2003). Soils that have been saturated for a long time have higher availability levels of Fe nutrients (Vose 1982). In saturated soils, Fe^{3+} is converted to the Fe^{2+} form, which is available to plants for their root uptake because oxygen is depleted in saturated soils. The soluble form of iron is again converted into the insoluble ferric form, which happens when water evaporates and soil dries out. When oxygen is reduced in the root zone, the growth and development rates of all field crops, particularly legume crops, slowly start to decrease and the crops eventually die off. One exception to how this condition plays out is rice crops, which can withstand the absence of oxygen at the root regions of plants (El-Shatnawi and Makhadmeh 2001). As a result, plants that grow in poorly drained soils can have their iron-deficiency chlorosis reduced. However, if oxygen deprivation inhibits root development and its biochemistry, the consequences of this reduction may stop or even be reversed.

The quality of water applied as irrigation may affect Fe deficiency. The water in soil that contains carbonates or bicarbonates may increase iron-deficiency chlorosis in plants (Hansen et al. 2003). The drainage of water under rain-fed conditions can reduce the formation of soil bicarbonates. In irrigated areas, applying acidic irrigation water to fields can decrease the concentration of bicarbonates and dissolves the carbonates of soils, which may improve the Fe status of soils. The high cost of these kinds of management methods is questionable given the low value crops. Soil salinity is another problem under the category of water management. Salt accumulation has mostly been linked to iron-deficiency chlorosis. As a result, managing soil salts through the proper management of irrigation and drainage may help to prevent iron-deficiency problems (Siebner-Freibach et al. 2004).

Iron-deficiency chlorosis can be improved or worsened by interactions with other nutrients. The type of nitrogen in the soil can affect Fe nutrition (Lucena 2003). A deficiency in iron can also be increased by applying N in the form of nitrate. The majority of ammoniacal fertilizers turned into the nitrate form of nitrogen within a few days of being in a soil that has high-aeration capacity. Managing iron-deficiency chlorosis by applying ammonium nitrogen is unnecessary when the total soil or more than 50% of it is applied as a basal dose. On the other hand, frequent applications of ammonium-based fertilizers during the growing season may help in overcoming the chlorosis problem caused by Fe deficiency. The presence of ammonium ions in soils or their accumulation by plants is a better strategy to treat Fe deficiency in plants, but this strategy alone is not sufficient to completely overcome Fe deficiency (Lucena 2003).

Fe nutrition may also be improved through a range of crop management strategies. Another method for reducing an Fe deficit is to increase the plant population by increasing the seeding rate or decreasing row spacing (Goos and Johnson 2000; Hansen et al. 2003), possibly due to improved Fe-stress-response systems or improved root growth. The late sowing of crops in soils having Fe deficiency problems is another agronomic practice that can overcome the Fe-deficiency stresses (Hansen et al. 2003). Fe deficiency mostly occurs in the early spring season, but as the root activity and the microbial activity increase, Fe deficiency decreases. The deficiency in Fe in plants may also increase if the plant roots are suffering from a mechanical or pest injury. However, the toxicity of some pesticides may also exacerbate iron-deficiency chlorosis (Franzen et al. 2004).

Growing Fe-efficient cultivars, varieties, hybrid varieties, or species is the only management tool for agriculture. The additional procedures listed above can still be necessary to decrease iron-deficiency chlorosis even if Fe-tolerant varieties are grown. However, while choosing any strategy to combat iron-deficiency chlorosis, the soil type, the conditions of the environment, and the cropping pattern should be kept in mind. Any strategy for combating Fe deficiency should first be verified through field research prior to applying.

5.2 *Breeding Methods to Improve Iron Content in Plants*

5.2.1 Selection of Varieties and Screening Them

Screening cultivars for resistance to iron-deficiency chlorosis can be conducted in two ways. Field screening, which also includes assessments of plants in greenhouses and environmental testing chambers, is the standard form of screening, where chlorosis is assessed among the varieties and compared with controls. Measuring the responses of plants to any particular physiological trait under iron-deficient conditions is the second way.

Screening genotypes for iron deficiency has been a principal method for isolating resistant cultivars in soybean. Through this method, various genotypes have been cultivated in lines in calcareous soils whose histories of iron accumulation are not yet known (Diers and Fehr 1989). The levels of resistance between lines are then quantified via visual inspection and an assessment of chlorosis severity. The visual evaluation of plants is performed mostly by an imaginary scale, consisting of 1 to 5. The different numbers refer to different levels of iron chlorosis. For example, plants that have no symptoms of chlorosis are ranked 1 on this scale; plants that have a few symptoms of chlorosis are ranked 2; plants that have moderate chlorosis symptoms are ranked 3; plants that have intense chlorosis symptoms are ranked 4; and, finally, plants that have severe symptoms of iron-deficiency chlorosis are ranked 5. SPAD chlorophyll meters have recently become popular as alternatives to visual evaluations (Frenkel et al. 2004). Successful field screening requires that the crop be cultivated in multiple locations and over several years, which would be expensive, time-consuming, and laborious. The expressions of iron-deficiency chlorosis change from time to time at different locations. There have been several attempts to address this heterogeneity issue, including managing the water in soil, the content of organic matter, and soil compaction and even utilizing pesticides (Ocumpaugh et al. 1992). Excess NO_3^- has been intentionally applied to some species to exhibit the symptoms of iron deficiency and encourage its homogeneity (Lucena 2000).

To hasten this selection process and to avoid any environmental or soil errors, research experts have now chosen to conduct their experiments in green houses and growth chambers by using either pots or nutrient solutions (Chaney et al. 1992a; Fairbanks et al. 1987; Ocumpaugh et al. 1992). In a few dicotyledonous species, nutrient solution screening has proven successful. Managing low levels of Fe (Jolley and Brown 1987) and adding bicarbonate (Dragonuk et al. 1989) are commonly used to exacerbate chlorosis. The variability of iron resistance between various chickpea and soybean cultivars was evaluated by using different levels of DPTA and bicarbonates (Chaney et al. 1992a, b). Some crop-breeding programs have used this approach to screen plants for Fe deficiency (Charlson et al. 2004).

Analyzing the physiological responses of plants that occur under iron-deficient conditions has been advised because it is a faster and simpler method than field screening. Differences in tolerance to iron-deficiency chlorosis can be discovered by using in situ measurements in the situations that exacerbate responses to low Fe.

Iron-deficiency responses are linked to a plant's efficiency in taking up Fe (Jessen et al. 1988). The iron-reduction-based screening approach takes advantage of the early response of resistant plants to an Fe shortage. When detecting an Fe reduction early in a plant's life cycle, this value correlates with chlorosis resistance in plants. Chlorosis in soybean roots is associated closely with the degree and time of Fe reduction (Jolley et al. 1992). They concluded that longer periods of measurement did not enhance the correlation, suggesting that a valid assessment could be achieved in a short amount of time.

The accuracy of the screening test for Fe reduction has been further enhanced by the addition of $CaCO_3$ as a nutrient solution buffer and by excluding Fe from the germination and growth solutions (Stevens et al. 1993). Measuring the quantity of reduced iron has become a practical tool in plant-breeding programs to screen for resistance to iron-deficiency chlorosis. A hydroponic experiment was conducted to analyze the reduced iron in five common bean cultivars. In the experiment, variations in all five cultivars were found when iron-deficiency chlorosis stress was induced (Krouma et al. 2003). However, this experiment has no relevance to field conditions.

Efficient dicot plants can be screened by the ability that measures the releasing capacity of H^+. Strong correlations between the releasing capacity of H^+ ions and iron-deficiency chlorosis and those between a decrease in Fe and the Fe-deficiency resistance of subclover cultivars were found (Wei et al. 1995). The cultivars that exhibited Fe resistance had an improved H^+ ion pump when exposed to an Fe shortage (Loeppert et al. 1995; Wei et al. 1995). An increase in H^+ extrusion under Fe-deficient conditions was evaluated in soybean and dry bean lines (Ellsworth et al. 1998).

The cereal crop species and the varieties within species are different in their tolerance levels. There was a clear correlation between the capacity of phytosiderophore release and the degree of susceptibility or tolerance to iron-deficiency chlorosis in several oat cultivars (Brown and Jolley, 1989). The level of iron sensitivity or iron tolerance and the releasing capacity of phytosiderophores in different species are closely connected (Römheld and Marschner 1990; Onyezili and Ross 1993). This is why measuring the release of phytosiderophore has been recognized as an effective method of screening Fe-efficient varieties of cereal crops (Hansen and Jolley 1995; Jolley and Hansen 1995; Hansen et al. 1996).

In plants, enormous improvements have been made in determining the genetic regulation and guidelines for iron-accumulating components (Bauer et al. 2004; Römheld and Schaaf 2004; Schaaf et al. 2004). In breeding programs, molecular and biochemical approaches can take the role of in vivo culture techniques that are employed to determine a plant's resistance to iron-deficiency chlorosis. However, these types of technologies have some limitations when practiced on large areas (Lin et al. 2000).

Instead of screening or employing breeding techniques, the most crucial method of minimizing the development of iron chlorosis in legume crops is the selection and growing of cultivars resistant to chlorosis. The only concern in growing cultivars resistant to iron-deficiency chlorosis is that they produce lower yields in the areas where iron-deficiency chlorosis is not an issue. In contrast, efforts to develop chlorosis-tolerant plants with higher yields continue. The planting of various cultivars in the same field can significantly enhance the yield and can at the same time combat iron-deficiency chlorosis by following precise farming methods. For example, the cultivars that are resistant to Fe deficiency but that have lower yields can be planted at the sites where iron-deficiency chlorosis is high, and the cultivars that have higher yields but that are susceptible to iron-deficiency chlorosis can be cultivated at the sites where iron deficiency is not a problem. This strategy may improve the yields of legume crops on a larger scale.

5.2.2 Molecular Breeding and Modern Biotechnological Tools to Improve Fe Status in Plants

Conventional breeding has always been engaged in improving various characteristics related to vegetative and reproductive development by focusing on different heredity values. It reflects the potential of each characteristic for genetic upgrading (Jacob et al. 2016). The plant genome has also been modified with the help of genetic-engineering (GE) technology. Any single plant character can be fully or partially enhanced or modified by using GE technology.

Approximately nineteen years ago, some breeding techniques were employed to achieve 14 μg per g dry weight (DW) of iron concentration in the grains of polished rice (Trijatmiko et al. 2016). However, this objective was accomplished later on by using the methods of genetic engineering, which demonstrated that high nutrition values cannot be achieved in crop plants by following only conventional breeding. The bioavailability of Fe takes precedence over the accumulation, transport, remobilization, and storage capacity of Fe in maximizing the Fe nutrition in legume plants (Carvalho and Vasconcelos 2013). According to Vasconcelos et al. (2003), rice's endosperm-specific promoter-controlled ferritin expression can enhance the iron content of grain. Additionally, genes that are engaged in the synthesis of phytosiderophores or encoding Fe carriers have brought high concentrations of iron to the grains of rice (Boonyaves et al. 2016). This strategy to enhance Fe content in plants has been employed by various researches. Fe content in the grains of rice lines was reached as high as 30% due to expressing the gene for *ferritin and the NAS2* gene (Trijatmiko et al. 2016). The population that consumed pearl millets, beans, and rice biofortified with iron displayed a high concentration of serum ferritin and complete Fe nutrition in individual bodies (Finkelstein et al. 2017). The individuals of that population were checked before conducting this experiment, and it was found that they were at high risk of Fe deficiency at the beginning, but the same individuals were found to have higher amounts of Fe in their bodies after consuming Fe-fortified food.

In Fe-biofortification efforts, new knowledge has been developed in genome editing, reverse breeding, oligo-directed mutagenesis, sequence-specific nuclease technology, and RNA-directed DNA methylation (Vasconcelos et al. 2017). Owing to some concerns about environmental safety and meeting human standards, the genome-editing technique has not been fully accepted by the public (Hefferon 2015; Tan et al. 2018). Before applying these strategies for Fe biofortification, the important allelic variation related to the genes of Fe metabolism should first be identified. The application of conventional breeding procedures for biofortification is restricted by the gene pool's diversity and the fertility of species (Tan et al. 2017). The genetic improvement of crops can be sped up by employing the latest molecular methods—such as clustered regularly interspaced short palindromic repeats (CRISPR) and CRISPR-associated (Cas) proteins—and by the actions of various genes—i.e., transcription activator-like effector nucleases (TALENs), meganucleases, and zinc-finger nucleases (ZIPs) (Jacob et al. 2016; Schaart et al. 2016; Vasconcelos et al. 2017). Although genetic engineering is a vital tool, it needs to overcome the challenges of public acceptance and regulations (Frewer et al. 2013).

5.2.3 Transgenic Crops Biofortified with Fe

Plant researchers have revealed that metal transporter proteins found in many crop species may use a variety of metal substrates for nutrient uptake from soil into roots, including iron, zinc, and even cadmium. Mutational analysis led researchers to discover that using mutants that lost the functions of these transporter proteins prevented the absorption of all three metals into plant cells (Morrissey and Guerinot 2009).

Moreover, iron accumulation in plant tissues can be performed through ferritin, a storage protein of iron. An experiment conducted by Masuda et al. (2012) found that Fe accumulation in rice was upregulated by ferritin and that the translocation of Fe within plant was enhanced by the overexpression of iron (II)-nicotinamide transporter *OsYSL2*. This method has been demonstrated to produce more iron (6.0 times more in greenhouses and 4.4 times more in field experiments) in transgenic lines by utilizing this methodology, implying that introducing numerous genes that regulate the homeostasis of Fe would be better than introducing only one gene in Fe-biofortification programs. A more recent study, conducted by Ali et al. (2020), reported that an increase in the transcriptional response of *OsYSL15* could lead the plants to have higher amounts of Fe content in roots under various pH levels in a hydroponic experiment.

Moreover, Masuda et al. (2013) reported that mugineic acid, a ferric iron chelator, considerably increased the accumulation of Fe via iron uptake and its transportation to other parts of plants. The authors used transgenic plants to express the soybean ferritin gene (*SoyferH2*)—which was driven by two endosperm-specific promoters—the barley nicotinamide synthase gene (*HvNAS1*), two nicotinamide aminotransferase genes (*HvNAAT*-A and B), and a mugineic acid synthase gene (*IDS3*). The mugineic acid synthase gene (*IDS3*) was used to increase the production of mugineic acid in plants. This showed a resistance to Fe deficiency and demonstrated a 2.5-fold increase in Fe accumulation by transgenic plants. Under conditions where Fe was sufficient in soil, these transgenic plants were able to accumulate 4.0 times the iron compared with the plants grown in either commercially supplied soil or calcareous soil. On transgenic lines grown in calcareous soil, the ferritin and mugineic acid biosynthetic genes showed signs of iron-deficiency resistance. Fe content in grains of polished rice was improved 4.0-fold in ferritin-expressed transgenic lines and 22.5-fold in biosynthetically expressed mugineic acid genes when compared with nontransgenic lines cultivated in normal soil and calcareous soil.

Because the same molecular pathway for Zn is followed to transport Fe into plants, the correction of Fe deficiency in crop plants also increases the accumulation of Zn. In an experiment, Aung et al. (2013) developed a line of transgenic rice that was consumed by the people of that area of Myanmar, where 70% of the population has an Fe deficiency. This line showed the overexpression of the *HvNAS1*, *OsYSL2*, and *SoyferH2* genes, which are responsible for enhancing Fe transport in plants, transporting Fe to the endosperm, and increasing the Fe accumulation in the endosperm, respectively. The study showed that Fe was increased 3.0–4.0-fold and zinc 1.0–2.0-fold when compared with conventional varieties. The authors concluded that transgenic lines of rice overexpressed with these genes not only overcame the Fe deficiencies in Myanmar's population but also increased their zinc contents.

Fe biofortification through biotechnological tools has been used not only in rice but in other crops too. Tan et al. (2018) developed transgenic lines of chickpeas for Fe biofortification. They used the combined action of two genes, namely chickpea nicotinamide synthase 2 (*CaNAS2*) and soybean ferritin (*GmFER*), to improve the accumulation and transport capacity of Fe in chickpeas. The overexpression of these genes significantly enhanced the levels of iron (Fe) and nicotianamine (NA). They also concluded that Fe bioavailability was enhanced by doubling the concentration of NA in plants. Likewise, Manwaring et al. (2016) analyzed the potential of Fe biofortification in pearl millets by further developing the presently available gene pool. The biofortification of Fe using the transgenic approach in crops such as pearl millets would be beneficial in areas where soil management is difficult or where supplementation programs are inadequate. In another experiment, it was observed that cassava plants accumulated a lot of Fe in their roots due to the overexpression of the iron-sequestering *Arabidopsis thaliana* gene *AtVIT1*. Because of this, most Fe was sequestered in the vacuoles of plant roots. An increase in Fe concentration in the upper parts of the plants was also noted (Narayanan et al. 2015). This shows that the biofortification strategy using molecular and biotechnological tools improves Fe contents in a number of crops grown in resource-poor countries.

5.3 *The Role of Microorganisms in Increasing the Uptake of Fe in Plants*

The growing understanding of the complex interactions between plants and microorganisms has piqued interest in using soil microorganisms to further improve the Fe-uptake processes of plants. In a number of studies, these microorganisms have been implicated in the uptake of micronutrients and macronutrients (Srinivasagam et al. 2013; Wang et al. 2014; Wu et al. 2015; Berruti et al. 2016), which suggests that these microorganisms can be used in the Fe-biofortification process to improve iron content in food crops. The following subsections list the microorganisms involved in improving iron content in plants.

5.3.1 Plant Growth–Promoting Bacteria (PGPB)

The importance of PGPB was first reported in 1980 by Kloepper et al. (1980). Since then, several experiments have been conducted to investigate the roles of these bacteria in the mobilization of micronutrients in plants. The growth of plants is impacted by these microorganisms either directly or indirectly through the production of growth hormones, such as indole-3-acidic acid (IAA), nitrogen fixation, 1-aminocy clopropane-1-carboxylate deaminase (ACC), the stimulation of systemic resistance, the production of siderophores, and mineral nutrient solubilization (Table 13.2.) (Zhang et al. 2011; Bhattacharyya and Jha 2012; Karthikeyan et al. 2012; Pereira and Castro 2014; Xu et al. 2014).

Siderophores, low-molecular-weight compounds, are synthesized through the activities of PGPB under Fe-deficient conditions. About five hundred varieties of siderophores have been reported in plants so far (Boukhalfa et al. 2003). These siderophores are divided into four categories: hydroxamate siderophores (the most common), pyoverdines, catecholates, and carboxylates. Siderophores such as ferribactin, pyoverdine, enterochelin, and rhizobactin are produced by the *Pseudomonas fluorescens*, *Pseudomonas aeruginosa*, *Escherichia coli*, and *Rhizobium meliloti* species of plant growth–promoting bacteria, respectively (Maurer et al. 1968; Smith and Neilands 1984; Schalk and Guillon 2013). With respect to Fe, the growth of plants is sped up by these siderophore-producing bacteria in two methods: directly, by increasing the absorption of Fe from soil, and indirectly, through the sequestration of iron within microbes, preventing their growth (Karthik et al. 2017).

Another strategy to increase the bioavailability of Fe is to have these bacteria develop hormonal compounds similar to plant hormones (Daly et al. 2017). Another example of using plant growth–promoting bacteria to increase Fe concentration in plants is the upregulation of *FIT1* and subsequent activation of *FRO2* and *IRT1* by

Table 13.2 Strategies to increase Fe absorption through PGPB and to enhance plant growth

Strategy	Description	References
Siderophore production	Directly: promotion of Fe absorption from soil Indirectly: sequestration of available Fe, preventing pathogen growth	Karthik et al. (2017)
Production of plant hormones	Production of IAA, gibberellins, and cytokinins	Hardoim et al. (2008)
Upregulation of *FIT1*	Activation of *FRO2* and *IRT1*	Zhang et al. (2009)

Bacillus subtilis GB03 (Zhang et al. 2009). Under Fe-deficient conditions, plants can choose the best microorganisms to interact with them in order to produce auxins and/or siderophores. The microorganisms that produce higher concentrations of auxin appear in the rhizosphere in iron-deficient plants when there is a higher concentration of phenolic root exudates. Above-ground applications of IAA (the most common naturally occurring plant hormone in the auxin class) have been shown to alleviate Fe-deficiency symptoms by decreasing Fe^{3+} production, increasing the expression of *FRO2* and *IRT1*, subsequently increasing the network of plant roots, creating more surface area for roots to absorb higher amounts of Fe, and increasing the porosity of the plant cell wall, thus enhancing root exudation (Wei Jin et al. 2008; Chen et al. 2010; Glick 2012; Wu et al. 2012).

The study conducted on chickpea (*Cicer arietinum* L.) crops revealed that the Fe content in plants was increased by up to 38% when the crops were inoculated with nineteen isolates of *Acinetobacter* (Sathya et al. 2016). When the same crop was inoculated with *Enterobacter ludwigii* SRI-229 and *Pseudomonas monteilii* SRI-360, the iron concentration increased by about 18% (Gopalakrishnan et al. 2016). Both teams of authors also reported that Fe content was reduced by up to 30% due to postharvest processing and increased by about 21% due to cooking the seed grains of chickpea crops.

5.3.2 Arbuscular Mycorrhizal Fungi (AMF)

AMF colonize plants, forming a symbiotic relationship in which the plant benefits from the increased surface area for water and nutrient adsorption while AMF are provided with carbohydrates necessary for their growth (Senés-Guerrero et al. 2014; Zhang et al. 2015).

The nutritional values of biofortified Fe and increased seed and biomass yields in chickpea crops were achieved when the soil was inoculated with AMF (Pellegrino and Bedini 2014). Additionally, these arbuscular mycorrhizal fungi could come into contact with beneficial rhizobacteria, which benefit these fungi by increasing productivity and plant fitness (Scheublin et al. 2010; Cruz and Ishii 2011; Moreira et al. 2016; Qin et al. 2016). The yield and Fe content was increased when wheat crops were inoculated with both AMF and various strains of *Pseudomonas* (Mäder et al. 2011). The soil structure and plant-soil system can be influenced by AMF through the secretion of high-molecular-weight glycoproteins such as glue gomalin (Srinivasagam et al. 2013; Wu et al. 2015), which act in the phytoremediation of heavy metals and trigger plant innate responses (He and Nara 2007; Gallou et al. 2011).

6 Conclusion

Iron is the fourth-most-abundant element on Earth and falls in the category of micronutrients—which are supplied in small quantities, as opposed to macronutrients. But they are as important as other essential elements. They are not always readily available to plants for uptake and absorption. According to research results, iron-deficiency chlorosis (IDC) occurs in most plants. Plants are considered as important sources of Fe for people in most countries. From plants, pulse legumes such as chickpeas, lentils, field peas, beans, soybeans, and pigeon peas are considered staple foods in most Asian and African countries and can provide large amounts of nutrients, including Fe, to the human diet. Because of Fe deficiency, most people in developing countries face serious health issues. Iron-deficiency anemia (IDA) is a well-known health problem that has affected two billion people around the world. Fe deficiency is usually observed in children and pregnant women. Fe deficiency can affect human health in several ways, including hampering the cognitive development and growth of children, lowering the physical performance and work efficiency of adults, weakening the body's immune system, and increasing the mortality rate during perinatal periods. Fe deficiency negatively affects the metabolism of some neurotransmitters and thyroid hormones and the activity of some iron-dependent enzymes.

In consideration of the importance of Fe in the human diet, the WHO has started several projects to improve Fe nutrition from plants to overcome hidden hunger. Among these, the Fe biofortification of legumes has garnered much attention because of the higher nutritional capacity of legume crops. *Biofortification* is defined as a series of strategies to enhance micronutrient concentrations and their bioavailability in edible food crops by following certain agronomic practices and plant-breeding methods.

In this chapter, we focused on the biofortification of legume grains to enhance Fe nutrition to overcome hidden hunger. The agronomic strategies that can increase Fe content and avoid its deficiency include fertilizer management, cultural practices, water management, the selection of varieties, seeding practices, and the time and method of sowing. The breeding methods include conventional breeding, mutation breeding, molecular breeding, and genetic engineering. The biotechnological means to improve Fe status in legumes uses transgenic plants and tissue cultures. The Fe contents of plants have also been increased through inoculation with various microorganisms such as plant growth–promoting bacteria and arbuscular mycorrhizal fungi.

References

Abadía J, Álvarez-Fernández A, Rombolaà AD, Sanz M, Tagliavini M, Abadía A (2004) Technologies for the diagnosis and remediation of Fe deficiency. Soil Sci Plant Nutr 50:965–971
Abbaspour N, Hurrell R, Kelishadi R (2014) Review on iron and its importance for human health. J Res Med Sci 19:164
Ali U, Zhong M, Shar T, Fiaz S, Xie L, Jiao G, Ahmad S, Sheng Z, Tang S, Wei X (2020) The influence of pH on cadmium accumulation in seedlings of rice (Oryza sativa L.). J Plant Growth Regul 39:930–940
Allen L, De Benoist B, Dary O, Hurrell R (2006) Guidelines on food fortification with micronutrients. JSTOR
Álvarez-Fernández A, Sierra MA, Lucena JJ (2002) Reactivity of synthetic Fe chelates with soils and soil components. Plant Soil 241:129–137
Amarakoon D, McPhee K, Thavarajah P (2012) Iron-, zinc-, and magnesium-rich field peas (Pisum sativum L.) with naturally low phytic acid: a potential food-based solution to global micronutrient malnutrition. J Food Compos Anal 27:8–13
Anderson WB (1982) Diagnosis and correction of iron deficiency in field crops-an overview. J Plant Nutr 5:785–795
Aung MS, Masuda H, Kobayashi T, Nakanishi H, Yamakawa T, Nishizawa NK (2013) Iron biofortification of Myanmar rice. Front Plant Sci 4:158
Ayers RS, Westcot DW (1985) Water quality for agriculture. Food and Agriculture Organization of the United Nations Rome
Balasundram N, Sundram K, Samman S (2006) Phenolic compounds in plants and Agri-industrial by-products: antioxidant activity, occurrence, and potential uses. Food Chem 99:191–203
Balk J, Schaedler TA (2014) Iron cofactor assembly in plants. Annu Rev Plant Biol 65:125–153
Bauer P, Bereczky Z, Brumbarova T, Klatte M, Wang H-Y (2004) Molecular regulation of iron uptake in the dicot species Lycopersicon esculentum and Arabidopsis thaliana. Soil Sci Plant Nutr 50:997–1001
Berruti A, Lumini E, Balestrini R, Bianciotto V (2016) Arbuscular mycorrhizal fungi as natural biofertilizers: let's benefit from past successes. Front Microbiol 6:1559
Bhattacharyya PN, Jha DK (2012) Plant growth–promoting rhizobacteria (PGPR): emergence in agriculture. World J Microbiol Biotechnol 28:1327–1350
Blair MW, Izquierdo P, Astudillo C, Grusak MA (2013) A legume biofortification quandary: variability and genetic control of seed coat micronutrient accumulation in common beans. Front Plant Sci 4:275
Boonyaves K, Gruissem W, Bhullar NK (2016) NOD promoter-controlled AtIRT1 expression functions synergistically with NAS and FERRITIN genes to increase iron in rice grains. Plant Mol Biol 90:207–215
Boukhalfa H, Lack J, Reilly SD, Hersman L, Neu MP (2003) Siderophore production and facilitated uptake of iron plutonium in p. putida. Los Alamos National Lab (LANL), Los Alamos
Brear EM, Day DA, Smith PM (2013) Iron: an essential micronutrient for the legume-rhizobium symbiosis. Front Plant Sci 4:359
Carvalho SM, Vasconcelos MW (2013) Producing more with less: strategies and novel technologies for plant-based food biofortification. Food Res Int 54:961–971
Chaney RL, Coulombe BA, Bell PF, Angle JS (1992a) Detailed method to screen dicot cultivars for resistance to Fe-chlorosis using FeDTPA and bicarbonate in nutrient solutions. J Plant Nutr 15:2063–2083
Chaney RL, Hamze MH, Bell PF (1992b) Screening chickpea for iron chlorosis resistance using bicarbonate in nutrient solution to simulate calcareous soils. J Plant Nutr 15:2045–2062
Charlson DV, Bailey TB, Cianzio SR, Shoemaker RC (2004) Breeding soybean for resistance to iron-deficiency chlorosis and soybean cyst nematode. Soil Sci Plant Nutr 50:1055–1062

Chen WW, Yang JL, Qin C, Jin CW, Mo JH, Ye T, Zheng SJ (2010) Nitric oxide acts downstream of auxin to trigger root ferric-chelate reductase activity in response to iron deficiency in Arabidopsis. Plant Physiol 154:810–819

Chungopast S, Duangkhet M, Tajima S, Ma JF, Nomura M (2017) Iron-induced nitric oxide leads to an increase in the expression of ferritin during the senescence of Lotus japonicus nodules. J Plant Physiol 208:40–46

Connorton JM, Balk J, Rodríguez-Celma J (2017) Iron homeostasis in plants–a brief overview. Metallomics 9:813–823

Coombs C, Lauzon JD, Deen B, Van Eerd LL (2017) Legume cover crop management on nitrogen dynamics and yield in grain corn systems. Field Crop Res 201:75–85

Cruz AF, Ishii T (2011) Arbuscular mycorrhizal fungal spores host bacteria that affect nutrient biodynamics and biocontrol of soil-borne plant pathogens. Biol Open 1:52–57

Daly DH, Velivelli SL, Prestwich BD (2017) The role of soil microbes in crop biofortification. In: Agriculturally important microbes for sustainable agriculture. Springer, pp 333–356

Dapaah H, Vyn T (1998) Nitrogen fertilization and cover crop effects on soil structural stability and corn performance. Commun Soil Sci Plant Anal 29:2557–2569

DellaValle DM, Thavarajah P, Vandenberg A, Glahn RP (2013) Lentil (Lens culinaris L.) as a candidate crop for iron biofortification: is there genetic potential for iron bioavailability? Field Crop Res 144:119–125

Denton MD, Phillips LA, Peoples MB, Pearce DJ, Swan AD, Mele PM, Brockwell J (2017) Legume inoculant application methods: effects on nodulation patterns, nitrogen fixation, crop growth and yield in narrow-leaf lupin and faba bean. Plant Soil 419:25–39

Díaz-Gómez J, Twyman RM, Zhu C, Farré G, Serrano JC, Portero-Otin M, Muñoz P, Sandmann G, Capell T, Christou P (2017) Biofortification of crops with nutrients: factors affecting utilization and storage. Curr Opin Biotechnol 44:115–123

Diers B, Fehr W (1989) Selection for iron efficiency of soybean in nutrient-solution and field tests. Crop Sci 29:86–90

Dragonuk M, Fehr W, Jessen H (1989) Effectiveness of nutrient-solution evaluation for recurrent selection for Fe efficiency of soybean. Crop Sci 29:952–955

Ellsworth JW, Jolley VD, Nuland DS, Blaylock AD (1998) Use of hydrogen release or a combination of hydrogen release and iron reduction for selecting iron-efficient dry bean and soybean cultivars. J Plant Nutr 21:2639–2651

El-Shatnawi M, Makhadmeh I (2001) Ecophysiology of the plant–rhizosphere system. J Agron Crop Sci 187:1–9

Eng B, Guerinot M, Eide D, Saier M (1998) Sequence analyses and phylogenetic characterization of the ZIP family of metal ion transport proteins. J Membr Biol 166:1–7

Erbersdobler H, Barth C, Jahreis G (2017) Legumes in human nutrition. Nutrient content and protein quality of pulses. Ernahrungs Umschau 64:134–139

Etcheverry P, Grusak MA, Fleige LE (2012) Application of in vitro bioaccessibility and bioavailability methods for calcium, carotenoids, folate, iron, magnesium, polyphenols, zinc, and vitamins B6, B12, D, and E. Front Physiol 3:317

Ezzati M, Lopez AD, Rodgers AA, Murray CJ (2004) Comparative quantification of health risks: global and regional burden of disease attributable to selected major risk factors. World Health Organization

Fabbri AD, Crosby GA (2016) A review of the impact of preparation and cooking on the nutritional quality of vegetables and legumes. Int J Gastron Food Sci 3:2–11

Fairbanks D, Orf J, Inskeep W, Bloom P (1987) Evaluation of soybean genotypes for iron-deficiency chlorosis in potted calcareous soil 1. Crop Sci 27:953–957

Finkelstein JL, Haas JD, Mehta S (2017) Iron-biofortified staple food crops for improving iron status: a review of the current evidence. Curr Opin Biotechnol 44:138–145

Franzen D, O'Barr J, Zollinger R (2004) Influence of certain postemergence broadleaf herbicides on soybean stressed from iron deficiency chlorosis. Agron J 96:1357–1363

Frenkel C, Hadar Y, Chen Y (2004) Peanut plants based bioassay for iron deficiency and its remediation. Soil Sci Plant Nutr 50:1063–1070
Frewer LJ, van der Lans IA, Fischer AR, Reinders MJ, Menozzi D, Zhang X, van den Berg I, Zimmermann KL (2013) Public perceptions of Agri-food applications of genetic modification–a systematic review and meta-analysis. Trends Food Sci Technol 30:142–152
Gallou A, Mosquera HPL, Cranenbrouck S, Suárez JP, Declerck S (2011) Mycorrhiza induced resistance in potato plantlets challenged by Phytophthora infestans. Physiol Mol Plant Pathol 76:20–26
Garcia Y, Gonzalez R, Menendez R, Gonzalez M, Bourg V (2008) Effect of supplementation with dehydrated formulation of heme iron+ iron ion in rats. J Food Nutr 18:204–212
Geissler C, Singh M (2011) Iron, meat and health. Nutrients 3:283–316
Gibson RS, Bailey KB, Gibbs M, Ferguson EL (2010) A review of phytate, iron, zinc, and calcium concentrations in plant-based complementary foods used in low-income countries and implications for bioavailability. Food Nutr Bull 31:S134–S146
Glick BR (2012) Plant growth–promoting bacteria: mechanisms and applications. Scientifica 2012:1
Godsey CB, Schmidt JP, Schlegel AJ, Taylor RK, Thompson CR, Gehl RJ (2003) Correcting iron deficiency in corn with seed row–applied iron sulfate. Agron J 95:160–166
Goos RJ, Johnson BE (2000) A comparison of three methods for reducing iron-deficiency chlorosis in soybean. Agron J 92:1135–1139
Goos RJ, Johnson B, Jackson G, Hargrove G (2004) Greenhouse evaluation of controlled-release iron fertilizers for soybean. J Plant Nutr 27:43–55
Gopalakrishnan S, Vadlamudi S, Samineni S, Sameer Kumar C (2016) Plant growth-promotion and biofortification of chickpea and pigeonpea through inoculation of biocontrol potential bacteria, isolated from organic soils. Springerplus 5:1–11
Guillamón E, García-Lafuente A, Lozano M, Rostagno MA, Villares A, Martínez JA (2010) Edible mushrooms: role in the prevention of cardiovascular diseases. Fitoterapia 81:715–723
Gupta RK, Gangoliya SS, Singh NK (2015) Reduction of phytic acid and enhancement of bioavailable micronutrients in food grains. J Food Sci Technol 52:676–684
Hansen NC, Jolley VD (1995) Phytosiderophore release as a criterion for genotypic evaluation of iron efficiency in oat. J Plant Nutr 18:455–465
Hansen N, Jolley V, Berg W, Hodges M, Krenzer E (1996) Phytosiderophore release related to susceptibility of wheat to iron deficiency. Crop Sci 36:1473–1476
Hansen N, Schmitt MA, Anderson J, Strock J (2003) Iron deficiency of soybean in the upper Midwest and associated soil properties. Agron J 95:1595–1601
Hardoim PR, van Overbeek LS, van Elsas JD (2008) Properties of bacterial endophytes and their proposed role in plant growth. Trends Microbiol 16:463–471
Hassan TH, Badr MA, Karam NA, Zkaria M, El Saadany HF, Rahman DMA, Shahbah DA, Al Morshedy SM, Fathy M, Esh AMH (2016) Impact of iron deficiency anemia on the function of the immune system in children. Medicine 95:e5395
He X, Nara K (2007) Element biofortification: can mycorrhizas potentially offer a more effective and sustainable pathway to curb human malnutrition? Trends Plant Sci 12:331–333
Hefferon KL (2015) Nutritionally enhanced food crops; progress and perspectives. Int J Mol Sci 16:3895–3914
Hopkins B, Ellsworth J, Jolley V, Hansen N, Horneck D (2005) Fe nutrition management. University of Idaho Agricultural Communication
Horneck D, Hart J, Stevens R, Petrie S (2005) Acidifying soil for crop production. Oregon State University. PNW
Horton S, Alderman H, Rivera JA (2009) Hunger and malnutrition. In: Global crises, global solutions: costs and benefits, pp 305–354
Jacob C, Carrasco B, Schwember AR (2016) Advances in breeding and biotechnology of legume crops. *Plant cell*. Tissue Organ Culture (PCTOC) 127:561–584
Jessen H, Fehr W, Cianzio S, d. (1988) Registration of germplasm lines of soybean, A11, A12, A13, A14, and A15. Crop Sci 28:204–204

Jolley VD, Brown JC (1987) Soybean response to iron-deficiency stress as related to iron supply in the growth medium. J Plant Nutr 10:637–651

Jolley V, Hansen N (1995) Using phytosiderophore release for the genotypic evaluation of iron efficiency in oat and wheat. In: Abstracts, 8th International Symposium on Iron Nutrition and Interactions in Plants (BioIron), pp. 164

Jolley VD, Fairbanks DJ, Stevens WB, Terry RE, Orf JH (1992) Root iron-reduction capacity for genotypic evaluation of iron efficiency in soybean. J Plant Nutr 15:1679–1690

Karthik M, Pushpakanth P, Krishnamoorthy R, Senthilkumar M (2017) Endophytic bacteria associated with banana cultivars and their inoculation effect on plant growth. J Hortic Sci Biotechnol 92:568–576

Karthikeyan B, Joe MM, Islam M, Sa T (2012) ACC deaminase containing diazotrophic endophytic bacteria ameliorate salt stress in Catharanthus roseus through reduced ethylene levels and induction of antioxidative defense systems. Symbiosis 56:77–86

Kloepper J, Schroth M, Miller T (1980) Effects of rhizosphere colonization by plant growth–promoting rhizobacteria on potato plant development and yield. Phytopathology 70:1078–1082

Kobayashi T, Itai RN, Senoura T, Oikawa T, Ishimaru Y, Ueda M, Nakanishi H, Nishizawa NK (2016) Jasmonate signaling is activated in the very early stages of iron deficiency responses in rice roots. Plant Mol Biol 91:533–547

Krouma A, Gharsalli M, Abdelly C (2003) Differences in response to iron deficiency among some lines of common bean. J Plant Nutr 26:2295–2305

Lin SF, Grant D, Cianzio S, Shoemaker R (2000) Molecular characterization of iron deficiency chlorosis in soybean. J Plant Nutr 23:1929–1939

Loeppert R, Wei L, Ocumpaugh W (1995) Variations in the strategy of iron uptake in plants-importance of the Fe-deficiency induced H+ pump. In: Abstracts, 8th International Symposium on Iron Nutrition and Interactions in Plants (BioIron), p. 22

Lucena JJ (2000) Effects of bicarbonate, nitrate and other environmental factors on iron deficiency chlorosis. A review. J Plant Nutr 23:1591–1606

Lucena JJ (2003) Fe chelates for remediation of Fe chlorosis in strategy I plants. J Plant Nutr 26:1969–1984

Lundgren JG, Fergen JK (2011) Enhancing predation of a subterranean insect pest: a conservation benefit of winter vegetation in agroecosystems. Appl Soil Ecol 51:9–16

Mäder P, Kaiser F, Adholeya A, Singh R, Uppal HS, Sharma AK, Srivastava R, Sahai V, Aragno M, Wiemken A (2011) Inoculation of root microorganisms for sustainable wheat–rice and wheat–black gram rotations in India. Soil Biol Biochem 43:609–619

Magalhães SC, Taveira M, Cabrita AR, Fonseca AJ, Valentão P, Andrade PB (2017) European marketable grain legume seeds: further insight into phenolic compounds profiles. Food Chem 215:177–184

Mallarino AP, Haq MU, Wittry D, Bermudez M (2001) Variation in soybean response to early season foliar fertilization among and within fields. Agron J 93:1220–1226

Mann J, Truswell AS (2017) Essentials of human nutrition. Oxford University Press

Manwaring HR, Bligh H, Yadav R (2016) The challenges and opportunities associated with biofortification of pearl millet (Pennisetum glaucum) with elevated levels of grain iron and zinc. Front Plant Sci 7:1944

Marschner H (1986) Effect of external and internal factors on root growth and development. In: The mineral nutrition of higher plants, pp 429–446

Martínez-Ballesta M, Dominguez-Perles R, Moreno D, Muries B, Alcaraz-López C, Bastías E, García-Viguera C, Carvajal M (2010) Minerals in plant food: effect of agricultural practices and role in human health. A review. Agron Sustain Dev 30:295–309

Martínez-Zavala M, Mora-Avilés MA, Anaya-Loyola MA, Guzmán-Maldonado H, Aguilera-Barreyro A, Blanco-Labra A, García-Gasca T (2016) Common bean leaves as a source of dietary iron: functional test in an iron-deficient rat model. Plant Foods Hum Nutr 71:259–264

Masuda H, Ishimaru Y, Aung MS, Kobayashi T, Kakei Y, Takahashi M, Higuchi K, Nakanishi H, Nishizawa NK (2012) Iron biofortification in rice by the introduction of multiple genes involved in iron nutrition. Sci Rep 2:1–7

Masuda H, Aung MS, Nishizawa NK (2013) Iron biofortification of rice using different transgenic approaches. Rice 6:1–12

Maurer B, Muller A, W, K., and Zahner, H. (1968) Ferribactin a siderochrome from Pseudomonas fluorescens migula. Archiv fur Mikrobiologie 60:326

Mengel K (1995) Iron availability in plant tissues—iron chlorosis on calcareous soils. In: Iron nutrition in soils and plants. Springer, pp 389–397

Moreira H, Pereira SI, Marques AP, Rangel AO, Castro PM (2016) Mine land valorization through energy maize production enhanced by the application of plant growth-promoting rhizobacteria and arbuscular mycorrhizal fungi. Environ Sci Pollut Res 23:6940–6950

Morikawa C (2004) Co-situs application of control release fertilizers as a new method to alleviate iron deficiency of paddy rice grown on calcareous soil. In: Abstracts of the XIIth International Symposium on Iron Nutrition and Interactions in Plants, ISINIP 2004, held April 11–15 in Tokyo

Morrissey J, Guerinot ML (2009) Iron uptake and transport in plants: the good, the bad, and the ionome. Chem Rev 109:4553–4567

Narayanan N, Beyene G, Chauhan RD, Gaitán-Solis E, Grusak MA, Taylor N, Anderson P (2015) Overexpression of Arabidopsis VIT1 increases accumulation of iron in cassava roots and stems. Plant Sci 240:170–181

Neudorf SJ (2015) Reaction of the hemochromogen test with the heme protein leghemoglobin obtained from soybean root nodules. J Can Soc Forensic Sci 48:36–45

Ocumpaugh W, Rahmes J, Bryn D, Wei L (1992) Field and greenhouse screening of oat seedlings for iron-nutrition efficiency. J Plant Nutr 15:1715–1725

Olson R (1950) Effects of acidification, iron oxide addition, and other soil treatments on sorghum chlorosis and iron absorption. Soil Sci Soc Am Proc (United States) 15

Onyezili FN, Ross JD (1993) Iron deficiency stress responses of five gram-inaceous monocots. J Plant Nutr 16:953–974

O'Reilly KA, Robinson DE, Vyn RJ, Van Eerd LL (2011) Weed populations, sweet corn yield, and economics following fall cover crops. Weed Technol 25:374–384

Ouf N, Jan M (2015) The impact of maternal iron deficiency and iron deficiency anemia. Saudi Med J 36:146–149

Pellegrino E, Bedini S (2014) Enhancing ecosystem services in sustainable agriculture: biofertilization and biofortification of chickpea (Cicer arietinum L.) by arbuscular mycorrhizal fungi. Soil Biol Biochem 68:429–439

Pereira SI, Castro PM (2014) Phosphate-solubilizing rhizobacteria enhance Zea mays growth in agricultural P-deficient soils. Ecol Eng 73:526–535

Popp C, Ott T (2011) Regulation of signal transduction and bacterial infection during root nodule symbiosis. Curr Opin Plant Biol 14:458–467

Proulx AK, Reddy MB (2006) Iron bioavailability of hemoglobin from soy root nodules using a Caco-2 cell culture model. J Agric Food Chem 54:1518–1522

Qin H, Brookes PC, Xu J (2016) Arbuscular mycorrhizal fungal hyphae alter soil bacterial community and enhance polychlorinated biphenyls dissipation. Front Microbiol 7:939

Randall G (1981) Correcting Fe chlorosis in soybeans. Minnesota Agric Ext Serv. Soils Fact Sheet 27

Romheld V (1986) Mobilization of iron in the rhizosphere of different plant species. Adv Plant Nutr 2:155–204

Römheld V, Marschner H (1990) Genotypical differences among graminaceous species in release of phytosiderophores and uptake of iron phytosiderophores. In: Genetic aspects of plant mineral nutrition. Springer, pp 77–83

Römheld V, Schaaf G (2004) Iron transport in plants: future research in view of a plant nutritionist and a molecular biologist. Soil Sci Plant Nutr 50:1003–1012

Rosado JL, Díaz M, Gonzalez K, Griffin I, Abrams SA, Preciado R (2005) The addition of milk or yogurt to a plant-based diet increases zinc bioavailability but does not affect iron bioavailability in women. J Nutr 135:465–468

Sathya A, Vijayabharathi R, Srinivas V, Gopalakrishnan S (2016) Plant growth–promoting actinobacteria on chickpea seed mineral density: an upcoming complementary tool for sustainable biofortification strategy. 3 Biotech 6:1–6

Schaaf G, Erenoglu BE, von Wirén N (2004) Physiological and biochemical characterization of metal-phytosiderophore transport in graminaceous species. Soil Sci Plant Nutr 50:989–995

Schaart JG, van de Wiel CC, Lotz LA, Smulders MJ (2016) Opportunities for products of new plant breeding techniques. Trends Plant Sci 21:438–449

Schalk IJ, Guillon L (2013) Pyoverdine biosynthesis and secretion in P seudomonas aeruginosa: implications for metal homeostasis. Environ Microbiol 15:1661–1673

Scheublin TR, Sanders IR, Keel C, Van Der Meer JR (2010) Characterisation of microbial communities colonising the hyphal surfaces of arbuscular mycorrhizal fungi. ISME J 4:752–763

Senés-Guerrero C, Torres-Cortés G, Pfeiffer S, Rojas M, Schüßler A (2014) Potato-associated arbuscular mycorrhizal fungal communities in the Peruvian Andes. Mycorrhiza 24:405–417

Shaikh S, Saraf M (2017) Biofortification of Triticum aestivum through the inoculation of zinc solubilizing plant growth–promoting rhizobacteria in field experiment. Biocatal Agric Biotechnol 9:120–126

Sharma P, Aggarwal P, Kaur A (2017) Biofortification: a new approach to eradicate hidden hunger. Food Rev Intl 33:1–21

Siebner-Freibach H, Hadar Y, Chen Y (2004) Interaction of iron chelating agents with clay minerals. Soil Sci Soc Am J 68:470–480

Singh K, Sharma H, Singh C, Singh Y, Nishizawa N, Mori S (2004) Effect of polyolefin resin coated slow release iron fertilizer and its methods of application on rice production in calcareous soil. Soil Sci Plant Nutr 50:1037–1042

Smith M, Neilands J (1984) Rhizobactin, a siderophore from Rhizobium meliloti. J Plant Nutr 7:449–458

Snapp SS, Swinton SM, Labarta R, Mutch D, Black JR, Leep R, Nyiraneza J, O'neil K (2005) Evaluating cover crops for benefits, costs and performance within cropping system niches. Agron J 97:322–332

Srinivasagam K, Natarajan B, Raju M, Selvan RK (2013) Myth and mystery of soil mycorrhiza: a review. Afr J Agric Res 8:4706–4717

Stein, J.A. (2010). Global impacts of human mineral malnutrition. Plant and Soil 335:133–154.

Stevens WB, Jolley VD, Hansen NC, Fairbanks DJ (1993) Modified procedures for commercial adaptation of root iron-reducing capacity as a screening technique. J Plant Nutr 16:2507–2519

Suttle NF (2010) Mineral nutrition of livestock. Cabi

Tan GZ, Das Bhowmik SS, Hoang TM, Karbaschi MR, Johnson AA, Williams B, Mundree SG (2017) Finger on the pulse: pumping iron into chickpea. Front Plant Sci 8:1755

Tan GZ, Das Bhowmik SS, Hoang TM, Karbaschi MR, Long H, Cheng A, Bonneau JP, Beasley JT, Johnson AA, Williams B (2018) Investigation of baseline iron levels in Australian chickpea and evaluation of a transgenic biofortification approach. Front Plant Sci 9:788

Thavarajah P (2012) Evaluation of chickpea (Cicer arietinum L.) micronutrient composition: biofortification opportunities to combat global micronutrient malnutrition. Food Res Int 49:99–104

Theil EC, Briat J-F (2004) Plant ferritin and non-heme iron nutrition in humans. International Food Policy Research Institute (IFPRI)

Tisdale SL, Nelson WL (1966) Soil fertility and fertilizers. Soil Sci 101:346

Toda M, Uchida Y (2017) Long-term use of green manure legume and chemical fertiliser affect soil bacterial community structures but not the rate of soil nitrate decrease when excess carbon and nitrogen are applied. Soil Res 55:524–533

Tonitto C, David M, Drinkwater L (2006) Replacing bare fallows with cover crops in fertilizer-intensive cropping systems: a meta-analysis of crop yield and N dynamics. Agric Ecosyst Environ 112:58–72

Trijatmiko KR, Dueñas C, Tsakirpaloglou N, Torrizo L, Arines FM, Adeva C, Balindong J, Oliva N, Sapasap MV, Borrero J (2016) Biofortified indica rice attains iron and zinc nutrition dietary targets in the field. Sci Rep 6:1–13

Valdes-Miramontes E, López-Espinoza A, Rodriguez-Macias R, Salcedo-Perez E, Ruiz-López M (2015) Effect of heat treatment on the chemical composition and minerals in wild lupine seeds. Rev Chil Nutr 42:186–190

Van Der Straeten D, Fitzpatrick TB, De Steur H (2017) Editorial overview: biofortification of crops: achievements, future challenges, socio-economic, health and ethical aspects. Curr Opin Biotechnol 44:VII-X

Vasconcelos M, Datta K, Oliva N, Khalekuzzaman M, Torrizo L, Krishnan S, Oliveira M, Goto F, Datta SK (2003) Enhanced iron and zinc accumulation in transgenic rice with the ferritin gene. Plant Sci 164:371–378

Vasconcelos MW, Gruissem W, Bhullar NK (2017) Iron biofortification in the 21st century: setting realistic targets, overcoming obstacles, and new strategies for healthy nutrition. Curr Opin Biotechnol 44:8–15

Vose P (1982) Iron nutrition in plants: a world overview. J Plant Nutr 5:233–249

Wallace A, Mueller R, Patel P, Soufi S (1976) Use of waste pyrites from mine operations on highly calcareous soil. Commun Soil Sci Plant Anal 7:57–60

Wang Y, Yang X, Zhang X, Dong L, Zhang J, Wei Y, Feng Y, Lu L (2014) Improved plant growth and Zn accumulation in grains of rice (Oryza sativa L.) by inoculation of endophytic microbes isolated from a Zn Hyperaccumulator, Sedum alfredii H. J Agric Food Chem 62:1783–1791

Wang GZ, Li HG, Christie P, Zhang FS, Zhang JL, Bever JD (2017) Plant-soil feedback contributes to intercropping overyielding by reducing the negative effect of take-all on wheat and compensating the growth of faba bean. Plant Soil 415:1–12

Wei Jin C, You GY, Zheng SJ (2008) The iron deficiency-induced phenolics secretion plays multiple important roles in plant iron acquisition underground. Plant Signal Behav 3:60–61

Wei L, Loeppert R, Ocumpaugh W (1995) Is root Fe3+ reduction capacity a determining factor for Fe-deficiency in dicots. In: Abstracts, 8th International Symposium on Iron Nutrition and Interactions in Plants (BioIron), p. 91

Welch RM, Graham RD (2004) Breeding for micronutrients in staple food crops from a human nutrition perspective. J Exp Bot 55:353–364

Whitney DA, Stone LR, Janssen KA, Long JH (1991) The effects of nitrogen fertilizer on soil. MF-Cooperative Extension Service

WHO (2015) The global prevalence of anaemia in 2011. World Health Organization

Wiersma JV (2005) High rates of Fe-EDDHA and seed iron concentration suggest partial solutions to iron deficiency in soybean. Agron J 97:924–934

Wu T, Zhang H-T, Wang Y, Jia W-S, Xu X-F, Zhang X-Z, Han ZH (2012) Induction of root Fe (lll) reductase activity and proton extrusion by iron deficiency is mediated by auxin-based systemic signalling in Malus xiaojinensis. J Exp Bot 63:859–870

Wu Z, Banuelos GS, Lin Z-Q, Liu Y, Yuan L, Yin X, Li M (2015) Biofortification and phytoremediation of selenium in China. Front Plant Sci 6:136

Xu M, Sheng J, Chen L, Men Y, Gan L, Guo S, Shen L (2014) Bacterial community compositions of tomato (Lycopersicum esculentum Mill.) seeds and plant growth–promoting activity of ACC deaminase producing Bacillus subtilis (HYT-12-1) on tomato seedlings. World J Microbiol Biotechnol 30:835–845

Yehuda Z, Hadar Y, Chen Y (2003) Immobilized EDDHA and DFOB as iron carriers to cucumber plants. J Plant Nutr 26:2043–2056

Yossef HEE-D (2010) Effect of calcium and phosphorus on nonhaeme iron absorption and haematogenic characteristics in rats. Food Nutr Sci 1:13

Zhang H, Sun Y, Xie X, Kim MS, Dowd SE, Paré PW (2009) A soil bacterium regulates plant acquisition of iron via deficiency-inducible mechanisms. Plant J 58:568–577

Zhang Y-F, He L-Y, Chen Z-J, Wang Q-Y, Qian M, Sheng X-F (2011) Characterization of ACC deaminase-producing endophytic bacteria isolated from copper-tolerant plants and their potential in promoting the growth and copper accumulation of Brassica napus. Chemosphere 83:57–62

Zhang Z-Z, Lou Y-G, Deng D-J, Rahman MM, Wu Q-S (2015) Effects of common mycorrhizal network on plant carbohydrates and soil properties in trifoliate orange–white clover association. PLoS One 10:e0142371

Zielińska-Dawidziak M (2015) Plant ferritin—a source of iron to prevent its deficiency. Nutrients 7:1184–1201

Zuo Y, Zhang F (2009) Iron and zinc biofortification strategies in dicot plants by intercropping with gramineous species. A review. Agron Sustain Dev 29:63–71

Chapter 14
The Biofortification of Zinc in Legumes to Alleviate Zinc Deficiency

Muhammad Talha Aslam, Imran Khan, Muhammad Umer Chattha, Muhammad Nawaz, Ayesha Mustafa, Fareeha Athar, Muhammad Umair Hassan, Hira Kanwal, and Adnan Noor Shah

Abstract Micronutrient malnutrition is a serious health problem in the world that requires serious attention. About two billion people around the globe are facing micronutrient malnutrition. The reduced zinc (Zn) availability in soil decreases the Zn concentration in dietary products, which is causing Zn deficiency in humans. Globally, different strategies are being used to improve the Zn concentration in grains to reduce Zn deficiency in humans. Fortification is considered an important strategy for enriching cereals with Zn and thus reducing Zn deficiency in humans. However, this strategy is costly, and low-income countries cannot afford it. In this context, agronomic and breeding approaches have emerged as excellent strategies to increase Zn concentration in grains in order to fulfill human needs. The breeding strategy is costly and time-consuming, and agronomic techniques (fertilizer application) are considered important strategies for increasing Zn contents in grains. Legumes are widely used around the world as food and the biofortification of legumes could be a promising approach to minimizing Zn deficiency in humans. Herein, we present information on the role of Zn in plants and humans and concept of Zn biofortification to mitigate Zn deficiency in humans. We also discuss the role of fertilization methods and breeding and molecular approaches to improve Zn concentrations in legumes.

M. T. Aslam · I. Khan · M. U. Chattha · A. Mustafa · F. Athar · H. Kanwal
Department of Agronomy, University of Agriculture Faisalabad, Faisalabad, Punjab, Pakistan

M. Nawaz · A. N. Shah (✉)
Department of Agricultural Engineering, Khwaja Fareed University of Engineering and Information Technology, Rahim Yar Khan, Punjab, Pakistan

M. U. Hassan
Research Center on Ecological Sciences, Jiangxi Agricultural University, Nanchang, China

M. A. Nadeem et al. (eds.), *Legumes Biofortification*,
https://doi.org/10.1007/978-3-031-33957-8_14

1 Introduction

Human health is directly affected by hidden hunger and malnutrition problems. Food insecurity from imbalanced diets or the consumption of poor-quality foods negatively affects human health (Onyango 2003; Gundersen and Ziliak 2015). The UN's Food and Agriculture Organization FAO (2020) has observed that malnutrition and obesity problems are associated with a lack of food or poor-quality food availability to humans. The approach of increasing the cultivation of cereals and cash crops in intensive cropping systems has negatively affected micronutrient concentrations in soils and diets (Cakmak et al. 2010). Adverse effects of this approach occur most commonly in developing nations but occur wherever the human diet comprises only the consumption of cereals.

The introduction of some highly responsive cereals into applied fertilizers has made farming profitable. Farmers adopted high-yielding cereals that boost the economy of the growers, and consequently, the areas dedicated to legume cultivation were significantly reduced (Cakmak et al. 2010). In addition, Cakmak et al. (2002) observed wide variation among dietary products: Only 30 of 7000 species are being used for human consumption in the daily diets of the world's population (Cakmak et al. 2010). Malnutrition and food security in developing nations have been evergreen problems for many centuries. Moreover, if the same attitude toward humans dietary habits remain, more than 840 million people will experience hunger as of 2030 (FAO 2020). The economic instability in developing countries plays a significant role in the choice of dietary foodstuff. In addition, the high consumption of cereals and the low consumption of fruits and vegetables are common practices among low-income people in developing countries. The long-term insufficiency of micronutrients in human food has resulted in micronutrient malnutrition, causing hidden hunger. Also, micronutrient deficiency has been found in more than two billion people in the world (Velu et al. 2014; Zaman et al. 2018). This was due to a high dependency on cereals or a diet containing low micronutrient concentrations (Welch 2005; Reddy 2010). Ruel et al. (2013) suggest that the risk of hidden hunger can be mitigated by adding specific nutrients to crops.

The nutrition-specific approach is also recognized as the direct approach, which requires changing dietary habits or taking nutrition supplements. The addition of nutrients to food can be attributed to the biofortification process (De Valença et al. 2017). Improving nutrient contents in grains or any other edible part of the plant is called biofortification, which is one of the promising approaches to meet the nutritional requirements that can stave off malnutrition (Cakmak 2008; Aciksoz et al. 2011; Das et al. 2019; Maqbool and Beshir 2019). Cereals are commonly being used to fulfill nutrient demands in most developing nations (Bouis and Welch 2010; Bouis and Saltzman 2017; Cakmak and Kutman 2018). Lacking micronutrients, especially Zn, causes serious health issues. Therefore, cereals are being treated with biofortification to increase Zn content in grains. Moreover, we can also combat the threat of hidden hunger through biofortification. Different scientists have found that zinc deficiency is most commonly found in plants and animals (Welch and Graham

2004; Cakmak 2008; Bouis and Welch 2010; Bouis and Saltzman 2017); Cakmak and Kutman 2018; Das et al. 2019; Maqbool and Beshir 2019), but many human health issues are related to Zn deficiency, so its inadequacy reduces human health (Welch and Graham 2004; Cakmak 2008; Bouis and Welch 2010; Cakmak and Kutman 2018; Das et al. 2019; Maqbool and Beshir 2019; Yaseen and Hussain 2021). Prasad et al. (2014) observed that 60–70% of the populations in Asian and African countries are susceptible to reduced Zn availability in their food. Moreover, Zn deficiency is one of the principal factors affecting the growth and development of children aged 5 years and under, and one-third of people in the world are at the risk of zinc deficiency (Wessells and Brown 2012; Cakmak and Kutman 2018). Meager physical growth, an underdeveloped immune system, uncontrolled cell division (cancer), infection, and difficulties in pregnancy, especially during birth, are the most common problems associated with a lack of Zn in the human body (Hotz and Brown 2004; Prasad 2007; Cakmak 2008; Gibson 2012; Cakmak and Kutman 2018). In addition, Zn deficiency in humans weakens the skeleton, epidermal, and nervous tissues, in addition to affecting the reproductive system (Roohani et al. 2013). Thus, a healthy diet comprising essential micronutrients, especially Zn, improves immunity and thus overall human health (Chatterjee et al. 2022). Therefore, improving Zn concentrations in dietary products is urgently needed to counteract malnourishment in humans due to zinc deficiency. Biofortification is an inexpensive and promising approach to increase Zn concentrations in food, so biofortifying staple crops can mitigate malnutrition from Zn deficiency.

2 Role of Zn in Human Health

Human health is heavily dependent on the percentage of Zn available in food. The importance of Zn for human health was first reported in 1961 (Roohani et al. 2013), an observation that has since been supported by Tapiero and Tew (2003): The average human body has approximately 2800–3000 Zn-containing proteins (Prasad et al. 2014; Cakmak and Kutman 2018). More than 300 enzymes need Zn to function properly (Zastrow and Pecoraro 2014). In addition, Maret (2013) observed the presence of Zn in enzymes such as hydrolases, lyases, ligases, isomerases, oxidoreductases, and transferases. Thus, zinc has secured a pivotal role in human growth, immunity development, and reproductive and neurobehavioral activities (Uriu-Adams and Keen 2010; Praharaj et al. 2021). Given all the abovementioned uses of Zn, the bioavailability of Zn should be improved. Because of its importance in the human body, Zn deficiency causes various illnesses and physiological disorders (Jurowski et al. 2014). The extent of Zn insufficiency and the age of the person experiencing it determine the number and severity of Zn-deficiency symptoms (Roohani et al. 2013). However, the symptoms of diarrhea appear at the beginning of Zn deficiency (Livingstone 2015).

Skin allergies and frequent infections are common if Zn deficiency occurs in school-age children (Hambidge 1997; Roohani et al. 2013; Prasad et al. 2014). In adults, zinc deficiency results in regular infections, hypogeusia, ulcers (nonhealing), and severe pregnancy pain during birth (Hotz and Brown 2004; Roohani et al. 2013; Prasad et al. 2014). That Zn deficiency hampers growth and development remains well understood. Zinc is most significant during the early growth stages and pregnancy (Roohani et al. 2013). Taking Zn supplements to fulfill Zn requirements is an alternative approach to reducing malaria (Veenemans et al. 2011), pneumonia (Bhutta et al. 1999), and diarrhea (Bhutta et al. 1999; Roohani et al. 2013) symptoms.

Given the significance of Zn in human health, researchers are taking interest in Zn to improve human nutrition. Many researchers have used different approaches to improving Zn contents in food in order to augment the bioavailability of Zn in humans (Moretti et al. 2013; Udechukwu et al. 2016; Saha et al. 2017). Sharma et al. (2013) defined *bioavailability* for humans as the proportion of Zn content absorbed by the human gut. And various agricultural researchers and dietary scientists are trying to improve Zn nutrition (Cakmak 2008; Zaman et al. 2018). In addition, Cakmak and Kutman (2018) observed that phytate hinder Zn improvement in grains, so by controlling the antinutrient (phytate), Zn nourishment can be increased (Abbas and Ahmad 2018). A high proportion of phytate is present in cereals, which leads to Zn malnutrition (Shahzad et al. 2014); this happens most commonly in developing countries where cereals are major sources of food (Gibson 2006; Praharaj et al. 2021). The bioavailability of zinc contents is reduced in the human body when it relies on only cereals for food (White and Broadley 2005; Sharma et al. 2013).

3 Role of Zn in Crops

Zinc is a highly important micronutrient. It exists either freely or in a complex by making a bond with low molecular-weight molecules in plants (Brown et al. 1993). Zn is a vital nutrient for plant growth, development, and yields and is required in low quantities (Sharma et al. 2013). Zn is involved in many biochemical processes, such as auxin metabolism (Brown et al. 1993); chlorophyll synthesis (Sharma et al. 2013), and enzyme activation (Tsonev and Cebola Lidon 2012). In addition, the functions of carbonic anhydrase, alcohol dehydrogenase, and SOD depend on the availability of zinc in plants (Tobin 1970; Sharma et al. 2013). Among these enzymes, carbonic anhydrase captures CO_2 during photosynthesis (Brown et al. 1993); therefore, the photosynthetic rate is negatively affected by reduced concentrations of zinc in plants. Alcohol dehydrogenase enzymes strengthen plants under anaerobic conditions, such as flooding (Du et al. 2018; Miro and Ismail 2013). Finally, SOD is an antioxidant enzyme that activates under stress conditions in response to reactive oxidant species, detoxifies them to promote plant survival (Alscher et al. 2002; Wang et al. 2018), and protects lipid and plant membranes from sustaining oxidative stress injuries.

Zn deficiency leads to membrane damage. Zn fingers control cell differentiation and proliferation (Sharma et al. 2013). Stromal processing peptide (SPP) bonds based on the Zn concentration in plants repairs photosystem-II, so chloroplast functioning is based on the Zn content in plants (Lu et al. 2011; Sharma et al. 2013). Zinc-deficient plants endure defects in chlorophyll functioning, abnormalities in their chloroplast bodies, and reduced photosynthetic efficiency (Brown et al. 1993). Moreover, Zn unavailability also affects the formation of carbohydrates, lipids, and nucleic acids in plants (Brown et al. 1993; Zaman et al. 2018). Many researchers have observed the significant role of zinc in plant water uptake, its transportation via xylem tissues, and its tolerance to heat stress (Kasim 2007; Peck and McDonald 2010; Tavallali et al. 2010; Disante et al. 2011; Hafeez et al. 2013). Additionally, the role of Zn fingers is well recognized by plant scientists (Guo et al. 2009; Li et al. 2010; Jan et al. 2013; Zhang et al. 2014). Lastly, zinc strengthens plants against pathogen infections and herbivore attacks, as reported by Cabot et al. (2019).

4 Zn Biofortification for Grains

The crop yields and nutritional contents of grain are heavily affected by Zn deficiency (Cakmak 2008; Khoshgoftarmanesh et al. 2010). In developing countries, cereals play prominent roles in meeting daily caloric requirements, but unfortunately, they have usually low Zn concentrations (De Valença et al. 2017; Cakmak and Kutman 2018). Soils with low zinc content consequently lower the amount of zinc in the cereals grown in them (Cakmak 2008). Globally, the regions that have low concentrations of Zn in their soils also have Zn deficiencies among the people living in them, thereby indicating robust interrelationships between plant crops, soils, and human health (Cakmak 2008). Several factors, such as intensive agriculture, resulting in the removal of huge numbers of essential micronutrients from soil; low organic matter (OM) content in soils; and a reduced use of micronutrient fertilizers, have contributed to the high rates of zinc deficiency in soils. Biofortification tries to recover the interrelationship between plant crops, soils, and human health to alleviate Zn deficiency in human beings (De Valença et al. 2017). Biofortification has elevated the concentration of Zn and Se in grains (Velu et al. 2014; De Valença et al. 2017; Bouis and Welch 2010). There are high rates of severe hidden hunger and micronutrient malnutrition among people in low-income countries, and they are inability to afford supplements and healthier diets.

Biofortification is a sustainable approach for alleviating micronutrient malnutrition and meeting daily nutrient requirements (Cakmak 2008; White and Broadley 2005; McDonald et al. 2008). An efficient biofortification approach should enhance crop yields, which eventually augments crop performance levels in all growing environments and increases Zn contents (Welch and Graham 2004; Zou et al. 2012). Zinc deficiency is frequently related to cereals, and people in developing regions—predominantly in rural areas—are greatly dependent on cereal-based diets because of high food prices and regional and cultural influences on food selection, so the biofortification of cereal grains with Zn may be a valuable solution to increasing Zn intake (Bouis and Welch 2010; Cakmak and Kutman 2018). The percentage of micronutrients in grains increases by increasing the fertilization of crops, and this bioavailability is governed by the absorption, translocation, and redistribution of micronutrients, itself controlled by a homeostatic mechanism in various plant parts that leads to the accumulation of adequate and nontoxic amounts of micronutrients (Welch and Graham 2004; De Valença et al. 2017).

Nutrient absorption in a plant is influenced by soil dynamics, including physicochemical and biological soil characteristics and agronomic practices, that increase micronutrient contents at the root zone and consequently in plant parts after uptake (Brown et al. 1993; Aciksoz et al. 2011; Prasad et al. 2014). All these factors are interrelated in plants and are adopted for micronutrient absorption. The absorption of Zn is differently affected by different agronomic approaches, such as adding organic matter, changing the fertilizer application method, and sustaining optimal soil moisture (Prasad et al. 2014). To improve Zn uptake in root cells, the micronutrient amount in the root–soil boundary should be increased. In this regard, root surface area can be increased by modifying root morphology. Moreover, the efflux of H^+ from roots, metal ions, and reductants also determine the level of Zn uptake in plants (Welch and Graham 2004). Making use of these facts can generate Zn-efficient cultivars, because after uptake, the micronutrients should efficiently move within the plant, eventually being added to the edible plant parts (De Valença et al. 2017). A straight translocation of micronutrients (from roots to grains) and retranslocation (from soil to vegetative parts to grains) were also identified (Cakmak 2008; De Valença et al. 2017). The translocation and retranslocation method for different genotypes and environmental conditions must be expanded on. For good human health, micronutrients should accumulate in bioavailable sources, namely biofortified food grains. Two strategies of biofortification, agronomic biofortification and genetic biofortification, are described in the next section.

5 Agronomic and Genetic Biofortification

The physiological and metabolic processes of plants require sufficient amounts of Zn. Approximately 50% of the agricultural land allocated for cereal cultivation is Zn deficient. Therefore, the plants grown there cannot uptake an adequate amount of Zn (Cakmak 2008). Eventually, this depletion in soils causes deficiencies in the grains of crops. To stave this off, Zn content in grains should be enhanced by increasing Zn availability to crops (Cakmak et al. 2010; Aciksoz et al. 2011; Velu et al. 2014; Zaman et al. 2018). Hence, the agronomic biofortification of grains with Zn has been found as a suitable approach to increasing Zn concentrations in crops. Different methods of applying fertilizer (soil/seed priming/foliar or their combined application), Zn sources, application times, genetic characteristics, and environmental factors greatly affect the various responses of crops. Through agronomic biofortification, the rate, time, source and method of fertilizer application must be adjusted to obtain desired results and potentially improve the efficacy of Zn concentrations in grains (Prasad et al. 2014; Zaman et al. 2018; Cakmak and Kutman 2018;

Yaseen and Hussian 2021). Agronomic biofortification was found to be a relatively inexpensive and economically viable method that offers multiple advantages to increasing yields with high Zn concentrations in grains. Cakmak (2008) found that genetic biofortification promises to establish new cultivars, which could take a lot of time and effort. Furthermore, genetic biofortification can be applied to reduce Zn deficiency in soil (Cakmak 2008).

6 Effect of Different Methods of Zn Application for Grain Zn Enrichment

Seed priming, soil and foliar applications, and combinations of the two are widely used methods to provide Zn to crops (Yilmaz et al. 1997; Khan et al. 2003; Mathpal et al. 2015; Hussain et al. 2012; Rehman et al. 2015). Diverse application methods produce various kinds of results in legumes. Each method of application has its benefits (Table 14.1). In addition, soil application is one of the predominant ways to provide zinc. The efficiency of soil-applied fertilizer depends mainly on soil properties (e.g., water-holding capacity (WHC) and pH), whereas after foliar sprays, Zn uptake relies primarily on the crop because the crop retains considerable influence on the uptake and translocation of Zn in its grains. Different crops respond to various Zn application methods in different ways.

7 Effect of Soil Application on Grain Zn Content of Legumes

The efficacy of soil-applied Zn fertilizer is determined mostly by soil pH. Zn availability is substantially high in acidic soils. With a one-unit increase in soil pH, the solubility of Zn drops by a hundredfold (Lindsay and Mortvedt 2018). It has been discovered that liming in acidic soils decreases Zn availability (Prasad et al. 2014). Zn deficiency in calcareous soils is caused by their high pH, which is due to the presence of $CaCO_3$ (Prasad 2007). Alkaline soil, which makes up around 30% of global croplands, has limited Zn availability for plants (Alloway 2009; Cakmak and Kutman 2018). Factors that are responsible for Zn fixation in soil, besides pH, are as follows: complexation with organic matter, diffusion into micropores, blocking mineral uptake, solid-phase diffusion, interparticle space, and coprecipitation with other metals (Tye et al. 2003; Prasad et al. 2014; Sparks 2015). Zn reaches the plant's roots primarily through diffusion (Wilkinson et al. 1968); therefore, the Zn availability in plants is hampered by limited soil moisture and limited organic matter content (Cakmak et al. 1996; Cakmak 2008; Cakmak and Kutman 2018; Rengel 2015). Poor water-holding capacity (WHC) and low organic matter (OM) content decrease soil fertility status, and such conditions are likely to impede the uptake of Zn by plant roots (Graham et al. 1992; Alloway 2009). Soil moisture is extremely

Table 14.1 Effect of different Zn application methods on grain biofortification of legumes

Crop	Method of Zn application	Rate of Zn application	Major effects	References
Cowpea	Soil application	25 kg/ha	The soil application of zinc (Zn) increased the Zn and protein contents in grains, which further decreased the concentration of phytic acid.	Silva et al. (2021)
Mungbean	Soil application	10 kg/ha	The yield and the yield component were significantly increased, as was the zinc concentration in grains, thanks to the zinc application in soil	Haider et al. (2021)
Chickpea	Foliar application	0.5%	Foliar applications of zinc (Zn) increased Zn contents, yields, and protein contents in grains.	Pal et al. (2021)
Chickpea	Foliar application	0.5% Zinc oxide nano-particles (Zno NPs)	Increases in zinc contents and grain yields were observed after foliar applications of zinc.	Dhaliwal et al. (2021)
Mungbean	Osmopriming + soil application + foliar application	2.01 M 10 kg ha^{-1} 0.5%	The maximum marginal net benefit, grain yield, and zinc concentration were observed.	Haider et al. (2020)
Lentil	Soil application	2 kg/ha	Soil applications of Zn significantly increased the Zn, nitrogen, and phosphorous contents in grains.	Islam et al. (2018)
Mungbean	Foliar spray	1%	Foliar applications of zinc increased Zn contents in grains and increased grain yields by 78–156% and 86–38%, respectively.	Haider et al. (2018)
Chickpea	Foliar spray	25 kg/ha	Foliar applications of zinc (Zn) increased the Zn contents of grains.	Hidoto et al. (2017)
Chickpea	Foliar application	0.5%	Foliar applications of zinc augmented nutrient (nitrogen and zinc) uptake. This resulted in the increased vegetative growth of plants, and high protein contents were recorded in grains.	Singh et al. (2015)

significant in soils that have low Zn availability. In the diffusion process, soil moisture serves as a conduit for Zn transport from soil to roots. Hence, the deleterious effects of Zn deficiency are more lethal under rain-fed conditions than under irrigated conditions (Cakmak 2008).

The interaction of Zn with other nutrients may alter the amount of Zn available to crops when using the soil application method. Many researchers have discovered a positive Zn–nitrogen relationship (Kutman et al. 2010; Erenoglu et al. 2011; Prasad et al. 2014). One of the well-studied nutritional interactions is the negative zinc–phosphorus interaction (Mousavi 2011; Prasad et al. 2014, 2016). The availability of Zn to plants is reduced when there is too much phosphorus in soil. Low concentrations of Zn in soil solutions and a decrease in vesicular arbuscular mycorrhizae (VAM) infections may be responsible for the decrease in Zn uptake caused by applying a high dose of phosphorus (Prasad et al. 2014). Zn has also been discovered to have negative interactions with Fe, Mn, and Cu (Prasad et al. 2014). These interactions must be taken into account when determining the Zn availability to plants.

The amount of moisture in the soil influences Zn availability by changing the pH, redox potential, and number of dissolved organic anions (Gao et al. 2012). The biological features of soil, especially its chemical characteristics, are important in determining Zn availability to plants (Gao et al. 2012; Prasad et al. 2014). Plant growth–promoting bacteria (PGPB) are a bacterial group that has several beneficial effects on the growth and development of plants (Aeron et al. 2011). They can improve mobility and nutrient absorption (Cakmakçi et al. 2006; Prasad et al. 2014; Vejan et al. 2016) and have been demonstrated to be effective in increasing plant Zn availability. Vesicular arbuscular mycorrhizae in the soil can also help plants by assisting in the mobilization of Zn (Cavagnaro 2008). The efficacy of soil-applied Zn, like any other technique of Zn application, is influenced by the plant's genetic characteristics. It has also been discovered that the source of zinc and the dose of zinc applied to the soil have impacts on yields and grain quality (Prasad et al. 2014; Zaman et al. 2018). The optimal dose of Zn to soil varies depending on the Zn in the soil solution and in the plant body. To prevent zinc toxicity in plants, care should be exercised when prescribing zinc dosages.

8 Effect of Foliar Application on Grain Zn Content of Legumes

The effectiveness of foliar applications of Zn highly depends on the type of fertilizer used, the crop's characteristics (especially its leaf features), and the crop's genetic potential (Zaman et al. 2018). Foliar applications have several benefits, including requiring less fertilizer and eliminating Zn fixation and antagonistic nutritional influences on Zn uptake (Prasad et al. 2014). In foliar applications, Zn was observed as a mobile nutrient in the phloem (Cakmak and Kutman 2018) and transferred to developing grains. When raising the grain Zn content in wheat, foliar applications of Zn are superior to soil applications (Yilmaz et al. 1997; Cakmak et al. 2010; Zou et al. 2012). The efficiency of foliar spray in increasing grain Zn concentration greatly varies depending on the time of its application (Cakmak 2008). A weak association has been discovered between soil DTPA-Zn content and Zn concentrations in grains (Zou et al. 2012). This discrepancy could be related to unfavorable soil conditions, which impede nutrient mobilization to plant roots and thus their absorption rate. They are not susceptible to any sort of fixation, because foliar applications do not interact with soil. Under adverse field conditions, this element remobilizes from vegetative tissues to grains (Zou et al. 2012), and maintaining a greater Zn concentration in vegetative tissues may help to increase Zn concentrations in grains (Pearson and Rengel 1994; Kutman et al. 2010). Unfavorable field conditions are common; thus, foliar Zn spray can be an effective way to boost Zn levels in grains.

The efficiency of foliar applications of Zn is based on when they are applied (Ozturk et al. 2006; Malesh et al. 2016). Zn grain augmentation was higher at the lateral growth stage of Zn application (Ozturk et al. 2006). Foliar applications of Zn were more effective when applied at the heading and early milk stages than when applied at the stem elongation and booting stages (Malesh et al. 2016). The greater efficiency of Zn supplied via foliar application at the milking stage could be due to the mobilization of micronutrients to the sink organs and active photoassimilation allocation to the sink. During the reproductive stage, Zn phloem mobility was shown to be higher after foliar applications (Haslett et al. 2001; Malesh et al. 2016). Foliar applications of 0.5% (w/v) $ZnSO_4.7H_2O$ at the heading and milking stages were beneficial for the augmentation of Zn in wheat grains, and this trend was observed in seven countries at seven sites in a 2-year-long experiment. According to the findings, foliar applications of Zn raise Zn content in grains by 83.5%, whereas soil applications of Zn increase it only by 12.3% (Zou et al. 2012). Under drought conditions, foliar Zn spray has also been shown to boost grain productivity (Karim et al. 2012), and this increase in yield under drought conditions could be attributed to better defense mechanisms (Cakmak 2000; Zou et al. 2012).

9 Effect of Seed Priming on Grain Zn Content of Legumes

In the field, high Zn concentrations in seeds have been demonstrated to boost seedling vigor and crop stand (Yilmaz et al. 1998). When seeds are grown in nutrient-deficient conditions, they lose vigor and reduce overall plant growth and yield (Yilmaz et al. 1998). Limited moisture impacts Zn availability to plant roots because Zn is transferred to plant roots via diffusion (Prasad et al. 2014), and soil-applied Zn has different efficiency levels for both irrigated and moisture deficit conditions. The consequence of Zn biofortification could be unreliability in rain-fed conditions. Seeds with higher Zn content can perform better even under rain-fed conditions, enhancing plant growth and productivity (Yilmaz et al. 1998). Wheat seed priming with Zn raised Zn concentration in grains by 12%, whereas Zn concentrations in chickpea grains and maize grains increased by 29% and 19%, respectively (Harris et al. 2008). In addition, the priming technique was observed to be cost-effective in wheat, maize, and chickpeas (Harris et al. 2008). Seed priming is less successful in enhancing the Zn contents of grains than foliar application is (Zaman et al. 2018); however, the former may play a significant role especially in a resource-constrained and stressed environment.

10 Additional Benefits of Zn Fertilization

Biofortification, along with improving Zn bioavailability, also provides some supplementary benefits. Zinc–phosphorous antagonism has been reported by various researchers (Aref 2007; Hussain et al. 2011). Therefore, soil-applied Zn significantly reduced the phosphorous (P) uptake concentration in plants (Cakmak 2008). Consequently, reduced P uptake helps in lowering the phytate contents in grains (Erdal et al. 2002; Cakmak 2008; Chattha et al. 2017). The uptake and storage availability of phosphorous increased by increasing the phloem mobility of phosphorous in plants growing in zinc-deficient soil conditions (Buerkert et al. 1998; Cakmak 2008). Inorganic P in grains transforms into an antinutritional factor—e.g., phytate or phytic acid, which further lowers Zn bioavailability (Egli et al. 2004; Hotz and Gibson 2007). Gibson (2006) used the phytate–Zn molar ratio to determine the bioavailability of zinc in plants. In addition to human health improvement thanks to high zinc concentrations, many additional agronomic benefits have been reported from Zn-enriched seeds. High-Zn seeds show batter tolerance to various abiotic stresses in the field (Welch 1999). Yilmaz et al. (1998) observed the poor stand establishment and seed vigor of crop that had low concentrations of Zn in their seeds. In addition, the high seedling vigor that is caused by an increased concentration of Zn in seeds can be attributed to reducing the seed rate (Braun 1999; Cakmak 2008). Lastly, a sufficient level of Zn in seeds provides a good defensive mechanism against soilborne pathogens (Cakmak 2008).

11 Strategies for the Genetic Biofortification of Legumes

Genetic biofortification is a practical, justifiable, and cost-effective method for ameliorating deficiencies in the micronutrient content of food (Saltzman et al. 2013). In genetic biofortification, the uptake, translocation, or distribution/accumulation of minerals in plant edibles can be increased through conventional breeding, mutation-breeding, or transgenic methods. To extract targeted nutritional traits, high genetic diversity is very important, and it can be made possible by implementing a conventional breeding program for biofortification. Breeders can easily predict the possible variable require to improve the concentration of minerals in high-yielding genotypes. For example, if significant diversity is not found for targeted characteristics, then plant breeders can take other approaches of breeding, such as distant hybridization and mutagenesis. Several advanced countries are governing a combined project, namely the HEALTHGRAIN project (2005–2010) and the HarvestPlus program, designed to biofortify agricultural crops with crop-breeding techniques to increase the nutritional status of malnourished people (Bouis and Welch 2010; Sarker and Agrawal 2015; Garg et al. 2018). Because of the more convenient use and desired results of conventional breeding, numerous crops are biofortified by using this breeding technique. However, pulses have not received much attention in biofortification, even though some pulses, such as cowpea, lentils, and common bean, have recently struggled.

Numerous lentil cultivars have been introduced with high Zn and Fe contents, thanks to the collaboration of the HarvestPlus program and the International Center for Agricultural Research in the Dry Areas (ICARDA) (HarvestPlus 2014; Shivay et al. 2016; Garg et al. 2018). Correspondingly, GB Pant University, in Pantnagar, Uttarakhand, India, in association with HarvestPlus, has developed four biofortified cowpea cultivars for high Fe content through the application of conventional breeding (Shivay et al. 2016; Garg et al. 2018). In relation to the biofortification of the common bean, high iron beans were introduced into six African and eight Latin American countries under the HarvestPlus program.

However, the correlations between mineral micronutrient status and yield/yield-contributing parameters still need to be studied. For instance, in many studies, scientists found a detrimental relationship between Zn content and the final yield of chickpeas (Diapari et al. 2014; Vandemark et al. 2018; Misra et al. 2020) and other crops (Banziger and Long 2000; Phuke et al. 2017; Naik et al. 2020), suggesting that Zn-rich cultivars with high-yielding characteristics need to bred. Consequently, this method favors evaluating the diversity of desirable traits and also helps in arranging a baseline for the preferred development of nutrient contents. Several factors, such as per capita consumption, that are relatively high in cereal crops compared with legume crops are included in decisions on the breeding targets of different crops. An inclusive analysis of various studies reaffirmed the significance of building baselines (Joshi-Saha et al. 2018; Misra et al. 2020), which can be used in conventional breeding and targeted mutagenesis techniques that reduce the destructive effects of low nutrients (Zn) and final yield counts.

12 Transgenic and Biotechnological Implements for Biofortification of Legumes

Studies have reported that the transgenic approach is being used in several biofortification projects to improve the status of Zn content in cereals and vegetable crops (Garg et al. 2018). In legume crops, this approach has been widely used in soybeans, and as a result, a few transgenic genotypes for oleic acid and stearidonic acid have been introduced (Garg et al. 2018). Studies have not yet reported any transgenically biofortified pulse crops. Additionally, the release of cultivars also faces several challenges, such as regulatory, legal, and sociopolitical restraints.

The transgenic research has been valuably considering the physiological activities in the uptake, transport, and translocation of micronutrients to the seeds, and it has significantly increased the nutrient contents in grains. Recently, gene assembling, with the combined use of conventional breeding and metabolic engineering, significantly improved nutrient uptake in targeted crops (Van Der Straeten et al. 2020). Further, genome editing with CRISPR/Cas9 is useful for cracking genes open (vacuolar iron transporter) to increase the bioavailability of Fe and Zn content (Ludwig and Slamet-Loedin 2019). In crop sciences, the transgenic strategy for biofortification is most beneficial when no natural change is present for the preferred trait that needs development. However, earlier studies have revealed that inconsistencies in Fe and Zn levels in most grain legumes can be remediated via conventional breeding.

13 Impact of Biofortified Legumes

Concerning the use of breeding tools in certain research areas, the benefits of using biofortified crops have been observed but still need to be analyzed. In more than forty countries, 340 biofortified genotypes of twelve crops have been introduced (Lusk 2020). Worldwide, Zn biofortification for lentils and beans has been successfully developed (Gaikwad et al. 2020). The main principle of introducing biofortified crops is to raise the mineral content in the crop. According to the Lusk (2020), no successful experiments for Zn biofortification in legume crops have been conducted, though there have been a few types of studies found relating to Fe biofortification in beans. From the summery of Fe biofortification, a significant variation was observed in the cultivars collected from breeders. Meanwhile, the cultivars already being used by farmers did not show a considerable increase in Fe from Fe biofortification that could not have been just as attributable to the agronomic and/or management practices followed in field (Glahn et al. 2020). Therefore, the future biofortification plan should be more precisely conducted to bring about significant outcomes. The Zn and Fe contents in cereals are higher than those in legumes (Hemalatha et al. 2007). The accessions with a wide range of Fe and Zn contents in various legumes are presented. As a result, the baselines to be adjusted for these crops should be planned in several broad and varied environments for the screening of designated genotypes in order to evaluate their potentials.

14 Conclusions

Zinc improvement in the human diet is required to combat Zn deficiency. For this, the biotechnological approach could prove a promising approach, and germplasms, characterized by its high Zn contents, are obvious targets. Moreover, germplasms should critically observe the bioavailability of the Zn micronutrient. To make genotypes agronomically superior, the breeding program must be formulated to fill research gaps. Zinc deficiencies in agricultural soils are common around the world. Plant and soil analyses have shown that 49% of the soils in Asian countries are zinc deficient (Singh 2008). Thus, in order to attain effective biofortification, farmers should follow appropriate agronomic practices such as timely supplies of Zn in the required concentration for the crops. Many antinutrients play significant roles in responding to various (biotic and abiotic) stresses and enhancing plant growth, though antinutritional factors also reduce the availability of micronutrients. Therefore, a baseline for such antinutrients should be carefully considered before setting a target, to avoid compromising crop yields. Extensive agronomic research studies should be carried out in order to evaluate biofortification on the basis of the bioavailability of Zn nutrients in legumes. The nutritional and health significance of pulses increases the demand for pulses to feed the global population (Rebello et al. 2014; Figueira et al. 2019; Perera et al. 2020). Designing an integrated biofortification approach remains a challenge for breeders, researchers, and growers. In addition, a paradigm shift in policies on investments and markets is required to encourage farmers to grow biologically enriched legumes.

References

Abbas Y, Ahmad A (2018) Impact of processing on nutritional and anti-nutritional factors of legumes: a review. Ann Food Sci Technol 19(2):199–215

Aciksoz SB, Yazici A, Ozturk L, Cakmak I (2011) Biofortification of wheat with iron through soil and foliar application of nitrogen and iron fertilizers. Plant Soil 349:215–225

Aeron A, Kumar S, Pandey P, Maheshwari DK (2011) Emerging role of plant growth promoting rhizobacteria in agrobiology. In: Bacteria in agrobiology: crop ecosystems. Springer, Berlin/Heidelberg, pp 1–36

Alloway BJ (2009) Soil factors associated with zinc deficiency in crops and humans. Environ Geochem Health 31:537–548

Alscher RG, Erturk N, Heath LS (2002) Role of superoxide dismutases (SODs) in controlling oxidative stress in plants. J Exp Bot 53:1331–1341

Aref F (2007) The effect of zinc and boron interaction on residual available phosphorous and zinc. Soil Environ 26(2):157–163

Banziger M, Long J (2000) The potential for increasing the iron and zinc density of maize through plant breeding. Food Nutr Bull 21:397–400

Bhutta Z, Black R, Brown K, Gardner J, Gore S, Hidayat A, Khatun F, Martorell R, Ninh N, Penny M et al (1999) Prevention of diarrhea and pneumonia by zinc supplementation in children in developing countries: pooled analysis of randomized controlled trials. J Pediatr 135:689–697

Bouis HE, Saltzman A (2017) Improving nutrition through biofortification: a review of evidence from HarvestPlus, 2003 through 2016. Glob Food Sec 12:49–58

Bouis HE, Welch RM (2010) Biofortification-a sustainable agricultural strategy for reducing micronutrient malnutrition in the global south. Crop Sci 50:S-20–S-32

Braun HJ (1999) Prospects of Turkey's wheat industry, breeding and biotechnology. In: Ekiz H (ed) Hububat Sempozyum. International Winter Cereal Research Center, Konya, pp 1–744

Brown PH, Cakmak I, Zhang Q (1993) Form and function of zinc plants. In: Zinc in soils and plants. Springer, Dordrecht, pp 93–106

Buerkert A, Haake C, Ruckwied M, Marschner H (1998) Phosphorus application affects the nutritional quality of millet grain in the Sahel. Field Crops Res 57:223–235

Cabot C, Martos S, Llugany M, Gallego B, Tolrà R, Poschenrieder C (2019) A role for zinc in plant defense against pathogens and herbivores. Front Plant Sci 10:1171

Cakmak I (2000) Tansley Review No. 111: possible roles of zinc in protecting plant cells from damage by reactive oxygen species. New Phytol 146:185–205

Cakmak I (2008) Enrichment of cereal grains with zinc: agronomic or genetic biofortification? Plant Soil 302:1–17

Cakmak I, Kutman UB (2018) Agronomic biofortification of cereals with zinc: a review. Eur J Soil Sci 69:172–180

Cakmak I, Yilmaz A, Ekiz H, Torun B, Erenoglu B, Braun HJ (1996) Zinc deficiency as a critical nutritional problem in wheat production in Central Anatolia. Plant Soil 180:165–172

Cakmak I, Graham R, Welch RM (2002) Agricultural and molecular genetic approaches to improving nutrition and preventing micronutrient malnutrition globally. Encycl Life Support Syst 1:1075–1099

Cakmak I, Pfeiffer WH, McClafferty B (2010) Review: biofortification of durum wheat with zinc and iron. Cereal Chem J 87:10–20

Cakmakçi R, Dönmez F, Aydın A, Sahin F (2006) Growth promotion of plants by plant growth-promoting rhizobacteria under greenhouse and two different field soil conditions. Soil Biol Biochem 38:1482–1487

Cavagnaro TR (2008) The role of arbuscular mycorrhizas in improving plant zinc nutrition under low soil zinc concentrations: a review. Plant Soil 304:315–325

Chatterjee P, Nirgude A, Chatterjee PK (2022) Healthy eating–a modifiable contributor to optimize healthy living in the COVID-19 pandemic: a review. J Sci Food Agric 102(5):1751–1758

Chattha MU, Hassan MU, Khan I, Chattha MB, Mahmood A, Chattha MU et al (2017) Biofortification of wheat cultivars to combat zinc deficiency. Front Plant Sci 8:281

Das S, Chaki AK, Hossain A (2019) Breeding and agronomic approaches for the biofortification of zinc in wheat (Triticum aestivum L.) to combat zinc deficiency in millions of a population: a Bangladesh perspective. Acta Agrobot 72:1–13

De Valença A, Bake A, Brouwer I, Giller K (2017) Agronomic biofortification of crops to fight hidden hunger in sub-Saharan Africa. Glob Food Secur 12:8–14

Dhaliwal SS, Sharma V, Shukla AK, Verma V, Behera SK, Singh P et al (2021) Comparative efficiency of mineral, chelated and nano forms of zinc and iron for improvement of zinc and iron in chickpea (Cicer arietinum L.) through biofortification. Agronomy 11:2436

Diapari M, Sindhu A, Bett K, Deokar A, Warkentin TD, Taran B (2014) Genetic diversity and association mapping of iron and zinc concentrations in chickpea (Cicer arietinum L.). Genome 57:459–468

Disante KB, Fuentes D, Cortina J (2011) Response to drought of Zn-stressed Quercus suber L. seedlings. Environ Exp Bot 70:96–103

Du HY, Liu DX, Liu GT, Liu HP, Kurtenbach R (2018) Relationship between polyamines and anaerobic respiration of wheat seedling root under water-logging stress. Russ J Plant Physiol 65:874–881

Egli I, Davidsson L, Zeder C, Walczyk T, Hurrell R (2004) Dephytinization of a complementary food based on wheat and soy increases zinc, but not copper, apparent absorption in adults. J Nutr 134:1077–1080

Erdal I, Yilmaz A, Taban S, Eker SELİM, Torun B, Cakmak I (2002) Phytic acid and phosphorus concentrations in seeds of wheat cultivars grown with and without zinc fertilization. J Plant Nutr 25(1):113–127

Erenoglu EB, Kutman UB, Ceylan Y, Yildiz B, Cakmak I (2011) Improved nitrogen nutrition enhances root uptake, root-to-shoot translocation and remobilization of zinc (65 Zn) in wheat. New Phytol 189:438–448

FAO; IFAD; UNICEF; WFP; WHO (2020) The state of food security and nutrition in the world 2020: transforming food systems for affordable healthy diets. FAO, Rome

Figueira N, Curtain F, Beck E, Grafenauer S (2019) Consumer understanding and culinary use of legumes in Australia. Nutrients 11:1575

Gaikwad KB, Rani S, Kumar M, Gupta V, Babu PH, Bainsla NK et al (2020) Enhancing the nutritional quality of major food crops through conventional and genomics-assisted breeding. Front Nutr 7:533453

Gao X, Hoffland E, Stomph T, Grant CA, Zou C, Zhang F (2012) Improving zinc bioavailability in transition from flooded to aerobic rice. A review. Agron Sustain Dev 32:465–478

Garg M, Sharma N, Sharma S, Kapoor P, Kumar A, Chunduri V et al (2018) Biofortified crops generated by breeding, agronomy, and transgenic approaches are improving lives of millions of people around the world. Front Nutr 5:12
Gibson RS (2006) Zinc: the missing link in combating micronutrient malnutrition in developing countries. Proc Nutr Soc 65(1):51–60
Gibson RS (2012) Zinc deficiency and human health: etiology, health consequences, and future solutions. Plant Soil 361:291–299
Glahn RP, Wiesinger JA, Lung'aho MG (2020) Iron concentrations in biofortified beans and non-biofortified marketplace varieties in East Africa are similar. J Nutr 150:3013–3023
Graham RD, Ascher JS, Hynes SC (1992) Selecting zinc-efficient cereal genotypes for soils of low zinc status. Plant Soil 146:241–250
Gundersen C, Ziliak JP (2015) Food insecurity and health outcomes. Health Aff 34:1830–1839
Guo Y, Yu Y, Wang D, Wu C, Yang G, Huang J, Zheng C (2009) GhZFP1, a novel CCCH-type zinc finger protein from cotton, enhances salt stress tolerance and fungal disease resistance in transgenic tobacco by interacting with GZIRD21A and GZIPR5. New Phytol 183:62–75
Hafeez B, Khanif YM, Saleem M (2013) Role of zinc in plant nutrition—a review. Am J Exp Agric 3:374–391
Haider MU, Farooq M, Nawaz A, Hussain M (2018) Foliage applied zinc ensures better growth, yield and grain biofortification of mungbean. Int J Agric Biol 20:2817–2822
Haider MU, Hussain M, Farooq M, Nawaz A (2020) Zinc nutrition for improving the productivity and grain biofortification of mungbean. J Soil Sci Plant Nutr 20(3):1321–1335
Haider MU, Hussain M, Farooq M, Ul-Allah S, Ansari MJ, Alwahibi MS et al (2021) Zinc biofortification potential of diverse mungbean [Vigna radiata (L.) Wilczek] genotypes under field conditions. PLoS One 16:e0253085
Hambidge KM (1997) Zinc deficiency in young children. Am J Clin Nutr 65:160–161
Harris D, Rashid A, Miraj G, Arif M, Yunas M (2008) 'On-farm' seed priming with zinc in chickpea and wheat in Pakistan. Plant Soil 306:3–10
HarvestPlus (2014) Biofortification progress briefs: iron and zinc lentils. ⟨https://www.harvestplus.org/sites/default/files/publications/Biofortification_Progress_Briefs_August2014_WEB_2_0.pdf⟩
Haslett BS, Reid RJ, Rengel Z (2001) Zinc mobility in wheat: uptake and distribution of zinc applied to leaves or roots. Ann Bot 87:379–386
Hemalatha S, Platel K, Srinivasan K (2007) Zinc and iron contents and their bioaccessibility in cereals and pulses consumed in India. Food Chem 102:1328–1336
Hidoto L, Worku W, Mohammed H, Taran B (2017) Effects of zinc application strategy on zinc content and productivity of chickpea grown under zinc deficient soils. J Soil Sci Plant Nutr 17:112–126
Hotz C, Brown KH (2004) Assessment of the risk of zinc deficiency in populations and options for its control. Food Nutr Bull 25:S91–S204
Hotz C, Gibson RS (2007) Traditional food-processing and preparation practices to enhance the bioavailability of micronutrients in plant-based diets. J Nutr 137:1097–1100
Hussain N, Khan MB, Ahmad R, Ali MA, Ahmed N, Saeed S (2011) Physiochemical traits, productivity and net return of wheat as affected by phosphorus and zinc requirements under arid climates. Pak J Bot 43(2):991–1002
Hussain S, Maqsood MA, Rengel Z, Aziz T (2012) Biofortification and estimated human bioavailability of zinc in wheat grains as influenced by methods of zinc application. Plant Soil 361:279–290
Islam MM, Karim MR, Oliver MMH, Urmi TA, Hossain MA, Haque MM (2018) Impacts of trace element addition on lentil (Lens culinaris L.) agronomy. Agronomy 8:100
Jan A, Maruyama K, Todaka D, Kidokoro S, Abo M, Yoshimura E, Shinozaki K, Nakashima K, Yamaguchi-Shinozaki K (2013) OsTZF1, a CCCH-tandem zinc finger protein, confers delayed senescence and stress tolerance in rice by regulating stress-related genes. Plant Physiol 161:1202–1216

Joshi-Saha A, Mehzabin P, Reddy KS, Ramachandran V (2018) Effect of seasonal variation on micronutrient content in chickpea (Cicer arietinum L.) and identification of accessions having high iron and zinc. J Food Legumes 31:1–4
Jurowski K, Szewczyk B, Nowak G, Piekoszewski W (2014) Biological consequences of zinc deficiency in the pathomechanisms of selected diseases. JBIC J Biol Inorg Chem 19(7):1069–1079
Karim R, Zhang Y-Q, Zhao R-R, Chen X-P, Zhang F-S, Zou C-Q (2012) Alleviation of drought stress in winter wheat by late foliar application of zinc, boron, and manganese. J Plant Nutr Soil Sci 175:142–151
Kasim WA (2007) Physiological consequences of structural and ultra-structural changes induced by Zn stress in Phaseolus vulgaris. I. Growth and photosynthetic apparatus. Int J Bot 3:15–22
Khan MU, Qasim M, Subhan M, Jamil M, Ahmad RD (2003) Response of rice to different methods of zinc application in calcareous soil. Pak J Appl Sci 3:524–529
Khoshgoftarmanesh AH, Schulin R, Chaney RL, Daneshbakhsh B, Afyuni M (2010) Micronutrient-efficient genotypes for crop yield and nutritional quality in sustainable agriculture—a review. Agron Sustain Dev 30:83–107
Kutman UB, Yildiz B, Ozturk L, Cakmak I (2010) Biofortification of durum wheat with zinc through soil and foliar applications of nitrogen. Cereal Chem 87:1–9
Li C, Lv J, Zhao X, Ai X, Zhu X, Wang M, Zhao S, Xia G (2010) TaCHP: a wheat zinc finger protein gene down-regulated by abscisic acid and salinity stress plays a positive role in stress tolerance. Plant Physiol 154:211–221
Lindsay WL, Mortvedt JJ (2018) Inorganic equilibria affecting micronutrients in soils. Micronutr Agric 4:89–112
Livingstone C (2015) Zinc: physiology, deficiency, and parenteral nutrition. Nutr Clin Pract 30(3):371–382
Lu Y, Hall DA, Last RL (2011) A small zinc finger thylakoid protein plays a role in maintenance of photosystem II in Arabidopsis thaliana. Plant Cell 23:1861–1875
Ludwig Y, Slamet-Loedin IH (2019) Genetic biofortification to enrich rice and wheat grain iron: from genes to product. Front Plant Sci 10:833
Lusk J (2020) Economic impacts of covid-19 on food and agricultural markets. CAST Commentary Council for Agricultural Science and Technology, 1–44
Malesh AA, Mengistu DK, Aberra DA (2016) Linking agriculture with health through genetic and agronomic biofortification. Agric Sci 2016(7):295–307
Maqbool MA, Beshir A (2019) Zinc biofortification of maize (Zea mays L.): status and challenges. Plant Breed 138:1–28
Maret W (2013) Zinc biochemistry: from a single zinc enzyme to a key element of life. Adv Nutr 4:82–91
Mathpal B, Srivastava PC, Shankhdhar D, Shankhdhar SC (2015) Zinc enrichment in wheat genotypes under various methods of zinc application. Plant Soil Environ 61:171–175
McDonald GK, Genc Y, Graham RD (2008) A simple method to evaluate genetic variation in grain zinc concentration by correcting for differences in grain yield. Plant Soil 306:49–55
Miro B, Ismail AM (2013) Tolerance of anaerobic conditions caused by flooding during germination and early growth in rice (Oryza sativa L.). Front Plant Sci 4:269
Misra G, Joshi-Saha A, Salaskar D, Reddy KS, Dixit GP, Srivastava AK et al (2020) Baseline status and effect of genotype, environment and genotype × environment interactions on iron and zinc content in Indian chickpeas (Cicer arietinum L.). Euphytica 216:137
Moretti D, Biebinger R, Bruins MJ, Hoeft B, Kraemer K (2013) Bioavailability of iron, zinc, folic acid, and vitamin A from fortified maize. Ann N Y Acad Sci 1312:54–65
Mousavi SR (2011) Zinc in crop production and interaction with phosphorus. Aust J Basic Appl Sci 5:1503–1509
Naik SM, Raman AK, Nagamallika M, Venkateshwarlu C, Singh SP, Kumar S et al (2020) Genotype × environment interactions for grain iron and zinc content in rice. J Sci Food Agric 100:4150–4164

Onyango AW (2003) Dietary diversity, child nutrition and health in contemporary African communities. Comp Biochem Physiol Part A Mol Integr Physiol 136:61–69
Ozturk L, Yazici MA, Yucel C, Torun A, Cekic C, Bagci A, Ozkan H, Braun HJ, Sayers Z, Cakmak I (2006) Concentration and localization of zinc during seed development and germination in wheat. Physiol Plant 128:144–152
Pal V, Singh G, Dhaliwal SS (2021) A new approach in agronomic biofortification for improving zinc and iron content in chickpea (Cicer arietinum L.) grain with simultaneous foliar application of zinc sulphate, ferrous sulphate and urea. Journal of. Soil Sci Plant Nutr 21(2):883–896
Pearson J, Rengel Z (1994) Distribution and remobilization of Zn and Mn during grain development in wheat. J Exp Bot 45:1829–1835
Peck AW, McDonald GK (2010) Adequate zinc nutrition alleviates the adverse effects of heat stress in bread wheat. Plant Soil 337:355–374
Perera T, Russo C, Takata Y, Bobe G (2020) Legume consumption patterns in US adults: national health and nutrition examination survey (NHANES) 2011–2014 and Beans, Lentils, Peas (BLP) 2017 survey. Nutrients 12:1237
Phuke RM, Anuradha K, Radhika K, Jabeen F, Anuradha G, Ramesh T et al (2017) Genetic variability, genotype × environment interaction, correlation, and GGE biplot analysis for grain iron and zinc concentration and other agronomic traits in RIL population of sorghum (Sorghum bicolor L. Moench). Front Plant Sci 8:712
Praharaj S, Skalicky M, Maitra S, Bhadra P, Shankar T, Brestic M et al (2021) Zinc biofortification in food crops could alleviate the zinc malnutrition in human health. Molecules 26(12):3509
Prasad R (2007) Crop nutrition—principles and practices. New Vishal Publications, Delhi, India, p 272
Prasad R, Shivay YS, Kumar D (2014) Agronomic biofortification of cereal grains with iron and zinc. Adv Agron 125:55–91
Prasad R, Shivay YS, Kumar D (2016) Interactions of zinc with other nutrients in soils and plants—a review. Indian J Fertil 12:16–26
Rebello CJ, Greenway FL, Finley JW (2014) A review of the nutritional value of legumes and their effects on obesity and its related co-morbidities. Obes Rev 15:392–407
Reddy A (2010) Regional disparities in food habits and nutritional intake in Andhra Pradesh, India. Regional disparities in food habits and nutritional intake in Andhra Pradesh, India (May 19, 2010). Reg Sect Econ Stud 10(2):125–134
Rehman A, Farooq M, Ahmad R, Basra S (2015) Seed priming with zinc improves the germination and early seedling growth of wheat. Seed Sci Technol 43:262–268
Rengel Z (2015) Availability of Mn, Zn and Fe in the rhizosphere. J Soil Sci Plant Nutr 15:397–409
Roohani N, Hurrell R, Kelishadi R, Schulin R (2013) Zinc and its importance for human health: an integrative review. J Res Med Sci 18:144–157
Ruel MT, Alderman H, Maternal and Child Nutrition Study Group (2013) Nutrition-sensitive interventions and programmes: how can they help to accelerate progress in improving maternal and child nutrition? Lancet 382:536–551
Saha S, Chakraborty M, Padhan D, Saha B, Murmu S, Batabyal K, Seth A, Hazra GC, Mandal B, Bell RW (2017) Agro-nomic biofortification of zinc in rice: influence of cultivars and zinc application methods on grain yield and zinc bioavailability. Field Crops Res 210:52–60
Saltzman A, Birol E, Bouis HE, Boy E, De Moura FF, Islam Y et al (2013) Biofortification: Progress toward a more nourishing future. Glob Food Sect 2:9–17
Sarker A, Agrawal SK (2015) Combating micronutrient malnutrition with biofortified lentils. CARDA Res Brief 1:1–4
Shahzad Z, Rouached H, Rakha A (2014) Combating mineral malnutrition through iron and zinc biofortification of cereals. Compr Rev Food Sci Food Saf 13(3):329–346
Sharma A, Patni B, Shankhdhar D, Shankhdhar SC (2013) Zinc—An indispensable micronutrient. Physiol Mol Biol Plants 19:11–20
Shivay YS, Singh U, Prasad R, Kaur R (2016) Agronomic interventions for micronutrient biofortification of pulses. Indian J Agron 61:161–172

Silva VM, Nardeli AJ, Mendes NAC, Rocha MM, Wilson L, Young SD et al (2021) Agronomic biofortification of cowpea with zinc: variation in primary metabolism responses and grain nutritional quality among 29 diverse genotypes. Plant Physiol Biochem 162:378–387
Singh MV (2008) Micronutrient deficiencies in crops and soils in India. Micronutrient Deficiencies in Global Crop Production, pp 93–125
Singh U, Kumar N, Praharaj CS, Singh SS, Kumar L (2015) Ferti-fortification: an easy approach for nutritional enrichment of chickpea. Ecoscan 9(3 & 4):731–736
Sparks DL (2015) Kinetics of soil chemical phenomena: future directions. Future Prospect Soil Chem 55:81–101
Tapiero H, Tew KD (2003) Trace elements in human physiology and pathology: zinc and metallothioneins. Biomed Pharmacother 57:399–411
Tavallali V, Rahemi M, Eshghi S, Kholdebarin B, Ramezanian A (2010) Zinc alleviates salt stress and increases antioxidant enzyme activity in the leaves of pistachio (Pistacia vera L. 'Badami') seedlings. Turk J Agric Forest 34:349–359
Tobin AJ (1970) Carbonic anhydrase from parsley leaves. J Biol Chem 245:2656–2666
Tsonev T, Cebola Lidon FJ (2012) Zinc in plants-an overview. Emir J Food Agric 24(4)
Tye A, Young S, Crout N, Zhang H, Preston S, Barbosa-Jefferson V, Davison W, McGrath S, Paton G, Kilham K et al (2003) Predicting the activity of Cd2+ and Zn2+ in soil pore water from the radio-labile metal fraction. Geochim Cosmochim Acta 67:375–385
Udechukwu MC, Collins SA, Udenigwe CC (2016) Prospects of enhancing dietary zinc bioavailability with food-derived zincchelating peptides. Food Funct 7:4137–4144
Uriu-Adams JY, Keen CL (2010) Zinc and reproduction: effects of zinc deficiency on prenatal and early postnatal development. Birth Defects Res B Dev Reprod Toxicol 89(4):313–325
Van Der Straeten D, Bhullar NK, De Steur H, Gruissem W, MacKenzie D, Pfeiffer W et al (2020) Multiplying the efficiency and impact of biofortification through metabolic engineering. Nat Commun 11:5203
Vandemark GJ, Grusak MA, McGee RJ (2018) Mineral concentrations of chickpea and lentil cultivars and breeding lines grown in the U.S. Pac Northwest Crop J 6 253–262
Veenemans J, Milligan P, Prentice AM, Schouten LR, Inja N, Van Der Heijden AC, De Boer LC, Jansen EJ, Koopmans AE, Enthoven WT et al (2011) Effect of supplementation with zinc and other micronutrients on malaria in Tanzanian children: a randomised trial. PLoS Med 8:e1001125
Vejan P, Abdullah R, Khadiran T, Ismail S, Nasrulhaq Boyce A (2016) Role of plant growth promoting rhizobacteria in agricultural sustainability—a review. Molecules 21:573
Velu G, Ortiz-Monasterio I, Cakmak I, Hao Y, Singh RP (2014) Biofortification strategies to increase grain zinc and iron concentrations in wheat. J Cereal Sci 59:365–372
Wang Y, Branicky R, Noë A, Hekimi S (2018) Superoxide dismutases: dual roles in controlling ROS damage and regulating ROS signaling. J Cell Biol 217:1915–1928
Welch RM (1999) Importance of seed mineral nutrient reserves in crop growth and development. In: Mineral nutrition of crops: fundamental mechanisms and implications. Food Products Press, New York, pp 205–226
Welch RM (2005) Biotechnology, biofortification, and global health. Food Nutr Bull 26(4_suppl3):S304–S306
Welch RM, Graham RD (2004) Breeding for micronutrients in staple food crops from a human nutrition perspective. J Exp Bot 55:353–364
Wessells KR, Brown KH (2012) Estimating the global prevalence of zinc deficiency: results based on zinc availability in national food supplies and the prevalence of stunting. PLoS One 7:e50568
White PJ, Broadley M (2005) Biofortifying crops with essential mineral elements. Trends Plant Sci 10:586–593
Wilkinson HF, Loneragan JF, Quirk JP (1968) The movement of zinc to plant roots. Soil Sci Soc Am J 32:831–833

Yaseen MK, Hussain S (2021) Zinc-biofortified wheat required only a medium rate of soil zinc application to attain the targets of zinc biofortification. Arch Agron Soil Sci 67:551–562

Yilmaz A, Ekiz H, Torun B, Gultekin I, Karanlik S, Bagci SA, Cakmak I (1997) Effect of different zinc application methods on grain yield and zinc concentration in wheat cultivars grown on zinc-deficient calcareous soils. J Plant Nutr 20:461–471

Yilmaz A, Ekiz H, Gültekin I, Torun B, Barut H, Karanlik S, Cakmak I (1998) Effect of seed zinc content on grain yield and zinc concentration of wheat grown in zinc-deficient calcareous soils. J Plant Nutr 21:2257–2264

Zaman QU, Aslam Z, Yaseen M, Ihsan MZ, Khaliq A, Fahad S, Bashir S, Ramzani PMA, Naeem M (2018) Zinc biofortification in rice: leveraging agriculture to moderate hidden hunger in developing countries. Arch Agron Soil Sci 64:147–161

Zastrow ML, Pecoraro VL (2014) Designing hydrolytic zinc metalloenzymes. Biochemistry 53:957–978

Zhang H, Liu Y, Wen F, Yao D, Wang L, Guo J, Ni L, Zhang A, Tan M, Jiang M (2014) A novel rice C2H2-type zinc finger protein, ZFP36, is a key player involved in abscisic acid-induced antioxidant defence and oxidative stress tolerance in rice. J Exp Bot 65:5795–5809

Zou CQ, Zhang YQ, Rashid A, Ram H, Savasli E, Arisoy RZ, Ortiz-Monasterio I, Simunji S, Wang ZH, Sohu VS et al (2012) Biofortification of wheat with zinc through zinc fertilization in seven countries. Plant Soil 361:119–130

Chapter 15
Organic Approaches Toward Iron and Zinc Biofortification in Legume Crops

Sehar Shahid, Aiman Hina, Asim Abbasi, Sumaira Maqsood, Madeeha Ansari, Inzamam Ul Haq, and Qamar uz Zaman

Abstract People all around the world especially in developing countries are at risk of malnutrition or micronutrient deficiencies. Moreover, the long-term yield and quality of crop plants, as well as the fertility status of soils, depend critically on the proper nutrition of crop plants. In the current era of pandemic, the dietary pattern in terms of quality has been largely compromised due to lack of diversity in diet and unavailability of nutrient-rich food material. One way to counter micronutrient deficiencies in crops and humans is the use of biofortification approaches. These approaches are cost-effective and eco-friendly in nature which not only improve nutritional content of the crop but also improve productivity by enhancing their resistance level against different biotic and abiotic stresses. Among different micronutrients, iron (Fe) and zinc (Zn) are essential plant cofactors, which play important roles in different plant processes such as respiration, photosynthesis, stress tolerance, chlorophyll synthesis, and auxin metabolism. Moreover, they are also pivotal for different human body functions, and their deficiencies cause serious health

S. Shahid
Department of Botany, Kohsar University Murree, Murree, Pakistan

A. Hina
Soybean Research Institute, Ministry of Agriculture (MOA) Key Laboratory of Biology and Genetic Improvement of Soybean (General), MOA National Centre for Soybean Improvement, State Key Laboratory for Crop Genetics and Germplasm Enhancement, Nanjing Agricultural University, Nanjing, China

A. Abbasi (✉) · S. Maqsood
Department of Environmental Sciences, Kohsar University Murree, Murree, Pakistan

M. Ansari
Institute of Botany, University of the Punjab, Lahore, Pakistan

I. U. Haq
College of Plant Protection, Gansu Agricultural University, Lanzhou, China

Qamar uz Zaman
Department of Environmental Sciences, The University of Lahore, Lahore, Pakistan

M. A. Nadeem et al. (eds.), *Legumes Biofortification*,
https://doi.org/10.1007/978-3-031-33957-8_15

issues particularly among young infants. This chapter contains detailed information regarding the adoption of different viable organic biofortification approaches, which enhance the micronutrient content of crops especially iron and zinc. Among these approaches, use of different organic manures, plant growth-promoting rhizobacteria, intercropping, phytohormones, biofertilizers, soil amendments, crop residues, genetic approaches, and transgenic approaches have been discussed in detail. These approaches provide useful insights for farmers, dietitians, food and genetic engineers, and policy makers regarding the potential of producing nutrient-rich food crops in field, which are readily available for humans in their natural form. However, synergism of different organic biofortification techniques will remain a challenge for our future researchers. In this regard, an integrated approach from farmers, extension workers, policy makers, dietitians, food engineers and educators will be needed to successfully implement organic biofortification techniques at the global level.

1 Introduction

Plants are the best example of autotrophic organisms that use a prime energy source via photosynthesis, providing food to nearly all living organisms. Photosynthesis requires water, essential nutrients, sunlight, and air to give plants the energy to grow, survive, and reproduce (Wiedenhoeft 2006). Food, whether it is bread, cereal, or cornflakes, is the major source of energy for poor and rich people and is derived from crops. Plants extract a considerable amount of essential nutrients from the soil as they cannot synthesize the required nutrients. The nutrients are taken up by the plants and then transferred to other organisms through the food chain. Several biotic and abiotic stressors affect the growth and productivity of crop plants. Among these stresses, nutrient stress is of prime importance as it affects not only crop yield but also the primary end-users, that is, humans or livestock. Subsistence agriculture provides a living for most of the people inhabiting semi-arid tropics (SAT). Low fertility and water scarcity are the characteristic indicators of these areas due to which their agricultural soils are referred as hungry and thirsty soils. Moreover, crop productivity in these areas is also lower as compared to irrigated agricultural areas. In order to feed rapidly increasing human population, such poor soils will have to be brought under cultivation in the future (Rego et al. 2006). Therefore, in order to upgrade the fertility status of these soils, micronutrients can play a central role. Hence, we have to develop a model strategy and create massive awareness, in order to get the maximum benefits of micronutrients on crop yield and quality, that can also be readily accepted by farmers, consumers, and scientific community.

2 Legumes

Legumes are considered as "poor man's meat," and the claim appears to be accurate based on their global consumption patterns (Messina 1999). After cereals, legumes are regarded as the second most important nutritionally valuable food source (Kouris-Blazos and Belski 2016) because they provide essential amino acids, proteins, complex carbohydrates, dietary fiber, unsaturated fats, vitamins, and minerals to humans (Rebello et al. 2014). Besides nutritional supremacy, legumes also produce certain beneficial bioactive compounds and possess cultural, physiological, and medicinal roles (Philips 1993). Various health benefits are linked with consuming legumes (Messina 1999) as these crops possess antiatherogenic, hypocholesterolemia, anticarcinogenic, and hypoglycemic properties. They contain a high amount of vitamin B-group such as folate, thiamine, and riboflavin and essential minerals such as zinc, iron, calcium, selenium, phosphorus, copper, potassium, magnesium, and chromium but low amounts of fat-soluble vitamins and vitamin C (Brigide et al. 2014; Kouris-Blazos and Belski 2016). These micronutrients are essential for bone health (calcium), protein synthesis, hemoglobin synthesis (iron), antioxidant activity, enzyme activity, carbohydrate, lipid (chromium and zinc), and iron metabolism (copper), as well as plasma membrane stabilization (zinc). Hence, it is necessary to escalate the legume yield keeping in view their nutritional quality traits (Naeem et al. 2017).

3 Role of Micronutrients in Crop Plants

Numerous factors influence crop production that has an impact on yield either directly or indirectly. Among these, certain soil factors, such as pH, soil texture, organic matter, soil–water relationships, and balanced nutrients, are typically highlighted in recent studies. The long-term yield and quality of legume plants, as well as the fertility status of soils, depend critically on the proper nutrition of crop plants. Mineral nutrition plays a crucial role in plant growth. Most crop plants exhibit a linear relationship between the amount of fertilizers taken up and the resulting harvest. Plants can successfully thrive to their full genetic potential with the right and steady supply of required minerals (Naeem et al. 2017).

The desire for greater crop productivity without balanced mineral nutrition created severe problems such as soil fertility depletion and plant nutrient imbalance of primary and secondary nutrients including micronutrients (Patel and Singh 2009). Plants cannot survive without micronutrients, even though they are required in a relatively low concentrations (Prasad et al. 2005). Although micronutrients are needed in trace amounts in a living system, they play a crucial role in maximizing the effectiveness of macronutrients and promoting plant growth and development (Shukla et al. 2009).

The group of micronutrients is comprised of eight essential elements, iron (Fe), sodium (Na), chlorine (Cl), boron (B), manganese (Mn), zinc (Zn), copper (Cu), and molybdenum (Mo). Among the scientific community, silicon (Si) has also been considered as a potential micronutrient. Although plants accumulate and use Si in relatively high concentrations even though it is not considered an essential nutrient. Some plant species with root nodules require cobalt (Co) as an essential micronutrient. Furthermore, nickel (Ni) is a micronutrient that is rarely in short supply or deficient in the natural world (Shukla et al. 2009).

4 Causes of Micronutrient Deficiency in Soils

Certain edaphic and ecological factors such as soil organic matter, pH, cation exchange capacity, and clay content affect the availability of micronutrients to crop plants. Moreover, water-retentive soils and calcareous/peat soils can also hinder the bioavailability of some micronutrients (Lindsay 1984; Ibrahim et al. 2011; Ramzan et al. 2014). In fertile soils, the critical limit of DTPA extractable Zn and Fe is 0.6 and 4.5 mg kg^{-1}, respectively (Alloway 2009). The soil's critical limit is the lowest possible soil test value that guarantees a maximum crop yield. The threshold below which insufficiency symptoms appear is the concentration at the bottom of the sufficiency range (Sillanpää 1982).

Soil contains a large amount of Fe, but only a small percentage is biologically available to crop plants. Most of the Fe on earth exists in Fe^{3+} form, which is readily not available to plants, making up a significant portion of the crust. The more soluble form of iron is Fe^{2+}; however, it is easily oxidized to a ferric (Fe^{3+}) form, which precipitates in the soil as oxide/hydroxide, phosphate, carbonate, and other unavailable complex forms (Lindsay and Schwab 1982). High soil pH reduces the zinc (Zn) bioavailability as it precipitates or adsorbs onto the surface of $CaCO_3$ and Fe oxides (Harter 1983; Chirwa and Yerokun 2012). The soil's cation exchange capacity (CEC) is negatively correlated with availability of Zn or Fe (Yoo and James 2002). The availability of Zn also decreases with increased clay content of soils. Similarly, a greater P content in soil and electrical conductivity (EC) also affects the bioavailability of Zn (Gao et al. 2011).

Soil micronutrient content is affected by various factors, including geochemical composition, soil type, macronutrient availability, micronutrient interactions, and vegetation type. Intrinsic properties of soil, such as pH, redox potential, soluble salt concentration, quantity and quality of soil organic matter, and trace element inputs, also play a role in determining the micronutrient content of soil (Fageria et al. 2002; Alloway 2008; Shukla et al. 2016). Additionally, agricultural intensification without sufficient micronutrient replenishment through fertilization is also a contributing factor toward low levels of micronutrient bioavailability. Furthermore, leaching, liming of agricultural soils, scarce manuring, and frequent application of micronutrient-deficient chemical fertilizers have also accelerated the depletion of available micronutrients in soils. Furthermore, as per the GPS-aided analysis of soil samples, element-wise deficiency is as follows: Zn 36.5%, B 23.4%, Fe 12.8%, Mn 7.1%, Cu 4.2% (Shukla et al. 2018).

Among these micronutrients, iron (Fe) and zinc (Zn) are essential plant cofactors which play important roles in different plant processes such as respiration, photosynthesis, and stress tolerance (Sharma et al. 2013; Rout and Sahoo 2015; Tripathi et al. 2018). Furthermore, in legume nodules, Fe- and Zn-dependent processes are crucial for establishing endosymbiotic associations between arbuscular mycorrhiza and soil rhizobia (González-Guerrero et al. 2016; Day and Smith 2021). Iron deficiency causes reduced vegetative development and chlorosis resulting in poor crop productivity. Although plants have a low affinity toward Zn but it is an important component of Cu and Zn superoxide dismutase enzymes actively involved in chlorophyll synthesis and auxin metabolism (Sharma et al. 2013). The detailed information regarding the significance of these two micronutrients is given below.

5 Importance of Iron (Fe)

Iron is a life-preserving micronutrient that is required for the survival of all living organisms including microorganisms, plants, and humans. The iron content of crops is a significant factor in determining human health in addition to its crucial role in promoting plant growth (Briat et al. 2015). Iron is instrumental in various metabolic processes including photosynthesis, respiratory functioning, and synthesis of DNA (Schmidt et al. 2020). Additionally, it triggers the activation of numerous metabolic pathways and is a crucial component in electron chains and a vital cofactor for many enzymes (Zuo and Zhang 2011). Plants require iron for various biological events, as it is the fourth plentiful element found in soil. Among all plants, legumes that participate in nitrogen-fixing symbiosis have a particularly high demand for iron (Brear et al. 2013). Iron is an indispensable micronutrient for plants being part of some antioxidant enzymes that protect chloroplasts from harmful free radicals. It also serves as an integral component of the heme group, which is a precursor of chlorophyll (Barker and Pilbeam 2015). Moreover, iron has a strong impact on phytohormonal regulation, specifically with regards to carbohydrate metabolic activities, auxin production, regulation of stress responses, and protein synthesis (Al-Amri et al. 2020). Iron also has a central role in formation of hemoglobin, body homeostasis, brain functioning and development, catecholamine metabolism, and muscle activity. Moreover, iron is also fundamental for the cellular respiration and transportation of oxygen (Sarkar et al. 2018). According to an estimate, adults need 14 mg of iron daily to maintain good health (Carvalho et al. 2012).

Approximately 80% of iron is found in photosynthetic cells which is crucial for the creation of various heme molecules such as cytochromes, chlorophyll, Fe-S clusters, and functioning of the electron transport system (Briat et al. 2007). Plants rely on iron for the production of chlorophyll and the preservation of chloroplast structure and its proper functioning. The porphyrin structure of chlorophyll is dependent on iron which makes it an intrinsic component of chloroplasts. In the photosynthetic process, iron atoms can be found in varying amounts in different components: 2–3 iron atoms in molecules linked to photosystem II, 12 atoms in photosystem I, 5 in the cytochrome complex, and 2 in the ferredoxin molecule. The distribution of iron in various components of the photosynthetic process demonstrates its direct involvement in plant photosynthesis and its impact on crop productivity (Varotto et al. 2002; Briat et al. 2007).

Despite the limited presence of iron in living organisms (50–100 μg per gram of dry matter), it is still a basic element for plant life. It plays a pivotal role in various metabolic processes through its involvement in key proteins and enzyme synthesis. Iron, being an integral component of proteins and enzymes, holds a prominent place in different fundamental biological procedures such as photosynthesis, DNA synthesis, chlorophyll synthesis, respiration and nitrogen fixation through the action of the ribonucleotide reductase (Asad and Rafique 2000; Rout and Sahoo 2015).

Legumes with an active symbiotic mechanism usually have a high demand for iron, as numerous symbiotic proteins need iron. The nitrogen-fixing enzyme, nitrogenase, as well as cytochromes, ferredoxin, and hydrogenase, which are synthesized by the abundant bacteroids, require iron (Dixon and Kahn 2004; Peters and Szilagyi 2006). The importance of iron in the symbiosis is emphasized by the high concentration of iron in the nodule relative to other plant tissues (Dixon and Kahn 2004; Peters and Szilagyi 2006; Brear et al. 2013).

6 Importance of Zinc (Zn)

Zinc is a life-sustaining micronutrient required for proper functioning of humans, crops, and livestock (Hussain et al. 2015). Although zinc is needed in limited quantities for the optimal growth of plants, it plays a significant role in plant metabolism. Its presence in the right concentration is necessary for various plant physiological activities such as photosynthesis, sugar production, seed and fertility formation, growth regulation, and resistance to diseases. Zinc is also considered indispensable for the proper functioning of cells in all living organisms and plays an important role in boosting the human immune system (Solanki et al. 2016).

Zinc is also involved in the synthesis of plant proteins, being an important constituent of ribosomes. Zinc deficiency causes significant decline in protein synthesis, which can be seen by the increased buildup of amino acids in plant tissues (Mousavi 2011). The synthesis of proteins usually occurs in pollen tubes and is disrupted by Zn deficiency, leading to negative effects on the pollination process (Outten and O'halloran 2001; Pandey et al. 2006). The protein synthesis process during transcription and translation is also impacted by a Zn deficiency. Moreover, Zn also safeguard rRNA from ribonuclease and helps in normal functioning of RNA polymerase (Kryvoruchko 2017). The decline in protein synthesis in Zn-deficient plants is due to the involvement of Zinc in nitrogen metabolism (Fageria 2002; Rehman et al. 2019). In addition, zinc plays a supporting role in the function of various enzymes such as metalloproteases, copper–zinc superoxide dismutase (SOD), nucleases, and aminopeptidases (Hänsch and Mendel 2009). Moreover, the transcriptionally regulated, MucR regulator of reactive oxygen species (ROS), responsible for maintaining cell integrity, balancing iron levels, synthesizing polysaccharides, regulating transcription, and promoting genome plasticity in numerous legume species is also maintained by specific concentration of Zn (Caswell et al. 2013). Zinc is also a major component of proteins that regulate gene expression by interacting with DNA (Liu et al. 2005).

Moreover, zinc plays its part in controlling the activity of enzymes involved in carbohydrate metabolism, such as aldolase, carbonic anhydrase, and fructose-1,6-bisphosphate aldolase, which operate in both the cytoplasm and chloroplasts. These enzymes aid in the transfer of sugar molecules during photosynthesis. Fructose-1,6-bisphosphate aldolase divides 6-carbon sugars between the cytoplasm and chloroplasts, and aldolase moves 3-carbon sugars from the cytoplasm to chloroplasts. Zinc deficiency inhibits the activity of these enzymes due to which carbohydrates starts to accumulate in plant foliage (Cakmak et al. 1989; Mousavi 2011). Legumes are susceptible to a variety of abiotic stressors, such as drought, salinity, high temperature, and chilling, leading to significant reductions in yield. This impact is further exacerbated when the plants are deficient in zinc. Being widely planted under rainfed conditions or marginal soils, legumes are often exposed to these environmental stresses at various stages of their growth. Under water scarce conditions, Zn deficiency results in stunted root development and hindered nutrient uptake (Broadley et al. 2007; Rehman et al. 2018; Ullah et al. 2019). The deficiency of zinc leads to a decrease in the functioning of multiple enzymes involved in plant metabolic and physiological processes (Salama et al. 2006; Ullah et al. 2019). Additionally, the active involvement of Zn in several enzymes helps plant to develop resistance against different abiotic stresses (Rehman et al. 2019). The Cu-Zn-SOD enzyme has zinc attached to copper and performs a catalytic function. When there is a shortage of zinc, the activity of Cu-Zn-SOD decreases and the production of free radicals increases, resulting in damage to plant cells (Marschner 1995; Rehman et al. 2019). Moreover, when plant is prone to a stressful condition, an excessive production of reactive oxygen species (ROS) can harm proteins, nucleic acids, and lipids, resulting in cell damage and eventual death (Gill and Tuteja 2010) By providing plants with zinc during periods of stress, the negative effects can be reduced through regulation of superoxide dismutase activity. This helps to counter the damaging impact of reactive oxygen species by detoxifying them (Cakmak 2000; Ullah et al. 2019). Studies have shown that zinc finger proteins also participate in the regulation of reactive oxygen species, conferring tolerance to various abiotic stresses (Davletova et al. 2005; Mittler et al. 2006; Ciftci-Yilmaz and Mittler 2008).

Zinc is deemed as a vital nutrient for plants and is involved in multiple plant regulatory, metabolic, and developmental processes (Broadley et al. 2007). Legumes rely heavily on zinc for various aspects of plant reproduction, including initiation of flowers, development of blooms, formation of male and female reproductive cells, fertilization, and seed growth (Pathak et al. 2012). Zinc scarcity leads to alterations in the shape, size of stigma, and secretions of black gram and interferes with the interaction between pollen and stigma (Pandey et al. 2009). Furthermore, Zn deficiency also triggers flower failure and infertility of pollen and ovules, leading to a decrease in yield due to limited seed formation, resulting in lower crop productivity (Pathak et al. 2012).

7 Biofortification

The deficiency of micronutrient mainly Zn and Fe remains one of the most serious agricultural and public health issues. Scientists have proposed numerous ecologically compatible approaches to combat micronutrient deficiencies in soils and to enhance the nutritional status of our food products (Marques et al. 2021). Among the approaches, biofortification is the most practical, eco-friendly, and cost-effective method in which nutritional profile of food products is upgraded using different agronomic practices, conventional breeding approaches, and modern biotechnological approaches (Sarkar et al. 2018). The agronomic procedures mainly enhance mineral uptake from soil to crop either through increased nutrient mobility or solubilization (White and Broadley 2009). The other two approaches improve the mineral content in edible portion of crop plants and increase their bioavailability for end-users (Carvalho and Vasconcelos 2013).

Biofortification aims to tackle the issues related to micronutrients deficiencies by enhancing the micronutrient levels in the edible portions of crops and enhancing their digestibility and absorption in the human body following ingestion (Carvalho and Vasconcelos 2013; Ramzani et al. 2016; Vasconcelos et al. 2017). In developing nations, over 20 million people are eating crops that have been biofortified (Rubiales and Mikić 2015; Soares et al. 2019). The United Nations emphasized the significance of biofortifying legumes in 2014, recognizing them as a crucial aspect in the battle against micronutrient deficiencies. Despite this recognition, progress in this area has been limited and the biofortification of legumes remains an underdeveloped area of focus (Foyer et al. 2019; Ghosh et al. 2019; Rehman et al. 2019).

7.1 Organic Agronomic Biofortification Approaches

Biofortification via agronomic methods is widely adopted globally due to its simplicity and efficiency. These methods refer to preharvest agricultural techniques that increase the nutritional value of food (Sarkar et al. 2018). A limitation of these agronomic biofortification methods is that they must take place before harvest in order for the food to be classified as biofortified. If the process is carried out after harvest, the food is considered fortified instead (Sarkar et al. 2018).

The green revolution saw a rise in technological advancements that led to an increase in crop production and an adequate supply of food to fulfill the caloric needs of the huge population worldwide (Smil 2000; Tilman et al. 2002). As a result, using organic methods instead of conventional synthetic fertilizers may be desirable. One commonly observed reason is to restore depleted soil organic matter, thereby taking advantage of the physical, chemical, and biological benefits that are associated with the soils high in organic matter (Loveland and Webb 2003; Diacono and Montemurro 2011; Murphy 2015). Moreover, by using organic methods, the uptake of plant nutrients can be enhanced, the efficiency of nutrient utilization can be maximized, and the environmental impact can be reduced when compared to the use of inorganic pesticides (Edmeades 2003; Quilty and Cattle 2011). In agriculture, particularly in small-scale subsistence farming, animal waste is utilized as fertilizer because it is the most accessible and economically viable source of nutrients for plants in many situations (Onduru et al. 2008). Adopting organic methods can raise the level of carbon in the soil and, through a series of interconnected processes, enhance the soil's biological activity, structure, ability to retain cations, water retention capacity, and other related factors (Lal 2006). As a result, these alterations can result in a rise in agricultural production (Diacono and Montemurro 2011). Moreover, organic methods can also supply substantial mineral-based nutrients to plants, which directly enhance crop yields through fertilization.

7.1.1 Organic Manures

The integrated application of different organic manures primarily gliricidia green leaf manure, composts, and vermicompost not only increase crop yield but also improve soil physical health (Babalad et al. 2009). Moreover, the nutrient requirement of crops has been fulfilled using liquid organic manures that ensure better nutrient use efficiency and counter the deficiency symptoms observed under an organic production system (Shwetha et al. 2009). As a legume crop, chickpeas obtain their larger nitrogen needs through biological nitrogen fixation, which can be aided by providing better soil chemical and physical conditions. Under the organic production system, the edaphic environment will be more promising for crop growth, and regular organic applications sustain its optimal level for greater time period. Studies have demonstrated that the organic production system can increase and sustain the productivity of the legume crop (Shwetha et al. 2009).

Foliar spray of liquid biofertilizers and soil application of organic manures had a better effect on growth, seed yield, and yield attributes of the targeted crop. The foliar application of panchagavya resulted in significantly increased plant height. The presence of sufficient amounts of N, P, and K, as well as additional amounts of the micronutrients Zn and Fe in vermicompost and FYM, provides favorable conditions for cell division, tissue growth, and improved plant growth through the development of a stronger and vigorous root system, thereby empowering plants to derive sufficient amounts of the nutrients available in the soil (Nekar et al. 2009; Deotale et al. 2011).

7.1.2 Plant Growth-Promoting Rhizobacteria

Actinomycetes, diazotrophic bacteria, mycorrhizal fungi, and rhizobia are useful soil microbes that live in symbiotic relationships with plant roots and guard plants from micronutrient deficiencies using different ways such as by producing certain plant growth hormones and promoting nutrient mineralization (Mekouar 2019). These microorganisms are naturally found in the soil; however, certain agricultural management practices or exogenous inoculation can increase their populations. A variety of different plant growth-promoting (PGP) soil microbes such as *Bacillus*, *Enterobacter*, and *Pseudomonas* can be used to enhance the bioavailability of micronutrients for crop plants. These microorganisms are mostly used in the form of seed inoculants, which promote growth of the targeted plants and induce resistance in them by producing different growth hormones, siderophores, chitinases, and antibiotics (Jha and Warkentin 2020). By producing siderophore compounds, plant growth-promoting (PGP) microorganisms avert the growth of pathogens, solubilize phosphorus, and chelate iron (Panhwar et al. 2012; Sreevidya et al. 2016), therefore playing an important role in iron fortification and fertility of the soil. PGP microorganisms are generally present in decomposing organic matter, compost, and soil and provide an ecologically compatible and economical method for improving environmental and soil health as well as increase crop production (Gopalakrishnan et al. 2016).

Several studies revealed that these microorganism-based inoculants via mycorrhizal associations increased the concentrations of Zn and Fe in different crop plants (Cavagnaro 2008; Brear et al. 2013; Jha and Warkentin 2020). Furthermore, the *Acinetobacter*, *Bacillus*, *Enterobacter*, and *Pseudomonas* species have been reported to ameliorate nitrogen fixation, plant growth, and yield of the grains in legumes including soybeans, peas, and chickpeas (Tokala et al. 2002; Valverde et al. 2007; Minorsky 2008; Soe et al. 2010; Gopalakrishnan et al. 2015). In chickpeas, PGP actinobacteria inoculation offered improved concentrations of seed minerals, that is, Zn (13%–30%) and Fe (10%–38%) as compared with uninoculated plants (Sathya et al. 2013). Likewise, field inoculation of these microorganisms and arbuscular mycorrhizal fungi enhanced the nutritional profile in chickpea grains along with protein content and yield by increasing the Zn and Fe concentrations (Pellegrino and Bedini 2014; Khalid et al. 2015; Gopalakrishnan et al. 2016).

Additionally, endophytic microorganisms can indirectly affect the regulation of metal transporters and contemplated as more promising agents to increase Zn and Fe uptake as well as their translocation within plant tissues (Reiter et al. 2002; Weyens et al. 2013). Rice and wheat grains have been biofortified with Zn and Fe using different fungal and bacterial endophytes (Ramesh et al. 2014; Abaid-Ullah et al. 2015).

7.1.3 Intercropping

Intercropping is an effective, economical, and environmentally friendly method, mostly adopted by small-scale farmers to overcome micronutrient deficiencies in crop plants (Gunes et al. 2007; Singh et al. 2016; Szerement et al. 2022). Achieving food security usually involves confirmation that everyone has access to adequate and nutritious food at all times, while also minimizing the environmental impact through the use of sustainable agriculture practices and maintaining a balance in the agroecosystem (Maitra and Ray 2019). Previous studies have demonstrated the benefits of intercropping in terms of enhanced crop quality, higher grain yield, increased protein content, and more efficient use of resources in a sustainable manner (Akhtar et al. 2013). Intercropping has also been shown to improve the iron and zinc content in grains, which can address the devastating consequences of micronutrient deficiencies on human health (Palmgren et al. 2008; Singh et al. 2016; De Valença et al. 2017). Moreover, Gunes et al. (2007) also found that intercropping of wheat and chickpea leads to higher levels of N, P, K, and Fe in wheat seeds and N, P, K, Fe, Zn, and Mn in chickpea seeds. Additionally, implementing multiple cropping systems can decrease the levels of anthropogenic disturbances associated with N and P contents while preserving soil fertility and lowering CO_2 emissions from the cropping system (Soares et al. 2019), potentially reducing the costs associated with their mitigation efforts. This approach is considered a key component of climate-smart agriculture and provides a comparatively cost-effective and comprehensive solution for mitigating micronutrient deficiencies particularly in those regions that are susceptible to climate change (Bouis et al. 2017; Maqbool et al. 2020). Therefore, intercropping represents a natural solution that can address malnutrition and simultaneously achieve sustainable crop production with reduced inputs, lower cultivation costs, preservation of land for nature, and a more holistic management of ecosystems (Hu et al. 2018; Kiwia et al. 2019). Similarly, Zuo and Zhang (2009) also reported that intercropping dicot and monocot plants leads to increased Fe and Zn content in seeds through interspecific root interaction and changes in the rhizosphere.

The principles of complementarity and facilitation are the key ecological concepts mostly seen in different intercropping systems and often results in improved resource utilization efficiencies (Li et al. 2020). The complementarity technique usually involves the division of resources and reduced competition between species, while the facilitation system refers to a positive interaction in which one species positively impact the growth, reproduction, or survival of another species by altering the biotic or abiotic environment (Duchene et al. 2017; Li et al. 2020).

7.1.4 Biofertilizers

Biofertilizers are eco-friendly and cost-effective components of organic farming which play a vital role in sustaining soil health and improving crop productivity. Biofertilizers are made up of certain living cells belonging to different microorganisms and are applied either directly to the soil, plant parts, or crop seeds. The applied nutrients usually colonize interior section of the crop plants or rhizosphere and enhance plant growth by converting unavailable nutrients forms into their available forms. This is usually done through activation of various biological processes such as solubilization of fixed soil phosphate, nitrogen fixation, and synthesis of growth-promoting substances (Vessey 2003). The microorganisms also play their part in building up soil organic matter and restoring natural nutrient cycle. Majority of the biofertilizers are specific to major nutrients. However, the information regarding the role of biofertilizers on the availability of micronutrients is still limited in literature. Azolla, a floating fern usually fixes atmospheric N and enhances the availability of certain secondary micronutrients (Ca, Fe, Zn, Mn, Cu, B, Co, and Ni) in rice crop. It has been reported that presence of few microorganisms significantly improved the bioavailability of secondary micronutrients such as Fe—*Thiobacillus ferroxidans* and *Ferrobacillus ferroxidans*; Zn—*Bacillus* spp.; and S—*Thiobacillus sulfoxidans* and *Beggiota* (Kc et al. 2016).

7.1.5 Soil Amendments

The application of different soil amendments mainly alters soil pH, which ultimately improves soil chemical and physical properties and crop productivity. Crop productivity in acidic soils can be improved with application of limestone (Foy 1992; Kochian 1995). Moreover, sodic and saline-sodic soils can be amended to a great extent with application of sulphur and gypsum. These elements also lower the pH of alkaline soils and ultimately increase contents of plant-available Mn, Fe, Zn, Cu, and Co. Gypsum is also used to mitigate bicarbonates from soil solution and to exchange Ca for Na on the soil cation-exchange complex. The decline in soil pH and increased bioavailability of micronutrients can be achieved with the removal of excessive bicarbonates from soil solution (Singh et al. 1989). Application of gypsum not only improves soil physical properties but also lowers soil pH and ultimately enhances the mobility of micronutrients from soil to crop plants (Singh et al. 1989).

7.1.6 Plant Residues

In many arid and semi-arid regions across the world, most crop residue is either utilized as animal feed or burned as fuel, leaving little to no residue in the field (Timsina and Connor 2001). It is believed that developing countries produce an estimated yearly total of over 1000 million tons of cereal residue. In 1998, the major crops that globally generated residues, including soybean, rice, corn, wheat, potato, barley, and rapeseed produced a total of 2956 million tons of residues (Mekouar 2019).

Crop residues are seen as significant providers of a variety of micronutrients. For instance, every ton of rice and wheat produces removals of Zn, Fe, Mn, Cu, B, and Mo of 96, 777, 745, 42, 55, and 4 gha − 1, respectively. Based on a total crop residue production estimate of 105 million tons in India, and the micronutrient content of the residues, it is estimated that the potential of these residues is to provide approximately 35,400 tons amounts of micronutrients (Prasad 1999). The recycling of crop residues has the potential to enhance soil availability of micronutrients, as it is estimated that 50%–80% of the Zn, Cu, and Mn was taken up by rice and wheat crops can be recovered through the incorporation of the residues (Prasad and Sinha 1995).

The addition of crop residues to flooded soils enhances microbial metabolism, leading to an increase in soil solution Fe and Mn concentrations as a result of a significant change in redox potential (Katyal 1977; Yodkeaw and De Datta 1989; Atta et al. 1996). In soils rich in calcium, the breakdown of crop residue produces organic acids that can enhance plant zinc absorption by loosening zinc from its solid form in the soil and making it more soluble in the soil solution (Prasad and Sinha 1995). The chelating agents released during the decay of crop residues boost the total concentration of zinc that is available for diffusion and enhance the diffusion rate of zinc. For instance, the use of rice straw has been discovered to enhance the zinc content of rice plants, which may be due to the improvement of soil pH and reduction of exchangeable sodium levels (Singh et al. 2005). Despite this, the successful implementation of crop residue management techniques in agricultural systems necessitates a thorough comprehension of the influence of crop residues on the movement of nutrients from the soil and fertilizers, as well as their effects on soil properties, both chemical and physical, and the production of crops.

7.2 Phytohormones

The role of phytohormones, such as gibberellic acid and cytokinin, in mitigating metal stress is crucial (Al-Hakimi 2007; Gangwar et al. 2010; Masood et al. 2016). Studies have shown that certain plant hormones have an impact on the expression of Fe uptake genes, namely, IRT1 and FRO2. Research has shown that auxin has a positive effect on FRO2 gene expression under iron-deficient conditions (Chen et al. 2010), while ethylene positively regulates both IRT1 and FRO2 in *Arabidopsis* and cucumber plants (Lucena et al. 2006). Strengthening Fe-deficiency-inducible responses in plants can improve their ability to acquire more iron from soil with limited iron contents.

In recent years, plant physiologists have been trying to discover the signals that trigger root responses to iron deficiency and have identified various hormonal compounds as signaling agents (Mori and Nishizawa 1987; Hindt and Guerinot 2012; Ivanov et al. 2012). Among these signaling agents, nitric oxide (NO) (Graziano and Lamattina 2007), auxins (Chen et al. 2010), cytokinin (Séguéla et al. 2008), brassinosteroids (Wang et al. 2012), and ethylene are of prime importance (García et al. 2011). Furthermore, nitric oxide, auxins, and ethylene are particularly noteworthy agents as they can be produced by various soil microorganisms. This highlights the important and potential connections between soil microorganisms and plant ability to uptake Fe. Research has shown that auxins play a key role as a chemical signal, boosting the response to iron deficiency. The exogenous application of synthetic auxin either in the form of α-naphthaleneacetic acid or of IAA has significantly enhanced expression of FRO2 and IRT1 gene and Fe-deficiency-induced reduction of ferric Fe. Moreover, auxins are also involved in stimulating the growth of root hairs and expansion of lateral roots for uptake of iron (Chen et al. 2010; Wu et al. 2012). Soil microorganisms that produce auxin-like compounds can positively impact plant iron uptake in situations where limited iron is available (Jin et al. 2006).

7.3 Genetic Approaches

Classical breeding techniques, gene discovery, and marker-assisted breeding are utilized in biofortification of different crops to increase their mineral contents (Grusak 2002). According to Hindu et al. (2018), genome-wide association studies (GWAS) were employed to identify various genomic regions in maize plants that are associated with Zn and Fe biofortification in kernels. It was suggested that the genomic selection (GS) could be a promising breeding approach for biofortifying wheat with Fe and Zn (Velu et al. 2016). Additionally, researchers have highlighted that crop breeding with a focus on nutrition has several advantages in terms of sustainability. While breeding typically involves a long and repetitive process of hybridization and selection, modern advancements have enabled us to control signaling pathways. However, this process can be time-consuming and labor-intensive. In recent years, the use of modern molecular tools such as DNA markers and marker-assisted selection (MAS) has accelerated the development of nutrient-rich varieties. According to the studies of Kumar et al. (2018), QTLs for Fe and Zn biofortification in pearl millet were identified using a combination of DArT and SSR markers. Literature has shown that numerous rice, wheat, and maize varieties enriched in Fe and/or Zn have been released in different parts of the globe. Ramesh et al. (2004) innovated a new method for enhancing seed zinc and iron content by overexpressing a zinc transporter in *Hordeum vulgare* cv. Golden Promise using a ubiquitin promoter. A threefold increase was achieved in iron content in rice grains through the *Agrobacterium*-mediated transfer of the complete coding sequence of the ferritin gene from soybean plants (Goto et al. 1999). Similarly, Lucca et al. (2002) created transgenic rice plants with elevated iron content, enhanced in phytase and cysteine-peptides, for improved iron intake and bioavailability. Similarly, Vasconcelos et al. (2003) engineered the expression of the soybean ferritin gene using the glutelin promoter in an elite Indica rice line with desirable agronomic traits. This led to an improvement in the nutritional levels of both brown and polished rice grains. Liu et al. (2004) created rice varieties that contained ferritin and had 64% more iron content after milling. The ferritin gene was expressed specifically and at a high level in the endosperm of the transgenic rice.

7.4 Transgenic Approaches

In addition to plant breeding and genetic approaches, several transgenic interventions have proven successful in biofortification of food crops. With transgenic techniques, genes can be transferred between completely different species or new genes can be introduced into food or cash crops. This approach can be an effective solution for developing biofortified crops when there is limited or no variation in nutrient content among plant varieties (Brinch-Pedersen et al. 2007; Zhu et al. 2007). The transgenic approach for biofortification takes advantage of the unlimited genetic pool for transferring and expressing desirable genes between plant species, regardless of their evolutionary or taxonomic relationship. When a specific micronutrient is not naturally present in crops, transgenic techniques are the only practical option for fortifying the crops with that nutrient (Pérez-Massot et al. 2013). The development of transgenic crops becomes possible with identification and understanding of gene function and use of these genes to manipulate plant metabolism (Christou and Twyman 2004). Additionally, alternative metabolic pathways from bacteria and other organisms can also be introduced into crops through transgenic techniques (Newell-Mcgloughlin 2008). Developing biofortified varieties through genetic approaches can be challenging for breeders in soils that are naturally low in iron and zinc micronutrients. To fully reap the benefits of these varieties, it is also important to consider other factors such as soil pH and organic matter, which can impact root exudation and enzyme activity in the rhizosphere and therefore affect micronutrient uptake and its accumulation in crop plants (Cakmak 2008).

8 Conclusion

Hidden hunger or malnutrition is a global phenomenon affecting lives of millions of people in both developing and developed nations. The deficiencies in our food system need to be addressed to make human immune system stronger in fight against global pandemics such as COVID-19. One way to combat malnutrition is the adoption of diverse organic biofortification approaches. Biofortification is a cost-effective method of producing nutrient-rich food crops which are readily available for humans in their natural form. Organic biofortification involves use of different organic manures, plant growth-promoting rhizobacteria, intercropping, phytohormones, biofertilizers, soil amendments, crop residues, genetic approaches, and transgenic approaches. However, synergism of different organic biofortification techniques and public acceptance of biofortified crops will remain a major challenge. In this regard, an integrated approach from farmers, extension workers, politicians, dietitians, food and genetic engineers, and educators will be needed to successfully implement biofortification techniques at the global level.

References

Abaid-Ullah M, Nadeem M, Hassan M, Ganter J, Muhammad B, Nawaz K, Shah AS, Hafeez FY (2015) Plant growth promoting rhizobacteria: an alternate way to improve yield and quality of wheat (*Triticum aestivum*). Int J Agric Biol 17:51–60

Akhtar M, Iqbal Z, Naeem A, Yaqub M, Lodhi AH, Tahir S, Qureshi TM (2013) Impact of chickpea and wheat co-cropping on crop growth, nutrient uptake and root induced biochemical changes. Soil Environ 32:152–157

Al-Amri N, Tombuloglu H, Slimani Y, Akhtar S, Barghouthi M, Almessiere M, Alshammari T, Baykal A, Sabit H, Ercan I (2020) Size effect of iron (III) oxide nanomaterials on the growth, and their uptake and translocation in common wheat (*Triticum aestivum* L.). Ecotoxicol Environ Saf 194:110377

Al-Hakimi A (2007) Modification of cadmium toxicity in pea seedlings by kinetin. Plant Soil Environ 53:129

Alloway BJ (2008) Zinc in soils and crop nutrition. International Zinc Association, Brussels

Alloway BJ (2009) Soil factors associated with zinc deficiency in crops and humans. Environ Geochem Health 31:537–548

Asad A, Rafique R (2000) Effect of zinc, copper, iron, manganese and boron on the yield and yield components of wheat crop in Tehsil Peshawar. Pak J Biol Sci 3:1615–1620

Atta SK, Mohammed S, Van Cleemput O, Zayed A (1996) Transformations of iron and manganese under controlled Eh, Eh-pH conditions and addition of organic matter. Soil Technol 9:223–237

Babalad H, Kambale A, Bhat S, Patil R, Math K, Shirnalli G, Palakshappa M (2009) Sustainable groundnut production through organic approach. J Oilseeds Res 26:365–367

Barker AV, Pilbeam DJ (2015) Handbook of plant nutrition. CRC Press, Boca Raton

Bouis H, Saltzman A, Low J, Ball A, Covic N (2017) An overview of the landscape and approach for biofortification in Africa. Afr J Food Agric Nutr Dev 17:11848–11864

Brear EM, Day DA, Smith PM (2013) Iron: an essential micronutrient for the legume-rhizobium symbiosis. Front Plant Sci 4:359

Briat J-F, Curie C, Gaymard F (2007) Iron utilization and metabolism in plants. Curr Opin Plant Biol 10:276–282

Briat J-F, Dubos C, Gaymard F (2015) Iron nutrition, biomass production, and plant product quality. Trends Plant Sci 20:33–40

Brigide P, Canniatt-Brazaca SG, Silva MO (2014) Nutritional characteristics of biofortified common beans. Food Sci Technol 34:493–500

Brinch-Pedersen H, Borg S, Tauris B, Holm PB (2007) Molecular genetic approaches to increasing mineral availability and vitamin content of cereals. J Cereal Sci 46:308–326

Broadley MR, White PJ, Hammond JP, Zelko I, Lux A (2007) Zinc in plants. New Phytol 173:677–702

Cakmak I (2000) Tansley Review No. 111 Possible roles of zinc in protecting plant cells from damage by reactive oxygen species. New Phytol 146:185–205

Cakmak I (2008) Enrichment of cereal grains with zinc: agronomic or genetic biofortification? Plant Soil 302:1–17

Cakmak I, Marschner H, Bangerth F (1989) Effect of zinc nutritional status on growth, protein metabolism and levels of indole-3-acetic acid and other phytohormones in bean (Phaseolus vulgaris L.). J Exp Bot 40:405–412

Carvalho SM, Vasconcelos MW (2013) Producing more with less: Strategies and novel technologies for plant-based food biofortification. Food Res Int 54:961–971

Carvalho LJ, Corrêa M, Pereira E, Nutti M, Carvalho JV, Ribeiro EG, Freitas S (2012) Iron and zinc retention in common beans (*Phaseolus vulgaris* L.) after home cooking. Food Nutr Res 56:15618

Caswell CC, Elhassanny AE, Planchin EE, Roux CM, Weeks-Gorospe JN, Ficht TA, Dunman PM, Roop RM (2013) Diverse genetic regulon of the virulence-associated transcriptional regulator MucR in Brucella abortus 2308. Infect Immun 81:1040–1051

Cavagnaro TR (2008) The role of arbuscular mycorrhizas in improving plant zinc nutrition under low soil zinc concentrations: a review. Plant Soil 304:315–325

Chen WW, Yang JL, Qin C, Jin CW, Mo JH, Ye T, Zheng SJ (2010) Nitric oxide acts downstream of auxin to trigger root ferric-chelate reductase activity in response to iron deficiency in Arabidopsis. Plant Physiol 154:810–819

Chirwa M, Yerokun OA (2012) The distribution of zinc fractions in surface samples of selected agricultural soils of Zambia. Int J Soil Sci 7:51

Christou P, Twyman RM (2004) The potential of genetically enhanced plants to address food insecurity. Nutr Res Rev 17:23–42

Ciftci-Yilmaz S, Mittler R (2008) The zinc finger network of plants. Cell Mol Life Sci 65:1150–1160

Davletova S, Schlauch K, Coutu J, Mittler R (2005) The zinc-finger protein Zat12 plays a central role in reactive oxygen and abiotic stress signaling in Arabidopsis. Plant Physiol 139:847–856

Day DA, Smith PM (2021) Iron transport across symbiotic membranes of nitrogen-fixing legumes. Int J Mol Sci 22:432

De Valença A, Bake A, Brouwer I, Giller K (2017) Agronomic biofortification of crops to fight hidden hunger in sub-Saharan Africa. Glob Food Sec 12:8–14

Deotale R, Kalamkar V, Banginwar A (2011) Effect of foliar sprays of cow urine and NAA on morpho-physiological, chemical, biochemical parameters and yield of soybean (*Glycine max* L.). J Soils Crops 21:332–337

Diacono M, Montemurro F (2011) Long-term effects of organic amendments on soil fertility. Sustain Agric 2:761–786

Dixon R, Kahn D (2004) Genetic regulation of biological nitrogen fixation. Nat Rev Microbiol 2:621–631

Duchene O, Vian J-F, Celette F (2017) Intercropping with legume for agroecological cropping systems: Complementarity and facilitation processes and the importance of soil microorganisms. A review. Agric Ecosyst Environ 240:148–161

Edmeades DC (2003) The long-term effects of manures and fertilisers on soil productivity and quality: a review. Nutr Cycl Agroecosyst 66:165–180

Fageria NK (2002) Influence of micronutrients on dry matter yield and interaction with other nutrients in annual crops. Pesq Agrop Brasileira 37:1765–1772

Fageria N, Baligar V, Clark R (2002) Micronutrients in crop production. Adv Agron 77:185–268

Foy CD (1992) Soil chemical factors limiting plant root growth. In: Limitations to plant root growth. Springer-Verlag New York Inc., pp 97–149

Foyer CH, Nguyen H, Lam HM (2019) Legumes—The art and science of environmentally sustainable agriculture, vol 42. Wiley Online Library, p 1

Gangwar S, Singh VP, Prasad SM, Maurya JN (2010) Modulation of manganese toxicity in Pisum sativum L. seedlings by kinetin. Sci Hortic 126:467–474

Gao X, Flaten DN, Tenuta M, Grimmett MG, Gawalko EJ, Grant CA (2011) Soil solution dynamics and plant uptake of cadmium and zinc by durum wheat following phosphate fertilization. Plant Soil 338:423–434

García MJ, Suárez V, Romera FJ, Alcántara E, Pérez-Vicente R (2011) A new model involving ethylene, nitric oxide and Fe to explain the regulation of Fe-acquisition genes in Strategy I plants. Plant Physiol Biochem 49:537–544

Ghosh A, Hasim Reja M, Nalia A, Kanthal S, Maji S, Venugopalan V, Nath R (2019) Micronutrient biofortification in pulses: an agricultural approach. Curr J Appl Sci Technol 35:1–12

Gill SS, Tuteja N (2010) Reactive oxygen species and antioxidant machinery in abiotic stress tolerance in crop plants. Plant Physiol Biochem 48:909–930

González-Guerrero M, Escudero V, Saéz Á, Tejada-Jiménez M (2016) Transition metal transport in plants and associated endosymbionts: arbuscular mycorrhizal fungi and rhizobia. Front Plant Sci 7:1088

Gopalakrishnan S, Srinivas V, Prakash B, Sathya A, Vijayabharathi R (2015) Plant growth-promoting traits of Pseudomonas geniculata isolated from chickpea nodules. 3 Biotech 5:653–661

Gopalakrishnan S, Vadlamudi S, Samineni S, Sameer Kumar C (2016) Plant growth-promotion and biofortification of chickpea and pigeonpea through inoculation of biocontrol potential bacteria, isolated from organic soils. Springerplus 5:1–11

Goto F, Yoshihara T, Shigemoto N, Toki S, Takaiwa F (1999) Iron fortification of rice seed by the soybean ferritin gene. Nat Biotechnol 17:282–286

Graziano M, Lamattina L (2007) Nitric oxide accumulation is required for molecular and physiological responses to iron deficiency in tomato roots. Plant J 52:949–960

Grusak MA (2002) Enhancing mineral content in plant food products. J Am Coll Nutr 21:178S–183S

Gunes A, Inal A, Adak MS, Alpaslan M, Bagci EG, Erol T, Pilbeam DJ (2007) Mineral nutrition of wheat, chickpea and lentil as affected by mixed cropping and soil moisture. Nutr Cycl Agroecosyst 78:83–96

Hänsch R, Mendel RR (2009) Physiological functions of mineral micronutrients (cu, Zn, Mn, Fe, Ni, Mo, B, cl). Curr Opin Plant Biol 12:259–266

Harter RD (1983) Effect of soil pH on adsorption of lead, copper, zinc, and nickel. Soil Sci Soc Am J 47:47–51

Hindt MN, Guerinot ML (2012) Getting a sense for signals: regulation of the plant iron deficiency response. Biochim Biophys Acta Mol Cell Res 1823:1521–1530

Hindu V, Palacios-Rojas N, Babu R, Suwarno WB, Rashid Z, Usha R, Saykhedkar GR, Nair SK (2018) Identification and validation of genomic regions influencing kernel zinc and iron in maize. Theor Appl Genet 131:1443–1457

Hu F, Tan Y, Yu A, Zhao C, Coulter JA, Fan Z, Yin W, Fan H, Chai Q (2018) Low N fertilizer application and intercropping increases N concentration in pea (*Pisum sativum* L.) grains. Front Plant Sci 9:1763

Hussain A, Arshad M, Zahir ZA, Asghar M (2015) Prospects of zinc solubilizing bacteria for enhancing growth of maize. Pak J Agric Sci 52:915–922

Ibrahim A, Usman A, Abubakar B, Aminu U (2011) Extractable micronutrients status in relation to other soil properties in Billiri Local Government Area. J Soil Sci Environ Manage 3:282–285

Ivanov R, Brumbarova T, Bauer P (2012) Fitting into the harsh reality: regulation of iron-deficiency responses in dicotyledonous plants. Mol Plant 5:27–42

Jha AB, Warkentin TD (2020) Biofortification of pulse crops: Status and future perspectives. Plan Theory 9:73

Jin C, He Y, Tang C, Wu P, Zheng S (2006) Mechanisms of microbial enhanced iron uptake in red clover. Plant Cell Environ 29:888–897

Katyal J (1977) Influence of organic matter on the chemical and electrochemical properties of some flooded soils. Soil Biol Biochem 9:259–266

Kc A, Yb R, Sb P, Shilpashree J (2016) Nutritive evaluation of Azolla as livestock feed. J Exp Biol Agric Sci 4:670–674

Khalid S, Asghar HN, Akhtar MJ, Aslam A, Zahir ZA (2015) Biofortification of iron in chickpea by plant growth promoting rhizobacteria. Pak J Bot 47:1191–1194

Kiwia A, Kimani D, Harawa R, Jama B, Sileshi GW (2019) Sustainable intensification with cereal-legume intercropping in Eastern and Southern Africa. Sustainability 11:2891

Kochian LV (1995) Cellular mechanisms of aluminum toxicity and resistance in plants. Annu Rev Plant Biol 46:237–260

Kouris-Blazos A, Belski R (2016) Health benefits of legumes and pulses with a focus on Australian sweet lupins. Asia Pac J Clin Nutr 25:1–17

Kryvoruchko I (2017) Zn-use efficiency for optimization of symbiotic nitrogenfixation in chickpea (*Cicer arietinum* L.). Turk J Bot 41:423–441

Kumar S, Hash CT, Nepolean T, Mahendrakar MD, Satyavathi CT, Singh G, Rathore A, Yadav RS, Gupta R, Srivastava RK (2018) Mapping grain iron and zinc content quantitative trait loci in an iniadi-derived immortal population of pearl millet. Genes 9:248

Lal R (2006) Enhancing crop yields in the developing countries through restoration of the soil organic carbon pool in agricultural lands. Land Degrad Dev 17:197–209

Li C, Hoffland E, Kuyper TW, Yu Y, Li H, Zhang C, Zhang F, Van Der Werf W (2020) Yield gain, complementarity and competitive dominance in intercropping in China: A meta-analysis of drivers of yield gain using additive partitioning. Eur J Agron 113:125987

Lindsay W (1984) Soil and plant relationships associated with iron deficiency with emphasis on nutrient interactions. J Plant Nutr 7:489–500

Lindsay WL, Schwab A (1982) The chemistry of iron in soils and its availability to plants. J Plant Nutr 5:821–840

Liu Q-Q, Yao Q-H, Wang H-M, Gu M-H (2004) Endosperm-specific expression of the ferritin gene in transgenic rice (*Oryza sativa* L.) results in increased iron content of milling rice. Yi Chuan xue bao = Acta Genet Sin 31:518–524

Liu PP, Koizuka N, Martin RC, Nonogaki H (2005) The BME3 (Blue Micropylar End 3) GATA zinc finger transcription factor is a positive regulator of Arabidopsis seed germination. Plant J 44:960–971

Loveland P, Webb J (2003) Is there a critical level of organic matter in the agricultural soils of temperate regions: a review. Soil Tillage Res 70:1–18

Lucca P, Hurrell R, Potrykus I (2002) Fighting iron deficiency anemia with iron-rich rice. J Am Coll Nutr 21:184S–190S

Lucena C, Waters BM, Romera FJ, García MJ, Morales M, Alcántara E, Pérez-Vicente R (2006) Ethylene could influence ferric reductase, iron transporter, and H+-ATPase gene expression by affecting FER (or FER-like) gene activity. J Exp Bot 57:4145–4154

Maitra S, Ray DP (2019) Enrichment of biodiversity, influence in microbial population dynamics of soil and nutrient utilization in cereal-legume intercropping systems: A Review. Int J Bioresour Sci 6:11–19

Maqbool A, Abrar M, Bakhsh A, Çalışkan S, Khan HZ, Aslam M, Aksoy E (2020) Biofortification under climate change: the fight between quality and quantity. In: Environment, climate, plant and vegetation growth. Springer, pp 173–227

Marques E, Darby HM, Kraft J (2021) Benefits and limitations of non-transgenic micronutrient biofortification approaches. Agronomy 11:464

Marschner H (1995) Mineral nutrition of higher plants, 2nd edn. Institute of Plant Nutrition University of Hohenheim, Germany

Masood A, Khan MIR, Fatma M, Asgher M, Per TS, Khan NA (2016) Involvement of ethylene in gibberellic acid-induced sulfur assimilation, photosynthetic responses, and alleviation of cadmium stress in mustard. Plant Physiol Biochem 104:1–10

Mekouar MA (2019) 15. Food and Agriculture Organization of the United Nations (FAO). Oxford University Press

Messina MJ (1999) Legumes and soybeans: overview of their nutritional profiles and health effects. Am J Clin Nutr 70:439s–450s

Minorsky PV (2008) On the inside. Plant Physiol 147:1–2

Mittler R, Kim Y, Song L, Coutu J, Coutu A, Ciftci-Yilmaz S, Lee H, Stevenson B, Zhu J-K (2006) Gain-and loss-of-function mutations in Zat10 enhance the tolerance of plants to abiotic stress. FEBS Lett 580:6537–6542

Mori S, Nishizawa N (1987) Methionine as a dominant precursor of phytosiderophores in Graminaceae plants. Plant Cell Physiol 28:1081–1092

Mousavi SR (2011) Zinc in crop production and interaction with phosphorus. Aust J Basic Appl Sci 5:1503–1509

Murphy B (2015) Impact of soil organic matter on soil properties—a review with emphasis on Australian soils. Soil Res 53:605–635

Naeem M, Ansari AA, Gill SS, Aftab T, Idrees M, Ali A, Khan MMA (2017) Regulatory role of mineral nutrients in nurturing of medicinal legumes under salt stress. In: Essential plant nutrients: uptake, use efficiency, and management, pp 309–334

Nekar N, Babalad H, Bhat S, Sreenivasa M (2009) Response of groundnut (Arachis hypogaea L.) to foliar application of liquid organic manures. J Oilseeds Res 26:390–392

Newell-Mcgloughlin M (2008) Nutritionally improved agricultural crops. Plant Physiol 147:939–953
Onduru DD, Snijders P, Muchena F, Wouters B, De Jager A, Gachimbi L, Gachini G (2008) Manure and soil fertility management in sub-humid and semi-arid farming systems of Sub-Saharan Africa: Experiences from Kenya. Int J Agric Res 3:166–187
Outten CE, O'halloran TV (2001) Femtomolar sensitivity of metalloregulatory proteins controlling zinc homeostasis. Science 292:2488–2492
Palmgren MG, Clemens S, Williams LE, Krämer U, Borg S, Schjørring JK, Sanders D (2008) Zinc biofortification of cereals: problems and solutions. Trends Plant Sci 13:464–473
Pandey N, Pathak GC, Sharma CP (2006) Zinc is critically required for pollen function and fertilisation in lentil. J Trace Elem Med Biol 20:89–96
Pandey N, Pathak GC, Sharma C (2009) Impairment in reproductive development is a major factor limiting yield of black gram under zinc deficiency. Biol Plant 53:723–727
Panhwar QA, Othman R, Rahman ZA, Meon S, Ismail MR (2012) Isolation and characterization of phosphate-solubilizing bacteria from aerobic rice. Afr J Biotechnol 11:2711–2719
Patel K, Singh M (2009) Scenario of micro-and secondary-nutrients deficiencies and their management in soils and crops of arid and semi-arid regions of Gujarat
Pathak GC, Gupta B, Pandey N (2012) Improving reproductive efficiency of chickpea by foliar application of zinc. Braz J Plant Physiol 24:173–180
Pellegrino E, Bedini S (2014) Enhancing ecosystem services in sustainable agriculture: biofertilization and biofortification of chickpea (*Cicer arietinum* L.) by arbuscular mycorrhizal fungi. Soil Biol Biochem 68:429–439
Pérez-Massot E, Banakar R, Gómez-Galera S, Zorrilla-López U, Sanahuja G, Arjó G, Miralpeix B, Vamvaka E, Farré G, Rivera SM (2013) The contribution of transgenic plants to better health through improved nutrition: opportunities and constraints. Genes Nutr 8:29–41
Peters JW, Szilagyi RK (2006) Exploring new frontiers of nitrogenase structure and mechanism. Curr Opin Chem Biol 10:101–108
Philips R (1993) Starchy legumes in human nutrition and culture. Plant Foods Hum Nutr 44:195–211
Prasad B (1999) Conjoint use of fertilisers with organics, crop residues and green manuring for their efficient use in sustainable crop production. Fertiliser News 44:67–73
Prasad B, Sinha S (1995) Nutrient recycling through crop residues management for sustainable rice and wheat production in calcareous soil. Fertiliser News 40:15–28
Prasad MNV, Sajwan KS, Naidu R (2005) Trace elements in the environment: biogeochemistry, biotechnology, and bioremediation. CRC Press
Quilty J, Cattle S (2011) Use and understanding of organic amendments in Australian agriculture: a review. Soil Res 49:1–26
Ramesh SA, Choimes S, Schachtman DP (2004) Over-expression of an Arabidopsis zinc transporter in Hordeum vulgare increases short-term zinc uptake after zinc deprivation and seed zinc content. Plant Mol Biol 54:373–385
Ramesh A, Sharma SK, Sharma MP, Yadav N, Joshi OP (2014) Inoculation of zinc solubilizing Bacillus aryabhattai strains for improved growth, mobilization and biofortification of zinc in soybean and wheat cultivated in Vertisols of central India. Appl Soil Ecol 73:87–96
Ramzan S, Bhat MA, Kirmani N, Rasool R (2014) Fractionation of zinc and their association with soil properties in soils of Kashmir Himalayas. Int Inv J Agric Soil Sci 2:132–142
Ramzani PMA, Khalid M, Naveed M, Ahmad R, Shahid M (2016) Iron biofortification of wheat grains through integrated use of organic and chemical fertilizers in pH affected calcareous soil. Plant Physiol Biochem 104:284–293
Rebello CJ, Greenway FL, Finley JW (2014) A review of the nutritional value of legumes and their effects on obesity and its related co-morbidities. Obes Rev 15:392–407
Rego T, Wani S, Sahrawat K, Pardhasaradhi G (2006) Macro-benefits from boron, zinc and sulfur application in Indian SAT: A step for Grey to Green Revolution in agriculture. J SAT Agric Res 2:1–21

Rehman A, Farooq M, Nawaz A, Al-Sadi AM, Al-Hashmi KS, Nadeem F, Ullah A (2018) Characterizing bread wheat genotypes of Pakistani origin for grain zinc biofortification potential. J Sci Food Agric 98:4824–4836

Rehman HM, Cooper JW, Lam HM, Yang SH (2019) Legume biofortification is an underexploited strategy for combatting hidden hunger. Plant Cell Environ 42:52–70

Reiter B, Pfeifer U, Schwab H, Sessitsch A (2002) Response of endophytic bacterial communities in potato plants to infection with Erwinia carotovora subsp. atroseptica. Appl Environ Microbiol 68:2261–2268

Rout GR, Sahoo S (2015) Role of iron in plant growth and metabolism. Rev Agric Sci 3:1–24

Rubiales D, Mikić A (2015) Introduction: legumes in sustainable agriculture

Salama ZA, El-Fouly MM, Lazova G, Popova LP (2006) Carboxylating enzymes and carbonic anhydrase functions were suppressed by zinc deficiency in maize and chickpea plants. Acta Physiol Plant 28:445–451

Sarkar J, Potdar AA, Saidel GM (2018) Whole-body iron transport and metabolism: Mechanistic, multi-scale model to improve treatment of anemia in chronic kidney disease. PLoS Comput Biol 14:e1006060

Sathya A, Vijayabharati R, Srinivas V, Gopalakrishnan S (2013) Plant growth-promoting actionbacteria on chickpea seed mineral density: an upcoming complementary tool for sustainable biofortification strategy. 3 Biotech 6(2):138

Schmidt W, Thomine S, Buckhout TJ (2020) Iron nutrition and interactions in plants, vol 10. Frontiers Media SA

Séguéla M, Briat JF, Vert G, Curie C (2008) Cytokinins negatively regulate the root iron uptake machinery in Arabidopsis through a growth-dependent pathway. Plant J 55:289–300

Sharma A, Patni B, Shankhdhar D, Shankhdhar S (2013) Zinc–an indispensable micronutrient. Physiol Mol Biol Plants 19:11–20

Shukla A, Dwivedi B, Singh V, Gill M (2009) Macro role of micronutrients. Indian J Fertil 5:11–30

Shukla AK, Behera SK, Lenka NK, Tiwari PK, Prakash C, Malik R, Sinha NK, Singh V, Patra AK, Chaudhary S (2016) Spatial variability of soil micronutrients in the intensively cultivated Trans-Gangetic Plains of India. Soil Tillage Res 163:282–289

Shukla AK, Behera SK, Pakhre A, Chaudhari S (2018) Micronutrients in soils, plants, animals and humans. Indian J Fertil 14:30–54

Shwetha B, Babalad H, Patil R (2009) Effect of combined use of organics in soybean-wheat cropping system. J Soils Crops 19:8–13

Sillanpää M (1982) Micronutrients and the nutrient status of soils: a global study. Food & Agriculture Organization

Singh G, Abrol I, Cheema S (1989) Effects of gympsum application on mesquite (*Prosopis juliflora*) and soil properties in an abandoned sodic soil. For Ecol Manag 29:1–14

Singh B, Natesan SKA, Singh B, Usha K (2005) Improving zinc efficiency of cereals under zinc deficiency. Curr Sci 88:36–44

Singh S, Hazra K, Praharaj C, Singh U (2016) Biofortification: pathway ahead and future challenges. In: Biofortification of food crops, pp 479–492

Smil V (2000) Feeding the world. A challenge for the twenty-first century. MIT Press, Cambridge, MA

Soares JC, Santos CS, Carvalho SM, Pintado MM, Vasconcelos MW (2019) Preserving the nutritional quality of crop plants under a changing climate: importance and strategies. Plant Soil 443:1–26

Soe KM, Bhromsiri A, Karladee D (2010) Effects of selected endophytic actinomycetes (Streptomyces sp.) and Bradyrhizobia from Myanmar on growth, nodulation, nitrogen fixation and yield of different soybean varieties. CMU J Nat Sci 9:95–109

Solanki M, Didwania N, Nandal V (2016) Potential of zinc solubilizing bacterial inoculants in fodder crops. momentum

Sreevidya M, Gopalakrishnan S, Kudapa H, Varshney R (2016) Exploring plant growth-promotion actinomycetes from vermicompost and rhizosphere soil for yield enhancement in chickpea. Braz J Microbiol 47:85–95

Szerement J, Szatanik-Kloc A, Mokrzycki J, Mierzwa-Hersztek M (2022) Agronomic biofortification with Se, Zn, and Fe: An effective strategy to enhance crop nutritional quality and stress defense—A Review. J Soil Sci Plant Nutr 22:1129–1159

Tilman D, Cassman KG, Matson PA, Naylor R, Polasky S (2002) Agricultural sustainability and intensive production practices. Nature 418:671–677

Timsina J, Connor DJ (2001) Productivity and management of rice–wheat cropping systems: issues and challenges. Field Crop Res 69:93–132

Tokala RK, Strap JL, Jung CM, Crawford DL, Salove MH, Deobald LA, Bailey JF, Morra M (2002) Novel plant-microbe rhizosphere interaction involving Streptomyces lydicus WYEC108 and the pea plant (*Pisum sativum*). Appl Environ Microbiol 68:2161–2171

Tripathi DK, Singh S, Gaur S, Singh S, Yadav V, Liu S, Singh VP, Sharma S, Srivastava P, Prasad SM (2018) Acquisition and homeostasis of iron in higher plants and their probable role in abiotic stress tolerance. Front Environ Sci 5:86

Ullah A, Romdhane L, Rehman A, Farooq M (2019) Adequate zinc nutrition improves the tolerance against drought and heat stresses in chickpea. Plant Physiol Biochem 143:11–18

Valverde A, Burgos A, Fiscella T, Rivas R, Velazquez E, Rodríguez-Barrueco C, Cervantes E, Chamber M, Igual J-M (2007) Differential effects of coinoculations with Pseudomonas jessenii PS06 (a phosphate-solubilizing bacterium) and Mesorhizobium ciceri C-2/2 strains on the growth and seed yield of chickpea under greenhouse and field conditions. In: First international meeting on microbial phosphate solubilization. Springer, pp 43–50

Varotto C, Maiwald D, Pesaresi P, Jahns P, Salamini F, Leister D (2002) The metal ion transporter IRT1 is necessary for iron homeostasis and efficient photosynthesis in Arabidopsis thaliana. Plant J 31:589–599

Vasconcelos M, Datta K, Oliva N, Khalekuzzaman M, Torrizo L, Krishnan S, Oliveira M, Goto F, Datta SK (2003) Enhanced iron and zinc accumulation in transgenic rice with the ferritin gene. Plant Sci 164:371–378

Vasconcelos MW, Gruissem W, Bhullar NK (2017) Iron biofortification in the 21st century: setting realistic targets, overcoming obstacles, and new strategies for healthy nutrition. Curr Opin Biotechnol 44:8–15

Velu G, Crossa J, Singh RP, Hao Y, Dreisigacker S, Perez-Rodriguez P, Joshi AK, Chatrath R, Gupta V, Balasubramaniam A (2016) Genomic prediction for grain zinc and iron concentrations in spring wheat. Theor Appl Genet 129:1595–1605

Vessey JK (2003) Plant growth promoting rhizobacteria as biofertilizers. Plant Soil 255:571–586

Wang Z-Y, Bai M-Y, Oh E, Zhu J-Y (2012) Brassinosteroid signaling network and regulation of photomorphogenesis. Annu Rev Genet 46:701–724

Weyens N, Beckers B, Schellingen K, Ceulemans R, Croes S, Janssen J, Haenen S, Witters N, Vangronsveld J (2013) Plant-associated bacteria and their role in the success or failure of metal phytoextraction projects: first observations of a field-related experiment. Microb Biotechnol 6:288–299

White PJ, Broadley MR (2009) Biofortification of crops with seven mineral elements often lacking in human diets–iron, zinc, copper, calcium, magnesium, selenium and iodine. New Phytol 182:49–84

Wiedenhoeft AC (2006) Plant nutrition. Infobase Publishing

Wu T, Zhang H-T, Wang Y, Jia W-S, Xu X-F, Zhang X-Z, Han ZH (2012) Induction of root Fe (lll) reductase activity and proton extrusion by iron deficiency is mediated by auxin-based systemic signalling in *Malus xiaojinensis*. J Exp Bot 63:859–870

Yodkeaw M, De Datta S (1989) Effects of organic matter and water regime on the kinetics of iron and manganese in two high pH rice soils. Soil Sci Plant Nutr 35:323–335

Yoo MS, James BR (2002) Zinc extractability as a function of pH in organic waste-amended soils. Soil Sci 167:246–259

Zhu C, Naqvi S, Gomez-Galera S, Pelacho AM, Capell T, Christou P (2007) Transgenic strategies for the nutritional enhancement of plants. Trends Plant Sci 12:548–555

Zuo Y, Zhang F (2009) Iron and zinc biofortification strategies in dicot plants by intercropping with gramineous species. A review. Agron Sustain Dev 29:63–71

Zuo Y, Zhang F (2011) Soil and crop management strategies to prevent iron deficiency in crops. Plant Soil 339:83–95

Chapter 16
Legume Biofortification to Increase Selenium Content

Roheela Ahmad, Aamir Hassan Mir, Tahir Ahmad Sheikh, Ayman Javed, Shakeel Ahmad Mir, and Javed Ahmad Bhat

Abstract Legumes carry out major and varied roles in the global agrarian system and global foodstuffs. They are considered ideal crops because they simultaneously achieve three developmental goals in a community: poverty reduction, enhanced human health and nutrition, and increased environmental resilience. Pulses are grown on land that is mostly used for low-input systems, and overall harvested area is around one-tenth that of all cereal crops. In 2008, cereal yields (3.54 t/ha) were over fourfold higher than pulse crop production (0.88 tonnes/ha). In contrast to population growth in emerging and industrialized countries, the worldwide pulse production has increased during the past 14 years at a faster rate.

Selenium is an essential element whose nutritional value is derived mostly from the actions of various selenoproteins and/or tiny selenium metabolites in the human body. Both inorganic (selenide, selenite, and selenate) and organic (selenomthionine, selenocysteine, and methylseleninic acid) forms of this metalloid element are in plants. Men with low dietary Se intake have been associated with oxidative stress, decreased reproductive and immunological functions, and an increased risk of cancer. The Institute of Medicine of the United States National Academy of Sciences states that the daily tolerable maximum consumption of Se for individuals is 400 grams, while the recommended dietary allowance (RDA) for adults is 55 grams. Factors such as parent material and atmospheric inputs, as well as soil parameters

R. Ahmad (✉) · A. Javed · J. A. Bhat
Division of Soil Science and Agricultural Chemistry, Faculty of Agriculture SKUAST-Kashmir, Sopore, Jammu and Kashmir, India

A. H. Mir
DARAS Rangreth, Srinagar, Jammu and Kashmir, India

T. A. Sheikh
Division of Agronomy, Faculty of Agriculture SKUAST-Kashmir, Sopore, Jammu and Kashmir, India

S. A. Mir
Division of Soil Science and Agricultural Chemistry, Faculty of Horticulture-Shalimar SKUAST-Kashmir, Srinagar, Jammu and Kashmir, India

M. A. Nadeem et al. (eds.), *Legumes Biofortification*,
https://doi.org/10.1007/978-3-031-33957-8_16

that influence Se sorption strength, such as pH, soil organic matter content, and Fe, Al, and Mn hydrous oxides, influence selenium bioavailability in soil. Se intake can be raised through crop selection or agronomic biofortification, which entails applying Se-containing fertilizers to soil or using foliar sprays. In Finland, selenium-enriched fertilizers were introduced in 1984. If the daily caloric intake were 10 MJ, this would lead to an increase in selenium consumption from 38 g d^{-1} to 80 g d^{-1} following fortification in 2000. The increased consumption of genetically predisposed crop species for enhanced accumulation of Se is a possibility, while dietary changes are also a possibility. Within-species genetic variation, on the other hand, could be used to boost Se delivery to human food, a process known as genetic biofortification. It might be able to first choose current crop varieties that accumulate more Se. In the long run, it might be possible to cultivate crops with greater Se levels. Traditional breeding programs as a replacement to agronomic biofortification could well be devised if there is sufficient genetic diversity in Se concentration for a particular crop species and if this variation is heritable, which would reduce the need for Se fertilizers other than at the minimum soil Se concentrations. To boost Se concentrations in crops, more breeding materials and faster breeding techniques are required.

The Se fertilization of crops intended for direct consumption by humans will be necessary to provide the highest number of people imaginable with sufficient amounts and varieties of Se. This method is environmentally friendly and biologically safe, according to a large body of evidence. The bioavailability of selenium is influenced by a number of parameters, in addition to its total quantity in soil, which must be taken into account. In the long run, it may be possible to choose or breed crop varieties with more selenium content by utilizing genotypic variation in Se buildup in crops, which would reduce the requirement for selenium fertilizers in all but the lowest soil selenium conditions. This method may be more cost-effective than one that focuses only on research.

1 Introduction

Food legume crops make up significant parts of the agricultural food crops used in developing nations, and they are regarded as critical crops for achieving food and nutritional security for both impoverished producers and consumers. Food legumes support cereal crops as a source of protein and minerals in the human diet. Additionally, they are essential crops for cereal replacement, lowering soil pathogens and providing nitrogen to the cereal crops (Beebe, n.d.). Nodulated pulses are increasingly being grown to increase farmer revenue because they are more productive than grain crops and may be used as feed crops in a variety of farming methods (Gowda et al. 1997). Food legumes are regarded as promising crops to achieve the development goals of the CGIAR, which include lowering levels of poverty and hunger, enhancing nutritional intake for people, and enhancing ecological resilience, thanks to their significant and varied roles in farming systems and the diets of low-income people.

Pulse crop seeds generally comprise 20–25% protein, whereas most cereal crops have 6–10% protein. Pulses are full of dietary fiber and usually have minimal oil content. Thanks to the high quantities of lysine and methionine in pulses' proteins, which are deficient in cereals, pulses have a similar nutritional value to cereals. The primary protein sources in vegetarian diets are pulses, and they are found in practically every traditional cuisine across the globe. Furthermore, in certain rich countries, pulse intake has changed in recent years, where pulses are considered as healthy foods (USDA-ERS 2011).

Pulse crops take up nearly one-tenth of the land area that cereal crops harvested globally do (Fig. 16.1). Since the middle of the 1990s, the area where pulse crops are cultivated has expanded at a pace of 0.4% annually, which compares to a nearly stable trend in the growth rate of cereals globally but is insufficient to elevate it from a supporting to a leading food crop.

Food legumes receive fewer land resources, as well as other inputs, than cereal crops (such as wheat, maize, rice, barley, sorghum, and millets). Additionally, unlike cereal grains, pulse crops are cultivated in dry, arid regions. Figure 16.2 shows the three systems' relative relevance in cultivating pulses and cereals in industrialized and developing countries as well as the rest of the world. Especially in comparison to more than 60% for cereal grains, only roughly 25% of the entire land in the developing world is planted for larger irrigated and rain-fed production systems. Even in industrialized areas, cereal crops are grown on a far larger percentage of land using rain-fed, high-input irrigation systems than pulses are. Whereas just 30% of cereals are grown globally in low-input rain-fed systems, 70% of all pulses are. The situation is substantially worse than the world average in areas such as South Asia,

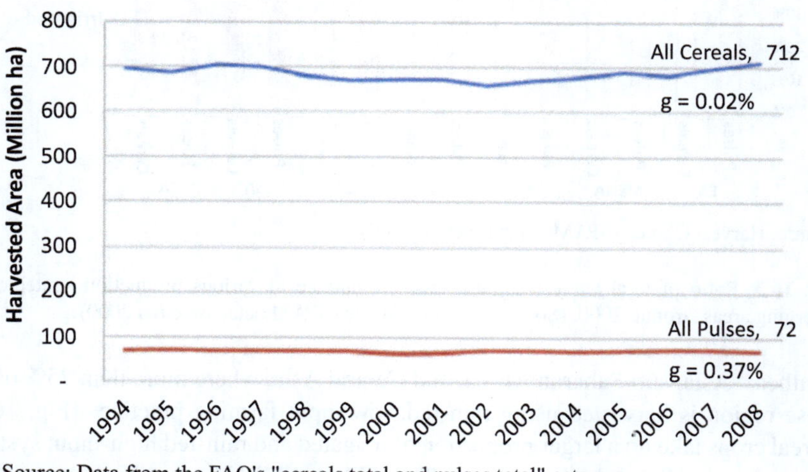

Fig. 16.1 Worldwide area used to harvest grain and pulse crops. (Source: Data from the FAO's "cereals total and pulses total" category)

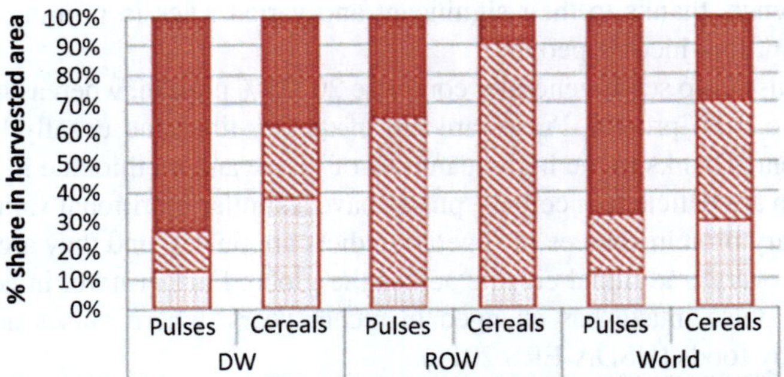

	DW Pulses	DW Cereals	ROW Pulses	ROW Cereals	World Pulses	World Cereals
Rainfed-Low input	43.15	178.03	3.08	18.09	46.24	196.13
Rainfed-High input	8.45	113.21	5.02	153.20	13.47	266.41
Irrigated	7.31	174.76	0.70	17.06	8.01	191.82

Source: Harvest Choice (SPAM database circa 2000)

Fig. 16.2 Comparing the total harvested area (m ha) of pulses and grains across various production methods in the Total Dry Weight (TDW), developing world (DW), the rest of the world (ROW), and the global market, in the year 2000. (Source: Harvest Choice (SPAM database circa 2000))

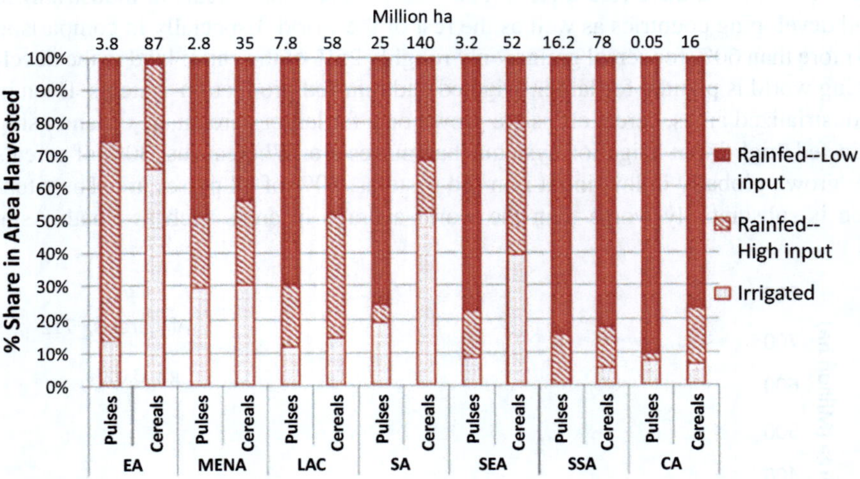

Source: Harvest Choice (SPAM database circa 2000)

Fig. 16.3 Ratio of total harvested pulse area attributable to various production strategies by emerging areas, around 2000. (Source: Harvest Choice (SPAM database circa 2000))

Southeast Asia, sub-Saharan Africa, and Central Asia, where more than 75% of the pulse region is harvested using rain-fed, low-input farming practices (Fig. 16.3). Cereal crops take up a larger proportion of irrigated and rain-fed high-input systems than pulses in all developing regions. The SSA and MENA areas have the smallest discrepancies in the proportions of the three production systems, while the South Asia (SA) and South East Asia (SEA) regions have the largest. World average pulse outputs (0.86 t/ha) are only about one-fourth of cereal yields (3.54 t/ha) because a

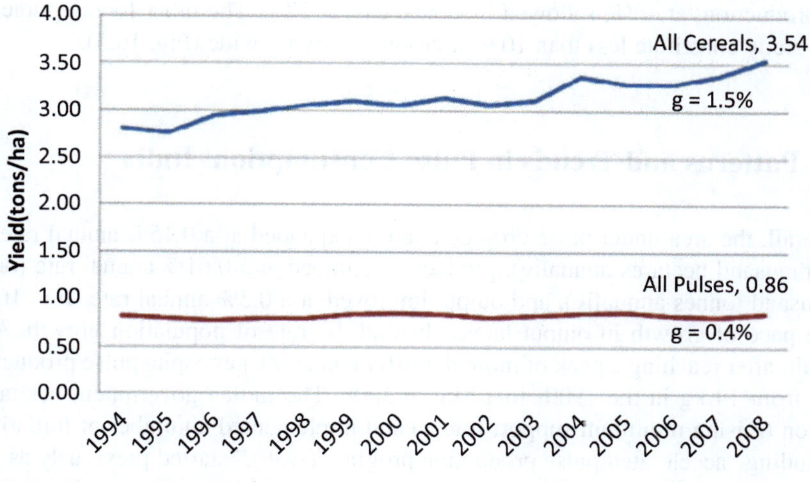

Fig. 16.4 Global cereal and pulse crop yield trends from 1994 to 2008

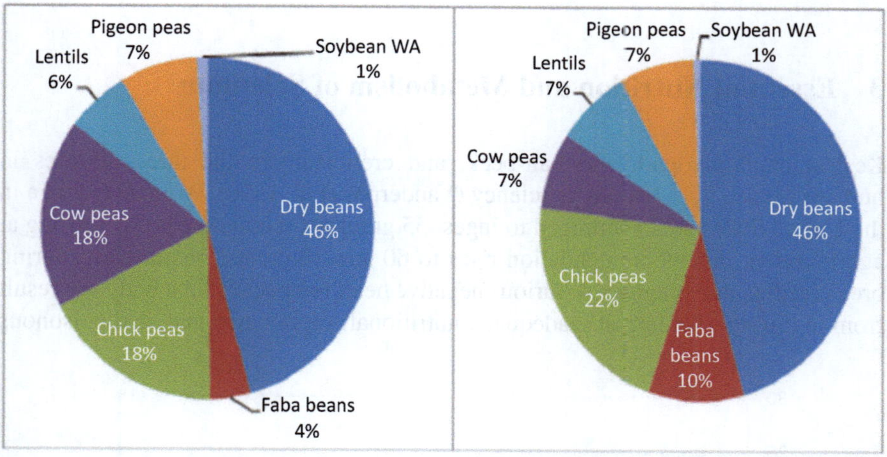

Total World's Area = 62 m ha

Total World's Production = 47 m tons

Fig. 16.5 Different legume crops' percentages of total world area and output from 2006 to 2008. (Source: SitouAkibode and MywishMaredia 2011)

higher percentage of pulse land is harvested under rain-fed minimal-input systems compared to cereal land (Fig. 16.4). From roughly 800 kg/ha to 840 kg/ha, average pulse yields have only marginally improved over the past 14 years, indicating a 0.4% annual growth rate. The yields of cereal crops, on the other hand, have risen at a rate of 1.5% yearly, or around 500 kg/ha, over the same time period.

The overall global prominence of diverse dietary legumes in terms of area and output between 2006 and 2008 is depicted in Fig. 16.5. Dry beans are the most popular food legume crop, accounting for 46% of total area, cowpeas and chickpeas are next, taking over 18% of the total area each. Dry beans continue to lead in terms of production, at 46%, followed by chickpeas, at 22%. The other food legumes in this study constitute less than 10% of production worldwide (Fig. 16.5).

2 Patterns and Trends in Pulse Consumption: India

Overall, the area under pulse crop cultivation expanded at a 0.15% annual rate (or 36 thousand hectares annually), production climbed at a 0.61% annual rate (or 93 thousand tonnes annually), and output improved at a 0.5% annual rate (Fig. 16.6). The pace of growth in output lagged behind the rate of population growth. As a result, after reaching a peak of more than 10 kg in 2000, per capita pulse production fell from 14 kg in the 1900s to 12 kg in 2008. The Indian government has taken action (raising minimum support prices) and implemented a number of initiatives, including "accelerated pulse production program (A3P)," started previously as part of the National Food Security Mission in response to the requirement to boost pulse crop profitability (Ministry of Agriculture and Cooperation 2010).

3 Essential Nutrition and Metabolism of Selenium

Keshan cardiomyopathy, kashan-beck, and cretinism are the three diseases in humans caused by selenium deficiency (Vanderpas et al. 1990). Women and men in the United States are encouraged to ingest 55 grams of selenium per day starting at age fourteen; this recommendation rises to 60 and 70 grams, respectively, during breastfeeding and pregnancy. Various negative health impacts on the body can result from ingesting selenium at inadequate, nutritional, supranutritional, and poisonous

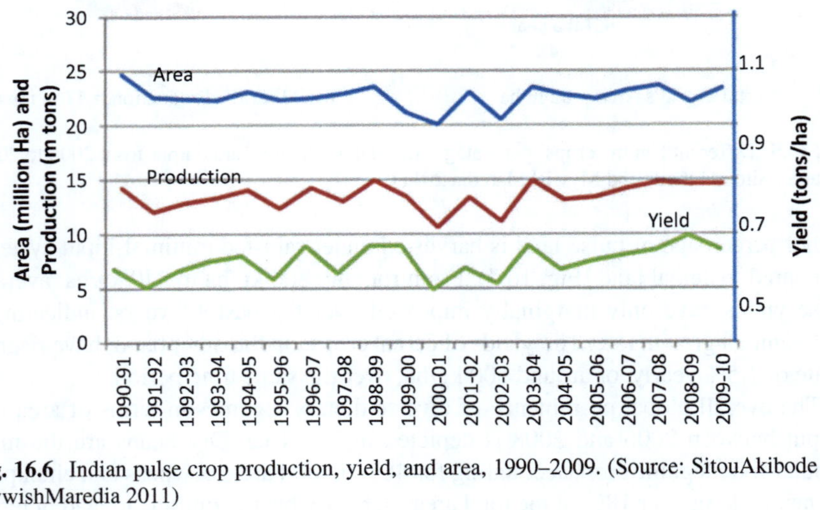

Fig. 16.6 Indian pulse crop production, yield, and area, 1990–2009. (Source: SitouAkibode and MywishMaredia 2011)

levels. Numerous life-forms may have their selenium levels measured using glutathione peroxidase (GPX) activity, plasma or serum selenium levels, and selenoprotein P (Sepp1) concentrations (Combs et al. 2011). The most common dietary form of selenium in the diets of people and both wild and domesticated animals is selenomethionine. It is believed that simple or carrier-mediated diffusion, as well as active transporters, mediate selenium uptake in the small intestine. Despite this, it is unknown what the exact mechanisms are. Selenomethionine interacts with methionine for intracellular transport in Caco-2 colorectal cells, and this interaction allows selenium to be used by methionine transporters (Thiry et al. 2013). The bioavailability of selenium is influenced by its chemical forms, which impacts subsequent metabolism and retention in addition to absorption (Wastney et al. 2011). In people who have a wide range of selenium statuses, selenomethionine consumption increases plasma selenium levels more than inorganic selenium does (Burk et al. 2006). The two elements' shared chemical structure explains a sizable portion of selenium's nonspecific incorporation into amino acids that include sulfur. While the body retains the bulk of the selenium that it consumes through food, a sizable quantity can be excreted in the urine and feces. According to Pedrosa et al. (2012), urinary loss replaces dietary loss as the main route for excretion when selenium levels are higher than the dietary requirement. The chemical forms of selenium in feces and urine are distinct; the latter metabolites enter the urine after methylating (Zeng and Combs 2008). The trimethylselenonium ion ($[CH_3]_3Se^+$) is the representation of urine selenium in rats, and selenosugar (1-methylseleno-N-acetyl-d-galactosamine) is the form of selenium in humans (Kobayashi et al. 2002). Modeling has been used to describe total selenium absorption, transport, and retention in light of the complexity of selenium metabolism and the various forms of ingested selenium (Wastney et al. 2011). By having subjects ingest both organic and inorganic stable isotope tracers of selenium, such as 74Se as selenomethionine and 76Se as sodium selenite, these can be assessed by taking blood, urine, and stool samples.

4 Selenium Inputs to Soils

Although seleniferous soils can have concentrations as high as 1200 mg of Se/kg, most soils have low Se concentrations (normal range: 0.01–2.0 mg of Se/kg; mean: 0.4 mg of Se/kg) (Fordyce 2013). On the other hand, tropical soils have slightly higher Se levels, often between 2 and 4.5 mg/kg^{-1} (Mehdi et al. 2013). Most of the selenium in soils comes from soils' original sources. Se composition is influenced by lithology, mineralogy, organic matter concentration, rainfall intensity, the type of soil and texture, predominant soil geogenic processes, and Se deposition (Mehdi et al. 2013). Volcanic plumes, sea spray, and organic and synthetic Se emissions all contribute to the deposition of Se, which is affected by wind speed and direction, altitude, and proximity to the coast (Saha et al. 2017).

Low Se soils are prevalent in areas with high rates of erosion and fewer atmospheric depositions because they are generated mostly from igneous rocks

(Christophersen et al. 2012). Granite and volcanic soils, which are often found in hilly Northern European countries, are abundant in Se (Mehdi et al. 2013). Selenites and selenides, as well as sulfide minerals, are found in sedimentary rocks; notably, Cretaceous deposits, such as black shales, are found in Se-rich soils. Ireland, China (such as Enshi in Hubei Province and Ziyang in Shaanxi Province), arid areas of the west and southwest of the United States (such as the San Joaquin Valley in California), and India (Punjab) are all rich in black shale (Winkel et al. 2015). Increased Se levels in soil can be attributed to human infusions (agricultural or industrial), dust depositions from nearby fossil sites, or emissions of hazardous amounts of geologically derived Se in soils and streams (Ohlendorf et al. 2020). Agricultural methods can potentially increase the amount of Se in soils through irrigation, protracted chemical fertilizer use, and farmyard manure (FYM). Irrigation may dramatically raise the soil levels of Se by assisting in the solubilization of Se-rich minerals or by transporting Se loads to soil when Se-rich waters are employed (He et al. 2018). However, when biofortification is carried out in Se-deficient locations, where soil Se is <0.6 mg/kg^{-1}, or even when ordinary soil conditions make Se accessibility to plants problematic, adding Se to chemical fertilizers is a regular technique (Bajaj et al. 2011).

Research on the precise effects and progression of various Se additions in agricultural soils hasn't been extensively conducted up to this point. More thorough mass balance analyses of selenium in soils are necessary to ascertain the percentage of selenium retained in soils and selenium lost via leaching, volatilization, and crop removal. This will allow us to accurately assess the contribution of each selenium source to the amount of selenium in agro-ecosystems. These investigations may be helpful in assessing current climate changes, which are projected to lead to future soil Se reductions (Jones et al. 2017).

Selenium is in both organic and inorganic forms in soil with varying oxidation states between II to VI. As the oxidizing conditions (redox potential) increase, selenium dissolution and movement increase in soil. Among the different inorganic forms of selenium, selenate (SeO_{42}) is highly accessible in oxic soils, because of its water soluble and mobile nature with low adsorption affinity on oxide surfaces (Hartikainen 2005). Under low oxidizing conditions, selenate (SeO_{42}) is transformed into selenite (SeO_{32}), which is less mobile and accessible than selenate because of its high inclination to be adsorbed on oxide surfaces at a low soil reaction, according to Hartikainen (2005). As a result, there is a minimal association between accessible soil selenium and total soil selenium. Winkel et al. in 2015 also cited the point when selenite is further converted to elemental selenium (SeO) or selenides (Se_2). Nanosize selenium (0) from selenate and selenate is also being produced by bacteria, according to reports by Jaurez-Maldonado et al. in 2019. According to Ni et al. (2015), selenium [II] can reside as H_2Se or Se^{2-}, which are exceedingly resistant forms of selenium.

A substantial fraction of organic compounds have yet to be undiscovered, but a few—such as methylated and unmethylated Se amino acids, along with volatile selenium forms such as dimethyl selenide (DMSe) and dimethyl diselenide (DMDSe)—are the organic selenium forms found in complexes and in conjugation with organic or polymeric colloids (Winkel et al. 2015). When crops are biofortified with Se, these compounds can be delivered via plant and microbial biomass degradation processes.

5 Selenium Intake, Absorption, and Accretion by Plants

According to Chauhan et al. (2019), plants may absorb both forms of selenium, such as selenite, elemental selenium, inorganic selenate, and organic selenium form (Se amino acids), but plants are not able to absorb H_2Se (selenides) or colloidal elemental Se. Plants are capable of adsorbing the organic form of selenium more than its inorganic form, as pointed by Kikkert et al. in (2013). Broad-specific amino acid transporters help in the penetration of Se amino acids into the plant cell, as cited by Lima et al. (2018). The most common inorganic form of selenium absorbed by plants is selenate, whose movement throughout the cellular membrane is regulated by the sulfate transport system, an energy-dependent pathway (White 2018). The selenate absorption by plants is significantly altered by competitive activities involving sulfate and selenite in soil, along with plant sulfate transporters (SULTRs) exhibiting distinct affinities for these two anions (Mehdawi et al. 2018). Selenite compounds are carried by phosphorus (P) and silicon (Si) transporters across cell membranes of plants, whereas distinct transporters take part in the transportation of hydrogen-selenite ion ($HSeO_3$), selenite anion (SeO_{32}), and selenous acid (H_2SeO_3). Ansporters (LS II) and aquaporins (OsNIP2:1) carry out the transportation of selenous acid (Wang et al. 2019), whereas, as cited by Zhang et al. (2014), hydrogen-selenite ion and a portion of selenite anion are transported largely via high- and low-preference P transporters (OsPT2).

Inorganic Se is transferred from the root cells to the plastids after entering the root cells, where it continues along the Se assimilation route to form SeCys and SeMet (Chauhan et al. 2019). Selenate must be activated before it can be transformed into selenite in order for plants to absorb it. The activation of selenate is controlled by the Adenosine triphosphate (ATP)-sulfurylase enzyme (APS), which joins ATP and selenate to form adenosine 5′-phosphoselenate (APSe) (Smits et al. 2009). Next, in order to convert APSe into selenite, the APS reductase enzyme (APR) accepts two electrons from glutathione (GSH). An alternative method involves phosphorylating APSe to produce 3′-phosphoadenosine 5′-phosphoselenate (PAPSe), a catalyst for the desulfation of desulfoglucosinolates. Serine acetyl transferase (SAT) and O-acetylserine (thiol) lyase (OAS-TL) are both components of the cysteine synthase enzyme complex, which transforms selenite into selenide and then incorporates it into SeCys (White 2018). In a nonenzymatic two-step process, selenodiglutathione (GS-Se-SG) is initially produced from selenonite in the presence of GSH. GS-Se-SG combines with O-acetylserine to make selenopersulfide/glutathionylselenol (GS-SeH), which then transforms into SeCys (OAS) (White 2016). As an alternative, the enzyme selenomethyltransferase (SMT) can create SeCys directly from selenite (Chauhan et al. 2019). In order to create SeMet from

SeCys, the cytosol must first produce the intermediates selenocystathionine and selenohomocysteine. The three enzymes that work together to increase the production of Se cystathionine are cystathionine-lyase (CBL), methionine-synthase (MS), and cystathionine-synthase (CGS), which increases the synthesis of Se cystathionine by condensing O-phosphohomoserine (OPH) and SeCys (White 2018).

Shoot Se concentrations vary widely among genera, species, and even ecotypes within species and are dependent on soils' phytoavailability of Se (Schiavon and Pilon-Smits 2017). In seleniferous soils, nonhyperaccumulator species, which typically acquire less than 100 g of Se/g^{-1} DW (dry weight), collect up to 1000 g of Se/g^{-1} DW, which can eventually be harmful to grazers (Schiavon and Pilon-Smits 2017). The two primary Se organic molecules found in hyperaccumulators are methylselenocysteine (MeSeCys) and selenocystathionine, whereas SeMet is the predominant Se organic molecule found in nonhyperaccumulators (Pilon-Smits 2019).

6 Biofortification of Legumes with Selenium: Agronomic and Genetic

For the biofortification of crops with Se, a thorough understanding of Se biogeochemistry, Se absorption processes, and plant assimilation is required. The practice where vital health-improving substances and micronutrients are added to crops with to improve the nutritional content of meals ingested by people is known as biofortification (Jha and Warkentin 2020). It is a novel and relatively simple approach to maintain, and it has had economical and long-term success in combating micronutrient deficiencies (Ros et al. 2016). The environmental and economic factors of local food systems, as well as producer and community acceptance of biofortified goods, affect positive outcomes. According to Miller and Welch (2013), for producers to adopt biofortified crops, they must be highly productive and profitable staples and the majority of consumers in certain places must consume enough biofortified food to significantly enhance these consumers' nutritional statuses.

Several biofortification systems have been developed and tentatively optimized in order to produce Se-enhanced functional foods by using plant breeding, genetic engineering, and agronomic methods (D'Amato et al. 2020). According to Haug et al. (2007), another potential substitute for agronomic biofortification using Se fertilizers is the inclusion of Se components in food during the manufacturing process. Fortification techniques are influenced by a number of variables, including the methods of selenium supplementation, the dosage, the selenium species, soil selenium, agronomic practices, the ecological and climatic conditions, crop varieties, and the level of cooperative in the implementation of adding micronutrients (Dall'Acqua et al. 2019; D'Amato et al. 2020). Up to this point, the majority of investigations into Se biofortification have focused on the use of Se alone or in combination with just one other element, often Si, iodine (I), or zinc (Zn) (Cakmak et al.

2020; Golob et al. 2020). Although micronutrient deficiencies are ubiquitous, only a few studies have concentrated on crop biofortification employing a variety of micronutrients (Zou et al. 2020).

7 Agronomic Biofortification

In low-Se locations, agronomic biofortification is commonly used mostly where the incorporation of Se to soil and its foliar fertilization is carried out, which is normally accomplished through the use of selenium-containing fertilizers, such as selenate or selenite (Alfthan et al. 2015). To meet biofortification goals, selenium fertilizers are typically used in small levels (10–20 g Se ha^{-1}). As a result, they are typically combined into the other industrial fertilizers (e.g., urea and calcium nitrate) to facilitate administration, serving as "carriers" of Se (Ramkissoon et al. 2019). Selenium may be more effectively chelated with organic molecules when organic acids are added to Se fertilizers, which increases Se plant uptake and Se fertilizer effectiveness. Broadley et al. (2010) found that applying Se fertilizers to soil increases total and bioavailable Se, resulting in elevated levels of Se in the consumable parts of crops (Poblaciones et al. 2014). Selenium leaching into groundwater does not appear to pose any environmental risks, as it is restricted via bonding with SOM and the cationic soil surfaces and limited to the volatilization loss (De Feudis et al. 2019). In the United Kingdom (Lyons 2010), Finland (Alfthan et al. 2015), and New Zealand (Lyons 2010), field tests using Se in soil have been successful (Hartikainen 2005). The effectiveness of this method, however, may be constrained by the uneven distribution of Se in soil and soil factors that influence Se speciation and absorption by plants, such as pH, organic matter, oxygenation, the presence of competing ions, the age of the soil, and chemical and biological changes to Se (Duncan et al. 2017). Only around 12% of the Se fertilizer supplied to soil is typically absorbed by plants because the majority of Se is bound to and maintained in soil, making it inaccessible (Broadley et al. 2010). Because of this, only a limited amount of residual Se is available for crops, necessitating that soil be fertilized with Se at the beginning of each growing season. The fastest selenium acquisition and absorption processes, the removal of selenium root-to-shoot transfer to edible sections of crops, and the avoidance of selenium losses owing to the soil immobilization of selenium compounds are the most likely reasons for this (Ramkissoon et al. 2019).

The ability of helpful rhizosphere bacteria to decrease the oxidized and methylated forms of selenium and expand the soil surface that plant roots can access to receive Se may improve the effectiveness of a specific fertilizer's use by plants and soils' phytoavailability (Mora et al. 2015; Yasin et al. 2015a,b). The Se biofortification of crops may be improved by introducing beneficial bacteria to the soil or inoculating plants with plant growth–promoting bacteria (PGPB). For example, inoculating wheat plants with certain microbes, either alone or in combination with arbuscular mycorrhizal fungi (AMF), will increase the Se levels of the wheat grains, as reported by Duran et al. (2014). The plants used for the remediation of Se-rich soils (Se hyperaccumulators) can also be utilized as sources of selenium in soils or as cattle feed because of their richness in selenium, particularly organic forms such as SeMet and MetSeCys (Banuelos et al. 2017; Wan et al. 2018). In this case, the plant matter must be thoroughly tested for harmful components before being included in the food chain because remediated soils are typically afflicted by a variety of metals/loids. The food-containing organic molecules of Se may be more rapidly used by enzymes compared with the inorganic molecules of Se, a fact that could have important consequences for optimal nutrition. The next Se biofortification option to research involves growing crops in Se-rich soils or watering them with Se-rich water. This procedure, known as natural biofortification, is gaining popularity across the world (Wu et al. 2015). It has been recognized in some parts of the United States (Banuelos et al. 2019), China (Dinh et al. 2018), and India (Dhillon and Dhillon 2009), but the proportion of research carried in this domain remains limited, because the majority of studies have been conducted in homogenous, controlled environments that do not accurately reflect the actual soil conditions of agro-ecosystems. Se might be more quickly released into the environment if seleniferous soils were used for agricultural purposes, especially when Se is concentrated via evapotranspiration, posing an ecotoxicological threat, which is one of the drawbacks of following this strategy.

8 Genetic Biofortification

Selenium biofortification by conventional breeding, also known as genetic biofortification, is considered to be the most efficient, enduring, and sustainable technique (White and Broadley 2009). It entails selecting varieties with high propensities to absorb and translocate selenium, particularly in its organic form, to edible parts (SeMet and/or MetSeCys) (Banuelos et al. 2017; Wu et al. 2015). Selenium phytoavailability is regarded as the most significant predictor of Se concentration in edible components, notably cereal grains and legumes. However, when selenium phytoavailability is high, genotypic variation in Se concentration may become significant (Zhu et al. 2009). Excessive selenium accumulation is associated with a number of chromosomal quantitative trait loci (QTLs) (Ates et al. 2016). By choosing edible cultivars with high selenium concentrations, it is possible to transfer these high-Se QTLs to high-yielding low-Se cultivars via marker-assisted breeding (MAB) (Wu et al. 2015). Plant breeding, whether traditional or marker assisted, has certain fundamental limitations, including the requirement to use Se fertilizer sources for agronomic biofortification when growing plants in soils with low Se contents.

The development of contemporary molecular techniques and analytical technology has resulted in advancements in Se biofortification research, allowing for the development of more-effective future methods. Molecular techniques for biofortification include reverse breeding, RNA-directed DNA methylation and editing, oligo-directed mutagenesis, and next-generation sequencing (NGS) at high speed and low cost. Analytical techniques include X-ray fluorescence and X-ray absorption near-edge spectroscopies (Carvalho and Vasconcelos 2013). Breeding and genetic engineering may benefit from such technologies and from functional genomics technology (Wang et al. 2018). Genetic engineering is not yet widely accepted and used compared to agronomic fortification and traditional breeding because many countries still have obligatory transgenic usage restrictions (Zhu et al. 2009).

Until now, there haven't been many transgenic varieties developed that have an improved tendency to collect and accumulate Se mostly in organic forms (White and Broadley 2009). The overexpression of sulfate transporters—enzymes that catalyze rate-limiting steps in selenium assimilation, like ATP-sulfurylase, and mechanisms that prevent selenium from being misincorporated into proteins, like selenocysteine lyase and selenocysteine methyltransferase—is a crucial characteristic of these transgenics, according to Zhu et al. (2009). Through the transfer of numerous genes targeted by genetic engineering, selenium biofortification outcomes in the past 10 years were favorable. For example, in *Arabidopsis thaliana*, the overexpression of the selenium-binding protein gene SBPI increased plant resistance to selenite through a GSH- dependent mechanism (Agalou et al. 2005). According to the same theory, *A. thaliana* exhibits higher selenium tolerance and accumulation in response to the overexpression of the ethylene response factor ERF96 or loss-of-function mutations in the APX1 gene, which codes for a cytosolic ascorbate peroxidase enzyme (Jiang et al. 2020). A novel enzyme, namely selenocysteine methyltransferase, was also found in a Se accumulator plant (*Brassica juncea* L.) with the capability of methylating both homocysteine and SeCys substrates (Chen et al. 2019). The total Se and MeSeCys accumulation increased as a result of this enzyme's overexpression in tobacco plants (Zhang et al. 2019).

In addition to selenate/sulfate transporters and one ATP-sulfurylase isoform (APS), *Stanleya pinnata*, a Se hyperaccumulator, has been found to possess new genetic traits. Because of the presence of the root high affinity sulfate/selenate transporter SULTR1;2, the plant is not suppressed when high sulfate is present, unlike nonhyperaccumulators (Wang et al. 2018). Any of the SULTR transporters can be used to achieve high levels of Se in transgenic crops. According to Jiang et al. (2018), *S. pinnata*'s APS2 isoform, which expresses itself only in the cytoplasm, is another intriguing enzyme. Lastly, the APS2 isoform, present in *A. thaliana* and *Stanleya elata*, two Se nonhyperaccumulators, has dual localization—i.e., plastidial and cytosolic—in contrast to the APS isoform, which typically operates in plastids for S/Se assimilation (Bohrer et al. 2015). The function of APS2 is now being studied because of the high expression of APS2 in *S. pinnata*, which exhibits hypertolerance to selenium.

9 Conclusions and Future Prospects

Se can be used in a variety of agrotechnological applications. Biofortification with Se fertilizers—i.e., agronomic biofortification and conventional breeding—is perhaps the most widely used and approved way to combat Se deficiencies around the world. Circular systems that incorporate crop genetic engineering, biofortification techniques, and Se phytoremediation are still in the early stages of development despite being promising in the context of sustainable agriculture. Scientists, on the other hand, are developing interest in nanosize Se biofortification. It is necessary to conduct more studies to ascertain whether Se nanoparticles (NPs) are safe for consumers and what chemical changes they could go through in the environment and during food preparation. The applications of PGPR in agronomic biofortification and agricultural biofortification with several micronutrients are still in the early stages of research. The biofortification of crops with selenium species and antioxidant chemicals has had a significant impact on mammalian nutrition and health. The effect of Se-enriched food on preventing or treating viral infections is crucial in the current viral epidemic and constitutes a study field that requires more examination. Selenium biofortification could be utilized to boost crop output in less-than-ideal circumstances, reducing the deleterious impacts of such conditions on plant physiology while enhancing antioxidant qualities and phytochemical content. On this basis, investigations seeking to optimize Se biofortification systems for enhancing food crop nutrition in difficult conditions are attracting a lot of attention on a global scale.

References

Agalou A, Roussis A, Spaink HP (2005) The Arabidopsis selenium-binding protein confers tolerance to toxic levels of selenium. Funct Plant Biol 32:881–890. https://doi.org/10.1071/FP05090

Alfthan G, Eurola M, Ekholm P, Venäläinen E-R, Root T, Korkalainen K, Hartikainen H, Salminen P, Hietaniemi V, Aspila P, Aro A, for the Selenium Working Group (2015) Effects of nationwide addition of selenium to fertilizers on foods, and animal and human health in Finland. From deficiency to optimal selenium status of the population. J Trace Elem Med Biol 31:142–147. https://doi.org/10.1016/j.jtemb.2014.04.009

Ates D, Sever T, Aldemir S, Yagmur B, Temel HY, Kaya HB, Alsaleh A, Kahraman A, Ozkan H, Vandenberg A, Tanyolac B (2016) Identification QTLs controlling genes for Se uptake in lentil seeds. PLoS One 11(3):e0149210. https://doi.org/10.1371/journal.pone.0149210

Banuelos GS, Lin Z-Q, Broadley MR (2017) Selenium biofortification. In: Pilon-Smits EAH, Winkel LHE, Lin ZQ (eds) Selenium in plants: molecular, physiological, ecological and evolutionary aspects. Springer International Publishing, Cham, pp 231–256. https://doi.org/10.1007/978-3-319-56249-0

Banuelos GS, Freeman J, Arroyo I (2019) Accumulation and speciation of selenium in biofortified vegetables grown under high boron and saline field conditions. Food Chem 5:100073. https://doi.org/10.1016/j.fochx.2019.100073

Bohrer AS, Yoshimoto N, Sekiguchi A, Rykulski N, Saito K, Takahashi H (2015) Alternative translational initiation of ATP sulfurylase underlying dual localization of sulfate assimilation pathways in plastids and cytosol in Arabidopsis thaliana. Front Plant Sci 5:750. https://doi.org/10.3389/fpls.2014.00750

Broadley MR, Alcock J, Alford J, Cartwright P, Foot I, Fairweather-Tait SJ, Hart DJ, Hurst R, Knott P, McGrath SP, Meacham MC (2010) Selenium biofortification of highyielding winter wheat (Triticum aestivum L.) by liquid or granular Se fertilisation. Plant Soil 332:5–18. https://doi.org/10.1007/s11104-009-0234-4

Burk RF, Norsworthy BK, Hill KE, Motley AK, Byrne DW (2006) Effects of chemical form of selenium on plasma biomarkers in a high-dose human supplementation trial. Cancer Epidemiol Biomarkers Prev 15:804–810

Cakmak I, Marzorati M, Van den Abbeele P, Hora K, Tjalling Holwerda H, Yazici MA, Savasli E, Neri J, Du Laing G (2020) Fate and bioaccessibility of iodine in food prepared from agronomically biofortified wheat and rice and impact of cofertilization with zinc and selenium. J Agr Food Chem 68(6):1525–1535. https://doi.org/10.1021/acs.jafc.9b05912

Carvalho SMP, Vasconcelos MW (2013) Producing more with less: strategies and novel technologies for plant-based food biofortification. Food Res Int 54:961–971. https://doi.org/10.1016/j.foodres.2012.12.021

Chauhan R, Awasthi S, Srivastava S, Dwivedi S, Pilon-Smits EAH, Dhankher OP, Rudra D (2019) Understanding selenium metabolism in plants and its role as a beneficial element. Crit Rev Environ Sci Tech 49:1937–1958. https://doi.org/10.1080/10643389.2019.1598240

Chen M, Zeng L, Luo X, Mehboob MZ, Ao T, Lang M (2019) Identification and functional characterization of a novel selenocysteine methyltransferase from Brassica juncea L. J Exp Bot 70(21):6401–6416. https://doi.org/10.1093/jxb/erz390

Christophersen OA, Lyons G, Haug A, Steinnes E (2012) Sources of heavy metals and metalloids in soils. In: Alloway BJ (ed) Heavy metals in soils: trace elements and metalloids in soils and their bioavailability, 3rd edn. Environ Pollut. Springer, Dordrecht, pp 11–50. https://doi.org/10.1007/978-94-007-4470-7

Combs GF Jr, Watts JC, Jackson MI, Johnson LK, Zeng H, Scheett AJ, Uthus EO, Schomburg L, Hoeg A, Hoefig CS, Davis CD, Milner JA (2011) Determinants of selenium status in healthy adults. Nutr J 10:75

D'Amato R, Regni L, Falcinelli B, Mattioli S, Benincasa P, Dal Bosco A, Pacheco P, Proietti P, Troni E, Santi C, Businelli D (2020) Current knowledge on selenium biofortification to improve the nutraceutical profile of food: a comprehensive review. J Agr Food Chem 68:4075–4097. https://doi.org/10.1021/acs.jafc.0c00172

Dall'Acqua S, Ertani A, Pilon-Smits EAH, Fabrega-Prats M, Schiavon M (2019) Selenium biofortification differentially affects sulfur metabolism and accumulation of phytochemicals in two rocket species (Eruca Sativa Mill. and Diplotaxis Tenuifolia) grown in hydroponics. Plants (Basel) 8(3):68. https://doi.org/10.3390/plants8030068

De Feudis M, D'Amato R, Businelli D, Guiducci M (2019) Fate of selenium in soil: a case study in a maize (Zea mays L.) field under two irrigation regimes and fertilized with sodium selenite. Sci Tot Environ 659:131–139. https://doi.org/10.1016/j.scitotenv.2018.12.200

Dhillon SK, Dhillon KS (2009) Phytoremediation of selenium contaminated soils: the efficiency of different cropping systems. Soil Use Manag 25:441–453. https://doi.org/10.1111/j.1475-2743.2009.00217.X

Dinh QT, Cui Z, Huang J, Tran TAT, Wang D, Yang W, Zhou F, Wang M, Yu D, Liang D (2018) Selenium distribution in the Chinese environment and its relationship with human health: a review. Environ Int 112:294–309. https://doi.org/10.1016/j.envint.2017.12.035

Duncan EG, Maher WA, Jagtap R, Krikowa F, Roper MM, O'Sullivan CA (2017) Selenium speciation in wheat grain varies in the presence of nitrogen and sulfurfertilisers. Environ Geochem Health 39:955–966. https://doi.org/10.1007/s10653-016-9857-6

Duran P, Acuña JJ, Jorquera MA, Azcón R, Paredes C, Rengel Z, Mora ML (2014) En- dophytic bacteria from seleniumsupplemented wheat plants could be useful for plant-growth promotion, biofortification and Gaeumanno- mycesgraminis biocontrol in wheat production. Biol Fert Soils 50:983–990

Golob A, Novak T, Maršić NK, Šircelj H, Stibilj V, Jerše A, Kroflič A, Germ M (2020) Biofortification with selenium and iodine changes morphological properties of Brassica oleracea L. var. gongylodes and increases their contents in tubers. Plant Physiol Biochem 150:234–243. https://doi.org/10.1016/j.plaphy.2020.02.044

Hartikainen H (2005) Biogeochemistry of selenium and its impact on food chain quality and zhuman health. J Trace Elem Med Biol 18:309–318. https://doi.org/10.1016/j.jtemb.2005.02.009

Haug A, Graham R, Christophersen O, Lyons G (2007) How to use the world's scarce selenium resources efficiently to increase the selenium concentration in food. Microb Ecol Health Dis 19(4):209–228. https://doi.org/10.1080/08910600701698986

He Y, Xiang Y, Zhou Y, Yang Y, Zhang J, Huang H, Shang C, Luo L, Gao J, Tang L (2018) Selenium contamination, consequences and remediation techniques in water and soils: a review. Environ Res 164:288–301. https://doi.org/10.1016/j.envres.2018.02.037

Jha AB, Warkentin TD (2020) Biofortification of pulse crops: status and future perspectives. Plants (Basel) 9(1):73. https://doi.org/10.3390/plants9010073

Jiang Y, Schiavon M, Lima LW, Tripti JRR, El Mehdawi AF, Royer S, Zeng Z, Hu Y, Pilon-Smits EAH, Pilon M (2018) Comparison of ATP sulfurylase 2 from selenium hyperaccumulator Stanleya pinnata and non-accumulator Stanleyaelata reveals differential intracellular localization and enzyme activity levels. Biochim Biophys Acta Gen Subj 1862(11):2363–2371. https://doi.org/10.1016/j.bbagen.2018.03.014

Jiang L, Yang J, Liu C, Chen Z, Yao Z, Cao S (2020) Overexpression of ethylene response factor ERF96 gene enhances selenium tolerance in Arabidopsis. Plant Physiol Biochem 149:294–300. https://doi.org/10.1016/j.plaphy.2020.02.024

Jones GD, Droz B, Greve P, Gottschalk P, Poffet D, McGrath SP, Seneviratne SI, Smith P, Winkel LH (2017) Selenium deficiency risk predicted to increase under future climate change. Proc Natl Acad Sci USA 114(11):2848–2853. https://doi.org/10.1073/pnas.1611576114

Kikkert J, Hale B, Berkelaar E (2013) Selenium accumulation in durum wheat and spring canola as a function of amending soils with selenite, selenate and or sulfate. Plant Soil 372:629–641. https://doi.org/10.1007/s11104-013-1773-2

Kobayashi Y, Ogra Y, Ishiwata K, Takayama H, Aimi N, Suzuki KT (2002) Selenosugars are key and urinary metabolites for selenium excretion within the required to low-toxic range. Proc Natl Acad Sci U S A 99:15932–15936

Lima LW, Pilon-Smits EAH, Schiavon M (2018) Mechanisms of selenium hyperaccumulation in plants: a survey of molecular, biochemical and ecological cues. Biochim Biophys Acta Gen Subj S0304-4165(18):30090–30094. https://doi.org/10.1016/j.bbagen.2018.03.028

Lyons G (2010) Selenium in cereals: improving the efficiency of agronomic biofortification in the UK. Plant Soil 332:1–4. https://doi.org/10.1007/s11104-010-0282-9

Mehdawi AF, Jiang Y, Guignardi ZS, Esmat A, Pilon M, PilonSmits EAH, Schiavon M (2018) Influence of sulfate supply on selenium uptake dynamics and expression of sulfate/ selenate transporters in selenium hyperaccumulator and nonhyperaccumulator Brassicaceae. New phytol 217:194–205. https://doi.org/10.1111/nph.14838

Mehdi Y, Hornick JL, Istasse L, Dufrasne I (2013) Selenium in the environment, metabolism and involvement in body functions. Molecules 18(3):3292–3311. https://doi.org/10.3390/molecules18033292

Miller DD, Welch RM (2013) Food system strategies for preventing micronutrient malnutrition. Food Policy 42:115–128. https://doi.org/10.1016/j.foodpol.2013.06.008

Ministry of Agriculture and Cooperation (2010) Operational guidelines for accelerated pulses production programme. National Food Security Mission. Government of India, Ministry of Agriculture Department of Agriculture & Cooperation, May, 2010

Mora ML, Durán P, Acuña J, Cartes P, Demanet R, Gianfreda L (2015) Improving selenium status in plant nutrition and quality. J Soil Sci Plant Nutr 15(2):486–503. https://doi.org/10.4067/S0718-95162015005000041

Ohlendorf HM, Santolo GM, Byron ER, Eisert MA (2020) Kesterson reservoir: 30 years of selenium risk assessment and management. Integr Environ Assess Manag 16(2):257–268. https://doi.org/10.1002/ieam.4222

Pedrosa LF, Motley AK, Stevenson TD, Hill KE, Burk RF (2012) Fecal selenium excretion is regulated by dietary selenium intake. Biol Trace Elem Res 149:377–381

Pilon-Smits EAH (2019) On the ecology of selenium accumulation in plants. Plants (Basel) 8(7):197. https://doi.org/10.3390/plants8070197

Poblaciones MJ, Rodrigo S, Santamaria O, Chen Y, McGrath SP (2014) Selenium accumulation and speciation in biofortified chickpea (Cicer arietinum L.) under Mediterranean conditions. J Sci Food Agric 94(6):1101–1106. https://doi.org/10.1002/jsfa.6372

Ramkissoon C, Degryse F, da Silva RC, Baird R, Young SD, Bailey EH, McLaughlin MJ (2019) Improving the efficacy of selenium fertilizers for wheat biofortification. Sci Rep 9(1):19520. https://doi.org/10.1038/s41598-019-55914-0

Ros G, van Rotterdam A, Bussink D, Bindraban P (2016) Selenium fertilization strategies for bio-fortification of food: an agro-ecosystem approach. Plant Soil 404:99–112. https://doi.org/10.1007/s11104-016-2830-4

Saha U, Fayiga A, Sonon L (2017) Selenium in the soil-plant environment: a review. Int J Appl Agr Sci 3(1):1–18. https://doi.org/10.11648/j.ijaas.20170301.11

Schiavon M, Pilon-Smits EA (2017) The fascinating facets of plant selenium accumulation - biochemistry, physiology, evolution and ecology. New Phytol 213(4):1582–1596. https://doi.org/10.1111/nph.14378

SitouAkibode and MywishMaredia (2011) Global and regional trends in production, trade and consumption of food legume crops. https://www.researchgate.net/publication/266353209

Thiry C, Ruttens A, Pussemier L, Schneider YJ (2013) An in vitro investigation of species-dependent intestinal transport of selenium and the impact of this process on selenium bioavailability. Br J Nutr 109:2126–2134

USDA-Economic Research Service (ERS) (2011) Vegetables and melons outlook/VGS-343/ February 17, 2011, pp 23–26. Economic Research Service, USDA. http://www.ers.usda.gov/briefing/drybeans/PDFs/DBnOutlook.pdf, Accessed March 2011

Vanderpas JB, Contempre B, Duale NL, Goossens W, Bebe N, Thorpe R, Ntambue K, Dumont J, Thilly CH, Diplock AT (1990) Iodine and selenium deficiency associated with cretinism in northern Zaire. Am J Clin Nutr 52:1087–1093

Wan J, Zhang M, Adhikari B (2018) Advances in seleniumenriched foods: from the farm to the fork. Trends Food Sci Tech 76:1–5. https://doi.org/10.1155/2018/8749345

Wang J, Cappa JJ, Harris JP, Edger PP, Zhou W, Pires JC, Adair M, Unruh SA, Simmons MP, Schiavon M, Pilon-Smits EAH (2018) Transcriptome-wide comparison of selenium hyperaccumulator and nonaccumulatorStanleya species provides new insight into key processes mediating the hyperaccumulation syndrome. Plant Biotechnol J 16:1582. https://doi.org/10.1111/pbi.12897

Wang M, Yang W, Zhou F, Du Z, Xue M, Chen T, Liang D (2019) Effect of phosphate and silicate on selenite uptake and phloem-mediated transport in tomato (Solanum lycopersicum L.). Environ Sci Pollut Res Int 26(20):20475–20484. https://doi.org/10.1007/s11356-019-04717-x

Wastney ME, Combs GF Jr, Canfield WK, Taylor PR, Patterson KY, Hill AD, Moler JE, Patterson BH (2011) A human model of selenium that integrates metabolism from selenite and selenomethionine. J Nutr 141:708–717

White PJ (2016) Selenium accumulation by plants. Ann Bot 117:217–235. https://doi.org/10.1093/aob/mcv180

White PJ (2018) Selenium metabolism in plants.Biochim Biophys Acta Gen Subj S0304-4165(18)30138-7. https://doi.org/10.1016/j.bbagen.2018.05.006

White PJ, Broadley MR (2009) Biofortification of crops with seven mineral elements often lacking in human diets–iron, zinc, copper, calcium, magnesium, selenium and iodine. New Phytol 182(1):49–84. https://doi.org/10.1111/j.1469-8137.2008.02738.x

Winkel LH, Vriens B, Jones GD, Schneider LS, Pilon-Smits E, Bañuelos GS (2015) Selenium cycling across soil-plant-atmosphere interfaces: a critical review. Nutrients 7(6):4199–4239. https://doi.org/10.3390/nu7064199

Wu Z, Bañuelos GS, Lin ZQ, Liu Y, Yuan L, Yin X, Li M (2015) Biofortification and phytoremediation of selenium in China. Front Plant Sci 6:136. https://doi.org/10.3389/fpls.2015.00136

Yasin M, El-Mehdawi AF, Jahn CE, Anwar A, Turner MFS, Faisal M, Pilon-Smits EAH (2015a) Seleniferous soils as a source for production of selenium-enriched foods and potential of bacteria to enhance plant selenium uptake. Plant Soil 386:385–394. https://doi.org/10.1007/s11104-014-2270-y

Yasin M, El-Mehdawi AF, Anwar A, Pilon-Smits EAH, Faisal M (2015b) Microbial-enhanced selenium and iron biofortification of wheat (Triticum aestivum L.): applications in phytoremediation and biofortification. Int J Phytorem 17:341–347. https://doi.org/10.1080/15226514.2014.922920

Zeng H, Combs GF Jr (2008) Selenium as an anticancer nutrient: roles in cell proliferation and tumor cell invasion. J Nutr Biochem 19:1–7

Zhang L, Hu B, Li W, Che R, Deng K, Li H, Yu F, Ling H, Li Y, Chu C (2014) OsPT2, a phosphate transporter, is involved in the active uptake of selenite in rice. New Phytol 201(4):1183–1191. https://doi.org/10.1111/nph.12596

Zhang L, Hu B, Deng K, Gao X, Sun G, Zhang Z, Li P, Wang W, Li H, Zhang Z, Fu Z, Yang J, Gao S, Li L, Yu F, Li Y, Ling H, Chu C (2019) NRT1.1B improves selenium concentrations in rice grains by facilitating selenomethinone translocation. Plant Biotechnol J 17(6):1058–1068. https://doi.org/10.1111/pbi.13037

Zhu YG, Pilon-Smits EA, Zhao FJ, Williams PN, Meharg AA (2009) Selenium in higher plants: understanding mechanisms for biofortification and phytoremediation. Trends Plant Sci 14(8):436–442. https://doi.org/10.1016/j.tplants.2009.06.006

Zou C, Du Y, Rashid A, Ram H, Savasli E, Pieterse PJ, OrtizMonasterio I, Yazici A, Kaur C, Mahmood K, Singh S, Le Roux MR, Kuang W, Onder O, Kalayci M, Cakmak I (2020) Simultaneous biofortification of wheat with zinc, iodine, selenium, and iron through foliar treatment of a micronutrient cocktail in six countries. J Agric Food Chem 67(29):8096–8106. https://doi.org/10.1021/acs.jafc.9b01829

Chapter 17
RETRACTED CHAPTER: Biofortifying Legumes with Iodine

Muhammad Aslam, Haroon Zaman Khan, Ramala Masood Ahmad, Ayesha Aslam, and Ahsan Raza Mallhi

1 Introduction

Legumes comprise nearly 19,500 species across 751 genera, which shows the level of diversity in this plant family (Christenhusz and Byng 2016). Globally, legumes provide basic sources of protein to developing economies. Negligence in the genetic improvement of legumes, which are good sources of calories and nutrients for developing economies, might exert a severe adverse effect on global food and nutritional security. Legumes are not considered a top priority in wider ecological contexts. With the help of next-generation sequencing, the opportunities for improving legumes such as chickpeas (*Cicer arietinum*), groundnuts (*Arachis hypogaea*), and pigeon peas (*Cajanus cajan*) have been documented, though the rate of improvement has not been comparable with the pace of improvement in cereals or other main staple crops (Varshney 2016).

Legumes provide a range of diverse micronutrients and amino acids, which exceeds or complements the profiles of cereals. The enrichment of legumes' nutritional profiles is an appropriate target to address the problem of hidden hunger from global micronutrient malnutrition. Biofortifying legumes to address the problem of

---

The original version of this chapter was retracted: The retraction note to this chapter is available at https://doi.org/10.1007/978-3-031-33957-8_24

M. Aslam (✉) · R. M. Ahmad · A. Aslam
Department of Plant Breeding and Genetics, University of Agriculture, Faisalabad, Pakistan

H. Z. Khan
Department of Agronomy, University of Agriculture, Faisalabad, Pakistan

A. R. Mallhi
Maize Research Station, Ayub Agricultural Research Institute, Faisalabad, Pakistan

M. A. Nadeem et al. (eds.), *Legumes Biofortification*,
https://doi.org/10.1007/978-3-031-33957-8_17

hidden hunger is a strategy that remains unexploited. Providing easy excess to nutritionally enriched legumes to malnourished and socioeconomically deprived populations will serve the dual purpose of hidden hunger amelioration and increasing legumes-related positive health impacts.

According to Godfray et al. (2010), at the current rate of crop improvement, by 2050, yields will not be sufficient to fulfill the demand of the global population. Although food security is a crucial global concern, nutritional security should also be considered equally important for the health of the world's population. Food insecurity is usually a major problem in developing and underdeveloped economies, but nutritional insecurity is prevailing in both developing and developed economies (Finn 2014). The International Food Policy Research Institute was developed to precisely measure and track the magnitude of hunger and malnutrition.

There are almost fifty well-known dietary nutrients that are essential for sustainable human health. This list comprises of nine amino acids, eight macroelements, thirteen vitamins, two fatty acids, sixteen microelements, water and carbohydrates (Welch and Graham 2005). The World Health Organization (WHO) recommends a daily intake of 150 micrograms (µg) of iodine for pregnant and lactating people and school-age children. Unfortunately, many countries have inadequate dietary intakes of iodine and many of the people among their populations are considered iodine deficient. Legumes are good sources of dietary iodine, containing upward of 25 µg of iodine per 100 grams of dry weight (WHO 2019a).

A chronic lack of micronutrients and vitamins is referred to hidden hunger. In developing regions of the world, it is a major problem for women and young children. Inadequate nutritional availability during infancy or a lack of sufficient nutritional intake by mothers during pregnancy causes stunted growth in children, which is the main indicator of malnutrition. Micronutrient malnutrition is adversely affecting the health of almost two billion people, with problems ranging from stunted growth to life-threatening illnesses (WHO 2019b). Over the past decade, the pace of decline in undernourishment at the global level has slowed down.

The biofortification of legumes with iodine is a method of increasing the iodine content of legumes, such as beans, peas, and lentils, through fertilization and other methods. Legumes are important sources of dietary iodine, and the biofortification of legumes has been shown to be beneficial in areas where iodine deficiency is an issue because it can help to improve dietary iodine intake. Increasing iodine content in legumes makes them more iodine dense and can help to reduce iodine deficiency, which can lead to impaired physical and mental development, especially in pregnant people and young children. Biofortifying legumes with iodine is a cost-effective and sustainable method of increasing iodine intake in populations because it does not require the fortification of processed foods, which can be costly and difficult to implement. Additionally, legumes are versatile and important parts of many diets, making them ideal vehicles for increasing iodine intake. The biofortification of legumes with iodine can help to ensure that people have access to a nutritious source of this important trace element. The contents of this chapter present the need for making nutritional improvements to legumes with a focus on iodine, in a wider ecological context.

2 Nutritional Benefits of Biofortifying Legumes with Iodine

Iodine is an essential trace element that helps to regulate metabolic processes and has a role in the formation of hormones. Legumes are rich sources of protein, iron, zinc, and other essential minerals; the biofortification of legumes can help to improve nutrition and dietary quality (Mukherjee et al. 2017). Biofortifying legumes with iodine is a process of increasing the iodine content of a legume crop, traditionally through natural means and modern technologies, including genetically modifying seeds. This process can improve the overall nutritional profiles of legumes by providing essential trace elements that are needed for the normal growth, development, and functioning of the human body. This is done to address the issue of iodine deficiency, which is a major public health concern in many parts of the world. Iodine is an essential mineral that is necessary for the proper functioning of the thyroid gland, which is responsible for the regulation of growth and metabolism. Biofortifying legumes with iodine is specifically important for populations living in iodine-deficient areas in that it can help to prevent iodine deficiency disorders. A deficiency in iodine can lead to a range of health problems, including goiters, hypothyroidism, and intellectual disabilities. Iodine biofortification in legumes has a number of important nutritional benefits.

Sufficient intake of iodine addresses a variety of health problems, including stunted growth and impaired cognitive development. Furthermore, sufficient dietary iodine plays an important role in hormone production, brain development, and metabolism. The biofortification of legumes with iodine helps to reduce the risk of iodine deficiency, which is a common cause of preventable mental disabilities and goiters.

Legumes are good sources of protein, fiber, and other essential vitamins and minerals. However, they can also be sources of antinutrients such as phytates and trypsin inhibitors. Iodine biofortification can reduce the antinutrient content of legumes, making them more nutrient dense and beneficial to health. It also helps to improve the taste, flavor, and shelf life of legumes because it reduces the rate at which the legumes spoil. This supports improving the societal acceptance of legumes in the human diet and is beneficial for those in developing countries who may not have access to refrigeration.

Additionally, studies have found that biofortified legumes are more nutritious than their nonbiofortified counterparts, providing higher levels of essential vitamins and minerals. This can help to improve the overall health of those consuming legumes as staple foods.

Iodine biofortification in legumes may also help to reduce the environmental impact of food production. Legumes require less water and fertilizer to produce than other crops do, making legumes more sustainable and environmentally friendly options. Additionally, by increasing the iodine contents of legumes, it may be possible to reduce the use of iodine-fortified salt, which can help to reduce environmental pollution.

3 Challenges in Biofortifying Legumes with Iodine

Iodine biofortification in legumes is a promising approach to reducing iodine deficiency, which affects over two billion people worldwide (WHO 2019c). Overall, iodine biofortification has great potential for improving the iodine status of populations. However, several challenges must be addressed in order to make this approach successful:

(a) There is a lack of knowledge about and expertise in how to best biofortify legumes with iodine. In addition, there is also a lack of awareness of the health benefits of iodine and the importance of including it in the human diet. Many legume species have not been well studied, and limited data are available on the genetic variability of iodine biofortification in legumes. This makes it difficult to develop and select high-yielding and iodine-fortified varieties.

(b) There is lack of access to appropriate technology for iodine biofortification in legumes. In order to produce iodine-fortified legumes, a variety of methods, such as legume breeding, soil amendment, tissue culture, and genetic engineering, must be employed. However, these methods are often not available or affordable in many of the countries where iodine insufficiency is a problem.

(c) Legumes generally have low levels of bioavailable iodine, so additional iodine needs to be added to soil in order to be effectively taken up by plants. This is an expensive and time-consuming process, and it is difficult to ensure that the right amount of iodine is added.

(d) Susceptibility to pests and diseases in legumes is prevailing, which can damage the plant and reduce the amount of iodine taken up. Controlling these pests and diseases can be costly, and it is not always possible to do so effectively.

(e) In some cultures, legumes are not traditionally eaten, making it difficult to encourage people to include them in their diets.

(f) Legumes are prone to iodine losses during processing, storage, and cooking. Therefore, biofortified legumes must be carefully handled to ensure that the iodine content is not compromised. Iodine is a volatile element that can easily be lost during storage or processing. This means that legumes biofortified with iodine may not retain their iodine levels over time.

(g) Quality control is essential to ensure that the final product contains the desired levels of iodine. This includes testing soils, seeds, and the harvested legumes.

(h) Higher amounts of iodine are required by legume crops than by other crops, and iodine is often in short supply in many areas. This makes it difficult to add enough iodine to soil to effectively biofortify the legume crop. Ensuring that the iodine added to soil is of high enough quality and effective enough to be absorbed by the plants is a challenge.

(i) Iodine is often expensive, making it difficult to purchase enough for effective biofortification.

(j) Legumes are sensitive to environmental conditions, and changes in temperature, humidity, or soil composition can affect the uptake of iodine. Therefore, careful monitoring is necessary to ensure that the right conditions are maintained.

(k) The bioavailability of iodine from legumes is not well understood, meaning that it is not known how much of the iodine is actually absorbed by the human body.
(l) Legumes are staple foods in many parts of the world, and their culture and preparation vary widely from region to region. This makes it difficult to ensure that iodine-biofortified legumes are being used in the most effective way to reduce iodine deficiencies.
(m) Research into the effectiveness of iodine biofortification in legume crops is still in its early stages, and more research is needed to understand how best to implement iodine biofortification in legume crops.
(n) The agronomic, economic, and social implications of iodine biofortification in legumes still need to be better understood. This includes understanding the impact of iodine biofortification on the yield and quality of legumes and the costs and benefits of such interventions.

Despite these challenges, iodine biofortification in legumes could reduce iodine deficiency and improve public health. With continued research and development, these challenges can be overcome. and the approach can be further improved to ensure its success.

4 Opportunities of Iodine Biofortification in Legumes

Biofortifying legumes with iodine is an important opportunity to improve nutrition and health and could provide a sustainable and cost-effective way to address iodine deficiency. Legumes contain high amounts of proteins, minerals, and vitamins, which makes them ideal sources for biofortification with iodine. Legumes are staple foods in many countries and are major sources of proteins and other essential nutrients. However, they are often low in iodine, a mineral essential for thyroid health and cognitive development. Legumes can be grown in a variety of environments and climates, including areas with limited access to iodized salt and other sources of iodine (Graham 2019).

Iodine biofortification can increase the iodine content of legumes and provide an effective and sustainable way to improve nutrition and health. It involves the breeding of legumes with higher levels of iodine. This can be carried out through traditional breeding or by using biotechnology. Biofortifying legumes with iodine can be achieved through a number of approaches. These include breeding for naturally occurring higher levels of iodine, enriching seeds with iodine, and using iodine-containing fertilizers. Breeding for higher levels of iodine could create a variety of legumes that contain higher levels of iodine than those currently available. Enriching seeds with iodine can increase the amount of iodine available in legumes, while using iodine-containing fertilizers can ensure that legumes are receiving adequate supplies of iodine. This strategy has been successfully implemented in several countries, where it has been shown to reduce the prevalence of iodine deficiency.

The aim is to produce legume varieties that are high in iodine but that still retain the beneficial nutritional and agronomic traits of the original variety. Iodine biofortification can improve nutrition, reduce the risk of iodine-deficiency disorders, and help to reduce the economic burden of ill-health. It can also help to reduce the need for costly and potentially harmful iodine supplementation programs. Biofortifying legumes with iodine is a cost-effective, sustainable, and safe way to improve nutrition and health.

This means that iodine-enriched legumes can be grown in iodine-deficient soils, providing a reliable source of iodine for the people who consume them. Opportunities for biofortifying legumes with iodine include the following:

(a) Cost-effectiveness—biofortification is much cheaper than supplementation with iodized salt or iodized oil, which are the two most common methods of supplementing iodine.
(b) Increased nutritional benefits—legumes are already nutrient dense, and biofortification adds the benefit of iodine.
(c) Widespread availability—legumes are widely available and can be grown in almost all parts of the world.
(d) Improved taste—biofortified legumes have been shown to have improved taste compared with nonbiofortified legumes.
(e) Sustainable solution—biofortification is a sustainable solution to address iodine deficiency because it does not rely on the availability of iodized salt or iodized oil.

5 Economic Impacts of Iodine Biofortification in Legumes

The global economic impact of biofortifying legumes with iodine is multifaceted. It can help to reduce malnutrition and its associated health costs, particularly in iodine-deficient areas. Similarly, it reduces the financial burden on households in those areas, who often must spend a disproportionate amount of their income on iodine-rich foods. Additionally, it can increase local production and trade in legumes, leading to a more stable and resilient global food system. Further, legume biofortification can help to reduce the environmental impact of food production because legumes are relatively low-input crops that require less fertilizer and water than other crops, such as cereals.

It has been estimated that the cost of iodine-deficiency disorder (IDD) is in the range of USD 6–7 billion annually, representing 0.2% to 0.3% of global gross domestic product (ICRISAT 2020). According to World Bank estimates, the cost of IDD can range from 1.7% to 4.4% of a country's gross domestic product. It has also been suggested that iodine-biofortified legume availability could lead to positive health outcomes and potential cost savings. For instance, a study conducted in India estimated that the introduction of iodine-biofortified legumes could reduce the cost of providing iodized salt by up to 28% (FAO 2020a).

The World Health Organization (WHO) has identified iodine deficiency as a major public health concern because it affects more than two billion people worldwide (WHO 2020a, b). Iodine biofortification has the potential to benefit millions of people in the world, particularly in developing countries, and improve economic conditions. It can reduce poverty, improve food security, and enhance economic development. It is a promising strategy to address the economic and developmental consequences of iodine deficiency. According to the WHO, the amount of iodine in food crops should be between 20 and 30 parts per million (ppm). However, many countries have not achieved this target (WHO 2018a).

Iodine is an essential micronutrient for human health, and its deficiency is a major public health problem in many parts of the world. Iodine deficiency has been linked to adverse health effects, including intellectual disabilities, neurological damage, and even death in severe cases (Cai 2011). Iodine biofortification could increase agricultural productivity and the incomes of smallholder farmers while reducing the cost of nutrition interventions. Moreover, it could create new markets and employment opportunities and reduce the cost of healthcare services. This could lead to improved economic outcomes for individuals, families, and communities.

The economic benefits of biofortifying legumes with iodine can be seen in a number of ways. For instance, it can help to reduce the costs of healthcare associated with iodine deficiency and reduce the prevalence of iodine-deficiency-related diseases (Mukherjee et al. 2017). Improved health can lead to savings in healthcare costs, allowing for more resources to be allocated to other sectors, such as education, infrastructure, and social programs, which can further contribute to economic development. Efficient iodine intake helps to improve the nutritional status of individuals in iodine-deficient areas, which in turn can lead to improved productivity and an increase in economic growth (Bhagavathula et al. 2018). The availability of iodine-biofortified legumes can help to reduce the amount of money spent on imported iodine because more of the population will be able to obtain the mineral through the consumption of biofortified legumes (Bhagavathula et al. 2018). Biofortifying legumes with iodine can help to address this issue because legumes are major sources of dietary iodine in many parts of the world.

Through the process of biofortification, legume crops can be genetically engineered to contain higher levels of iodine, allowing people to consume more iodine through the foods they eat (Kaur 2018). This approach could provide a cost-effective, sustainable, and scalable solution to iodine deficiency and could improve nutrition and health outcomes in many parts of the world (Zimmermann 2013). By improving the health of children and adults, iodine biofortification could increase economic productivity, improve educational outcomes, and multiply incomes (Hoddinott and Yisehac 2008).

The economic benefits of iodine biofortification are not limited to individuals. With the improvement in the health and productivity of the population, iodine biofortification can bring benefits to entire countries. For example, improved health can lead to increased worker productivity (Wilkinson and Maret 2019). This, in turn, could lead to increased economic growth and development, potentially resulting in improved standards of living in many parts of the world (O'Donnell 2011; Wilkinson and Maret 2019). People in iodine-deficient countries may experience improved economic outcomes when they have access to iodine-rich foods (Vijayaraghavan 2020).

Finally, iodine biofortification could contribute to poverty alleviation and sustainable economic growth. Therefore, it is important that more research be done to promote and implement iodine biofortification in legumes, in order to maximize its potential benefits.

In conclusion, the economic benefits of biofortifying legumes with iodine should not be overlooked; biofortifying legumes with iodine has the potential to bring economic benefits to individuals, families, and countries. It can reduce poverty, improve food security, and contribute to economic development. As a result, it can be an important contributor to improve world economic conditions.

6 Role of Genetic Modification in Biofortifying Legumes with Iodine

Genetically modifying legumes involves the introduction of a gene that enables the targeted species to produce a compound that can increase the bioavailability of iodine from those legumes. Examples of such genes include those from iodine-accumulating plant species, such as *Thlaspi caerulescens*, which can be introduced into legumes to increase their iodine content and bioavailability (Gonzalez et al. 2009). This can be carried out either through direct transformation or through *Agrobacterium*-mediated transformation, a process that uses a bacterium to transfer genes from one organism to another (Kumar and Verma 2014). The introduction of an iodine-accumulating gene into legumes has been shown to significantly increase the amount of iodine available in those legumes (Kumar and Verma 2014). This can be beneficial for those who rely on legumes as major sources of dietary iodine because it can increase the overall dietary intake of iodine. This is especially important for populations in areas with limited access to fish and other iodine-rich foods (Delange and Glinoer 2001). Additionally, genetic modification can improve the nutritional quality of legumes in that the increased iodine content can help to prevent iodine-deficiency-related diseases (WHO 2019b, c). Overall, genetic modification can provide an effective and sustainable solution to biofortifying legumes with iodine.

Genetic modification (GM) technology has enabled researchers to introduce genes that can increase the synthesis of thyroid hormones, which are essential for the absorption of iodine. For example, researchers have successfully introduced a gene from *Arabidopsis thaliana* into soybean and fava bean plants, which significantly increased the content of iodine in those species (Chen et al. 2016). GM technology has also enabled researchers to modify genes that are involved in the biosynthesis of phytic acid, which is known to reduce the absorption of iodine (Chen et al. 2017b). This technology can be used to increase the expression of genes that are involved in the absorption of iodine, thereby increasing its bioavailability (Bresciani et al. 2018). Genetic modification can be achieved by inserting genes that code for the production of iodine-containing compounds, such as iodothyronine and iodine-rich proteins. The resulting plants can then be used to produce foods with higher levels of iodine than their non-GM counterparts (Bhatia et al. 2014). The concentration, uptake, transport, and storage of iodine in legumes can be enhanced by using GM technology; for example, GM has been used to introduce a sulfate transporter gene from maize into soybeans, thus increasing the uptake of iodide by the plant and increasing the iodine content of the soybeans (Nair 2018). GM could increase the amount of available iodine in legumes by manipulating the genes that are responsible for iodine uptake and metabolism. For instance, the gene MYB12 has been identified as a potential target for GM-based iodine biofortification in soybeans; it encodes a transcription factor that controls the expression of the genes involved in iodine uptake and metabolism (Zhang et al. 2018). Similarly, other genes, such as ICL1, ICL2, and ICL3, have also been identified as potential targets for GM-based iodine biofortification in legumes (Wang et al. 2017). Researchers have used GM technology to introduce a gene from the kelp species *Laminaria japonica* into soybean plants, resulting in a 26-fold increase in iodine content compared to the non-GM control (Ito et al. 2011). Additionally, GM technology can be used to reduce the number of antinutrients in legumes, further increasing the bioavailability of iodine (Hashimoto and Takahashi 2018). One study has shown that the introduction of genes into soybeans increased the levels of iodothyronine in the plants by almost tenfold. This technology has also been used to increase the levels of iodine-rich proteins, such as glutathione-S-transferase, in lentils, increasing its iodine content by up to 50% (Sadhu and Jain 2017).

In addition to GM, there are other methods of biofortifying legumes with iodine, such as selecting and breeding plants with naturally higher levels of iodine and supplementing the soil with iodine-rich fertilizers. However, GM technology provides an efficient, cost-effective, and sustainable way of biofortifying legumes with iodine and thus increases the iodine content of food crops.

GM can significantly improve the iodine content of legumes while reducing the need for other external inputs such as fertilizers, pesticides, and water (Zhang et al. 2018). By introducing an iodine-accumulating gene into a legume, it is possible to significantly improve the bioavailability of iodine in that legume, enabling populations with limited access to iodine-rich foods to obtain the iodine they need for optimal health. GM could also improve the nutritional quality of legumes in a much shorter period of time than traditional breeding methods can. However, it must be used responsibly and with caution because potential risks are associated with the use of GM technology.

7 Strategies for Implementing Iodine Biofortification in Legumes

Iodine is an essential micronutrient that is needed for normal growth and development and is necessary for the production of hormones in the thyroid gland. Biofortifying legumes with iodine is an effective strategy for addressing the global iodine-deficiency crisis. There are different workable strategies for the implementation of iodine biofortification in legumes.

Crop Breeding Crop breeding is an effective strategy for biofortifying legumes with iodine. By selecting for desirable traits through crossbreeding, scientists can create varieties of legumes that have higher levels of iodine. For example, in 2017, researchers at the International Crops Research Institute for the Semi-Arid Tropics (ICRISAT) successfully developed a chickpea variety with higher levels of iodine compared with those of traditional varieties (Singh et al. 2017). Plant breeders can cross existing varieties of legumes with naturally occurring high-iodine variants, such as wild species of *Phaseolus vulgaris* and *Vigna unguiculata*. This technique has been shown to be effective in increasing the iodine content of common bean varieties (Black et al. 2008).

(a) Hybridization is a strategy being used to develop plant varieties with improved traits, such as higher iodine content. This could be achieved by crossing high-iodine varieties of legumes with popular varieties that have high yields and other desirable characteristics. For example, the development of a high-iodine variety of chickpeas (*Cicer arietinum* L.) through the hybridization of two separate varieties has been documented (Singh et al. 2018).
(b) Selective breeding is a method of plant breeding used to develop varieties with desired traits. Through this process, legume varieties with higher iodine contents could be developed from existing varieties. For example, a chickpea variety with a high iodine content was developed in India through selective breeding and was found to contain more than three times the amount of iodine than the current varieties on the market (Gupta et al. 2017).

Fertilization Fertilizing legumes with iodine-containing compounds, such as potassium iodide, can be an effective way to increase the iodine content in these crops. For example, in 2011, a study showed that applying potassium iodide to soybean plants increased the iodine levels in the crop by up to 35% (Sharma et al. 2011). Iodine-enriched fertilizers are composed of potassium iodide, which can be readily taken up by plants and stored in their seeds. The use of iodine-enriched fertilizers can effectively increase the iodine content of legumes, such as kidney beans and chickpeas (Naeem et al. 2013).

Foliar Spray Foliar spraying is another method for biofortifying legumes with iodine. This involves spraying a solution containing iodine onto the leaves of plants. Foliar spraying has been found to be effective in increasing the iodine content of several legume species, including lentils, chickpeas, mungbeans, and peas (Saseendran and Antony 2013).

Seed Priming Seed priming can be used as an effective and cost-efficient way to increase the amount of iodine biofortified in legumes. Seed priming involves pre-treating the seeds with a solution containing a micronutrient, such as iodine, and then growing the seeds in a soil environment. This method has been used to successfully increase the iodine contents of several legume species, including chickpeas, lentils, mungbeans, and field peas (Singh and Prasad 2011).

Irrigation Irrigating legumes with iodine-rich water sources, such as seawater, can increase the iodine content in the plants. For example, in 2015, researchers in Mexico found that irrigating cowpeas with seawater increased the iodine content of the crop by up to 19% (Muñoz-Gutiérrez et al. 2015).

Inoculation Rhizobia are soil-dwelling bacteria that are capable of fixing nitrogen from the atmosphere and can also take up and store iodine from the soil. Legumes are known to form symbiotic relationships with nitrogen-fixing bacteria, which can be used to increase the iodine content in the plant. This involves introducing beneficial microorganisms, such as nitrogen-fixing bacteria, into soil to improve the overall nutrient status of the soil. This method has been found to be effective in increasing the iodine content of several legume species, including chickpeas, lentils, and mungbeans (Prasad and Singh 2010). A study by Yadav et al. (2013) showed that inoculating legumes with nitrogen-fixing bacteria can increase the levels of iodine in legumes. Studies have shown that the use of rhizobial inoculants can effectively increase the iodine content of legumes, such as kidney beans and chickpeas (Mahajan and Datta 2017; Smulders et al. 2006).

Soil Amendment Soil amendment is another method for biofortifying legumes with iodine. This involves adding iodine-containing compounds to soil to increase the overall iodine content of the soil. This method has been found to be effective in improving the iodine content of chickpeas and mungbeans (Prasad and Singh 2012). However, soil amendments should be used judiciously because their overuse can lead to nutrient imbalances in soil.

Fortification Legumes can be fortified with iodine in a variety of ways:

(a) Fortifying the raw material is a strategy that involves the direct addition of iodine to the raw legume material prior to processing. This approach is advantageous because the iodine is added directly to the legume material, and there is no need for special fortification equipment. However, this approach is not often used, because of increased labor costs, the need for regular testing, quality control, and the difficulty of ensuring uniform distribution of iodine throughout the raw legume material (Karmakar and Saha 2013).

(b) Fortification during processing involves the addition of iodine to the legume material during processing. This approach is advantageous because the iodine is added directly to the legume material, and there is no need for extra fortification equipment (Vanstone and Rae 2003). This method has been used in several countries, including Brazil, Mexico, Peru, and the Dominican Republic, to fortify beans and lentils with iodine.

Potassium iodate can be added directly to legumes during processing. This approach is simple and cost-effective, as it does not require any additional steps or equipment. The amount of iodate added depends on the type and form of the legume and on the desired level of iodine fortification. This method has been used to fortify legumes such as lentils, chickpeas, and mungbeans (Kaur and Singh 2016). Iodized salt can be added to legumes during processing to provide iodine fortification. This method is cost-effective and can be easily scaled up to reach larger populations. However, the level of iodine fortification is limited by the amount of salt that can be added without impacting the taste and texture of the legume (Singh et al. 2015b). Legumes can be coated with iodized oil to provide iodine fortification. This approach is simple and cost-effective, and it allows for the fortification of a wide range of legumes. The amount of iodized oil used depends on the type and form of the legume and on the desired level of iodine fortification (Singh et al. 2015a, c). Legumes can be sprayed with iodized solutions to provide iodine fortification. This approach is simple and cost-effective, and it allows for the fortification of a wide range of legumes. The amount of iodized solution used depends on the type and form of the legume and on the desired level of iodine fortification (Singh et al. 2015a, b, c).

(c) Fortifying finished products involves the addition of iodine to the finished legume product. This approach is advantageous because the iodine is added directly to the finished product, and there is no need for extra fortification equipment. However, this approach is not often used, because of the need for regular testing and quality control and the difficulty of ensuring uniform distribution of iodine throughout the finished product (Bressani and Moncada 2012).

(d) Fortification through fortifiers is another strategy for fortifying legumes with iodine. This method involves using fortifiers, such as iodized salt or iodized oil, to increase the iodine content of legumes. This approach has been used in several countries, including India, Nigeria, Egypt, and Pakistan, to fortify legumes with iodine.

(e) Biofortification is another strategy for fortifying legumes with iodine. This approach involves breeding crops to increase their iodine content. This method has been used successfully to increase the iodine content of chickpeas. Several biofortification technologies, such as nanotechnology and genetic engineering, can be used to increase the iodine content of legumes. For example, researchers in China used genetic engineering to create a variety of soybean with higher levels of iodine compared with those of traditional varieties (Chen et al., 2017b).

Intercropping Legumes can be intercropped with other iodine-rich crops, such as onions and garlic. This can help to increase the iodine content in soil and in legumes (Guimaraesn et al. 2017).

In conclusion, strategies for implementing iodine biofortification in legumes include all the strategies mentioned in this part of chapter. These strategies have been shown to be effective in increasing the iodine content of legumes and can help to address the global iodine-deficiency crisis.

8 Environmental Considerations for Iodine Biofortification in Legumes

Environmental considerations for iodine biofortification in legumes are important for ensuring the long-term sustainability of this dietary intervention, and biofortification is an important strategy for addressing iodine deficiency, especially in developing countries where soil iodine levels are low and access to iodized salt is limited. However, the environmental considerations must be taken into account before implementing iodine biofortification. The main environmental considerations for iodine biofortification in legumes include the following:

(a) Selection of appropriate legumes—Different legumes have different levels of iodine bioavailability, and some may be more suitable than others for iodine biofortification. In addition, some legumes may be better suited to particular soil and climate conditions, making them more suitable for biofortification for that particular area (Bressani et al. 2020). Farmers should select legume varieties that are well-adapted to the local climate and soil type, as these varieties are more likely to absorb and retain iodine from the soil. Similarly, farmers should monitor soil and water levels for iodine to ensure that levels are not too high.

(b) Climate change—Legumes are known to be efficient nitrogen fixers, which can reduce the need for nitrogen fertilizers. In addition, legumes can reduce global warming potential through their ability to sequester carbon in their biomass, contributing to climate change mitigation (Singh and Singh 2020). Climate change can also affect the success of iodine biofortification in legumes. Changes in precipitation and temperature can alter the uptake of nutrients and the growth of legumes.

(c) Nutrient balance—Iodine biofortification may lead to an imbalance in other essential nutrients, such as iron, zinc, or calcium. Therefore, it is important to consider the overall nutrient balance when selecting legumes for iodine biofortification (Patil et al. 2017).

(d) Interactions with other soil components—iodine biofortification may interact with other soil components, such as other micronutrients, organic matter, and soil pH. Therefore, it is important to consider the potential interactions of iodine biofortification with other soil components (Sandhu et al. 2013).

(e) Cost and sustainability—the cost of iodine biofortification must be taken into account when considering its sustainability. In addition, the long-term sustainability of iodine biofortification must also be considered because the effects of iodine biofortification may not be permanent (Bressani et al. 2020).
(f) Soil fertility—biofortified legumes may require higher levels of soil fertility to maximize yields and nutrient content. To ensure sustainability, soil fertility should be regularly tested and fertilizers should be applied in accordance with soil requirements (Bielinski and Seebauer 2020). In addition, increased evaporation in warmer climates can lead to greater losses of iodine from soils, reducing the availability of this nutrient to crops.
(g) Soil salinity—biofortified legumes may be sensitive to high levels of soil salinity, which can reduce crop yields and nutrient content. Therefore, soil salinity levels should be monitored and, where necessary, measures taken to reduce soil salinity (Buresh and Lal 2009).
(h) Soil quality—legumes can improve soil quality by improving soil structure, fertility, and water-holding capacity. Legumes can also increase microbial biomass, organic matter, and nitrogen availability (Delate and Cambardella 2004).
(i) Water management—biofortified legumes require adequate water for optimal growth and yields. Therefore, proper water management, including irrigation and drainage, should be implemented to ensure adequate water availability (Bar-Yosef et al. 2006).
(j) Water use—legumes are generally more water efficient than other crops and can reduce the need for irrigation, particularly in dry climates (Avila and Pereira 2016).
(k) Pesticide use—legumes are susceptible to a range of pests and diseases, and pesticide use may be necessary to ensure that maximum yields and nutrient contents are produced. However, pesticides can have negative environmental impacts. Therefore, it is important to ensure that pesticides are used responsibly and in accordance with best practices (Smith and Gilbert 2005).
(l) Pest control—legumes can reduce the need for chemical pesticides because their nitrogen-fixing ability can act as a natural pest repellent (Gahukar 2004).
(m) Habitat—legumes can create habitat for wildlife, such as providing food and shelter for pollinators and other beneficial insects (Landis et al. 2000).

Fertilizer use and pesticide use can have negative impacts on the environment. Therefore, strategies should be implemented to reduce the use of these chemicals while increasing the efficiency of their applications. This includes the use of slow-release fertilizers, precision in the timing and location of the application, and the use of biological controls when possible. Careful consideration should be given to the types of fertilizers and pesticides used and their potential to contaminate the local water supply.

Environmental considerations for iodine biofortification in legumes include the potential for increased levels of iodine in the environment, the potential leaching of the iodine from the legumes, the potential for iodine to accumulate in soils, and the potential for changes in the nutritional profiles of the legumes. The potential for increased levels of iodine in the environment is a concern because iodine is a non-biodegradable substance and can accumulate in the environment over time. The potential for the leaching of iodine from legumes is a concern in that this could lead to increase levels of iodine in groundwater and surface water. Similarly, the potential for iodine to accumulate in soils is a concern because it could lead to increased levels of soilborne contamination and could lead to changes in soil fertility. Lastly, the potential for changes in the nutritional profiles of the legumes is a concern because some nutrients could be lost or gained when adding iodine (Jones et al. 2017; Ritchey 2017; Thorpe 2009; Donovan and Jones 2015).

Environmental considerations for iodine biofortification in legumes also include reducing the environmental footprint of iodine-fortified legumes, utilizing natural sources of iodine, and minimizing contamination with toxic metals. Reducing the environmental footprint of iodine-fortified legumes can be accomplished through sustainable agricultural practices such as conservation tillage and the use of cover crops. Utilizing natural sources of iodine can help reduce the amount of synthetic iodine used in biofortification. Natural sources of iodine include seaweed, fish, and shellfish. These sources can be incorporated into soil through composting or by spraying them directly onto crops. This method reduces the amount of synthetic iodine that is released into the environment. Minimizing contamination with toxic metals is also an important environmental consideration. These practices can help ensure the safety and sustainability of iodine-fortified legumes (Ahmed et al. 2016; Eisen and Saffery 2013; Hazell and Wood 2009; Khan and Shukla 2018). Furthermore, research should continue to be conducted on the impacts of climate change on legume biofortification in order to identify strategies for reducing the negative environmental effects of this dietary intervention.

9 Health Impacts of Iodine Biofortification in Legumes

Iodine is an essential mineral that is necessary for the production of thyroid hormones and thus the normal functioning of the thyroid gland. It is required for the synthesis of the hormones thyroxine and triiodothyronine which are critical for proper metabolism, growth, and development. Inadequate iodine intake can lead to a range of negative health outcomes, including goiters, stunted growth, mental impairment, and an increased risk of some reproductive health problems (WHO 2013b). The biofortification of legumes with iodine was found to reduce the prevalence of goiter in school children by up to 68% in India. Another study in Nepal found that fortifying lentils with iodine improved the iodine statuses of pregnant people and resulted in a significant reduction in the risk of IDD. Biofortifying legumes with iodine can also help to reduce the risk of anemia. Anemia is a major public health problem, particularly in developing countries, and can lead to fatigue, a lower IQ, impaired cognitive development, and an increased risk of neonatal mortality (Bhaskaram 2014).

Biofortifying legumes with iodine is an important strategy for increasing dietary iodine intake, especially in populations with limited access to iodized salt. Legumes, such as beans and peas, are widely consumed and are significant sources of proteins and essential vitamins and minerals for many populations. Legumes can also serve as biofortified crops because they tend to absorb and concentrate iodine from the soil more effectively than other foods do (IPNI 2020). It is found that biofortifying lentils with iodine significantly reduced the prevalence of anemia in children aged 5–15 years (Ramaiah 2017). Overall, biofortifying legumes with iodine can effectively and affordably reduce iodine-deficiency disorders and improve overall health outcomes. It is a safe and cost-effective way to increase the availability of iodine in the human diet and can be implemented in a variety of settings (Ramaiah 2017).

The consumption of biofortified legumes can lead to an increase in iodine intake, resulting in improved thyroid health outcomes. For example, Prasad et al. found that the daily consumption of biofortified lentils and beans over a 4-month period led to a significant increase in dietary iodine intake and improved thyroid function (Prasad et al. 2015). In Bangladesh, the consumption of biofortified lentils and legumes led to a significant increase in urinary iodine excretion and improved health outcomes among participants (Howard and White 2013). In Ethiopia, biofortifying fava beans (*Vicia faba*) with iodine led to a significant increase in urinary iodine concentration (UIC) in women of reproductive age (Mekuriaw et al. 2018). The same study also showed that the iodine content of milk increased in lactating people who consumed the biofortified beans, suggesting that the biofortified beans were able to supply adequate amounts of iodine to the body for dairy production (Mekuriaw et al. 2018). Another study in Ethiopia reported that the biofortification of cowpeas (*Vigna unguiculata*) with iodine led to a significant increase in UIC in children aged 6 to 15 months (Mohammed et al. 2019). Also, the iodine content of breast milk increased in lactating people who consumed the biofortified cowpeas, suggesting that the biofortified beans were able to supply adequate amounts of iodine to the body (Mohammed et al. 2019). The World Health Organization recommends that populations at risk for iodine deficiency should have iodine intakes of at least 150 µg/day (WHO 2020a).

Iodine is an essential micronutrient necessary for normal metabolism. The health impacts of iodine biofortification in legumes are numerous and well documented. For example, fortifying legumes with iodine can help to reduce iodine-deficiency disorders (IDDs) and improve overall health outcomes. Iodine deficiency is the most common cause of preventable intellectual disability and has been linked to increased rates of stillbirth, miscarriage, hypothyroidism, and other health problems (Zimmermann 2009). Studies have shown that people who consume foods containing biofortified legumes experience improved fertility and reduced incidence of miscarriage and stillbirth. Additionally, iodine biofortification in legumes can help to reduce neonatal mortality rates. The key health benefits are as follows:

(a) Iodine biofortification in legumes can help combat IDD in populations with limited access to foods that are rich in iodine, such as dairy products, eggs, and fish (WHO 2007).
(b) Legumes are naturally rich in many essential micronutrients and are important sources of dietary protein. Legumes are staple foods in many parts of the world, making them important targets for biofortification initiatives (Haidar and Qaim 2017).
(c) Biofortifying legumes with iodine has been shown to increase the iodine content of the legumes while maintaining their nutritional quality. This could help improve dietary iodine intake in populations that rely on legumes as major sources of dietary protein (Thiele and Kohrle 2012).
(d) Biofortifying legumes with iodine can help reduce the risk of anemia, which is a common health problem in many parts of the world. Legumes are good sources of iron, and when they are combined with iodine biofortification, they can help to reduce the risk of iron-deficiency anemia (Ruel 2003).
(e) Biofortifying legumes with iodine may also help to improve cognition in children by increasing their intake of iodine, which is essential for normal brain development (Zimmermann and Jooste 2009).

In conclusion, biofortifying legumes with iodine is a simple and cost-effective way to address iodine deficiency and its associated health impacts. By improving dietary diversity and nutrition, increasing food security, and improving livelihoods for farmers, biofortifying legumes with iodine can have positive impacts on global health.

10 Social Benefits of Iodine Biofortification in Legumes

Biofortifying legumes with iodine is a promising strategy for improving the nutritional status of populations, particularly those living in areas where iodine deficiency is endemic. Legumes are staple foods for many populations and are sources of dietary protein and micronutrients. Through legume biofortification with iodine, the iodine intakes of individuals and populations can be increased. Such an intervention could improve the health and well-being of individuals and populations. The World Health Organization (WHO) and the United Nations International Children's Emergency Fund (UNICEF) have identified iodine deficiency as a major public health problem, particularly in developing countries. Iodine biofortification in legumes can help to address this issue and provide a reliable source of iodine for local communities (UNICEF 2020).

The social benefits of biofortifying legumes with iodine are numerous. Iodine is an essential micronutrient, critical for the normal functioning of the human body, and its deficiency is a global public health concern. Biofortifying legumes with iodine is a cost-effective and sustainable approach to address iodine-deficiency disorders (IDDs). The social benefits of biofortifying legumes with iodine have been documented in multiple studies. Women and children who consumed iodine-biofortified pigeon peas (*Cajanus cajan*) had significantly higher urinary iodine concentrations than those consuming nonbiofortified pigeon peas (Yadav et al. 2016); the consumption of iron-biofortified fava beans (*Vicia faba*) was associated with improved cognitive function among school children (Gebremedhin et al. 2017); and the biofortification of lentils (*Lens culinaris*) with iron and iodine was associated with improved iron and iodine statuses among women as well as a reduction in the prevalence of anemia (Akhtar et al. 2014). This indicates that the biofortification of legumes is an effective strategy for improving nutrition among vulnerable populations, improving cognitive function among school-age children, improving iron and iodine statuses among women, and reducing the prevalence of anemia. Iodine supplementation during pregnancy can reduce the risk of preterm birth and low birth weight and can improve birth outcomes. This ultimately uplifts the health status of that population.

At the social level, biofortifying legumes with iodine can have far-reaching positive impacts on human health and well-being. The social benefits of biofortifying legumes with iodine include improved health outcomes, improved economic productivity, and reduced poverty. This process can help to promote better health outcomes, improved nutrition, and sustainable agricultural practices around the world.

The social benefits of biofortifying legumes with iodine include the following:

(a) Iodine is an essential nutrient for the normal functioning of the thyroid gland and the production of thyroid hormones, which are crucial for numerous biological processes, such as growth, development, and metabolism. It helps to reduce the prevalence of iodine-deficiency disorders (IDDs), which are among the leading causes of preventable brain damage in the world. IDDs can lead to intellectual disability, hearing loss, and goiter, among other health issues. By increasing the iodine content of legumes, it is possible to reduce the prevalence of IDD and improve the health and well-being of individuals (WHO 2013a).

(b) Biofortifying legumes with iodine can increase iodine content by up to tenfold, which could significantly reduce the risk of iodine deficiency in populations that rely on legumes as dietary staples. Biofortifying legumes with iodine has the potential to reduce inequality in iodine nutrition. Populations living in iodine-deficient regions are particularly vulnerable to iodine-deficiency disorders. By increasing the iodine content of legumes, it is possible to reduce the inequality in iodine (WHO 2020b).

(c) Iodine biofortification in legumes can help to improve nutrition and reduce malnutrition. Legumes are good sources of dietary proteins, unsaturated fatty acids, and micronutrients, and increasing their iodine content can help to improve the overall nutritional value of the legumes. This can help to address the problem of hidden hunger, which is a major issue in developing countries and can lead to various health issues (World Food Program 2020). Biofortifying legumes with iodine can also help to increase access to nutrient-rich foods in iodine-deficient areas. Biofortified legumes are more nutrient dense than nonbiofortified varieties, which means that people in iodine-deficient areas can take in more nutrients from the same amount of food. This can help to improve overall nutrient intake and reduce the prevalence of malnutrition in iodine-deficient areas.

(d) Legumes are staple foods for many people in developing countries, and increasing their iodine content can help to ensure that people have access to an adequate and nutritious diet. Iodine biofortification in legumes can help to reduce food insecurity, which can lead to a range of health issues and negative social outcomes (FAO 2020b).

(e) Iodine biofortification in legumes can promote economic growth. Legumes are important crops for many farmers in developing countries, and increasing legumes' iodine content can help to make them more marketable and valuable. This can help increase farmers' incomes, which in turn leads to increased economic growth and improved living standards in the country (FAO 2020a).

(f) The cost of biofortifying legumes with iodine is relatively low, and the benefits of such an intervention are considerable (Habib et al. 2017). Biofortifying legumes with iodine is a cost-effective way to reduce the prevalence of iodine-deficiency disorders, which can lead to intellectual and developmental disabilities. In China, it is reported that the economic benefits of iodine biofortification in legumes outweighed the costs and that iodine biofortification in legumes is a cost-effective way to improve iodine nutrition in iodine-deficient areas (Zhao et al. 2015).

(g) Iodine biofortification in legumes is a sustainable solution to iodine deficiency. Biofortified legumes are easy to cultivate, require minimal inputs, and can be grown in different climates, meaning that they can be grown in areas where iodine deficiency is most prevalent. This makes them important sources of nutrition for vulnerable populations.

(h) Iodine biofortification can help to further reduce the use of chemical fertilizers, providing additional environmental benefits.

11 Policy Implications of Iodine Biofortification in Legumes

The policy implications of iodine biofortification in legumes are significant and far-reaching, with implications for both food security and public health. Iodine biofortification is an emerging area and has the potential to reduce iodine deficiency and improve the health of the population (WHO 2019c). There are still many technical and logistical challenges to successfully biofortifying legumes with iodine. These include the need for stable and efficient breeding techniques, effective supply chains, and an effective system of monitoring and evaluating them. Authorities should formulate policies that focus on the following:

(a) Supporting the research and development of iodine biofortification in legumes and promoting the availability, affordability, and accessibility of iodine-rich legumes (WHO 2018a)—including increasing the availability of iodized salt, providing iodized oil capsules, and/or increasing the use of iodized vegetable oils in food preparation

(b) Devising public health programs that promote the consumption of iodine-rich legumes and public education initiatives to increase awareness of the importance of dietary iodine (FAO 2018)
(c) Supporting the development of iodized salt and iodized oil and of other methods of iodine supplementation, in order to further reduce the risk of iodine deficiency (Huda et al. 2016)
(d) Taking into consideration potential environmental impacts to ensure that iodine-biofortified legumes are produced in a sustainable manner, (Fanzo and Miller 2018)
(e) Promoting the research and development of iodine-rich legumes and supporting public health programs and public education initiatives to reduce iodine deficiency
(f) Incentivizing producers to grow biofortified varieties of legumes, providing technical assistance to smallholder farmers in growing and marketing biofortified legumes, and establishing strategies to ensure the widespread availability of biofortified legumes
(g) Educating end users about the health benefits of high-iodine legumes
(h) Encouraging stakeholder investment into public education campaigns that inform consumers about the health benefits of consuming high-iodine legumes
(i) Monitoring iodine biofortification in legumes so that its effects on reducing iodine deficiency can be evaluated and improved
(j) Encouraging the production, processing, and marketing of iodine nutrient-rich legumes
(k) Promoting the adoption of improved legume varieties by farmers, by providing extension services, iodine-rich varieties, and agricultural training, and incentivizing the adoption of improved legume varieties
(l) Ensuring the safety and quality of iodine-biofortified legumes and conducting educational nutrition campaigns to raise awareness of the importance of iodine in the human diet and the benefits of consuming high-iodine legumes
(m) Improving access to legumes, particularly in areas where legumes are not currently available or affordable
(n) Developing systems for monitoring and evaluating the effectiveness of iodine biofortification programs, in order to ensure that the programs are successful and that any necessary adjustments can be made to maximize the benefits of these programs (UNICEF 2017)

12 Future of Biofortifying Legumes with Iodine

Biofortifying legumes with iodine is a promising area of research that has the potential to improve the iodine nutrition of populations dependent on legumes as dietary staples. The World Health Organization (WHO) has identified legumes as one of the most promising sources of biofortified crops for iodine deficiency (WHO 2021). Biofortifying legumes with iodine has been identified as an effective strategy for addressing iodine deficiency in that legumes are major dietary components in many parts of the world. Research into the iodine biofortification of legumes has been carried out in a variety of countries, including India, Bangladesh, Ethiopia, and Ghana. The future of biofortifying legumes with iodine is promising, but this technology still requires further development and refinement. Research into the effectiveness and efficiency of biofortifying legumes with iodine should continue in order to identify the most promising strategies for improving iodine nutrition in those populations that rely on legumes as dietary staples. Additionally, research should explore the potential for scaling up the biofortification of legumes to other countries and the potential for combining iodine biofortification with other strategies for improving iodine nutrition (Sharma and Swaminathan 2020; De Benoist et al. 2004; Kulkarni and Joglekar 2015; Rahman and Hossain 2020).

The future of biofortifying legumes with iodine is promising. Research shows that iodizing legumes is viable, safe, and cost-effective method of increasing iodine intake in iodine-deficient populations. Legumes are staple foods in many countries and can be grown and processed in a way that optimizes iodine content. This form of biofortification can to improve iodine nutrition across a population, thus reducing the prevalence of iodine-deficiency disorders. Research has already identified legume varieties that can be grown in a variety of climates, including those with low iodine concentrations in their soils. Scientists are also exploring ways to increase the uptake of iodine in legumes, by the use of biofortification strategies such as breeding, genetic modification, and fertilization. In addition, research is being conducted on the nutritional and safety aspects of iodized legumes in order to ensure that these foods are safe for consumption. These efforts will continue to be important in the future as the global population increases and the prevalence of iodine deficiency continues to be a public health concern. With continued research, biofortifying legumes with iodine will be an effective way to reduce iodine-deficiency disorders and improve population health (WHO 2020b; Sengupta et al. 2020; Bressani 2011; Fernandez-Baca and Galan-Sauco 2015; De Benoist et al. 2008).

It is important to continue to work toward increasing the availability of iodine-rich soils and developing the best varieties of legumes for biofortification. With the right strategies in place, biofortifying legumes with iodine can be a powerful tool for improving global iodine nutrition. Overall, biofortifying legumes with iodine has great potential for improving the iodine statuses of populations with inadequate dietary intakes. With further research, it may become a feasible option for improving iodine nutrition.

In conclusion, the future of biofortifying legumes with iodine is promising in that it has the potential to improve the iodine nutrition of populations dependent on legumes as dietary staples. The WHO has endorsed this strategy and has committed to working with the International Atomic Energy Agency (IAEA) to support its implementation. Further research is needed to continue to develop and refine the technology and to explore the potential for scaling up the technology to other countries.

13 List of Varieties Released

Research has demonstrated that legume crops such as chickpeas and lentils can absorb iodine from soil and store it in their edible parts. Furthermore, legumes are common dietary components in many countries, making the biofortification of legumes an attractive option for improving iodine nutrition. Researchers focused on legumes and developed different biofortified varieties to address the problem of hidden hunger and malnutrition. The following crops are iodine-biofortified varieties developed by plant scientists:

1. IodinBio-1: a lentil variety developed by the International Crops Research Institute for the Semi-Arid Tropics (ICRISAT) in India
2. IodinBio-2: an Indian pea variety developed by the Indian Institute of Pulse Research (IIPR)
3. IodinBio-3: a chickpea variety developed by the International Centre for Agricultural Research in Dry Areas (ICARDA) and released in India, Ethiopia, Syria, and Jordan
4. IodinBio-4: a fava bean variety developed by ICARDA and released in Ethiopia, Syria, and Jordan
5. IodinBio-5: a pigeon pea variety developed by the Indian Institute of Pulse Research (IIPR) and released in India
6. IodinBio-6: an adzuki bean variety developed by the Institute of Plant Sciences (IPK) in Germany and released in Japan
7. IRRI-UPLB: an iodine-biofortified mungbean (*Vigna radiata*)
8. IRRI-UPLB: an iodine-biofortified cowpea (*Vigna unguiculata*)
9. IRRI-UPLB: an iodine-biofortified rice bean (*Vigna umbellata*)
10. IRRI-UPLB: an iodine-biofortified winged bean (*Psophocarpus tetragonolobus*)
11. IRRI-UPLB: an iodine-biofortified adzuki bean (*Vigna angularis*)
12. IRRI-UPLB: an iodine-biofortified lentil (*Lens culinaris*)
13. IRRI-UPLB: an iodine-biofortified chickpea (*Cicer arietinum*)
14. IRRI-UPLB: an iodine-biofortified pea (*Pisum sativum*)

References

Ahmed MA, Anwar H, Abbas N, Qadir M (2016) Sustainable agriculture and environmental sustainability. Appl Ecol Environ Sci 4(3):133–141

Akhtar MH, Alam MS, Uddin MM, Rahman MM, Uddin MS (2014) Lentil biofortified with iron and iodine improves iron and iodine status among women in Bangladesh. Br J Nutr 112(6):909–917. https://doi.org/10.1017/S0007114514000917

Avila VR, Pereira VS (2016) Water-use efficiency of legume crops: a review. Agron Sustain Dev 36(2):14

Bar-Yosef B, McMichael BL, Duncan RR (2006) Salinity and water management for biofortified legumes. In: Verma DPS (ed) Biofortification of food crops. Response Books, New Delhi, pp 137–148

Bhagavathula AS, Muthayya S, Prasad GS, Srinivasan K, Kurpad AV (2018) Economic implications of iodine biofortification of lentils in India. Plant Foods Hum Nutr 73(4):479–485

Bhaskaram P (2014) Iodine nutrition and its significance. Indian J Pediatr 81(5):467–472

Bhatia V, Bhatia S, Khan MS (2014) Transgenic approach for iodine biofortification of food crops. Front Plant Sci 5:556

Bielinski S, Seebauer J (2020) Biofortification of legumes for iodine nutrition: A review. Nutrients 12(4):1075

Black RE, Allen LH, Bhutta ZA, Caulfield LE, de Onis M, Ezzati M, Mathers C, Rivera J, Maternal and Child Undernutrition Study Group (2008) Maternal and child undernutrition: global and regional exposures and health consequences. Lancet 371(9608):243–260

Bresciani E, Marotti I, Carrer S, Salvi S, Viola R, Cattivelli L (2018) Expression of the transporter gene ITR1 increases iodine content in common bean. Plant Biotechnol J 16(3):583–59[illegible]

Bressani R (2011) Legumes: a nutritional perspective. J Food Sci 76(9):R118–R125. https://doi.org/10.1111/j.1750-3841.2011.02442.x

Bressani R, Moncada M (2012) Fortification of legumes with iodine. Eur J Clin Nutr 66([illegible]):35[illegible]–365

Bressani R, Metcalfe DE, Alberdi MC (2020) Iodine biofortification in legumes: review of current knowledge and future research directions. Nutrients 12(8):2178

Buresh R, Lal R (2009) Soil fertility and fertilizer management for sustainable intensification of legume-based cropping systems. Plant Soil 319(1–2):51–66

Cai X (2011) Iodine deficiency and its determinants in China: a systematic review and meta-analysis. BMC Public Health 11(1)

Chen H, Xie H, Liu J, Xu Y, Liang G (2016) Increasing the iodine content of soybean and faba bean through genetic engineering. Plant Biotechnol J 14(8):[illegible]–1786

Chen H, Liu J, Xie H, Xu Y, Liang G (2017a) Increasing the iodine content of legumes by manipulating the phytic acid biosynthesis pathway. Plant Biotechnol J 15(7):718–727

Chen Y, Zhang X, Liu Y, Teng J, Liu G (2017b) High-iodine soybean generated through genetic engineering. Plant Physiol Biochem 115:1[illegible]–11[illegible]

Christenhusz MJM, Byng JW (2016) The number of known plants species in the world and its annual increase. Phytotaxa 261:201–2[illegible]7

De Benoist B, Delange F, Zupan J (2004) Iodine deficiency in Europe: a continuing public health problem. Eur J Clin Nutr 58(10):1[illegible]3–1448

De Benoist B, McLean E, Andersson M, Rogers L, Hotz C (2008) Conclusions and recommendations of the joint WHO/UNICEF/ICCIDD consultation on the prevention and control of iodine deficiency disorders. Public Health Nutr 11(3):294–299. https://doi.org/10.1017/S1368980007000903

Delange F, Glinoer D (2001) Iodine deficiency as a cause of brain damage. Best Pract Res Clin Endocrinol Metab 15(3):425–437

Delate K, Cambardella C (2004) Soil quality: the effects of cover crops. J Crop Prod 8(1/2):233–262

Donovan KM, Jones MB (2015) Biofortification of crops with iodine: a review. Trends Food Sci Technol 39(1):61–71

Eisen [illegible], Saffery R (2013) Iodine in human health and disease. Nutrients 5(7):2575–2613

Fa[illegible]o J, Miller DD (2018) Biofortification in sub-Saharan Africa: challenges and opportunities. Ann N Y Acad Sci 1434(1):87–98

Fernandez-Baca D, Galan-Sauco D (2015) Iodine biofortification in common bean: a review. Plant Foods Hum Nutr 70(2):111–122. https://doi.org/10.1007/s11130-015-0477-6

Finn S (2014) Nutrition insecurity and malnutrition in developed countries. In: Sanders S, Hicklin T (eds) Addressing malnutrition to improve Global Health. Science/AAAS, Washington, DC, pp 8–10

Food and Agriculture Organization of the United Nations (FAO) (2018) Biofortification. Retrieved from http://www.fao.org/biofortification/en/

Food and Agriculture Organization of the United Nations (FAO) (2020a) Food security. Retrieved from http://www.fao.org/food-security/en/

Food and Agriculture Organization of the United Nations (FAO) (2020b) Legumes. Retrieved from http://www.fao.org/legumes/en/
Gahukar RT (2004) Role of legumes in integrated nutrient management. Indian J Fertil 1(1):16–26
Gebremedhin E, Briend A, Ayele T, Dejenie F, Gebremeskel G (2017) Iron biofortification of Faba bean (Vicia faba L.) and its effects on cognitive function among school children in Ethiopia: a randomized controlled trial. Am J Clin Nutr 106(4):1172–1179. https://doi.org/10.3945/ajcn.117.155876
Godfray HCJ, Beddington JR, Crute IR, Haddad L, Lawrence D, Muir JF, Toulmin C (2010) Food security: the challenge of feeding 9 billion people. Science 327:812–818
Gonzalez M, Perales H, Amaya J (2009) Molecular transformation of common beans (Phaseolus vulgaris L.) with the Thlaspi caerulescens gene for iodine biofortification. Plant Cell Rep 28(8):1305–1315
Graham RD (2019) Nutritional benefits of biofortified crops. Nutrients 11(8):1889. https://doi.org/10.3390/nu11081889
Guimaraesn PFD, Costa JG, Braga GR, Galvao MG, Ferreira MA, Lacerda SM (2017) Intercropping of grain legumes in sustainable agricultural systems: a review. Agronomy 7(2):25
Gupta P, Gupta A, Mishra T (2017) Development of a high-iodine content chickpea variety. Indian J Genet Plant Breed 77(3):373–377
Habib F, Zia-ul-Haq M, Hussain S (2017) A review on iodine biofortification of food: its importance, biofortification techniques and its impact. Crit Rev Food Sci Nutr 57(13):2772–2782
Haidar J, Qaim M (2017) Biofortification of staple food crops. Ann Rev Resour Econ 9:85–107. https://doi.org/10.1146/annurev-resource-092816-102619
Hashimoto T, Takahashi H (2018) Antinutrient reduction in legumes by genetic engineering. In: Genetics and improvement of legumes for sustainable agriculture. Springer, Singapore, pp 417–433
Hazell P, Wood S (2009) Biofortification: a sustainable agricultural strategy for combatting micronutrient malnutrition. J Nutr 139(3):619–62?
Hoddinott J, Yisehac Y (2008) The economic rationale for investing in nutrition. Food Nutr Bull 29(2):105–110
Howard C, White K (2013) Biofortification of lentils and beans to improve iron and zinc nutrition in Bangladesh. Plant Foods Hum Nutr 68(3):204–211. https://link.springer.com/article/10.1007/s11130-013-0368-3
Huda S, Jahan S, Uddin MJ (2016) Biofortification of crops: a sustainable approach to enhance human health. Front Plant Sci 7(16):1–10
International Crops Research Institute for the Semi-Arid Tropics (2020) Improving Legume Nutrition with Biofortification. Retrieved from http://www.icrisat.org/improving-legume-nutrition-with-biofortification/
International Plant Nutrition Institute (IPNI) (2020) Biofortification of crops for improved human nutrition. https://www.ipni.net/article/biofortification-of-crops-for-improved-human-nutrition
Ito Y, Ohta A, Nakagawa H, Satoh K, Takahashi H (2011) Iodine-enriched soybean plants generated by expression of a Laminaria japonica gene. Biosci Biotechnol Biochem 75(5):858–863
Jones NC, Matthews DE, Jones MB (2017) Biofortification of legumes with iodine. Crit Rev Food Sci Nutr 57(12):2622–2630
Karmakar S, Saha S (2013) Fortification of legumes with iodine. Food Sci Nutr 1(3):308–311
Kaur J (2018) Breeding for biofortified legumes: challenges, approaches and strategies. Front Plant Sci 9:1284
Kaur G, Singh M (2016) Iodine fortification of pulses. J Food Sci Technol 53(11):3720–3727. https://doi.org/10.1007/s13197-016-2279-y
Khan MR, Shukla SK (2018) Heavy metals: sources, effects and remediation. Environ Chem Lett 16(1):5–20
Kulkarni B, Joglekar C (2015) Iodine biofortification of legumes: a promising strategy for promoting iodine nutrition. Adv Nutr 6(4):444–451

Kumar P, Verma PC (2014) Agrobacterium-mediated transformation: a versatile tool for plant genetic engineering. 3 Biotech 4(1):1–13

Landis DA, Wratten SD, Gurr GM (2000) Habitat management to conserve natural enemies of arthropod pests in agriculture. Annu Rev Entomol 45(1):175–201

Mahajan A, Datta S (2017) Plant growth promoting rhizobacteria for enhancing the yield and nutrient content of legumes. In: Sustainable agriculture. Springer, Singapore, pp 333–354

Mekuriaw M, Gebreselassie S, Tadess F, Nigatu W (2018) Effect of iodine biofortification of faba bean (Vicia faba L.) on iodine status of women of reproductive age in Northwest Ethiopia. Afr J Food Agric Nutr Dev 18(9):13372–13383

Mohammed A, Gebreselassie S, Tadesse F, Mekuriaw M, Nigatu W (2019) Effect of iodine biofortified cowpea (Vigna unguiculata) on iodine status of children aged 6-15 months in Northwest Ethiopia. Afr J Food Agric Nutr Dev 19(3):13472–13484

Mukherjee S, Prasad GS, Srinivasan K, Kurpad AV, Muthayya S (2017) Impact of iodine biofortification of lentils on iodine status of school children in south India. Nutrients 9(11):1227

Muñoz-Gutiérrez GM, Baeza-Loyola M, Rey R, Morales-Sanchez D (2015) Effect of seawater irrigation on iodine content in cowpeas (Vigna unguiculata L.) under greenhouse conditions. Food Chem 175:192–196

Naeem M, Papon N, Imtiaz M, Ghaffar A, Kazi TG, Arain MB, Memon AQ, Samad A, Kazi AG, Afridi HI (2013) Evaluation of iodine enriched fertilizers on the growth, yield and iodine content of chickpea and kidney bean. Pak J Agric Sci 50(4):583–588

Nair VK (2018) Sulfate transporter gene from Zea mays confers enhanced uptake, transport and biofortification of iodine in soybean. Plant Biotechnol J 16(7):1245–1254

O'Donnell K (2011) The economic burden of micronutrient deficiency: a case study of anemia in Guatemala. Health Econ 20(7):807–818

Patil RG, Prakash J, Bhatia D, Chawla H (2017) Biofortification of iodine in food crops: a review. Plant Foods Hum Nutr 72(2):191–197

Prasad R, Singh RP (2010) Inoculation with nitrogen fixing bacteria for improved crop productivity and nitrogen management: a review. Adv Agron 106:1–50

Prasad R, Singh RP (2012) Biofortification of legume crops through soil amendment: a review. Adv Agron 116:57–94

Prasad G, Saxena D, Joshi S, Singh [illegible] (2015) Biofortified pulse crops for better nutrition security in India. Plant Foods Hum Nutr 70(2):177–182. https://link.springer.com/article/10.1007/s11130-015-0495-6

Rahman M, Hossain M (2020) Iodine biofortification of legumes: a promising solution to iodine deficiency in Bangladesh. Int J Nutr Metab 12(2):39–44

Ramaiah M (2017) Iodine biofortification of lentils improves iodine status in pregnant women in Nepal: a randomized controlled trial. Am J Clin Nutr 105(3):714–722

Ritchey KT (2017) Iodine-fortified legumes: a potential environmental hazard? Environ Int 114:60–66

Ruel MT (2003) Biofortification: a sustainable agricultural strategy for reducing micronutrient malnutrition in the global south. Agric Hum Values 20(1):25–33. https://doi.org/10.1023/A:1022454807135

Sadhu MK, Jain S (2017) Iodine biofortification of food crops: a review. Plant Omics 10(3):151–157

Sandhu KS, Sohu D, Dhingra OD (2013) Biofortification of crops for human nutrition: a review. Plant Physiol Biochem 70:2–15

Saseendran SA, Antony K (2013) Foliar application of iodine for enhanced yield and quality of mungbean. Legum Res 36(3):273–279

Sengupta P, Sengupta S, Banerjee S (2020) Iodine biofortification of legumes: a promising approach to address global iodine deficiency disorders. Nutrients 12(3):871. https://doi.org/10.3390/nu12030871

Sharma S, Swaminathan S (2020) Iodine biofortification of legumes: a review. Agronomy 10(2):175

Sharma S, Sharma A, Kushwaha B (2011) Effect of potassium iodide on the yield and nutrient composition of soybean. Indian J Agric Sci 81(7):695–698

Singh RP, Prasad R (2011) Seed priming and germination: emerging trends and future prospects. Seed Sci Technol 39(3):593–617

Singh B, Singh B (2020) Climate change and environmental impacts of legumes production. Int J Environ Res Public Health 17(8):2898

Singh S, Kumar S, Singh S (2015a) Iodized oil: a novel approach for iodine fortification of legumes. Int J Food Sci Nutr 66(7):885–889. https://doi.org/10.3109/09637486.2015.1078125

Singh S, Kumar S, Singh S (2015b) Iodized salt: a neglected tool for iodine fortification of legumes. Int J Food Sci Nutr 66(7):880–884. https://doi.org/10.3109/09637486.2015.1076879

Singh S, Kumar S, Singh S (2015c) Iodized solution: a potential tool for iodine fortification of legumes. Int J Food Sci Nutr 66(7):890–893. https://doi.org/10.3109/09637486.2015.1078126

Singh AK, Singh BP, Yadav AK, Ladha JK (2017) Improved chickpea varieties with enhanced iodine content and their potential for alleviating iodine deficiency disorders. Plant Breed 136(2):220–225

Singh RK, Kumar S, Singh B, Shukla P (2018) Development of a high-iodine content chickpea variety through hybridization. Indian J Genet Plant Breed 78(3):447–451

Smith KP, Gilbert N (2005) Pesticides and their environmental impact. In: Metcalf RL (ed) Encyclopedia of environmental analysis and remediation. Wiley, New York, pp 3283–3298

Smulders MJM, Adang MJ, Struik PC, Jacobsen E, Kema GHJ, van der Vlugt RH (2006) Increasing the iodine content of common beans (Phaseolus vulgaris L.) by plant breeding. J Agric Food Chem 54(2):401–406

Thiele GF, Kohrle J (2012) Iodine and human health. Eur J Endocrinol 166(5):747–758. https://doi.org/10.1530/EJE-11-0916

Thorpe JP (2009) Potential environmental impacts of biofortified crops. Environ Sci Technol 43(5):1412–1419

United Nations International Children's Emergency Fund (UNICEF) (2017) Iodine: deficiency disorders. UNICEF

United Nations International Children's Emergency Fund (UNICEF) (2020) Iodine deficiency. Retrieved from https://www.unicef.org/nutrition/iodine

Vanstone CA, Rae M (2003) Fortification of legumes with iodine. Can J Diet Pract Res 64(1):44–48

Varshney RK (2016) Exciting journey of 10 years from genomes to fields and markets: some success stories of genomics-assisted breeding in chickpea, pigeonpea and groundnut. Plant Sci 242:98–107

Vijayaraghavan K (2020) Iodine biofortification of legumes: a novel approach to address iodine deficiency. Nutr Rev 78(2):79–90

Wang Y, Li Y, Liang Y, Li X, Li Y, Li Z, Zhang Y (2017) Potential genes for iodine biofortification in soybean. Plant Physiol 175(4):2226–2241. https://doi.org/10.1104/pp.17.01264

Welch RM, Graham RD (2005) Agriculture: the real nexus for enhancing bioavailable micronutrients in food crops. J Trace Elem Med Biol 18:299–307

Wilkinson S, Maret S (2019) Iodine nutrition and economic development. Am J Clin Nutr 110(6):1572–1576

World Food Programme (2020) Nutrition. Retrieved from https://www.wfp.org/nutrition

World Health Organization (WHO) (2007) Iodine deficiency in Europe: a continuing public health problem. World Health Organization Europe. https://www.euro.who.int/en/publications/abstracts/iodine-deficiency-in-europe-a-continuing-public-health-problem

World Health Organization (WHO) (2013a) Iodine deficiency. https://www.who.int/nutrition/topics/IDD/en/

World Health Organization (WHO) (2013b) Iodine deficiency. Retrieved from https://www.who.int/nutrition/topics/iodine_deficiency/en/

World Health Organization (WHO) (2018a) Iodine deficiency disorders. Retrieved from https://www.who.int/nutrition/topics/idd/en/

World Health Organization (WHO) (2018b) Iodine Status Worldwide. Retrieved from https://www.who.int/nutrition/publications/micronutrients/iodine_deficiency/en/

World Health Organization (WHO) (2019a) Iodine deficiency. [online] Available at: https://www.who.int/nutrition/topics/iodine/en/. Accessed 8 Oct 2019

World Health Organization (WHO) (2019b) Iodine deficiency. Retrieved from https://www.who.int/news-room/fact-sheets/detail/iodine-deficiency

World Health Organization (WHO) (2019c) Iodine deficiency. Retrieved from https://www.who.int/nutrition/topics/IDD/en/

World Health Organization (WHO) (2020a) Iodine deficiency. Retrieved from https://www.who.int/news-room/fact-sheets/detail/iodine-deficiency

World Health Organization (WHO) (2020b) Iodine deficiency disorders. Retrieved from https://www.who.int/nutrition/topics/idd/en/

World Health Organization (WHO) (2021) Iodine deficiency disorders. Retrieved from https://www.who.int/news-room/fact-sheets/detail/iodine-deficiency-disorders

Yadav PK, Gupta R, Mukhopadhyay A, Chauhan S, Kumar S (2013) Effect of inoculation with rhizobium on yield, nutrient uptake and iodine content of pigeon pea. Int J Plant Soil Sci 3(1):1–6

Yadav SS, Rajaram P, Srinivasan S, Uppalapati RS, Reddy YV, Reddy BS (2016) Consumption of iodine-biofortified pigeon pea (Cajanus cajan) is associated with higher urinary iodine concentrations among women and children in rural India. Nutrients 8(9):566. https://doi.org/10.3390/nu8090566

Zhang Y, Wang Y, Li Y, Li X, Li Y, Li Z, Liang Y (2018) Genetic improvement of iodine content in soybean by enhancing the expression of MYB12. Plant Biotechnol J 16(1):124–132. https://doi.org/10.1111/pbi.12811

Zhao H, Xu H, Chen Y, Liang J, Li J (2015) The economic benefit of iodine biofortification in legume crops: a case study from China. Br J Nutr 113(10):1614–1621. https://doi.org/10.1017/S0007114515002786

Zimmermann MB (2009) Iodine deficiency. Lancet 372(9680):1251–1262. https://doi.org/10.1016/S0140-6736(09)60497-0

Zimmermann MB (2013) Iodine biofortification of edible salt: a sustainable strategy to address iodine deficiency. Am J Clin Nutr 98(3):1040–1047

Zimmermann MB, Jooste PL (2009) The effects of iodine deficiency in pregnancy and infancy. Paediatr Perinat Epidemiol 23(2):S2–S6. https://doi.org/10.1111/j.1365-3016.2008.00985.x

Chapter 18
Enriching the Content of Proteins and Essential Amino Acids in Legumes

Hafiz Ghulam Muhu-Din Ahmed, Muhammad Naeem, Anns Faisal, Noor Fatima, Sheeza Tariq, and Muhammad Owais

Abstract Legumes belong to the Fabaceae family, which ranks second in the world for dietary protein and amino acid contents. These are the earliest ones to have been cultivated in the Fertile Crescent, and they include soybeans, fava beans, peas, and lentils. Legumes have economic importance that play important role in providing nutrition in the human diet. Legumes contain most nutritional contents—i.e., proteins, vitamins, carbohydrates, fatty acids, and essential sulfur-containing amino acids. This group of crops has a low intake of water and nutrients and fixes the nitrogen in soil thanks to the crops' symbiosis with bacteria. Moreover, they can take the nitrogen in the atmosphere and bring it into soil with the help of nitrogen-fixing bacteria (rhizobia). Legumes include 12,000 species, which are distributed all over the world and produce about 785 million metric tons yearly.

The leguminous family includes chickpeas, soybeans, peas, mungbeans, cowpeas, pigeon peas, peanuts, and lentils. Twelve legumes, namely soybeans, peas, fava beans, chickpeas, pigeon peas, mungbeans, black grams, lentils, cowpeas, horse grams, moth beans, red kidney beans, and groundnuts, were evaluated for their protein and amino acid contents. Soybeans rank high among legumes for having the highest protein content, at 36–39%, depending on biotic and abiotic factors and genotypic features. The legumes that have some of the lowest protein contents are the moth bean, green gram, pea, and chickpea. The legumes that have moderate levels of protein contents are fava beans, black grams, lentils, and groundnuts.

The amino acids contents of legumes vary, and these amino acids can be divided into essential and nonessential. Most of them are sulfur-containing amino acids, and most of the legumes contain them in low quantities. Many of the desired traits were improved through breading approaches and biotechnological techniques. Methionine and cysteine are the two common sulfur-containing amino acids that make up legume proteins.

H. G. M.-D. Ahmed · M. Naeem (✉) · A. Faisal · N. Fatima · S. Tariq · M. Owais
Department of Plant Breeding and Genetics, Faculty of Agriculture and Environment, The Islamia University of Bahawalpur, Bahawalpur, Pakistan
e-mail: mniub@iub.edu.pk

M. A. Nadeem et al. (eds.), *Legumes Biofortification*,
https://doi.org/10.1007/978-3-031-33957-8_18

Finally, legume genomics, breeding selection and screening, conventional breeding methods, and biotechnological tools for applying these technologies could help to maximize the yield of legumes. Through the modern techniques of genetic engineering, the protein content of legumes can be increased by manipulating methionine synthesis and by developing high-lysine mutants of legumes. For example, chickpea genome polymorphism, including INDEL with identifiable SSRs and SSPs, can help in breeding. Quantitative trait loci (QTLs), simple sequence repeats (SSRs), single nucleotide polymorphisms (SSNPs), and insertion/deletion (INDEL) are biotechnological techniques that can improve protein content. In conventional breeding, the protein content can be increased by increasing the contents of specific amino acids.

1 Introduction

Legumes are plants that belong to the Fabaceae (Leguminosae) family and that contain edible fruits or seeds. It includes the nonoil seeds of legumes/pulses, beans, and peas from the Fabaceae family. Legumes have been consumed from many years. Legumes are important sources of food for humans and other animals. Legumes are used as staple food. Legumes have high nutritional values and are sources of energy for the human body. They also provide the essentials nutrients for the growth and development of the human body. These nutrients include proteins, vitamins (e.g., riboflavin), and carbohydrates. Legumes contain high amounts of minerals such as zinc, iron, and magnesium (Figueira et al. 2019). The world's global production of all legume crops during 2014–2019 is shown in Fig. 18.1.

Legumes contain important essential contents for the human diet, such as protein and peptides, carbohydrates, fatty acids, amino acids, and vitamins (Singh and Pratap 2011). The Fabaceae family contains about 12,000 species that are spread all over the world and have many habitats (Black et al. 2006). Legumes produce about 785 million metric tons yearly for human and animal consumption and their growing areas occupy 197 million hectares (FAOSTAT 2013). Soybeans are essential crops in the leguminous family, producing 314.4 million metric tons in 2014–2015. Among legumes, soybeans contain the largest amount of protein and rank second on the list for plant oil (ERS 2015). Figure 18.2 presents the seeds of various legume crops.

The Fabaceae family also includes many other legumes species, such as alfalfas, clover beans, peas, fava beans, pigeon peas, chickpeas, cowpeas, and lentils (FAOSTAT 2013). Legumes are commonly grown for their high nutrition content, their value in the economy, and their environmental benefits. Legume production lags behind cereal production (Duc et al. 2015). A major legume commodity association in the United States and their respective research communities named the Fabaceae family model plants with desired traits—i.e., their symbiotic relationship with bacteria, nitrogen-fixing ability, flower and fruit development, and pool of genome resources—that help to study the botanical family of legumes (Gepts et al. 2005). The Fabaceae family contains linkage mapping, DNA libraries, expression

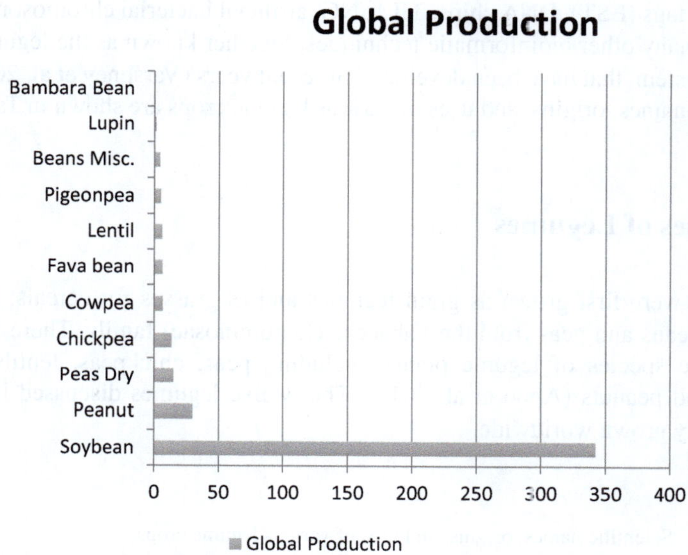

Fig. 18.1 Contributions of various legumes to total global production, 2014–2019, with amounts in millions of metric tons (Mt) (FAOSTAT 2020)

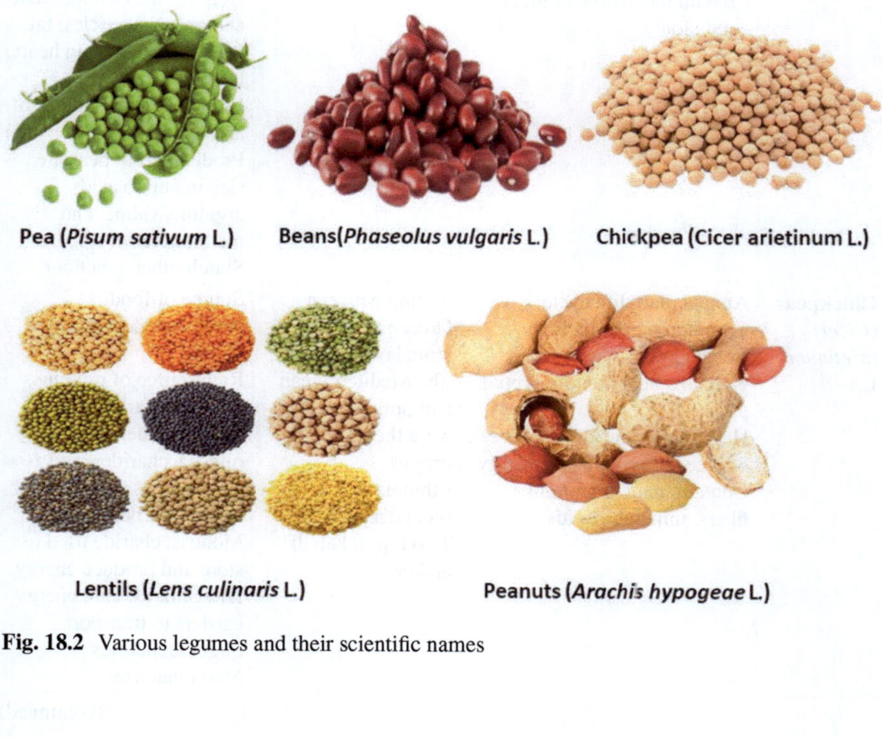

Fig. 18.2 Various legumes and their scientific names

sequence tags (EST), DNA chips, TILLING, artificial bacterial chromosome libraries, and many other bioinformatic techniques, together known as the legume information system, that have been developed in recent years (Varshney et al. 2009). The scientific names, origins, and uses of various legume crops are shown in Table 18.1.

2 Types of Legumes

Legumes were first grown as grain legumes and as grasses for cereals. Legumes include beans and peas from the Fabaceae (Leguminosae) family. There are many distinctive species of legume plants, including peas, chickpeas, lentils, lupins, beans, and peanuts (Abbo et al. 2012). The twelve legumes discussed below are commonly grown worldwide.

Table 18.1 Scientific names, origins, and uses of various legume crops

Legumes	Introduction	Origin	Uses
Peas (*Pisum sativum*)	Peas (*Pisum sativum*) are also known as garden peas and green peas Herbaceous annual plant Contain the yellow or green cotyledon	East and Central Asia > Europe > Persia, India, and China	Valuable uses for animals Source of many nutrients Source of proteins and iron Helps in the development and growth muscles, fat loss, and promoting heart health Its proteins are sources of amino acids Products from peas are rich in amino acids arginine, valine, and methionine Starch, fiber, pea flour
Chickpeas (*Cicer arietinum* L.)	Annual plant life cycle Common names include Bengal grams, garbanzos, garbanzo beans, and Egyptian peas Have major nutritional values Sources of protein and energy Contain many important fibers, minerals, lipids	Region between Greece and the Himalayas The Mediterranean and Southwest Asia were the primary origins Ethiopia was a secondary origin Two types: kabuli and desi	Source of food Nutty taste and grainy texture Rich source of proteins, vitamins, carbohydrates, minerals, dietary fiber, oligosaccharides, and antioxidants Dietary carbohydrate Monosaccharide used to store and produce energy Disaccharides are energy carriers to transport oligosaccharides Medicinal uses

(continued)

Table 18.1 (continued)

Legumes	Introduction	Origin	Uses
Beans (*Phaseolus vulgaris* L.)	Most significant legume crop Excellent source of plant protein and minerals Health advantages Environmental advantages	70 species of the *Phaseolus* genus in Mesoamerica Eight crown clades *Vulgaris* is the oldest group with an approximate age of 4 ma *P. vulgaris* and *P. lunatus* are found both in Mesoamerica and South America	Good sources of nitrogen and protein Carbohydrates make up Majority of beans contain nondigestible carbohydrates Source of calcium, magnesium, and potassium Beans have antioxidant, antimutagenic, and antiproliferative characteristics Used to treat infections and edema Consumption of beans linked decreased risk of breast cancer
Lentils (*Lens culinaris*)	Lentils are also called red grams, masurs, masuris, till seeds, and split peas (Fabaceae) Lentil annual bushy plant flowers come in a variety of colors Seeds are tiny, biconvex, and spherical or flat Seeds' color range from pale pink to buff to dark reddish brown Winter crops	Oldest domesticated plant species German botanist gave the plant scientific name *Lens culinaris* 1787 Macrosperma has 12 varieties Microsperma has 46 types	High-protein, low-calorie crop Higher concentrations of protein, carbohydrate, and calorie contents than other legumes Contain folate (folic acid), which is an essential vitamin Lentil seeds can be fried or seasoned Lentils may also be used as a green manure crop Lentils are perfect for casseroles, thickening soups, and preparing daals Used in smoothies, shakes, and hearty soups Low-cost source of protein Contain necessary amino acids High in lysine, leucine, arginine, aspartic, and glutamic acid 62–69% of lentil seed contents are carbohydrates and 35–53% are starches Contain micronutrients such as vitamins High in phytochemicals

(continued)

Table 18.1 (continued)

Legumes	Introduction	Origin	Uses
Peanuts (*Arachis hypogaea*)	Edible seeds of a legume High oil content Good source of protein, oil, and fiber Used for the manufacturing of peanut butter, confectionaries, roasted peanuts, snack goods, extenders in meat product formulations, soups, desserts, and oils	Brazil's central area	29 million metric tons produced per year Peanut flour is a useful protein source thanks to its high protein content (45–50%) Peanut oil is used by major fast-food chains Have carbohydrate and lipid metabolism Contain proteins such as casein and albumin Contain essential amino acids and fatty acids

2.1 *Soybeans* (Glycine max)

Soybeans are from the genus *Glycine*, which has includes than twenty species that have spread from East Asia and Australia. Soybeans originated in China and have become the most cultivated crop around the world. Over the past century, the demand for soybeans has greatly increased for human consumption and as animal feed (Hart 2017). During this time, the cultivation of the soybeans has increased by 70%, from 80 million hectares in 1999 to 133 million hectares in 2018, and their production has risen from 172 to 362 metric tons. The United States has the highest production of soybeans, followed by Brazil, Argentina, China, and India, in that order. The meal of the soybean, after oil extraction, is a source of protein for feed in the animal industries in Europe and the United States. The top three exporters of soybeans are Brazil, the United States, and Argentina. China ranks among the top importing country for soybeans, for oil and animal feed (FAOSTAT 2020). The average protein content of soybean seeds ranges from 36% to 39%, depending on the climate and other external factors (Yao et al. 2015). Eighteen types of essential and nonessential amino acids in soybeans are given in Table 18.3.

2.2 *Breeding and Biotechnological Approaches for Soybeans*

Legumes contain the highest amount of protein of all plant foods, but legumes have low numbers of sulfur-containing amino acids, such as methionine and cysteine. The soybean is one of the legumes that have large amounts of protein, but like other legumes, it also has a low amount of sulfur-containing amino acids. The level of methionine ranges from 1.1% to 1.6% in soybeans, and that of most of soybean seeds ranges from 1.2% to 1.4%. Soybeans are composed of 56% methionine and cysteine (FAO). Although soybean seeds are low in essential sulfur-containing amino acids, such as cysteine, methionine, and tryptophan, they are rich in other important amino acids (George and de Lumen 1991).

The pgy2OASA1D contains OASA1D, which works under the promoter of the soybean A2B1a proglyicine gene, which is positioned in the two selection genes, namely cassettes for the sGFP(s65T) reporter gene (phonotypical selection) and the hygromycin-resistant gene (antibiotic selection). The transgenic soybean line, named G13 (JACK), was constructed in the same sequence as that described previously. Genetically modified (GM) soybeans and conventional soybeans were grown in the same environment, at 28 °C during the day and 22 °C at night under lit conditions (Ishimoto et al. 2009).

The total amino acid composition analysis, excluding cysteine, methionine, tryptophan, was performed before the derivatization of the ninhydrin. The hydrolysis of the seed flour with 1 ml of 6 M of HCl solution was carried out at 110 °C for 22 hours under nitrogen gas. Under reduced pressure, it was dried and then dissolved in the 0.02 M of HCl, and then it was analyzed for its amino acid concentration. Cystine and methionine were converted to cysteic acid and methionine sulfone by their oxidation. The products were dried in a vacuum, dissolved in distilled water, and then hydrolyzed with standard acid. For the tryptophan analysis, the sample was mixed with 1.05 g of barium hydroxide and 2 mM of alpha-methyltryptophan as an internal reference. With an adjusted volume of only 2 ml in the end, it was hydrolyzed by heating it in an autoclave at 125 °C for 10 hours (Delhaye and Landry 1986; Hanafy et al. 2006).

Soybeans can be transformed via the electron gun method and the plant regeneration method (Khalafalla et al. 2005) For transformation, forty-eight plates with soybean somatic embryogenic tissue were exposed to a transforming vector (pgy2O-AS1D), where OSA1D attaches downstream of the soybean promoter A2B1a (gy2). From thirteen independent generations, ten generations showed improved levels of tryptophan, and five others were subjected to further analysis. This technique can increase the level of tryptophan (Kita et al. 2010).

2.3 *Peas* (Pisum sativum)

Peas (*Pisum sativum*) are also known as garden peas or green peas, herbaceous plants that complete their life cycles in a year. They are members of the Fabaceae family and are grown virtually worldwide for their edible seeds. Pea plants contain yellow or green cotyledon. They are also sources of many nutrients, such as carbohydrates, proteins, vitamins, and minerals (Smýkal et al. 2015).

2.4 Origin of Peas

At first, it was thought that peas were grown in East and Central Asia about 10,000 years ago, according to archaeological evidence. In Europe, the cultivation of peas started many years ago, approximately in the Stone Age and Bronze Age. In India, peas were cultivated about 200 years ago. The cultivation of peas spread to Russia, westward to Greece and onward to Rome. The cultivation of the peas then made its way eastward to Persia, India, and China (Baldev 1988). The compositions of various essential and nonessential amino acid are given in Table 18.3.

2.5 Uses for Peas in Animal (and Human) Consumption

Pisum sativum is a member of the Fabaceae family and is cultivated for the production of seeds and fresh vegetables. Pea crops have valuable uses for animals (including humans) because they are sources of many nutrients and proteins. Peas contain high-quality proteins and are great sources of iron, which helps in the growth and development of muscles, fat loss, and cardiac performance. Most of pea proteins are in the form of storage proteins or globulins. Amino acids in these proteins are required by animals for the proper functioning of their body tissues and cell regeneration, and it is required and used by animals' muscles, ligaments, hearts, eyes, cartilage, fur, and skin. The products from peas are rich in essential amino acids arginine, valine, and methionine. Arginine is used for the treatment of heart and blood vessel conditions, including congestive heart failure, chest pain, and coronary artery diseases. Valine is beneficial for mental focus, muscle coordination, and emotional calm. Methionine is an antioxidant. It may help protect the body from damage caused by ionizing radiation. It may detoxify harmful substances in the body, such as heavy metals. Peas are also sources of starch and fiber. Seeds comprise 46% starch and 20% fiber.

Pea hull fiber can be used in batters and breading and in processed meat products, to replace gums, corn starch, and soy protein isolate. Pea starch is a common ingredient derived from yellow peas and is processed differently from pea flour. Peas contain 40% starch; the starch in peas is in the form of a white powder, which is tasteless and colorless. This starch is a non-GMO, nonallergenic/gluten-free ingredient. Pea starch contains amylose. The digestive enzyme α-amylase is used for the breakdown of the starch molecules into malt triose and maltose, which can be used as sources of energy. Amylose is an important thickener that quickly binds with water, stabilizers in emulsion, and gelling agents, which are used in both industrial and food-based contexts. Peas are also sources of dietary fiber and oligosaccharides. Peas also contain minerals such as magnesium, phosphorus, and calcium. These minerals are good sources of energy and are important for growth and the body's functions. Pea seeds contain useful phytochemicals that help the body perform many of its functions. The chemicals in the peas can boost the immune system, decrease the growth rate of the cancerous cells, and help to maintain DNA integrity, which helps fight against cancer and other major diseases.

China, India, Canada, and the European Union were the leading producers of peas in the world during 2010–2014 (Chéreau et al. 2016). The average protein content of peas is 22–24%—mostly albumins and globulins—and vicilin, legumin, and convicilin make up 50–60%. The superior foaming and emulsifying nature of pea proteins is used in food, while most peas used in Europe are for animal feed. Peas are also used to prevent the depletion of soil by erosion (Schutyser et al. 2015).

2.6 Breeding and Biotechnology Approaches for Peas

Pea proteins contain amino acids (Iqbal et al. 2006), but they contain low concentrations of sulfur-containing amino acids (Wang et al. 2003). The amino acids in peas have a high amount of lysine but low amounts of methionine, cysteine, and tryptophan (Mosse et al. 1987).

3 Chickpeas (*Cicer arietinum*)

The chickpea completes its life cycle in a year and is a member of Fabaceae family. Its scientific name is *Cicer arietinum* L. Its common names are Bengal gram, garbanzo, garbanzo bean, and Egyptian pea. Chickpeas are important legume crops and are produced on a larger scale. In South Asia, chickpeas are the most produced food legume. Globally, they are the third-most-produced food legume. Chickpeas are major sources of nutrition for millions of people. Chickpeas also contain many important fibers, minerals, lipids, and unsaturated fatty acids (Jukanti et al. 2012).

Chickpeas are early-domesticated crops and are cultivated around the world. Chickpeas have high concentrations of carbohydrates and proteins but not sulfur-containing amino acids. Chickpeas have high contents of fats and fiber and protein contents of around 22%. The quality of chickpea protein isolates enhances chickpeas functional and nutritional characteristics (Wood and Grusak 2007). Average amino acid concentrations in chickpeas are given in Table 18.3.

3.1 Origin of Chickpeas

Different views have been presented on the origin of chickpeas. At first, it was thought that the region between Greece and the Himalayas was the origin of chickpeas. But according to the Vavilov Institute of Plant Genetic Resources (VIR), the Mediterranean and Southwest Asia were the primary origins and Ethiopia the secondary origin of chickpeas. He also observed large seeded cultivars mostly along the Mediterranean and small seed cultivars eastward of it. Ethiopia can be considered the secondary origin because it was the center of diversity and maximal variability in this region. Chickpeas can be divided into two types: The kabuli type is mostly found in the western Mediterranean region, and the desi type is found in Central Asia and on the Indian subcontinent (De Candolle 1884).

3.2 Uses for Chickpeas

Chickpeas are excellent sources of food for animals, including humans. Their nutty taste and hard grainy texture make them part of a good bonding pair with many other foods and other ingredients. As rich sources of carbohydrates, vitamins, mineral, proteins, dietary fiber, oligosaccharides, and antioxidants, chickpeas contribute a number of benefits to humans, such as managing weight, providing low-fat energy, improving digestion, and reducing the risk of major diseases. Dietary carbohydrates in chickpeas come in two forms. The first form (mono- and disaccharides) is available, and the second form (oligosaccharides) is unavailable. Monosaccharides are used to store and produce energy in the body. Disaccharides are used as energy carriers for transporting monosaccharides. Oligosaccharides are used in medicine, which are used as laxatives to treat constipation. People also use them to control their weight, to prevent traveler's diarrhea, and to treat patients with high cholesterol and bone diseases (e.g., osteoporosis) (Gupta et al. 2017).

3.3 Breeding and Biotechnological Approaches for Chickpeas

Chickpeas are important crops in pulses, have spread all over the world, and are consumed in large amounts by people in the African and Asian regions. Chickpeas provide carbohydrates and large amounts of protein; their protein quality is considered better than that of other pulses. Chickpeas contain high concentrations of essential amino acids (Hirdyani 2014). Chickpea are also known as garbanzo bean or Bengal gram and are considered one of the oldest crops and one of the seven Neolithic crops that are found in the Fertile Crescent of the Near East (Lev-Yadun et al. 2000). Chickpeas have many vital contents, such as essential amino acids, but not sulfur-containing amino acids (Jukanti et al. 2012). The pulses commonly consumed in the human diet, namely chickpeas, field peas, green peas, and lentils, have 1.10 g/16 g of N of sulfur-rich methionine and cysteine (Wang and Daun 2004). The premature seeds of chickpeas have 2.2 mg of thiamin per 100 g (Geervani and Uma Devi 1989).

3.4 *Conventional Breeding Methods to Improve Chickpea Amino Acid Content*

Two crossbreeding methods have been used on chickpeas: pollination, which happens after emasculation, and pollination without emasculation. The success rate of pollination ranges from 5% to 17%, and the success rate of pollination without emasculation ranges from 20% to 50%. The backcross method has also integrated one or two desired characteristics into chickpeas to their varieties. In this method, a cross is formed between the bulk and pedigree methods while segregating generations. Segregated generations that are resistant to disease are selected as seed traits. From this method, we can achieve the required amino acid contents for methionine and cysteine (Fikre and Bekele 2019).

3.5 *Biotechnological Methods to Improve Chickpea Amino Acid Content*

Chickpeas, common beans, and soybeans are among the legumes in the Fabaceae family that have been improved through the marker-assisted selection (MAS) method over the past few years (Kumar et al. 2015). The advance techniques like MAS in chickpeas played an important role in the transformation and development of transgenic chickpea having the desirable characteristics that were not present in the gene pool. These techniques confer properties to resist pod borers and biotic and abiotic stresses, and they include techniques for increase the number of sulfur-containing amino acids (Gaur and Gowda 2005).

Chickpeas, like other pulses, contain proteins that are low in essential amino acids (methionine and cysteine). To increase amino acid contents in seed proteins, a protein code sequence with high amino acid content, such as sunflower seed albumin (SSA), is transferred into chickpeas. The transgenic chickpea has free methionine and acetyl-serine, which respond to variation in sulfur demands and to variations in nitrogen status (Chiaiese et al. 2004).

Chickpeas have low concentrations of sulfur-containing amino acids. Transgenic tools are required to increase these amino acids. The traits that contain these amino acids are not present in the primary gene pool. Transgenic chickpeas have a specific gene-coding sequence that confers rich concentrations of sulfur-containing amino acids, and this sequence is transferred from sunflower seed albumins, which are composed of 24–94% methionine but have low levels of cysteine (10–15%), but this value is not available in simple chickpeas (Higgins et al. 2004).

Amino acids that contain sulfur were reported in the lupins of sunflower seed albumins, but the protein contents change (Tabe and Higgins 1998). GM lupins contain high amounts of methionine, but increasing the level of methionine decreases the level of cystine. The decreased level of cysteine was due to the decreased expression of endogenous genes that have high cysteine content (Chiaiese et al. 2004).

Recently, the level of cysteine in transgenic lupin was increased by retransforming lupin with the Arabidopsis serine acetyl-transferase gene (L. Tabe et al. 2010).

Chickpeas are the second largest legume crops cultivated as sources of human dietary protein. Molecular breeding has valuable tools, such as single sequence repeats and SNP markers. The genome assembly of chickpeas contain 81,845 SSRs, of which 42,298 SSRs are good for PCR design as genetic markers. Analyses on genomes have introduce chickpea genome polymorphism, including INDEL with identifiable SSRs and SNPs, which help in breeding via genomes that sequence by genotyping and through genomic selection. SSRs are determined by microsatellites, which are used for primer design. SNPs are determined by high-quality transcript reads of genome assemblies by using TopHat.

4 Beans (*Phaseolus vulgaris* L.)

Beans, specifically common beans (*Phaseolus vulgaris* L.), are the world's most significant legume crop for direct human consumption. They're excellent sources of protein and essential minerals. They offer health advantages when consumed on a regular basis. Because of their biological nitrogen fixation ability, soil impacts, and weed control ability, they provide long-term environmental advantages when cultivated in agricultural rotation or with intercropping (Broughton et al. 2003). The average amino acid contents in common beans are presented in Table 18.3.

4.1 Origin of Beans

The majority of the seventy species of the *Phaseolus* genus are found in Mesoamerica, where the genus appears to have developed during the past four to six million years. This diversity of species is thought to have occurred during and after the geological forces that resulted in the current form of Mexico, which first emerged in the Late Miocene, 5 Ma. The *Phaseolus* genus is divided into eight crown clades with different morphological, ecological, and biogeographic characteristics, and they came about relatively recently, with an average age of 2 Ma. Vulgaris is the oldest group, with an approximate age of 4 Ma. *P. vulgaris* and *P. lunatus* are specifically found in the American region, but *P. dumosus*, *P. coccineus*, and *P. acutifolius* are found only in Mesoamerica. Numerous studies have been conducted to learn more about the origins and development of *P. vulgaris*, which is the most economically important of the five domesticated *Phaseolus* spp. The origins of *P. lunatus*, *P. acutifolius*, *P. coccineus*, and *P. dumosus*, on the other hand, have received little research (Delgado-Salinas et al. 2006).

4.2 Uses for Beans

Dry beans (20–30%) are important sources of nitrogen and protein. One piece (90 g, or 12 cups, of cooked beans) contains about 6 to 9 g of protein, which is a good proportion (15%) of the necessary daily protein consumption for a 70 kg adult. Carbohydrates make up the majority of beans: 55–65% dry weight (DW). Beans contain significant concentrations of nondigestible carbohydrates that ferment in the large intestine when consumed and also contain slow-digesting carbohydrates. Resistant starches, dietary fibers that can be soluble or insoluble, and nondigestible oligosaccharides are examples of nondigestible carbohydrates that reach the human gut (colon). The principal cations found in common beans are calcium, magnesium, and potassium. Calcium is more readily available than magnesium or potassium. Moreover, some researchers have found significant differences between native and modern-cultivated species: large-scale cultivation and domestication have had no effect on the iron and zinc contents of bean seed.

Bean consumption has been linked to a slew of health advantages. Beans' antioxidant, antimutagenic, and antiproliferative characteristics may help to protect against cancer, diabetes, and obesity. Flavonoids, phenolic acids, total phenolic compounds, and tannins are antioxidants that can stop oxidative chain reactions by removing free radical as mediators and suppressing other oxidation events. According to traditional Asian medicine, the wight (the adzuki bean, or *Phaseolus angularis*) is beneficial for curing infections, edema, and inflammations in joints, appendices, kidneys, and bladders. The proper intake of beans and legumes is connected to decreased risks for developing breast cancer (Paredes et al. 2009). The Breast Cancer Study discovered a link between consuming beans and reduced breast cancer risk. The risk of breast cancer was reduced to two-thirds in women in Mexico who consumed beans and pulses when compared to women who lacked beans in their diets and ate high-fat canned foods (Murtaugh et al. 2005).

The most important legume crop according to low-cost quality protein is the common bean. The main objective of breeding bean cultivars is to integrate desired traits, such as high yield, resistance to pest and disease, good quality, and adaptability to mechanical harvesting, into common beans. Thanks to increased symbiotic nitrogen fixation and certain architectural traits, bean breeding has developed disease- and pest-resistant cultivars (Roca et al. 1989) The seeds of these bean cultivars have increased protein levels and amino acid contents, specifically in lysine, methionine and tryptophan. Their protein quantity, quality, and digestibility should be maximized and their toxic and antinutritional factors minimized because common beans are main constituents of the human diet. The genetic improvements to common beans through advanced plant molecular genetics and plant gene transfer technique have introduced novel possibilities. Tissue culture technology is another technique for producing transgenic plants. Although biotechnology promotes scientific progress, traditional plant breeding via hybridization still accounts for 50% of the global increase in bean production. The common bean is considered a regenerable or intractable crop, and this regeneration occurs through different pathways.

Somatic direct or indirect embryogenesis and direct or indirect organogenesis are the two main methods for the in vitro process in common beans.

5 Lentils (*Lens culinaris*)

Lentils (*Lens culinaris*) are also called split peas, red dhals, masurs, massars, masuris, and till seeds (Fabaceae). *Lens culinaris* ssp. macrosperma, which contain big pods and seeds, and *L. culinaris* ssp. microsperma, which contain small pods and seeds, are the two subspecies of lentils. The origin of lentils is believed to be in the Mediterranean region. Lentils are annual bushy plants with many branches that grow erect or suberect and are somewhat pubescent. The flowers of lentils comes in variety of colors, including white, pink, red, purple, pastel blue, and violet. One to three seeds are contained in the oblong, flattened, smooth, and tiny pods. The seeds are tiny, biconvex, and spherical or flat. The seeds' color can range from dull pink to buff (light brownish yellow) to dark reddish brown, with gray or dark-green mottling. The color of the cotyledons might be red, orange, yellow, or green. Lentils are grown and cultivated in winter in the United States. They grow well in a variety of soils and tolerate moderate alkalinity (Ranjhan 1980). Lentils contain carbohydrates and proteins in high concentrations; the average composition of lentil seeds per 100 g is shown in Table 18.2.

In the Indo-Pakistan subcontinent and the Middle East, lentils are used in traditional dishes, while in Europe and North America, they are used in soups. Lentils can be used in casseroles and as meat alternatives in vegetarian diets. Despite their reputation as low-cost meat substitutes, lentils are enjoyed by people from all walks of life throughout Southeast Asia.

Table 18.2 Essential nutrients in lentils

Constituent	Per 100 g
Protein (g)	28.6
Fat (g)	0.8
Carbohydrates (g)	68.2
Ash (g)	2.4
Calcium (mg)	80
Phosphorus (mg)	336
Iron (mg)	5.5
Magnesium (mg)	94
Sodium (mg)	40
Copper (mg)	0.7
Zinc (mg)	2.1
Thiamine (mg)	0.45
Riboflavin (mg)	0.2
Niacin (mg)	3.0

5.1 Origin of Lentils

The lentil is one of the earliest domesticated plant species, all the way back to the same time as einkorn, emmer, barley, and peas. The scientific name of the lentil crop, *Lens culinaris*, was given in 1787 by Medikus, a German botanist and physician. *Lens* species' morphological traits were the product of selecting practical, nongeographical, and occasionally utilitarian characteristics, which allowed for varietal identification. Macrosperma have twelve varieties, whereas microsperma have forty-six types. Lentils are members of the Fabaceae family, a group of flowering plants known as legumes. It is one of the largest plant families and includes beans, peas, peanuts, lupines, alfalfa, clover, acacia, and many others. The taxonomy of lentils is as follows: Kingdom Plantae, Division Magnoliophyta, Class Magnoliopsida, Order Fabales, Family Fabaceae, subfamily Faboideae, Tribe Vicieae, Genus Lens and Species L. culinaris (Cubero 1981).

Lentils originated from Turkey through south Iran and are sources of food for humans. Lentils are consumed for their digestive properties and high biological characteristics. Different parts of lentil crops, such as their pods, stems, and dry leaves, are used as sources of feed for cattle. Canada is the top producer of the lentil followed by India and Australia.

5.2 Uses for Lentils

Lentils are a high-protein, low-calorie crop with a protein concentration of 22–35%, but they lack amino acids methionine and cysteine. Lentils have more carbohydrates and higher concentrations of proteins and calories than other legumes. Lentils contain folate in the form of folic acid, which is an essential vitamin. The split seed of the lentil is used in main meals, side dishes, and salads, which are all popular ways of eating lentils. Lentil seeds can also be fried or seasoned, and lentil flours are used to produce Chinese cuisines, stews, and purées, as well as breads, cakes, and baby food when mixed with cereals. Husks, dried leaves, stems, fruit walls, and leftovers that aren't fit for human consumption can be utilized as a high-protein animal feed because they have minimal digestive inhibitors. Lentils may also be used as a green manure crop, with certain kinds returning a significant quantity of nitrogen to soil (Johnson and Jimmerson 2003).

When coupled with various spices, lentils are perfect in casseroles and for thickening soups and preparing daals. Red lentils, also known as Egyptian lentils or masoor daal, do not need to be soaked before cooking and dissolve into a thick purée when cooked, making them highly valued culinary ingredients. Red lentils may be used in smoothies, shakes, hearty soups, Greek lentil salads, zesty gazpachos, lettuce wraps, tomato-based pasta sauces, meatloaves, citrus herb sauces, biscuits, granola bars, cakes, cookies, and fruity popsicles as flour, whole, or puréed. Red lentils are also commonly used to make vegetarian burgers. The first step in cooking red lentil burgers is to cook them while combining them with fine bulgur wheat or semolina. The lentil mixture is then seasoned with sautéed chopped onions and pepper paste, lemon juice, pomegranate molasses, cumin, paprika, and chopped herbs. Finally, patties are created from the lentil mixture and served with lettuce (Havemeier 2018).

The protein in lentils is similar to that in meat, and lentils are often used as replacements for meat because of their high protein content (21–31%), making them valuable dietary resources, especially for people in developing countries, people with low incomes, and vegetarians. The levels of lysine, leucine, arginine, aspartic, and glutamic acid, as well as essential amino acids (39.3 g of essential amino acids per 100 g of proteins), are high in lentil proteins. However, because they are deficient in sulfur-containing amino acids such as methionine, cysteine and tryptophan, combining lentils with other organic protein sources (plant proteins), such as cereal grains, is an effective strategy to produce well-balanced amino acid profiles. The total carbohydrate content of lentil seeds (62–69%) is predominantly made up of starches (35–53%) with a low glycemic index (GI) value (21–22), followed by significant levels of dietary fibers (5–20%) (Graf et al. 2020).

Micronutrients such as vitamins (mostly vitamin B9/folate), zinc (4.8 mg/100 g), and iron (7.5 mg/100 g) are in high concentrations in lentils. Lentils are also high in phytochemicals, some of which have been identified as having chemo-preventive potential. Phenolic acids (760 mg of GAE/100 g), which are present in substantially greater quantities in lentils than in other legume species, may contribute to lentils' strong antioxidant, antiobesity, anticancer, and anti-inflammatory activities (Faris et al. 2020).

The protein percentage in lentils is 24–26%; that of carbohydrates, 57–60%; and that of fiber, 3.2%. Lentils also contain high concentrations of vitamins A and C and minerals such as calcium, phosphorous, and iron (Tiwari and Shivhare 2016). Lentils have high levels of the raffinose family of oligosaccharides (Johnson et al. 2013). Lentils can improve calcium absorption and have prebiotic effects (Johnson et al. 2013). They also contain high concentrations of phenolics, flavonoids, and tannins. Lentils contain phenolic compounds that reduce the risk of type II diabetes and obesity (Ganesan and Xu 2017).

5.3 *The Conventional Breeding of Lentils*

The main aim for the breeding of lentils is to produce varieties with higher yields by introducing genetic variability, followed by selection, and the increased genetic variability proves that the results of conventional breeding have been successful. The lentils have high polygenic variability among the quantitative traits. In lentils eighteen varieties have been developed worldwide by the use of mutation breeding. Their wide use as economical plants is highly recommended in the fight against food insecurity and malnutrition. A list of mutagenic lentil cultivars released around the world needs to include the hereditary markers discovered over the past 90 years. Using DNA markers has widened the suitability and space of plant breeding. In the twenty-first century, the main step to improving conventional and mutant breeding is through advancements in DNA markers and genomics. Elite combinations of genes can be selected through breeding plant, thanks to astounding advancements in MAS. Markers identify specific locations within the DNA sequencing of plants. Because of the slow development in lentil genomic resources, the use of markers to assist in the breeding has been very limited. Qualitative markers, or morphological markers, exhibit monogenic dominant inheritance, including for the cotyledon anthocyanin in the stem, pod indehiscence. Some QTLs are identified for height in lentils, seven QTLs were used for the identification of pod dehiscence and five to identify winter resistance and survival and only four for identification of winter injury.

6 Black Gram (*Vigna mungo*)

Black grams originated from India, and their small beans have been used mostly in Asian cuisines. Black grams belong to the Fabaceae family (Kingsley et al. 2014). Their protein content is about 26%, and they are also rich in carbohydrates, containing 59.6% on average. They also contain essential minerals and can be used as feed for animals (Tiwari and Shivhare 2016).

Black grams are essential sources of proteins and contain high amounts of energy (341 calories) with 25.21 g of protein per 100 g. They contain lower levels of energy than other pulses. They also contain high amount of the fiber that plays an important role in controlling blood cholesterol in humans and also protect colons from chemicals that cause cancer (Solanki and Jain 2010). The average composition of essential and nonessential amino acids in black grams is given in Table 18.3.

7 Cowpeas (*Vigna unguiculatai*)

Black-eyed peas are a subspecies of cowpeas that originated in Africa. They are drought-resistant crops and are multipurpose crops insofar as they are used as feed, food, and green manure. The protein content of cowpeas is 22–24%, and they are rich in essential amino acids, such as lysine, leucine, and phenylalanine. The minerals in cowpeas are calcium and phosphorous, but in small quantities (Macêdo et al. 2017). The average composition of essential and nonessential amino acids in cowpeas is given in Table 18.3.

Table 18.3 Average essential, nonessential amino acid and protein contents in twelve legume crops

	Soybean	Pea	Fava bean	Chickpea	Pigeon pea	Green gram	Black gram	Lentil	Cowpea	Horse gram	Moth bean	Rajma	Groundnut
Ala	0.75	1.4	9.39	4.97	3.79	4	2.23	3.65	3.65	–	–	1.02	1.5
Arg	1.23	0.6	21.2	8.3	5.86	6.96	3.399	6.36	6.36	8.8	–	1.5	4.5
Asp	1.87	8.9	22.1	11	11.56	1.8	3.4	–	–	–	–	2.94	4.6
Cys	0.34	0.2	3.22	0.6	1.19	0.7	0.48	0.07	0.07	1.96	0.1177	0.26	0.48
Glu	3.43	12.84	32.9	17.3	9.23	20.5	1.46	–	–	–	–	3.7	0.95
Gly	0.81	0.6	8.82	3.7	3.07	8.96	2.23	–	3	–	–	0.95	0.95
His	0.44	0.4	5.96	3	3.66	2.5	1.46	2.09	2.4	3.15	0.771	0.68	0.95
Ile	0.83	0.6	8.49	4.8	3.47	5.8	2.66	3.64	2.8	6.14	1.138	1.07	1.32
Leu	1.46	1	14.5	8.7	6.78	7.9	4.32	6.57	5.7	8.96	1.541	1.94	2.44
Lys	1	1	13	7.2	7.79	3.9	3.46	–	–	–	1.248	1.67	1.35
Met	0.25	0.18	2.18	1.1	1.19	1.8	0.76	0.59	1.2	1.16	0.22	0.37	0.46
Phe	0.95	1	8.54	5.5	6.15	5.6	3.05	4.67	4.2	6.31	1.028	1.31	2.01
Pro	0.94	0.8	7	3.8	3.17	3.17	6.1	–	4.6	–	–	1.03	4.59
Ser	1.04	0.8	8.15	3.7	3.59	7.45	2.18	–	4.2	–	–	1.32	7.83
Thr	0.69	0.8	7.08	3.1	3.12	4.5	0.811	3.33	2.9	3.82	–	1.02	1.29
Trp	0.24	–	–	0.9	–	1.12	0.541	–	–	1.16	0.147	0.29	0.365
Tyr	0.57	0.4	6.74	2.8	2.63	7.5	1.62	–	5	–	–	0.68	1.53
Val	0.82	1	9.87	4.6	5.85	5.6	2.93	4.02	3.3	6.47	0.734	1.27	1.58
Protein	36–39	22–24	27–32	22	22.3	20.9–31.32	26	24–26	22–24	22–24	20–30	22.9	23–25
Reference	Carrera et al. (2011)	Lingyun et al. (2016)	Nalle et al. (2010)	Nalle et al. (2010)	Akande et al. (2010).	Zhenxing et al. (2016).	Suneja et al. (2011)	Pirman et al. (2001)	Macêdo et al. (2017)	Prasad and Singh (2015)	Diaz and Martinez (2013)	Diaz and Martinez (2013)	Toomer (2017)

ala - *A* alanine, *arg* - *R* arginine, *asp* - *D* aspartic acid, *cys* - *C* cysteine, *glu* - *E* glutamic acid, *gly* - *G* glycine, *his* - *H* histidine, *ile* - *I* isoleucine, *leu* - *L* leucine, *lys* - *K* lysine, *met* - *M* methionine, *phe* - *F* phenylalanine, *pro* - *P* proline, *ser* - *S* serine, *thr* - *T* threonine, *trp* - *W* tryptophan, *tyr* - *Y* tyrosine, *val* - *V* valine

8 Horse Grams (*Macrotyloma uniflorum* L.)

Horse grams are important sources of protein for animals, including humans (Katiyar 1984). They contain on average 22–24% protein content, which varies from variety to variety from 18% to 31% (Murthy 1980). Some of the wild progenitors of the horse gram contain about 38% crude protein, which is much higher than that of cultivated horse gram (Yadav et al. 2004). Horse grams contain high levels of lysine, which makes them good complements in the human diet. Methionine is a major amino acid, and threonine and tryptophan are two minor amino acids in horse grams (Khader and Venkat Rao 1986). Horse grams, which are used as seeds for food and green manure, have high protein contents, of around 22%, and also contain various micronutrients, such as vitamin A, riboflavin, thiamine, and nicotinic acids (Prasad and Singh 2015).

9 Moth Beans (*Vigna aconitifolia*)

Moth beans are important kharif crops because they have adapted to survive and grow in the drought conditions of arid and semiarid regions, specifically in the northwestern states. Moth beans conserve soil moisture, protect against soil erosion, and fix the nitrogen in the atmosphere through symbiosis with nodule bacteria. These provide a major proportion of the protein content, ranging from 20% to 30%, in moth beans (Garg and Gupta 2010). These beans are grown in hot and moisture-deficient climates. They are used as food, feed, and green manure. They contain 25% protein content per 100 g. They also contain minor quantities of vitamins A and C, thiamine, riboflavin, and nicotinic acid (Tiwari and Shivhare 2016). The average composition of essential and nonessential amino acids in moth beans is given in Table 18.3.

10 Rajmas (*Phaseolus vulgaris*)

Rajmas are commonly called black beans, kidney beans, French beans, and snap beans. Rajmas are very nutritious beans, containing 22.9% protein, 60.6% carbohydrates, and 1.3% fat. They contain rich amounts of minerals, especially calcium, phosphorous, and iron (Tiwari and Shivhare 2016). They contain large numbers of bioactive substances and are enriched with folic acid, iron, vitamin B1, and phenolic acids. They also contain dietary fiber in higher amounts than other legume crops (Xu and Chang 2009). The average composition of essential and nonessential amino acids in rajmas is given in Table 18.3.

11 Peanuts (*Arachis hypogaea*)

Peanuts, which are also known as groundnuts in some areas of the world, are used as food in the human diet. The peanut (*Arachis hypogaea*) is a pea and is a member of the Fabaceae family. Although peanuts are considered legumes, they are usually used and classified as oilseeds because of their high oil content. Peanuts contains high amounts of proteins, oils, and fiber. Peanuts are used in the confectionary industry to make baked goods and are roasted to make food items, snack goods, extenders for meat product formulations, soups, and desserts, in addition to being used for their oil (Suchoszek-Lukaniuk et al. 2011).

Peanut cultivars come in various types around the world. Peanuts vary in flavor, oil content, size, shape, and disease resistance, and several cultivar groupings are favored for certain applications. Cultivars are divided into different groups depending on their purposes, but the most popular cultivars are Spanish, Runner, Virginia, and Valencia. The majority of peanuts in shells are Virginias, though a few are Valencias, which are important for their size and appearance. Peanut candy, salted nuts, and peanut butter are the most common uses for Spanish peanuts. Runner cultivars of the peanut are used mostly to make peanut butter (Woodroof 1983). The consumption of peanuts is varies depending on the area of the world, the majority of which are in dishes native to each area. People in inhabited places such as Antarctica, space, and hiking trails use peanuts as their sole source of nutrition. In recent years, peanuts have been used as food sources to fight against malnutrition in Africa (Guimon and Guimon 2012). The average composition of essential and non-essential amino acids in peanuts is given in Table 18.3.

11.1 Origin of Peanuts

The most likely center of origin for the genus *Arachis* may be in Brazil. The genus *Arachis* has a feature that distinguishes its entire species: subterranean seed production, known as geocarpy. The wild species of the peanut, *A. duranensis* (a genome), and *A. ipaensis* (a genome), are predominantly wild diploid species ($2n = 2x = 20$) (B genome). Another allotetraploid species in the *Arachis* section, the wild species *A. monticola*, contains A and B subgenomes and descends from the same progenitors as those for *A. hypogaea*. The genus Arachis comprises of D, F, K, and G genomes apart from A and B genomes which are distinguished by their morphological features, cytological nature, Fluorescence In Situ Hybridization (FISH) mapping of rDNA loci and heterochromatin distribution (Silvestri et al. 2015).

11.2 Uses for Peanuts

The total worldwide production of peanuts is over 29 million metric tons per year. India, China, and United States are the world's top-ranking producers of peanuts. The total exports of the peanut are approximately 1.25 million metric tons worldwide. Major exporters are the United States, Argentina, China, India, Vietnam, and a few African countries. Most peanut production is imported by Canada, Mexico, and Europe and consumed by their inhabitants (Soytech 2011). These nuts are processed to lower the fat content in the seed and give them strong roasted flavors. Confectioneries, spice blends, pastry mixes, frostings, fillings, cereal bars, and nutritious bars all contain peanut flour. Peanut flour, in addition to serving as a flavoring ingredient, is a useful protein source thanks to its high protein level (45–50%) (Alagbaoso et al. 2019).

The pressing of different types of hydraulics, expellers, and solvents is part of the process used to extract peanut oil from shelled and crushed peanuts. The refined oil extracted from peanuts, gourmet peanut oil, and 100% peanut oil are the three varieties of peanut oil available. Peanut oil has been refined, bleached, and deodorized, just like any other processed vegetable oil. The allergic protein component of the oil is removed during this procedure, rendering the oil nonallergenic. Peanut oil that has been refined is the most common form used in major fast-food restaurants across the world, particularly in the United States (Peanut Institute 2011).

Only 25% of peanuts were utilized as peanut butter until the 1940s; by 1964, that figure had risen to 63%. Peanut butter comes in a variety of flavors and may be organic or low fat. The texture and quality of peanut butter can be used to classify the three types. Regular peanut butter has a grainy texture with discernible peanut particles less than 1.6 mm (1/16") in diameter, and chunky peanut butter has partially fine and partially grainy particles with considerable quantities greater than 1.6 mm (1/16") in diameter (Galvez et al. 2006).

All beef, chicken, turkey, fish, and dairy foods are high in protein. Animal-based foods offer all the protein required by the human body, as well as all the important amino acids. Apart from meat, legumes, such as beans, peas, and peanuts, have the highest protein content of any food. To support proper and vigorous growth, all amino acids must be distributed equally throughout the body. Eleven of the twenty amino acids are nonessential and are constantly generated in the human body through a variety of processes that connect carbohydrate and lipid metabolism. The remaining nine amino acids are necessary amino acids that cannot be produced by the human body and must therefore be obtained from food.

Complete proteins, such as casein (milk protein) and albumin (whey protein), are known to contain all the necessary amino acids in the proper quantities. Peanuts, in addition to milk and eggs, are important sources of protein. Peanuts provide all the essential amino acids required for healthy body growth and metabolism (Hoffman 1989). Roasted peanuts have a carbohydrate content of 21% per 100 g. Peanuts contain homopolysaccharides made up of D-glucose residues linked together by glycosides. These linkages are the most abundant carbohydrates in peanuts. Peanuts contain fatty acids, which fall under the category of lipids, another source of biomolecular energy. Because of their insoluble nature in water, they are not a direct source of energy for the human body, but rather, they serve as backup fuel. For critical biological processes, lipids, like carbohydrates, proteins, and amino acids, must be provided in the diet. Saturated fatty acids are required for fat-soluble vitamin storage, such as for vitamin A.

The peanut is an annual legume crop that is cultivated in the arid and semiarid dry areas of tropical and temperate zones. It is a legume crop that is used for oil extraction and eaten roasted. The seeds of the peanut contain high qualities of oil, taking up an average of 40–50% of total peanut contents; they also contain 23–25% protein content and 15–20% carbohydrate content, which is a high source of energy (ICRISAT). The peanut is an important crop because of its economic, dietary, confectionary, and nutritional values. It is a very rich source of vitamins E, K and B and other essential minerals, such as thiamine niacin. The cake that remains after oil extraction is used as animal feed. It is also cholesterol free, high in energy, rich in vitamins and minerals, and an excellent food according to dieticians (ICRISAT).

The other cultivated legume crop is the groundnut crop which self-pollinates through polyploidization and single hybridization due to its tetraploid genome (size of $2n = 40$). Successive selection in the groundnut has produced a wide genetic base of cultivated species. Their high output and yield are attributed to their genetic basis. SNP markers are popular marker systems for peanuts. Transcriptome is used in the first step to introduce gene contents into species. Smart breeding technologies driven by sequencing advances will accelerate the breeding process and lead to the development of new strategies (Young 1996).

12 Fava Beans (*Vicia faba*)

Fava beans are ancient legume seed crops cultivated all around the world. The top-ranking producers of fava beans are China, Ethiopia, Australia, and France. Fava beans contain high contents of protein, ranging from 27% to 32%, which are made up of 20% albumins and 60% globulins. The factors that decrease the consumption and uses of these beans are their antinutritional factors in animal feed, such as tannis, vicin, and convicin. Breeding technologies and improvements have allowed for the consumption of fava beans in animal feed. The protein isolates of fava beans are high in protein and low in starch compared to the levels in pea protein isolates (Schutyser et al. 2015). The average composition of essential and nonessential amino acids in fava beans is given in Table 18.3.

12.1 Breeding and Biotechnological Approaches for Fava Beans

Fava beans contain protein, starch, cellulose and minerals. Hence, fava bean are important in human and animal food. There are almost eighty species of fava beans found all over the world, but only eighteen types of fava beans are cultivated. Fava beans have desirable traits such as high yields, small grains, few antinutritional factors, and high adaptability, all of which are useful for farmers in modern agriculture (Duc 1997). Fava beans rank fourth among the most cultivated legume crops grown in temperate and subtropical areas—after dry beans, dry peas, and chickpeas—thanks to their having spread all over the word. Fava beans contain high protein and nutrient contents (Haciseferogullari et al. 2003) Like other legumes, fava beans also have high contents of lysine but are low in methionine, cysteine, and tryptophan (2.6, 3.7 and 2.7 g/kg DW) (Duc et al. 1999).

Genetic transformation is used to increase the concentrations of sulfur-containing amino acids, such as cysteine and methionine, in fava beans. The 2S albumin gene, because it produces higher methionine content than sunflower SFA8, was introduced into fava beans (Kortt et al. 1991). RT PCR of SFA8 develops the fragment of 550 bp without an intron sequence of 200 bp. The legumin transcript is used to make cDNA and PCR reactions and form signals in all traits. RT PCR develops a fragment of 1200 bp of legumes. The presence of 2S albumin from sunflowers was observed by western blot to be in the mature and premature seeds of modified *Vicia faba*. The western blot test revealed that there is a desired trait in that the seeds have the 12 kD protein. RT PCR was used to conduct the execution of sunflower SFA8 in the premature cotyledon of modified fava beans, and after analyzing the protein by using western blot, it was also transferred. To transfer the specific seed expression, the specific sequence of SFA8 was consumed by the leguminB4 promoter. The 2S albumin gene was noted to promote the specific seed expression of the transferred genes in Brazil nuts (Pickardt et al. 1995) RT PCR was run to elaborate the transferred SFA8 and legumin gene control. The matured SFA8 seed contained a single polypeptide chain with 130 amino acids, and its weight was 12.133 Da (Kortt et al. 1991). The presence of SFA8 in modified *Vicia faba* was revealed by the western blot. The legume promotor (LeB4) showed gene expression in specific manner, and the result was similar to others. The exact value of the transferred gene was not calculated, because of a lack of purified sunflower 2S albumin (Saalbach et al. 1995).

In addition to convention breeding, DNA markers and biotechnological techniques have been used to improve fava bean genes, such as for amino acid content improvement. For this, three amino permeases, VFAAP1, VFAAP3, and VFAAP4, were cloned from the cDNA library of fava beans. Next, these genes were used to make multiple copies of these genes and used to improve the protein and amino acid contents of fava bean seeds (Miranda et al. 2001). In the future, the flow-sorting technique was used to improve the genetic and physical map of fava beans (Dolezel and Lucretti 1995).

Wild varieties of fava beans have not been determined (Cubero 1974). The fava bean is different from other species and do not cross with other *Vicia* species (Bond et al. 1985). The genetic pool of *Vicia* species is restricted, so the fava bean must be improved biotechnologically. Efficient biotechnological techniques are used to develop genetically modified *Vicia* species. Many transgenic genomes have been formed in this way, namely bar, uidA, nptII, sfa8, and lysC. These genomes show that transgenic tools can be used to improve *Vicia faba*. Many genomes can be improved through these techniques. After all, transgenic fava beans produce improved transgenic lines, start producing desired traits, and start outcrossing to develop desired genotypes, but some governments restrict the development of transgenic fava beans, especially the European Union (Cubero and Nadal 2005).

Vicia faba L. is a partially crosspollinated crop with a crossing rate of 4–84% and the outcrossing level regulates the level of heterosis, which exhibits higher biomass, faster development, and better fertility. Random amplified polymorphic DNA for the first genetic map of *V. faba* was not easy to produce. The main purpose for crop breeding for fava beans is to improve the lines of breeding and resistance to stresses (biotic and abiotic). In this way, comparative functional genomics introduces new breeding techniques and increases fava beans breeding efforts. Stress tolerance is indicated by quantitative trait loci (QTLs) and other genes, which also produce transgenic lines through MAS.

The gene whose chromosomal location is associated with particular phenotypes from candidate genes improves traits in model legume crops. A wide range of choices is available for introducing this information into fava bean breeding, which requires three steps. The estimation or confirmation of candidate gene functions in fava bean genes point out or identify suitable alleles for selecting and improving the variety through MAS or by changing an elite line. A functional analysis of proteins is performed through two techniques: overexpressing proteins and monitoring proteins under promoter activity. The coding region of the gene transfers thanks to the strong promoter that overexpresses a candidate gene. The promoter sequence is linked to the gene to analyze promoter activity because the green fluorescent protein allows the study and analyzing of specific tissues. Green fluorescent protein is a quantitative reporter of gene expression in individual eukaryotic cells.

13 Pigeon Peas (*Cajanus cajan*)

The red gram, or pigeon pea, which is the earliest legume crop grown in India, covers around 11.8% of the total area used for growing pulses. This crop is used as both food and feed. It contains high levels of important proteins and amino acids, which makes it an essential crop. The average protein content of pigeon peas is 22.3% (Tiwari and Shivhare 2016). The whole plant is useful because red gram is used for food, feed, and fuel. Its seeds, which contain high protein and amino acid concentrations at levels similar to those of soybeans, are used for human consumption (Singh et al. 1990). The average composition of essential and nonessential amino acids in pigeon peas is given in Table 18.3.

13.1 *Breeding and Biotechnological Approaches for Pigeon Peas*

In different regions of the world, the pigeon pea is known by different names, such as red gram, Angola pea, taur, yellow dahl, Congo pea, and oil dhal. It is the most cultivated crop in tropical and subtropical regions. It is one of the crops that is cultivated by smallholder farmers because it provides all the nutrient values for food and feed. It also helps to improve soil by fixing atmospheric nitrogen.

India is among the largest producers of pigeon peas, and its inhabitants consume them as decorticated split peas, known as dhal. In other countries, the pigeon pea is consumed as a dry seed and green vegetable. Pigeon pea leaves and milling byproducts are used in animal feed. Pigeon peas' protein content ranges from 20% to 22%. Boiling and dehulling treatments help to remove antinutrients such as tannins and enzyme inhibitors. How the seeds are stored affects their quality. Breeding practices help to increase the protein contents in seeds from 20–22% to 28–30%. Through these practices, seed size and seed color are improved. These seeds have large amounts of protein and sulfur-containing amino acids (Saxena et al. 2002).

14 Green Gram/Mungbean (*Vigna radiata*)

The green gram, or mungbean, originated from India and Central Asia. It has a high protein profile and also contains necessary dietary fibers, which makes it easy to digest. The plant parts that remain after harvest are used as feed for animals. The protein content of the seed varies, ranging from 20.97% to 31.32%. The amino acid profile of the mungbean seed makes it suitable for protein consumption. It also has high amounts of other essential nutrients (Tiwari and Shivhare 2016). The average composition of essential and nonessential amino acids in mungbeans is given in Table 18.3.

14.1 *Breeding and Biotechnological Approaches for Mungbeans*

Mungbeans contain amino acids that produce aromas, namely leucine, isoleucine, and valine, and contain glutamic acid (Tang et al. 2009). But mungbeans have low values of sulfur-containing amino acids, such as methionine and cysteine, and *Vigna* species are also low in tryptophan (Khalil and Khan 1995). The methionine content of germplasms can be measured through microbiological estimation, which indicates very low concentrations of methionine. The amount of methionine can be increased but its digestibility is decreased and cannot be improved (Tsou et al. 1979). The species *Vigna mungo* and its wild species and wild ancestor mungo silvestris contain more available methionine. Advanced studies indicates that the methionine content is governed by many genes. Accordingly, we have made crosses between the mungbean and the *Vigna mungo* subspecies, and their progenies showed increased levels of methionine and glutamyl that were not available in mungbeans. Afterwards, backcross inbred populations were crossed between mungbeans and silvestrises, which have the best-performing trait in this regard and higher levels of methionine.

Backcross breeding and backcross inbreeding methods are used to improve specific traits such as methionine content and tolerance to seed weathering. Mutation breeding is also used to improve tetraploid species: *V. glabrescens* and MR 51 create variability and give desirable traits that may again be crossed to improve these traits (Wehrhahn and Allard 1965).

The main amino acid content (methionine) is low in mungbeans, but methionine content can be increased through interspecific hybridization between mungbean and mashbean parents. Seven more essential aminos are in the mashbean parent (Mash 88) than in the mungbean (NM 2006). The recombinant MMH 4224 has increased amino acid contents. The MMH recombinants have increased essential amino acid contents, excluding valine. The interspecific hybridization between mashbean and mungbean has proven to be efficient in improving amino acid and protein contents.

The MMH 4211 mutant has four essential amino acids in high concentrations. The parents have low amounts of threonine (121.150 ng), methionine (18.062 ng), leucine (14.787 ng), and histidine (17.730 ng). The mutant MMH 2225 has an increased value of methionine (22.438 ng). For the most part, both parents have high concentrations of amino acids, but these improved traits are expressed only in female parents. Interspecific hybridization was used to improve the contents of mungbeans. The six varieties, namely MMH 2112, MMH 4255, MMH 1115, MMH 4295, MMH 4224, and MMH 7124, have improved essential amino acid contents. The variety MMH 2333 has improved contents for only three essential amino acids. The mutant MMH 7142 has also increased amino acid contents, excluding that for arginine.

Compared to the recombinant variety of NM2006 (16.739 ng) and mashbean (16.482 ng), MMH 1125 has higher arginine content. The tryptophan contents were improved in all the recombinant varieties except MMH 2333 and MMH 16425, and the MMH 1125 has a high concentration of tryptophan (Abbas et al. 2019).

References

Abbas G, Ahsan M, Asghar MJ, Rizwan M, Hussain M (2019) Genetic improvement in nutritional quality of mungbean through interspecific hybridization with mashbean. Iran J Sci Technol Trans A Sci 43(5):2173–2184

Abbo S, Lev-Yadun S, Gopher A (2012) Plant domestication and crop evolution in the near east: on events and processes. Crit Rev Plant Sci 31:241–257

Akande KE, Abubakar MM, Adegbola TA, Bogoro SE, Doma UD (2010) Chemical evaluation of the nutritive quality of pigeon pea [Cajanus cajan (L.) Millsp.]. Int J Poult Sci 9(1):63–65

Alagbaoso SO, Olawuni IA, Ibeabuchi JC, Ofoedum AF, Onyeneke, E. N. (2019) Functional and sensory properties of biscuit produced from Peanut butter substituted with shortening. IOSR J Environ Sci Toxicol Food Technol 13:1–10

Baldev B (1988) Origin, distribution, taxonomy, and morphology. In: Baldev B, Ramanujam S, Jain HK (eds) Pulse crops. Oxford and IBH Publishing Co, New Delhi, pp 3–51

Black M, Bewley JD, Hlamer P (2006) The encyclopedia of seeds: science, technology and uses. CAB International, Wallingford, p 828

Bond DA, Lawes DA, Hawtin GC, Saxena MC, Stephens JS (1985) Faba bean (Vicia faba L.). In: Summerfield RJ, Roberts EH (eds) Grain legume crops. William Collins Sons, London, pp 199–265

Broughton WJ, Hernández G, Blair M, Beebe S, Gepts P, Vanderleyden J (2003) Beans (*Phaseolus* spp.): model food legumes. Plant Soil 252:55–128. https://doi.org/10.1023/A:1024146710611

Carrera CS, Reynoso CM, Funes GJ, Martínez MJ, Dardanelli J, Resnik SL (2011) Amino acid composition of soybean seeds as affected by climatic variables. Pesq Agropecu Bras 46:1579–1587

Chéreau D, Videcoq P, Ruffieux C, Pichon L, Motte JC, Belaid S et al (2016) Combination of existing and alternative technologies to promote oilseeds and pulses proteins in food applications. Oil seeds and fats crops and lipids. 23:D406. https://doi.org/10.1051/ocl/2016020

Chiaiese P, Ohkama-Ohtsu N, Molving L, Godfree R, Dove D, Hocart C, Fujiwara T, Higgins TJV, Tabe LM (2004) Sulphur and nitrogen nutrition influence the response of chickpea seeds to an added, transgenic sink for organic Sulphur. J Exp Bot 55(404):1889–1901

Cubero JI (1974) On evolution of Vicia faba L. Theor Appl Genet 45:47–51

Cubero JI (1981) Origin, domestication and evolution. In: Webb C, Hawtin GC (eds) Lentils. Commonwealth Agricultural Bureau, Slough, UK, pp 15–38

Cubero JI, Nadal S (2005) Genetic resources, chromosome engineering, and crop improvement. In: Ram JS, Prem PJ (eds) Grain Legumes, vol 1. CRC Press, Boca Raton, pp 163–186

De Candolle A (1884) Origin of cultivated plants. Kesinger Publishing LCC, Whitefish, p 2006

Delgado-Salinas A, Bibler R, Lavin M (2006) Phylogeny of the genus *Phaseolus* (Leguminosae): a recent diversification in an ancient landscape. Syst Bot 31:779–791. https://doi.org/10.1600/036364406779695960

Delhaye S, Landry J (1986) High-performance liquid chromatography and ultraviolet spectrophotometry for quantification of tryptophan in barytic hydrolysates. Anal Biochem 159:175–178

Díaz I, Martinez M (2013) Chapter 418 – Plant C1A cysteine peptidases in germination and senescence. In: Salvesen G (ed) Handbook of proteolytic enzymes (Third Edition). Academic Press, pp 1852–1858. ISBN 9780123822192, https://doi.org/10.1016/B978-0-12-382219-2.00417-8

Dolezel J, Lucretti S (1995) High-resolution flow karyotyping and chromosome sorting in Vicia faba lines with standard and reconstructed karyotypes. Theor Appl Genet 90:797–802

Duc G (1997) Faba bean (Vicia faba L.). Field Crop Res 53:99–109

Duc G, Marget P, Esnault R, Le Guen J, Bastianelli D (1999) Genetic variability for feeding value of faba bean seeds (Vicia faba L.): comparative chemical composition of isogenics involving zero-tannin and zero-vicine genes. J Agric Sci Camb 133:185–196

Duc G, Agrama H, Bao S, Berger J, Bourion V, De Rone AM, Gowda CLL, Mikic A, Millot D, Singh KB, Tullu A, Vandenberg A, Patto MCV, Warkentin TD, Zongj X (2015) Breeding annual grain legumes for sustainable agriculture: new methods to approach complex traits and target new cultivar ideotypes. Crit Rev Plant Sci 34:381–411

ERS (2015) Economic research service, United States Department of Agriculture (USDA). In: Ash M (ed) Oil crops outlook, p 11

FAOSTAT (2013) Food and agriculture organization of the United Nations, statistics division. http://faostat3.fao.org/home/E. Accessed 28 Feb 2015

FAOSTAT (2020) Data on legumes and their distribution around the world

Faris MAIE, Mohammad MG, Soliman S (2020) Lentils (Lens culinaris L.): a candidate chemopreventive and antitumor functional food. Funct Foods Cancer Prevent Ther 6:99–120. Edited by Yearul Kabir

Figueira N, Curtain F, Beck E, Grafenauer S (2019) Consumer understanding and culinary use of legumes in Australia. Nutrients 11(7):1575

Fikre A, Bekele D (2019) Chickpea breeding and crop improvement in Ethiopia: past, present and the future. Univers J Agric Res 8(2):33–40

Galvez FCF, Aquino MB, Villarino BJ, Francisco MLL, Lustre AO, Resurreccion AVA (2006) Development and optimization of chocopeanut spread. In: Lustre AO, Francisco MLL, Palomar LS, Resurreccion AVA (eds) Peanut butter and spreads. United States Agency for International Development, Tbilisi

Ganesan K, Xu B (2017) Polyphenol-rich lentils and their health promoting effects. Int J Mol Sci 18(11):2390

Garg N, Gupta H (2010) Isolation and purification of fungal pathogen (Macrophomina phaseolina) induced chitinase from moth beans (Phaseolus aconitifolius). J Pharm Bioallied Sci 2(1):38–43

Gaur PM, Gowda CLL (2005) Trends in world chickpea production, research and development. In: Knights EJ, Merrill R (eds) Proceedings of focus 2005: Chickpea in the farming systems. 21–23 September 2005, Goondiwindi, Qld, Australia. Pulse Australia, Toowoomba, Australia, pp 8–15

Geervani P, Uma Devi T (1989) Effect of maturation on nutrient composition of selected vegetable legumes. J Sci Food Agric 46(2):243–248

George AA, de Lumen BO (1991) A novel methionine-rich protein in soybean seed: identification, amino acid composition, and n-terminal sequence. J Agric Food Chem 39:224–227

Gepts P, Beavis WD, Brummer EC, Shoemaker RC, Stalker HT, Weeden NF, Young ND (2005) Legumes as a model plant family. Genomics for food and feed report of the cross-legume advances through genomics conference. Plant Physiol 137:1228–1235

Graf D, Monk JM, Wu W, Wellings HR, Robinson LE, Power KA (2020) Red lentil supplementation reduces the severity of dextran sodium sulfate-induced colitis in C57BL/6 male mice. J Funct Foods 64:103625

Guimon J, Guimon P (2012) How ready-to-use therapeutic food shapes a new technological regime to treat child malnutrition. Technol Forecast Soc Chang 79(7):1319–1327

Gupta RK, Gupta K, Sharma A, Das M, Ansari IA, Dwivedi PD (2017) Health risks and benefits of chickpea (Cicer arietinum) consumption. J Agric Food Chem 65(1):6–22

Haciseferogullari H, Gezer I, Bahtiyarca Y, Menges HO (2003) Determination of some chemical and physical properties of sakis faba bean (Vicia faba L. var. major). J Food Eng 60:475–479

Hanafy MS, Rahman MS, Khalafalla MM, El-Shemy HA, Nakamoto Y, Ishimoto M, Wakasa K (2006) Accumulation of free tryptophan in azuki bean (Vigna angularis) induced by expression of a gene (OASA1D) for a modified a-subunit of rice anthranilate synthase. Plant Sci 171:670–676

Hart C (2017) The economic evolution of the soybean industry. In: Nguyen HT, Bhattacharyya MK (eds) The Soybean Genome. Springer, Cham, pp 1–9

Havemeier SM (2018) Evaluating the agricultural, historical, nutritional, and sustainable uses of pulse grains and legumes (Doctoral dissertation, University of Minnesota). https://hdl.handle.net/11299/216081

Higgins TJ, Chiaiese P, Sarmah B, Molvig L, Tabe L (2004) Genetic improvement of chickpeas: seed composition and protection. In: Proceedings 5th European conference on grain legumes and 2nd international conference on legume genomics and genetics. 7–11 June 2004, Dijon, France, pp 21–22

Hirdyani H (2014) Nutritional composition of Chickpea (Cicerarietinum-L) and value added products – a review. Indian J Commun Health 26(Suppl S2):102–106

Hoffman G (1989) The chemistry and technology of edible oils and fats and their high fat products. Academic Press Inc., San Diego

http://usda.mannlib.cornell.edu/usda/current/OCS/OCS-02-12-2015.pdf. Accessed 28 Feb 2015

Iqbal A, Khalil IA, Ateeq N, Sayyar Khan M (2006) Nutritional quality of important food legumes. Food Chem 97:331–335

Ishimoto M, Rahman SM, Hanafy MS, Khalafalla MM, El-Shemy HA, Nakamoto Y, Kita A, Takahashi K, Matusda F, Murano Y, Funabashi T, Miyagawa H, Wakasa K (2009) Evaluation of amino acid content and nutritional quality of transgenic soybean seeds with high-level tryptophan accumulation. Mol Breed 25:313. https://doi.org/10.1007/s11032-009-9334-3

Johnson JB, Jimmerson J (2003) Lentils (No. 1119-2016-91868)

Johnson CR, Thavarajah D, Combs GF Jr, Thavarajah P (2013) Lentil (Lens culinaris L.): A prebiotic-rich whole food legume. Food Res Int 51:107–113

Jukanti AK, Gaur PM, Gowda CLL, Chibbar RN (2012) Nutritional quality and health benefits of chickpea (Cicer arietinum L.): a review. Br J Nutr 108(S1):S11–S26. https://doi.org/10.1017/S0007114512000797

Katiyar RP (1984) 'Kulthi' a promising crop for Himachal hills. Indian Farming 34(9):31–35

Khader V, Venkat Rao S (1986) Limiting amino acids in horsegram (Dolichos biflorus). Indian J Nutr Diet 23(6):158–164

Khalafalla MM, Rhaman MS, El-Shemy HA et al (2005) Optimization of particle bombardment conditions by monitoring of transient sGFP(S65T) expression in transformed soybean. Breed Sci 55:257–263

Khalil IA, Khan S (1995) Protein quality of Asian beans and their wild progenitor, Vigna sublobata (Roxb). Food Chem 52:327–330

Kingsley D, Ravikumar G, Chauhan R, Abraham J (2014) Extraction and screening of bioactive metabolites from Vigna mungo against various pathogens. Int J Pharm Sci Res 5:428–431. https://doi.org/10.13040/IJPSR.0975-8232.5(2).428-31

Kita Y, Nakamoto Y, Takahashi M, Kitamura K, Wakasa K, Ishimoto M (2010) Manipulation of amino acid composition in soybean seeds by the combination of deregulated tryptophan biosynthesis and storage protein deficiency. Plant Cell Rep 29(1):87–95

Kortt AA, Caldwell JB, Lilley GG, Higgins TJV (1991) Amino acid and cDNA sequences of a methionine-rich 2S protein from sunflower seed (Helianthus annuus L). Eur J Biochem 195:329–334

Kumar S, Rajendran K, Kumar J, Hamwieh A, Baum M (2015) Current knowledge in lentil genomics and its application for crop improvement. Front Plant Sci 6. https://doi.org/10.3389/fpls.2015.00078

Lev-Yadun S, Gopher A, Abbo S (2000) The cradle of agriculture. Science 288:1062–1063

Lingyun Y, Jian W, Chenggang W, Shan L, Shidong Z (2016) Effect of zinc enrichment on growth and nutritional quality in Pea sprouts. J Food Nutr Res 4:100–107

Macêdo KD, Frota G, Lopes LAR, Silva ICV, Arêas JAG (2017) Nutritional quality of the protein of Vigna unguiculata L. Walp and its protein isolate. Rev Cienc Agron 48:792–798. https://doi.org/10.5935/1806-6690.20170092

Miranda M, Borisjuk L, Tewes A, Heim U, Sauer N, Ulrich W, Hans W (2001) Amino acid permeases in developing seeds of Vicia faba L.: expression precedes storage protein synthesis and is regulated by amino acid supply. Plant J 28:61–71

Mosse J, Huet J, Baudet A (1987) Changement de la composition en acides amines des grains de pois en fonction de leur taux d'azote. Sci Aliments 7:301–324

Murtaugh MA, Sweeney C, Giuliano AR, Herrick JS, Hines L, Byers T et al (2005) Diet patterns and breast cancer risk in Hispanic and non-Hispanic white women: the four-corners breast cancer study. Am J Clin Nutr 87:978–984

Murthy NRS (1980) In vitro protein digestibility and dry matter disappearance in relation to different levels of tannin and fibre in horsegram (Dolichos biflorus L.). Mysore J Agri Sci 14(3):466

Nalle CL, Ravindran V, Ravindran G (2010) Nutritional value of faba beans (Vicia faba L.) for broilers: apparent metabolisable energy, ileal amino acid digestibility and production performance. Anim Feed Sci Technol 156(3-4):104–111. https://doi.org/10.1094/CCHEM-87-4-0342

Paredes M, Becerra V, Tay J (2009) Inorganic nutritional composition of common bean (Phaseolus vulgaris L.) genotypes race Chile. Chil J Agric Res 69:486–495. https://doi.org/10.4067/S0718-58392009000400002

Peanut Institute (2011) Types of Peanut oil. Available from http://wwwpeanutinstituteorg/health-and-nutrition/. Accessed 20 Oct 2011

Pickardt T, Saalbach I, Waddell D, Meixner M, Muntz K, Schieder O (1995) Seed specific expression of the 2S albumin gene from Brazil nut (Bertholletia excelsa) in transgenic Vicia narbonensis. Mol Breed 1:295–301

Prasad SK, Singh MK (2015) Horse gram- an underutilized nutraceutical pulse crop: a review. J Food Sci Technol 52:2489–2499. https://doi.org/10.1007/s13197-014-1312-z

Ranjhan SK (1980) Animal nutrition and feeding practices in India, 2nd edn. Vikas Publishing House, New Delhi

Roca WM, Munoz LC, Chavarriaga P, Ramirez H (1989) The potential role of biotechnology in the improvement of Phaseolus beans

Saalbach I, Pickardt T, Waddell D, Hillmer S, Schieder O, Muntz K (1995) The Sulphur-rich Brazil nut 2S albumin is specifically formed in transgenic seeds of the grain legume Vicia narbonensis. Euphytica 85:181–192

Saxena KB, Kumar RV, Rao PV (2002) Pigeonpea nutrition and its improvement. J Crop Prod 5(1–2):227–260

Schutyser MAI, Pelgrom PJM, van der Goot AJ, Boom RM (2015) Dry fractionation for sustainable production of functional legume protein concentrates. Trends Food Sci Technol 45(2):327–335

Silvestri MC, Ortiz AM, Lavia GI (2015) rDNA loci and heterochromatin positions support a distinct genome type for 'x = 9 species' of section Arachis (Arachis, Leguminosae). Plant Syst Evol 301:555–562

Singh NP, Pratap A (2011) Food legumes for nutritional security and health benefits. In: Singh U, Praharaj SC, Singh SS, Singh PN (eds) Biofortification of food crops. Springer, New Delhi, pp 41–50

Singh U, Jambunathan R, Saxena KB, Subrahmanyam N (1990) Nutritional quality evaluation of newly developed high-protein genotypes of pigeon pea (Cajanus cajan L.). J Sci Food Agric 50:201–209

Smýkal P, Coyne CJ, Ambrose MJ, Maxted N, Schaefer H, Blair MW et al (2015) Legume crops phylogeny and genetic diversity for science and breeding. Crit Rev Plant Sci 34(1–3):43–104

Solanki YB, Jain SM (2010) Immunostimulatory activities of Vigna mungo L. extract in male Sprague-Dawley rats. J Immunotoxicol 7:213–218

Soytech (2011) Peanut facts and industrial overview. Available from www.soytech.com/peanut facts.htm. Accessed 2 Oct 2011

Suchoszek-Lukaniuk K, Jaromin A, Korycińska M, Kozubek A (2011) Nuts and seeds in health and disease prevention. Elsevier

Suneja Y, Kaur S, Anil K, Kaur GN (2011) Levels of nutritional constituents and antinutritional factors in black gram (Vigna mungo L. Hepper). Food Res Int 44:621–628. https://doi.org/10.1016/j.foodres.2010.12.020

Tabe L, Higgins TJV (1998) Engineering plant protein composition for improved nutrition. Trends Plant Sci 3(7):282–286

Tabe L, Wirtz M, Molvig L, Droux M, Hell R (2010) Overexpression of serine acetlytransferase produced large increases in O-acetylserine and free cysteine in developing seeds of a grain legume. J Exp Bot 61(3):721–733

Tang CH, Chen L, Ma CY (2009) Thermal aggregation, amino acid composition and in vitro digestibility of vicilin-rich protein isolates from three Phaseolus legumes: a comparative study. Food Chem 113:957–963

Tiwari AK, Shivhare AK (2016) Pulses in India: retrospect and prospects. Published by director, Govt. of India, Ministry of Agri. and Farmers Welfare (DACandFW). Publication No.: DPD/Pub.1/Vol. 2/2016, 2, pp 1–21 https://farmer.gov.in/imagedefault/prospects_2017.pdf

Toomer OT (2017) Nutritional chemistry of the peanut (Arachis hypogaea). Crit Rev Food Sci Nutr 29:1–12. https://doi.org/10.1080/10408398.2017.1339015

Tsou CS, Hsu MS, Tan ST, Park HG (1979) The protein quality of mungbean and its improvement. Acta Hortic 93:279–287

Varshney RK, Close TJ, Singh NK, Hoisington DA, Cook DR (2009) Orphan legume crops enter the genomics era! Curr Opin Plant Biol 12:202–210

Wang N, Daun JK (2004) The chemical composition and nutritive value of Canadian pulses. Canadian Grain Commission Report, pp 19–29

Wang TL, Domoney C, Hedley CL, Casey R, Grusak MA (2003) Can we improve the nutritional quality of legume seeds? Plant Physiol 131:886–891

Wehrhahn C, Allard RW (1965) The detection and measurement of the effects of the individual genes involved in the inheritance of a quantitative character in wheat. Genetics 51:109–119

Wood JA, Grusak MA (2007) Nutritional value of chickpea. In: Yadav SS, Redden R, Chen W, Sharma B (eds) Chickpea breeding and management. Centre for Agriculture and Bioscience International, Wallingford, pp 101–142. https://doi.org/10.1079/9781845932138.005

Woodroof JG (1983) Historical background in peanuts: production, processing, products, vol 181. AVI Westport CT

Xu B, Chang SK (2009) Total phenolic, phenolic acid, anthocyanin, flavan-3-ol, and flavonol profiles and antioxidant properties of pinto and black beans (Phaseolus vulgaris L.) as affected by thermal processing. J Agric Food Chem 57:4754–4764. https://doi.org/10.1021/jf900695s

Yadav S, Negi KS, Mandal S (2004) Protein and oil rich wild horsegram. Genet Resour Crop Evol 51:629–633

Yao DN, Kouassi KN, Erba D, Scazzina F, Pellegrini N, Casiraghi MC (2015) Nutritive evaluation of the bambara groundnut Ci12 landrace [Vigna subterranea (L.) Verdc. (Fabaceae)] produced in Côte d'Ivoire. Int J Mol Sci 16(9):21,428–21,441

Young SN (1996) Behavioral effects of dietary neurotransmitter precursors: basic and clinical aspects. Neurosci Biobehav Rev 20(2):313–323

Zhenxing S, Yang Y, Yingying Z, Guixing R (2016) Nutritional composition and antioxidant activity of twenty mung bean cultivars in China. Crop J 4:398–406. https://doi.org/10.1016/j.cj.2016.06.011

Chapter 19
Enriching Legume Protein Contents

Amreena Sultan, Abdel Rahman Mohammad Said Al Tawaha, and Abdel Razzaq M. Altawaha

Abstract Leguminous plants, which include beans, peas, and lentils, are vital to long-term agricultural sustainability because of their ability to thrive in regions with changing weather patterns and diminished precipitation. Active research is going on to generate new types of legumes that can thrive in temperatures 4 or 5 degrees higher than average. The discovery of novel bioactive chemicals and functional food ingredients for disease prevention has made significant strides in recent years. In the past, legumes were considered foods that were good for people's health. Legumes are plants in the family Fabaceae. Since antiquity, their seeds have been essential to human nutrition. Legumes have high contents of proteins and complex carbohydrates such as dietary fiber and resistant starch, as well as low levels of lipids. Also, legumes are an essential source of minerals such as iron, zinc, and calcium. Legumes are rich in vitamins, including folate, which lessens the chance of neural tube abnormalities. In addition, they contribute to the prevention of chronic diseases by providing bioactive substances. In this chapter, we discuss several topics related to enriching legume protein contents.

1 Introduction

The family Leguminaceae, commonly known as the Fabaceae, are known as legumes because they produce seeds inside pods. Humans consume only a small percentage of the more than 18,000 species of climbers, herbs, shrubs, and trees in the vast family. Oil seeds and pulses are the two categories into which food legumes

A. Sultan
SKUAST-K, Shalimar, Srinagar, Jammu and Kashmir, India

A. R. M. S. Al Tawaha (✉)
Department of Biological Sciences, Al Hussein Bin Talal University, Ma'an, Jordan

A. R. M. Altawaha
Department of Crop Science, Faculty of Agriculture, Universiti Putra Malaysia, Serdang, Selangor, Malaysia

M. A. Nadeem et al. (eds.), *Legumes Biofortification*,
https://doi.org/10.1007/978-3-031-33957-8_19

are split. The former are the dried seeds of cultivated legumes like soybeans and peanuts, which have a high oil content and are consumed as a traditional cuisine. The Food and Agriculture Organization of the United Nations recognizes 11 basic types of leguminous plants (FAO 2016a, b). Legumes are still a typical food in many cultures worldwide and are said to have been one of the first crops humans ever domesticated. As a low-cost substitute for meat, these seeds have gained widespread acclaim worldwide (Kouris-Blazos and Belski 2016). Development and harvest are affected by several biotic and abiotic factors. Plant physiology techniques, particularly those applied to grain legumes, help to describe plant responses and discover stress-tolerant features or elite cultivars.

Several physiological variables, such as nutrient uptake and translocation, shoot architecture, yield, seed quality, leaf gas exchange, and soil biota interaction, have been evaluated and recommended in the past decade to understand better how environmental stress affects grain legumes. Legumes are nutritionally valuable as they contain essential amino acids, complex carbohydrates, unsaturated fats, vitamins, and minerals for human nutrition (Rebello et al. 2014). In addition, legumes contain advantageous bioactive components. Also, legumes have been given responsibilities in economy, culture, physiology, and medicine in addition to being superior in terms of nutrition. The intake of legumes has also been linked to several positive health effects, including hypocholesterolemic, antiatherogenic, anticarcinogenic, and hypoglycemic qualities (Ndidi et al. 2014; Messina 2016). In addition to being a cheap source of nutrition, legumes can generate cash for subsistence farmers who grow them in their own homes. Rural farmers who cannot afford pricey irrigation systems and fertilizers are wonderful crops. This is because legumes are extremely disease and pest resistant, serve as cover crops, and grow well in unfavorable climates. Thus, they minimize soil erosion. Legumes and legume-based products must be made more available to low-income people to fight poverty and malnutrition. Nearly 170 million preschoolers and nursing moms in poor African and Asian nations suffer from protein-energy malnutrition (PEM), a severe nutritional disease. PEM is too expensive for lower-income people because of the high cost of animal protein (eggs, meat, and milk), the common cereal-based diet, and the rising price of essential commodities. High-protein legumes like soybeans and cowpeas are in high demand due to a shortage.

2 Protein Content of Legumes

Legumes are an excellent source of high-quality protein, containing 20–45% of it and are often rich in the essential amino acid lysine (Philips 1993) (Table 19.1).. With proteins ranging from 17% to 20% in peas and beans and 38% to 45% in lupins and soybeans, respectively, peas and beans fall on the lower end of the spectrum (Mlyneková et al. 2014). Legumes provide more protein than most plant diets, around twice as much as cereals (Nedumaran et al. 2015). Legumes are high in protein because nitrogen-fixing bacteria in their roots convert inert nitrogen gas into

Table 19.1 Seed protein content and amino acid deficiencies in major grain legumes

Species	Botanical name	Protein content of grain seeds (%)	Deficit amino acid
Chickpea	*Cicererietinum* L.	17–22 before dehulling 25.3–28.9 after dehulling	Methionine, cysteine, threonine, and valine
Lentil	*Lens culinaris* Medik	20.6–31.4	Methionine, cysteine
Lupin	*LupLiunpuisn ualsbaulsb uLs.* L.	44–44	Alanine, tryptophan
Soybean	*Glycine max*(L.)*Merr*	Up to 40	Methionine, cysteine, threonine, and lysine
Common bean	*Phaseolus vulgaris* L	20–30	Methionine, cysteine
Pigeon pea	*Cajanus cajan* (L.)	20–22	Methionine, cysteine, valine
Faba bean	*Vicia faba* L.	26–41	Methionine
Mung bean	*Vigna radiata* L.	20.97–31.32	Methionine, cysteine
Cowpea	*Vigna unguiculata* (L. *Walp.*)	14.8–25	Methionine
Pea	*Pisum sativum* L.	13.7–30.7	Methionine, cysteine, and tryptophan
Urad bean	*Vigna mungo* L. *Hepper*	25–28	Methionine, cysteine
Lathyrus	*Lathyrus sativus* L.	8.6–34.6	Methionine, cysteine

ammonium, which the plant uses for protein synthesis. However, legume proteins, except soy protein (Table 19.2), are deficient in tryptophan, methionine, cystine, and other critical sulfur-containing amino acids (SCAA). Therefore, they lack protein. Vialin and legumin include albumins and globulins. Because vialin, the main protein group in most legumes, has a low concentration of SCAA, this explains why legumes have relatively low levels of SCAA (FAO 2016a, b). As a result, eating legumes and cereals considerably improves protein quality. Legumes and grains should be taken in a ratio of 35:65 to maintain nutritional balance. The low amount of SCAA in legumes is not entirely insufficient because it improves calcium retention. The hydrogen ions produced when SCAA degrades cause the demineralization of bone and other body tissues. As a result, legume protein can improve calcium retention compared to proteins high in SCAA derived from animals or grains. Low-density lipoproteins, known to contribute to the development of coronary artery disease, have also been associated with decreases in legume protein (Philips 1993). Regarding protein, legumes and cereals complement each other because grains are high in SCAA (which legumes are low in SCAA) and low in lysine (high in legumes). Legumes are crucial in vegetarian diets as the main source of protein, vitamins, and minerals. Therefore, vegetarians must include legumes and grains to balance amino acids properly (Leonard 2012).

Table 19.2 List of legume genotypes with higher seed protein content

Crop	Genotypes	Seed protein content	Source
Chickpea	ICC 5912, LEGCA608, LEGCA609, LEGCA614, LEGCA619, LEGCA716	29.2%	ICRISAT, Patancheru, India
Common Bean	J-216, FJIP-43, LR05, 6-EX, Accession 4049	25.23%	Mexican state of Jalisco and Durango
Cowpea	HC-6, HC-5, CP-21, LST-II-C-12, CP-16, TVu-2723, TVu-3638, TVu-2508	26.7–27.9% 32.50%	India Minjibir, Kano State, Nigeria
Faba bean	25 genotypes	28.43–29.68%	Manitoba and Saskatchewan
Grass pea lentil	IC127616	32.20%	Canada
Mung bean	MGG330, Nagpuri	29.9% and 29.3%	India
Pea	PI206793, PI206801, PI206838, PI210619, PI210644, PI210675, PI210678, PI210684	>30%	Manitoba and Ontario, Canada
Soybean	TN03–350, TN04–5321 Lines developed from Kwangan- kong _ Samnamkong and Danbaegkong _ Samnam-kong	High-protein content 34.3–44.4% and 35.8–49.6%	Tennessee Agricultural Experiment Station, Tennessee, USA, USDA–ARS and the North Carolina Agricultural Research Service Yeongnam Agricultural Research Institute (YARI), Milyang, Republic of Korea

3 Grain Legumes: A Rich Protein

Due to seed viability requirements, grain legume protein compositions vary. Many grains and legumes have an SPC of 25–40%, and raising that proportion over 40% may be problematic. The SPC of chickpeas varies between 17% and 22% before shelling and 25.3% and 28.9% after shelling. Chickpea seed has trace amounts of glutelins, prolamin, albumin, and globulin but lacks cysteine and methionine (Saharan and Khetarpaul 1994). Although chickpeas are a rich source of several essential amino acids, the main limiting amino acids are cysteine, methionine, valine, and threonine. Despite having no variations in crucial amino acids, the Desi form of chickpea has a greater SPC than the Kabali type (Khan et al. 1995). Common beans (*Phaseolus vulgaris* L.) have an SPC ranging from 20% to 30% (Shellie-Dessert and Bliss 1991) and are essential for reducing protein-related malnutrition, particularly in impoverished countries. Phaseolin, which accounts for 50–60% of total seed protein content in common beans and 36% to 46% storage protein content, is deficient in the essential amino acids methionine, cysteine, and tryptophan. Cowpea is a grain legume with multiple uses in human nutrition. Cowpea SPC ranges between 15% and 25%. Cowpea storage proteins contain tryptophan and lysine but not cysteine or methionine. After globulins, albumins, glutelins, and prolamin are cowpea grain's most common storage protein fractions (Jayathilake et al. 2018). The SPC of field beans ranges from 26% to 41%, with tryptophan, cysteine, and methionine being the only three necessary amino acids missing. Globulins comprise over 80% of seed proteins (vicilin and legume). The amino acid lysine is most commonly found in broad bean seeds. SPC for lenses varies between 20% and 30%. Like other legumes, lentil seed contains significant amounts of albumin (2661%) and globulin (4470% storage protein; 11S-leguminin, 7S-vicilin, and convicilin). However, prolamin and glutelin levels are low. Two primary protein classes—albumin (15%) and globulin (85%)—make up the majority of the protein found in white lupin (*Lupinus albus* L.) seed, which can contain up to 44% of it. White lupin seed protein has fewer antinutritive qualities than other grain legumes like peas and soybean, albeit having some allergenic effects. Additionally, lysine, phenylalanine, arginine, and leucine are all present in greater concentrations in white lupin seed than in soybean, making it a highly sought-after grain legume from a dietary perspective. A soybean, *Glycine max (*L.*)* Merr., has 35–45% protein. It lacks methionine but has enough lysine to make up for grains' lack of lysine. It was projected that in 2018, soybeans would account for 70% of the world's protein meal. According to Kollarova et al. (2010), mung beans (*Vigna radiata* L.) are a good source of various necessary micronutrients and easily digestible proteins. It is rich in proteins, except for methionine and cysteine, which contain sulfur. Mung bean, which is simpler to digest than other legumes and mildly hypoallergenic, is used to wean newborns. Furthermore, mung bean is an excellent meat substitute for vegans. The protein content of peas (*Pisum sativum* L.) ranges from 13.7% to 30.7% (Tzitzikas et al. 2006). Legumin, vicilin, convicilin, and proteins related to globulin make up the pea seed protein. Pea protein extract contains 26.3–52.0% vicilin. Gluten-free and low-allergenic pea protein is also popular in the food industry. Pigeon pea seeds are an essential source of plant-based dietary protein for India's vegetarian population, ensuring protein-based food security (Varshney et al. 2010). The broad bean (*Vigna mungo* L. Hepper) is a critical grain legume that contains globulin (63%), albumin (12%), and glutelin (21%). It is also high in protein (up to 25%). While urad bean seeds lack methionine and cysteine, they are high in glutamic and aspartic acid as well as lysine.

4 Genetic Variability in Grain Legumes to Increase Seed Protein

SPC in grain legume crops can be improved economically by utilizing crop germplasm variety (Gottschalk et al. 1975). Grain legume high-SPC genotypes are studied using agricultural genetic resources. Chickpeas show significant genetic variation for SPC, including 12.4–31.5%, 17–22% (Jukanti et al. 2012), and

14.6–23.2% (Serrano et al. 2017). Several high-SPC genotypes such as LEGCA608 and LEGCA716 were identified in these studies and could be used for cowpeas, an inexpensive protein source to improve human nutrition. Boukar et al. (2011) studied a collection of 1541 cowpea lines for genetic diversity in grain protein content and mineral profiles. Weng et al. (2019) evaluated 173 cowpea accessions from two global regions (Fayetteville and Alma, Arkansas). They also observed considerable genetic diversity for SPC (22.8–28.9%) that could be exploited to transfer the high SPC trait into top-yielding cowpea cultivars (Weng et al. 2019). Nutritional profiles of 22 foods, including grain protein content. The protein content range of the seed, determined by combustion, was 23.16–28.13% (Gerrano et al. 2019).

Scientists found strong positive associations between SPC and different mineral compositions, suggesting that both grades could be selected together. Gerrano et al. (2019) found significant SPC in COVU-702 and HC-98-64 in 21 cowpea genotypes. In another study, Paulistinha had 29.2% SPC among 18 cowpea genotypes. MNC01-649F-2, BRS-Cauam, BRS-Paraguacu, BRS-Marataoa, Canapuzinho, and BRS-Tumucumaque had high SPC among 30 Brazilian cowpea lines. Grass pea is a grain legume that is naturally climate adaptable and high in SPC. 37 grass pea genotypes were examined, and IC127616 turned out to be very SPC-rich (32.2%) (Kumari et al. 2018). The micro-Kjeldahl technique was used to evaluate 27 local mung bean landraces and found considerable genetic variation for SPC (17.2–29.9%). Vigna radiata var. sublobata, a wild mung bean species, exhibits a genetic diversity of 15.2–21% SPC and is rich in lysine, isoleucine, and phenylalanine amino acids (Babu et al. 1988). The genetic variability of SPC in lenses is 20–30% such as *L. orientalis* and *L. ervoides* elite lentil cultivars can aid in SPC breeding programs.

5 Legumes' Mendelian Seed Protein Inheritance

Numerous studies already use Mendelian genetics to determine the genetics of SPC in several crop legumes (Hynes 1968). Matta and Gatehouse (1982) consider pea storage proteins (legumin and convicilin). Mahmoud and Gatehouse (1984) used an F2 hybrid between 360 and 611 to explain the monogenic inheritance of another SPC vicilin (Vc-1) gene in the pea. Using seed size genetics, Perez et al. (1993) identified the genes responsible for producing peas with high and low SPC (round vs. wrinkled). Low SPC and albumin were found in round-seeded pea plants (RR/RbRb), but high SPC and albumin levels were found in recessive "alleles" (rr/rbrb) (Perez et al. 1993). Since the protein content of cowpeas is highly heritable and regulated by a small number of genes, it may be possible to increase their protein content (Ravelombola et al. 2021). There was a strong inverse relationship between yield and protein content in the high-protein population of six populations created via diallel crosses between two high-yielding soybean lines and two high-protein lines, and a significant positive association in the high-yield population.

6 QTL Mapping for Seed Protein Content

The use of biparental mapping populations from the genomics of different grain legumes has made it easier to identify the underlying QTLs affecting SPC (Diers et al. 1992). Few studies have found the QTLs that control SPC in chickpeas. A study used genotyping by sequencing to characterize ICC995–ICC5912 RILs in four environments and discovered a large influence of QTL (Quantitative Trait Locus) on chromosome 3 (LG3) for SPC, accounting for 44.3% of the phenotypic variance (PV) (Wang et al. 2019). Irzykowska and Wolko (2004) mapped five SPC-QTLs to LG2, LG5, and LG7 in peas, accounting for 13.1–25.8% PV using an F2-derived Wt10245-Wt11238 mapping population. Wt11238 found a protein content QTL on LGVb surrounded by cp, gp, and te markers in two F5 mapping populations, Wt3557 and Wt10245 (Krajewski et al. 2012). Genotyping of an orb-mapping population revealed two SPC-QTLs on LG1b, accounting for 16% of the PV, and two on LG4a, accounting for 10.2% (Gali et al. 2018). The LGV-QTL has two candidate genes, Ls (GA biosynthesis) and Rbcs4 (small Rubisco subunit), while the LGI-QTL has Rgp (cell wall synthesis) (Burstin et al. 2007). On the other hand, Obala et al. (2020) gained an understanding of the genetic factors controlling SPC in pigeon peas based on the results of five F2 populations that segregated for SPC. CcLG02, CcLG03, CcLG06, and CcLG11 main effect QTLs were found to account for 14.5% of PV (Obala et al. 2020). Several alleles in soybean regulate the SPC trait, which is strongly influenced by G-E interactions. More than 300 QTLs affecting soybean SPC have been identified (Van and McHale 2017).

Nevertheless, the three chromosomes with the largest SPC QTLs are 5, 15, and 20. In a population created by mating cultivated and wild soybeans, (Diers et al. 1992) initially described a significant QTL regulating high SPC on chromosome 20, which was later mapped to a 3 cM on LGI (Nichols et al. 2006). This QTL's position was subsequently pinpointed to 8.4 Mb (Bolon et al. 2010), 1 MB, and 77.4 kb (Fliege et al. 2022). Similar to the previous large SPC QTL, qSeedPro 15, which overlapped the previously found genomic area on chromosome 15, was reduced to 4 Mb (Zhang et al. 2020) (Table 19.3). On the other hand, Zhang et al. (2020) identified the QTL's potential candidate gene Glyma.15G049200. Using Illumina Infinium Bead Chip sequencing technology, we identified five SPC QTLs on chromosomes that collectively accounted for 4.6–19.6% of the observed FV in recombinant inbred lines derived from Williams 82 *G. soja* (PI 483460B) (Patil et al. 2017). The qPro 20 QTL was the only one of them that remained stable throughout all four situations.

7 High-Protein Genomic Region/Haplotypes GWAS

Several researchers (Sonah et al. 2015). Jadhav et al. (2015) performed SPC association mapping on 187 chickpea genotypes using SSR markers (desi, kabuli, and exotic). Nine SPC MTAs on LG1 to LG5 accounted for 16.85% PV. A GWAS

Table 19.3 Proteomic approach for investigating novel proteins for improving seed protein content in grain legumes

Crop	Protein identified	Approach used	Genotype
Chickpea	High-amino acid content, 454 protein spots	Two-dimensional electrophoresis and mass spectrometry	Flip97-171C, Elite
Common bean	Sulfur-containing amino acids, S-methylcysteine accumulation	Chromatography-tandem mass spectrometry	Flip97-171C, Elite
Faba bean	Sulfur-containing amino acids; enhanced concentration of cysteine and methionine	Mass spectrometry	SARC1 and SMARC1N-PN1
Narrow-leafed lupin	2760 protein identifications	LC-MS	P27255, Tanjil
Pea	156 proteins	2-D gels, MALDI-TOF MS	Unicrop
Soybean	High arginine content in Nepalese	MALDI-TOF; two-dimensional gel electrophoresis	Caméor

employing multilocation and multiyear phenotyping of a large global sample of pea germplasm identified significant MTAs for agronomic and qualitative parameters (Gali et al. 2019). Two important MTAs, Chr3LG5 138253621 and 194530376, impacted SPC. In a large-scale association and linkage study, 985 soybean accessions were analyzed and haplotypic variations at the high SPC locus cqProt-003 on chromosome 20 were identified (wild, landrace, historical, and current cultivars). As reported by Marsh et al. (2022), Glyma20G084500, Glyma20G085250, and Glyma20G085100 are three causative candidate genes identified within a 173 kb linkage block. In addition, research revealed large trait-associated genomic regions (Marsh et al. 2022). One of these, Glyma.20G085100, which has trinucleotide insertions and a 304 bp deletion, is strongly associated with the high-protein phenotype (Marsh et al. 2022).

8 Crude Protein of Grain Legumes' Seeds: Functional Genomics Perspectives

In the previous decade, RNA sequencing has advanced functional genomics research, notably transcriptome evaluation for grain legume trait gene(s) identification (Wang et al. 2021). Numerous studies have revealed candidate gene functions for grain legume SPC. The functional characterization of storage protein genes like pea seed albumin ("PA1," "PA1b") and the conglutin family in thin leaf lupin using cDNA cloning is remarkable (Foley et al. 2011). By sequencing cDNA clones from growing seeds, researchers were able to characterize the function of 11 novel storage proteins (conglutin family)-encoding genes in narrow-leaf lupin (Foley et al. 2011). Sixteen conglutin genes producing storage proteins were discovered using

RNA-seq transcriptome analysis in the Tanjil cultivar. Using a QTL-seq method, the functional genomic study of the soybean genome revealed 329 differentially expressed genes underlying the QTL regions qSPC 20-1 and qSPC 20-2 and responsible for SPC (Wang et al. 2021). The soybean genome's SPC-controlling region's transposon insertion Glyma.20g085100's regulatory function was revealed using RNAi technology (Fliege et al. 2022). RNA interference inhibited Glyma.20g085100 expression in the low-protein Thorne soybean genotype, increasing protein (Fliege et al. 2022).

9 Metabolomics/Proteomics Show the Genetics of Legumes' High Seed Protein Content

Proteomics enables researchers to examine all proteins generated in a given organism or cell at a given time and under certain conditions. This approach might identify previously unknown seed storage proteins and shed light on the molecular drivers of enhanced SPC in a range of legumes (George and De Lumen 1991). Two-dimensional gel electrophoresis was used to identify soybean's methionine-rich protein (George and De Lumen 1991). Two-dimensional polyacrylamide gel electrophoresis (2D-PAGE) proteomics analysis later discriminated between wild soybean (*G. soja*) and farmed soybean by identifying 44 protein spots in the former and 34 in the latter based on their content of high-density storage proteins (beta-conglycinin and glycinin). Soybean genotypes with high-protein content have been found. SDSPAGEN, SDS-PAGE, and MALDI-TOF-MS studies of nine soybean accessions and William 82 revealed significant changes in protein content in seed 11S. The pea's broad spectrum of 156 proteins was identified in its mature seed by Bourgeois et al. (2011), and novel storage proteins for improved SPC storage globulins were found. Sulfur amino acid concentrations are elevated in kidney beans lacking phaseolin and lectins due to increased concentrations of legumin, albumin 2, and defensin (Marsolais et al. 2010). Santos et al. (2017) used proteomics to analyze the protein composition of 24 chickpea genotypes to study the heritable variation of chickpea storage proteins. HPLC analysis verified the inherited variability of SPC. Positively correlated with seed protein content (SPC) and inversely correlated with seed oil content (SOC) were the metabolites asparagine, aspartic acid, glutamic acid and free 3-cyanoalanine in GC-TOF/MS metabolomics of seed protein soybean lines (Wang et al. 2019). However, there were negative correlations between the amount of seed protein and oil and various sugars (sucrose, fructose, glucose, and mannose) (Wang et al. 2019). Common bean genotypes with different concentrations of S-methylcysteine in seeds were metabolite-profiled (Saboori-Robat et al. 2019). They discovered that S-methylcysteine accumulates primarily as gamma-glutamyl-S-methylcysteine during seed maturation and that free methylcysteine accumulates much less. The nutritionally rich Valle Agricola chickpea genotype from southern Italy included 66% glutamic acid, glutamine, aspartic acid, phenylalanine, and glutamate. Metabolomics developments might help us comprehend legume SPC-related cellular metabolic networks and mechanisms.

10 Genetic Engineering and Genome Editing for Grain Legumes SPC Improvement

Numerous genetic engineering methods have increased the content of essential amino acids in grain legumes (Aragão et al. 1999). Transgenic technology can modify essential amino acids in legumes, especially high-sulfur amino acids. Chiaiese et al. (2004) introduced a sunflower albumin transgene into chickpeas to increase methionine content. Cysteine and methionine dominate this protein. Transgenic chickpea seeds have more methionine than controls. In thin-leaved lupins, the albumin transgene from sunflower seeds increased seed methionine levels (Molvig et al. 1997).

11 Selecting High SPC Lines for Greater Genetic Gain Through Genomic Selection and Rapid Generation

Genomic selection (GS) can predict the genetic merit of progenies with complicated characteristics and no phenotypic values from "large target populations" by constructing a prediction model and calculating genomic-assisted breeding values in a "training population" with known phenotypes. Breeders should increase selection accuracy (I), selection intensity (I), and breeding cycle length to maximize the GS advantage for genetic gain: an I r/L = DG = R = h2S. (L). Heredity (H2), selective pressure (R), and additive genetic variation (a) One area where GS has supplanted phenotypic selection is in chickpea grain production in the face of moisture stress. They use the optimal light intensity, photoperiod, and temperature to speed up photosynthesis, which shortens the breeding cycle (Watson et al. 2018). There are recognized speed breeding procedures for faba bean, chickpea, lupin, lentil, pea, and soybean (Crosser et al. 2016). To maintain global food security, further speed breeding procedure modification could expedite advances in breeding-relevant traits, particularly SPC, in grain legumes.

12 Constraints on Seed Protein Content

Seeds are the progeny of plants and are susceptible to several fundamental compromises that affect their structure and size. To develop a healthy embryo, seeds require essential nutrients, including cell walls and a small number of lipids, carbohydrates, and nucleic acids. As a result, there are limitations in selecting the possible protein content. In addition, the selective power has limits, as shown by the long-term selection of the corn seed oil content. "Ecologists have conceived these trade-offs as a component of an economic spectrum over the past two or more decades, which affects the variety of features seen in leaves, stems and roots. Smaller seeds can go farther as a dispersal unit, while larger seeds can establish more quickly. The genetic diversity and architecture of many specific legume crops have likely been constrained by millennia of selection on these trade-offs in seed size and composition." Few researchers have examined how breeding efforts affect seed protein content or related ecological and evolutionary restrictions on seed composition to those efforts. Although it plays a slightly different role in grain legumes than in other legumes, seed size is often a significant covariate in seed protein content.

13 Conclusions and Future Perspective

Malnutrition-related issues, such as dietary protein deficiencies, are becoming more prevalent due to the growing human population, particularly in underdeveloped and emerging nations. The growing problem of hunger and malnutrition-related issues could be reduced by providing protein-rich legumes enhanced by plant breeding and molecular breeding techniques. Improved grain and legume dietary proteins may also provide a significant and cost-effective replacement for expensive animal-based nutritional proteins. Major grain legumes can be protein biofortified to help to develop, and underdeveloped nations meet their daily dietary protein needs. Accurate characterization of different culture gene pools and landrace haplotypes with genetic variation is urgently needed to accelerate the improvement of SPC in legumes. Prebreeding of high-yielding, low-protein top legume varieties with CWR genes or QTLs can be crucial. Recent genomics, genome-wide association mapping, whole-genome resequencing, and complete genome and pangenome sequences in a range of legumes may facilitate genomics-assisted selection of causal alleles/QTLs/haplotypes/candidate genes that influence high-protein concentration at the genome level–grain legume protein increase. Functional genomics, proteomics, and metabolomics may explain complex molecular networks that enhance protein concentration in legumes with various grains. However, most protein assessment methods are damaging, making it challenging to choose grain legume genotypes with high-protein content in germplasm or offspring. Thus, protein-rich legume genotype selection needs high-throughput, nondestructive methods. Genetic selection and fast generational advances may help select and create high-protein legumes. Genome editing will allow us to change the gene(s) regulating high-protein concentration at select legume genome regions to enhance SPC, overcoming transgenic technology's limitations. These modern breeding methods should help us generate grain legumes with higher protein content, as yield and protein content are inversely related. The combination of approaches might satisfy battle protein-based malnutrition and hunger to ensure global human growth.

References

Al-Tawaha AM, Seguin P, Smith DL et al (2006) Foliar application of elicitors alters isoflavone concentrations and other seed characteristics of field-grown soybean. Can J Plant Sci 86:677–684

Aragão FJL, Barros LMG, De Sousa MV, Grossi de Sá MF, Almeida ERP, Gander ES, Rech EL (1999) Expression of a methionine-rich storage albumin from the Brazil nut (Bertholletia excelsa HBK, Lecythidaceae) in transgenic bean plants (*Phaseolus vulgaris* L., Fabaceae). Genet Mol Biol 22:445–449

Babu CR, Sharma SK, Chatterjee SR, Abrol YP (1988) Seed protein and amino acid composition of wild Vigna radiata var. sublobata (Fabaceae) and two cultigens, V. mungo and V. radiata. Econ Bot 42:54–61.

Bolon YT, Joseph B, Cannon SB, Graham MA, Diers BW, Farmer AD, May GD, Muehlbauer GJ, Specht JE, Tu ZJ et al (2010) Complementary genetic and genomic approaches help characterise the linkage group I seed protein QTL in soybean. BMC Plant Biol 10:41.

Boukar O, Massawe F, Muranaka S, Franco J, Maziya-Dixon B, Singh B, Fatokun C (2011) Evaluation of cowpea germplasm lines for protein and mineral concentrations in grains. Plant Genet Resour 9:515–522.

Bourgeois F, Ekholm T, Eliashberg Y (2011) Symplectic homology product via Legendrian surgery. PNAS 108(20):8114–8121

Burstin J, Marget P, Huart M, Moessner A, Mangin B, Duchene C, Desprez B, Munier-Jolain N, Duc G (2007) Developmental genes have pleiotropic effects on plant morphology and source capacity, eventually impacting on seed protein content and productivity in pea. Plant Physiol 144:768–781.

Chiaiese P, Ohkama-Ohtsu N, Molvig L, Godfree R, Dove H, Hocart C, Fujiwara T, Higgins TJV, Tabe LM (2004) Sulphur and nitrogen nutrition influence the response of chickpea seeds to an added, transgenic sink for organic sulphur. J Exp Bot 55:1889–1901.

Crosser JS, Pazos-Navarro M, Bennett RG, Tschirren S, Edwards K, Erskine W, Creasy R, Ribalta FM (2016) Time to flowering of temperate pulses in vivo and generation turnover in vivo–in vitro of narrow-leaf lupin accelerated by low red to far-red ratio and high intensity in the far-red region. Plant Cell Tissue Organ Cult 127:591–599.

Diers BW, Keim P, Fehr WR, Shoemaker RC (1992) RFLP analysis of soybean seed protein and oil content. Theor Appl Genet 83:608–612.

FAO (2016a) International year of legumes: nutritious seeds for a sustainable future. Food and Agriculture Organisation of the United Nations and World Health Organisation. FAO, Rome. www.fao.org/pulses-2016

FAO (2016b) Legumes can help fight climate change, hunger and obesity in Latin America and the Caribbean. Food and Agriculture Organisation of the United Nations and World Health Organisation. FAO, Santiago de Chile

Fliege C, Ward RA, Vogel P, Nguyen H, Quach T, Guo M, Viana JPG, Dos Santos LB, Specht JE, Clemente TE et al (2022) Fine mapping and cloning of the major seed protein QTL on soybean chromosome 20. Plant J 110:114–128.

Foley RC, Gao LL, Spriggs A, Soo LY, Goggin DE, Smith P, Atkins CA, Singh KB (2011) Identification and characterisation of seed storage protein transcripts from Lupinus angustifolius. BMC Plant Biol 11:59.

Gali KK, Liu Y, Sindhu A, Diapari M, Shunmugam ASK, Arganosa G, Daba K, Caron C, Lachagari RVB, Tar'An B et al (2018) Construction of high-density linkage maps for mapping quantitative trait loci for multiple traits in field pea (*Pisum sativum* L.). BMC Plant Biol 18:1–25.

Gali KK, Sackville A, Tafesse EG, Lachagari VR, McPhee K, Hybl M, Miki'c A, Smýkal P, McGee R, Burstin J et al (2019) Genome-wide association mapping for agronomic and seed quality traits of field pea (*Pisum sativum* L.). Front Plant Sci 10:1538.

George AA, De Lumen BO (1991) A novel methionine-rich protein in soybean seed: identification, amino acid composition, and N-terminal sequence. J Agric Food Chem 39:224–227.

Gerrano AS, Jansen van Rensburg WS, Venter SL, Shargie NG, Amelework BA, Shimelis HA, Labuschagne MT (2019) Selection of cowpea genotypes based on grain mineral and total protein content. Acta Agric Scand Sect B Soil Plant Sci 69:155–166
Gottschalk W, Mueller HP, Wolff G (1975) The genetic control of seed protein production and composition (peas). Egypt J Gene Cytol 4:453–468
Hynes M (1968) Genetically controlled variants of a storage protein in *Pisum sativum*. Aust J Biol Sci 21:827–829.
Irzykowska L, Wolko B (2004) Interval mapping of QTLs controlling yield-related traits and seed protein content in Pisum sativum. J Appl Gen 45(3):297–306
Jadhav AA, Rayate SJ, Mhase LB, Thudi M, Chitikineni A, Harer PN, Jadhav AS, Varshney RK, Kulwal PL (2015) Marker-trait association study for protein content in chickpea (*Cicer arietinum* L.). J Genet 94:279–286.
Jayathilake C, Visvanathan R, Deen A, Bangamuwage R, Jayawardana BC, Nammi S, Liyanage R (2018) Cowpea: An overview on its nutritional facts and health benefits. J Sci Food Agric 98:4793–4806.
Jukanti AK, Gaur PM, Gowda CLL, Chibbar RN (2012) Nutritional quality and health benefits of chickpea (*Cicer arietinum* L.): a review. Br J Nutr 108:S11–S26.
Khan MA, Akhtar N, Ullah I, Jaffery S (1995) Nutritional evaluation of desi and kabuli chickpeas and their products commonly consumed in Pakistan. Int J Food Sci Nutr 46:215–223. [CrossRef]
Kollárová K, Vatehová Z, Slováková LU, Lišková D (2010) Interaction of galactoglucomannan oligosaccharides with auxin in mung bean primary root. Plant Physiol Biochem 48(6):401–406
Kouris-Blazos A, Belski R (2016) Health benefits of legumes and pulses with a focus on Australian sweet lupins. Asia Pac J Clin Nutr 21(1):1–17. https://doi.org/10.6133/apjcn.2016.25.1.23
Krajewski P, Bocianowski J, Gawłowska M, Kaczmarek Z, Pniewski T, Święcicki W, Wolko B (2012) QTL for yield components and protein content: a multienvironment study of two pea (*Pisum sativum* L.) populations. Euphytica 183:323–336.
Kumari S, Jha VK, Kumari D, Ranjan R, Nimmy MS, Kumar A, Kishore C, Kumar V (2018) Protein content of Lathyrus sativus collected from diverse locations. J Pharmacogn Phytochem 7(1S):1610–1611
Leonard E (2012) Cultivating good health. In: Grains and legumes nutrition council. Cadillac Printing, Adelaide
Mahmoud SH, Gatehouse JA (1984) Inheritance and mapping of vicilin storage protein genes in *Pisum sativum* L. Heredity 53:185–191.
Marsh JI, Hu H, Petereit J, Bayer PE, Valliyodan B, Batley J, Nguyen HT, Edwards D (2022) Haplotype mapping uncovers unexplored variation in wild and domesticated soybean at the major protein locus cqProt-003. Theor Appl Genet 135:1443–1455.
Marsolais F, Pajak A, Yin F, Taylor M, Gabriel M, Merino DM, Ma V, Kameka A, Vijayan P, Pham H et al (2010) Proteomic analysis of common bean seed with storage protein deficiency reveals up-regulation of sulfur-rich proteins and starch and raffinose metabolic enzymes, and down-regulation of the secretory pathway. J Proteome 73:1587–1600.
Matta NK, Gatehouse JA (1982) Inheritance and mapping of storage protein genes in *Pisum sativum* L. Heredity 48:383–392.
Messina MJ (2016) Legumes and soybeans: overview of their nutritional profiles and health effects. Asia Pac J Clin Nutr 25(1):1–17. https://doi.org/10.1.1.847.8636
Mlyneková Z, Chrenková M, Formelová Z (2014) Cereals and legumes in nutrition of people with celiac disease. Int J Celiac Dis 2(3):105–109. https://doi.org/10.12691/ijcd-2-3-3
Molvig L, Tabe LM, Eggum BO, Moore AE, Craig S, Spencer D, Higgins TJV (1997) Enhanced methionine levels and increased nutritive value of seeds of transgenic lupins (*Lupinus angustifolius* L.) expressing a sunflower seed albumin gene. Proc Natl Acad Sci U S A 94:8393–8398.
Ndidi US, Ndidi CU, Aimola IA, Bassa OY, Mankilik M, Adamu Z (2014) Effects of processing (Boiling and roasting) on the nutritional and antinutritional properties of Bambara groundnuts

(*Vigna subterranean* [L.] Verdc.) from Southern Kaduna, Nigeria. J Food Processing 2014:1–9. https://doi.org/10.1155/20172129

Nedumaran S, Abinaya P, Jyosthnaa P, Shraavya B, Parthasarathy R, Bantilan C (2015) Grain legumes production, consumption and trade trends in developing countries, Working Paper Series No 60. International Crops Research Institute for the Semi-Arid Tropics (ICRISAT). International Crops Research Institute for the Semi-Arid Tropics, Telangana, pp 1–57. https://doi.org/10.7910/DVN/V61SNB

Nichols DM, Glover KD, Carlson SR, Specht JE, Diers BW (2006) Fine mapping of a seed protein QTL on soybean linkage group I and its correlated effects on agronomic traits. Crop Sci 46:834–839. [CrossRef]

Obala J, Saxena RK, Singh VK, Kale SM, Garg V, Kumar CV, Saxena KB, Tongoona P, Sibiya J, Varshney RK (2020) Seed protein content and its relationships with agronomic traits in pigeonpea is controlled by both main and epistatic effects QTLs. Sci Rep 10:1–17.

Patil G, Mian R, Vuong T, Pantalone V, Song Q, Chen P, Shannon GJ, Carter TC, Nguyen HT (2017) Molecular mapping and genomics of soybean seed protein: a review and perspective for the future. Theor Appl Genet 130:1975–1991.

Perez MD, Chambers SJ, Bacon JR, Lambert N, Hedley CL, Wang T (1993) Seed protein content and composition of near-isogenic and induced mutant pea lines. Seed Sci Res 3:187–194.

Philips RD (1993) Starchy legumes in human nutrition and culture. Plant Foods Hum Nutr 44(3):195–211. https://doi.org/10.1007/BF01088314

Ravelombola W, Shi A, Huynh BL (2021) Loci discovery, network-guided approach, and genomic prediction for drought tolerance index in a multi-parent advanced generation intercross (MAGIC) cowpea population. Hort Res 8:24.

Rebello CJ, Greenway FL, Finley JW (2014) A review of the nutritional value of legumes and their effects on obesity and its related co-morbidities. Obes Rev 15(5):392–407. https://doi.org/10.1111/obr.12144

Saboori-Robat E, Joshi J, Pajak A, Solouki M, Mohsenpour M, Renaud J, Marsolais F (2019) Common bean (*Phaseolus vulgaris* L.) accumulates most S-methylcysteine as its -glutamyl dipeptide. Plan Theory 8:126.

Saharan K, Khetarpaul N (1994) Protein quality traits of vegetable and field peas: varietal differences. Plant Foods Hum Nutr 45:11–22.

Santos A, Giráldez FJ, Groenen MAM, Madsen O, Frutos J, Valdés C, Andrés S (2017) Modifications in liver transcriptomic profile of fattening lambs by early suckled milk intake level. The 68th Annual Meeting of the European Federation of Animal Science (EAAP), Tallinn, Estonia.

Serrano C, Carbas B, Castanho A, Soares A, Patto MC, Brites C (2017) Characterisation of nutritional quality traits of a chickpea (*Cicer arietinum*) germplasm collection exploited in chickpea breeding in Europe. Crop Pasture Sci 68:1031–1040.

Shellie-Dessert K, Bliss F (1991) Genetic improvement of food quality factors. Common beans. In: van Schoonhoven A, Voyset O (eds) Research for crop improvement. CAB International, Wallingford, pp 649–677

Sonah H, O'donoughue L, Cober E, Rajcan I, Belzile F (2015) Identification of loci governing eight agronomic traits using a GBS–GWAS approach and validation by QTL mapping in soya bean. *Plant Biotech. J*. 3:10. https://doi.org/10.1111/pbi.12249

Tzitzikas EN, Vincken JP, de Groot J, Gruppen H, Visser RG (2006) Genetic variation in pea seed globulin composition. J Agric Food Chem 54:425–433.

Van K, McHale LK (2017) Meta-analyses of QTLs associated with protein and oil contents and compositions in soybean (*Glycine max* (L.) Merr.) seed. Int J Mol Sci 18:1180.

Varshney RK, Penmetsa RV, Dutta S, Kulwal PL, Saxena RK, Datta S, Sharma TR, Rosen B, Carrasquilla-Garcia N, Farmer AD et al (2010) Pigeonpea genomics initiative (PGI): An international effort to improve crop productivity of pigeonpea (*Cajanus cajan* L.). Mol Breed 26:393–408.

Wang N, Daun JK (2004) Effect of variety and crude protein content on nutrients and certain antinutrients in field peas (*Pisum sativum*). J Sci Food Agric 84:1021–1029

Wang J, Zhou P, Shi X, Yang N, Yan L, Zhao Q, Yang C, Guan Y (2019) Primary metabolite contents are correlated with seed protein and oil traits in near-isogenic lines of soybean. Crop J 7:651–659

Wang J, Mao L, Zeng Z, Yu X, Lian J, Feng J, Yang W, An J, Wu H, Zhang M et al (2021) Genetic mapping high protein content QTL from soybean 'Nanxiadou 25'and candidate gene analysis. BMC Plant Biol 21:388.

Watson A, Ghosh S, Williams MJ, Cuddy WS, Simmonds J, Rey MD, Asyraf Md Hatta M, Hinchliffe A, Steed A, Reynolds D et al (2018) Speed breeding is a powerful tool to accelerate crop research and breeding. Nat Plants 4:23–29.

Weng Y, Qin J, Eaton S, Yang Y, Ravelombola WS, Shi A (2019) Evaluation of seed protein content in USDA cowpea germplasm. Hort Sci 54:814–817.

Zhang H, Goettel W, Song Q, Jiang H, Hu Z, Wang ML, An Y-QC (2020) Selection of GmSWEET39 for oil and protein improvement in soybean. PLoS Genet 16:e1009114

Chapter 20
Biomolecular Aspects of Plant Nutrition Related to Food Biofortification

Zahid Manzoor, Maria Batool, Muhammad Fraz Ali, Zeshan Hassan, and Umbreen Shahzad

Abstract Depleting the nutrients in soil and continuously growing crops cause nutrient deficiencies in soils. When crops are grown in nutrient-deficient soils, agricultural food also becomes nutrient deficient. Further food demand is rapidly increasing while production area is decreasing day by day. Therefore, it is crucial to develop highly nutritious crops to fight against malnutrition and meet global food demands. Biofortification is the most efficient method to provide highly nutritious food to the world within a short amount of time. There are different methods to develop highly efficient crops for food biofortification, but using biomolecular markers is the best way to quickly achieve this goal. Molecular breeding and transgenic breeding are the advanced forms of using biomolecules in modern breeding to develop highly nutritious food. This chapter discusses plant nutrition through use of molecular markers for food biofortification.

Z. Manzoor
National Key Laboratory of Crop Genetic Improvement, Hubei Hongshan Laboratory, College of Plant Science and Technology, Huazhong Agricultural University, Wuhan, China

M. Batool
MOA Key Laboratory of Crop Ecophysiology and Farming System in the Middle Reaches of the Yangtze River, College of Plant Science and Technology, Huazhong Agricultural University, Wuhan, China

M. F. Ali (✉)
College of Agronomy, Northwest A & F University, Yangling, Shaanxi, China

Z. Hassan · U. Shahzad
College of Agriculture, Bahuddin Zakariya University, Bahadur Sub-Campus Layyah, Layyah, Punjab, Pakistan

M. A. Nadeem et al. (eds.), *Legumes Biofortification*,
https://doi.org/10.1007/978-3-031-33957-8_20

1 Introduction

Biofortification is commonly defined as a food strategy to produce micronutrient-enhanced food crops to address nutrient-deficiency issues that prevail in especially low-income countries (Ali et al. 2021a; Dhaliwal et al. 2021). It is a sustainable and cost-effective approach that reduces malnutrition and regulates balanced nutrient metabolism (Koç and Karayiğit 2021). Dietary inadequacies in one or more micronutrients, including iron, vitamin A, and zinc, exert far-reaching adverse impacts on human health and nutrition, which are global public health problems, especially in developing countries (Msungu et al. 2022). Commonly used strategies for increasing the supply of these micronutrients include micronutrient supplementation as drug preparations, but this method is not readily available or accessible to low-income rural populations. Alternatively, several other strategies can be used, including dietary diversification through biofortification using different genetic and agronomic approaches (Ali et al. 2021b; Foley et al. 2021).

Micronutrients and several beneficial or health-related elements can supplement plants via different methods that enrich edible products (Kumar et al. 2021). In addition to enhancing the nutritional statuses of food crops, plant enrichment leads to positive outcomes for physiological and metabolic processes, along with higher tolerance to stresses, such as abiotic or biotic factors (Mittal et al. 2022). Many populations in developing countries depend mainly on staple crops for their nutrition (Mustafa et al. 2019), an essential subject of biofortification, which works through several biotechnological, agronomic, and conventional breeding approaches (Cominelli et al. 2020).

Fourteen essential macro- and micronutrients are needed for the adequate nutrition of plants (El-Ramady et al. 2022). These nutrients include nitrogen (N), potassium (K), phosphorus (P), calcium (Ca), sulfur (S), magnesium (Mg), chlorine (Cl), molybdenum (Mo), boron (B), copper (Cu), nickel (Ni), iron (Fe), zinc (Zn), and manganese (Mn) (Romera et al. 2021). These are taken mainly from soil, which is why crop productivity is related to soil physiochemical characteristics that regulate nutrient availability and cause the nutrient deficiency in nutrient-deficient soil environments (Das et al. 2021).

Soil nutrient availability depends mainly on soil pH, which is highly sensitive to excessive nutrients in soil (Pahalvi et al. 2021). For instance, higher aluminum (Al^{3+}) and iron (Fe^{3+}) levels in acidic soils significantly reduce the inorganic P availability in the roots of plants (Penn and Camberato 2019). Thus, nutrient deficiencies should be considered to attain the optimal growth and yields of crops. The level of any nutrient beyond the optimum threshold level (deficiency or toxicity) leads to a reduction in the yields and quality of crops.

Plants take up minerals and water mainly through their roots (Stachiw et al. 2019); meanwhile, foliage uptake processes are also important. The foliar application of nutrients leads to nutrient entry via cuticle, stomata, or lenticels into different cells and organelles to participate in various metabolic processes (Malhotra et al. 2020). Foliar applications of nutrients are used for the rapid recovery of plants from deficiency (Pandey et al. 2021). Mineral nutrients as inorganic elements are essential for proper plant functioning during the life cycle; several yield-related mechanisms, including biomass accumulation and nutrient partitioning, are directly regulated by nutrient availability in the external solution (Burgess et al. 2022). Plants mainly require seventeen essential elements for higher biomass and yield formation. Meanwhile, several other elements are also beneficial for their survival under unfavorable cues and/or for the enhancement of product quality (Teklić et al. 2021). The relation of nutrient supply and yield is known as the law of diminishing returns, which states that if the supply of a particular nutrient is high, it will reduce the availability of other nutrient elements and limit the genetic potential of crops, leading to reductions in crop yield (Pandey et al. 2021). Moreover, it has been reported that an excessive supply of specific nutrients leads to a decrement in crop yield, which might be due to their toxicity or to the deficiency of other nutrients in plants or soils (García-Sánchez et al. 2020).

Micronutrients are essential in cellular metabolism and ion homeostasis (Afzal et al. 2020); however, their deficiency leads to severe disorders and metabolic perturbations (Shafi and Zahoor 2020). Nutrient deficiencies, including those in iron (Fe), zinc (Zn) and iodine (I), are increasing worldwide, with adverse effects on public health (Prom-U-Thai et al. 2020). Previously, food supplementation has been the major approach used to produce vitamin- and mineral-enriched food (Srivastav et al. 2022). There are also certain limitations in food supplementation, including the lower availability of micronutrients after food processing.

The current global challenge in agricultural production is to produce biofortified crops (Maqbool et al. 2020) that enhance consumers' nutritional levels while permitting producers to offer micronutrient-rich products (Marques et al. 2021). Furthermore, the speedy identification and characterization of plant metabolism–related genes have been driving forces in biofortification strategies (Wang et al. 2021), through metabolic profiling, genome sequencing, high-throughput physical mapping, and the gene expression analysis of crops that can provide impetuses for plant biofortification and nutrient enrichment (Vasconcelos et al. 2017).

The objective of this chapter is to cover the biomolecular techniques used in developing biofortified crops with better nutrition to provide nutrified agricultural food to the world, to fight against malnutrition. Molecular strategies are needed at the time to make nutrient-enriched, readily available crops to all people around the globe. Thus, we are compiling previous literature related to biomolecular techniques for developing biofortified food with enhanced nutrition.

2 Molecular Mechanisms of Nutrient Uptake in Plants

Several reports documented the plant molecular mechanisms related to nutrient and water uptake at the cell and organ levels (Hossain et al. 2012). The subsequent research was conducted on isolated membrane vesicles and electrophysiological mechanisms regulated by a specific group of genes, involving transportation processes in the cellular membrane (Gill et al. 2021). Plants have specific proteins known as transporters that facilitate the transportation of molecules across the plasma membrane, leading to membrane impermeability (Horemans et al. 2000; Vishwakarma et al. 2019). Recently, the gene cloning and knockout mutants of numerous transport proteins have aided in deciphering the detailed transport mechanisms at the organ level (Kailasam and Peiter 2021; Tang et al. 2020).

Additionally, phospholipids are selectively permeable to gaseous molecules, including CO_2 and O_2 but impermeable to hydrophilic solutes and various inorganic ions and selectively permeable to water (Kochhar and Gujral 2020). Transporter proteins assist in transporting different ions—inorganic and organic solutes—across the cell and vacuolar membrane to regulate cellular homeostasis (Mansour 2022; Yadav et al. 2021). Membranes consist of various transporter proteins: ATPases or ATP pumps, channel proteins, and cotransporters (Babla et al. 2019; Michalak et al. 2022) (Fig. 20.1).

Phosphorus (P) is one of the vital elements and essential for optimal development and crop yields (Poirier et al. 2022). P in the soil is not limited but reduced in the plants because of bioavailability. However, plants have shown adaptation under P stress through improved acquisition or efficient utilization and activation of a tolerance mechanism that actives corresponding genes (Chen et al. 2022). These physiological processes in the plants include root structural variations, the secretion of carboxylates, protons and root secretory phosphohydrolases, and mycorrhizal associations that are regulated by many genes in all plants (Vengavasi et al. 2021). The molecular studies have reported several candidate genes related to high-affinity Pi transporters, organic anion efflux transporters, and purple acid phosphatases (PAPs) that have led to improved P uptake (Elanchezhian et al. 2015). The identified quantitative trait loci (QTLs) for P stress tolerance might assist in the development of P-efficient plants by using marker-assisted breeding in different crops (Wang et al. 2019). Under P-deficient condition, plants adapt to the environment by

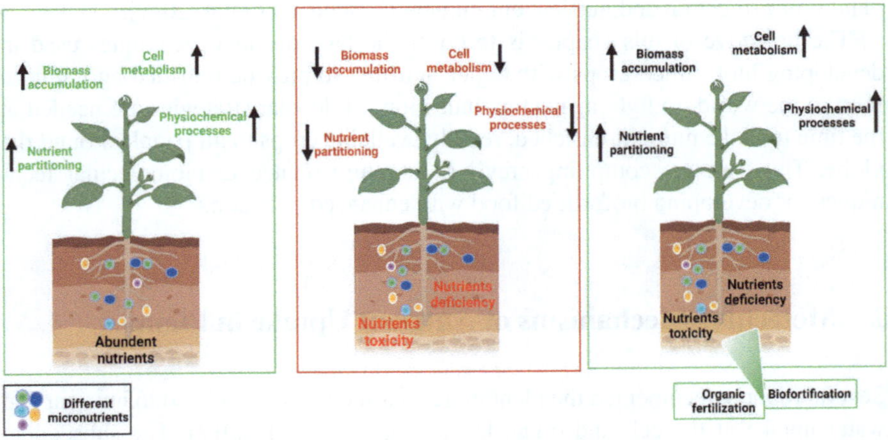

Fig. 20.1 Different mechanistic approaches adopted by plants to take up and absorb nutrients from different sources

maintaining cellular Pi homeostasis in their cells (Poirier et al. 2022). Physiological adaptation includes a variation of lipid composition, the enhancement of anthocyanin biosynthesis, the lower accumulation of P in pectin, the activities of RNase and PAPs for remobilizing P during senescence and facilitating various processes of glycolysis that require adenylate (Soumya et al. 2022).

Calcium (Ca^{2+}) under low nitrogen supply is absorbed in the form of nitrate through different transporters such as *NRT2.1* (Guan et al. 2021). Ca^{2+} acts as a secondary messenger and works to regulate nitrate transporters, so it has nitrate-signaling and primary nitrate responses (Zhang et al. 2020). Cytoplasmic Ca^{2+} concentration regulates nitrate starvation in a dose-dependent manner (Adavi and Sathee 2021). Nitrogen-efficient plants show a higher Ca^{2+} level that increased the expression pattern of the negative regulators of nitrate in the lateral organ boundary domain family (*LBD37* and *LBD38*) (Ahn et al. 2021; Song et al. 2020). The abundance of *LBD37* and *LBD38* lowers the expression level of high-affinity nitrate transporter *NRT2.1*, nitrate reductase gene, and nitrogen related enzyme activities (Adavi and Sathee 2021).

Micronutrients such as Fe and Zn are important in optimal plant growth, and they are popular for their use in biofortification to enrich edible crops (Jaiswal et al. 2022). Fe is one of the important mineral nutrients found in all living organisms, and *IRT1* is an Fe-coding transporter that was first cloned in *Arabidopsis thaliana* (Rehman et al. 2021; Vélez-Bermúdez and Schmidt 2022). Field plants use strategies such as secreting DMA and absorbing Fe^{2+}, which is more highly abundant than Fe^{3+}, under submerged environments (Pereira 2014). Additionally, homologs of ferric-chelate reductase have been found in crop plants, but they could not express under Fe deficiency or adequacy (Schmidt 2003); moreover, the lowered activity of Fe^{3+}-chelate reductase is also associated with this mechanism (Celletti et al. 2020; Shi et al. 2022). Zinc is an important element, and its mechanisms related to zinc acquisition garner special attention in research (Bouain et al. 2019). Metal transporters (*GmZIP1*) belonging to the *ZIP* family have been observed in soybeans to take up and transport Zn (Zeng et al. 2019). Genetically modified *A. thaliana* with the overexpression of the *ZAT* gene contained increased Zn contents in its root cells under zinc toxicity, and this leads to zinc resistance by sequestration in the roots (Gu et al. 2021). A lot of research has reported insights into the various zinc transportation processes of plants at the molecular level (Vijayalakshmi 2022). As a result, different useful methods have been identified for the engineering of zinc-efficient plants to enhance the zinc content in edible their parts (Praharaj et al. 2021). Those targets include membrane transporters of divalent cations, including the *ZIP* family members in cell membranes and tonoplasts (Barzana et al. 2021). Another class of vacuolar membrane transporters, *ZAT* and *MHX*, have also been identified (Yağız et al. 2022). Further research on gene overexpression and the important engineering membrane transporters in plants could verify the efficiency in enhancing zinc uptake.

Generally, soil nitrogen availability is highly varied in space and time because of different environmental factors; therefore, the preferred form of N for uptake depends on the adaptation of plants to soil conditions (Briat et al. 2020; Zuluaga and Sonnante 2019). The uptake of nitrate ions in root cells via two transport systems has been found to coexist in plants, and it act coordinately to transport nitrates from a soil solution and distribute them to different plant parts (Islam 2022). In addition, the *NRT1* gene family facilitates the low-affinity transport system (LATS) in roots, excluding *AtNRT1.1*, which is a dual affinity transporter and a nitrate sensor (Li et al. 2022; Zhao et al. 2022). A previous study has shown that fifty-one genes are responsible for different expression patterns in different plant tissues, indicating their specific and unique functions (Zuluaga and Sonnante 2019).

Macronutrients, particularly cations, including ammonium (NH_4^+), potassium (K^+), calcium (Ca^{2+}), and magnesium (Mg^{2+}), are vital in the normal functioning of plant growth and development and are required in large amounts (Abbas et al. 2021). Other positively charged micronutrients (iron, manganese, zinc, copper, and nickel) act as important cofactors and activators for different enzymes (Adhikary et al. n.d.). Magnesium (Mg^{2+}) is one of the abundant divalent cation cells that possess essential functions in several of the physiological processes of plants (Chaudhry et al. 2021). Ten important members of a gene family (*AtMGT*) that encode putative Mg^{2+} transport proteins have been found in *Arabidopsis thaliana*, which were expressed mostly in different Arabidopsis plant tissues (Ishijima 2021). *AtMGT1* functionally complemented a bacterial mutant lacking a Mg^{2+} transport ability; another member, *AtMGT10*, complemented a yeast mutant defective in Mg^{2+} uptake and enhanced the cellular Mg^{2+} level in deficient cells threefold during an hour of uptake time (Chaudhry et al. 2021). However, a radioactive isotope of K^+ has not yet been reported; therefore, K^+ transport could be measured with radioactive Rb^+ (Dai et al. 2021). Earlier reports on K^+ uptake from root cells revealed two important transport components indicating high-affinity (mechanism I) and low-affinity (mechanism II) transportation mechanisms, respectively (Xie et al. 2020).

3 Biomolecular Techniques for Food Biofortification

Transgenic breeding and molecular breeding are advanced techniques to develop biofortified crops (Adlak et al. 2019). RNA interference (RNAi) and genome editing are required to determine the breeding for the biofortification of field crops (Marques et al. 2021). These breeding techniques are helping scientists to generate novel types of plants and sources of generating genetic variations (Fig. 20.2).

3.1 Transgenic Breeding

Transgenic breeding is considered the most effective and time-saving technique to be adopted for the biofortification of crops to meet the demands of world hunger (Kiran 2020). Transgenic crops were developed at a critical time, when there was a shortage of genetic variations in wild types and in gene pools (Singer et al. 2021). Through transgenic breeding, scientists achieved the goal of developing genetically

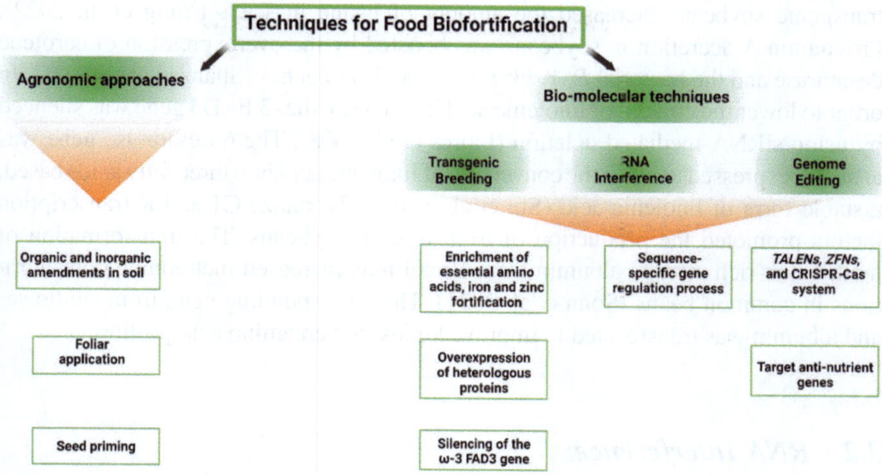

Fig. 20.2 Different techniques to produce nutrient-enriched crops for food biofortification

modified crops by inserting genes of interest (Turnbull et al. 2021). Either gene of interest could reduce the absorption of toxic compounds or increase nutrient availability in agricultural food products.

Transgenic breeding has already been introduced in several legume crops, such as chickpeas, lupines, soybeans, garden peas, and common beans, to combat malnutrition (Jha et al. 2022). The primary goals of pulse biofortification through transgenic breeding are the enrichment of important amino acids, the fortification of iron and zinc, and the decrease of antinutrient chemicals (Bouis 2003; Shahzad et al. 2021). The overexpression of heterologous proteins rich in these amino acids has been shown to remedy the lack of sulfur-rich amino acids (Kim et al. 2020). A cysteine-rich protein derived from maize called 27 kDa zein was introduced and overexpressed in various pulse crops to improve the nutritional value of cysteine (Garg 2022; Kim and Krishnan 2019; Shahzad et al. 2021). The overexpression of S-rich proteins also increased the methionine concentration in Narbon beans and lupins (Whitcomb et al. 2020). By using a seed-specific promotor to overexpress the 2S albumin storage protein and aspartate kinase, methionine concentration in Brazil nuts was increased, resulting in a fourfold increase in the accumulation of methionine in their seeds (Girija et al. 2020; Sharma et al. 2022). Following transformation, the rice *OASA1D* transgene increased free tryptophan accumulation in adzuki beans (Uppal et al. 2021).

To improve iron bioavailability in chickpeas, the glycine max ferritin and chickpea *NAS2* genes were added and overexpressed (Shahzad et al. 2021). Cystathionine-synthase gene overexpression increased methionine content in soybeans. Increased cysteine and methionine contents were achieved in soybeans by overexpressing the maize zein protein (Song et al. 2013). When the O acetyl serine sulfhydrylase gene was overexpressed, the cysteine content of seeds increased (Kim et al. 2012). Dihydrodipicolinic acid synthase and aspartokinase gene overexpression in transgenic soybeans increased the amount of lysine in seeds (Yang et al. 2022). Provitamin A accretion in soybeans was boosted by the overexpression of carotene desaturase and the bacterial *PSY*, *bkt1*, *crtW*, and *crtB* genes (Shahzad et al. 2021). In order to lower the content of linolenic acid in soybean, the -3 FAD3 gene was silenced by using siRNA-mediated deletion (Flores et al. 2008). The 6-desaturase gene was also overexpressed to boost the conversion of linolenic acid into three fatty acids based, a stable form of linolenic acid (Shi et al. 2016). The maize C1 and R transcription factors promoted the production of isoflavone in soybeans. The transformation of methionine-rich storage albumin from Brazil nuts increased methionine concentrations in common beans (Sohn et al. 2021). The corresponding gene from sunflower and albumin was transformed to improve lupins' S-rich amino acid profiles.

3.2 RNA Interference

A double-stranded RNA (dsRNA) molecule powers the sequence-specific gene regulatory mechanism known as RNA interference (RNAi), which prevents a particular gene from being translated or transcribed (Hashiro and Yasueda 2022). RNAi has opened new possibilities for agricultural enhancement even since it was discovered. Compared to antisense technology, it is a more accurate, reliable, effective tool. RNAi makes the incorporation of tolerance to biotic and abiotic stresses and the delivery of high-quality food through biofortification and bioelimination possible (Shahzad et al. 2021). It is frequently used to improve the nutritional value of crops and remove contaminants and allergens from food.

Even if the function of RNAi in pulses biofortification has not been investigated to a great extent, there is still room for development in the field of pulses. The bioavailability of micronutrients in pulses is limited by phytate. Studies have shown that phytic acid forms complexes with calcium, magnesium, copper, and iron, making them less soluble (Amat et al. 2022). The bioavailability of micronutrients could be improved by overexpressing the genes encoding these molecules. Additionally, it has been suggested that selenium increases the bioavailability of iodine in lentils, peas, and chickpeas (Jha et al. 2022). However, more research is needed on a subset of bioavailability inhibitors. To generate pulse crops that are richer in nutrients, it is necessary to investigate the biochemical pathways involved in the creation of antinutrients and use RNAi to mute or knock out the genes that play crucial roles in the absorption of toxic compounds.

3.3 Genome Editing

In plant genome editing (GE), sequence-specific nucleases (SSNs) are utilized to produce transgene-free plants through the inheritance of stable gene modifications in the desired crop (Bilichak et al. 2020). Plant genome editing uses a variety of

SSNs, including TALENs, ZFNs, and the cluster regulatory interspaced short palindromic repeats (CRISPR)-Cas9 system (Karmakar et al. 2022). Using CRISPR, targeted gene editing is possible by using Cas9/13, RNA-guided DNA endonucleases guided by a short-guided RNA (sgRNA) (Marzec and Hensel 2019). The use of genome editing for crop plant biofortification was less common, although several notable cases are detailed below.

In the current state of the research, there is a lack of examples that show the function of genome editing for biofortifying pulses. Yet there is potential for using genome editing to improve the uptake of iron and zinc in all pulse crops by manipulating the iron-regulated transporters (IRTs) such as protein (ZIP) (Ram et al. 2021), ferric-chelate reductase oxidases (FRO) (Ebbisa 2022), Yellow Stripe 1-likes (YSLs) (Joshi-Saha et al. 2022), natural resistance-associated macrophage proteins (NRAMPs) (Urwat et al. 2021), and zinc-regulated transporters (Liao et al. 2022).

When modifying the genome of pulses for biofortification purposes, scientists also aim to disable the antinutrient genes responsible for the lower uptake of micronutrients. Saponins are antinutrient substances that are useful in lower concentrations but can act as antinutrients when ingested in greater quantities (Gadanya et al. 2022). The genes *OSC*, *P450* and *UGT* are involved in saponin production in *A. thaliana*. The production of saponins can be lowered by locating critical regulatory genes in the biosynthesis pathway and then silencing them (Melini and Melini 2021). It is also necessary to identify and subsequently modify the genes that produce other antinutrients, such as lathyrogens, protease inhibitors, and amylases, to decrease the levels of these substances (Table 20.1).

4 Conclusion and Future Prospects

We discussed plant improvement through modern techniques for food biofortification. Mainly, we focused on legume crops in our chapter. Using different modern breeding techniques to develop nutrient-enriched crops is the best solution to deliver nutrients according to demands. We reviewed various recent studies, focusing on the use of biomolecules for nutrient biofortification. Molecular pathways regulate the system of plants and are responsible for all physiological and biochemical events that happen inside plants. Therefore, molecular breeding is the best way to change the targeted pathway and achieve desired results. Many scientists have worked on bimolecular aspects of food biofortification, but a gap remains to be addressed in legume crops in the future. By using advanced CRISPR technology and transgenic techniques, many accomplishments have been achieved. Still, work on legume crops using genome editing, RNA interference, and transgenic breeding techniques is still needed. Plant nutrition through biomolecular techniques is the best way to defeat global hunger. Legumes must be focused on in the future because they are staple foods in many countries in the world.

Table 20.1 Different techniques recently used for improving legume crops to enhnace nutrient uptake and accumulation in grains for food biofortification

Crop	Technology	Output	Reference
Chickpea	Genome editing	Enhanced abiotic stress tolerance through biofortification for cytokinin dehydrogenase	Mahto et al. (2022)
Soybean	*Agrobacterium*-mediated transformation	Enhanced β-carotene content	Kim et al. (2012)
Mungbean	Genome-wide single nucleutide polymorphism (SNP) identification and association mapping	Variation among genotypes for seed nutrition and future molecular breeding resources	Wu et al. (2020)
Soybean	Gene expression, subcellular localization, mutant phenotype	A pair of vacuolar iron transporter like (VTL) genes, GmVTL1a/b facilitate nitrogen fixation	Liu et al. (2020)
Wild-type chickpea	SNP marker identification	16 associated SNP markers related to Fe and Zn concentrations to be used in marker-assisted selection (MAS)	Karaca et al. (2020)
Soybean	RNAi-mediated gene silencing	*GmMIPS1* silencing restricts phytate accumulation, which leads to mineral bioavailability	Kumar et al. (2019)
Soybean	CRISPR-Cas9-mediated gene disruption	Modulation in oleic acid content, lowest level of linoleic acid	Al Amin et al. (2019)

References

Abbas S, Javed MT, Ali Q, Azeem M, Ali S (2021) Nutrient deficiency stress and relation with plant growth and development. In: Engineering tolerance in crop plants against abiotic stress. CRC Press, pp 239–262

Adavi SB, Sathee L (2021) Influence of calcium on nitrate starvation response of bread wheat. Plant Physiol Rep 26(4):661–670

Adhikary S, Das P, Paramanik B, Roy D (n.d.) Chapter functions of micronutrients in the soil-plant system and their management towards sustainable agriculture. Insights Agric Sci 127

Adlak T, Tiwari S, Tripathi M, Gupta N, Sahu VK, Bhawar P, Kandalkar V (2019) Biotechnology: an advanced tool for crop improvement. Curr J Appl Sci Technol 33(1):1–11

Afzal S, Sirohi P, Sharma D, Singh NK (2020) Micronutrient movement and signalling in plants from a biofortification perspective. In: Plant micronutrients. Springer, Cham, pp 129–171

Ahn JY, Jung YH, Song H, Yi H, Hur Y (2021) Alleles disrupting LBD37-like gene by an 136 bp insertion show different distributions between green and purple cabbages (Brassica oleracea var. capitata). Genes Genomics 43(6):679–688

Al Amin N, Ahmad N, Wu N, Pu X, Ma T, Du Y et al (2019) CRISPR-Cas9 mediated targeted disruption of FAD2–2 microsomal omega-6 desaturase in soybean (Glycine max. L). BMC Biotechnol 19(1):1–10

Ali MF, Adnan M, Khan M, Asif M, Perveen S, Ammar A et al (2021a) Agronomic zinc biofortification in rice production systems. Plant Cell Biotechnol Mol Biol 22(25–26):103–113

Ali MF, Ali Ammar SB, Ali U, Huma N, Adnan M (2021b) Mitigating zinc deficiency in plants and soils through agronomic techniques: a review. J Environ Agric Sci 23(1 & 2):1–10

Amat T, Assifaoui A, Schmitt C, Saurel R (2022) Importance of binary and ternary complex formation on the functional and nutritional properties of legume proteins in presence of phytic acid and calcium. Crit Rev Food Sci Nutr:1–23

Babla M, Cai S, Chen G, Tissue DT, Cazzonelli CI, Chen Z-H (2019) Molecular evolution and interaction of membrane transport and photoreception in plants. Front Genet 10:956

Barzana G, Rios JJ, Lopez-Zaplana A, Nicolas-Espinosa J, Yepes-Molina L, Garcia-Ibañez P, Carvajal M (2021) Interrelations of nutrient and water transporters in plants under abiotic stress. Physiol Plant 171(4):595–619

Bilichak A, Gaudet D, Laurie J (2020) Emerging genome engineering tools in crop research and breeding. In: Cereal genomics. Springer, pp 165–181

Bouain N, Krouk G, Lacombe B, Rouached H (2019) Getting to the root of plant mineral nutrition: combinatorial nutrient stresses reveal emergent properties. Trends Plant Sci 24(6):542–552

Bouis HE (2003) Micronutrient fortification of plants through plant breeding: can it improve nutrition in man at low cost? Proc Nutr Soc 62(2):403–411

Briat J-F, Gojon A, Plassard C, Rouached H, Lemaire G (2020) Reappraisal of the central role of soil nutrient availability in nutrient management in light of recent advances in plant nutrition at crop and molecular levels. Eur J Agron 116:126069

Burgess AJ, Masclaux-Daubresse C, Strittmatter G, Weber AP, Taylor SH, Harbinson J et al (2022) Improving crop yield potential: underlying biological processes and future prospects. Food Energy Secur 12:e435

Celletti S, Astolfi S, Guglielmo N, Colla G, Cesco S, Mimmo T (2020) Evaluation of a legume-derived protein hydrolysate to mitigate iron deficiency in plants. Agronomy 10(12):1942

Chaudhry AH, Nayab S, Hussain SB, Ali M, Pan Z (2021) Current understandings on magnesium deficiency and future outlooks for sustainable agriculture. Int J Mol Sci 22(4):1819

Chen W, Tang L, Wang J, Zhu H, Jin J, Yang J, Fan W (2022) Research advances in the mutual mechanisms regulating response of plant roots to phosphate deficiency and aluminum toxicity. Int J Mol Sci 23(3):1137

Cominelli E, Pilu R, Sparvoli F (2020) Phytic acid and mineral biofortification strategies: from plant science to breeding and biotechnological approaches. In: Plants, vol 9. MDPI, p 553

Dai H, Wang Y, Ren C, Ji X, Zhou Y, Zhang X et al (2021) A new method for detecting Na+, K+-ATPase activity by ICP–MS: quantitative analysis on the inhibitory effect of rhein on Na+, K+-ATPase activity by ICP–MS in HCT116 cells. Biomed Chromatogr 35(12):e5199

Das P, Gogoi N, Sarkar S, Patil SA, Hussain N, Barman S et al (2021) Nano-based soil conditioners eradicate micronutrient deficiency: soil physicochemical properties and plant molecular responses. Environ Sci Nano 8(10):2824–2843

Dhaliwal SS, Sharma V, Shukla AK, Verma V, Behera SK, Singh P et al (2021) Comparative efficiency of mineral, chelated and nano forms of zinc and iron for improvement of zinc and iron in chickpea (Cicer arietinum L.) through biofortification. Agronomy 11(12):2436

Ebbisa A (2022) Mechanisms underlying cereal/legume intercropping as nature-based biofortification: a review. Food Prod Process Nutr 4(1):1–17

Elanchezhian R, Krishnapriya V, Pandey R, Rao AS, Abrol YP (2015) Physiological and molecular approaches for improving phosphorus uptake efficiency of crops Curr Sci:1271–1279

El-Ramady H, Hajdú P, Törős G, Badgar K, Llanaj X, Kiss A et al (2022) Plant nutrition for human health: a pictorial review on plant bioactive compounds for sustainable agriculture. Sustainability 14(14):8329

Flores T, Karpova O, Su X, Zeng P, Bilyeu K, Sleper DA et al (2008) Silencing of GmFAD3 gene by siRNA leads to low α-linolenic acids (18: 3) of fad3-mutant phenotype in soybean [Glycine max (Merr.)]. Transgenic Res 17(5):839–850

Foley JK, Michaux KD, Mudyahoto B, Kyazike L, Cherian B, Kalejaiye O et al (2021) Scaling up delivery of biofortified staple food crops globally: paths to nourishing millions. Food Nutr Bull 42(1):116–132

Gadanya M, Ismail R, Lawal M (2022) A comparative study on phytochemical, anti-nutrient and mineral composition of some frutis. Biosci J Fudma 3(1):45–51

García-Sánchez F, Simón-Grao S, Martínez-Nicolás JJ, Alfosea-Simón M, Liu C, Chatzissavvidis C et al (2020) Multiple stresses occurring with boron toxicity and deficiency in plants. J Hazard Mater 397:122713

Garg M (2022) Genetically modified crops and their applications. In: Plant genomics for sustainable agriculture. Springer, Singapore, pp 71–98
Gill RA, Ahmar S, Ali B, Saleem MH, Khan MU, Zhou W, Liu S (2021) The role of membrane transporters in plant growth and development, and abiotic stress tolerance. Int J Mol Sci 22(23):12792
Girija A, Shotan D, Hacham Y, Amir R (2020) The level of methionine residues in storage proteins is the main limiting factor of protein-bound-methionine accumulation in arabidopsis seeds. Front Plant Sci 11:1136
Gu D, Zhou X, Ma Y, Xu E, Yu Y, Liu Y et al (2021) Expression of a Brassica napus metal transport protein (BnMTP3) in Arabidopsis thaliana confers tolerance to Zn and Mn. Plant Sci 304:110754
Guan M, Chen M, Cao Z (2021) NRT2. 1, a major contributor to cadmium uptake controlled by high-affinity nitrate transporters. Ecotoxicol Environ Saf 218:112269
Hashiro S, Yasueda H (2022) RNA interference-based pesticides and antiviral agents: microbial overproduction systems for double-stranded RNA for applications in agriculture and aquaculture. Appl Sci 12(6):2954
Horemans N, Foyer CH, Asard H (2000) Transport and action of ascorbate at the plant plasma membrane. Trends Plant Sci 5(6):263–267
Hossain MA, Piyatida P, da Silva JAT, Fujita M (2012) Molecular mechanism of heavy metal toxicity and tolerance in plants: central role of glutathione in detoxification of reactive oxygen species and methylglyoxal and in heavy metal chelation. J Bot 2012:1
Ishijima S (2021) Magnesium and Mg2+ transport proteins in cells magnesium and Mg2+ transport proteins in cells. J Biol Macromol 21(2):55–67
Islam MS (2022) Nitrate transport in plant through soil-root-shoot systems: a molecular view. J Plant Nutr 45(11):1748–1763
Jaiswal DK, Krishna R, Chouhan GK, de Araujo Pereira AP, Ade AB, Prakash S et al (2022) Biofortification of minerals in crops: current scenario and future prospects for sustainable agriculture and human health. Plant Growth Regul 98:1–18
Jha UC, Nayyar H, Parida SK, Deshmukh R, von Wettberg EJ, Siddique KH (2022) Ensuring global food security by improving protein content in major grain legumes using breeding and 'omics' tools. Int J Mol Sci 23(14):7710
Joshi-Saha A, Sethy SK, Misra G, Dixit G, Srivastava A, Sarker A (2022) Biofortified legumes: present scenario, possibilities and challenges. Field Crop Res 279:108467
Kailasam S, Peiter E (2021) A path toward concurrent biofortification and cadmium mitigation in plant-based foods. New Phytol 232(1):17–24
Karaca N, Ates D, Nemli S, Ozkuru E, Yilmaz H, Yagmur B et al (2020) Identification of SNP markers associated with iron and zinc concentrations in Cicer seeds. Curr Genomics 21(3):212–223
Karmakar A, Taufiqa S, Baig MJ, Molla KA (2022) Increasing disease resistance in host plants through genome editing. Proc Indian Natl Sci Acad 88:1–13
Kim W-S, Krishnan HB (2019) Impact of co-expression of maize 11 and 18 kDa δ-zeins and 27 kDa γ-zein in transgenic soybeans on protein body structure and sulfur amino acid content. Plant Sci 280:340–347
Kim W-S, Chronis D, Juergens M, Schroeder AC, Hyun SW, Jez JM, Krishnan HB (2012) Transgenic soybean plants overexpressing O-acetylserine sulfhydrylase accumulate enhanced levels of cysteine and Bowman–Birk protease inhibitor in seeds. Planta 235(1):13–23
Kim W-S, Sun-Hyung J, Oehrle NW, Jez JM, Krishnan HB (2020) Overexpression of ATP sulfurylase improves the sulfur amino acid content, enhances the accumulation of Bowman–Birk protease inhibitor and suppresses the accumulation of the β-subunit of β-conglycinin in soybean seeds. Sci Rep 10(1):1–13
Kiran K (2020) Advanced approaches for biofortification. In: Advances in agri-food biotechnology. Springer, Singapore, pp 29–55
Koç E, Karayiğit B (2021) Assessment of biofortification approaches used to improve micronutrient-dense plants that are a sustainable solution to combat hidden hunger. J Soil Sci Plant Nutr:1–26

Kochhar S, Gujral SK (2020) Plant physiology: theory and applications. Cambridge University Press, Cambridge
Kumar A, Kumar V, Krishnan V, Hada A, Marathe A, Jolly M, Sachdev A (2019) Seed targeted RNAi-mediated silencing of GmMIPS1 limits phytate accumulation and improves mineral bioavailability in soybean. Sci Rep 9(1):1–13
Kumar A, Anju T, Kumar S, Chhapekar SS, Sreedharan S, Singh S et al (2021) Integrating omics and gene editing tools for rapid improvement of traditional food plants for diversified and sustainable food security. Int J Mol Sci 22(15):8093
Li M, Wang T, Zhang H, Liu S, Li W, Abou Elwafa SF, Tian H (2022) Tanrt2. 1-6b is a dual-affinity nitrate transporter contributing to nitrogen uptake in bread wheat under both nitrogen deficiency and sufficiency. Crop J 10(4):993–1005
Liao F, Lilay GH, Castro PH, Azevedo H, Assunção AG (2022) Regulation of the zinc deficiency response in the legume model Medicago truncatula. Front Plant Sci 13:916168
Liu S, Liao LL, Nie MM, Peng WT, Zhang MS, Lei JN et al (2020) A VIT-like transporter facilitates iron transport into nodule symbiosomes for nitrogen fixation in soybean. New Phytol 226(5):1413–1428
Mahto RK, Singh C, Chandana B, Singh RK, Verma S, Gahlaut V et al (2022) Chickpea biofortification for cytokinin dehydrogenase via genome editing to enhance abiotic-biotic stress tolerance and food security. Front Genet 13:900324
Malhotra H, Pandey R, Sharma S, Bindraban PS (2020) Foliar fertilization: possible routes of iron transport from leaf surface to cell organelles. Arch Agron Soil Sci 66(3):279–300
Mansour MMF (2022) Role of vacuolar membrane transport systems in plant salinity tolerance. J Plant Growth Regul:1–38
Maqbool A, Abrar M, Bakhsh A, Çalışkan S, Khan HZ, Aslam M, Aksoy E (2020) Biofortification under climate change: the fight between quality and quantity. In: Environment, climate, plant and vegetation growth. Springer, pp 173–227
Marques E, Darby HM, Kraft J (2021) Benefits and limitations of non-transgenic micronutrient biofortification approaches. Agronomy 11(3):464
Marzec M, Hensel G (2019) More precise, more universal and more specific–the next generation of RNA-guided endonucleases for genome editing. FEBS J 286(23):4657–4660
Melini V, Melini F (2021) Functional components and anti-nutritional factors in gluten-free grains: a focus on quinoa seeds. Foods 10(2):351
Michalak A, Wdowikowska A, Janicka M (2022) Plant plasma membrane proton pump: one protein with multiple functions. Cell 11(24):4052
Mittal U, Kumar V, Kukreja S, Singh B, Pandey NK, Goutam U (2022) Role of beneficial elements in developing resilience to abiotic and biotic stresses in plants: present status and future prospects. J Plant Growth Regul:1–25
Msungu SD, Mushongi AA, Venkataramana PB, Mbega ER (2022) A review on the trends of maize biofortification in alleviating hidden hunger in sub-Sahara Africa. Sci Hortic 299:111029
Mustafa M, Mayes S, Massawe F (2019) Crop diversification through a wider use of underutilised crops: a strategy to ensure food and nutrition security in the face of climate change. In: Sustainable solutions for food security. Springer, pp 125–149
Pahalvi HN, Rafiya L, Rashid S, Nisar B, Kamili AN (2021) Chemical fertilizers and their impact on soil health. In: Microbiota and biofertilizers, vol 2. Springer, pp 1–20
Pandey R, Vengavasi K, Hawkesford MJ (2021) Plant adaptation to nutrient stress, vol 26. Springer, pp 1–4
Penn CJ, Camberato JJ (2019) A critical review on soil chemical processes that control how soil pH affects phosphorus availability to plants. Agriculture 9(6):120
Pereira AMP (2014) Study of the Fe uptake systems in two different rice cultivars
Poirier Y, Jaskolowski A, Clúa J (2022) Phosphate acquisition and metabolism in plants. Curr Biol 32(12):R623–R629
Praharaj S, Skalicky M, Maitra S, Bhadra P, Shankar T, Brestic M et al (2021) Zinc biofortification in food crops could alleviate the zinc malnutrition in human health. Molecules 26(12):3509

Prom-U-Thai C, Rashid A, Ram H, Zou C, Guilherme LRG, Corguinha APB et al (2020) Simultaneous biofortification of rice with zinc, iodine, iron and selenium through foliar treatment of a micronutrient cocktail in five countries. Front Plant Sci 11:589835

Ram H, Sardar S, Gandass N (2021) Vacuolar iron transporter (like) proteins: regulators of cellular iron accumulation in plants. Physiol Plant 171(4):823–832

Rehman AU, Masood S, Khan NU, Abbasi ME, Hussain Z, Ali I (2021) Molecular basis of iron biofortification in crop plants; A step towards sustainability. Plant Breed 140(1):12–22

Romera FJ, Lan P, Rodríguez-Celma J, Pérez-Vicente R (2021) Nutrient interactions in plants, vol 12. Frontiers Media SA, p 782505

Schmidt W (2003) Iron solutions: acquisition strategies and signaling pathways in plants. Trends Plant Sci 8(4):188–193

Shafi A, Zahoor I (2020) Abiotic and biotic stress-induced alterations in the micronutrient status of plants. In: Plant micronutrients. Springer, pp 285–309

Shahzad R, Jamil S, Ahmad S, Nisar A, Khan S, Amina Z et al (2021) Biofortification of cereals and pulses using new breeding techniques: current and future perspectives. Front Nutr 665

Sharma JK, Santal AR, Singh NP (2022) Improvement of seed protein quality in some important food crops using genetic engineering approaches. In: Biotechnology and crop improvement. CRC Press, pp 107–119

Shi H, Chen H, Gu Z, Zhang H, Chen W, Chen YQ (2016) Application of a delta-6 desaturase with α-linolenic acid preference on eicosapentaenoic acid production in Mortierella alpina. Microb Cell Factories 15(1):1–14

Shi Y, Guo S, Zhao X, Xu M, Xu J, Xing G et al (2022) Comparative physiological and transcriptomics analysis revealed crucial mechanisms of silicon-mediated tolerance to iron deficiency in tomato. Front Plant Sci 13:5288

Singer SD, Laurie JD, Bilichak A, Kumar S, Singh J (2021) Genetic variation and unintended risk in the context of old and new breeding techniques. Crit Rev Plant Sci 40(1):68–108

Sohn SI, Pandian S, Oh YJ, Kang HJ, Cho WS, Cho YS (2021) Metabolic engineering of isoflavones: an updated overview. Front Plant Sci 12:670103

Song S, Hou W, Godo I, Wu C, Yu Y, Matityahu I et al (2013) Soybean seeds expressing feedback-insensitive cystathionine γ-synthase exhibit a higher content of methionine. J Exp Bot 64(7):1917–1926

Song B, Tang Z, Li X, Li J, Zhang M, Zhao K et al (2020) Mining and evolution analysis of lateral organ boundaries domain (LBD) genes in Chinese white pear (Pyrus bretschneideri). BMC Genomics 21(1):1–17

Soumya PR, Vengavasi K, Pandey R (2022) Adaptive strategies of plants to conserve internal phosphorus under P deficient condition to improve P utilization efficiency. Physiol Mol Biol Plants 28:1–13

Srivastav P, Vutukuru M, Ravindran G, Awad MM (2022) Biofortification—present scenario, possibilities and challenges: a scientometric approach. Sustainability 14(18):11632

Stachiw S, Bicalho B, Grant-Weaver I, Noernberg T, Shotyk W (2019) Trace elements in berries collected near upgraders and open pit mines in the Athabasca Bituminous Sands Region (ABSR): distinguishing atmospheric dust deposition from plant uptake. Sci Total Environ 670:849–864

Tang R-J, Luan M, Wang C, Lhamo D, Yang Y, Zhao F-G et al (2020) Plant membrane transport research in the post-genomic era. Plant Commun 1(1):100013

Teklić T, Parađiković N, Špoljarević M, Zeljković S, Lončarić Z, Lisjak M (2021) Linking abiotic stress, plant metabolites, biostimulants and functional food. Ann Appl Biol 178(2):169–191

Turnbull C, Lillemo M, Hvoslef-Eide TA (2021) Global regulation of genetically modified crops amid the gene edited crop boom–a review. Front Plant Sci 12:630396

Uppal C, Kaur A, Sharma C (2021) Genome engineering for nutritional improvement in pulses. In: Genome engineering for crop improvement. John Wiley & Sons, Inc, Hoboken, pp 157–180

Urwat U, Ahmad SM, Masi A, Ganai NA, Murtaza I, Khan I, Zargar SM (2021) Fe and Zn stress induced gene expression analysis unraveled mechanisms of mineral homeostasis in common bean (Phaseolus vulgaris L.). Sci Rep 11(1):1–17
Vasconcelos MW, Gruissem W, Bhullar NK (2017) Iron biofortification in the 21st century: setting realistic targets, overcoming obstacles, and new strategies for healthy nutrition. Curr Opin Biotechnol 44:8–15
Vélez-Bermúdez IC, Schmidt W (2022) Plant strategies to mine iron from alkaline substrates. Plant Soil:1–25
Vengavasi K, Pandey R, Soumya P, Hawkesford MJ, Siddique KH (2021) Below-ground physiological processes enhancing phosphorus acquisition in plants. Plant Physiol Rep 26(4):600–613
Vijayalakshmi U (2022) Insights into zinc and silver co-substitution over cytotoxicity, antibacterial efficacy, and bone regeneration capabilities of strontium phosphosilicate. Ceram Int 48(4):5054–5065
Vishwakarma K, Mishra M, Patil G, Mulkey S, Ramawat N, Pratap Singh V et al (2019) Avenues of the membrane transport system in adaptation of plants to abiotic stresses. Crit Rev Biotechnol 39(7):861–883
Wang W, Ding G-D, White PJ, Wang X-H, Jin K-M, Xu F-S, Shi L (2019) Mapping and cloning of quantitative trait loci for phosphorus efficiency in crops: opportunities and challenges. Plant Soil 439(1):91–112
Wang X, He Y, Gao Q, Yang D, Liang J (2021) Approaches to evaluate nutrition of minerals in food. Food Sci Human Wellness 10(2):141–148
Whitcomb SJ, Rakpenthai A, Brückner F, Fischer A, Parmar S, Erban A et al (2020) Cysteine and methionine biosynthetic enzymes have distinct effects on seed nutritional quality and on molecular phenotypes associated with accumulation of a methionine-rich seed storage protein in rice. Front Plant Sci 11:1118
Wu X, Islam AF, Limpot N, Mackasmiel L, Mierzwa J, Cortés AJ, Blair MW (2020) Genome-wide Snp identification and association mapping for seed mineral concentration in mung bean (Vigna Radiata L.). Front Genet 11:656
Xie Q, Ma L, Tan P, Deng W, Huang C, Liu D et al (2020) Multiple high-affinity K+ transporters and ABC transporters involved in K+ uptake/transport in the potassium-hyperaccumulator plant Phytolacca acinosa Roxb. Plan Theory 9(4):470
Yadav B, Jogawat A, Lal SK, Lakra N, Mehta S, Shabek N, Narayan OP (2021) Plant mineral transport systems and the potential for crop improvement. Planta 253(2):1–30
Yağız AK, Yavuz C, Naeem M, Dangol SD, Aksoy E (2022) Genome editing for nutrient use efficiency in crops. In: Principles and practices of OMICS and genome editing for crop improvement. Springer, pp 347–383
Yang Q, Zhao D, Zhang C, Sreenivasulu N, Sun SS-M, Liu Q (2022) Lysine biofortification of crops to promote sustained human health in the 21st century. J Exp Bot 73(5):1258–1267
Zeng H, Zhang X, Ding M, Zhang X, Zhu Y (2019) Transcriptome profiles of soybean leaves and roots in response to zinc deficiency. Physiol Plant 167(3):330–351
Zhang Z, Hu B, Chu C (2020) Towards understanding the hierarchical nitrogen signalling network in plants. Curr Opin Plant Biol 55:60–65
Zhao Z, Li M, Xu W, Liu J-H, Li C (2022) Genome-wide identification of NRT gene family and expression analysis of nitrate transporters in response to salt stress in Poncirus trifoliata. Genes 13(7):1115
Zuluaga DL, Sonnante G (2019) The use of nitrogen and its regulation in cereals: structural genes, transcription factors, and the role of miRNAs. Plan Theory 8(8):294

Chapter 21
Medicinal Legumes in Turkey: A Gift of Nature for Bios

Muzaffer Barut, Asiye Sena Cavdar, and Leyla Sezen Tansi

Abstract Leguminosae is one of the largest families of ethnopharmacological importance. Most of the legumes are biochemical factories for the production of a variety of secondary metabolites (SMs) that are effective in treating various diseases. These important SMs produced by legumes are directly or indirectly of great importance for bios. SMs consisting of terpenoids, phenols, and nitrogen-containing compounds exhibit various biological activities such as anti-inflammatory, antitumor, antiviral, and antibacterial. In the meantime, the biofortification of plant foods has recently garnered a great deal of consideration, as it attempts to improve the consumption of vitamins, minerals, and antioxidants, which are essential because of their importance to human health and disease prevention. Moreover, the definition of biofortification has been expanded to include not just the micronutrient enhancement but also the crop nutritional content enhancement by the addition of SMs. In this perspective, the investigation of a plant's SMs is crucial to the improvement of a new breed of plants that partially compensate for Western diets' inferior nutritional quality with improved quality food sources. This chapter articulates different aspects of medicinal legumes including plant breeding, biotechnology techniques, and new breakthroughs in genomic approaches as a useful implement for biofortification of SMs and the improvement of healthful food options.

1 Introduction

Leguminosae is one of the most diverse flowering plants family containing over 900 genera and 26,800 species around the world (Anonymous 2013). The family is widespread and the third largest land plant family with regard to the number of species, after the Orchidaceae and the Compositae. It includes herbs, shrubs, trees, and perennial or annual plants that are simply identified by their fruits (legume). The

M. Barut (✉) · A. S. Cavdar · L. S. Tansi
Department of Field Crops, Faculty of Agriculture, Çukurova University, Adana, Turkey
e-mail: mbarut@cu.edu.tr

M. A. Nadeem et al. (eds.), *Legumes Biofortification*,
https://doi.org/10.1007/978-3-031-33957-8_21

largest genera are Astragalus (over 2700 species), Acacia (around 1500 species), Mimosa (over 750 species), Crotalaria (over 750 species), and Racosperma (over 700 species). Leguminous plants are not only used as food, green manures, and forages, but they are also used to make a wide range of natural products, such as poisons, drugs, flavors, and dyes. Legumes are important in traditional medicine around the world. China uses more than 50 leguminous species as sources of medicines for the treatment of about 40 ailments (Quattrocchi 2012). Likewise, India uses about 55% of species of Leguminosae as medicines (Quattrocchi 2012). Antithrombotic warfarin, derived from coumarin of sweet clover, and antidiabetic metformin, derived from guanidine of sainfoin, were produced from legumes as medicine.

The potential of legumes to form root nodules and fix nitrogen in symbiosis with suitable rhizobium bacteria is well documented. This feature makes legumes favorable for nitrogen-deficient soils, to provide high-protein plant material (Raj et al. 2020). Aside from their use as fodder, forage legumes have gained attention for nonfood applications such as phytoremediation, biofuel, and manure (Stoddard 2008). Moreover, legumes produce a diverse range of secondary metabolites (SMs), including alkaloids, terpenoids, and phenolics. These phytochemicals defend legumes from competitive plants, oxidative stress, and herbivores, and a bunch of them have livestock-friendly properties (Wink 2013). Hence, considering their widespread distribution and extensive cultivation, these plants appear to be an underutilized, potentially valuable resource for the production of SMs.

Food fortification with various biological and chemical additions, as well as changes to the food processing system, is required. Several ways have been implemented in the current biotechnology era to increase the nutritional quality of food through value-addition and biofortification of plants (Zhu et al. 2020). These emerging trends in plant biotechnology have contributed to the biofortification of SMs with therapeutic properties in plants. Plants biofortified for SMs can help strengthen the immune system by increasing the adequacy of daily intake of beneficial compounds such as antioxidants among individuals throughout the life cycle. Thus, biofortified plants are a sustainable option that can be extremely advantageous for people who have limited access to a variety of medication supplies.

2 Plant Genetic Resources of Legumes in Turkey

There are 11,707 plant taxa belonging to 167 families exist in the flora of Turkey, and 3649 of these taxa are endemic (Güner et al. 2012). The level of endemism, as well as the total number of plant species, is thought to affect the richness of a region's flora. Floristic records show that more than 3000 plant species are endemic to Turkey, accounting for around 30% of the overall flora (Özhatay et al. 2017).

There are 1145 species belonging to the Leguminosae family in Turkey, and 39% of these species are endemic. The share of the genera belonging to the Leguminosae family in Turkey is given in Fig. 21.1. Legume genetic resources in Turkey and their distribution based on the grid squares are presented (Table 21.1 and Fig. 21.2).

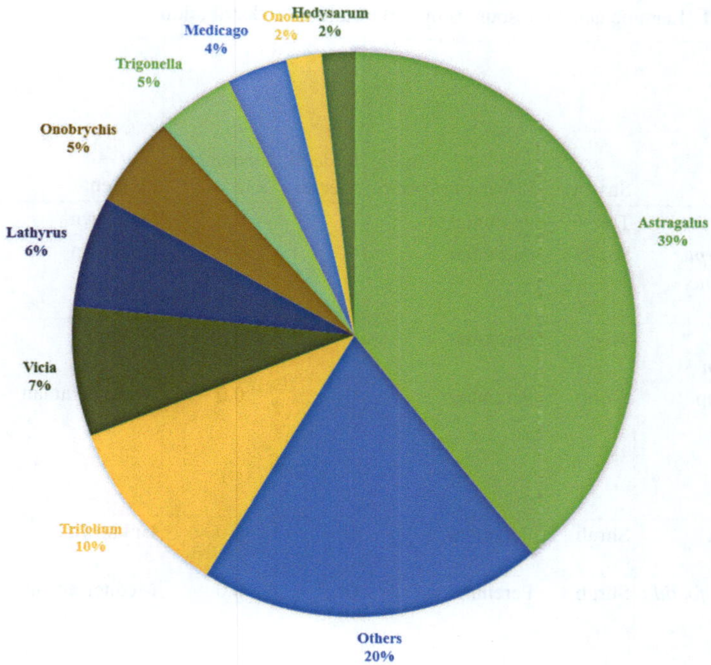

Fig. 21.1 The share of the genera belonging to the Leguminosae family in Turkey

Thirty-six of the genera belonging to the family are herbs, 21 of them are shrubs, 6 of them are herbs and shrubs, 4 of them are trees, and 2 of them are trees or shrubs. The leading genera by the number of species are *Astragalus* spp. (447), *Trifolium* spp. (119), *Vicia* spp. (85), *Lathyrus* spp. (73), *Onobrychis* spp. (60), *Trigonella* spp. (51), *Medicago* spp. (40), *Ononis* spp. (23), *Hedysarum* spp. (22), and *Lotus* spp. (20).

3 Health Benefits of Medicinal Legumes

Various plant parts, such as the root, the leaf, and the stem, are the source of so many valuable medications. Several local residents continue to rely on medicinal plants for primary medical care and the treatment of a variety of illnesses (Samydurai et al. 2012). Some legumes are possible sources of glycosides, biologics, antibiotics, and alkaloids, which are used in medicine manufactured by the pharmaceutical industry (Table 21.2). Large amounts of SMs of traditional and therapeutic importance are found in legumes. For example, isoflavones, phytosterols, saponins, and other basic nutritive ingredients are all found in soybeans and have strong therapeutic properties. *Astragalus* is a common plant that is used as a medicine and is a popular herbal product around the world. Flavonoids, saponins, amino acids, polysaccharides,

Table 21.1 Legume genetic resources in Turkey and their distribution

Species	Structure	Life cycle	Total species	Endemic species	Endemism rate	Element	Distribution of the species based on grids in Turkey
Acacia karroo	Tree	Perennial	1	0	0.0	Unknown	"C4, C5"
Adenocarpus complicatus	Shrub	Perennial	1	0	0.0	Unknown	"B1, B2, C2, C3, C4, C6"
Albizia julibrissin	Tree	Perennial	1	0	0.0	Hirkan	"A8"
Alhagi spp.	Shrub	Perennial	2	0	0.0	Irano-Turanian	"A3, A5, A9, B4, B5, B10, C1, C2, C4, C5, C6, C8"
Amorpha fruticosa	Shrub	Perennial	1	0	0.0	Unknown	"A2"
Anagyris foetida	Shrub	Perennial	1	0	0.0	Mediterranean	"A1, B1, B8, C1, C2, C3, C4, C5, C6, C9"
Anthyllis	Herb and shrub	Annual or perennial	8	1	12.5	Mediterranean, Euro-Siberian	"A1, A2, A3, A4, A5, A6, A7, A8, A9, B1, B2, B3, B4, B5, B6, B7, B8, B9, B10, C1, C2, C3, C4, C5, C6, C7, C8, C9, C10"
Arachis hypogaea	Herb	Annual	1	0	0.0	Unknown	"C2, C3, C4, C5, C6"
Argyrolobium spp.	Herb	Perennial	3	0	0.0	Sahra-Arabistan	"A2, A4, A5, A6, A7, A8, A9, B2, B4, C5, C6"

(continued)

Table 21.1 (continued)

Species	Structure	Life cycle	Total species	Endemic species	Endemism rate	Element	Distribution of the species based on grids in Turkey
Astragalus spp.	Herb and small shrub	Annual or perennial	447	277	62.0	Mediterranean, Euro-Siberian, Irano-Turanian	"A1, A2, A3, A4, A5, A6, A7, A8, A9, B1, B2, B3, B4, B5, B6, B7, B8, B9, B10, C1, C2, C3, C4, C5, C6, C7, C8, C9, C10"
Biserrula pelecinus	Herb	Annual	1	0	0.0	Mediterranean	"B1, C1, C2, C3, C5, C6"
Caesalpinia gilliesii	Shrub	Perennial	1	0	0.0	Unknown	"A2, B1"
Calicotome villosa	Shrub	Perennial	1	0	0.0	Mediterranean	"A1, A2, B1, C1, C2, C3, C4, C5, C6"
Caragana spp.	Shrub	Perennial	2	1	50.0	Irano-Turanian	"A8, C4"
Ceratonia siliqua	Tree or shrub	Perennial	1	0	0.0	Mediterranean	"A2, C2, C3, C4, C6"
Cercis spp.	Tree or shrub	Perennial	2	0	0.0	Unknown	"A1, A2, A3, A4, A5, B1, C2, C3, C5, C6, C8, C9"
Chamaecytisus spp.	Shrub	Perennial	11	4	36.4	E. Mediterranean, Euro-Siberian	"A1, A2, A3, A4, A5, A6, A7, A8, A9, B1, B2, B3, B5, C2, C3, C5, C6"
Chesneya spp.	Herb	Perennial	2	1	50.0	Irano-Turanian	"A7, B7, A8, A9"
Chronanthus orientalis	Shrub	Perennial	1	1	100.0	E. Mediterranean	"B1, B2"

(continued)

Table 21.1 (continued)

Species	Structure	Life cycle	Total species	Endemic species	Endemism rate	Element	Distribution of the species based on grids in Turkey
Cicer spp.	Herb	Annual or perennial	10	4	40.0	E. Mediterranean, Irano-Turanian	"A1, A2, A4, A5, A6, A7, A8, B1, B2, B4, B5, B6, B7, B8, B9, C2, C3, C4, C5, C6, C7, C8, C9"
Colutea spp.	Shrub	Perennial	5	1	20.0	E. Mediterranean	"A1, A2, A3, A4, A5, A6, A7, A8, A9, B1, B2, B3, B4, B5, B6, B7, B8, B9, C2, C3, C5, C6, C9"
Coronilla spp.	Shrub	Annual or perennial	11	1	9.1	E. Mediterranean	"A1, A2, A3, A4, A5, A6, A7, A8, A9, B1, B2, B3, B4, B5, B6, B7, B8, B9, B10, C1, C2, C3, C4, C5, C6, C8, C9, C10"
Cytisopsis spp.	Shrub	Perennial	2	1	50.0	E. Mediterranean	"B2, C1, C2, C5, C6"
Cytisus spp.	Shrub	Perennial	4	1	25.0	Unknown	"A1, A3, A5, C1, C3"
Dorycnium spp.	Herb and shrub	Perennial	9	4	44.4	Mediterranean, Irano-Turanian, Karadeniz	"A1, A2, A3, A4, A5, A6, A7, A8, A9, B1, B2, B3, B4, B5, B7, B8, C1, C2, C3, C4, C5, C6, C9"

(continued)

Table 21.1 (continued)

Species	Structure	Life cycle	Total species	Endemic species	Endemism rate	Element	Distribution of the species based on grids in Turkey
Ebenus spp.	Shrub	Perennial	17	17	100.0	E. Mediterranean, Irano-Turanian	"A5, B2, B3, B4, B5, B6, B7, C2, C3, C4, C5"
Factorovskya aschersoniana	Herb	Annual	1	0	0.0	E. Mediterranean	"C5"
Galega officinalis	Herb	Perennial	1	0	0.0	Euro-Siberian	"A1, A2, A3, A4, A5, A6, B1, B2, B3, B4, B5, C3, C8"
Genista spp.	Shrub	Perennial	13	4	30.8	E. Mediterranean, Euro-Siberian, Irano-Turanian	"A1, A2, A3, A4, A5, A6, A7, A8, A9, B1, B2, B3, B4, B5, B6, B7, B8, C1, C2, C3, C4, C5, C6"
Gleditsia triancanthos	Tree	Perennial	1	0	0.0	Unknown	"A2"
Glycyrrhiza spp.	Herb	Perennial	7	3	42.9	E. Mediterranean, Irano-Turanian	"A1, A2, A4, A5, A6, A7, A9, B1, B2, B3, B4, B5, B6, B7, B8, B9, B10, C1, C2, C3, C4, C5, C6, C7, C8, C9, C10"
Gonocytisus spp.	Shrub	Perennial	3	1	33.3	E. Mediterranean	"A1, A2, B1, B2, C2, C3, C4, C5, C6"
Halimodendron halodendron	Shrub	Perennial	1	0	0.0	Unknown	"A9"
Hammatolobium lotoides	Herb	Perennial	1	0	0.0	E. Mediterranean	"C5, C6"

(continued)

Table 21.1 (continued)

Species	Structure	Life cycle	Total species	Endemic species	Endemism rate	Element	Distribution of the species based on grids in Turkey
Hedysarum spp.	Herb	Annual or perennial	22	12	54.5	E. Mediterranean, Irano-Turanian	"A1, A2, A3, A4, A5, A7, A8, A9, B1, B2, B3, B4, B5, B6, B7, B8, B9, B10, C2, C3, C4, C5, C6, C7, C8, C9, C10"
Hippocrepis spp.	Herb	Annual	4	0	0.0	Mediterranean	"A1, A2, A5, A6, B1, B2, C1, C2, C3, C4, C5, C6, C7, C8, C9"
Hymenocarpus circinnatus	Herb	Annual	1	0	0.0	Mediterranean	"A1, A2, A5, B1, C1, C2, C3, C4, C5, C6, C9"
Lathyrus spp.	Herb	Annual, biennial, or perennial	73	22	30.1	Mediterranean, Euro-Siberian, Irano-Turanian	"A1, A2, A3, A4, A5, A6, A7, A8, A9, B1, B2, B3, B4, B5, B6, B7, B8, B9, B10, C1, C2, C3, C4, C5, C6, C7, C8, C9, C10"
Lens spp.	Herb	Annual or perennial	7	1	14.3	Mediterranean, Irano-Turanian	"A1, A2, A5, A6, B1, B2, B4, B6, B7, B8, C2, C3, C4, C5, C6, C7, C8, C9"
Lotononis genistoides	Herb	Perennial	1	0	0.0	Irano-Turanian	"A2, B1, B2, B3, B4, B5, C2, C3, C4, C5"

(continued)

Table 21.1 (continued)

Species	Structure	Life cycle	Total species	Endemic species	Endemism rate	Element	Distribution of the species based on grids in Turkey
Lotus spp.	Herb	Annual or perennial	20	2	10.0	Mediterranean, Irano-Turanian	"A1, A2, A3, A4, A5, A6, A7, A8, A9, B1, B2, B3, B4, B5, B6, B7, B8, B9, B10, C1, C2, C3, C4, C5, C6, C7, C8, C9, C10"
Lupinus spp.	Herb	Annual	7	1	14.3	E. Mediterranean	"A1, A2, A3, A5, B1, C1, C2, C3, C4, C6"
Medicago spp.	Herb and shrub	Annual or perennial	40	1	2.5	Mediterranean, Irano-Turanian	"A1, A2, A3, A4, A5, A6, A7, A8, A9, B1, B2, B3, B4, B5, B6, B7, B8, B9, C1, C2, C3, C4, C5, C6, C7, C8, C9, C10"
Melilotus spp.	Herb	Annual	10	1	10.0	Mediterranean, Irano-Turanian	"A1, A2, A3, A4, A5, A6, A8, A9, B1, B2, B3, B4, B5, B6, B7, B8, B9, B10, C1, C2, C3, C4, C5, C6, C8, C9, C10"
Onobrychis spp.	Herb and shrub	Annual or perennial	60	35	58.3	Mediterranean, Irano-Turanian	"A1, A2, A3, A4, A5, A6, A7, A8, A9, B1, B2, B3, B4, B5, B6, B7, B8, B9, B10, C1, C2, C3, C4, C5, C6, C7, C8, C9, C10"

(continued)

Table 21.1 (continued)

Species	Structure	Life cycle	Total species	Endemic species	Endemism rate	Element	Distribution of the species based on grids in Turkey
Ononis spp.	Herb and shrub	Annual or perennial	23	4	17.4	Mediterranean, Euro-Siberian, Irano-Turanian	"A1, A2, A3, A4, A5, A6, A7, A8, B1, B2, B3, B4, B5, B6, B7, B8, B9, C1, C2, C3, C4, C5, C6, C7, C8, C9"
Ornithopus spp.	Herb	Annual	3	0	0.0	Mediterranean	"A1, A2, A5, A8, B1, C1, C2, C3, C5, C6"
Oxytropis spp.	Herb	Perennial	13	3	23.1	Irano-Turanian	"A4, A7, A8, A9, B3, B5, B6, B7, B8, B9, B10, C5, C6, C9, C10"
Phaseolus spp.	Herb	Annual or perennial	2	0	0.0	Unknown	Unknown
Pisum spp.	Herb	Annual	6	0	0.0	Mediterranean	"A1, A2, A4, A5, A8, A9, B1, B4, B5, B6, B7, B8, C1, C2, C3, C4, C5, C6, C7, C8, C9, C10"
Podocytisus caramanicus	Herb	Perennial	1	0	0.0	E. Mediterranean	"C4, C5, C6"
Prosopis farcta	Shrub	Perennial	1	0	0.0	Unknown	"B7, B8, C5, C6, C8"
Psoralea spp.	Herb	Perennial	3	0	0.0	Mediterranean, Karadeniz, Irano-Turanian	"A1, A2, A3, A5, A6, A7, A8, A9, B1, B7, B8, C2, C3, C5, C6, C7, C8"

(continued)

Table 21.1 (continued)

Species	Structure	Life cycle	Total species	Endemic species	Endemism rate	Element	Distribution of the species based on grids in Turkey
Robinia pseudoacacia	Tree	Perennial	1	0	0.0	Unknown	"A2, A3, A9"
Sartoria hedysaroides	Herb	Perennial	1	1	100.0	Unknown	"C4"
Scorpiurus muricatus varyete subvillosus	Herb	Annual	1	0	0.0	Mediterranean	"A1, A2, A5, A6, B1, B2, B4, B8, C1, C2, C3, C4, C5, C6, C8, C9"
Securigera spp.	Herb	Annual	2	0	0.0	E. Mediterranean	"A1, A2, A6, B1, C1, C3, C4, C5, C6, C9"
Sophora spp.	Herb	Perennial	3	0	0.0	Karadeniz	"A2, A3, A4, A5, A6, A8, A9, B4, B5, B6, B7, B9, B10, C5, C6"
Spartium junceum	Shrub	Perennial	1	0	0.0	Mediterranean	"A1, A2, A4, A5, A6, A7, C1, C2, C3, C5, C6"
Sphaerophysa kotschyana	Herb	Perennial	1	1	100.0	Irano-Turanian	"B5"
Teline monspessulana	Shrub	Perennial	1	0	0.0	Mediterranean	"A1, A2, C1"
Tetragonolobus spp.	Herb	Annual or Perennial	3	0	0.0	Unknown	"A2, A3, A5, A7, B1, B2, B4, B6, B7, C2, C5, C6"
Thermopsis turcica	Herb	Perennial	1	1	100.0	E. Mediterranean	"B3"

(continued)

Table 21.1 (continued)

Species	Structure	Life cycle	Total species	Endemic species	Endemism rate	Element	Distribution of the species based on grids in Turkey
Trifolium spp.	Herb	Annual, biennial, or perennial	119	12	10.1	Mediterranean, Euro-Siberian, Irano-Turanian	"A1, A2, A3, A4, A5, A6, A7, A8, A9, B1, B2, B3, B4, B5, B6, B7, B8, B9, B10C1, C2, C3, C4, C5, C6, C7, C8, C9, C10"
Trigonella spp.	Herb	Annual	51	18	35.3	Mediterranean, Irano-Turanian	"A1, A2, A3, A4, A5, A6, A7, A8, A9, B1, B2, B3, B4, B5, B6, B7, B8, B9, B10, C1, C2, C3, C4, C5, C6, C7, C8, C9, C10"
Ulex europaeus	Shrub	Perennial	1	0	0.0	Unknown	"A2, A7"
Vavilovia formosa	Herb	Perennial	1	0	0.0	Unknown	"B3, B7, B8, B9, C2, C3, C5, C9"
Vicia spp.	Herb	Annual, biennial, or perennial	85	11	12.9	E. Mediterranean, Irano-Turanian, Karadeniz	"A1, A2, A3, A4, A5, A6, A7, A8, A9, B1, B2, B3, B4, B5, B6, B7, B8, B9, B10C1, C2, C3, C4, C5, C6, C7, C8, C9, C10"
Vigna spp.	Herb	Annual	2	0	0.0	Unknown	"A2, C3"
Total			1145	448			

Reproduced from TÜBİVES (Bakış et al. 2011)

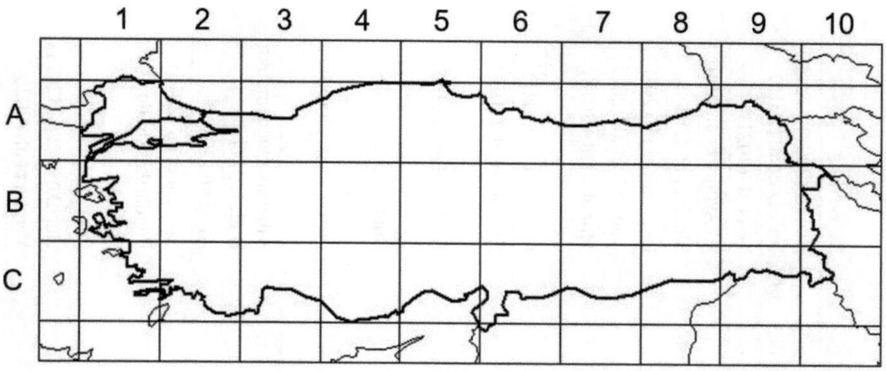

Fig. 21.2 Map of Turkey in grid squares

glycosides, alkaloids, organic acids, and trace elements are the principal types of chemical compounds found in *Astragalus* species. The main claimed benefits of the therapeutic use of medicinal plants in many diseases are their safety as well as being affordable, effective, and readily available (Hosseinzadeh et al. 2015).

4 Biofortification Techniques for the Development of Secondary Metabolites in Medicinal Legumes

Biofortification refers to a variety of techniques for enhancing the nutritious value of food crops, ranging from agronomical methods and conventional breeding to cutting-edge biotechnology. Shortly, the agronomic approaches for biofortification involve the applications to improve the concentration of minerals and vitamins. Recently, the definition of biofortification has been expanded to include not only the enhancement of micronutrients but also the improvement of plant nutritive content by the addition of SMs. Several agronomic practices, such as nitrogen fertilization, have been documented to influence the levels of SMs, such as polyphenols (Nguyen and Niemeyer 2008). Environmental conditions and the application of stressors can also influence accumulation of SMs and offer an alternate biofortification method. For example, because biosynthesis can be modified by light intensity, light response or UV exposure in a controlled environment can boost the biosynthesis of flavonoids or anthocyanins (Scarano et al. 2020).

The most popular method for selecting agronomic traits with nutritional characteristics of relevance to customers is classical breeding techniques. Genetic improvement can be accomplished by selecting plants with favorable traits, or by crossing individuals that are closely related to improve new hybrids. In both instances, genetic variety is a prerequisite for the achievement of breeding strategies designed to advance nutritional qualities. In addition, new developments in metabolomics and genomics have contributed to marker-assisted selection (MAS), and the

Table 21.2 Prominent medicinal species of the family Leguminosae native to Turkey and their medicinal uses

Scientific name	English name	Local name	Part(s)/extract	Biological activity	References
Alhagi pseudalhagi	Camelthorn	Aguldikeni	Stem gum	Antibacterial activity	Bonjar (2004) Abdul-Hafeez et al. (2015)
			Aerial parts	Osteogenic activity	Ye et al. (2021)
			Aerial parts	Antidiabetic activity	Al Ayoubi et al. (2018)
			Ethanol extract	Antiarthritic activity	Singh et al. (2018)
Anthyllis vulneraria	Woundwort	Çobangülü	Leaves and flowers	Antioxidant activity	Ouerfelli et al. (2018)
			Leaves and flowers	Antibacterial activity	Ouerfelli et al. (2021)
			Hydroalcoholic extract	Antiviral activity	Moradi et al. (2018)
			Herbal parts	Hemolytic activity	Lorenz et al. (2021)
Astragalus cicer	Chickpea milkvetch	Nohut geveni	Whole plants	Antioxidant activity	Butkutė et al. (2018)
Astragalus diphtherites	–	Yamaç geveni	Shoot and root extracts	Antioxidant and antimicrobial activity	Keskin et al. (2018)
Astragalus glycyphyllos	Liquorice milkvetch	Dev geven	Whole plants	Antioxidant activity	Butkutė et al. (2018)
Astragalus macrocephalus	–	Topaç geveni	Methanol extract	Antimutagenic activity	Ozbek et al. (2008)
Ceratonia siliqua	Carob	Keçiboynuzu	Ethanol extract	Antioxidant and anticancer effects.	Amessis-Ouchemoukh et al. (2017)
Coronilla scorpioides	Annual scorpion vetch	Akrep burçağı	Aerial parts	Antitumor activity	Al-Snafi (2016)
			Aerial parts	Antioxidant activity	El-Hela et al. (2013)
Galega officinalis	Goat's rue	Keçisedefi	Alcoholic extracts	Antibacterial activity	Pundarikakshudu et al. (2001)
			Dried aerial parts	Weight-reducing activity	Palit et al. (1999)
			Alcoholic extracts	Antidiabetic activity	Hachkova et al. (2021)
			Alcoholic extracts	Antioxidant activity	Vergun et al. (2020)

Glycyrrhiza glabra	Liquorice	Meyan	Roots	Antiulcer activity	Malek and Ghazvini (2007)
			Aqueous extract	Memory-enhancing activity	Dhingra et al. (2004)
			Isolated components	Antiallergic activity	Shin et al. (2007)
			Isolated components	Hepatoprotective activity	Wan et al. (2009)
			Roots	Anticancer activity	Sheela et al. (2006)
			Isolated components	Antiviral activity	De Simone et al. (2001)
			Crude extracts	Antitussive activity	Kuang et al. (2018)
			Alcoholic extract	Antimalarial activity	Schwikkard and van Heerden (2002)
			Methanol extract	Antioxidant activity	Latcef et al. (2012)
Halimodendron halodendron	Salt tree	Pembesalkım	Aerial parts	Antimicrobial and antioxidant activities	Wang et al. (2012)
Lotus corniculatus	Common bird's-foot trefoil	Gazal boynuzu	Alcoholic extracts	Antioxidant, antiprotozoal and, immunostimulant activities	Abdallah et al. (2021)
			Isolated components	Antibacterial activity	Dalmarco et al. (2010)
			Ethyl acetate, methanol, and water extracts	Antiamylase activity	Yerlikaya et al. (2019)
			Aerial parts	Antidepressant activity	Gürağaç Dereli et al. (2020)
			Isolated components	Anti-inflammatory	Koelzer et al. (2009)

(continued)

Table 21.2 (continued)

Scientific name	English name	Local name	Part(s)/extract	Biological activity	References
Ononis spinosa	Spiny restharrow	Kayışkıran, demirdelen, acram, şıbrık	Flowering plants, root	Antioxidant, anticancer, antidiabetic, antiparasitic, gastrointestinal, antimicrobial, respiratory, dermatological, hepato and reno-protective, antipyretic, anti-inflammatory, analgesic, immunological	Al-Snafi (2020)
Prosopis farcta	Syrian mesquite	Çediotu	Roots, leaf, pods, and seeds extract	Antibacterial activity	Sharifi-Rad et al. (2014)
			Aqueous extract of fruit	Antifungal activity	Miri et al. (2020)
			Aqueous extract of fruit	Apoptotic effects	Miri et al. (2018)
Psoralea bituminosa	Pitch trefoil	Katran yoncası,	Aqueous leaves extract	Antihyperglycemic activity	Lemouchi et al. (2015)
			The essential oil and hydrosol extracts	Antioxidant activity	Lemouchi et al. (2017)
Thermopsis turcica	Yellow licorice	Sarı meyan, piyan	Alcoholic extracts	Antioxidant, antimicrobial, and genotoxic activities	Aksoy et al. (2013) Ali and Ciğerci (2019) Korcan et al. (2009)

Trigonella foenum-graecum	Fenugreek	Çemen otu	Seeds	Antidiabetic activity	Geberemeskel et al. (2019) Skalli et al. (2019)
			Seeds	Antibacterial activity	Marzougui et al. (2012)
			Seeds	Dermatological disorders	Gul et al. (2012)
			Seeds and leaves	Kidney stones	Al-Ramahi et al. (2015)
			Plant extract	Anti-inflammatory activity	Subhashini et al. (2011), Liu et al. (2012), Pal and Mukherjee (2020)
			Seeds	Antiasthmatic activity	Emtiazy et al. (2018)
			Seeds	Antianemic activity	Chourasiya et al. (2019)
			Seeds	Antifungal activity	Sudan et al. (2020)
			Seeds	Antinociceptive activity	Kaviarasan et al. (2007)
			Seeds	Antilipidemic activity	Bordia et al. (1997) Belaïd-Nouira et al. (2012)

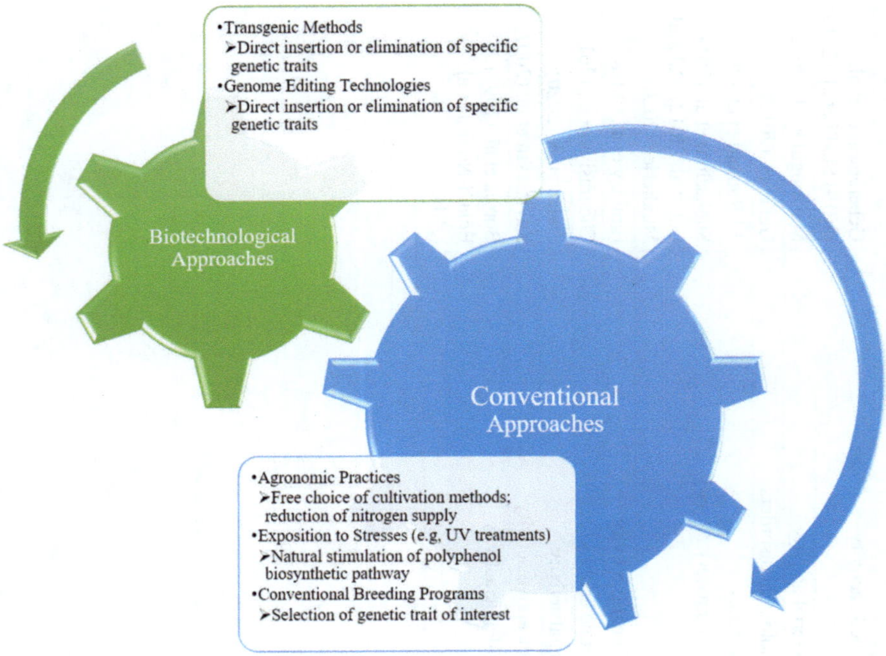

Fig. 21.3 Common biofortification techniques for the enhancement of secondary metabolites in plants

identification and mapping of the key quantitative trait loci (QTL) hold enormous promise in this field. This strategy incorporates both traditional transgenic changes and innovative breeding approaches, which are currently undergoing rapid development and are considered as the flywheel for the future generation of functional meals (Fig. 21.3).

In many cases, bioprocessing in the form of transformed hairy roots, suspension culture, or organ culture has proven to be effective in increasing the production of SMs. For instance, Shabani et al. (2009) reported that glycyrrhizin synthesis is more likely to occur by in vitro cultured *Glycyrrhiza glabra* in response to methyl jasmonate (MeJA) and salicylic acid. Moreover, the levels of trigonelline and 4-hydroxyisoleucine in cell suspension cultures treated with 100 M MeJA increased significantly in *Trigonella foenum-graecum* (MA Abd-El Mawla and Osman 2011). Furthermore, MeJA enhanced the production of camptothecin, an alkaloid, by 1.6-fold in *Astragalus flavus* (El-Sayed et al. 2022). Similarly, Feng et al. (2021) stated that MeJA accumulation stimulates calycosin-7-O-β-D-glucoside biosynthesis in adventitious roots of *Astragalus membranaceus*. In the same way, treatment of *Astragalus gossypinus* cells with Se at various doses significantly increased the amount of phenolic acids compared to controls (Maassoumi et al. 2022). Additionally, Kowalczyk et al. (2021) revealed that the accumulation of betulinic acid was enhanced in transgenic hairy roots of *Senna obtusifolia*. Saviranta et al. (2010) indicated that the total phenolic content of leaves was enhanced by elevated

Table 21.3 Biofortification of secondary metabolites in medicinal legumes

Plant Species	Application	Effect(s)	References
Arachis hypogaea	Hairy root cultures	Increased resveratrol, arachidin-1,m, and arachidin-3	Yang et al. (2015)
Astragalus flavus	MeJA	Increased camptothecin	El-Sayed et al. (2022)
Astragalus gossypinus	Se application	Increased phenolic acids	Maassoumi et al. (2022)
Astragalus membranaceus	MeJA	Increased calycosin-7-O-β-D-glucoside	Feng et al. (2021)
Crotalaria spp.	In an in vitro propagated is more than in field-cultivated medicinal preparations	Increased quantity of SMs	Nakka and Devendra (2012)
Glycyrrhiza glabra	Methyl jasmonate (MeJA) and salicylic acid	Increased glycyrrhizin synthesis	Shabani et al. (2009)
Glycyrrhiza glabra	Transgenic roots	Increased glycyrrhizin content	Shirazi et al. (2018)
Lupinus polyphyllus	Alkaloids such as coniine, papaverine, and other chemicals such as cAMP and polyamines in cell suspension cultures	Increased quinolizidine alkaloid synthesis	Wink and Witte (1983)
Lupinus polyphyllus	Cutting up the leaves	Increased in quinolizidine alkaloid synthesis	
Lupinus succulentus	High nitrogen content	Increased alkaloids were up to five times	Johnson et al. (1987).
Macrotyloma uniflorum and *Trigonella foenum-graecum*	Silver nanoparticles treated seed	Increased alkaloid, flavonoid, and tannin content	Priya et al. (2018)
Medicago sativa	Transgenic alfalfa	Increased isoflavonoid synthesis	Deavours and Dixon (2005)
Medicago truncatula, Trigonella foenum-graecum, Cicer arietinum, and *Trifolium pratense*	Seedlings, intact roots, or cell suspension cultures (in reaction to electro-elicitation)	Increased quantities of SMs	Kaimoyo et al. (2008)
Psoralea corylifolia	Jasmonic and acetyl salicylic acid in hairy root cultures	Increased daidzin synthesis	Zaheer et al. (2016)
Senna obtusifolia	In transgenic hairy roots	Increased betulinic acid	Kowalczyk et al. (2021)
Trifolium pratense	Elevated ozone	Increased total phenolic content	Saviranta et al. (2010)

(continued)

Table 21.3 (continued)

Plant Species	Application	Effect(s)	References
Trigonella foenum-graecum	MeJA	Increased levels of trigonelline and 4-hydroxyisoleucine	MA Abd-El Mawla and Osman (2011)
Trigonella foenum-graecum	Foliar application of gamma-irradiated sodium alginate and phosphorus	Increased the carotenoid and trigonelline content	Dar et al. (2016)
Trigonella foenum-graecum	Hairy roots	The expression of the Δ^{24}-reductase gene was highly upregulated in hairy roots that produced more diosgenin (eightfold)	Zolfaghari et al. (2020)
Trigonella foenum-graecum	Comparative transcriptome analysis	Increased diosgenin synthesis	Zhou et al. (2019)
Trigonella foenum-graecum	MeJA increased the expression of two key genes involved in the mevalonate pathway	Increased diosgenin synthesis	Chaudhary et al. (2015)
Trigonella foenum-graecum	Hairy-root transformation	Specialized SMs pathway elucidation	Garagounis et al. (2020)
Vicia faba	Tryptophan, potassium silicate, chitosan	Increased total phenols synthesis	Fouda et al. (2022)
Vicia faba	Underwater deficiency conditions	Increased phenolic and vicine contents of the produced seeds	Bakhoum et al.(2022)

ozone, as well as the quantities of particular phenolic components in *Trifolium pratense*. The hairy root cultures of *Arachis hypogaea* were shown to produce pharmacologically significant stilbenoids, including increases in resveratrol, arachidin-1,m, and arachidin-3 (Yang et al. 2015). Kaimoyo et al. (2008) reported that seedlings, intact roots, or cell suspension cultures of *Medicago truncatula*, *Trigonella foenum-graecum*, *Cicer arietinum*, and *Trifolium pratense* also produced higher quantities of SMs in reaction to electro-elicitation. What's more, Zaheer et al. (2016) indicated that daidzin production increased with jasmonic and acetyl salicylic acid in hairy root cultures of *Psoralea corylifolia.* In cell suspension cultures of *Lupinus polyphyllus*, foreign alkaloids such as coniine, papaverine, and other chemicals such as cAMP and polyamines can cause a short-term rise in quinolizidine alkaloid accumulation (Wink and Witte 1983). Alkaloid, flavonoid, and tannin content increased in silver nanoparticles treated seed compared to control in *Macrotyloma uniflorum* and *Trigonella foenum-graecum*, but alkaloid, flavonoid, and tannin content reduced in copper nanoparticles treated seed compared to control (Priya et al. 2018). Besides, total phenols were significantly increased in *Vicia faba* by application of 75 ppm tryptophan, 100 ppm potassium silicate, and 750 ppm chitosan (Fouda et al. 2022).

The biosynthesis of SMs can increase in some cultural treatments as well as in all various elicitor applications. Alkaloids were up to five times more abundant in plants with a high nitrogen content compared to plants with low nitrogen content in *Lupinus succulentus* (Johnson et al. 1987). Cutting up the leaves of *Lupinus polyphyllus* causes a 400% increase in quinolizidine alkaloid concentration (Wink 1983). Nakka and Devendra (2012) reported that the quantity of SMs in an in vitro propagated *Crotalaria* species is more than in field-cultivated medicinal preparations. The combination of gamma-irradiated sodium alginate and phosphorus increased the carotenoid and trigonelline content of *Trigonella foenum-graecum* through the foliar application (Dar et al. 2016). Underwater deficiency conditions, yield, and its constituents decreased, along with total carbohydrate and protein contents, whereas phenolic and vicine contents of the produced seeds increased in *Vicia faba* (Bakhoum et al. 2022).

One of the most prevalent metabolic engineering strategies is the use of essential structural genes of the SMs pathway, mostly for over-expression, to promote the new biochemical pathways that lead to the biosynthesis of SMs. For example, transgenic alfalfa was produced by constitutively expressing IFS, which is associated with its enhanced isoflavonoid content (Deavours and Dixon 2005). The expression of the Δ^{24}-reductase gene was highly upregulated in hairy roots that produced more diosgenin (eightfold) in *Trigonella foenum-graecum* (Zolfaghari et al. 2020). Zhou et al. (2019) used comparative transcriptome analysis to identify the genes involved in diosgenin production in *Trigonella foenum-graecum* MeJA increased the expression of two key genes involved in the mevalonate pathway which leads to diosgenin and could be a promising elicitor for *Trigonella foenum-graecum* plants to produce diosgenin (Chaudhary et al. 2015). Garagounis et al. (2020) reported a hairy-root transformation protocol for *Trigonella foenum-graecum* as a tool for specialized SMs pathway elucidation. Shirazi et al. (2018) conducted research on *Glycyrrhiza glabra* transgenic roots and suggested that many genes involved in the glycyrrhizin biosynthesis pathway could be simultaneously transferred to the hairy root to increase glycyrrhizin content (Table 21.3).

5 Conclusion

Efforts should be intensified to maximize the utilization of potentially valuable medicinal legume species. Plant biotechnology can enhance the exploitation of these leguminous genetic resources by allowing the isolation of very promising genetic traits. However, characterization of the key groups of SMs is currently lacking for variety of these species, and further labor will be required to fill this gap. All the biofortification solutions that have been established thus far constitute a second intriguing method for the enhancement of beneficial components in main crops. Selecting genetic traits of interest and introducing them into cultivars require a significant investment of time using standard breeding approaches. Stress application is an appealing alternative, but it necessitates the careful selection of those that have the optimum impact on the quality of plants. Among the biofortification options for enhancing SMs, genomic strategies are a sophisticated method for carefully regulating and inducing new target genes. In addition, given the diversity of genome-editing applications, selecting the best techniques for each intervention type (insertion/deletion) and plant species will be critical. Despite the numerous technological obstacles (e.g., the performance of genome-editing systems), other crucial issues, such as public approval and the adoption of more lenient regulatory policies, must be studied and overcome prior to the widespread deployment of these techniques in agriculture.

References

Abdallah RM, Hammoda HM, Radwan MM et al (2021) Phytochemical and pharmacological appraisal of the aerial parts of *Lotus corniculatus* L. growing in Egypt. Nat Prod Res 35:5914–5917

Abdul-Hafeez EY, Mahmoud AF, Ibrahim OH (2015) Antibacterial activities and phytochemical screening of *Alhagi pseudalhagi*. Assiut J Agric Sci 46:33–47

Aksoy L, Kolay E, Ağılönü Y et al (2013) Free radical scavenging activity, total phenolic content, total antioxidant status, and total oxidant status of endemic *Thermopsis turcica*. Saudi J Biol Sci 20:235–239

Al Ayoubi S, Raafat K, El-Lakany A et al (2018) Phytochemical investigation of *Psoralea bituminosa* L. and its anti-diabetic potentials. Pharm J 10:841–853

Ali MM, Ciğerci İH (2019) Genotoxic evaluation of an endemic plant *Thermopsis turcica* extracts on liver cancer cell Line. Pak J Zool 51:355–357

Al-Ramahi R, Jaradat N, Shalalfeh R et al (2015) Evaluation of potential drug-herb interactions among a group of Palestinian patients with chronic diseases. BMC Complement Altern Med 15:1–5

Al-Snafi AE (2016) The pharmacological and toxicological effects of *Coronilla varia* and *Coronilla scorpioides*: A review. Pharm Chem J 3:105–114

Al-Snafi AE (2020) The traditional uses, constituents and pharmacological effects of *Ononis spinosa*. IOSR J Pharm 10:53–59

Amessis-Ouchemoukh N, Ouchemoukh S, Meziant N et al (2017) Bioactive metabolites involved in the antioxidant, anticancer and anticalpain activities of *Ficus carica* L., *Ceratonia siliqua* L. and *Quercus ilex* L. extracts. Ind Crop Prod 95:6–17

Anonymous (2013) The plant list. Version 1.1. Published on the Internet. http://www.theplantlist.org/. Accessed 1 May

Bakhoum GS, Amin G, Sadak MS (2022) Biochemical study of some faba bean (*Vicia faba* L.) cultivars under different water regimes in sandy soil. Egypt J Chem 0

Bakış Y, Babac MT, Uslu E (2011) Updates and improvements of Turkish Plants Data Service (TÜBİVES). In: Health Informatics and Bioinformatics (HIBIT), 2011 6th international symposium on. IEEE, pp 136–140

Belaïd-Nouira Y, Bakhta H, Bouaziz M et al (2012) Study of lipid profile and parieto-temporal lipid peroxidation in $AlCl_3$ mediated neurotoxicity. Modulatory effect of fenugreek seeds. Lipids Health Dis 11:1–8

Bonjar S (2004) Evaluation of antibacterial properties of some medicinal plants used in Iran. J Ethnopharmacol 94:301–305

Bordia A, Verma S, Srivastava K (1997) Effect of ginger (*Zingiber officinale* Rosc.) and fenugreek (*Trigonella foenum-graecum* L.) on blood lipids, blood sugar and platelet aggregation in patients with coronary artery disease. Prostaglandins Leukot Essent Fat Acids 56:379–384

Butkutė B, Dagilytė A, Benetis R et al (2018) Mineral and phytochemical profiles and antioxidant activity of herbal material from two temperate *Astragalus* species. Biomed Res Int 2018:6318630

Chaudhary S, Chikara SK, Sharma MC et al (2015) Elicitation of diosgenin production in *Trigonella foenum-graecum* (fenugreek) seedlings by methyl jasmonate. Int J Mol Sci 16:29889–29899

Chourasiya A, Sahu RK, Khan MA (2019) Anti-anemic and haemopoietic evaluation of *Trigonella foenum-graecum* (Fenugreek) in rodent model. J Drug Deliv Ther 9:332–337

Dalmarco JB, Dalmarco EM, Koelzer J et al (2010) Isolation and identification of bioactive compounds responsible for the anti-bacterial efficacy of *Lotus corniculatus* var. São Gabriel. Int J Green Pharm 4:108

Dar TA, Uddin M, Khan MM et al (2016) Modulation of alkaloid content, growth and productivity of *Trigonella foenum-graecum* L. using irradiated sodium alginate in combination with soil applied phosphorus. J Appl Res Med Aromat Plants 3:200–210

De Simone F, Aquino R, De Tommasi N et al (2001) Anti-HIV aromatic compounds from higher-plants. Bioactive Compounds from Natural Sources 305

Deavours BE, Dixon RA (2005) Metabolic engineering of isoflavonoid biosynthesis in alfalfa. Plant Physiol 138:2245–2259

Dhingra D, Parle M, Kulkarni SK (2004) Memory enhancing activity of *Glycyrrhiza glabra* in mice. J Ethnopharmacol 91:361–365

El-Hela AA, Abdel-Hady NM, Dawoud G et al (2013) Phenolic content, antioxidant potential and Aedes aegyptii ecological friend larvicidal activity of some selected Egyptian plants. J Egypt Soc Parasitol 43:215–234

El-Sayed AS, Zayed RA, El-Baz AF et al (2022) Bioprocesses optimization and anticancer activity of camptothecin from *Aspergillus flavus*, an endophyte of in vitro cultured *Astragalus fruticosus*. Mol Biol Rep 49:1–16

Emtiazy M, Oveidzadeh L, Habibi M et al (2018) Investigating the effectiveness of the *Trigonella foenum-graecum* L. (fenugreek) seeds in mild asthma: a randomized controlled trial. Allergy, Asthma Clin Immunol 14:1–8

Feng Y, Zhao Y, Ha Y et al (2021) Drought stress-induced methyl jasmonate accumulation promotes calycosin-7-O-β-d-glucoside production in *Astragalus membranaceus* adventitious roots. Plant Cell Tiss Org Cult 147:561–568

Fouda SE, El-Saadony FM, Saad AM et al (2022) Improving growth and productivity of faba bean (*Vicia faba* L.) using chitosan, tryptophan, and potassium silicate anti-transpirants under different irrigation regimes. Saudi J Biol Sci 29:955–962

Garagounis C, Beritza K, Georgopoulou ME et al (2020) A hairy-root transformation protocol for *Trigonella foenum-graecum* L. as a tool for metabolic engineering and specialised metabolite pathway elucidation. Plant Physiol Biochem 154:451–462

Geberemeskel GA, Debebe YG, Nguse NA et al (2019) Antidiabetic effect of fenugreek seed powder solution (*Trigonella foenum-graecum* L.) on hyperlipidemia in diabetic patients. J Diabetes Res 2019:1

Gul F, Shinwari ZK, Afzal I (2012) Screening of indigenous knowledge of herbal remedies for skin diseases among local communities of North West Punjab, Pakistan. Pakistan J Bot 5:1609–1616

Güner A, Aslan S, Ekim T et al (2012) Türkiye Bitkileri Listesi. Nezahat Gökyiğit Botanik Bahçesi Yayınları. Flora Dizisi, IBSN, 978-605

Gürağaç Dereli FT, Khan H, Sobarzo-Sánchez E et al (2020) Antidepressant potential of *Lotus corniculatus* L. subsp. *corniculatus*: an ethnobotany based approach. Molecules 25:1299

Hachkova H, Nagalievska M, Soliljak Z et al (2021) Medicinal plants *Galega officinalis* L. and yacon leaves as potential sources of antidiabetic drugs. Antioxidants 10:1362

Hosseinzadeh S, Jafarikukhdan A, Hosseini A et al (2015) The application of medicinal plants in traditional and modern medicine: a review of *Thymus vulgaris*. Int J Clin Med 6(09):635

Johnson ND, Liu B, Bentley BL et al (1987) The effects of nitrogen fixation, soil nitrate, and defoliation on the growth, alkaloids, and nitrogen levels of *Lupinus succulentus* (Fabaceae). Oecologia 74:425–431

Kaimoyo E, Farag MA, Sumner LW et al (2008) Sub-lethal levels of electric current elicit the biosynthesis of plant secondary metabolites. Biotechnol Prog 24:377–384

Kaviarasan S, Sundarapandiyan R, Anuradha CV (2007) Protective action of fenugreek (*Trigonella foenum-graecum*) seed polyphenols against alcohol-induced protein and lipid damage in rat liver. In: Proceedings of the VIIIth Conference of the International Society for Trace Element Research in Humans (ISTERH), the IXth Conference of the Nordic Trace Element Society (NTES), and the VIth Conference of the Hellenic Trace Element Society (HTES), 2007. Springer, pp 111–120

Keskin C, Özen HÇ, Toker Z et al (2018) Determination of in vitro antioxidant and antimicrobial properties of shoot and root extracts of *Astragalus diphtherites* FENZL var. *diphtherites* and *Astragalus gymnalopecias* RECH. FIL. obtained by different solvents. KSÜ Tarim ve Doga Dergisi 21:157–166

Koelzer J, Pereira DA, Dalmarco JB et al (2009) Evaluation of the anti-inflammatory efficacy of *Lotus corniculatus*. Food Chem 117:444–450

Korcan SE, Cigerci IH, Dilek M et al (2009) Antimicrobial activity of an endemic species, *Thermopsis turcica*, Turkey. Kuwait J Sci Eng 36:101–112

Kowalczyk T, Sitarek P, Toma M (2021) Enhanced accumulation of betulinic acid in transgenic hairy roots of *Senna obtusifolia* growing in the sprinkle bioreactor and evaluation of their biological properties in various biological models. Chem Biodivers 18:e2100455

Kuang Y, Li B, Fan J et al (2018) Antitussive and expectorant activities of licorice and its major compounds. Bioorg Med Chem 26:278–284

Latcef M, Iqbal L, Fatima N et al (2012) Evaluation of antioxidant and urease inhibition activities of roots of *Glycyrrhiza glabra*. Pak J Pharm Sci 25:99–102

Lemouchi R, Selles C, Medjdoub H et al (2015) Assessment of possible efficacy of aqueous leaves extract of *Psoralea bituminosa* L. for anti-hyperglycaemic activity. Asian Pac J Trop Dis 5:575–578

Lemouchi R, Selles C, Dib ME et al (2017) Chemical composition and antioxidant activity of essential oil and hydrosol extract obtained by hydrodistillation (HY) and liquid–liquid extraction (LLE) of *Psoralea bituminosa*. J Herbs Spices Med Plants 23:299–307

Liu Y, Kakani R, Nair MG (2012) Compounds in functional food fenugreek spice exhibit anti-inflammatory and antioxidant activities. Food Chem 131:1187–1192

Lorenz P, Conrad J, Klaiber I et al (2021) Monodesmosidic oleanene-type saponins from kidney vetch (*Anthyllis vulneraria* L.) with hemolytic activity. Phytochem Lett 46:21–28

MA Abd-El Mawla A, Osman EH (2011) Elicitation of Trigonelline and 4-Hydroxyisoleucine with Hypoglycemic Activity in Cell Suspension Cultures of *Trigonella foenum-graecum* L. Open Conf Proc J 2:80

Maassoumi N, Ghanati F, Zare-Maivan H et al (2022) Metabolic changes network in selenium-treated *Astragalus* cells derived by glutathione as a core component. Plant Cell Tissue Organ Cult 149:455–465

Malek JM, Ghazvini K (2007) In vitro susceptibility of *Helicobacter pylori* to licorice extract. Iran J Pharm Res 6(1):69–72

Marzougui N, Boubaya A, Thabti I et al (2012) Antibacterial activity of extracts of diploid and induced autotetraploid Tunisian populations of *Trigonella foenum-graecum* L. J Med Plant Res 6:5166–5172

Miri A, Darroudi M, Entezari R (2018) Biosynthesis of gold nanoparticles using *Prosopis farcta* extract and its in vitro toxicity on colon cancer cells. Res Chem Intermed 44:3169–3177

Miri A, Khatami M, Ebrahimy O (2020) Cytotoxic and antifungal studies of biosynthesized zinc oxide nanoparticles using extract of *Prosopis farcta* fruit. Green Chem Lett Rev 13:27–33

Moradi MT, Karimi A, Alidadi S (2018) In vitro anti-herpes simplex virus activity, antioxidant potential and total phenolic compounds of selected Iranian medicinal plant extracts

Nakka S, Devendra BN (2012) A rapid in vitro propagation and estimation of secondary metabolites for in vivo and in vitro propagated *Crotalaria* species, a Fabaceae member. J Microbiol Biotechnol Food Sci 2(3):897–916

Nguyen PM, Niemeyer ED (2008) Effects of nitrogen fertilization on the phenolic composition and antioxidant properties of basil (*Ocimum basilicum* L.). J Agric Food Chem 56:8685–8691

Ouerfelli M, Bettaieb Ben Kâab L, Almajano MP (2018) Radical scavenging and antioxidant activity of *Anthyllis vulneraria* leaves and flowers. Molecules 23:1657

Ouerfelli M, Majdoub N, Aroussi J et al (2021) Phytochemical screening and evaluation of the antioxidant and anti-bacterial activity of Woundwort (*Anthyllis vulneraria* L.). Rev Bras Bot 44:549–559

Ozbek T, Gulluce M, Agar G (2008) Investigation of the antimutagenic effects of methanol extract of *Astragalus* L. species growing in eastern Anatolia region of Turkey. Fresenius Environ Bull 17:2052–2058

Özhatay N, Kültür Ş, Gürdal B (2017) Check-list of additional taxa to the supplement flora of Turkey VIII. İstanbul J Pharm 47(1):30–44

Pal D, Mukherjee S (2020) Fenugreek (*Trigonella foenum*) seeds in health and nutrition. In: Nuts and seeds in health and disease prevention. Elsevier, pp 161–170

Palit P, Furman BL, Gray AI (1999) Novel weight-reducing activity of *Galega officinalis* in mice. J Pharm Pharmacol 51:1313–1319

Priya T, Kumari P, Kumari M (2018) Influence of nanoparticles on the seedlings of the plants belonging to the Fabaceae family

Pundarikakshudu K, Patel JK, Bodar MS et al (2001) Anti-bacterial activity of *Galega officinalis* L. (Goat's Rue). J Ethnopharmacol 77:111–112

Quattrocchi U (2012) Dictionary of medicinal and poisonous plants, common names, scientific names, eponyms, synonyms and etymology. CRC Press, Taylor and Francis Group, Boca Raton London New York

Raj A, Jhariya MK, Banerjee A et al (2020) Multifunctional role of legumes in agroforestry. In: Environmental and sustainable development through forestry and other resources. Apple Academic Press, pp 51–74

Samydurai P, Thangapandian V, Aravinthan V (2012) Wild habits of Kolli Hills being staple food of inhabitant tribes of Eastern Ghats, Tamil Nadu, India

Saviranta NM, Julkunen-Tiitto R, Oksanen E et al (2010) Leaf phenolic compounds in red clover (*Trifolium pratense* L.) induced by exposure to moderately elevated ozone. Environ Pollut 158:440–446

Scarano A, Chieppa M, Santino A (2020) Plant polyphenols-biofortified foods as a novel tool for the prevention of human gut diseases. Antioxidants 9:1225

Schwikkard S, van Heerden FR (2002) Antimalarial activity of plant metabolites. Nat Prod Rep 19:675–692

Shabani L, Ehsanpour AA, Asghari G et al (2009) Glycyrrhizin production by in vitro cultured *Glycyrrhiza glabra* elicited by methyl jasmonate and salicylic acid. Russ J Plant Physiol 56:621–626

Sharifi-Rad J, Hoseini-Alfatemi S, Sharifi-Rad M et al (2014) Phytochemical screening and antibacterial activity of different parts of the *Prosopis farcta* extracts against methicillin-resistant *Staphylococcus aureus* (MRSA). Minerva Biotecnol 26:287–293

Sheela ML, Ramakrishna MK, Salimath BP (2006) Angiogenic and proliferative effects of the cytokine VEGF in *Ehrlich ascites* tumor cells is inhibited by *Glycyrrhiza glabra*. Int Immunopharmacol 6:494–498

Shin YW, Bae EA, Lee B et al (2007) In vitro and in vivo antiallergic effects of *Glycyrrhiza glabra* and its components. Planta Med 73:257–261

Shirazi Z, Aalami A, Tohidfar M (2018) Metabolic engineering of glycyrrhizin pathway by over-expression of beta-amyrin 11-oxidase in transgenic roots of *Glycyrrhiza glabra*. Mol Biotechnol 60:412–419

Singh S, Mukerjee A, Mishra SB et al (2018) Assessment of anti-arthritic potential of ethanolic extract and its fractions of aerial parts of *Alhagi Pseudalhagi* (Bieb.) Desv. in animal models. Orient Pharm Exp Med 18:199–208

Skalli S, Hassikou R, Arahou M (2019) An ethnobotanical survey of medicinal plants used for diabetes treatment in Rabat, Morocco. Heliyon 5:e01421

Stoddard F (2008) Bioenergy, legumes and biotechnology. In: Recent advances in biotechnology. Excel India Publishers, pp 63–66

Subhashini N, Thangathirupathi A, Lavanya N (2011) Antioxidant activity of *Trigonella foenum-graecum* using various in vitro and ex vivo models. Int J Pharm Pharm Sci 3:96–102

Sudan P, Goswami M, Singh J (2020) Antifungal potential of fenugreek seeds (*Trigonella foenum-graecum*) crude extracts against *Microsporum gypseum*

Vergun O, Shymanska O, Rakhmetov D (2020) Parameters of antioxidant activity of *Galega officinalis* L. and *Galega orientalis* Lam.(Fabaceae Lindl.) plant raw material. Slovak J Food Sci 14:125–134

Wan XY, Luo M, Li XD (2009) Hepatoprotective and anti-hepatocarcinogenic effects of glycyrrhizin and matrine. Chem Biol Interact 181:15–19

Wang J, Lou J, Luo C et al (2012) Phenolic compounds from *Halimodendron halodendron* (Pall.) Voss and their antimicrobial and antioxidant activities. Int J Mol Sci 13:11349–11364

Wink M (1983) Wounding-induced increase of quinolizidine alkaloid accumulation in lupin leaves. Zeitschrift für Naturforschung C 38:905–909

Wink M (2013) Evolution of secondary metabolites in legumes (Fabaceae). S Afr J Bot 89:164–175

Wink M, Witte L (1983) Evidence for a wide-spread occurrence of the genes of quinolizidine alkaloid biosynthesis: induction of alkaloid accumulation in cell suspension cultures of alkaloid-'free'species. FEBS Lett 159:196–200

Yang T, Fang L, Nopo-Olazabal C et al (2015) Enhanced production of resveratrol, piceatannol, arachidin-1, and arachidin-3 in hairy root cultures of peanut co-treated with methyl jasmonate and cyclodextrin. J Agric Food Chem 63:3942–3950

Ye Z, Li T, Qing D et al (2021) Structural elucidation and osteogenic activity of a novel heteropolysaccharide from *Alhagi pseudalhagi*. Int J Biol Macromol 171:185–197

Yerlikaya S, Baloglu MC, Diuzheva A (2019) Investigation of chemical profile, biological properties of *Lotus corniculatus* L. extracts and their apoptotic-autophagic effects on breast cancer cells. J Pharm Biomed Anal 174:286–299

Zaheer M, Reddy VD, Giri CC (2016) Enhanced daidzin production from jasmonic and acetyl salicylic acid elicited hairy root cultures of *Psoralea corylifolia* L. (Fabaceae). Nat Prod Res 30:1542–1547

Zhou C, Li X, Zhou Z et al (2019) Comparative transcriptome analysis identifies genes involved in diosgenin biosynthesis in *Trigonella foenum-graecum* L. Molecules 24:140

Zhu Q, Wang B, Tan J et al (2020) Plant synthetic metabolic engineering for enhancing crop nutritional quality. Plant Commun 1(1):100017

Zolfaghari F, Rashidi-Monfared S, Moieni A et al (2020) Improving diosgenin production and its biosynthesis in *Trigonella foenum-graecum* L. hairy root cultures. Ind Crop Prod 145:112075

Chapter 22
Variability in the Biofortification Properties of Fenugreek (*Trigonella foenum-graecum* L.)

Mahmut Camlica and Gulsum Yaldiz

Abstract Fenugreek (*Trigonella foenum-graecum* L.) is one of the most important annual herbaceous plants belonging to Fabaceae family. It has high concentrations of proteins, fiber, and nutrients, as well as medicinal properties thanks to sapogenins, alkaloids, and other compounds. Also, fenugreek seeds include many critical chemical properties, such as fatty acids, vitamins, and nicotinic acid. Fenugreek has also been used in pharmaceutical and nutraceutical industries because it contains valuable metabolites. In this context, the biofortification of fenugreeks can be used as sustainable and cost-effective materials to combat malnutrition in developing countries. The biofortification of fenugreeks differs depending on the genotype, the level of genetic variability, and environmental factors. For this reason, this review highlights the genetic diversity of fenugreeks together with different culture studies, such as those on breeding, transgenic techniques, and/or agronomic applications, in order to reveal how to biofortify fenugreeks.

This chapter will deal with topics associated with the biofortification of fenugreeks grown in conventional agricultural systems and other cultivation systems.

1 Introduction

Malnutrition is one of the main problems in the world, especially in most developing countries. Millions of people die because of malnutrition, and most of them are children (Briend et al. 2013). Today, 8.0 billion people are alive in the world, but it is predicted that the population of the world will increase to 8.6 and 9.7 billion by 2030 and 2050, respectively (United Nations 2016). More than two billion people (representing more than one-fourth of the world population) experience deficiency in one or more essential elements (Wakeel et al. 2018). Malnutrition problems can be eliminated by enriching the nutrients and yield properties of plants. To meet the

M. Camlica (✉) · G. Yaldiz
Department of Field Crops, Faculty of Agriculture, Bolu Abant İzzet Baysal University, Bolu, Türkiye

M. A. Nadeem et al. (eds.), *Legumes Biofortification*,
https://doi.org/10.1007/978-3-031-33957-8_22

food consumption of this increasing population, both agricultural areas and the characteristics of agricultural products, such as yield and quality (primary and secondary metabolites), must be increased.

Biofortification (biological fortification) is nutritionally defined as increasing the bioavailability of edible plants for human consumption. It is developed and cultivated with advanced biotechnological methods, traditional growing systems, and agricultural practices (McGuire 2015).

Biofortification can be carried out by using conventional selective breeding or genetic engineering to increase nutrients in food plants during their growing periods. Plant biofortification is an important part in taking a sustainable and cost-effective approach to eliminating malnutrition in developing countries. It was reported that agronomic biofortification is simple, cost-effective, sustainable, and cheap and includes more micronutrients. However, nutrient sources, application methods, and environmental effects require special attention during biofortification (Bouis et al. 2011; Garg et al. 2018).

As plant bioenrichment increases, people will be able to sustainably obtain products rich in micronutrients. In addition, biofortified plants, which contain essential micronutrients, can be consumed by undernourished and low-income families (Garg et al. 2018). More specifically, fenugreek nutritional content can be increased by using biofortification to feed people in low-income countries.

Fenugreeks (*Trigonella foenum-graecum* L.) are important legumes and medicinal and aromatic plants belonging to the Fabaceae family. The fenugreek plant has been used in the food industry as a spice and in medicine because it has phytochemical medicinal benefits (Camlica and Yaldiz 2021; Yaldiz and Camlica 2021). Fenugreek seeds are used externally to heal broken bones and treat skin infections and used internally to increase milk production for infants and treat cough and other respiratory problems (Meghwal and Goswami 2012).

The leaves and seeds of this plant are used for both human nutrients and animal feeding. Previous studies reported that moisture content was the main component of the fenugreek leaves at 86%, followed by protein, lipid, and fiber contents at 4.4%, 1%, and 1%, respectively. Also, each 100 g of leaves contains 395 mg Ca, 67 mg Mg, 51 mg P, 16.5 mg Fe, 76 mg Na, 31 mg K, 260 mg Cu, 167 mg S, 165 mg Cl, 2.3 mg carotene, 40 mg thiamine, 310 mg riboflavin, 800 mg nicotinic acid, and 52 mg vitamin C; with traces of vitamin K and high amounts of choline (13.5 mg/g). Fenugreek seeds also contain carbohydrates (45–60%), proteins (20–30%), fatty oils (5–10%), pyridine alkaloids, flavonoids, free amino acids, calcium, iron, saponins (0.6–1.7%), essential oils, etc. (Budavari 1996; Newall et al. 1996; Mehrafarin et al. 2010). Some of the chemical properties of fenugreek seeds and leaves are given in Fig. 22.1.

The yield properties, quality criteria, and secondary metabolites of fenugreek are affected by biotic and abiotic stress factors. These stress factors affect its physiology and reduce the plant growth, nutrient content, and primary and secondary metabolites of fenugreeks. To overcome these factors, the application of breeding methods or cultures should be applied to obtain high yields and rich nutrient contents in fenugreeks (Camlica and Yaldiz 2021).

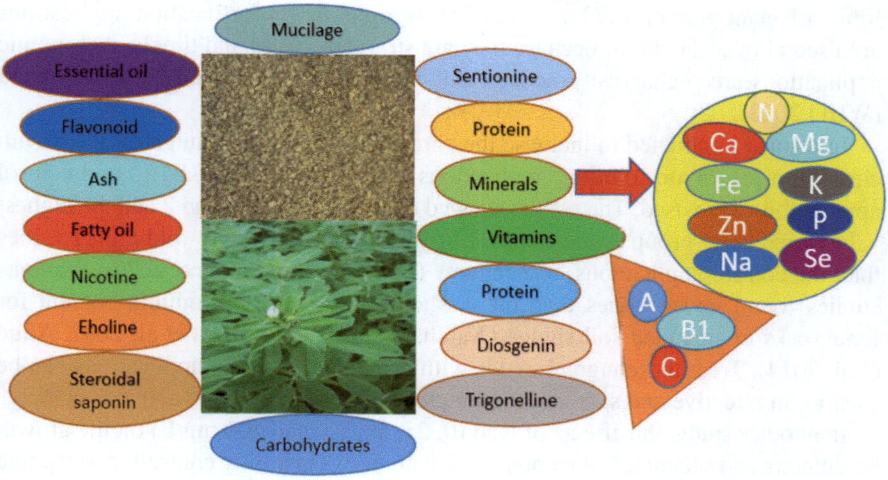

Fig. 22.1 Chemical properties of fenugreek seeds

A previous study reported that fenugreeks can be used to overcome micronutrient deficiencies (Singh et al. 2014). For this reason, studies should be carried out on increasing the micro- and macronutrients in fenugreeks, and appropriate biotechnological or traditional agricultural applications should be found.

In this context, biofortified fenugreeks can be used as sustainable and cost-effective materials to avoid or combat malnutrition in developing countries. The biofortification method for fenugreeks differs depending on the genotype, the level of genetic variability, and environmental factors. For this reason, this review highlights the genetic diversity of fenugreeks together with different culture studies, such as those on breeding or agronomic applications, in order to reveal how to biofortify fenugreeks. Therefore, in this chapter, the issues related biofortifying fenugreeks grown in conventional farming and other growing systems are discussed, emphasizing the malnutrition caused by deficiencies in macro- and microelements.

2 Macro and Microelement Variability in the Fenugreek Biofortification

Among the all biofortification methods, the simplest involves applying fertilizers with micronutrients. However, mineral accumulation and variability, which differ among plant species, as well as ecological conditions and soil properties, also affect agricultural biofortification (Ismail et al. 2007). Briefly, principal food crop biofortification can be divided into two methods: conventional breeding and agricultural applications.

Classical plant breeding improves the main nutrients of foods through the development of varieties by intervening in a controlled manner within the natural genetic limits of plant genomes (Acquaah 2015). Agronomic biofortification applications are used al over the world because they are straightforward and timely. Agronomic applications are preharvest practices that enhance the nutritional content of food (WHO 2022).

In a study conducted to increase the chromium content of fenugreek seeds with chromium picolinate solution, applications 1 and 2 (0.02 and 0.04 g) and control applications were used. The results showed that applications 1 and 2 had the highest values, and these groups were found at fifty-five-fold and eighty-fold higher values than the control applications, respectively (Priyadarshini and Brar 2020). Previous studies found lower values than these, showing increased chromium content for fenugreeks in amended soil sludge (Xanthate et al. 2012; Sinha et al. 2007; Allué et al. 2014). Treating fenugreek seeds with chromium solution applications can be used as an effective and safe method for chromium accumulation and inheritance.

In another study, the effects of lead (0, 25, 50, 75, and 100 mg/L) on the growth parameters and biochemical responses of fenugreeks in a fully controlled aeroponic climate chamber were evaluated. The results showed variability under heavy metal stress. The mineral contents in this study decreased in fenugreek leaves, stems, and roots by increasing Pb application doses (Tuntürk et al. 2022).

Water-stress conditions affect the mineral compositions of fenugreek varieties. It was noted that N, P, Zn, and Cu concentrations decreased as affected by drought, while K, Fe, and Mn concentrations increased by drought with missing of the 1 irrigation and sharply decreased with the missing of 2 irrigation (Hussein and Camilia 2011).

Thanks to Rezaei-Chiyaneh et al. (2021a), the effect of the combined use of bacteria and mycorrhizal fungi on the sowing systems of black cumin and fenugreeks was investigated. It was reported that different intercropping quantities affect N and P concentrations after a biofertilization application, and it was reported that these concentrations were higher in the seeds of fenugreeks compared to those in a monocropping system without an application of biofertilizer. Also, the highest increases in N and P concentrations were noted in the intercropping pattern of black cumin (34%) and fenugreeks (66%) after using bacteria.

Different organic manures and biofertilizers showed no positive effects on the N, P, and K uptakes of fenugreek straws and seeds. However, the removal of more nutrients from seeds and straws occurred under the combined use of organic sources of nitrogen and chemical fertilizer (Malav et al. 2018).

Previous studies reported that fenugreek leaves contained 1.94% protein, 12.21% carbohydrate, and 3.60% ash (Malghani et al. 2022) and that the seeds of fenugreeks contained 1286.6 mg/100 g of Ca, 197.1 mg/100 g of Mg, 83.2 mg/100 g of Fe, 4.79 mg/100 g of Zn, and 3.16 mg/100 g of Mn (Issa et al. 2014).

The micro- and macroelements in fenugreeks can be increased via traditional or biotechnological methods and can be consumed as food, especially in countries whose inhabitants are facing hunger.

3 Effect of Various Factors on Metabolite Production in Fenugreek Cultivation

Many studies have been carried out to produce high secondary metabolites by using various systems in fenugreeks because of their use in traditional treatments until the fifteenth century. These researches revealed the focusing on medicinal uses and chemical properties. Fenugreeks are grown to obtain different properties, such as their polyphenol compounds and their anti-inflammatory and antimicrobial properties.

Irankhah et al. (2020) reported that fenugreek plants were grown for their trigonelline and diosgenin contents by applying arbuscular mycorrhizal (AM) fungal inoculum, with methyl jasmonate chemical treatment combinations under mild water limitation. Different treatments (e.g., full watered and water scarcity (WS) featured different methyl jasmonate applications. The results of the study showed that trigonelline and diosgenin components increased the application of AM fungal or methyl jasmonate treatment combinations under water-deficient conditions. In addition, difference levels of bioactive molecules in plant production were observed in all application and treatment combinations. High protein levels were found after a MeJA treatment applied to fenugreeks under various stress conditions, confirming its role as an elicitor (Ciura et al. 2017). Previous studies reported that methyl jasmonate applications reduced the root content. As a result, MeJA provides some significant ingredients that account for the plant growth parameters and biochemical contents under stress conditions in the literature (Ludwig-Muller et al. 2002; Herrera-Medina et al. 2008).

Similarly, Sánchez-Rodriguez et al. (2012) reported that mild water stress had a positive effect on improving the quality of fenugreek plants. Unlike those studies, a different study stated that mild water stress did not show a significant positive effect on the accumulation of trigonelline and diosgenin in fenugreeks (Irankhan et al. 2020). Differences between previous studies can be attributed to factors such as agricultural practices and ecological and genotypical differences.

It was reported that balances among the growth and development of primary and secondary metabolites were based on the lack of water intensity, the duration, and the repetition together during the plant growth stage under other stress conditions (González-Chavira et al. 2018). Similarly, different fenugreek genotypes (eight genotypes) were exposed to drought stress to determine their respective drought tolerance values. It was reported that drought stress affected the genotype and yield values at flowering time and during pod-linking periods (Chauhan et al. 2017).

The legume–rhizobium symbiosis is a significant aspect of symbiotic nitrogen fixation used in agriculture for sustainability. On the other hand, fungi increase product yield. Seven fungi genera behave in an arbuscular mycorrhizal symbiosis with plants. The most abundant fungi in agricultural soils are AM fungi (Gaur et al. 1998).

Turrini et al. (2018) investigated how fungal inoculation affects the formation of bacterial nodules, it was reported that AMF (AM fungi)-based inoculum and AMF spores were sources of rhizobacteria. Previous studies have revealed that secondary metabolite production is promoted by microorganism applications in medicinal and aromatic plants (Chamkhi et al. 2021). Plants grown under stress conditions produce effective substances, such as antioxidants, phenolics, etc., to respond to reactive oxygen species (ROS) (Vaishnav et al. 2016).

Many studies have shown that plant growth–promoting rhizobacteria (PGPR) improve salt-stress tolerance in crops (Dung et al. 2021; Vaishnav et al. 2020). In addition, inoculating legumes with rhizobacteria and rhizobia is an effective strategy to increase the nodule formation and N_2 fixation capacity of plants under salt-stress conditions (Nohwar et al. 2019; Noori et al. 2018).

The useful rhizobacteria application promotes crop production under stress conditions by compacting the number of potential rhizobium plantation sites and suppressing pathogens through antibiotic production while lowering ethylene levels and promoting root system elongation through hormone excretion for good nutrition uptake (Jain and Pandey 2016; Rizvi et al. 2022).

Sharma et al. (2022) also found that 100 mM of NaCl stress significantly affected the N content, decreasing it by up to 65% in uninoculated fenugreek, while N content was reduced by 25% in bacteria-inoculated plants. In addition, the trigonelline content increased by approximately 54% compared to the uninoculated control under salt-stress conditions. Similarly, Alagna et al. (2020), Chiappero et al. (2019), and Desoky et al. (2020) reported that plants inoculated with PGPR strains under salt stress experienced a significant reduction in ROS levels and an increase in osmolyte content.

Dung et al. (2021) reported that PGPR inoculation under stress conditions can eliminate stress conditions in plants and plant parts at different growth stages by synthesizing antioxidants and by producing phytohormones, ACC-deaminase, and exopolysaccharides.

Many studies have shown that vermicompost and bacterial applications increase seed yield, fixed oil content, and fatty acid concentrations. Additionally, phosphate solvents, N stabilizers, and other effective microorganisms increased secondary metabolite production, ensuring optimal availability to nutrients (Rezaei-Chiyaneh et al. 2020; Ostadi et al. 2020).

Many studies have shown that applying bacterial biofertilizer to fenugreek plants grown in intercropping systems increases the content of trigonelline and diosgenin. For example, with organic manure increases these contents in fenugreek buckwheat (*Fagopyrum esculentum*) according to Salehi et al. (2018); these contents increase in cocultivation of safflower and fenugreeks according to Abdelkader and Hamad (2015); and when fertilized with vermicompost, these contents increase in fenugreek monocrops according to Baghbani-Arani et al. (2017). The co-implementation of vermicompost and organic manures enhance the substrate utilization capacity of microorganisms and soil water holding. Thus, this application contributes to the gradual utilization of nutrients throughout plant development (Goswami et al. 2017; Rezaei-Chiyaneh et al. 2021b). Many researchers have shown that biofertilizers and organic manures improve the nutrition uptake of medicinal and aromatic plants grown under intercropping cultivation with limited P in soil (Weisany et al. 2016; Amani Machiani et al. 2019).

In another study, the effects of emeninethylenediaminetetraacetic acid (EDTA), Indole-3-acetic acid (IAA), and $FeSO_4$ applications on fenugreek plants grown under lead excess (2000 mg/L PbC12) were evaluated in terms of the plants' phytochemical mechanisms, and it was determined that fenugreek germination parameters improved under EDTA, IAA, and $FeSO_4$ applications. In terms of radicle length and amylase activity, better results were obtained after the IAA application compared to the EDTA and $FeSO_4$ applications. It was determined that exogenous $FeSO_4$ was more efficient in the plant development process. In addition, reductions in hydrogen peroxide (H_2O_2) and malondialdehyde (MDA) levels were found after all applications, especially after the IAA application.

In particular, it was revealed that the main phenol produced after the addition of all chelators with IAA was gallic. Moreover, it was reported that quercetin was detected only after the addition of EDTA, while syringe acid was produced only after the exogenous IAA application. As a result, fenugreeks exhibited higher activity after IAA than after EDTA and $FeSO_4$ in reducing Pb stress through the upregulated mechanisms of the antioxidant system, which decreased ROS activities and increased specific phenols (Mnafgui et al. 2022).

Motevalli et al. (2021) detected the highest amount of diosgenin in roots and shoots in a microcobalt-free Murashige and Skoog (MS) medium containing only 300 μM L^{-1} of cobalt nanoparticle. In addition, it has been reported that diosgenin increased in fenugreeks following the application of high concentrations of cobalt nanoparticles (NPs). Similarly, De and De (2011) determined the production of diosgenin in fenugreeks via heavy metals, including cobalt chloride, and found that the use of cobalt in MS culture media could increase diosgenin production. Unlike these findings, it was stated that silver nanoparticles reduced diosgenin production (Jasim et al. 2017).

Many researches have stated that the root of the fenugreek is an acceptable plant stabilizer and can accumulate high Pb contents. Also, it was noted that low Pb content can be found in the aerial part of fenugreeks (Cataldo et al. 1975; Sharma and Ubey (2005); Pourrut et al. 2011). In another study, the root tip cell proteins of fenugreeks can be damage by Cd content, and this Cd content shortens the root lengths of fenugreek plants (Bhat et al. 2012).

Zulfiqar et al. (2019) reported that the seedling, development, and metabolism phases of fenugreek plants grown in high-lead soil conditions were adversely affected. Furthermore, high Pb in soil has been reported to cause multiple effects on fenugreek morphology and its biochemical properties, such as reducing seedling rate, rootlet height, and crop biomass, as well as reducing ROS activity induction in response to the detrimental effects of lead on fenugreeks Zhang et al. (2020a, b) showed that chemical agent chelators and plant growth regulators have important roles in alleviating heavy metal stress. Previous studies noted that an application of auxin (IAA) improved the growth of plants under lead- and cadmium-stress conditions (Bashri and Prasad 2016; Tammam et al. 2021). Similarly, Ostrowski et al. (2016) stated that the IAA application can decrease the phytotoxicity caused by high cadmium concentrations during the germination phase. It was also revealed that auxin stimulates amylase activity during starch disruption to provide the energy required for embryo cell proliferation, resulting in an increase in root length, which was already known (Liu et al. 2005; Wang et al. 2021).

Similarly, it was stated by Kaur (2018) that an EDTA application under Pb-stress conditions increased the germination rate of fenugreeks. Chen et al. (2017) demonstrated the effects of EDTA, IAA, and $FeSO_4$ on gallic acid, syringing acid, chlorogenic acid, and quercetin production in fenugreeks grown under 2000 mg/L of Pb.

Xanthate et al. (2012) reported that in their study on the chromium (Cr) biofortification of fenugreeks (80 kg/ha of potassium dichromate) in soil, a 24.83-fold higher Cr concentration was found in fenugreek leaves and an over fivefold higher Cr concentration in fenugreeks compared to the controls. A previous study reported that Cr concentration (3.61 μg/g DW) was found in fenugreek seeds grown under 100% tannery-sludge-reclaimed soil. In addition, poor translocation of chromium from roots to shoots and thus to leaves and seeds causes low chromium concentration in fenugreek (Sinha et al. 2007). In general, crops have Cr^{3+} from an inactive mechanism and block the stem cell vacuoles, so transportation is limited (Shanker et al. 2005; Sinha et al. 2007). Therefore, chromium picolinate solution can be used to realize a more feasible and productive process for direct fenugreek seed treatments compared to using Cr biosupplementation in soil.

Although an increase in zinc content after germination has been reported in studies in the literature, at 5.27 mg/100 g (Sharara 2017) and 4.43 mg/100 g (Atlaw and Kumar 2018), Pandey and Awasthi (2013) reported a slight decrease (5.5 mg/100 g) compared to the zinc content of raw fenugreek seeds, and the latter autors attributed this to mineral leaching into the wetting medium. They also reported that the manganese content in 100 grams of fenugreek seed flour was 0.95 mg for raw seeds, 1.23 mg for germinated seeds, and 1.18 mg for chromium-treated seeds. Ali et al. (2012), Pajak et al. (2018), and Khorshidian et al. (2016) found the manganese content in raw fenugreek seeds to be slightly higher than in germinated ones. In contrast, El-Shimi et al. (1984) and El-Mahdy and El-Sebaiy (2003) reported an increase in manganese content after germination.

Thus, the increases in zinc, manganese, and chromium contents after germination (Atlaw and Kumar 2018) have been reported to be related to decreases in antinutrient contents, particularly phytate, which ultimately led to the increased bioavailability of these minerals. However, the differences between the results found by the researchers are due to environmental, climatic, and soil differences and to the different methods and cultivation systems (Naga Raju et al. 2006). It was stated that the treatment of fenugreek seeds with chromium did not make any difference in the mineral contents of seeds when compared to the controls.

Baghbani-Arani et al. (2017) reported that the maximum Chl fluorescence (Fm), variable Chl fluorescence (Fv), photochemical activity of PSII (Fv/Fm), Chl (a, b and a + b), and leaf area index (LAI) decreased in fenugreek plants grown under water-deficiency stress, but at minimum fluorescence (F0), the biological yield and seed yield and the carotenoid, anthocyanin, and trigonellin concentrations increased. They stated that both vermicompost and zeolite applications under water-deficiency stress increased the LAI, biological yield, seed yield, and trigonellin concentration in fenugreeks. Lastly, the application of fish oil increased the polyunsaturated essential fatty acid contents in the sprouts of fenugreeks and flaxes (Holub and Nagpurkar 2008).

4 Conclusion

Biofortification has been successfully applied to fenugreeks by using conventional breeding and agronomical applications. These applications increased the nutrient concentrations of fenugreeks. The results showed that biofortified fenugreeks can be used to promote human health. In addition, the influence and application of biofortification in fenugreeks and other crops should be increased throughout the world to avoid and combat hidden hunger.

References

Abdelkader MAI, Hamad EHA (2015) Evaluation of productivity and competition indices of safflower and fenugreek as affected by intercropping pattern and foliar fertilization rate. Middle East J Agric Res 4:956–966

Acquaah G (2015) Conventional plant breeding principles and techniques. In: Al-Khayri J, Jain S, Johnson D (eds) Advances in plant breeding strategies: breeding, biotechnology and molecular tools, vol 1. Springer, Cham, pp 115–158

Alagna F, Balestrini R, Chitarra W, Marsico AD, Nerva L (2020) Getting ready with the priming: innovative weapons against biotic and abiotic crop enemies in a global changing scenario. In: Hossain MA, Liu F, Burritt DJ, Fujita M, Huang B (eds) Primingmediated stress and cross-stress tolerance in crop plants. Academic Press, Cambridge, pp 35–56

Ali MA, Sayeed MAA, Yeasmin MS, Khan MS, Muhamad AM, I.I. (2012) Characteristics of oils and nutrient contents of *Nigella sativa* L. and *Trigonella foenum-graecum* seeds. Bull Chem Soc Ethiop 26:55–64. https://doi.org/10.4314/bcse.v26il.6

Allué J, Garcés AM, Bech J, Barceló J, Poschenrieder C (2014) Fractionation of chromium in tannery sludge-amended soil and its availability to fenugreek plants. J Soils Sediments 14:697–702. https://doi.org/10.1007/s11368-013-0776-1

Amani Machiani M, Rezaei-Chiyaneh E, Javanmard A, Maggi F, Morshedloo MR (2019) Evaluation of common bean (*Phaseolus vulgaris* L.) seed yield and quali-quantitative production of the essential oils from fennel (*Foeniculum vulgare* mill.) and dragonhead (*Dracocephalum moldavica* L.) in intercropping system under humic acid application. J Clean Prod 235:112–122. https://doi.org/10.1016/j.jclepro.2019.06.241

Atlaw TK, Kumar JY (2018) Effect of germination on nutritional composition and functional properties of fenugreek (*Trigonella foenum-graecum* Linn) seed flour. J Food Process Technol 09. https://doi.org/10.4172/2157-7110.1000738

Baghbani-Arani A, Modarres-Sanavy SAM, Mashhadi-Akbar-Boojar M, Mokhtassi-Bidgoli A (2017) Towards improving the agronomic performance, chlorophyll fluorescence parameters and pigments in fenugreek using zeolite and vermicompost under deficit water stress. Ind Crop Prod 109:346–357. https://doi.org/10.1016/j.indcrop.2017.08.049

Bashri G, Prasad SM (2016) Exogenous IAA differentially affects growth, oxidative stress and antioxidants system in Cd stressed *Trigonella foenum-graecum* L. seedlings: toxicity alleviation by up-regulation of ascorbate-glutathione cycle. Ecotoxicol Environ Saf 132:329–338. https://doi.org/10.1016/j.ecoenv.2016.06.015

Bhat TM, Choudhary S, Ansari M, Aslam R (2012) Genotoxicity and molecular screening of proteins in root tip cells of Trigonella foenum-graecum (Fenugreek var-Azad) under cadmium stress condition. Int J Plant Physiol Biochem 4(3):46–51

Bouis HE, Hotz C, McClafferty B, Meenakshi JV, Pfeiffer WH (2011) Biofortification: a new tool to reduce micronutrient malnutrition. Food Nutr Bull 32(suppl. 1):31–40

Briend A, Collins S, Golden M, Manary M, Myatt M (2013) Maternal and child nutrition. Lancet 382(9904):1549

Budavari S (1996) The merck index: an encyclopedia of chemicals, drugs, and biological, 12th edn. N.J. Merk & Co, Inc., Whitehouse Station

Camlica M, Yaldiz G (2021) Employing modern technologies in the cultivation and production of fenugreek (*Trigonella foenum-graecum* L.). In: Naeem M, Aftab T, Khan MMA (eds) Fenugreek. Springer, Singapore. https://doi.org/10.1007/978-981-16-1197-1_3

Cataldo DA, Maroon M, Schrader LE, Youngs VL (1975) Rapid colorimetric determination of nitrate in plant tissue by nitration of salicylic acid. Commun Soil Sci Plant Anal 6(1):71–80. https://doi.org/10.1080/00103627509366547

Chamkhi I, Benali T, Aanniz T, El Menyiy N, Guaouguaou FE, El Omari N et al (2021) Plant-microbial interaction: the mechanism and the application of microbial elicitor induced secondary metabolites biosynthesis in medicinal plants. Plant Physiol Biochem 167:269–295

Chauhan J, Singhal RK, Kakralya BL, Kumar S, Sodani R (2017) Evaluation of yield and yield attributes of fenugreek (*Trigonella foenum graecum*) genotypes under drought conditions. Int J Pure Appl Biosci 5:477–484

Chen S, Zhao H, Zou C, Li Y, Chen Y, Wang Z, Jiang Y, Liu A, Zhao P, Wang M, Ahammed GJ (2017) Combined inoculation with multiple arbuscular mycorrhizal fungi improves growth, nutrient uptake and photosynthesis in cucumber seedlings. Front Microbiol 8:2516. https://doi.org/10.3389/fmicb.2017.02516

Chiappero J, del Rosario Cappellari L, Alderete LGS, Palermo TB, Banchio E (2019) Plant growth promoting rhizobacteria improve the antioxidant status in *Mentha piperita* grown under drought stress leading to an enhancement of plant growth and total phenolic content. Ind Crop Prod 139:111553

Ciura J, Szeliga M, Grzesik M, Tyrka M (2017) Next-generation sequencing of representational difference analysis products for identification of genes involved in diosgenin biosynthesis in fenugreek (*Trigonella foenum-graecum*). Planta 245(5):977–991

De D, De B (2011) Elicitation of diosgenin production in *Trigonella foenum-graecum* L. seedlings by heavy metals and signaling molecules. Acta Physiol Plant 33:1585–1590

Desoky ESM, Saad AM, El-Saadony MT, Merwad ARM, Rady MM (2020) Plant growth-promoting rhizobacteria: potential improvement in antioxidant defense system and suppression of oxidative stress for alleviating salinity stress in *Triticum aestivum* (L.) plants. Biocatal Agric Biotechnol 30:101878

Dung MHT, Nguyen TTM, Hung SH, Huang E, Huang CC (2021) Role of plant growth promoting rhizobacteria (PGPR) in stimulating salinity stress defense in plants: a review. Int J Mol Sci 22(6):3154

El-Mahdy AR, El-Sebaiy LA (2003) Changes in phytate and minerals during germination and cooking of fenugreek seeds. Food Chem 9:149–158. https://doi.org/10.1016/0308-8146(82)90092-9

El-Shimi NM, Damir AA, Ragab M (1984) Changes in some nutrients of fenugreek seeds during germination. Food Chem 14:11–19. https://doi.org/10.1016/0308-8146(84)90013-X

Garg M, Sharma N, Sharma S, Kapoor P, Kumar A, Chunduri V, Arora P (2018) Biofortified crops generated by breeding, agronomy, and transgenic approaches are improving lives of millions of people around the world. Front Nutr 5:12. https://doi.org/10.3389/fnut.2018.00012

Gaur A, Adholeya A, Mukerji KG (1998) A comparison of AM fungi inoculants using capsicum and Polianthes in marginal soil amended with organic matter. Mycorrhiza 7:307–312

Gonzalez-Chavira MM, Herrera-Hernandez MG, Guzman-Maldonado H, Pons-Hernandez JL (2018) Controlled water deficit as abiotic stress factor for enhancing the phytochemical content and adding-value of crops. Sci Hort 234:354–360

Goswami L, Nath A, Sutradhar S, Bhattacharya SS, Kalamdhad A, Vellingiri K, Kim KH (2017) Application of drum compost and vermicompost to improve soil health, growth, and yield parameters for tomato and cabbage plants. J Environ Manag 200:243–252. https://doi.org/10.1016/j.jenvman.2017.05.073

Herrera-Medina MJ, Tamayo MI, Vierheilig H, Ocampo JA, Garcia-Garrido JM (2008) The jasmonic acid signalling pathway restricts the development of the arbuscular mycorrhizal association in tomato. J Plant Growth Regul 27:221–230

Holub BJ, Nagpurkar A (2008) Method of fortifying seeds with an essential fatty acid, fortified seed and food product. U.S. patent 7416752, 26 August 2008

Hussein MM, Camilia YD (2011) Mineral constituents of fenugreek varieties grown under water stress condition. Aust J Basic Appl Sci 5(12):2904–2909

Irankhah S, Sillo F, Nerva L, Ganjeali A, Balestrini R, Chitarra W (2020) Combined effects of water deficit, exogenous ethylene application and root symbioses on trigonelline and ABA accumulation in fenugreek. Appl Sci (Basel) 10:2338

Ismail AM, Heuer S, Thomson MJ, Wissuwa M (2007) Genetic and genomic approaches to develop rice germplasm for problem soils. Plant Soil 65:547–570. https://doi.org/10.1007/s11103-007-9215-2

Issa MY, El Zalabani SM, El-Askary HI, El-Dine RS, Yousif MF (2014) Microscopical, physicochemical and nutritional characterisation of three herbal hepatoprotectives. World J Pharm Pharm Sci 3(5):1430–1446

Jain R, Pandey A (2016) A phenazine-1-carboxylic acid producing polyextremophilic pseudomonas chlororaphis (MCC2693) strain, isolated from mountain ecosystem, possesses biocontrol and plant growth promotion abilities. Microbiol Res 190:63–71

Jasim B, Thomas R, Mathew J, Radhakrishnan EK (2017) Plant growth and diosgenin enhancement effect of silver nanoparticles in fenugreek (*Trigonella foenum-graecum* L.). Saudi Pharmaceut J 25(3):443–447. https://doi.org/10.1016/j.jsps.2016.09.012

Kaur L (2018) Accumulation potential of Indian mustard (*Brassica juncea* var. *arawali*) and fenugreek (*Trigonella foenum-graecum* L.) planted on lead and nickel contaminated soil. Trop Plant Res 5:217–223. https://doi.org/10.22271/tpr.2018.v5.i2.027

Khorshidian N, Asli MY, Arab M, Mortazavian AM, Mirzaie AA (2016) Fenugreek: potential applications as a functional food and nutraceutical. Nutr Food Sci Res 3:5–16. https://doi.org/10.18869/acadpub.Nfsr.3.1.5

Liu X, Zhang S, Shan X, Zhu YG (2005) Toxicity of arsenate and arsenite on germination, seedling growth and amylolytic activity of wheat. Chemosphere 61:293–301. https://doi.org/10.1016/j.chemosphere.2005.01.088

Ludwig-Muller J, Bennett RN, Garcia-Garrido JM, Piche Y, Vierheilig H (2002) Reduced arbuscular mycorrhizal root colonization in *Tropaeolum majus* and *Carica papaya* after jasmonic acid application can not be attributed to increased glucosinolate levels. J Plant Phys 159:517–523

Malav JK, Patel JK, Pavaya RP, Patel BB, Patel VR (2018) Effect of different organic sources on fenugreek (*Trigonella foenum-graecum* L.) under organic farming module. Int J Curr Microbiol App Sci 7(2):17–25

Malghani NA, Mahesar SA, Baig JA, Talpur FN, Hussain Sherazi ST, Junaid M (2022) Nutritional assessment and proximate analysis of selected vegetables grown in Larkana, Sindh, Pakistan. J Turk Chem Soc 9(4):985–998

McGuire S (2015) FAO, IFAD, and WFP. The state of food insecurity in the world 2015: meeting the 2015 international hunger targets: taking stock of uneven progress. Rome: FAO Adv Nutr 6(5):623–624. https://doi.org/10.3945/an.115.009936

Meghwal M, Goswami TK (2012) A review on the functional properties, nutritional content, medicinal utilization and potential application of fenugreek. Food Process Technol 3. https://doi.org/10.4172/2157-7110.1000181

Mehrafarin A, Qaderi A, Rezazadeh S, Naghdibadi H, Noormohammadi G, Zand E (2010) Bioengineering of important secondary metabolites and metabolic pathways in fenugreek (*Trigonella foenum graecum* L.). J Med Plants 9(35):1–18

Mnafgui W, Hajlaoui H, Rizzo V, Muratore G, Elleuch A (2022) Priming with EDTA, IAA and Fe alleviates Pb toxicity in *Trigonella Foneum graecum* L growth: phytochemicals and secondary metabolites. J Biotechnol 356:42–50

Motevalli S, Batool Hassani R, Ghalamboran M, Rezadoost Chahardeh H (2021) Increase of diosgenin in fenugreek seedlings by cobalt nanoparticles. Rhizosphere 18:100335

Naga Raju GJ, Sarita P, Ramana Murty GA, Kumar MR, Reddy BS, Charles MJ, Lakshminarayanaa S, Reddy TS, Reddy SB, Vijayan V (2006) Estimation of trace elements in some antidiabetic medicinal plants using PIXE technique. Appl Radiat Isot 64:893–900. https://doi.org/10.1016/j.apradiso.2006.02.085

Newall CA, Anderson LA, Phillipson JD (1996) Herbal medicines: a guide for healthcare professionals. The Pharmaceutical Press, London

Ostadi A, Javanmard A, Amani Machiani M, Morshedloo MR, Nouraein M, Rasouli F, Maggi F (2020) Effect of different fertilizer sources and harvesting time on the growth characteristics, nutrient uptakes, essential oil productivity and composition of *Mentha x piperita* L. Ind Crop Prod 148:112290. https://doi.org/10.1016/j.indcrop.2020.112290

Ostrowski M, Ciarkowska A, Jakubowska A (2016) The auxin conjugate indole-3-acetyl-aspartate affects responses to cadmium and salt stress in *Pisum sativum* L. J Plant Physiol 191:63–72

Pajak P, Socha R, Broniek J, Królikowska K, Fortuna T (2018) Antioxidant properties, phenolic and mineral composition of germinated chia, golden flax, evening primrose, phacelia and fenugreek. Food Chem 275:69–76

Pandey H, Awasthi P (2013) Effect of processing techniques on nutritional composition and antioxidant activity of fenugreek (*Trigonella foenum-graecum*) seed flour. J Food Sci Technol 52:1054–1060. https://doi.org/10.1007/s13197-013-1057-0

Pourrut B, Shahid M, Dumat C, Winterton P, Pinelli E (2011) Lead uptake, toxicity, and detoxification in plants. Rev Environ Contam Toxicol 213:113–136. https://doi.org/10.1007/978-1-4419-9860-6_4

Priyadarshini S, Brar JK (2020) Biofortification of chromium in fenugreek seeds. J Trace Elem Med Biol 61:126521

Rezaei-Chiyaneh E, Amani Machiani M, Javanmard A, Maggi F, Morshedloo MR (2020) Vermicompost application in different intercropping patterns improves the mineral nutrient uptake and essential oil compositions of sweet basil (*Ocimum basilicum* L.). J Soil Sci Plant Nutr 21:450–466. https://doi.org/10.1007/s42729-020-00373-0

Rezaei-Chiyaneh E, Battaglia ML, Sadeghpour A, Shokrani F, Nasab ADM, Raza MA, von Cossel M (2021a) Optimizing intercropping systems of black cumin (*Nigella sativa* L.) and fenugreek (*Trigonella foenum-graecum* L.) through inoculation with bacteria and mycorrhizal fungi. Adv Sustain Syst 5:2000269

Rezaei-Chiyaneh E, Amirnia R, Satar Fotohi Chiyaneh S, Maggi F, Barin M, Razavi BS (2021b) Improvement of dragonhead (*Dracocephalum moldavica* L.) yield quality through a coupled intercropping system and vermicompost application along with maintenance of soil microbial activity. Land Degrad Dev 32:2833. https://doi.org/10.1002/ldr.3957

Rizvi A, Ahmed B, Khan MS, El-Beltagi HS, Umar S, Lee J (2022) Bioprospecting plant growth promoting rhizobacteria for enhancing the biological properties and phytochemical composition of medicinally important crops. Molecules 27(4):1407

Salehi A, Mehdi B, Fallah S, Kaul HP, Neugschwandtner RW (2018) Productivity and nutrient use efficiency with integrated fertilization of buckwheat–fenugreek intercrops. Productivity and nutrient use efficiency with integrated fertilization of buckwheat–fenugreek intercrops. Nutr Cycl Agroecosyst 110:407–425. https://doi.org/10.1007/s10705-018-9906-x

Sanchez-Rodriguez E, Leyva R, Constan-Aguilar C, Romero L, Ruiz J (2012) Grafting under water stress in tomato cherry: improving the fruit yield and quality. Ann Appl Biol 161:302–312

Shanker AK, Cervantes C, Loza-Tavera H, Avudainayagam S (2005) Chromium toxicity in plants. Environ Int 31:739–753. https://doi.org/10.1016/j.envint.2005.02.003
Sharara MS (2017) Effect of germination and heat treatment on chemical composition and bioactive components of fenugreek seeds. World J Diary Food Sci 12:33–41. https://doi.org/10.5829/idosi.wjdfs.2017.33.41
Sharma P, Ubey RSD (2005) Lead toxicity in plants. Braz J Plant Physiol 17:35–52. https://doi.org/10.1590/S1677-04202005000100004
Sharma K, Sharma S, Vaishnav A, Jain R, Singh D, Bahadur Singh H, Goel A, Sing S (2022) Salt-tolerant PGPR strain Priestia endophyticaSK1 promotes fenugreek growth under salt stress by inducing nitrogen assimilation and secondary metabolites. J Appl Microbiol 133:2802–2813
Singh KP, Singh BN, Jain PK (2014) Role of fenugreek (*Trigonella foenum graecum* L.) in crop biofortification. Int J Plant Res 27(1):51–59
Sinha S, Gupta AK, Bhatt K (2007) Uptake and translocation of metals in fenugreek grown on soil amended with tannery sludge: involvement of antioxidants. Ecotoxicol Environ Saf 67:267–277. https://doi.org/10.1016/j.ecoenv.2006.07.005
Tammam A, El-Aggan W, Abou-Shanab R, Mubarak M (2021) Improved of growth and phytostabilization potential of lead (Pb) in *Glebionis coronaria* L. under the effect of IAA and GA3 alone and in combination with EDTA by altering biochemical attributes of stressed plants. Int J Phytoremediation 23:958–968
Tuntürk M, Tunçtürk R, Oral E, Nohutcu L (2022) Effects of lead doses on the mineral content and antioxidant capacity of fenugreek (*Trigonella foenum-graecum* L.). J Elem 27(2):317–333
Turrini A, Avio L, Giovannetti M, Agnolucci M (2018) Functional complementarity of *Arbuscular mycorrhizal* fungi and associated microbiota: the challenge of translational research. Front Plant Sci 9:1407
United Nations (2016) Sustainable Development Goal 2-Zero Hunger. Available at: www.undp.org/content/undp/en/home/sustainable-development-goals/goal-2-zero-hunger/targets. Accessed 8 December 2022
Vaishnav A, Varma A, Tuteja N, Choudhary DK (2016) PGPR mediated amelioration of crops under salt stress. In: Plantmicrobe interaction: an approach to sustainable agriculture. Springer, Singapore, pp 205–226
Vaishnav A, Singh J, Singh P, Rajput RS, Singh HB, Sarma BK (2020) Sphingobacterium sp. BHU-AV3 induces salt tolerance in tomato by enhancing antioxidant activities and energy metabolism. Front Microbiol 11:443
Wakeel A, Farooq M, Bashir K, Ozturk L (2018) Micronutrient malnutrition and biofortification: recent advances and future perspectives: molecular and genomic perspectives in crop plants. Elsevier, Amsterdam, pp 225–243. https://doi.org/10.1016/B978-0-12-812104-7.00017-4
Wang J, Moeen-ud-din M, Yang S (2021) Dose-dependent responses of Arabidopsis thaliana to zinc are mediated by auxin homeostasis and transport. Environ Exp Bot 189:104554. https://doi.org/10.1016/j.envexpbot.2021.104554
Weisany W, Raei Y, Salmasi SZ, Sohrabi Y, Ghassemi-Golezani K (2016) Arbuscular mycorrhizal fungi induced changes in rhizosphere, essential oil and mineral nutrients uptake in dill/common bean intercropping system. Ann Appl Biol 169:384–397. https://doi.org/10.1111/aab.12309
WHO (2022) World Health Organization. Biofortification of Staple Crops. Available online: http://www.who.int/elena/titles/complementary_feeding/en/ (accessed on 11 December 2022)
Xanthate DS, Murthy CVR, Rajan SCS, Sekhar DMR (2012) Chromium uptake by fenugreek. Nat Proc 4-6
Yaldiz G, Camlica M (2021) Impact of various environmental stress factors on productivity, quality, and secondary metabolites of fenugreek (*Trigonella foenum-graecum* L.). In: Naeem M, Aftab T, Khan MMA (eds) Fenugreek. Springer, Singapore. https://doi.org/10.1007/978-981-16-1197-1_14
Zhang C, He Q, Wang M, Gao X, Chen J, Shen C (2020a) Exogenous indole acetic acid alleviates Cd toxicity in tea (*Camellia sinensis*). Ecotoxicol Environ Saf 190:110090. https://doi.org/10.1016/j.ecoenv.2019.110090

Zhang J, Yang R, Li YC, Peng Y, Wen X, Ni X (2020b) Distribution, accumulation, and potential risks of heavy metals in soil and tea leaves from geologically different plantations. Ecotoxicol Environ Saf 195:1–39. https://doi.org/10.1016/j.ecoenv.2020.110475

Zulfiqar U, Farooq M, Hussain S, Maqsood M, Hussain M, Ishfaq M, Ahmad M, Anjum MZ (2019) Lead toxicity in plants: impacts and remediation. J Environ Manag 250:109557. https://doi.org/10.1016/j.jenvman.2019.109557

Chapter 23
Biofortification: Lessons from the Past and Strategies for Future Food Security

Shumaila Ijaz, Javed Iqbal, Banzeer Ahsan Abbasi, Aasma Tufail, Zakir Ullah, Javad Sharifi-Rad, Tabassum Yaseen, and Siraj Uddin

Abstract Micronutrient delivery can be made more sustainable, affordable, and long lasting through the process of biofortification, which involves breeding nutrients into food crops. The purpose of the biofortification strategy is to introduce a nutrient-dense micronutrient trait into cultivars that already have other desirable agronomic and dietary features, such as high yield and disease resistance. Crop surpluses may find their way into retail stores, where they will likely reach first rural consumers and later urban consumers. Here, we review the experiences from the past and different strategies for future food security, such as maintaining the physical, chemical, and biological characteristics of healthy soil and following proper cultivation practices, including tillage, water management, integrated and balanced nutritional management, the application of organic matter, the application of synthetic fertilizers, micronutrient and bioavailability applications, the application of microorganisms, crop rotations, intercropping, and proper pest management. This chapter also discusses how to build awareness, policy-supported research interventions, crop development, transgenic approaches, and low-cost, high-throughput

---

Shumaila Ijaz and Javed Iqbal contributed equally with all other contributors.

S. Ijaz · A. Tufail
Department of Botany, Division of Science and Technology, University of Education, Lahore, Pakistan

J. Iqbal (✉) · T. Yaseen
Department of Botany, Bacha Khan University, Charsadda, Khyber Pakhtunkhwa, Pakistan

B. A. Abbasi
Department of Botany, Rawalpindi Women University, Rawalpindi, Punjab, Pakistan

Z. Ullah · S. Uddin
Department of Plant Sciences, Faculty of Biological Sciences, Quaid-i-Azam University, Islamabad, Pakistan

J. Sharifi-Rad
Facultad de Medicina, Universidad del Azuay, Cuenca, Ecuador

M. A. Nadeem et al. (eds.), *Legumes Biofortification*,
https://doi.org/10.1007/978-3-031-33957-8_23

methods. Although biofortification is not yet being fully scaled up in a single nation, a wealth of data and many experiences point to its potential usefulness. Biofortification will become a cost-effective more-nourishing investment thanks to policies that enable cross-sectoral implementation at all levels.

1 Introduction

One of the biggest issues confronting humanity is providing food for the 9.7 billion people who will be alive in 2050. A rise in manufacturing might worsen environmental harms. Biofortification is a new agricultural sustainability technique that might serve as a paradigm for sustainable intensification by increasing food security without using more farmland or water. Adding micronutrients and/or nutraceuticals to food, such as vitamin A or iron, that have been associated with better health outcomes is known as biofortification. Increased productivity and profit from biofortified crops can help the economy while also reducing hunger and infectious illnesses. This chapter provides a summary of the century-long effects of biofortification on crop nutrition and provides a list of future food security strategies.

The United Nations' (UN's) Sustainable Development Goals (SDGs) were adopted after the Millennium Development Goals. Malnutrition, the cause of hidden hunger, is already affecting every region in the world. Around two billion individuals in the world have experienced malnutrition (Tulchinsky 2010). The United Nations' Millennium Development Goals stated in 2006 that around 850 million of them are affected by the negative impacts of undernourishment (United Nations Millennium Development Goals Report 2006). In low-income nations in Africa, the risk for developing a micronutrient deficiency are as follows: for Fe, 5%; Zn, 40%; Ca, 54%; Se, 28%; and I, 19% (Joy et al. 2014).

In poor nations, malnutrition mostly affects mothers and small children. According to one estimate, fifty-one million children under the age of five are "going to waste," and 151 million show "inhibited growth," meaning that they do not have appropriate weights for their heights (Ramadas et al. 2020). According to Krishnaswamy (2009), in India, 79.1% of children between the ages of 3 and 6 years and 56.2% of married women between the ages of 15 and 49 are anemic. Vitamin A deficiency affects 104 million preschoolers (32% of all preschoolers) in sub-Saharan Africa and 169 million preschoolers in Southeast Asia and South Asia. Malnutrition can be caused by a number of factors, but the main one is an unbalanced diet. Malnutrition is exacerbated by the declining levels of important nutrients such as minerals and vitamins in food owing to the growing micronutrient shortage in soil (IRRI 2006).

Micronutrient deficits, even moderate to mild ones, might have major effects on human health, including an impaired metabolism, a weakened immunity, a higher risk of illness, an inability to grow, deterioration in brain performance, and eventually decreased output (deficiency of iron, vitamin D, vitamin C, vitamin A, iodine, B vitamins, and zinc). Direct methods and indirect methods, such as

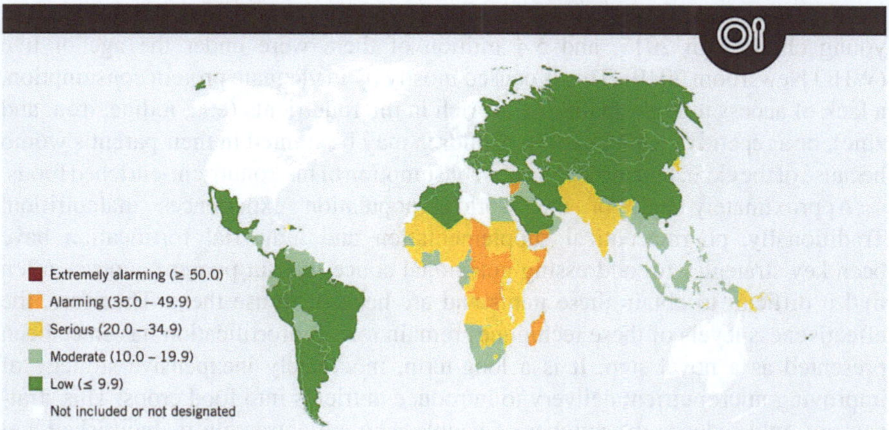

Fig. 23.1 Undernourished proportion of each country's population. (Source: Global Hunger Index 2022, Wikipedia)

nutrition-specific and nutrition-sensitive therapies, can alleviate hidden hunger (Fig. 23.1) (Ruel et al. 2013).

Fortification, the process of adding essential vitamins, minerals, and other trace elements to food, enhances a crop's nutritional content and reduces the risk of developing public health problems. Fortification is a realistic, affordable, and sustainable technique. Contrarily, enhancing the nutritional content of food crops by the application of modern biotechnology, conventional plant breeding, or agronomic methods is known as biofortification. In contrast to traditional fortification, which increases nutrient levels in crops after harvest, biofortification aims to increase nutrient levels while plants are still growing. Thus, biofortification may be an effective strategy for reaching people when traditional and/or supplemental fortification approaches are restricted or difficult to use (World Health Organization 2019).

Because the bulk of meals worldwide consists of starchy staple foods, including sorghum, rice, millets, wheat, sweet potatoes, maize, and legumes, biofortification focuses primarily on these crops. Reaching undernourished people that have little access to a variety of diets, commercially fortified foods, and many supplements is made possible by following this practical strategy (Saltzman et al. 2013; Kumar 2012).

Our bodies require trace amounts of vitamins and minerals, which are referred to as micronutrients. However, they have significant effects on us, and deficiencies in them lead to major health problems, such chronic illnesses, stunted growth, weakened immune systems, and a decline in our physical and mental capabilities (WHO) (Kennedy et al. 2007; WHO 2003). Globally, there are more than two billion people who are micronutrient deficient, which results in more than twenty million yearly fatalities. Micronutrient deficiencies are also known as hidden hunger. Zn and Fe deficiencies are ranked fifth and sixth, respectively, and are most prevalent in low-income nations (World Health Organization 2002). Women and children are especially at risk of nutritional deficits. According to WHO estimates, malnutrition, particularly a lack of micronutrients, killed approximately 6.3 million infants and young children in 2017, and 5.4 million of them were under the age of five (WHO Newsroom 2019). This is caused mostly by inadequate protein consumption, a lack of access to high-quality meals rich in micronutrients (e.g., iodine, iron, and zinc), or a repetitive eating pattern. Children may be stunted in their parent's womb because of the expectant mother's poor consumption of micronutrient-enriched foods.

Approximately half of the world's population experiences malnutrition. Traditionally, pharmaceutical supplementation and industrial fortification have been key strategies for addressing nutritional concerns. But poorer countries often find it difficult to obtain these items and are hesitant to use them. Therefore, the effectiveness levels of these techniques remain low. Biofortification has since been presented as a novel step. It is a long-term, moderately inexpensive strategy of improving micronutrient delivery to introduce nutrients into food crops. This strategy not only reduces the number of people who are extremely malnourished and require supplemental therapy but also helps those people retain their improved nutritional statuses.

Additionally, biofortification is a workable solution for rural residents who live in poverty and perhaps lack access to fortified foods available on the market. The soils in this region are depleted of zinc, iron, and iodine, where the bulk of the soil has been deteriorated by alkalinity and salt issues. Many people prefer cereal-based diets, which are lower in proteins and vitamins (Fuge and Johnson 2015). Micronutrient deficits have impacts on crop quality, productivity, and other metabolic processes, including seed development and blooming. Playing roles in cell wall integrity and strength, micronutrients, particularly B, Mg, and Cu, increase plant resistance to disease penetration. Therefore, biofortification also has a lot of interest in agronomic fortification. Fortification prioritizes increasing the number or concentration of micronutrients in foods while also boosting their bioavailability and lowering the number of antinutritional agents. To ensure the effectiveness of biofortification, the following three issues must be resolved (Hotz and McClafferty 2007):

1. The biofortified crop must be efficient and successful in eliminating micronutrient deficiencies in people.
2. The biofortified crop must produce a substantial amount of food and be lucrative for farmers.
3. The target regions' farmers and consumers must embrace the biofortified crop.

As of 2018, 6.7 million farming families around the world produce biofortified crops, and these products undoubtedly end up in meals. More than three hundred cultivars of crops, including sweet potatoes, wheat, cassavas, maize, bananas, oranges, lentils, beans, potatoes, cowpeas, rice, and plantains, have so far been planted in thirty nations (HarvestPlus 2013).

2 Agronomic Biofortification Restrictions

The following difficulties arise while improving crop characteristics through agronomic biofortification (Fig. 23.2):

- When farmers are unable to access micronutrient fertilizers at the right time, they frequently fail to apply them to their crops, which further contributes to widespread deficiencies.
- Plants are unable to absorb applied micronutrients owing to the low utilization efficiency of micronutrients such as copper, iron, zinc, etc.
- Genetics restricts the effects of agronomic biofortification because protein content is genetically determined.
- Deficiencies in zinc and iron are common across the globe. The public has to be made aware of the adverse effects of iron and zinc deficiencies because these deficiencies may not come with visible symptoms, so they often go unnoticed.
- How minerals move from soil to plants remains unclear for the majority of crops. Therefore, more research is required on the processes that control the rate of micronutrient absorption and translocation in the soil–plant system.
- Before releasing a crop to the market, any safety concerns must be thoroughly examined. The extensive information gaps in the patterns governing mineral distributions in plant systems and in the bioavailability of micronutrients in dietary grains need to be filled.

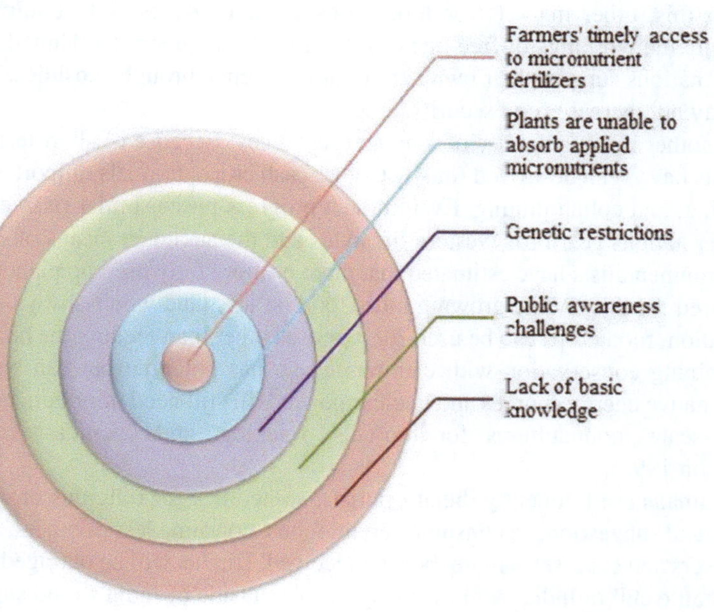

Fig. 23.2 Difficulties arise while improving crop characteristics through agronomic biofortification

- For the majority of crops, the losses in micronutrients because of postharvest procedures that remove external tissues during washing and preparation have not been studied but must be taken into account.

All sorts of food are susceptible to physical dangers, such as bad weather, natural catastrophes, fires (such as the recent ones in Australia), pests, and illnesses. The grassy stunt virus (qv) and the Irish potato famine (qv) have demonstrated the risks of relying on only one crop, and locusts recently wreaked havoc in East Africa, in 2020. Several crops are affected by *Xylella fastidiosa*, and numerous livestock illnesses and threats from aphids, viruses, bacteria, fungi, and pigeons, rats, and other animals can wreak havoc. Up to a point, such dangers can be controlled, but by whom?

No one wants the task, the expense, or the blame if supplies run out; hence, the responsibility for ensuring food security has not been clearly defined. But it won't happen on a national or international scale without consensus. Ricardo's law on specialization has been lauded by economists, but nations that trade cash crops for food take a risk. Grain exports from Russia were banned in 2010 because of the heat wave and drought, and future panic purchasing and hoarding might result from such acts. China's famine (1958–1962) demonstrates the limitations of central planning, but free markets far from immune to such issues. Inelastic demand for dietary energy has meant that intake doesn't increase if costs decrease (Dikotter 2010). As a result, producers are penalized by food surpluses because unit prices drop, and local agriculture can be destroyed when Western surpluses are dumped on underdeveloped nations without providing famine assistance (Fowler 2000). Instead of doing this, other markets for food crops consume excesses that might be kept. Despite the workings of free markets, the European Union, the United States, and other nations support their own agricultural systems through subsidies, which may or may not increase food security.

Another issue is the use of land and resources. Because of short-termism, rainforests have been destroyed for soya, other cash crops, biofuels, imported livestock, logging, and coltan mining. Even though progress poses similar risks everywhere, poorer nations yearn for Western lifestyles and the attendant means of subsistence. Environmentalists have estimated that three or more fully used up planets would be required for this. While growing more food on less land won't always ensure conservation, food crops can be used for purposes other than feeding the hungry world; combining conservation with cautious usage has potential but can be overdone. Alternative energy sources and nuclear power will still need minimum acreage than large-scale monocultures for biofuels, fracking, and open-cast coal mining (Wilson 1992).

Humans are frequently dietary purists, sometimes for religious or cultural reasons, and suggestions to consume certain foods, consume less of others, and prevent leftovers and excesses will not be well received. Hindus will be outraged by the idea of a cattle cull in India, but how many people outside of India would support it and eat more beef and dairy products today than they would later?

Finally, even efforts that appear logical can backfire. Many people have criticized traditional livestock, notably cattle and sheep. Impact per head might be reduced, but let's cut back on population. The present supplies of beef, lamb, and dairy being consumed require upfront investments, suitable targeting, alternatives for byproducts (particularly manure), and a commitment to conservation. If a cull had taken place, the meat and offal that resulted would obviously not be wasted. But take into account that Shetland is 110 miles north of Scotland and not well adapted for growing crops. In 1998, 20,000 healthy sheep would have been mostly squandered had they been culled for economic reasons. Others in 2007 had less good fortune (Capper et al. 2009; Poore and Nemecek 2018).

While several people will defend their specific preferences, convenience, and vested interests, theoretical solutions may easily fail in actuality. What can be done, then?

3 Biofortification Strategies for Future Food Security

The long-term replacement for conventional techniques for enhancing mineral nutrition has been presented as biofortification. In the parts of staple crops that can be eaten, biofortification boosts the mineral content and bioavailability. The first can be accomplished by agronomic intervention, plant breeding, or genetic engineering, but the latter can be altered only through plant breeding and genetic engineering (Fig. 23.3) (Zhu et al. 2007).

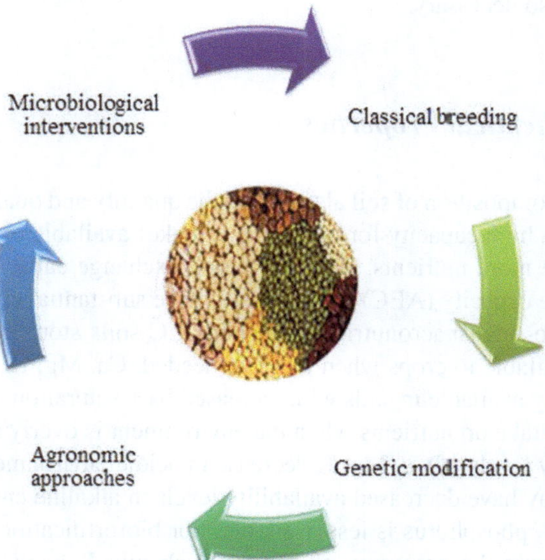

Fig. 23.3 Various biofortification approaches

3.1 Maintaining the Physical, Chemical, and Biological Characteristics of Healthy Soil

One of the key aspects affecting plant strength is the health of soil. Healthy soil promotes proper root development, increased nutrient availability, biological activity, optimal aeration, water-holding capacity, etc. Nutrients are more concentrated in the edible parts of plants as a result of the need for water for plant growth.

3.2 Physical Properties of Soil

The texture of soil has a considerable impact on its micronutrient availability. Because of their high water and nutrient retention capacity, sandy soils have fewer micronutrients than fine-textured soils. Reduced moisture conditions result in slower rates of nutrient transport and dissolution and in slower root activity. In contrast to blocky structured soils, excellent soils have high nutrient contents, good porosity, and high water-holding capacity; boost nutrient absorption; decrease soil compaction; and promote appropriate root development, which together results in higher-quality produce. Applying amendments, organic matter, tank silt, pressed mud, and bentonite clay can help enhance soil structure and increase the soil's capacity to hold water and minerals, which increases the concentrations of vitamins and nutrients in crops. Although submersion can boost Mn and Fe absorption, it decreases the absorption of other macro- and micronutrients. In order for nutrients to be absorbed and transferred to grains, bulbs, stems, fruits, etc., good drainage facilities are also necessary.

3.3 Soil Chemical Properties

The chemical composition of soil also affects the quantity and quality of crop yields. If the soil has a high capacity for buffering, it makes available or replenishes those crops that have more nutrients. Both the cation exchange capacity (CEC) and the anion exchange capacity (AEC) of the topsoil have substantial effects on the availability of micro- and macronutrients. Higher-CEC soils store more nutrients and make them available to crops when they are needed. Ca, Mg, K, and other cations are more readily available in soils with increased base saturation. In addition, crops are less able to take up nutrients when the environment is overly acidic or alkaline. The availability levels of Ca, Mg, K decrease in acidic environments, although Zn, Cu, and Mn may have decreased availability levels in alkaline environments. At an unbalanced pH, phosphorus is less available. For biofortification to be successful, the soil must have the necessary physical and chemical characteristics. Gypsum, sulfur compounds in alkaline soil and lime in acidic soil can all be used to preserve the chemical qualities of the soil.

3.4 Soil Biological Properties

The rapid mineralization of nutrients requires optimal biological activities. The number of various microorganisms that may mobilize or solubilize both immobile and mobile nutrients and make them accessible to plants is growing, including plant growth–promoting rhizobacteria (PGPR), arbuscular mycorrhiza fungi (AMF), and vesicular arbuscular mycorrhiza (VAM). Some macroorganisms, such as earthworms, mole crickets, and ants, increase soil porosity and facilitate deep root penetration. Therefore, biological activity is essential for biofortification. Increased biological activity in soil can be achieved by adding organic matter to soil, including legume crops during agricultural rotation, and using less pesticide.

3.5 Proper Cultivation Practices

3.5.1 Tillage

The majority of crops strongly benefit from tillage. The best soil conditions for crops to develop sprouts and complete their life cycle may be achieved by proper tillage. Tillage gives the crops with a competitive advantage by removing weeds and disease inoculants. Tillage must take place at ideal moisture levels (50–75%) called major histocompatibility complex (MHC) because too much or too little humidity can cause hardpan in subsoil, which inhibits the development of roots and, as a result, lowers nutrient uptake and crop yields. Nowadays, zero tillage or reduced tillage is becoming more popular, but for them, soil compaction is the key issue because it hinders root multiplication (Busscher and Sojka 1987). The Zn concentration in the winter wheat plant tissue at the beginning of heading was unaffected by the tillage treatments during the first 2 years, but in the third year, it was 11.7 mg/kg in the conventional tillage plots and only 6.4 mg/kg in the zero-till plots. For them, a subsoil plowing every 3 to 4 years is a solution. The nutrient absorption by crops is also increased by some enhanced tillage techniques, such as planting in furrow-irrigation-raised beds (FIRBs) and ridge and furrow planting (Stipesevic et al. 2009).

3.5.2 Water Management

Because mass flow and diffusion account for the majority of nutrient absorption, the main factor affecting the concentration of nutrients in agricultural products is soil moisture. The right amount of moisture promotes stronger root development and enhances the availability of nutrients to plants by increasing their solubility. By limiting mobilization or reducing leaching, both water surpluses and water deficits decrease the concentration of nutrients in the root zone. Sometimes, a little stress may boost the amount of nutrients in grains. Wheat grains' lipid content can be reduced by imposing a water deficit during grain filling, although the grain fills and starch compositions might benefit from a modest water shortage, greatly enhancing the quality of the bread (Zhao et al. 2009). After harvest, the wheat field's water was appropriately controlled, which boosted the protein concentration and grain composition. However, P, Zn, K, Mg, and Ca levels can rise at the preanthesis stage in response to water stress or a water deficit. If the product's quality is to be raised, water management must be properly carried out at all essential stages. In both limed and unlimed acid laterite and alluvial soils, continuous flooding throughout the rice-growing season reduces the availability of Zn and Cu to plants while increasing the availability of B, Fe, and Mn. Alternately flooding and drying under or above ground soil conditions were shown to be better for rice than continuous flooding because this process makes more Zn, B, and Cu nutrients available to plants while making fewer Mn and Fe nutrients available.

3.5.3 Integrated and Balanced Nutritional Management

The most crucial stage in methods of biofortification is the administration of nutrients. The best technique in a sustainable method of biofortification is the comprehensive use of microorganisms, inorganic and organic fertilizers, manure, and compost.

3.5.3.1 Application of Organic Matter

The organic matter in soil has a significant impact on soil's physical, biological, and chemical qualities. It enhances the physical characteristics of soil, such as its structure, porosity, bulk density, and ability to stabilize soil aggregates. It also functions as a reclamation agent for saline and alkaline soils. It can provide plants with all other nutrients, in addition to enhancing the health of soil (Lindsay 1991). Humic and fulvic acids, which are created during the breakdown of organic materials, aid in making iron more soluble and available to plants. Other nutrients, including Cu and Ni, are less accessible because they are strongly bound to soil's organic materials. Zinc and other nutrients are taken up more readily by plants thanks to the addition of bio solids, compost, green manure, and biochar, which also increase the amount of zinc that is available to plants (Watson et al. 2012).

Additionally, these modifications decrease the absorption of heavy metals (Liu et al. 2009). The variety in the microbial community and microbial biomass carbon significantly increase with the addition of organic materials. These biological characteristics of soils may contribute to the preservation of soil quality and to nutrient cycling. Foods cultivated under organic circumstances have a higher nutritious content such as vitamins and minerals. From an agronomic standpoint, organic matter is the only method that can produce viable biofortified agricultural products (Woese et al. 1997).

3.5.3.2 Application of Synthetic Fertilizers

According to the results of the soil test, macronutrients such N, P, and K should be applied; nitrogen, in specific, should be applied in two applications. These nutrients enhance plant absorption of all nutrients and encourage the growth of roots and shoots. Micronutrients are occasionally provided by intensive macronutrient fertilizers because they are sometimes added to these during production or are present as contaminants. The absorption of nutrients with poor phloem mobility, such as calcium, which is susceptible to dilution effects, is decreased by adding high dosages of minerals such N, P, and K (Marcelle 1995). By boosting root respiration, an over-reliance on ammonium-based fertilizers restricts cation nutrient absorption and lowers root vegetable glucose content. More phytate is produced by excess P in soil, which can increase Zn deficiency, and excessive K intake interferes with the absorption of Ca and Mg (Marschner 1995). Therefore, the careful use of macronutrients is crucial to ensuring the correct absorption of additional nutrients.

3.5.3.3 Micronutrient Application and Bioavailability

Micronutrients are simply transferred from soil to the human body by consuming crops. Numerous crucial aspects influence whether agronomic biofortification is effective in addressing human micronutrient deficiencies. Nutritional bioavailability at different stages affects most of these parameters (Fig. 23.4).

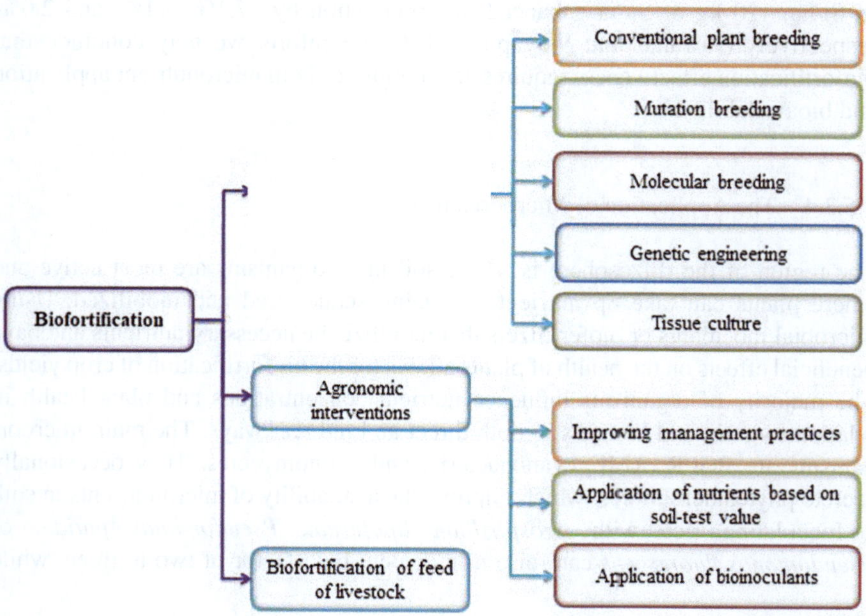

Fig. 23.4 Major methods of biofortification to improve crop production

Depending on the crop species and crop genotype (Rashid and Fox 1992; Graham et al. 1992), the application of micronutrients in soil can improve the nutritional contents in grains. For example, adding zinc to soil can increase the amount of zinc in grain crops twofold to threefold. Green manure and fertilizers with Zn coatings improve the nutritional content and absorption of basmati rice grain and straw. Zn-coated urea (ZCU) contains $ZnSO_4$ and foliar fertilization with 0.2% zinc sulfate led to greater Zn content in rice. The highest overall Zn absorption was recorded by Pooniya and Shivay (2013). Following the application of 100% P, 10 kg of Zn, and 1 kg of B per hectare, Kumar et al. (2012) produced a substantial increase in the absorption of micronutrients (Zn, Mn, Fe, Cu, and B) by wheat. The highest value was obtained when treated with boron at a lower level (0.5 kg/ha), while the lowest value was recorded (4.2%) with boron applied at 1.5 kg/ha (Kumar et al. 2012). Boron application levels should be increased from 0.5 to 1.5 kg/ha, which should reduce boron consumption efficiency. Foliar sprays speed up the transfer of nutrients from the beginning of the implementation to the consuming site. When sprayed in chelated form, Fe, Zn, and Mn were shown to have higher translocation inside plants (Ferrandon and Chamel 1988). The production of fruits and vegetables has used foliar fertilization and chelates, including EDTA, $ZnSO_4$, Zn, and others. Nutrients will move from these vegetative portions to the edible parts. When rice is processed (dehusked and milled), the grains' outermost layer, where Fe and Zn are in protein bodies, is frequently removed, leaving less Zn and Fe in the finished product (Zimmermann et al. 2004; Haas et al. 2005). When the water receives micronutrients before heat processing, rice can be effectively increased in nutrients because the procedure transfers nutrients to the endosperm from the germ layer and bran (Hotz et al. 2015; Oli et al. 2016). In comparison to applications of 0, 40, and 80 kg of Si/ha, 120 kg of Si/ha enhanced rice production by 17.1%, 7.1%, and 2.0%, respectively (Jawahar and Vaiyapuri 2013). Therefore, we may conclude that biofortification effectiveness requires accounting for both micronutrient application and bioavailability.

3.5.3.4 The Application of Microorganisms

The region of the rhizosphere is where soil microorganisms are most active and where plants can take up nutrients by being sequestered and mobilized. Using microbial inoculants or biofertilizers that mobilize the necessary nutrients and have beneficial effects on the health of plants allows for the biofortification of crop yields. The majority of organisms influence nutrient concentrations and plant health in relation to grain and biomass in both direct and indirect ways. The main microorganisms are PGPR, AMF, cyanobacteria, and actinomycetes. They occasionally secrete phytosiderophores, which improve the availability of micronutrients in soil.

Inoculating rice with *Azospirillum lipoferum*, *Pseudomonas putida*, or *Pseudomonas fluorescens* can raise iron content by a factor of two to three, while treating wheat with *Providencia* sp. bacteria can rise copper and zinc concentrations by a factor of two to three (Rana et al. 2012). When a certain strain of *Trichoderma asperellum* is applied to the plot, it produces siderophores, thanks to which the Fe content in wheat biomass may increase by up to 1.5 times (De Santiago et al. 2011). The commercial use of *Pseudomonas* sp. in rice soil increases Zn content in rice by up to 157% (Tariq et al. 2007). Crop plants' absorption levels for K, P, Ca, Fe, and Mg are improved by inoculating them with *Pseudomonas* species and actinobacteria species (Khan 2005). Because the majority of fungi are heterotrophs (saprotrophs, biotrophs, and necrotrophs) in nature, they are crucial in controlling soil fertility because they break down and cycle minerals and organic materials. The large hyphal network of arbuscular mycorrhiza both internally and externally extends into plant roots. Because their hyphae have distinct qualities, including a quicker rate of development and thin and widespread branches, they can more effectively investigate the soil. According to a comparison of the foraging area of the crop to the crop's root length, AMF may increase it up to a hundredfold. The supplies of Zn, N, K, Fe, Ni, P, Cu, Mn, Ca, B, etc. can be improved by using AMS (Clark and Zeto 2000). In addition to producing low-molecular-weight organic acids, certain ectomycorrhizal fungi also aid in increasing nutrient mobilization. Total Ca, N, B, P, and Mo absorption rates in okra were risen by the use of AMF + P + proper irrigation by 8%, 24%, 5%, 14%, and 40%, respectively. In the case of pea, the absorption rate increases for elements were as follows: Ca (22%), N (8%), Zn (22%), B (11%), P (19%), Mg (12%), Mo (38%), K (12%), Mn (10%), Cu (28%), and Fe (10%) (Kumar 2012). The agents that encourage plant development are cyanobacteria, or blue-green algae, which also play significant roles in nutrient intake and increasing user effectiveness. They raise the level of nutrients in plants in the following ways:

- They prevent harmful pathogenic activity and preserve the health of plants.
- They create extracellular polysaccharides called indole-3-acetic acid (IAA) and allelochemicals that stabilize the soil and boost N and C levels in the rhizosphere.
- They facilitate the mobilization of nutrients into plants and aid in nutrient sequestration.

In flooded rice production systems, the application of *Anabaena*-based biofilm inoculants to rice soils increases the amount of both zinc and iron in the rice grains by 13–46% and 15–41%, respectively. Treatments with biofilms made of *Anabaena–Pseudomonas* resulted in an increase in copper buildup in rice grains. Particularly in South Asia and certain other regions of the world, cyanobacterial inoculation aids in boosting grain yields from rice crops by 10–24% (Karthikeyan et al. 2007). Actinomycetes can play significant roles in dissolving fundamental rock-forming minerals to acquire essential nutrients; they also serve as nucleation sites for the precipitation of secondary minerals. By doing so, they aid plants in absorbing nutrients (Fig. 23.5).

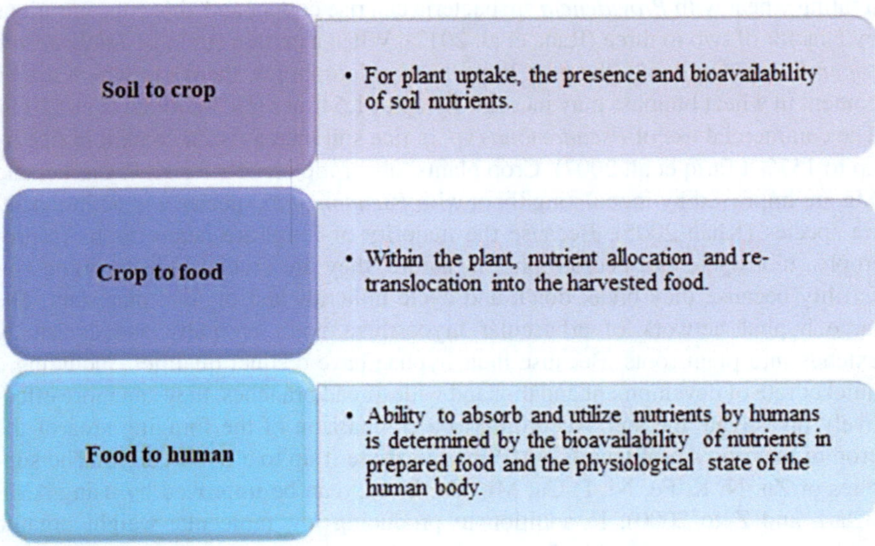

Fig. 23.5 Agronomic biofortification simply adds mineral fertilizer containing micronutrients to the soil and/or plant leaves in order to increase the micronutrient content of food crops' edible parts

3.6 Crop Rotation

Crop rotation has positive impacts on soil's physical and chemical fertility as well as on weed and disease infestation. Crop rotation and cover crops may increase the availability of Fe, Cu, and Zn. In comparison to the same rotation fed with inorganic fertilizer alone, the application of farmyard manure (FYM) and green manure in rice–wheat rotations maintained an accessible percentage of soil micronutrients, including Mn, Cu, Zn, and Fe (Karlen 1994; Kumar and Yadav 1995). The greatest choice after cereals for enhancing food quality is to include pulse crops in the cropping system because of their significance for people and animals as well as their benefits for soils and their capacity to survive in challenging and vulnerable locations.

3.6.1 Intercropping

In order to protect soil from nutrient mining, prevent weed and disease infestation, preserve soil physical and biological health, and enhance nutrient density, intercropping between crops that deplete soil can help to promote regrowth.

3.6.2 Proper Pest Management

Pests, weeds, and inoculants for diseases and insects have significant influences on both the quantity and quality of crop outputs. They limit crop development and can even cause plant death. They also produce certain poisons that give plants a harsh flavor. The greatest solution to managing their infestation and keeping the product's quality high is integrated pest management.

3.6.3 Proper Drying and Storage

It is important to thoroughly dry grains during the postharvest season because improperly dried grains can occasionally grow mold and produce dangerous compounds such as aflatoxins and ochratoxin A. Grain crops such as wheat and rice are susceptible to nutritional losses and are subject to pollutants, pests, and diseases. Therefore, following harvest, adequate storage is essential (Fig. 23.6).

Institutions in the public sector must work hard to develop promotional strategies and policies that will help agronomic methods for biofortification become much more widely accepted. These agronomic approaches can spread quickly if micronutrient fertilizers and other bioinoculants such as AMF, PGPR, and cyanobacteria are made available. Farmers are encouraged to plant more biofortified crops thanks to guaranteed premium remunerative pricing for the crops on the marketplace. Growers' industries, customers, and the general public would become more

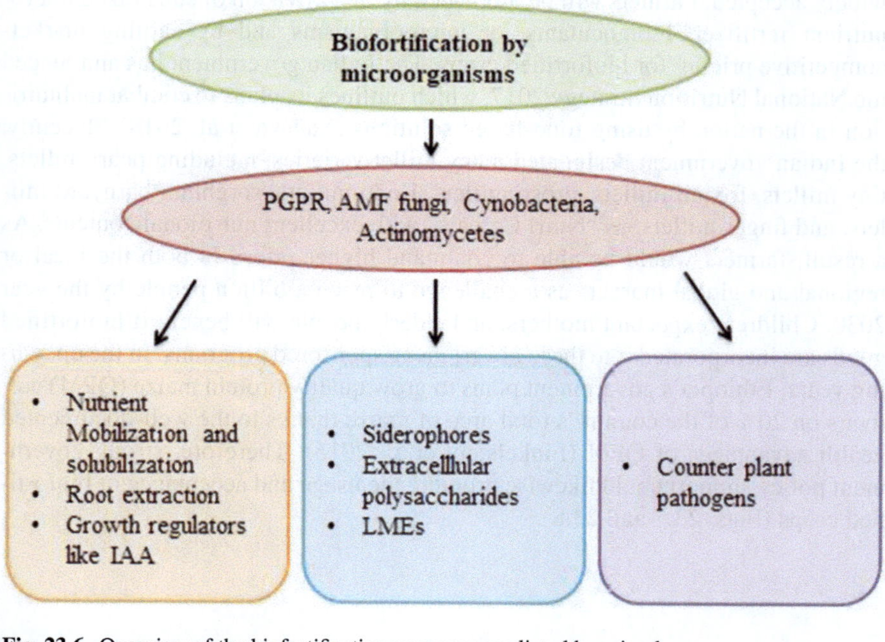

Fig. 23.6 Overview of the biofortification processes mediated by microbes

conscious of the possibilities and advantages of these biofortified crops through active investment in extension initiatives. The widespread adoption of biofortified crops has to be conditioned by a few fundamental measures.

3.7 Awareness Generation

The main cause of the sluggish adoption is incomplete awareness of the health advantages of biofortified crops. To ensure that farmers actively engage in this initiative, several demonstration experiments should be carried out on the farmer's field. Another reason for the slow adoption is lower yields compared to conventional methods, but farmers' interest will be kept if they receive premium prices. Strong ties should be established through agroprocessing sectors that offer farmers assurance. Agriculturalists, industry, and consumers would become aware of the presence and advantages of biofortified crop yields through vigorous promotional extension operations such field demonstrations, live theaters, radio programs, and message transmissions through TV discussion.

3.8 Policy Support

Enhancing input supply is a significant step in making biofortified crops more widely accepted. Farmers will be motivated by the provision of subsidized micronutrient fertilizer, bioinoculants, or microorganisms and by earning market-competitive pricing for biofortified crops. The Indian government has announced the National Nutrition Strategy 2017, which outlines its plans to combat malnutrition in the nation by using food-based solutions (Yadava et al. 2018). Recently, the Indian government designated many millet varieties, including pearl millets, tiny millets, foxtail millets, proso millets, Kodo millets, sorghum, barnyard millets, and finger millets, as "Nutri Cereals" with excellent nutritional contents. As a result, farmers would be able to command higher prices in both the local or regional and global markets as a challenge to reach a billion people by the year 2030. Children, expectant mothers, and elderly people will benefit if biofortified goods are incorporated into these government-sponsored programs. In the upcoming years, Ethiopia's government plans to grow quality protein maize (QPM) cultivars on 20% of the country's total area of maize thanks to the well-documented health advantages of QPM (Finkelstein et al. 2015). Therefore, strong government policy support could likewise promote the usage and acceptance of biofortified crops (Figs. 23.7 and 23.8).

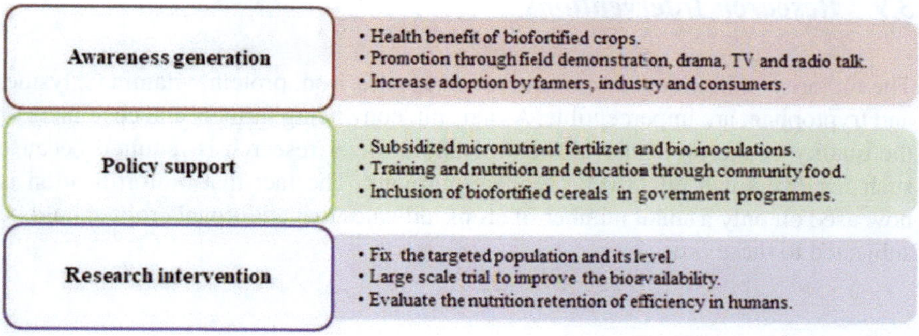

Fig. 23.7 Theoretical illustration of the potential for agricultural biofortification

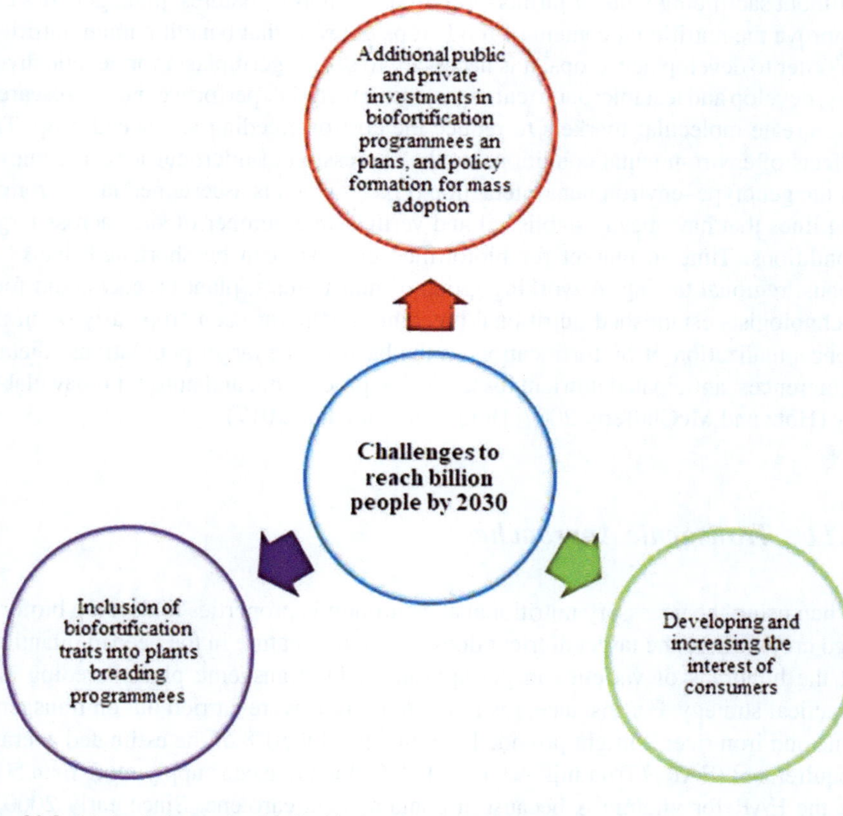

Fig. 23.8 Methods for getting biofortified goods to consumers

3.9 Research Interventions

The majority of nutrient components, such as zinc, iron, protein, vitamin C, lysine, and tryptophan, are imperceptible. As a result, convincing farmers and customers of the quality of the product is difficult. Nanofertilizer research is required because such fertilizers can effectively regulate nutrients. The fact that biofortification is now used on only a small number of crops indicates that additional crops should be subjected to these procedures.

3.10 Crop Development

Without sacrificing yield or farmer-preferred agronomic features, plant genetics can improve the nutritional content of food crops to levels that benefit human nutrition. In order to develop new crops, it is necessary to screen germplasm for genetic diversity, develop and test micronutrient-rich genetic material, perform genomic research, and create molecular markers to reduce the cost of breeding and speed it up. The effects of environmental conditions on the expression of micronutrients are known as the genotype–environment interaction (GxE), which is ascertained later, promising lines that have been established and verified in a number of sites across target conditions. Time to market for biofortified cultivars can be shortened thanks to robust regional testing. A working group of nutritionists, plant breeders, and food technologists established nutritional breeding targets for each crop early on in the conceptualization of biofortification on the basis of the target populations' dietary preferences, anticipated nutrient losses during processing, and nutrient bioavailability (Hotz and McClafferty 2007; Bouis and Saltzman 2017).

3.11 Transgenic Approaches

When using the necessary nutritional and agronomic properties to develop biofortified crops where the target nutrient does not exist in nature in the needed quantities in the hundreds of varieties in germplasm banks, transgenic plant breeding is a practical strategy. For instance, restricted field trials were carried out on transgenic zinc and iron rice; it might provide both metals with 30% of the estimated average requirement (EAR) (Trijatmiko et al. 2016). Golden rice can supply more than 50% of the EAR for vitamin A because it contains beta carotene. Since early 2000, a prototype of golden rice has been known, although it hasn't yet been commercialized in any nation, largely because regulatory approval procedures are so risk averse (Wesseler and Zilberman 2014). However, the distribution of such genetically altered cultivars to farmers is some years away, and pending certification under national biosafety and regulatory procedures, they have significant nutritional potential. All the crops that have been released or are soon to be released for the HarvestPlus programs use conventional breeding rather than transgenic breeding. HarvestPlus believes that conventional breeding is the quickest way to giving farmers and consumers access to more nutrient-dense crops because it does not face the same regulatory barriers and is universally recognized.

3.12 *Low-Cost, High-Throughput Methods*

Biofortification breeding requires the development of or adoption of rapid, low-cost testing techniques for micronutrients because of the requirements of the hundreds of samples that are analyzed for their mineral and vitamin contents each season. These diagnostic traits include near-infrared spectroscopy (NIRS) and colorimetric methods for measuring carotenoid levels. Because it involves minimal preanalytical preparation and allows for nondestructive inspection, X-ray fluorescence spectroscopy (XRF) has become the method of choice for mineral analysis (Paltridge et al. 2012a, b).

3.13 *Releases of Biofortified Crops*

More than 150 biofortified cultivars of ten agricultural crops have been sent to thirty countries overall. Twelve additional crop types with potential for biofortification are being evaluated for distribution in twenty-five other countries. The orange sweet potato has been promoted by the International Potato Center (CIP) in the countries depicted on this map.

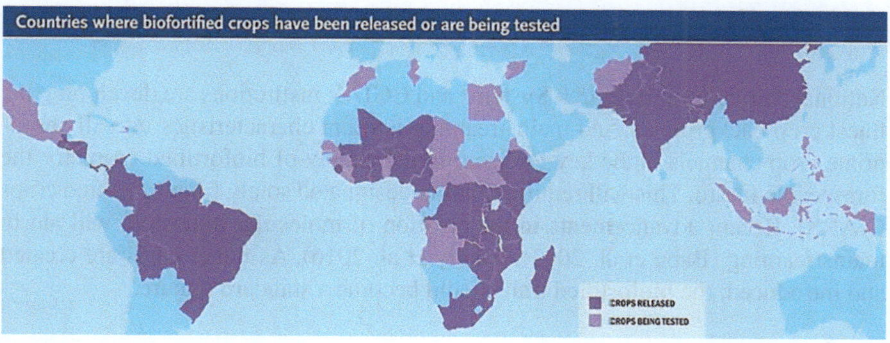

3.13.1 Vegetatively Propagated Crops

Farmers who plant stems, tubers, or vines instead of seeds are those whose seed systems are frequently characterized by tiny, informal participants. Planting materials must be replaced shortly after harvesting because they are fragile, costly, and cumbersome to transport over large distances. A lack of commercial private sector involvement offers a challenge and an opportunity for the development of seed varieties for biofortified crops such as orange sweet potato (Low et al. 2017).

3.13.2 Hybrid Crops

The largest opportunity for private sector commercialization can be found in hybrid crops, which require new seeds every year to maintain the same crop yields and agronomic features. The percentage of private sector acceptance relies on how they evaluate the need, even if employing the private industry for service may lead to long-term sustainability. Therefore, proponents of biofortification must focus their efforts on generating targeted demand from farmers and consumers (Bouis and Saltzman 2017).

4 Institutional Leadership Will Promote and Direct Mainstreaming in the Future

Biofortification needs to be incorporated as a primary activity within a variety of international organizations in order to realize its full potential. Three essential components are needed for this (Fig. 23.9) (Bouis and Saltzman 2017).

4.1 Supply

National Agricultural Research Systems and CGIAR institutions are developing the finest crop backgrounds, and mainstreaming nutrient characteristics into all appropriate crop channels is the key to ensuring the supply of biofortified crops for the foreseeable future. This will require funding designated solely for biofortified crops (NARS). Recent advancements in the creation of molecular indicators will aid in mainstreaming (Babu et al. 2013; Swamy et al. 2016). As novel kinds are created and introduced, the biofortified trait should become a standard feature.

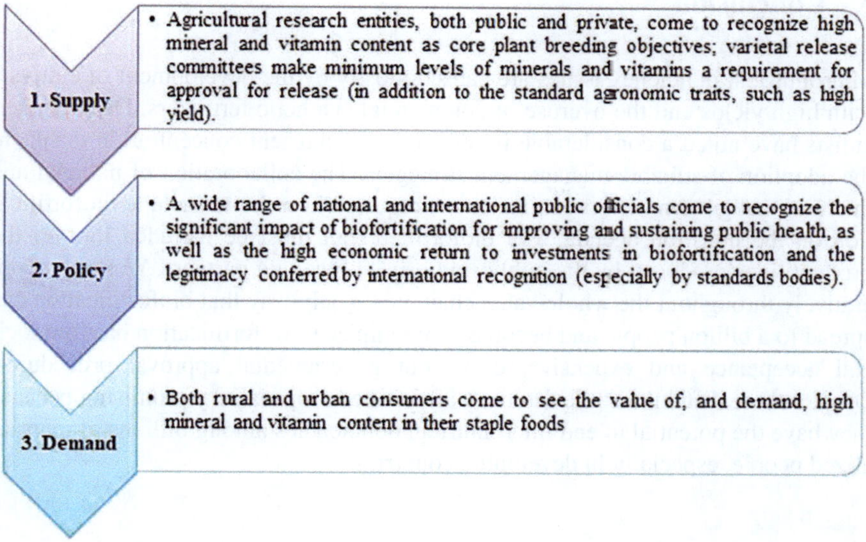

Fig. 23.9 Three critical elements of biofortification for global institutions

4.2 Policy

A supportive environment for biofortification is being created by regional and international initiatives such as the Scaling Up Nutrition movement and the Comprehensive Africa Agriculture Development Program (CAADP) of the African Union. Governments have asked for the introduction of new biofortified crops because they are so pleased with the results of the lead-biofortified plant species that have been accessible in their nations. These initiatives need to keep including biofortification in all kinds of policy (Bouis and Saltzman 2017).

4.3 Demand

In order for biofortified foods to reach communities that are micronutrient deficient, it is important to weigh the possible benefits of raising market demand for biofortified crops and making them more desirable to farmers. To maintain biofortified crops, customers in both rural and urban areas must demand high mineral and vitamin content in their staple foods. Rural smallholder demand is driven by superior agronomic features and nutrition messages, as was covered in the delivery section (Bouis and Saltzman 2017).

5 Conclusion

Micronutrient deficiency issues are caused mostly by the development of cultivars with high yields and the overuse of commercial synthetic fertilizers. Different scientists have noted a considerable improvement in nutrient concentration thanks to the adoption of suitable management strategies. The collaboration of national and international government organizations is required in order to include biofortification on the nutrition agenda. The biofortified trait must be included into all the product lines produced by the public and private breeding partners. Working cooperatively throughout the whole value chain is the only way that biofortification can spread to a billion people and become commonplace. Biofortification has low societal acceptance and expensive, drawn-out governmental approval procedures. Despite these difficulties, the future of biofortified crops is very promising because they have the potential to end micronutrient deficiencies among billions of marginalized people, especially in developing countries.

References

Babu R, Rojas NP, Gao S, Yan J, Pixley K (2013) Validation of the effects of molecular marker polymorphisms in LcyE and CrtRB1 on provitamin A concentrations for 26 tropical maize populations. Theor Appl Genet 126(2):389–399. https://doi.org/10.1007/s00122-012-1987-3

Bouis HE, Saltzman A (2017) Improving nutrition through biofortification: a review of evidence from HarvestPlus, 2003 through 2016. Glob Food Sec 12:49–58. https://doi.org/10.1016/j.gfs.2017.01.009

Busscher WJ, Sojka RE (1987) Enhancement of subsoiling effect on soil strength by conservation tillage. Trans ASAE 30(4):888-0892. https://doi.org/10.13031/2013.30493

Capper JL, Cady RA, Bauman DE (2009) Demystifying the environmental sustainability of food production. In: Proceedings of Cornell Nutrition Conference, October, pp 187–203

Clark RÁ, Zeto SK (2000) Mineral acquisition by arbuscular mycorrhizal plants. J Plant Nutr 23(7):867–902. https://doi.org/10.1080/01904160009382068

De Santiago A, Quintero JM, Avilés M, Delgado A (2011) Effect of Trichoderma asperellum strain T34 on iron, copper, manganese, and zinc uptake by wheat grown on a calcareous medium. Plant Soil 342(1):97–104. https://doi.org/10.1007/s11104-010-0670-1

Dikötter F (2010) Mao's great famine: the history of China's most devastating catastrophe, 1958–62. A & C Black

Ferrandon M, Chamel AR (1988) Cuticular retention, foliar absorption and translocation of Fe, Mn and Zn supplied in organic and inorganic form. J Plant Nutr 11(3):247–263. https://doi.org/10.1080/01904168809363800

Finkelstein JL, Mehta S, Udipi SA, Ghugre PS, Luna SV, Wenger MJ et al (2015) A randomized trial of iron-biofortified pearl millet in school children in India. J Nutr 145(7):1576–1581. https://doi.org/10.3945/jn.114.208009

Fowler, P. (2000). Harnessing Trade for Development. Oxfam International

Fuge R, Johnson CC (2015) Iodine and human health, the role of environmental geochemistry and diet, a review. Appl Geochem 63:282–302. https://doi.org/10.1016/j.apgeochem.2015.09.013

Graham RD, Ascher JS, Hynes SC (1992) Selecting zinc-efficient cereal genotypes for soils of low zinc status. Plant Soil 146(1):241–250

Haas JD, Beard JL, Murray-Kolb LE, Del Mundo AM, Felix A, Gregorio GB (2005) Iron-biofortified rice improves the iron stores of nonanemic Filipino women. J Nutr 135(12):2823–2830. https://doi.org/10.1093/jn/135.12.2823

HarvestPlus. Biofortification Prioritization Index (BPI). Washington, DC: International Food Policy Research Institute (IFPRI); 2013. Available from: http://www.harvestplus.org/content/worldfood-day-new-ranking-toolguideinvestment-biofortified-crops-launched

Hotz C, McClafferty B (2007) From harvest to health: challenges for developing biofortified staple foods and determining their impact on micronutrient status. Food Nutr Bull 28(2_suppl2):S271–S279. https://doi.org/10.1177/15648265070282S206

Hotz C, Kabir KA, Dipti SS, Arsenault JE, Bipul M (2015) Rice fortification with zinc during parboiling may improve the adequacy of zinc intakes in Bangladesh. J Sci Food Agric 95(2):379–385. https://doi.org/10.1002/jsfa.6730

IRRI (2006) Bringing hope, improving lives: strategic plan 2007–2015. Vol. 61. International Rice Research Institute, Manila, Philippines

Jawahar S, Vaiyapuri V (2013) 7. Effect of sulphur and silicon fertilization on yield_ nutrient uptake and economics of rice by S. Jawahar and V. Vaiyapuri. Int Res J Chem 1:34

Joy EJ, Ander EL, Young SD, Black CR, Watts MJ, Chilimba AD et al (2014) Dietary mineral supplies in Africa. Physiol Plant 151(3):208–229. https://doi.org/10.1111/ppl.12144

Karlen DL (1994) Crop rotations for the 21st century. Adv Agron 53:1–45

Karthikeyan N, Prasanna R, Nain L, Kaushik BD (2007) Evaluating the potential of plant growth promoting cyanobacteria as inoculants for wheat. Eur J Soil Biol 43(1):23–30. https://doi.org/10.1016/j.ejsobi.2006.11.001

Kennedy GL, Pedro MR, Seghieri C, Nantel G, Brouwer I (2007) Dietary diversity score is a useful indicator of micronutrient intake in non-breast-feeding Filipino children. J Nutr 137(2):472–477. https://doi.org/10.1016/j.jtemb.2005.02.006

Khan AG (2005) Role of soil microbes in the rhizospheres of plants growing on trace metal contaminated soils in phytoremediation. J Trace Elem Med Biol 18(4):355–364. https://doi.org/10.1016/j.jtemb.2005.02.006

Krishnaswamy K (2009) The problem and consequences of the double burden-a brief overview. Programme and abstracts. In: Symposium on nutritional security for India-issues and way forward, Indian National Science Academy. New Delhi, August 2009, pp 3–4

Kumar A (2012) Phosphorus and rain-harvested water economy through vesicular arbuscular mycorrhizae (VAM) in okra–pea sequence [PhD thesis]. Palampur: CSK HPKV

Kumar A, Yadav DS (1995) Use of organic manure and fertilizer in rice (Oryza sativa)-wheat (Triticum aestivum) cropping system for sustainability. Indian J Agric Sci (India)

Kumar A, Parmar DK, Suri VK (2012) Effect of boron fertilizers and organic manure on autumn cauliflower in western Himalayas. Ann Hortic 5(1):17–24

Lindsay WL (1991) Iron oxide solubilization by organic matter and its effect on iron availability. In: Iron nutrition and interactions in plants. Springer, Dordrecht, pp 29–36

Liu L, Chen H, Cai P, Liang W, Huang Q (2009) Immobilization and phytotoxicity of cd in contaminated soil amended with chicken manure compost. J Hazard Mater 163(2–3):563–567. https://doi.org/10.1016/j.jhazmat.2008.07.004

Low J, Ball A, Magezi S, Njoku J, Mwanga R, Andrade M et al (2017) Sweet potato development and delivery in sub-Saharan Africa. Afr J Food Agric Nutr Dev 17(2):11955–11972. https://doi.org/10.18697/ajfand.78.HarvestPlus07

Marcelle R (1995) Mineral nutrition and fruit quality. Acta Hortic 383:219–226

Marschner H (1995) Mineral nutrition of higher plants, 2nd edn. Germany, Institute of Plant Nutrition University of Hohenheim

Oli P, Ward R, Adhikari B, Mawson AJ, Adhikari R, Wess T et al (2016) Synchrotron X-ray fluorescence microscopy study of the diffusion of iron, manganese, potassium and zinc in parboiled rice kernels. LWT-Food Sci Technol 71:138–148. https://doi.org/10.1016/j.lwt.2016.03.034

Paltridge NG, Milham PJ, Ortiz-Monasterio JI, Velu G, Yasmin Z, Palmer LJ et al (2012a) Energy-dispersive X-ray fluorescence spectrometry as a tool for zinc, iron and selenium analysis in whole grain wheat. Plant Soil 361(1):261–269. https://doi.org/10.1007/s11104-012-1423-0

Paltridge NG, Palmer LJ, Milham PJ, Guild GE, Stangoulis JC (2012b) Energy-dispersive X-ray fluorescence analysis of zinc and iron concentration in rice and pearl millet grain. Plant Soil 361(1):251–260. https://doi.org/10.1007/s11104-011-1104-4

Pooniya V, Shivay YS (2013) Enrichment of basmati rice grain and straw with zinc and nitrogen through ferti-fortification and summer green manuring under indo-Gangetic plains of India. J Plant Nutr 36(1):91–117. https://doi.org/10.1080/01904167.2012.733052

Poore J, Nemecek T (2018) Reducing food's environmental impacts through producers and consumers. Science 360(6392):987–992. https://doi.org/10.1126/science.aaq0216

Ramadas S, Vellaichamy S, Ramasundaram P, Kumar A, Singh S (2020) Biofortification for enhancing nutritional outcomes and policy imperatives. In: Wheat and barley grain biofortification. Woodhead Publishing, pp 309–327. https://doi.org/10.1016/B978-0-12-818444-8.00013-4

Rana A, Joshi M, Prasanna R, Shivay YS, Nain L (2012) Biofortification of wheat through inoculation of plant growth promoting rhizobacteria and cyanobacteria. Eur J Soil Biol 50:118–126. https://doi.org/10.1016/j.ejsobi.2012.01.005

Rashid A, Fox RL (1992) Evaluating internal zinc requirements of grain crops by seed analysis. Agron J 84(3):469–474. https://doi.org/10.2134/agronj1992.00021962008400030022x

Ruel MT, Alderman H, Maternal and Child Nutrition Study Group (2013) Nutrition-sensitive interventions and programmes: how can they help to accelerate progress in improving maternal and child nutrition? Lancet 382(9891):536–551. https://doi.org/10.1016/S0140-6736(13)60843-0

Saltzman A, Birol E, Bouis HE, Boy E, De Moura FF, Islam Y, Pfeiffer WH (2013) Biofortification: progress toward a more nourishing future. Glob Food Sec 2(1):9–17. https://doi.org/10.1016/j.gfs.2012.12.003

Stipesevic B, Jug D, Jug I, Tolimir M, Cvijovic M (2009) Winter wheat and soybean zinc uptake in different soil tillage systems. Cereal Res Commun 37(2):305–310. https://doi.org/10.1556/crc.37.2009.2.19

Swamy BPM, Rahman MA, Inabangan-Asilo MA et al (2016) Advances in breeding for high grain zinc in rice. Rice 9:49. https://doi.org/10.1186/s12284-016-0122-5

Tariq M, Hameed S, Malik KA, Hafeez FY (2007) Plant root associated bacteria for zinc mobilization in rice. Pak J Bot 39(1):245

Trijatmiko KR, Dueñas C, Tsakirpaloglou N, Torrizo L, Arines FM, Adeva C et al (2016) Biofortified indica rice attains iron and zinc nutrition dietary targets in the field. Sci Rep 6(1):1–13

Tulchinsky TH (2010) Micronutrient deficiency conditions: global health issues. Public Health Rev 32(1):243–255

United Nations Millennium Development Goals Report (2006). Available from: https://www.un.org/zh/millenniumgoals/pdf/MDGReport2006

Watson CA, Öborn I, Edwards AC, Dahlin AS, Eriksson J, Lindström BEM et al (2012) Using soil and plant properties and farm management practices to improve the micronutrient composition of food and feed. J Geochem Explor 121:15–24. https://doi.org/10.1016/j.gexplo.2012.06.015

Wesseler J, Zilberman D (2014) The economic power of the Golden Rice opposition. Environ Dev Econ 19(6):724–742. https://doi.org/10.1017/S1355770X1300065X

WHO Newsroom (2019) Factsheet "children: reducing mortality". Available from: https://www.who.int/newsroom/fact-sheets/detail/children-reducing-mortality

Wilson EO (1992) The diversity of life–Belknap Press of Harvard University Press

Woese K, Lange D, Boess C, Bögl KW (1997) A comparison of organically and conventionally grown foods—results of a review of the relevant literature. J Sci Food Agric 74(3):281–293. https://doi.org/10.1002/(SICI)1097-0010(199707)74:3<281::AID-JSFA794>3.0.CO;2-Z

World Health Organization (2002) The world health report 2002: reducing risks, promoting healthy life. World Health Organization

World Health Organization (2003) Diet, nutrition, and the prevention of chronic diseases: report of a joint WHO/FAO expert consultation, vol 916. World Health Organization

World Health Organization (2019) Biofortification of Staple Crops. Available from: https://www.who.int/elena/titles/biofortification/en/

Yadava DK, Hossain F, Choudhury PR, Kumar D, Singh AK (2013) Biofortification of crops: a sustainable tool for nutritional security in India

Zhao FJ, Su YH, Dunham SJ, Rakszegi M, Bedo Z, McGrath SP, Shewry PR (2009) Variation in mineral micronutrient concentrations in grain of wheat lines of diverse origin. J Cereal Sci 49(2):290–295. https://doi.org/10.1016/j.jcs.2008.11.007

Zhu C, Naqvi S, Gomez-Galera S, Pelacho AM, Capell T, Christou P (2007) Transgenic strategies for the nutritional enhancement of plants. Trends Plant Sci 2007(12):548–555

Zimmermann MB, Wegmueller R, Zeder C, Chaouki N, Biebinger R, Hurrell RF, Windhab E (2004) Triple fortification of salt with microcapsules of iodine, iron, and vitamin A. Am J Clin Nutr 80(5):1283–1290. https://doi.org/10.1093/ajcn/80.5.1283

Retraction Note to: Biofortifying Legumes with Iodine

Muhammad Aslam, Haroon Zaman Khan, Ramala Masood Ahmad, Ayesha Aslam, and Ahsan Raza Mallhi

Retraction Note to:
Chapter 17 in: M. A. Nadeem et al. (eds.), *Legumes Biofortification*, https://doi.org/10.1007/978-3-031-33957-8_17

The Editors have retracted this chapter because the majority of the references cited could not be verified or contain errors. Muhammad Aslam, Haroon Zaman Khan, Ramala Masood Ahmad and Ahsan Raza Mallhi have not responded to correspondence from the Publisher about this retraction. The Publisher was unable to find a current email address for Ayesha Aslam.

The retracted version of this chapter can be found at https://doi.org/10.1007/978-3-031-33957-8_17

M. A. Nadeem et al. (eds.), *Legumes Biofortification*, https://doi.org/10.1007/978-3-031-33957-8_24

Index

M. A. Nadeem et al. (eds.), *Legumes Biofortification*,
https://doi.org/10.1007/978-3-031-33957-8

MIX
Papier aus verantwortungsvollen Quellen
Paper from responsible sources
FSC® C105338

If you have any concerns about our products,
you can contact us on
ProductSafety@springernature.com

In case Publisher is established outside the EU,
the EU authorized representative is:
Springer Nature Customer Service Center GmbH
Europaplatz 3, 69115 Heidelberg, Germany

Printed by Libri Plureos GmbH
in Hamburg, Germany